Introduction to CRIMINOLOGY

A Text/Reader

Anthony Walsh • Craig Hemmens
Boise State University

Los Angeles • London • New Delhi • Singapore

For information:

Sage Publications, Inc.
2455 Teller Road
Thousand Oaks, California 91320
E-mail: order@sagepub.com

Sage Publications Ltd.
1 Oliver's Yard
55 City Road
London EC1Y 1SP
United Kingdom

Sage Publications India Pvt. Ltd.
B 1/I 1 Mohan Cooperative Industrial Area
Mathura Road, New Delhi 110 044
India

Sage Publications Asia-Pacific Pte Ltd
33 Pekin Street #02-01
Far East Square
Singapore 048763

Printed in the United States of America

Library of Congress Cataloging-in-Publication Data

Walsh, Anthony, 1941-
Introduction to criminology: a text/reader/Anthony Walsh, Craig
Hemmens.
p. cm.
Includes bibliographical references and index.
ISBN 978-1-4129-5683-3 (pbk.)
1. Criminology. 2. Crime. 3. Criminal behavior. I. Hemmens, Craig. II. Title.

HV6025.W3655 2008
364—dc22 2007044493

This book is printed on acid-free paper.

09 10 11 12 10 9 8 7 6 5 4 3 2

Acquisitions Editor: Jerry Westby
Developmental Editor: Denise Simon
Editorial Assistant: Eve Oettinger
Copy Editor: Diana Breti
Typesetter: C&M Digitals (P) Ltd.
Proofreader: Doris Hus
Indexer: Holly Day
Cover Designer: Edgar Abarca
Marketing Manager: Jennifer Reed Banando

Brief Contents

Detailed Contents

Preface

There are a number of excellent criminological textbooks and criminological readers available to students and professors, so why this one? The reason is that stand-alone textbooks and readers (often assigned as an expensive addition to a textbook) have a pedagogical fault that we seek to rectify with the present book. Textbooks focus on providing a broad overview of the criminological research and lack depth, while readers often feature in-depth articles about a single topic, with little or no text to unify the readings and little in the way of pedagogy. This book provides more in the way of text and pedagogy and will use recent research-based articles to help students understand criminology. This book is unique in that it is a hybrid text/reader offering the best of both worlds. It includes a collection of articles on criminology that have previously appeared in a number of leading criminology journals along with original textual material that serves to explain and synthesize the readings. We have selected some of the best recent research and literature reviews and assembled them into this text/reader for an undergraduate or graduate criminology class.

Journal articles selected for inclusion have been selected based primarily on how they add to and complement the textual material and how interesting we perceive them to be for students. In our opinion, these articles are the best contemporary work on the issues they address. However, journal articles are written for professional audiences, not for students, and thus often contain quantitative material students are not expected to understand. They also often contain concepts and hair-splitting arguments over minutiae that tend to turn students glassy-eyed. Mindful of this, we have substantially edited and abridged the articles contained in this text/reader to make them as student friendly as possible. We have done this without doing injustice to the core points raised by the authors or detracting from the authors' key findings and conclusions. Those wishing to read these articles (and others) in their entirety are able to do so by accessing the Sage Web site provided for users of this book. Research-based quantitative articles are balanced by review/overview essay-type articles and qualitative articles providing subjective insight into criminal behavior from the points of view of offenders.

This book can serve as a supplemental reader or the primary text for an undergraduate course in criminology, or as the primary text for a graduate course. In a graduate course in criminology, it would serve as both an introduction to the extant literature and a sourcebook for additional reading, as well as a springboard for enhanced class discussion. When used as a supplement to an undergraduate course, this book can serve to provide greater depth than the standard textbook. It is important to note that the readings and/or the introductory textual material in this book provide a comprehensive survey of the current state of the existing scientific literature in virtually all areas of criminology, and give a history of how we got to this point in each topic area.

Structure of the Book

We use the typical outline for criminology textbook topics/sections, beginning with the definitions of crime and criminology and measuring crime, proceeding into theories of crime and criminality, and then into typologies. We depart from this typicality in one way only, that of the ordering of the theory chapters. The typical criminology textbook begins with a discussion of biological and psychological theories and proceeds to demolish concepts that others demolished decades ago, such as atavism and the XYY syndrome. Having shown how wrong these concepts were, and leaving the impression that those concepts exhaust the content of modern biological and psychological theories, they proceed to sociological theories.

Unfortunately, this is the exact opposite of the way that normal science operates. Normal science begins with observations and descriptions of phenomena and then asks a series of "why" questions that systematically take it down to lower levels of analysis. Wholes are wonderful meaningful things, and holistic explanations are fine, as far as they go. But they only go so far before they exhaust their explanatory power and require a more elementary look. Philosophers of science agree that holistic accounts describe phenomena, whereas reductionist (examining a phenomenon at a more fundamental level) accounts explain them. Scientists typically observe and describe what is on the surface of a phenomenon and then seek to dig deeper to find the fundamental mechanisms that drive the phenomenon.

In the natural sciences, useful observations go in both holistic and reductionist directions, such as from quarks to the cosmos in physics and from nucleotides to ecological systems in biology. There is no zero-sum competition between levels of analysis in these sciences, nor should there be in ours. Thus, following our discussion of the early schools, we begin with the most holistic (social structural) theories. These theories describe elements of whole societies that are supposedly conducive to high rates of criminal behavior, such as capitalism or racial heterogeneity. Because only a small proportion of people exposed to these alleged criminogenic forces commit crimes, we must move down to social process theories that talk about how individuals interpret and respond to structural forces. We then have to move to more individualistic (psychosocial) theories that focus on the traits and abilities of individuals that would lead them to arrive at different interpretations than other individuals, and finally to theories (biosocial) that try to pin down the exact mechanisms underlying these predilections.

This text/reader is divided into 14 sections that mirror the sections in a typical criminology textbook, each dealing with a particular type of subject matter in criminology. Each of the theory sections concludes with an evaluation of the theories and the policy implications derivable from them. These sections are as follows:

1. **Introduction:** We first provide an introductory section dealing with the discipline of criminology. We describe crime and criminality and introduce the concept, function, and pitfalls of criminological theory. We also offer a brief history of the discipline, which will allow readers to understand how the science of criminology started. The social context in which various perspectives began at key times in certain periods of cultural and political development is discussed, as well as the technology available for criminologists seeking to understand the

quicksilver of criminal behavior. We also discuss the policy implications of criminological theory in this section.

2. **Measuring crime:** This section describes the various ways data on the prevalence and incidence of crime is collected. We describe the strengths and weaknesses of the Uniform Crime Reports, the National Incident-Based Reporting System, and the National Crime Victimization Survey, which are measures collected by government agencies, and self-report studies, which are collected by criminologists. The three journal articles in this section illustrate how the different measures are used to explore issues of importance to criminologists.

3. **Early schools of criminology and their modern counterparts:** This section explores a basic dichotomy in criminology: classical and positivist schools. The classical school emphasizes human rationality, free will, and choice; the positivist school emphasizes the scientific search for factors that influence how these human attributes are exercised. The articles selected for inclusion in this section are about how modern thinkers view the argument about "free will" and "determinism" in the context of criminal behavior.

4. **Social structural theories:** Social structural theories are "macro" theories that explore the behavioral effects of how society is structured on criminal behavior. They look at such things as culture, neighborhood, and social practices and how these things serve to generate crime. They do not seek to explain individual criminal behavior, but rather aggregate crime rates of different groups who are exposed to these factors.

5. **Social process theories:** Social process theories are "micro" theories that explore how individuals subjectively perceive the kinds of factors social structural theories identify; all people do not react similarly to similar situations. Theorists in this tradition concentrate on exploring the influence of smaller social groupings (such as peer groups and the family) on the behavior of individuals.

6. **Critical theories:** Critical theories are also structural theories, but they differ on the critical stance that they have on society and on their emphasis on social conflict rather than social consensus. The capitalist mode of production is *the* cause of crime for many theorists in this tradition, although many others do not share this extreme view. For feminist theorists, the beast is patriarchy rather than capitalism.

7. **Psychosocial theories:** Although some writers have classified social process theories as "psychological" because of their emphasis on subjective interpretation, the primary difference between them and the material in this chapter is that psychological traits such as IQ, impulsiveness, and empathy are emphasized more than influences outside of the actor.

8. **Biosocial theories:** Biosocial perspectives are having an ever-increasing impact on criminology. This chapter examines what the disciplines of behavior genetics, evolutionary psychology, and neuroscience have to offer our discipline. Theorists in these disciplines go to great pains to convince us that we cannot understand the role of genes, hormones, and brain structures without understanding the complementary role of the environment—there is no nature versus nurture argument here; only nature via nurture.

9. **Developmental theories:** Developmental theories bring the disciplines of biology, psychology, and sociology together to offer a more complete understanding of antisocial

behavior. This chapter looks at the onset, acceleration, deceleration, and desistance from offending along with all the risk and protective factors for it.

10. **Violent crimes:** This section examines the UCR Part I violent crimes (murder, rape, robbery, and aggravated assault). It also features multiple murder (mass, spree, and serial killing) and the very contemporary problem of terrorism. The causes and context of these crimes are addressed. These are the topics that capture student interest the most, and thus more space is allotted to them.

11. **Property Crime:** This section examines the UCR Part I property crimes (burglary, larceny/theft, motor vehicle theft, and arson). It concentrates primarily on the subjective reasons for engaging in property crime by looking at what offenders themselves have to say about why they engage in crime. Additionally, the growing area of cybercrime is discussed.

12. **Public Order Crime:** Public order crimes can be more harmful than many other types of crimes, although they may be legal at some times and at some places. This section looks at the links between alcohol, drugs, and crime. It also examines prostitution, drunk driving, and gambling.

13. **White-collar and organized crime:** White-collar crime is more costly to American society than common street crime. In this section, we differentiate between occupational and corporate crime and look at such issues as the similarities and differences among white-collar and street criminals. We then examine organized crime and the reasons that it exists, and where it is most likely to exist.

14. **Victimology:** We then present a section on the neglected topic of victimology. We examine who is most likely to be victimized in terms of gender, race, age, and socioeconomic class, and find that those most likely to be perpetrators of crime are those most likely to be victims of crime. We also explore the consequences of victimization.

Ancillaries

To enhance the use of this text/reader and to assist those using this book as a core text, we have developed high-quality ancillaries for instructors and students.

Instructor's Resource CD A variety of instructor's materials are available. For each chapter, this includes summaries, PowerPoint slides, chapter activities, Web resources, and a complete set of test questions.

Student Study Sites This comprehensive student study site features chapter outlines students can print for class, flashcards, interactive quizzes, Web exercises, links to additional journal articles, links to Frontline videos, NPR and PBS radio shows, and more.

Acknowledgments

We would first of all like to thank executive editor Jerry Westby. Jerry's faith in and commitment to the project is greatly appreciated, as are those of his very able developmental editor, Denise Simon. This tireless two kept up a most useful three-way dialogue between authors, publisher, and a parade of excellent reviewers, making this text the best that it could possibly be. Our copy editor, Diana Breti, spotted every errant comma, dangling participle, and missing reference in the manuscript, for which we are truly thankful. We also want to express our gratitude to Erin Conley for organizing and writing an excellent draft manuscript for the "How to Read a Research Article" guide for students. It should prove to be very useful! Thank you one and all.

We are also most grateful for the many reviewers who spent considerable time providing us with the benefit of their expertise during the writing/rewriting phase of the text's production. Trying to please so many individuals is a trying task, but one that is ultimately satisfying and one that undoubtedly made the book better than it would otherwise have been. These expert criminologists were

James David Ballard
California State University Northridge

Ashley Blackburn
University of North Texas

Dennis R. Brewster
Oklahoma State University

Tammy Castle
University of West Florida

Sue Cote
California State University Sacramento

Addrain Conyers
Southern Illinois University

Roger Cunningham
Eastern Illinois University

Julie Globokar
Kaplan University

Julia Glover Hall
Drexel University

Patricia K. Hendrickson
Tarleton State University

Heath C. Hoffmann
College of Charleston

Art Jipson
University of Dayton

John McMullen
Frostburg State University

Allison Ann Payne
Villanova University

Craig T. Robertson
University of North Alabama

Bonnie Semora
University of Georgia

Ronald E. Severtis, Jr.
Ohio State University

Tracey Steele
Wright State University

Rebecca Stevens
Kent State University

Staci Strobl
John Jay College of Criminal Justice

Brian J. Stults
Florida State University

Finally, Anthony Walsh would like to acknowledge his most wonderful wife, Grace Jean, for her love and support during this and numerous other projects. She is a real treasure and the center of his universe. Craig Hemmens would like to acknowledge Tony Walsh, who did the heavy lifting, and Mary Stohr, who puts up with so much.

Dedication

Anthony Walsh dedicates this book to his ever pleasant, ever gorgeous wife Grace (AKA "the face") and to all his children and grandchildren.

Craig Hemmens dedicates this book, all the books in this series, and everything of value that he has ever done to his father, George Hemmens, who showed him the way; James Marquart and Rolando Del Carmen, who taught him how; and Mary and Emily, for giving him something he loves even more than his work.

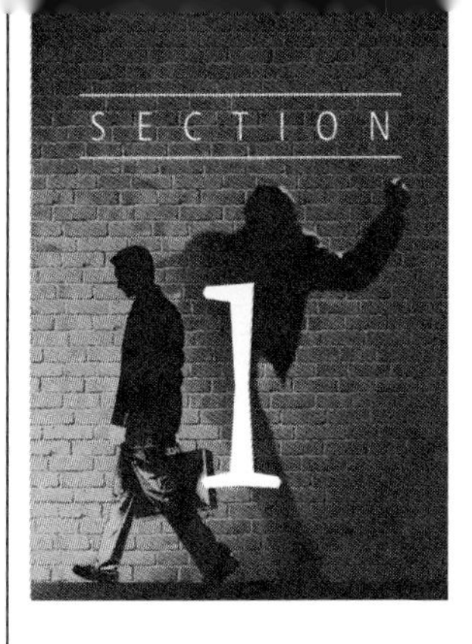

INTRODUCTION AND OVERVIEW OF CRIME AND CRIMINOLOGY

In 1996, Iraqi refugees Majed Al-Timimy, 28, and Latif Al-Husani, 34, married the daughters, aged 13 and 14, of a fellow Iraqi refugee in Lincoln, Nebraska. The marriages took place according to Muslim custom, and everything seemed to be going well for awhile until one of the girls ran away and the concerned father and her husband reported it to the police. It was at this point that American and Iraqi norms of legality and morality clashed head on. Under Nebraska law, people under 17 years old cannot marry, so both grooms and the father and mother of the girls were arrested and charged with a variety of crimes from child endangerment to rape.

According to an Iraqi woman interviewed by the police (herself married at 12 in Iraq), both girls were excited and happy about the wedding. Suddenly these men faced up to 50 years in prison for their actions, and the Iraqi community was shocked, as would have been earlier generations of Americans who were legally permitted to marry girls of this age. The men were sentenced to 4 to 6 years in prison and paroled in 2000 with conditions that they have no contact with their "wives." Thus, something that is legally and morally permissible in one culture can be severely punished in another. Were the actions of these men child sex abuse or simply unremarkable marital sex? Which culture is right? Can we really ask such a question? Is Iraqi culture "more right" than American culture, given that marrying girls of that age was permissible here, too, at one time? Most important, how can criminologists hope to study crime scientifically if what constitutes a crime is relative to time and place?

What Is Criminology?

Criminology is an interdisciplinary science that gathers and analyzes data on various aspects of crime and criminal behavior. As with all scientific disciplines, its goal is to understand its subject matter and to determine how that understanding can benefit humankind. In pursuit of this understanding, criminology asks questions such as the following:

- Why do crime rates vary from time to time and from culture to culture?
- Why are some individuals more prone to committing crime than others?
- Why do crime rates vary across different ages, genders, and racial/ethnic groups?
- Why are some harmful acts criminalized and not others?
- What can we do to prevent crime?

By a *scientific* study of crime and criminal behavior we mean that criminologists use the scientific method to try to answer the questions they ask, rather than simply philosophizing about them. The scientific method is a tool for winnowing truth from error by demanding evidence for one's conclusions. Evidence is obtained by formulating hypotheses derived from theory that are rigorously tested with data. How this is accomplished will be addressed later in this section, after we discuss the nature of crime.

What Is Crime?

The term *criminal* can and has been applied to many types of behavior, some of which nearly all of us have been guilty of at some time in our lives. We can all think of acts that we feel *ought* to be criminal but are not, or acts that should not be criminal but are. The list of acts that someone or another at different times and at different places may consider to be crimes is very large, and only a few are defined as criminal by the United States law at this time. Despite these difficulties, we need a definition of crime in order to proceed. The most often-quoted definition is that of Paul Tappan (1947), who defined **crime** as "an intentional act in violation of the criminal law committed without defense or excuse, and penalized by the state" (p. 100). A crime is thus an *act* in violation of a *criminal law* for which a *punishment* is prescribed; the person committing it must have *intended* to do so and must have done so without legally acceptable *defense* or *justification.*

Tappan's definition is strictly a legal one that reminds us that the state, and only the state, has the power to define crime. Hypothetically, a society could eradicate crime tomorrow simply by rescinding all of its criminal statutes. Of course, this would not eliminate the behavior specified by the laws; in fact, the behavior would doubtless increase because the behavior could no longer be officially punished. While it is absurd to think that any society would try to solve its crime problem by eliminating its criminal statutes, legislative bodies are continually revising, adding to, and deleting from their criminal statutes.

Crime as a Moving Target

Every vice is somewhere and at some time a virtue. There are numerous examples, such as the vignette at the beginning of this chapter, of acts defined as crimes in one country being tolerated and even expected in another. Laws also vary within the same culture from time to time as well as across different cultures. Until the Harrison Narcotics Act of 1914, there were few legal restrictions in the United States on the sale, possession, or use of most drugs, including heroin and cocaine. Following the Harrison Act, many drugs became controlled substances, their possession became a crime, and a brand new class of criminals was created overnight.

Crimes pass out of existence also, even acts that had been considered crimes for centuries. Until the United States Supreme Court invalidated sodomy statutes in *Lawrence v. Texas* (2003), sodomy was legally punishable in many states. Likewise, burning the American flag had serious legal consequences until 1989, when the Supreme Court in *Texas v. Johnson* invalidated anti-flag-burning statutes as unconstitutional. What constitutes a crime, then, can be defined in and out of existence by the courts or by legislators. As long as human societies remain diverse and dynamic, there will always be a moving target of activities with the potential for nomination as crimes, as well as illegal activities nominated for decriminalization.

If what constitutes crime differs across time and place, how can criminologists hope to agree upon a scientific explanation for crime and criminal behavior? Science is about making universal statements about stable or homogeneous phenomena. Atoms, the gas laws, the laws of thermodynamics, photosynthesis, and so on are not defined or evaluated differently by scientists around the globe according to local customs or ideological preferences. But the phenomenon we call "crime keeps moving around, and because it does some criminologists have declared it impossible to generalize about what is and is not 'real' crime" (Hawkins, 1995, p. 41).

What these criminologists are saying is that crime is a socially constructed phenomenon that lacks any "real" objective essence and is defined into existence rather than discovered. At one level, of course, everything is socially constructed; nature does not reveal herself to us sorted into ready-labeled packages; humans must do it for her. *Social construction* means nothing more than that humans have perceived a phenomenon, named it, and categorized it according to some classificatory rule that makes note of the similarities and differences among the things being classified. Most classification schemes are not arbitrary; if they were, we would not be able to make sense of anything. Categories have empirically meaningful referents and are used to impose order on the diversity of human experience, although arguments exist about just how coherent that order is.

Crime as a Subcategory of Social Harms

So, what *can* we say about crime? How *can* we conceive of it in ways that at least most people would agree are coherent and correspond with their view of reality? When all is said and done, crime is a subcategory of all harmful acts that range from simple acts like smoking to very serious acts like murder. Harmful acts are thus arrayed on a continuum in terms of the seriousness of the harm involved. Some harmful acts, such as smoking tobacco and drinking to excess, are not considered anyone's business other than the actor's if they take place in private or even in public if the person creates no annoyance.

Socially (as opposed to private) harmful acts are acts deemed to be in need of regulation (health standards, air pollution, etc.), but not by the criminal law except under exceptional circumstance. Private wrongs (such as someone reneging on a contract) are socially harmful, but not sufficiently so to require the heavy hand of the criminal law. Such wrongs are regulated by the civil law, in which the wronged party (the plaintiff), rather than the state, initiates legal action, and the defendant does not risk deprivation of his or her liberty if the plaintiff prevails.

Further along the continuum we find a category of harmful acts considered so socially harmful that they come under the purview of the coercive power of the criminal justice system. Even here, however, we are still confronted with the problem of human judgment in determining what goes into this subcategory. But this is true all along the line; smoking was once actually considered rather healthy, and air pollution and unhealthy conditions were simply facts of life about which nothing could be done. Categorization always requires a series of human judgments, but that does not render the categorizations arbitrary.

The harm wrought by criminal activity exacts a huge financial and emotional price. The emotional pain and suffering borne by crime victims is obviously impossible to quantify, but many estimates of the financial harm are available. Most estimates focus on the costs of running the criminal justice system, which includes the salaries and benefits of personnel and the maintenance costs of buildings (offices, jails, prisons, stations) and equipment (vehicles, weapons, uniforms, etc.). Added to these costs are the costs associated with each crime (the average cost per incident multiplied by the number of incidents as reported to the police). All these costs combined are estimates of the *direct* costs of crime.

The *indirect* costs of crime must also be considered as part of the burden. These costs include all manner of surveillance and security devices, protective devices (guns, alarms, security guards) and insurance costs, medical services, and the lost productivity and taxes of incarcerated individuals. Economist David Anderson (1999) lists a cascade of direct and indirect costs of crime and concludes that the aggregate burden of crime in the United States (in 1997 dollars) is about $1,102 *billion*, or a per capita burden of $4,118. Crime thus places a huge financial burden on everyone's shoulders, as well as a deep psychological burden on its specific victims.

Beyond Social Construction: The Stationary Core Crimes

Few people would argue that an act is not arbitrarily categorized or is not seriously harmful if it is universally condemned. That is, there is a core of offenses defined as wrong at almost all times and in almost all cultures. Some of the strongest evidence in support of the stationary core perspective comes from the International Criminal Police Organization (Interpol) (1992), headquartered in Lyon, France. Interpol serves as a repository for crime statistics from each of its 125 member nations. Interpol's data show that such acts as murder, assault, rape, and theft are considered serious crimes in every country.

Criminologists call these universally condemned crimes **mala in se** ("inherently bad"). Crimes that are time and culture bound are described as **mala prohibita** ("bad because they are prohibited"). But how can we be sure that an act is inherently bad? We would say that the litmus test for determining a mala in se crime is that no one, except under the most bizarre of circumstances, would want to be victimized by it. While millions of people seek to be "victimized" by prostitutes, drug dealers, bookies, or any of a number of other providers of illegal goods and services, no one wants to be murdered, raped, robbed, or have his or her property stolen. Being victimized by such actions evokes physiological reactions (anger, helplessness, sadness, depression, a desire for revenge) in all cultures, and would do so even if the acts were not punishable by law or custom. Mala in se crimes engage these emotions not because some legislative body has defined them as wrong, but because they hammer at our

▲ **Photo 1.1** Group portrait of a police department liquor squad posing with cases of confiscated alcohol and distilling equipment during Prohibition.

deepest primordial instincts. Evolutionary biologists propose that these built-in emotional mechanisms exist because mala in se crimes threatened the survival and reproductive success of our distant ancestors, and that they function to strongly motivate people to try to prevent such acts from occurring and punish the perpetrators if they do (Daly & Wilson, 1988; O'Manique, 2003; Walsh, 2000).

Figure 1.1 illustrates the relationship of core crimes (mala in se) to acts that have been arbitrarily defined (mala prohibita) as crimes and all harmful acts that may potentially be criminalized. The Figure is inspired by John Hagan's (1985) effort to distinguish between "real" crimes and "socially constructed" arbitrary crimes by examining the three highly interrelated concepts of *consensus* (the degree of public agreement on the seriousness of an act), the *severity* of penalties attached to an act, and the level of *harm* attached to an act.

Criminality

Perhaps we can avoid altogether the problem of defining crimes by studying individuals who commit predatory *harmful* acts, regardless of the legal status of the acts. Criminologists do this when they study criminality. **Criminality** is a clinical or scientific, rather than legal, term, and one that can be defined independently of legal definitions of crimes. Crime is an intentional act of commission or omission contrary to the law; *criminality* is a property of individuals that signals the willingness to commit those and other harmful acts (Gottfredson & Hirschi, 1990). Criminality is a continuously distributed trait that is a combination of other

Figure 1.1 Mala in Se and Mala Prohibita Crimes as Subsets of All Harms

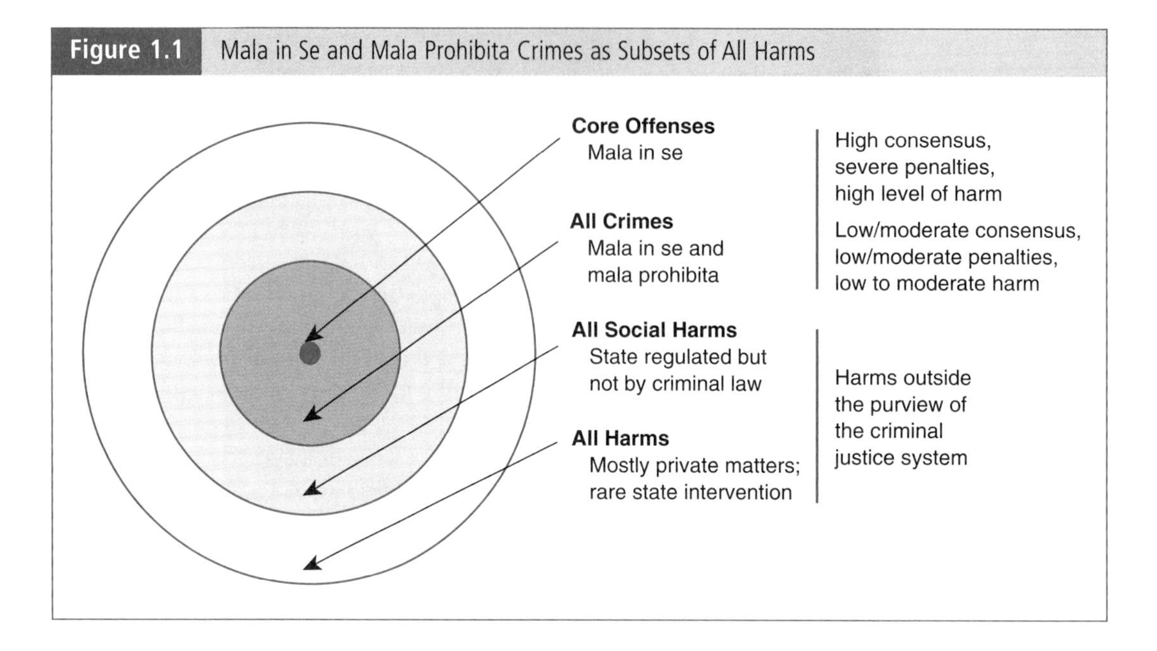

continuously distributed traits and that signals the willingness to use force, fraud, or guile to deprive others of their lives, limbs, or property for personal gain. People can use and abuse others for personal gain regardless of whether the means used have been defined as criminal; it is the propensity to do this that defines criminality, independent of the labeling of an act as a crime or of the person being legally defined as a criminal.

Defining criminality as a continuous trait acknowledges that there is no sharp line separating individuals with respect to this trait—it is not a trait that one has or has not. Just about everyone at some point in life has committed an act or two in violation of the law. But that doesn't make us all criminals; if it did, the term would become virtually synonymous with the word *human*! The point is, we are all situated somewhere on the criminality continuum, which ranges from saint to sociopath, just as our heights range from the truly short to the truly tall. Some are so extreme in height that any reasonable person would call them "tall." Likewise, a small number of individuals have violated so many criminal statutes over such a long period of time that few would question the appropriateness of calling them "criminals." Thus, both height and criminality can be thought of as existing along a continuum, even though the words we use often imply that people's heights and criminal tendencies come in more or less discrete categories (tall/short, criminal/noncriminal). In other words, just as height varies in fine gradations, so, too, does involvement in crime.

A Short History of Criminology

Criminology is a young discipline, although humans have probably been theorizing about crime and its causes ever since they first made rules and observed others breaking them. What and how people thought about crime and criminals (as well as all other things) in the past was strongly influenced by the social and intellectual currents of their time. This is no less true of what and how modern professional criminologists think about crime and criminals. In prescientific days, explanations for bad behavior were often of a religious or spiritual nature, such as demonic possession or the abuse of free will. Because of the legacy of Original Sin, all human beings were considered born sinners. The gift of the grace of God kept men and women on the straight and narrow, and if they deviated from this line it was because God was no longer their guide and compass.

Other more intellectual types believed that the human character and personality are observable in physical appearance. Consider Shakespeare's Julius Caesar's distrust of Cassius because he "has a lean and hungry look." Such folk wisdom was systematized by an Italian physician named Giambattista della Porta, who developed a theory of human personality called *physiognomy* in 1558. Porta claimed that the study of physical appearance, particularly of the face, could reveal much about a person's personality and character. Thieves, for instance, were said to have large lips and sharp vision.

▲ **Photo 1.2** Charles Darwin (1809–1882) heavily influenced biological positivism in criminology.

Porta was writing during a historical period known as the Renaissance, a period between approximately 1450 and 1600, which saw a change in thinking from the pure God-centered supernaturalism and relative barbarism of the Middle Ages to more human-centered naturalism. *Renaissance* means "rebirth" and refers to the rediscovery of the thinking traditions of the ancient Greeks. The sciences (primitive as they were) and arts were becoming important, the printing press was invented, and Christopher Columbus "discovered" America during this period. In short, the Renaissance began to move human thinking away from the absolute authority of received opinion and toward a way that would eventually lead to the modern scientific method.

Another major demarcation in the emergence of the modern world was the Enlightenment, or Age of Reason. The Enlightenment was the period approximately between 1650 and 1800. It might be said that the Renaissance provided a key to the human mind and the Enlightenment opened the door. Whereas the Renaissance is associated with advances in art, literature, music, and philosophy, the Enlightenment is associated with advances in mathematics, science, and the dignity and worth of the individual as exemplified by a concern for human rights. This concern led to reforms in criminal justice systems throughout Europe, a process given a major push by Cesare Beccaria's work *On Crimes and Punishments,* which ushered in the so-called *classical school.* The classical school emphasized human rationality and free will in its explanations for criminal behavior. Beccaria and other classical thinkers will be discussed at length in Section 3.

Modern criminology really began to take shape with the increasing faith among intellectuals that science could provide answers for everything. These individuals witnessed the harnessing of the forces of nature to build and operate the great machines and mechanisms that drove the Industrial Revolution. They also witnessed the strides made in biology after

Charles Darwin's works on the evolution of species. Criminology saw the beginning of the so-called *positivist school* during this period. Theories of character, such as Franz Josef Gall's system of *phrenology*—assessing character from physical features of the skull—abounded. The basic idea behind phrenology was that cognitive functions are localized in the brain, and that the parts regulating the most dominant functions were bigger than parts regulating the less dominant ones. Criminals were said to have large protuberances in parts of the brain thought to regulate craftiness, brutishness, moral insensibility, and so on, and small bumps in such "localities" as intelligence, honor, and piety.

The biggest impact during this period, however, was made by Cesare Lombroso's theory of *atavism,* or the born criminal. Criminologists from this point on were obsessed with measuring, sorting, and sifting all kinds of data (mostly physical) about criminal behavior. The main stumbling block to criminological advancement during this period was the inadequacy of its research. The intricacies of scientifically valid research design and measurement were not appreciated, and statistical techniques were truly primitive by today's standards. The early positivist thinkers will be discussed at length in Section 3.

The so-called Progressive Era (about 1890 to 1920) ushered in new social ideologies and new ways of thinking about crime. The era was one of liberal efforts to bring about social reform as unions, women, and other disadvantaged groups struggled for recognition. Criminology largely turned away from what was disparagingly termed "biological determinism," which implied that nothing could be done to reform criminals, to cultural determinism. If behavior is caused by what people experience in their environments, so the optimistic argument went, then we can change their behavior by changing their environment. It was during this period that sociology became the disciplinary home of criminology. Criminology became less interested in why individuals commit crime from biological or psychological points of view and more concerned with aggregate level (social structures, neighborhoods, subcultures, etc.) data. It was during this period that the so-called structural theories of crime, such as the Chicago school of social ecology, were formulated. Anomie strain theory was another structural/cultural theory that emerged somewhat later (1938). This theory was doubtless influenced strongly by the American experience of the Great Depression and of the exclusion of blacks from many areas of American society.

The period from the 1950s through the early 1970s saw considerable dissatisfaction with the strong structural approach, which many viewed as proceeding as if individuals were almost irrelevant to explaining criminal behavior. Criminological theory moved toward integrating psychology and sociology during this period and strongly emphasized the importance of socialization. Control theories were highly popular at this time, as was labeling theory; these are addressed in Section 5.

Because the latter part of this period was a time of great tumult in the United States (the anti-war, civil rights, women's, and gay rights movements), it also saw the emergence of several theories, such as conflict theory, that were highly critical of American society. These theories extended to earlier works of Marxist criminologists, who tended to believe that the only real cause of crime was capitalism. These theories provided little new in terms of our understanding of "street" criminal behavior, but they did spark an interest in white-collar crime and how laws were made by the powerful and applied against the powerless. These theories are addressed in Section 6.

Perhaps because of a new conservative mood in the United States, theories with the classical taste for free will and rationality (albeit modified) embedded in them reemerged in the

1980s. These were rational choice, deterrence, and routine activities theories, all of which had strong implications for criminal justice policy. These are discussed in Section 3.

In the late 1990s and early 2000s, we witnessed a resurgence of biosocial theories. These theories view all behavior as the result of various biological factors interacting with each other and with the past and present environments of the actors involved. Biosocial theories have been on the periphery of criminology since its beginning but have been hampered by perceptions of them as driven by an illiberal agenda and by their inability to "get inside" the mysteries of heredity and the workings of the brain. The truly spectacular advances in the observational techniques (brain scan methods, $10 cheek swabs to test DNA, etc.) in the genomic and neurosciences over the past two decades have made these things less of a mystery today, and social scientists are increasingly realizing that there is nothing illiberal about recognizing the biology of human nature.

No science advances without the technology at its disposal to plumb its depths. For instance, the existence of atoms was first proposed by Greek philosophers more than 2,500 years ago. This was dismissed as merely philosophical speculation until the early 19th century, when English chemist John Dalton proposed his atomic theory of chemistry, which asserted that all chemical reactions are the rearrangements of atoms. Dalton was heavily criticized by chemists who wanted a "pure" chemistry uncontaminated by physics. Yet chemists everywhere soon adopted the idea of atoms, but still debated whether they were an actual physical reality or just a useful concept. Using scanning tunneling microscopes, today we can see individual atoms, and the argument has been put to rest.

Criminologists are in a position similar to that of chemists 100 years ago. The concepts, methods, and measuring devices available to geneticists, neuroscientists, endocrinologists, and other biological scientists may do for the progress of criminology what physics did for chemistry, what chemistry did for biology, and what biology is increasingly doing for psychology. Exceptionally ambitious longitudinal studies carried out over decades in concert with medical and biological scientists, such as the Dunedin Multidisciplinary Health and Development Study (Moffitt, 1993), the National Longitudinal Study of Adolescent Health Study (Udry, 2003), and the National Youth Survey (Menard & Mihalic, 2001), are able to gather a wealth of genetic, neurological, and physiological data. Such studies are being conducted with increasing frequency. Integrating these hard science disciplines into criminology will no more rob it of its autonomy than physics robbed chemistry or chemistry robbed biology. On the contrary, physics made possible huge advances in chemistry, and chemistry did the same for biology. These advances would not have happened had scientists maintained their call for the "purity" of their disciplines.

The Role of Theory in Criminology

When an FBI agent asked the depression-era bank robber Willie Sutton why he robbed banks, Sutton replied, "Because that's where the money is." In his own way, Sutton was offering a theory explaining the behavior of bank robbers. Behind his witty answer is a model of a kind of person who has learned how to take advantage of opportunities provided by convenient targets flush with a valued commodity. Thus, if we put a certain kind of personality and learning together with opportunity and coveted resources, we get bank robbery. This is what theory making is all about: trying to grasp how all the known correlates of a phenomenon are linked together in noncoincidental ways to produce an effect.

Just as medical scientists want to find out what causes disease, criminologists are interested in finding factors that cause crime and criminality. As is the case with disease, there are a variety of risk factors to be considered when searching for causes of criminal behavior. The first step in detected causes is to discover **correlates,** which are factors that are related to the phenomenon of interest. To discover whether two factors are related, we must see whether they vary together; that is, if one variable goes up or down, the other goes up or down as well.

Establishing causality requires much more than simply establishing a correlation. Take gender, the most thoroughly documented correlate of criminal behavior ever identified. Literally thousands of studies throughout the world, including European studies going back five or six centuries, have consistently reported strong gender differences in all sorts of antisocial behavior, including crime, and the more serious the crime the stronger that difference is. All studies are unanimous in indicating that males are more criminal than females. Establishing *why* gender is such a strong correlate of crime is the real challenge, as it is with any other correlate. Trying to establish causes is the business of theory.

What Is Theory?

A **theory** is a set of logically interconnected propositions explaining how phenomena are related and from which a number of hypotheses can be derived and tested. Theories should provide coherent explanations of the phenomena they address, they should correspond with the relevant empirical facts, and they should provide practical guidance for researchers looking for further facts. This guidance takes the form of a series of statements that can be logically deduced from the assertions of the theory. We called these statements **hypotheses,** which are statements about relationships between and among factors we expect to find based on the logic of our theories. Hypotheses and theories support one another in the sense that theories provided the raw material (the ideas) for generating hypotheses, and hypotheses support or fail to support theories by exposing them to empirical testing.

Theories are devised to explain how a number of different correlates may actually be causally related to crime and criminality rather than simply associated with them. We emphasize that when we talk of causes we do not mean that when *X* is present *Y will* occur in a completely prescribed way. We mean that when *X* is present *Y* has a certain *probability* of occurring, and perhaps only then if *X* is present along with factors *A, B,* and *C.* In many ways, crime is like illness because there may be as many routes to becoming criminal as there are to becoming ill. In other words, criminologists have never uncovered a necessary cause (a factor that *must* be present for criminal behavior to occur and in the absence of which criminal behavior has never occurred) or a sufficient cause (a factor that is able to produce criminal behavior without being augmented by some other factor).

There is a lot of confusion among laypersons about the term *theory.* We often hear statements such as, "That's just theory" or hear it negatively contrasted with practice: "That's all right in theory, but it won't work in the real world." Such statements imply that a theory is a poor relative of a fact, something impractical we grasp at in the absence of solid, practical evidence. Nothing could be further from the truth. Theories help us to make sense of a diversity of seemingly unrelated facts and propositions, and they even tell us where to look for more facts, which make theories very practical things indeed.

We all use theory every day to fit facts together. A detective confronted with a number of facts about a mysterious murder must fit them together, even though their meaning and relationship to one another is ambiguous and perhaps even contradictory. Using years of experience, training, and good common sense, the detective constructs a theory linking those facts together so that they begin to make some sense and begin to tell their story. An initial theory derived from the available facts then guides the detective in the search for additional facts in a series of "*if* this is true, *then* this should be true" statements. There may be many false starts as our detective misinterprets some facts, fails to uncover others, and considers some to be relevant when they are not. Good detectives, like good scientists, will adjust their theory as new facts warrant; poor detectives and poor scientists will stand by their favored theory by not looking for more facts or by ignoring, downplaying, or hiding contrary facts that come to their attention. When detectives do this, innocent people suffer and guilty people remain undiscovered; when scientists do this, the progress of science suffers.

The physical and natural sciences enjoy a great deal of agreement about what constitutes the core body of knowledge within their disciplines; thus, they have few competing theories. Within criminology there is little agreement about the nature of the phenomena we study, and so we suffer an embarrassment of theoretical riches. Given the number of criminological theories, students may be forgiven for asking which one is true. Scientists never use the term "truth" in scientific discourse; rather, they tend to ask which theory is most useful. Criteria for judging the merits of a theory are summarized below:

1. *Predictive Accuracy:* A theory has merit and is useful to the extent that it accurately predicts what is observed. That is, the theory has generated a large number of research hypotheses that have supported it. This is the most important criterion.
2. *Predictive Scope:* Predictive scope is the scope or range of the theory and thus the scope or range of the hypotheses that can be derived from it. That is, how much of the empirical world falls under the explanatory umbrella of theory A compared to how much falls under theory B.
3. *Simplicity:* If two competing theories are essentially equal in terms of the first two criteria, then the less complicated one is considered more "elegant."
4. *Falsifiability:* A theory is never proven true, but it must have the quality of being falsifiable or disprovable. If a theory is formulated in such a way that no amount of evidence could possibly falsify it, then the theory is of little use (Ellis, 1994, pp. 202–205).

How to Think About Theories

You will be a lot less concerned about the numerous theories in criminology if you realize that different theories deal with different levels of analysis. A **level of analysis** is that segment of the phenomenon of interest that is measured and analyzed. We can ask about causes of crime at the levels of whole societies, subcultures, neighborhoods, families, or individuals. Answers to the question of crime causation at one level do not generally answer the same question at another level. For instance, suppose that at the individual level there is strong evidence to support the notion that crime is linked to impulsiveness and low IQ. Do you think that this evidence would help us to understand why the crime rate in society A is 2.5 times that of society B, or why the crime rate in society C last year was only 75% as high as it was 20 years ago? It would do so only in the extremely unlikely event that society A has 2.5 times as many impulsive low-IQ people as society B, or that society C has lost 25% of its people with those characteristics in the last 20 years. If the question posed asks about crime rates in whole societies, the answers must address sociocultural differences among different societies or in the same society at different times.

Conversely, if crime rates are found to be quite strongly related to the degree of industrialization or racial/ethnic diversity in societies, this tells us nothing about why some people in an industrialized, heterogeneous society commit crimes and others in the same society do not. To answer questions about individuals we need theories about individuals. Generally speaking, questions of cause and effect must be answered at the same level of analysis at which they were posed; thus, different theories are required at different levels.

The second reason we have so many theories is that causal explanations are also offered at different temporal levels: *ultimate* (distant in time) and *proximate* (close in time) explanations. If we define a criminal act as the result of a person who is psychologically prepared to commit it meeting a situation conducive to its commission (such as Willie Sutton and banks), the possible levels of explanation range from the ultimate (the evolutionary history of the species) to the most proximate level (the immediate precipitating situation). Between these extreme levels are genetic, temperamental, developmental, personality, familial, experiential, and social environmental explanations. We will be discussing theories offering explanations for crime at all levels, but you should realize that in reality these levels describe an integrated whole as people interact with their environments.

We know that crime rates change in society, sometimes drastically, without any corresponding change in the gene pool or personalities of the people in them. Because causes are sought only among factors that vary, changing sociocultural environments must be the only causes of changing crime rates. What environmental changes do, however, is raise or lower individual thresholds for engaging in crime, and some people have lower thresholds than others. People with weak criminal propensities (or high prosocial propensities) require high levels of environmental instigation to commit crimes, but some individuals would engage in criminal behavior in the most benign of environments. When—or whether—individuals cross the threshold to commit criminal acts depends on the interaction between their personal thresholds and the environmental thresholds.

Interpreting the meaning of research findings is not as simple as documenting correlates of crime. There is little room for error when contrasting rates of crime between and among the various demographic variables such as age, gender, and race/ethnicity. Nor is there much difficulty (unless one wants to split fine hairs) in defining and classifying people into those categories. But theory testing looks for causal explanations rather than simple descriptions, and that's where our problems begin. For example, when we consistently find positive correlations between criminal behavior and some other factor it is tempting to assume that something causal is going on, but as we have said previously, correlations merely *suggest* causes, they do not demonstrate them. Resisting the tendency to jump to causal conclusions from correlations is the first lesson of statistics.

Ideology in Criminological Theory

We have seen how criminological theorizing is linked to the social and intellectual climate of the times. It is also essential that we understand the role of ideology in criminology. **Ideology** is a way of looking at the world, a general emotional picture of "how things should be."

This implies a selective interpretation and understanding of evidence that comes to our senses rather than an objective and rational evaluation of the evidence. Ideology forms, shapes, and colors our concepts of crime and its causes in ways that lead to a tendency to accept or reject new evidence according to how well or poorly it fits our ideology. We rarely see a discussion of ideology in criminology textbooks, leading students to believe that criminological arguments are settled with data in the same manner as natural science arguments typically are settled. Unfortunately, this is not always the case in criminology.

According to Thomas Sowell (1987), two contrasting visions have shaped thoughts about human nature throughout history, and these visions are in constant conflict with each other. The first of these visions is the **constrained vision**, so called because believers in this vision view human activities as constrained by an innate human nature that is self-centered and largely unalterable. The **unconstrained vision** denies an innate human nature, viewing it as formed anew in each different culture. The unconstrained vision also believes that human nature is perfectible, a view scoffed at by those who profess the constrained vision. A major difference between the two visions is that the constrained vision says, "This is how the world *is*"; the unconstrained vision says, "This is how the world *should be*." These visions are what sociologists call *ideal types*, which are conceptual tools that accentuate differences between competing positions for purposes of guiding the exploration of them. There are many "visions" that are hybrids of the two extremes; Sowell lists Marxism, for instance, as a prominent hybrid of the two visions.

The two contrasting ways of approaching a social problem such as crime are aptly summed up by Sowell (1987): "While believers in the unconstrained vision seek the special causes of war, poverty, and crime, believers in the constrained vision seek the special causes of peace, wealth, or a law-abiding society" (p. 31). Note that this implies that unconstrained visionaries (mostly liberals) believe that war, poverty, and crime are aberrations to be explained, while constrained visionaries (mostly conservatives) see these things as historically normal and inevitable, although regrettable, and believe that what has to be understood are the conditions that prevent them. We will see the tension between visions constantly as we discuss the various theories in this book.

Given this, it should be no surprise to discover that criminological theories differ in how they approach the "crime problem." A theory of criminal behavior is at least partly shaped by the ideological vision of the person who formulated it, and that, in turn, is partly due to the ideological atmosphere prevailing in society. Sowell (1987) avers that a vision "is what we sense or feel *before* we have constructed any systematic reasoning that could be called a theory, much less deduced any specific consequences as hypotheses to be tested against evidence" (p. 14). Those who feel drawn to a particular theory likewise owe a great deal of their attraction to it to the fact that they share the same vision as its formulator. In other words, "visions," more so than hard evidence, all too often lead criminologists to favor one theory over another more strongly than most care to acknowledge (Cullen, 2005, p. 57).

Orlando Patterson (1998) views ideology as a major barrier to advancement in the human sciences. He states that conservatives believe only "the proximate internal cultural and behavioral factors are important ('So stop whining and pull up your socks, man!')," and "liberals and mechanistic radicals" believe that "only the proximate and external factors are worth considering ('Stop blaming the victim, racist!')" (p. ix). Patterson's observation reminds us of the ancient Indian parable of the nine blind men feeling different part of an elephant. Each man described the elephant according to the part of its anatomy he had felt, but each failed to appreciate the descriptions of the others who felt different parts. The men fell into dispute and departed in anger, each convinced of the utter stupidity, and perhaps the malevolence, of the others. The point is that ideology often leads criminologists to "feel" only part of the criminological elephant and then to confuse the parts with the whole. As with the blind men, criminologists sometimes question the intelligence and motives (e.g., having some kind of political agenda) of other criminologists who have examined different parts of the criminological elephant. Needless to say, such criticisms have no place in scientific criminology.

There is abundant evidence that political ideology is linked to favored theories among contemporary criminologists. Walsh and Ellis (2004) asked 137 criminologists which theory they considered to be "most viable with respect to explaining variations in serious and persistent criminal behavior." Twenty-three different theories were represented in their responses, but obviously they cannot all be the "most viable," so something other than hard evidence was instrumental in their choices. The researchers found that the best predictor of a favored theory was the criminologists' stated ideology (conservative, moderate, liberal, or radical), and the second best predictor was the discipline in which criminologists received the bulk of their training. Ideology and the lack of interdisciplinary training will no doubt continue to plague the development of a theory of crime and criminality that is acceptable to all criminologists. When reading this text, try to understand where the originators, supporters, and detractors of any particular theory being discussed are "coming from" ideologically as well as theoretically.

Connecting Criminological Theory and Social Policy

Theories of crime causation imply that changing the conditions the theory holds responsible for causing crime can reduce crime and even prevent it. We say "imply" because few theorists are explicit about the public policy implications of their work. Scientists are primarily concerned with gaining knowledge for its own sake; they are only secondarily concerned with how useful that knowledge may be to practitioners and policymakers. Conversely, policymakers are less concerned with hypothetical "causes" of a problem and more concerned with what strategies are both politically and financially feasible.

Policy is simply a course of action designed to solve some problem that has been selected from among alternative courses of action. Solving a social problem means attempting to reduce the level of the problem currently being experienced or to enact strategies that try to prevent it from occurring in the first place. Social science findings can and have been used to help policymakers determine which course of action to follow to "do something" about the crime problem, but there are many other concerns that policymakers must consider that go beyond maintaining consistency with social science theory and data. The question of "what to do about crime" involves political and financial considerations, the urgency of other problems competing for scarce financial resources (schools, highways, environmental protection, public housing, national defense), and a host of other major and minor considerations.

Policy choices are, at bottom, value choices, and as such only those policy recommendations that are ideologically palatable are likely to be implemented. Given all of these extratheoretical considerations, it would be unfair to base our judgment of a theory's power solely, or even primarily, on its impact on public policy. Even if some aspects of policy are theory based, unless all recommendations of the theory are fully implemented, the success or failure of the policy cannot be considered evidence of theoretical failure any more than a baker can blame a recipe for a lousy cake if he or she neglects to include all the ingredients it calls for.

Connecting problems with solutions is a tricky business in all areas of government policy making, but nowhere is it more difficult than in the area of criminal justice. No single strategy can be expected to produce significant results, and it may sometimes make matters worse. For example, President Johnson's "War on Poverty" was supposed to have a significant impact on the crime problem by attacking what informed opinion of the time considered its "root cause." Programs and policies developed to reduce poverty did so, but reducing poverty had no effect on reducing crime; in fact, crime rose as poverty was falling. Another high-profile example of failed policy is the Volstead Act of 1919 that prohibited the manufacture and sale of alcohol in the United States. Although based on a true premise (alcohol is a major factor in facilitating violent crime), it failed because it ushered in a wild period of crime as gangs fought over control of the illegal alcohol market. Policies often have effects that are unanticipated by policymakers, and these effects can be positive or negative.

Nevertheless, every theory has policy implications deducible from its primary assumptions and propositions The deep and lasting effects of the classical theories on legal systems around the world has long been noted, but the broad generalities about human nature contained in those theories offer little specific advice on ways to change criminals or to reduce their numbers. Although we caution against using the performance of a theory's public policy recommendations as a major criterion to evaluate its power, the fact remains that a good theory *should* offer useful practical recommendations, and we will discuss a theory's policy implications when appropriate.

A Brief Word About the Section Readings

Because this book is a hybrid text/reader, a few words are warranted about the rationale behind our choice of articles. The readings in each section are meant to provide further depth in the material covered in the text. The theoretical sections (3 through 9) contain a mixture of "classical" readings by the old masters and modern quantitative or qualitative readings. One may wonder why we bother presenting classical pieces; after all, the great philosopher/mathematician Alfred North Whitehead once opined that "A science that hesitates to forget its founders is lost" (in Kuhn, 1970, p. 138). Whitehead's warning is apt if taken to mean that the reverence and reputation attached to the founders should never stand in the way of evidence of better explanations. However, as Kuhn (1970) notes, a science needs its heroes: "Fortunately, instead of forgetting these heroes, scientists have been able to forget or revise their works" (p. 139). If science forgets its founders completely, it risks repeating some of their overly dogmatic errors. Additionally, we should not be asked to forget them before we get to know them because much of what they wrote still has relevance and has served as foundation material for subsequent researchers.

Lawrence Sherman's article, "The Use and Usefulness of Criminology 1751–2005: Enlightened Justice and Its Failures," serves a number of purposes for us. First, it adds a little more to the history of criminology, especially its beginnings in the Enlightenment. Of particular interest is his discussion of English magistrate Henry Fielding, who Sherman believes is more entitled to the mantle of Father of Criminology than Beccaria or Lombroso because, unlike those two, Fielding put his ideas to a real-world test. This is something about a founding figure that we should never forget. Sherman's article also illustrates our point about tying theory to policy; indeed, the whole piece is a plea to more closely tie criminology to policy.

Sherman argues that criminology has been, and is, overwhelmingly analytical (theory-generating and testing) rather than experimental ("show me evidence from the real world"). Although he maintains that the strength of experimental criminology will rest on the strength of analytic criminology, he believes that the growth and acceptance of criminology will rest more on its experimental results than on advances in its basic science.

Summary

- Criminology is the scientific study of crime and criminals. It is an interdisciplinary/multidisciplinary study, although criminology has yet to integrate these disciplines in any comprehensive way.

- The definition of crime is problematic because acts that are defined as criminal vary across time and culture. Many criminologists believe that because crimes are defined into existence we cannot determine what real crimes are and criminals are. However, there is a stationary core of crimes that are universally condemned and always have been. These crimes are predatory crimes that cause serious harm and are defined as *mala in se*, or "inherently bad" crimes, as opposed to *mala prohibita*—"bad because they are forbidden" crimes.

- The history of criminology shows that the cultural and intellectual climate of the time strongly influences how scholars think about and study crime and criminality. The Renaissance brought more secular thinking, the Enlightenment more humane and rational thinking, the Industrial Revolution brought with it more scientific thinking, and the Progressive Era saw a reform-oriented criminology reminiscent of the classical school.

- Advances in any science are also constrained by the tools available to test theories. The ever-improving concepts, methods, and techniques available from modern genetics, neuroscience, and other biological sciences should add immeasurably to criminology's knowledge base in the near future.

- Theory is the "bread and butter" of any science, including criminology. There are many contending theories seeking to explain crime and criminality. Although we do not observe such theoretical disagreement in the more established sciences, the social/behavioral sciences are young, and human behavior is extremely difficult to study.

- When judging among the various theories, we have to keep certain things in mind, including predictive accuracy, scope, simplicity, and falsifiability. We must also remember that crime and criminality can be discussed at many levels (social, subcultural, family, or individual) and that a theory that may do a good job of predicting crime at one level may do a poor job at another level.

- Theories can also be offered at different temporal levels. They may focus on the evolutionary history of the species (the most ultimate level), the individual's subjective appraisal of a situation (the most proximate level), or any other temporal level in between. A full account of an individual's behavior may have to take all these levels into consideration because any behavior arises from an individual's propensities interacting with the current environmental situation as that individual perceives it. This is why we approach the study of crime and criminality from social, psychosocial, and biosocial perspectives.

- Criminologists have not traditionally done this, preferring instead to examine only aspects of criminal behavior that they find congenial to their ideology and, unfortunately, often maligning those who focus on other aspects. The main dividing line in criminology has separated conservatives (who tend to favor explanations of behavior that focus on the individual) and liberals (who tend to favor structural or cultural explanations). The theories favored by criminologists are strongly correlated with sociopolitical ideology.

- All theories have explicit or implicit recommendations for policy because they posit causes of crime or criminality. Removing those alleged causes should reduce crime, if the theory is correct, but the complex nature of crime and criminality makes policy decisions based on theory very risky indeed. Policymakers must consider many other issues demanding scarce resources, so the policy content of a theory should never be used to pass judgment on the usefulness of theory for criminologists.

EXERCISES AND DISCUSSION QUESTIONS

1. Which of the following 10 acts do you consider mala in se crimes, mala prohibita crimes, or no crime at all? Defend your choices.
   A. drug possession
   B. vandalism
   C. drunk driving
   D. collaborating with the enemy
   E. sale of alcohol to minors
   F. fraud
   G. spouse abuse
   H. adult male having consensual sex with underaged person
   I. prostitution

2. Why is it important to consider ideology when evaluating criminologists' work? Is it possible for them to divorce their ideology from their work?

3. The following table presents a list of seven acts that are considered criminal offenses. Add three more offenses that interest you to this list. Then, rate each of the 10 acts on a scale from 1 to 10 in terms of your perception of each one's seriousness (with 10 being the most serious). Give your list to a member of the opposite gender without letting him or her see your ratings, and ask him or her to rate the offenses on the same 10-point scale. After he or she is finished, compare the two ratings with the other person present, and discuss each inconsistency of 2 or more ranking points. Write a one- to two-page double-spaced report on what you learned from this exercise about how you and the other person differ and resemble one another in your thoughts about the seriousness of crime. Is there a gender difference?

Offense	Ranking by Someone Else	Your Ranking
Alcohol consumption by a minor		
Assassinating an unpopular political leader		
Killing a repeatedly abusive spouse		
Raping a stranger with threats to use a deadly weapon		
Committing rape on a date by threatening bodily harm		
Driving while extremely drunk		
Molesting a young child		
Total of all rankings		

4. Go to http://www.lsus.edu/la/journals/ideology/ for the online journal *Quarterly Journal of Ideology*. Click on *Archives* and find and read "Ideology: Criminology's Achilles' Heel." What does this article say about the "conflict of visions" in criminology?

USEFUL WEB SITES

Anderson, K. Social constructionism and belief causation. http://www.stanford.edu/group/dualist/vol8/pdfs/anderson.pdf.

Critical Criminology. http://www.critcrim.org.

Conflict Criminology. http://faculty.ncwc.edu/toconnor/301/301lect13.htm.

Learning Theories of Crime. http://faculty.ncwc.edu/TOCONNOR/301/301lect10.htm.

Links to Criminological Theory. http://www.acs.appstate.edu/dept/ps-cj/cj-sour.html#Theory.

CHAPTER GLOSSARY

Constrained vision: One of the two so-called ideological *visions* of the world. The constrained vision views human activities as constrained by an innate human nature that is self-centered and largely unalterable.

Correlates: Factors that are related to the phenomenon of interest.

Crime: An intentional act in violation of the criminal law committed without defense or excuse and penalized by the state.

Criminality: A continuously distributed trait composed of a combination of other continuously distributed traits that signals the willingness to use force, fraud, or guile to deprive others of their lives, limbs, or property for personal gain.

Criminology: An interdisciplinary science that gathers and analyzes data on crime and criminal behavior.

Hypotheses: Statements about relationships between and among factors we expect to find based on the logic of our theories.

Ideology: A way of looking at the world; a general emotional picture of "how things should be" that forms, shapes, and colors our concepts of the phenomena we study.

Level of analysis: That segment of the phenomenon of interest that is measured and analyzed, i.e., individuals, families, neighborhoods, states, etc.

Mala in se: Universally condemned crimes that are "inherently bad."

Mala prohibita: Crimes that are "bad" simply because they are prohibited.

Policy: A course of action designed to solve some problem that has been selected from among alternative courses of action.

Theory: A set of logically interconnected propositions explaining how phenomena are related and from which a number of hypotheses can be derived and tested.

Unconstrained vision. One of the two so-called ideological *visions* of the world. The unconstrained vision denies an innate human nature, viewing it as formed anew in each different culture.

How to Read a Research Article

As you travel through your criminal justice/criminology studies, you will soon learn that some of the best known and/or emerging explanations of crime and criminal behavior come from research articles in academic journals. This book has research articles throughout the book, but you may be asking yourself, "How do I read a research article?" It is my hope to answer this question with a quick summary of the key elements of any research article, followed by the questions you should be answering as you read through the assigned sections.

Every research article published in a social science journal will have the following elements: (1) introduction, (2) literature review, (3) methodology, (4) results, and (5) discussion/conclusion.

In the introduction, you will find an overview of the purpose of the research. Within the introduction, you will also find the hypothesis or hypotheses. A hypothesis is most easily defined as an educated statement or guess. In most hypotheses, you will find that the format usually followed is: If X, Y will occur. For example, a simple hypothesis may be: "If the price of gas increases, more people will ride bikes." This is a testable statement that the researcher wants to address in his/her study. Usually, authors will state the hypothesis directly, but not always. Therefore, you must be aware of what the author is actually testing in the research project. If you are unable to find the hypothesis, ask yourself what is being tested and/or manipulated, and what are the expected results?

The next section of the research article is the literature review. At times the literature review will be separated from the text in its own section, and at other times it will be found within the introduction. In any case, the literature review is an examination of what other researchers have already produced in terms of the research question or hypothesis. For example, returning to my hypothesis on the relationship between gas prices and bike riding, we may find that five researchers have previously conducted studies on the increase of gas prices. In the literature review, I will discuss their findings and then discuss what my study will add to the existing research. The literature review may also be used as a platform of support for my hypothesis. For example, one researcher may have already determined that an increase in gas prices causes more people to roller-blade to work. I can use this study as evidence to support my hypothesis that increased gas prices will lead to more bike riding.

The methods used in the research design are found in the next section of the research article. In the methodology section you will find the following: who/what was studied, how many subjects were studied, the research tool (e.g., interview, survey, observation), how long the subjects were studied, and how the data that was collected was processed. The methods section is usually very concise, with every step of the research project recorded. This is important because a major goal of the researcher is "reliability," or, if the research is done over again the same way, will the results be the same?

The results section is an analysis of the researcher's findings. If the researcher conducted a quantitative study (using numbers or statistics to explain the research), you will find statistical tables and analyses that explain whether or not the researcher's hypothesis is supported. If the researcher conducted a qualitative study (non-numerical research for the purpose of theory construction), the results will usually be displayed as a theoretical analysis or interpretation of the research question.

Finally, the research article will conclude with a discussion and summary of the study. In the discussion, you will find that the hypothesis is usually restated, and perhaps a small discussion of why this is the hypothesis. You will also find a brief overview of the methodology and results. Finally, the discussion section will end with a discussion of the implications of the research and what future research is still needed.

Now that you know the key elements of a research article, let us examine a sample article from your text.

The Use and Usefulness of Criminology, 1751–2005: Enlightened Justice and Its Failures

By Lawrence W. Sherman

1. What is the thesis or main idea from this article?

   - The thesis or main idea is found in the introductory paragraph of this article. Although Sherman does not point out the main idea directly, you may read the introduction and summarize the main idea in your own words. For example: The thesis or main idea is that criminology should move away from strict analysis and towards scientific experimentation in order to improve the criminal justice system and crime control practices.

2. What is the hypothesis?

   - The hypothesis is found in the introduction of this article. It is first stated in the beginning paragraph: "As experimental criminology provides more comprehensive evidence about responses to crime, the prospects for better basic science—and better policy—will improve accordingly." The hypothesis is also restated in the middle of the second section of the article. Here, Sherman actually distinguishes the hypothesis by stating: "The history of criminology . . . provides an experimental test of this hypothesis about analytic versus experimental social science: *that social science has been most useful, if not most used, when it has been most experimental, with visibly demonstrable benefits (or harm avoidance) from new inventions.*"

3. Is there any prior literature related to the hypothesis?

   - As you may have noticed, this article does not have a separate section for a literature review. However, you will see that Sherman devotes attention to prior literature under the heading Enlightenment, Criminology, and Justice. Here, he offers literature regarding the analytical and experimental history of criminology. This is a brief overview to help the reader understand the prior research that explains why social science became primarily analytic.

4. What methods are used to support the hypothesis?

   - Sherman's methodology is known as a historical analysis. In other words, rather than conducting his own experiment, Sherman is using evidence from history to support his hypothesis regarding analytic and experimental criminology. When conducting a historical analysis, most researchers use archival material from books, newspapers, journals, etc. Although Sherman does not directly state his source of information, we can see that he is basing his argument on historical essays and books, beginning with Henry Fielding's *An Enquiry Into the Causes of the Late Increase of Robbers* (1751) to the social experiments of the 1980s by the National Institute of Justice. Throughout his methodology, Sherman continues to emphasize his hypothesis of the usefulness of experimental criminology, yet how experiments have also been hidden in the shadows of analytic criminology throughout history.

5. Is this a qualitative study or quantitative study?

   ◆ To determine whether a study is qualitative or quantitative, you must look at the results. Is Sherman using numbers to support his hypothesis (quantitative) or is he developing a non-numerical theoretical argument (qualitative)? Because Sherman does not utilize statistics in this study, we can safely conclude that this is a qualitative study.

6. What are the results and how does the author present the results?

   ◆ Because this is a qualitative study, as we earlier determined, Sherman offers the results as a discussion of his findings from the historical analysis. The results may be found in the section titled Criminology: Analytic, Useful, and Used. Here, Sherman explains that "*the vast majority of published criminology remains analytic and nonexperimental.*" He goes on to say that although experimental criminology has been shown to be useful, it has not always been used, or has not been used correctly. Because of the misuse of experimental criminology, criminologists have steered towards the safety of analysis rather than experimentation. Therefore, Sherman concludes that "[a]nalytic social science still dominates field experiments by 100 to 1 or better in criminology. . . . Future success of the field may depend upon a growing public image based on experimental results."

7. Do you believe that the author/s provided a persuasive argument? Why or why not?

   ◆ This answer is ultimately up to the reader, but looking at this article, I believe that it is safe to assume that the readers will agree that Sherman offered a persuasive argument. Let us return to his major premise: The advancement of theory may depend on better experimental evidence, but, as history has illustrated, the vast majority of criminology remains analytical. Sherman supports this proposition with a historical analysis of the great thinkers of criminology and the absence of experimental research throughout a major portion of history.

8. Who is the intended audience of this article?

   ◆ A final question that will be useful for the reader deals with the intended audience. As you read the article, ask yourself, to whom is the author wanting to speak? After you read this article, you will see that Sherman is writing for not only students, but also professors, criminologists, historians, and/or criminal justice personnel. The target audience may most easily be identified if you ask yourself, "Who will benefit from reading this article?"

9. What does the article add to your knowledge of the subject?

   ◆ This answer is best left up to the reader because the question is asking how the article improved your knowledge. However, one way to answer the question is as follows: This article helps the reader to understand that criminology is not just about theoretical construction. Criminology is both an analytical and experimental social science, and in order to improve the criminal justice system, as well as criminal justice policies, more attention needs to be given to the usefulness of experimental criminology.

10. What are the implications for criminal justice policy that can be derived from this article?

◆ Implications for criminal justice policy are most likely to be found in the conclusion or the discussion sections of the article. This article, however, emphasizes the implications throughout the article. From this article, we are able to derive that crime prevention programs will improve greatly if they are embedded in well-funded experimental-driven data rather than strictly analytical data. Therefore, it is in the hands of policymakers to fund criminological research and apply the findings in a productive manner to criminal justice policy.

Now that we have gone through the elements of a research article, it is your turn to continue through your text, reading the various articles and answering the same questions. You may find that some articles are easier to follow than others, but do not be dissuaded. Remember that each article will follow the same format: introduction, literature review, methods, results, and discussion. If you have any problems, refer to this introduction for guidance.

READING

The Use and Usefulness of Criminology, 1751–2005

Enlightened Justice and Its Failures

Lawrence W. Sherman

In this article, Lawrence Sherman adds to our knowledge about the history of criminology. His premise is that after a useful beginning in the eighteenth-century Enlightenment as both an experimental and analytic social science, criminology sank into two centuries of inactivity. Its resurrection in the late twentieth-century crime wave successfully returned criminology to the forefront of discovering useful, if not always used, facts about prevailing crime patterns and responses to crime. Criminology's failures of "use" in creating justice more enlightened by knowledge of its effects is linked to the still-limited usefulness of criminology, which lacks a comprehensive body of evidence to guide sanctioning decisions. Yet that knowledge is rapidly growing, with experimental (as distinct from analytic) criminology now more prominent than at any time since Henry Fielding founded criminology while inventing the police. In short, Sherman wants us to put criminology to use by experimenting with different replicable crime control practices, using experimental and control groups when possible, rather than simply being a theory-testing science.

Criminology was born in a crime wave, raised on a crusade against torture and execution, and then hibernated for two centuries of speculation. Awakened by the rising crime rates of the latter twentieth century, most of its scholars chose to pursue analysis over experiment. The twenty-first century now offers more policy-relevant science than ever, even if basic science still occupies center stage. Its prospects for integrating basic and "clinical" science are growing, with more scholars using multiple tools rather than pursuing single-method work. Criminology contributes only a few drops of science in an ocean of decision making, but the number of drops is growing steadily. As experimental criminology provides more comprehensive evidence about responses to crime, the prospects for better basic science—and better policy—will improve accordingly.

Enlightenment, Criminology, and Justice

The entire history of social science has been shaped by key choices scholars made in that transformative era, choices that are still made today. For criminology more than most disciplines, those Enlightenment choices have had enormous consequences for the use and usefulness of its social science. The most important of these consequences is that justice still remains largely un-Enlightened by empirical evidence about the effects of its actions on public safety and public trust.

Historians may despair at defining a coherent intellectual or philosophical content in the Age of Enlightenment, but one idea seems paramount: "that we understand nature and man best through

Editors' Note: This article is from The ANNALS of the American Academy of Political and Social Science, Vol. 600, No. 1, 115–135 (2005).

the use of our natural faculties" (May 1976, xiv) by systematic empirical methods, rather than through ideology, abstract reasoning, common sense, or claims of divine principles made by competing religious authorities. Kant, in contrast, stressed the receiving end of empirical science in his definition of Enlightenment: the time when human beings regained the courage to "use one's own mind without another's guidance" (Gay 1969, 384).

Rather than becoming *experimental* in method, social science became primarily *analytic*. This distinction between experimental manipulation of some aspect of social behavior versus detached (if systematic) observation of behavioral patterns is crucial to all social science (even though not all questions for social science offer a realistic potential for experiment). The decision to cast social science primarily in the role of critic, rather than of inventor, has had lasting consequences for the enterprise, especially for the credibility of its conclusions. There may be nothing so practical as a good theory, but it is hard to visibly—or convincingly—demonstrate the benefits of social analysis for the reduction of human misery. The absence of "show-and-tell" benefits of analytic social science blurred its boundaries with ideology, philosophy, and even emotion. This problem has plagued analytic social science ever since, with the possible exception of times (like the Progressive Era and the 1960s) when the social order itself was in crisis. As sociologist E. Digby Baltzell (1979) suggested about cities and other social institutions, "as the twig is bent, so grows the tree." Social science may have been forged in the same kind of salon discussions as natural science, but without some kind of empirical reports from factories, clinics, or farm fields. Social science has thus famously "smelled too much of the lamp" of the library (Gay 1969). Even when analytic social science has been most often used, it is rarely praised as useful.

That is not to say that theories (with or without evidence) have lacked influence in criminology, or in any social science. The theory of deterrent effects of sanctions was widely used to reduce the severity of punishment long before the theory could be tested with any evidence. The theories of "anomie" and "differential association" were used to plan the 1960s "War on Poverty" without any clear evidence that opportunity structures could be changed. Psychological theories of personality transformation were used to develop rehabilitation programs in prisons long before any of them were subject to empirical evaluation. Similarly, evidence (without theory) of a high concentration of crime among a small proportion of criminal offenders was used to justify more severe punishment for repeat offenders, also without empirical testing of those policies.

The criminologists' general preference for analysis over experiment has not been universal in social science. Enlightenment political science was, in an important—if revolutionary—sense, experimental, developing and testing new forms of government soon after they were suggested in print. The Federalist Papers, for example, led directly to the "experiment" of the Bill of Rights.

Perhaps the clearest exception to the dominance of analytic social science was within criminology itself in its very first work during the Enlightenment. The fact that criminologists do not remember it this way says more about its subsequent dominance by analytic methods than about the true history of the field. Criminology was born twice in the eighteenth century, first (and forgotten) as an experimental science and then (remembered) as an analytic one. And though experimental criminology in the Enlightenment had an enormous impact on institutions of justice, it was analytic criminology that was preserved by law professors and twentieth-century scholars as the foundation of the field.

The history of criminology thus provides an experimental test of this hypothesis about analytic versus experimental social science: *that social science has been most useful, if not most used, when it has been most experimental, with visibly demonstrable benefits (or harm avoidance) from new inventions.* The evidence for this claim in eighteenth-century criminology is echoed by the facts of criminology in the twentieth century.

In both centuries, the fraternal twins of analysis and experiment pursued different pathways through life, while communicating closely with each other. One twin was critical, the other imaginative; one systematically observational, the other actively experimental; one detached with its integrity intact, the other engaged with its integrity under threat. Both twins needed each other to advance their mutual field of inquiry. But it has been experiments in every age that made criminology most useful, as measured by unbiased estimates of the effects of various responses to crime.

The greatest disappointment across these centuries has been the limited usefulness of experimental criminology in achieving "geometric precision" (Beccaria 1764/1964) in the pursuit of "Enlightened Justice," defined as "the administration of sanctions under criminal law guided by (1) inviolate principles protecting human rights of suspects and convicts while seeking (2) consequences reducing human misery, through means known from (3) unbiased empirical evidence of what works best" (Sherman 2005). While some progress has been made, most justice remains unencumbered by empirical evidence on its effects. To understand why this disappointment persists amid great success, we must begin with the Enlightenment itself.

Inventing Criminology: Fielding, Beccaria, and Bentham

The standard account of the origin of criminology locates it as a branch of moral philosophy: part of an aristocratic crusade against torture, the death penalty, and arbitrary punishment, fought with reason, rhetoric, and analysis. This account is true but incomplete. Criminology's forgotten beginnings preceded Cesare Beccaria's famous 1764 essay in the form of Henry Fielding's 1753 experiments with justice in London. Inventing the modern institutions of a salaried police force and prosecutors, of crime reporting, crime records, employee background investigations, liquor licensing, and social welfare policies as crime prevention strategies, Fielding provided the viable preventive alternatives to the cruel excesses of retribution that Beccaria denounced—before Beccaria ever published a word.

The standard account hails a treatise on "the science of justice" (Gay 1969, 440) that was based on Beccaria's occasional visits to courts and prisons, followed by many discussions in a salon. The present alternative account cites a far less famous treatise based on more than a thousand days of Fielding conducting trials and sentencing convicts in the world's (then) largest city, supplemented by his on-site inspections of tenements, gin joints, brothels, and public hangings. The standard account thus chooses a criminology of analytic detachment over a criminology of clinical engagement.

The standard account in twentieth-century criminology textbooks traced the origin of the field to this "classical school" of criminal law and criminology, with Cesare Beccaria's (1738–1794) treatise *On Crimes and Punishments* (1764) as the first treatise in scientific criminology. (Beccaria is also given credit [incorrectly], even by Enlightenment scholars, for first proposing that utility be measured by "the greatest happiness divided among the greatest number"—which Frances Hutcheson, a mentor to Adam Smith, had published in Glasgow in 1725 before Beccaria was born [Buchan 2003, 68–71].) Beccaria, and later Bentham, contributed the central claims of the deterrence hypothesis on which almost all systems of criminal law now rely: that punishment is more likely to prevent future crime to the extent that it is certain, swift, and proportionate to the offense (Beccaria) or more costly than the benefit derived from the offense (Bentham).

Fielding

This standard account of Beccaria as the *first* criminologist is, on the evidence, simply wrong. Criminology did not begin in a Milanese salon among the group of aristocrats who helped Beccaria formulate and publish his epigrams but more than a decade earlier in a London magistrate's courtroom full of gin-soaked robbery defendants. The first social scientist of crime to publish in the English—and perhaps any—language was Henry Fielding, Esq. (1707–1754).

Fielding was appointed by the government as magistrate at the Bow Street Court in London. His years on that bench, supplemented by his visits to the homes of London labor and London poor, provided him with ample qualitative data for his 1751 treatise titled *An Enquiry Into the Causes of the Late Increase of Robbers.*

Fielding's treatise is a remarkable analysis of what would today be called the "environmental criminology" of robbery. Focused on the reasons for a crime wave and the policy alternatives to hanging as the only means of combating crime, Fielding singles out the wave of "that poison called gin" that hit mid-century London like crack hit New York in the 1980s. He theorizes that a drastic price increase (or tax) would make gin too expensive for most people to consume, thereby reducing violent crime. He also proposes more regulation of gambling, based on his interviews with arrested robbers who said they had to rob to pay their gambling debts. Observing the large numbers of poor and homeless people committing crime, he suggests a wider "safety net" of free housing and food. His emphasis is clearly on prevention without punishment as the best policy approach to crime reduction.

Fielding then goes on to document the failures of punishment in three ways. First, the system of compulsory "voluntary policing" by each citizen imposed after the Norman Conquest had become useless: "what is the business of every man is the business of no man." Second, the contemporary system of requiring crime victims to prosecute their own cases (or hire a lawyer at their own expense) was failing to bring many identified offenders to justice. Third, witnesses were intimidated and often unwilling to provide evidence needed for conviction. All this leads him to hint at, but not spell out, a modern system of "socialized" justice in which the state, rather than crime victims, pays for police to investigate and catch criminals, prosecutors to bring evidence to court, and even support for witnesses and crime victims.

His chance to present his new "invention" to the government came two years after he published his treatise on robbery. In August, 1753, five different robbery-murders were committed in London in one week. An impatient cabinet secretary summoned Fielding twice from his sickbed and asked him to propose a plan for stopping the murders. In four days, Fielding submitted a "grant proposal" for an experiment in policing that would cost £600 (about £70,000 or $140,000 in current value). The purpose of the money was to retain, on salary, the band of detectives Fielding worked with, and to pay a reward to informants who would provide evidence against the murderers.

Within two weeks, the robberies stopped, and for two months not one murder or robbery was reported in Westminster (Fielding 1755/1964, 191–193). Fielding managed to obtain a "no-cost extension" to the grant, which kept the detectives on salary for several years. After Henry's death, his brother John obtained new funding, so that the small team of "Bow Street Runners" stayed in operation until the foundation of the much larger—and uniformed—Metropolitan Police in 1829.

The birth of the Bow Street Runners was a turning point in the English paradigm of justice. The crime wave accompanying the penny-a-quart gin epidemic of the mid-eighteenth century had demonstrated the failure of relying solely on the *severity* of punishment, so excessive that many juries refused to convict people who were clearly guilty of offenses punishable by death—such as shoplifting. As Bentham would later write, there was good reason to think that the *certainty* of punishment was too low for crime to be deterrable. As Fielding said in his treatise on robbery, "The utmost severity to offenders [will not] be justifiable unless we take every possible method of preventing the offence." Fielding was not the only inventor to propose the idea of a salaried police force to patrol and arrest criminals, but he was the first to conduct an *experiment* testing that invention. While Fielding's police experiment would take decades to be judged successful (seventy-six years for the "Bobbies" to be founded at Scotland Yard in 1829), the role of experimental evidence proved central to changing the paradigm of practice.

Beccaria

In sharp contrast, Beccaria had no clinical practice with offenders, nor was he ever asked to stop a crime wave. Instead, he took aim at a wave of torture and execution that characterized European justice. Arguing the same ideology of prevention as Fielding (whose treatise he did not cite), Beccaria urged abolition of torture, the death penalty, and secret trials. Within two centuries, almost all Europe had adopted his proposals. While many other causes of that result can be cited, there is clear evidence of Beccaria's 1764 treatise creating a "tipping point" of public opinion on justice.

What Beccaria did not do, however, was to supply a shred of scientific evidence in support of his theories of the deterrent effects of non-capital penalties proportionate to the severity of the offense. Nor did he state his theories in a clearly falsifiable way, as Fielding had done. In his method, Beccaria varies little from law professors or judges (then and now) who argue a blend of opinion and factual assumptions they find reasonable, deeming it enlightened truth *ipse dixit* ("because I say so myself"). What he lacked by the light of systematic analysis of data, he made up for by eloquence and "stickiness" of his aphorisms. Criminology by slogan may be more readily communicated than criminology by experiment in terms of fame. But it is worth noting that the founding of the British police appears much more directly linked to Fielding's experiments than the steady abolition of the death penalty was linked to Beccaria's book.

Bentham

Beccaria the moral-empirical theorist stands in sharp contrast to his fellow Utilitarian Jeremy Bentham, who devoted twelve years of his life (and some £10,000) to an invention in prison administration. Working from a book he wrote on a "Panopticon" design for punishment by incarceration (rather than hanging), Bentham successfully lobbied for a 1794 law authorizing such a prison to be built. He was later promised a contract to build and manage such a prison, but landed interests opposed his use of the site he had selected. We can classify Bentham as an experimentalist on the grounds that he invested much of his life in "trying" as well as thinking. Even though he did not build the prison he designed, similar prisons (for better or worse) were built in the United States and elsewhere. Prison design may justifiably be classified as a form of invention and experimental criminology, as distinct from the analytic social science approach Bentham used in his writings—thereby making him as "integrated" as Fielding in terms of theory and practice. The demise of Bentham's plans during the Napoleonic Wars marked the end of an era in criminology, just as the Enlightenment itself went into retreat after the French Revolution and the rise of Napoleon. By 1815, experimentalism in criminology was in hibernation, along with most of criminology itself, not to stir until the 1920s or spring fully to life until the 1960s.

Two Torpid Centuries—With Exceptions

Analytic criminology continued to develop slowly even while experimental criminology slumbered deeply, but neither had any demonstrable utility to the societies that fostered them. One major development was the idea of involuntary causes of crime "determined" by either social (Quetelet 1835/2004) or biological (Lombroso 1876/1918) factors that called into question the legal doctrines of criminal responsibility. The empirical evidence for these claims, however, was weak (and in Lombroso's case, wrong), leaving the theoretical approach to criminology largely unused until President Johnson's War on Poverty in the 1960s.

Cambridge-Somerville

The first fully randomized controlled trial in American criminology appears to have been the Cambridge-Somerville experiment, launched in Massachusetts in the 1930s by Dr. Richard Clark Cabot. This project offered high-risk young males "friendly guidance and social support, healthful activities after school, tutoring when necessary, and medical assistance as needed" (McCord 2001). It also included a long-term "big brother" mentoring relationship that was abruptly terminated in most cases during World War II. While the long-term effects of the program would not be known until the 1970s, the critical importance of the experimental design was recognized at the outset. It was for that reason that the outcomes test could reach its startling conclusion: "The results showed that as compared with members of the control group, those who had been in the treatment program were more likely to have been convicted for crimes indexed by the Federal Bureau of Investigation as serious street crimes; they had died an average of five years younger; and they were more likely to have received a medical diagnosis as alcoholic, schizophrenic, or manic-depressive" (McCord 2001, 188). In short, the boys offered the program would have been far better off if they had been "deprived" of the program services in the randomly assigned control group.

No study in the history of criminology has ever demonstrated such clear, unintended, criminogenic effects of a program intended to prevent crime. To this day, it is "exhibit A" in discussions with legislators, students, and others skeptical of the value of evaluating government programs of any sort, let alone crime prevention programs. Its early reports in the 1950s also set the stage for a renaissance in experimental criminology, independently of the growth of analytic criminology.

Renaissance: 1950–1982

Amidst growing concern about juvenile delinquency, the Eisenhower administration provided the first federal funding for research on delinquency prevention. Many of the studies funded in that era, with both federal and nonfederal support, adopted experimental designs. What follows is merely a highlighting of the renaissance of experimental criminology in the long twilight of the FDR coalition prior to the advent of the Reagan revolution.

Martinson and Wilson

While experimental evidence was on the rise in policing, it was on the decline in corrections. The comprehensive review of rehabilitation strategies undertaken by Lipton, Martinson, and Wilks (1975) initially focused on the internal validity of the research designs in rehabilitation experiments within prisons. Concluding that these designs were too weak to offer unbiased estimates of treatment effects, the authors essentially said "we don't know" what works to rehabilitate criminals. In a series of less scientific and more popular publications, the summary of the study was transformed into saying that there is no evidence that criminals can be rehabilitated. Even the title "What Works" was widely repeated in 1975 by word of mouth as "nothing works."

The Martinson review soon became the basis for a major change in correctional policies. While the per capita rates of incarceration had been dropping throughout the 1960s and early 1970s, the trend was rapidly reversed after 1975 (Ruth and Reitz 2003). Coinciding with the publication of Wilson's (1975) first edition of *Thinking About Crime*, the Martinson review arguably helped fuel a sea change from treating criminals as victims of society to treating society as the victim of criminals. That, in turn, may have helped to feed a three-decade increase in prisoners (Laub 2004) to more than 2.2 million, the highest incarceration rate in the world.

Warp Speed: 1982–2005

Stewart

In September, 1982, a former Oakland Police captain named James K. Stewart was appointed director of the National Institute of Justice (NIJ). Formerly a White House Fellow who had attended a National Academy of Sciences discussion of the work of NIJ, Stewart had been convinced by James Q. Wilson and others that NIJ needed to invest more of its budget in experimental criminology. He acted immediately by canceling existing plans to award many research grants for analytic criminology, transferring the funds to support experimental work. This work included experiments in policing, probation, drug market disruption, drunk-driving sentences, investigative practices, and shoplifting arrests.

Schools

The 1980s also witnessed the expansion of experimental criminology into the many school-based prevention programs. Extensive experimental and quasi-experimental evidence on their effects—good and bad—has now been published. In one test, for example, a popular peer guidance group that was found effective as an alternative to incarceration was found to increase crime in a high school setting. Gottfredson (1987) found that high-risk students who were not grouped with other high-risk students in high school group discussions did better than those who were.

Drug Courts

The advent of (diversion from prosecution to medically supervised treatments administered by) "drug courts" during the rapid increase in experimental criminology has led to a large and growing volume of tests of drug court effects on recidivism. Perhaps no other innovation in criminal justice has had so many controlled field tests conducted by so many different independent researchers. The compilations of these findings into meta-analyses will shed increasing light on the questions of when, and how, to divert drug-abusing offenders from prison.

Boot Camps

Much the same can be said about boot camps. The major difference is that boot camp evaluations started off as primarily quasi-experimental in their designs (with matched comparisons or worse), but increasing numbers of fully randomized tests have been conducted in recent years (Mitchell, MacKenzie, and Perez 2005). Many states persist in using boot camps for thousands of offenders, despite fairly consistent evidence that they are no more effective than regular correctional programs.

Child Raising

Criminology has also claimed a major experiment in child raising as one of its own. Beginning at the start of the "warp speed" era, the program of nurse home visits to at-risk first mothers designed by Dr. David Olds and his colleagues (1986) has now been found to have long-term crime prevention effects. Both mothers and children show these effects, which may be linked to lower levels of child abuse or better anger management practices in child raising.

Criminology: Analytic, Useful, and Used

This recitation of a selected list of experiments in criminology must be labeled with a consumer warning: *the vast majority of published criminology remains analytic and nonexperimental.* While criminology was attracting funding and students during the period of rising crime of the 1960s to 1990s, criminologists put most of their efforts into the basic science of crime patterns and theories of criminality. Studies of the natural life course of crime among cohorts of males became the central focus of the field, as measured by citation patterns (Wolfgang, Figlio, and Thornberry 1978). Despite standing concerns that criminology would be "captured" by governments to become a tool for developing repressive policies, the evidence suggests that the greatest (or largest) generation of criminologists in history captured the field away from policymakers.

The renaissance in experimental criminology therefore addressed very intense debates over many key issues in crime and justice, providing the first unbiased empirical guidance available to inform those debates. That much made criminology increasingly useful, at least potentially. Usefulness alone, of course, does not guarantee that the information will be *used.* Police agencies today do make extensive use of the research on concentrating patrols in crime hot spots, yet they have few repeat offender units, despite two successful tests of the "invention." Correctional agencies make increasing use of the "what works" literature in the United States and United Kingdom, yet prison populations are still fed by people returned to prison on the unevaluated policy of incarcerating "technical" violators of the conditions of their release (who have not committed new crimes). Good evidence alone is not enough to change policy in any context. Yet absent good evidence, there is a far greater danger that bad policies will win out. Analytic criminology—well or badly done—poses fewer risks for society than badly done experimental criminology. It is not clear that another descriptive test of differential association theory will have any effect on policy making, unless it is embedded in a program evaluation. But misleading or biased evidence from poor-quality research designs—or even unreplicated experiments—may well cause the adoption of policies that ultimately prove harmful.

This danger is, in turn, reduced by the lack of influence criminology usually has on policy making or operational decisions. That, in turn, is linked to the absence of clear conclusions about the vast majority of criminal justice policies and decisions. Until experimental criminology can develop a more comprehensive basis of evidence for guiding operations, practitioners are unlikely to develop the habit of checking the literature before making a decision. The possibility of improving the quality of both primary evidence and systematic reviews offers hope for a future in which criminology itself may entail less risk of causing harm.

This is by no means a suggestion that analytic criminology be abandoned; the strength of experimental criminology may depend heavily on the continued advancement of basic (analytic) criminology. Yet the full partnership between the two has yet to be realized. Analytic social science still dominates field experiments by 100 to 1 or better in criminology, just as in any other field of research on human behavior. Future success of the field may depend upon a growing public image based on experimental results, just as advances in treatment attract funding for basic science in medicine.

Conclusion

Theoretical criminology will hold center stage for many years to come. But as Farrington (2000) has argued, the advancement of theory may depend on better experimental evidence. And that, in turn, may depend on a revival in the federal funding that has recently dropped to its lowest level in four decades. Such a revival may well depend on exciting public interest in the practical value of research, as perhaps only experiments can do.

"Show and tell" is hard to do while it is happening. Yet it is not impossible. Whether anyone ever sees a crime prevention program delivered, it is at least possible to embed an experimental design into every long-term analytic study of crime in the life course. As Joan McCord (2003) said in her final words to the American Society of Criminology, the era of purely observational criminology should come to an end. Given what we now know about the basic life-course patterns, McCord suggested, "all longitudinal studies should now have experiments embedded within them."

Doing what McCord proposed would become an experiment *in* social science as well as *of* social science. That experiment is already under way, in a larger sense. Criminology is rapidly becoming more multi-method, as well as multi-level and multi-theoretical. Criminology may soon resemble medicine more than economics, with analysts closely integrated with clinical researchers to develop basic science as well as treatment. The integration of diverse forms and levels of knowledge in "consilience" with each other, rather than a hegemony of any one approach, is within our grasp. It awaits only a generation of broadly educated criminologists prepared to do many things, or at least prepared to work in collaboration with other people who bring diverse talents to science.

References

Baltzell, D. (1979). *Puritan Boston and Quaker Philadelphia: Two protestant ethics and the spirit of class authority and leadership.* New York: Free Press.

Beccaria, C. (1964). *On crimes and punishments* (J. Grigson, Trans.). Milan, Italy: Oxford University Press. (Original work published 1764)

Buchan, J. (2003). *Crowded with genius: The Scottish Enlightenment: Edinburgh's moment of the mind.* New York: HarperCollins.

Farrington, D. (2000). Explaining and preventing crime: The globalization of knowledge. The American Society of Criminology 1999 Presidential Address. *Criminology, 38,* 1–24.

Fielding, H. (1964). *The journal of a voyage to Lisbon.* London: Dent. (Original work published 1755)

Gay, P. (1969). *The Enlightenment: An interpretation. Vol. 2, The science of freedom.* New York: Knopf.

Gottfredson, G. (1987). Peer group interventions to reduce the risk of delinquent behavior: A selective review and a new evaluation. *Criminology, 25,* 671–714.

Laub, J. (2004). The life course of criminology in the United States: The American Society of Criminology 2003 Presidential Address. *Criminology, 42,* 1–26.

Lipton, D., Martinson, R., & Wilks, J. (1975). *The effectiveness of correctional treatment: A survey of treatment evaluation studies.* New York: Praeger.

Lombroso, C. (1918). *Crime, its causes and remedies.* Boston: Little, Brown. (Original work published 1876)

May, H. (1976). *The Enlightenment in America.* New York: Oxford University Press.

McCord, J. (2001). *Crime prevention: A cautionary tale.* Proceedings of the Third International, Inter-Disciplinary Evidence-Based Policies and Indicator Systems Conference, University of Durham. Retrieved April 22, 2005, from http://cem.dur.ac.uk

McCord, J. (2003). *Discussing age, crime, and human development. The future of life-course criminology.* Denver, CO: American Society of Criminology.

Mitchell, O., MacKenzie, D., & Perez, D. (2005). A randomized evaluation of the Maryland correctional boot camp for adults: Effects on offender anti-social attitudes and cognitions. *Journal of Offender Rehabilitation, 40*(4).

Olds, D., Henderson, C, Chamberlin, R., & Tatelbaum, R. (1986). *Pediatrics, 78,* 65–78.

Quetelet, A. (2004). A treatise on man. As cited in F. Adler, G. O. W. Mueller, and W. S. Laufer, *Criminology and the criminal justice system* (5th ed., p. N-6). New York: McGraw-Hill. (Original work published 1835)

Ruth, H., & Reitz, K. (2003). *The challenge of crime: Rethinking our response.* Cambridge, MA: Harvard University Press.

Sherman, L. (2005). *Enlightened justice: Consequentialism and empiricism from Beccaria to Braithwaite.* Address to the 14th World Congress of Criminology, International Society of Criminology, Philadelphia, August 8.

Wilson, J. (1975). *Thinking about crime.* New York: Basic Books.

Wolfgang, M., Figlio, R., & Thornberry, T. (1978). *Evaluating criminology.* New York: Elsevier.

DISCUSSION QUESTIONS

1. What is the main point of Sherman's discussion of analytic versus experimental criminology?
2. Why should Fielding, rather than Beccaria, be considered the father of criminology, according to Sherman?
3. Sherman wants an integration of criminology with other sciences and wants criminology to be multi-method, multi-level, and multi-theoretical. What might be the ideological barriers to such integration?

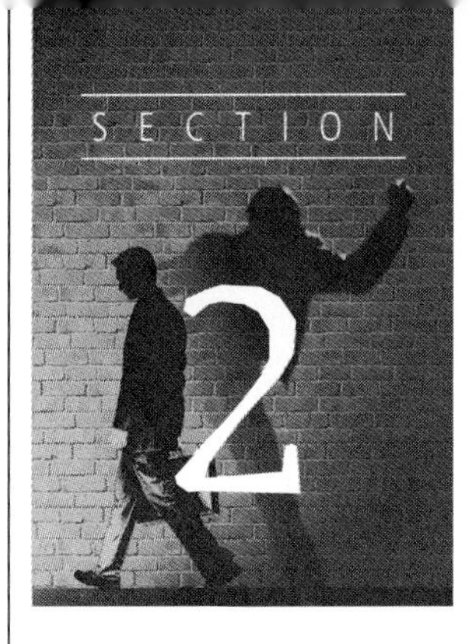

MEASURING CRIME AND CRIMINAL BEHAVIOR

A weary English bobby (a popular nickname for British police officers) patrolling his foot beat on a chilly November night hears the unmistakable sounds of sexual activity from the dark entranceway of a closed greengrocer's shop. He smiles to himself and tiptoes toward the sound. When he reaches the entranceway he switches on his flashlight and booms out the favorite line of the stereotypical bobby: "What's goin' on 'ere then?" The squeaking couple immediately come to attention and adjust their dress before the young man—obviously still in a state of arousal—stammers, "Why, nothing, constable." The officer recognizes the woman as a local "slapper" (prostitute) and he vaguely recognizes the man (more of a boy of around 17, really) as a local supermarket worker. The constable reasons that he should arrest both parties for public indecency but that would entail about an hour of paperwork (an hour in the warm police station with a nice cup of tea sounds good, though) and lead to the profound embarrassment of the poor boy. He finally decides to give some sound advice about sexually transmitted diseases to the boy and a stern warning to the woman and sends them both on their way.

This short story illustrates that official statistics are measuring police behavior as much as they are measuring crime. Sir Josiah Stamp, director of the Bank of England in the 1920s, cynically stated this criticism: "The government are very keen on amassing statistics. They collect them, raise them to the nth power, take the cube root and prepare wonderful diagrams. But you must never forget that every one of these figures comes in the first instance from the village watchman, who just puts down what he damn pleases" (in Nettler, 1984, p. 39). We don't recommend this kind of cynicism, but we do counsel that you keep a healthy skepticism about statistics as you read this chapter.

Categorizing and Measuring Crime and Criminal Behavior

When attempting to understand, predict, and control any social problem, including the crime problem, the first step is to determine its extent. Gauging the extent of the problem means discovering how much of it there is, where and when it occurs most often, and among what social categories it occurs most frequently. It also helps our endeavors if we have knowledge of the patterns and trends of the problem over time. Note that we did not address "why" questions (why does crime occur; why is it increasing/decreasing, who commits it and why, and so on); such questions can only be adequately addressed after we have reliable data about the extent of the problem. However, all social statistics are suspect to some extent, and crime statistics are perhaps the most suspect of all. They have been collected from many different sources in many different ways and have passed through many sieves of judgment before being recorded.

There is a wide variety of data provided by government and private sources to help us to come to grips with America's crime problem, all with their particular strengths and weaknesses. The major data sources can be grouped into three categories: official statistics, victimization survey data, and self-reported data. Official statistics are those derived from the routine functioning of the criminal justice system. The most basic category of official statistics comes from the calls made to police by victims or witnesses and the crimes that the police discover on patrol. Other major categories of official crime data consist of information about arrests, about convictions, and about correctional (prison, probation/parole) populations.

The Uniform Crime Reports: Counting Crime Officially

The primary source of official crime statistics in the United States is the annual **Uniform Crime Reports (UCR)** compiled by the Federal Bureau of Investigation (FBI). The UCR reports crimes known to the nation's police and sheriff's departments and the number of arrests made by these agencies; federal crimes are not included. Offenses known to the police are recorded whether or not an arrest is made or an arrested person is subsequently prosecuted and convicted. Participation in the UCR reporting program is voluntary, and thus not all agencies participate. This is unfortunate for anyone hoping for comprehensive crime data. In 2005, law enforcement agencies participating in the program represented approximately 281 million United States residents, or about 94% of the population (FBI, 2006). This means that crimes committed by about 6% of the American population (about 17 million people) were not included in the UCR data.

The UCR reports the number of each crime reported to the police as well as their rate of occurrence. The rate of a given crime is the actual number of reported crimes standardized by some unit of the population. We expect the raw number of crimes to increase as the population increases, so comparing the number of crimes reported today with the number reported 30 years ago, or the number of crimes reported in New York with the number reported in Wyoming, tells us little unless we consider population differences. For instance, California

▲ **Photo 2.1** The annual Uniform Crime Reports are produced by the FBI. To the J. Edgar Hoover Building in Washington, D.C., local, county, and state criminal justice agencies send their annual crime data. UCR data are, by their nature, incomplete, as many crimes are never reported to the police at all. This "dark figure of crime" might be as high as 50% of all crime incidents.

reported 2,407 murders to the FBI in 2003, and Louisiana reported 586. These figures don't provide an accurate image of the comparative murder picture in these states unless we take their respective populations into consideration. To obtain a **crime rate**, we divide the number of reported crimes in a state by its population and multiply the quotient by 100,000, as in the following comparison of California and Louisiana rates:

$$\text{California Murder Rate: } (2{,}407 \div 35{,}484{,}453) \times 100{,}000 = 6.8$$

$$\text{Louisiana Murder Rate: } (586 \div 4{,}490{,}334) \times 100{,}000 = 13.0$$

Thus, a person in Louisiana is almost at twice the risk of being murdered compared to a person in California. This statement is based on statewide averages; the actual risk will vary widely from person to person based on such factors as age, race, sex, socioeconomic status (SES), and place of residence.

The UCR separates crimes into two categories: **Part I offenses** (or **Index Crimes**), and **Part II offenses**. Part I offenses include four violent (homicide, aggravated assault, forcible rape, and robbery) and four property offenses (larceny/theft, burglary, motor vehicle theft, and arson). Notice that these are all universally condemned mala in se offenses. Part I offenses correspond with what most people think of as "serious" crime. Part II offenses are treated as less serious offenses and are recorded based on arrests made, rather than cases reported to the police. Part II offense figures understate the extent of criminal offending far more than is the case with Part I figures because only a very small proportion of these crimes result in arrest.

Table 2.1 is a page from the 2005 UCR listing all Part I and II crimes broken down by sex and age for the years 2000 and 2004.

Table 2.1 Estimated Number of Arrests for Part I and Part II Crimes by Sex and Age in 2000 and 2004

	Total			Males			Total			Females		
Offense charged	**2000**	**2004**	**Percent change**	**[illegible]00**	**2004**	**Percent change**	**2000**	**2004**	**Percent change**	**2000**	**2004**	**Percent change**
TOTAL[1]	6,345,009	6,185,599	−2.5	1,0[illegible]1,721	893,547	−11.7	1,822,827	1,935,212	+6.2	386,462	377,182	−2.4
Murder and nonnegligent manslaughter	6,655	6,568	−1.3	572	546	−4.5	842	860	+2.1	79	51	−35.4
Forcible rape	16,256	15,264	−6.1	[illegible],687	2,402	−10.6	194	214	+10.3	26	43	+65.4
Robbery	54,937	55,713	+1.4	1[illegible],789	13,185	−4.4	6,164	6,894	+11.8	1,438	1,353	−5.9
Aggravated assault	221,440	209,180	−5.5	[illegible],998	28,162	−6.1	54,852	54,004	−1.5	8,992	8,677	−3.5
Burglary	149,363	152,527	+2.1	[illegible],288	43,509	−15.2	22,843	26,343	+15.3	6,650	5,902	−11.2
Larceny-theft	453,659	432,974	−4.6	1[illegible],354	113,048	−20.6	256,556	269,413	+5.0	83,723	81,004	−3.2
Motor vehicle theft	68,765	69,452	+1.0	[illegible]211	18,352	−20.9	12,964	14,290	+10.2	4,760	3,777	−20.7
Arson	8,440	7,811	−7.5	[illegible]680	4,123	−11.9	1,449	1,442	−0.5	605	622	+2.8
Violent crime[1]	299,288	286,725	−4.2	4[illegible]046	44,295	−5.8	62,052	61,972	−0.1	10,535	10,124	−3.9
Property crime[2]	680,228	662,764	−2.6	22[illegible]533	179,032	−19.2	293,812	311,488	+6.0	95,738	91,305	−4.6
Other assaults	596,709	575,276	−3.6	97[illegible]400	100,699	+3.4	180,093	190,397	+5.7	43,021	49,988	+16.2
Forgery and counterfeiting	39,848	41,893	+5.1	[illegible]524	1,774	−32.4	25,687	27,883	+8.5	1,335	949	−28.9
Fraud	115,825	102,342	−11.6	[illegible]97	2,803	−33.2	98,865	87,840	−11.2	2,015	1,610	−20.1
Embezzlement	6,173	5,786	−6.3	[illegible]60	429	−35.0	6,177	5,897	−4.5	617	257	−58.3
Stolen property; buying, receiving, possessing	59,328	62,201	+4.8	14,[illegible]51	11,556	−18.9	12,508	14,600	+16.7	2,641	2,258	−14.5

Offense charged	Total 2000	Total 2004	Total Percent change	Males 2000	Males 2004	Males Percent change	Total 2000	Total 2004	Total Percent change	Females 2000	Females 2004	Females Percent change
Vandalism	140,960	132,323	−6.1	60,138	52,484	−12.7	25,651	26,867	+4.7	8,488	8,720	+2.7
Weapons; carrying, possessing, etc.	85,855	94,323	+9.9	19,383	21,156	+9.1	7,434	8,299	+11.6	2,150	2,721	+26.6
Prostitution and commercialized vice	17,448	13,197	−24.4	294	281	−4.4	28,561	31,586	+10.6	323	610	+88.9
Sex offenses (except forcible rape and prostitution)	48,738	47,721	−2.1	9,418	9,153	−2.8	3,522	3,444	−2.2	668	662	−0.9
Drug abuse violations	754,883	811,879	+7.6	101,885	92,553	−9.2	161,597	190,667	+18.0	17,909	19,710	+10.1
Gambling	5,320	5,745	+8.0	820	1,046	+27.6	657	611	−7.0	31	35	+12.9
Offenses against the family and children	65,253	56,050	−14.1	3,193	2,161	−32.3	18,717	17,500	−6.5	1,823	1,329	−27.1
Driving under the influence	704,473	655,849	−6.9	10,380	8,866	−14.6	138,345	150,005	+8.4	2,115	2,417	+14.3
Liquor laws	308,870	260,310	−15.7	65,951	48,720	−26.1	93,917	89,966	−4.2	30,254	26,410	−12.7
Drunkenness	356,057	302,896	−14.9	11 128	8,227	−26.1	54,762	53,548	−2.2	2,700	2,469	−8.6
Disorderly conduct	298,583	266,694	−10.7	72,806	73,238	+0.6	88,404	91,876	+3.9	28,558	35,112	+22.9
Vagrancy	11,984	13,636	+13.8	1,189	2,204	+85.4	3,364	3,426	+1.8	338	924	+173.4
All other offenses (except traffic)	1,658,025	1,709,488	+3.1	176,264	154,369	−12.4	444,471	505,447	+13.7	60,972	57,679	−5.4
Suspicion	1,952	1,080	−44.7	529	254	−52.0	463	302	−34.8	167	91	−45.5
Curfew and loitering law violations	56,414	50,151	−11.1	56,414	50,151	−11.1	24,806	21,346	−13.9	24,806	21,346	−13.9
Runaways	34,747	28,350	−18.4	34,747	28,350	−18.4	49,425	40,547	−18.0	49,425	40,547	−18.0

SOURCE: Federal Bureau of Investigation (2005).

1. Does not include suspicion.

2. Violent crimes are offenses of murder, forcible rape, robbery, and aggravated assault. Property crimes are offenses of burglary, larceny-theft, motor vehicle theft, and arson.

Cleared Offenses

If a person is arrested and charged for a Part I offense, the UCR records the crime as **cleared** ***by arrest.*** A crime may also be cleared by *exceptional means* when the police have identified a suspect and have enough evidence to support arrest but the suspect could not be taken into custody immediately, or at all. Such circumstances exist when the suspect dies or is in a location where the police cannot presently gain custody. For instance, he or she is in custody on other charges in another jurisdiction or is residing in a country with no extradition treaty with the United States. As can be seen in Figure 2.1, violent crimes are more likely to be cleared than property crimes because violent crime investigations are pursued more vigorously and because victims of such crimes may be able to identify the perpetrator(s).

Figure 2.1 Percentage of Crimes Cleared by Arrest or Exceptional Means in 2005

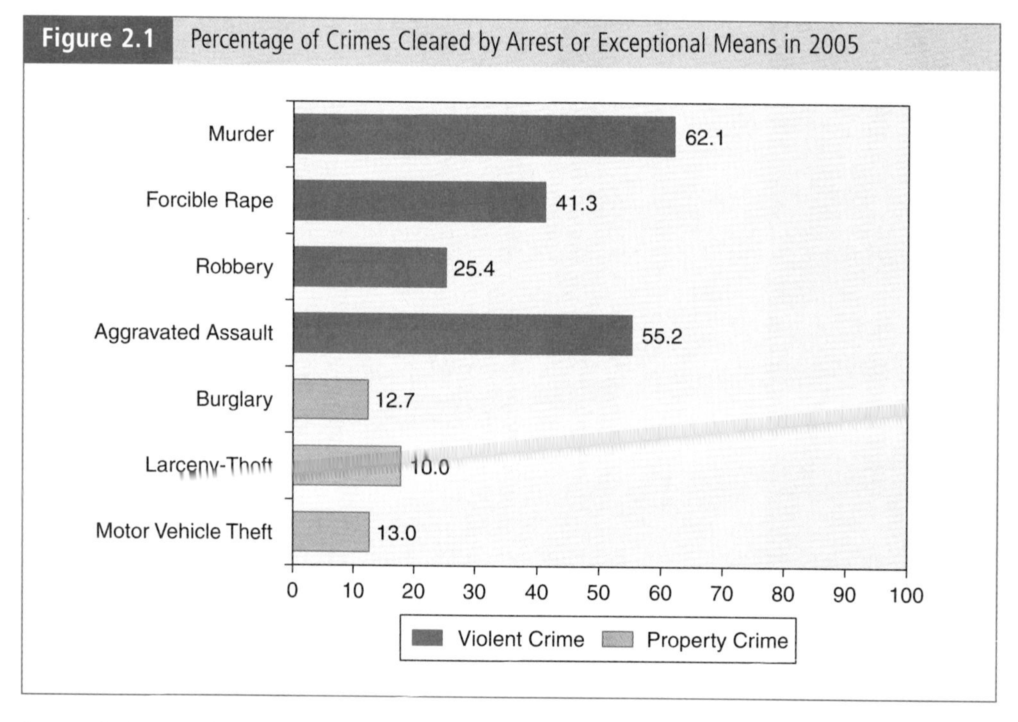

SOURCE: Federal Bureau of Investigation (2006).

Problems With the UCR

UCR data have serious limitations that restrict their usefulness for criminological research, particularly research seeking to uncover causes of crime. Some of the more serious of these limitations are outlined below.

- The UCR data significantly underrepresents the actual number of criminal events in the United States each year. According to a nationwide victim survey, only 47% of victims of violent crime and 40% of victims of property crime indicated that they reported their victimization to the police (Catalano, 2006). Victims are more likely to report violent crimes if injuries are serious and are more likely to report property crimes when losses are high. Females (54.6%) are more likely than males (47%) to report violent victimization; males and females are about equally as likely (39%) to report property victimization.

- Federal crimes, such as highly costly white-collar crimes like stock market fraud, hazardous waste dumping, tax evasion, and false claims for professional services, are not included.

- Crimes committed in the jurisdictions of nonparticipating law enforcement agencies are not included in the data. Even with full voluntary compliance, all departments would not be equally as efficient and thorough (or honest) in their record keeping.

- Crime data may be falsified by police departments for political reasons. The National Center for Policy Analysis (1998b) reports that police departments in Philadelphia, New York, Atlanta, and Boca Raton, Florida had underreported and/or downgraded crimes in their localities (and these are just the departments we know about).

- Because of the FBI's hierarchy rule, the UCR even underreports crimes that are known to the police. The **hierarchy rule** requires police to report only the highest (most serious) offense committed in a multiple offense-single incident to the FBI and to ignore the others. For instance, if a man robs five patrons in a bar, pistol whips one patron who tried to resist, locks the victims in the beer cooler, and then rapes the female bartender, only the rape is reported to the FBI.

NIBRS: The "New and Improved" UCR

Efforts to improve the reliability and validity of official statistics are occurring all the time, with the most ambitious being the **National Incident-Based Reporting System** (NIBRS). NIBRS began in 1982 and is designed for the collection of more detailed and more comprehensive crime statistics than is the UCR (which it is supposed to replace). As opposed to the current UCR, which monitors only a relatively few crimes and gathers few details associated with them, NIBRS collects data on 46 "Group A" offenses and 11 "Group B" offenses. There is no hierarchy rule under the NIBRS system; it reports multiple victims, multiple offenders, and multiple crimes that may be part of the same incident. It also provides information about the circumstances of the offense and about victim and offender characteristics, such as offender/victim relationship, age, sex, and race of victims and perpetrators (if known). Unfortunately, only 19 states and three cities with populations greater than 500,000 (Austin, Memphis, and Nashville) were reporting crime incidents to NIBRS as of 2004 (Finkelhor & Ormrod, 2004). Many police departments lack the manpower and technical expertise to collect and process the wide and detailed range of information that is part of each crime incident their officers deal with, and administrators see little benefit to their department to justify the effort (Dunworth, 2001).

Because NIBRS data provides information about the offender and the victim (victims can identify physical characteristics of perpetrators), it can be used to try to resolve certain criminological issues. One issue is the disproportionately high rate of arrest for blacks in the United States. Is this the result of disproportionately high involvement in crime or the result of discriminatory arrest patterns of police? This issue was explored by Stewart D'Alessio and

▲ **Photo 2.2** The use of technology by police has been credited in part for crime reduction in the 1990s.

Lisa Stolzenberg in the article presented in this section. Using data on 335,619 incidents in 1999 from 17 participating states, they found that whites were significantly more likely to be arrested than blacks for all reported violent crimes except rape, in which case there was no significant difference. D'Alessio and Stolzenberg conclude that the disproportionately high rate of arrest rate for blacks in the United States is the result of disproportionately high involvement of blacks in crime (at least in violent crime).

Crime Victimization Survey Data and Their Problems

Crime victimization surveys involve asking large numbers of people if they have been criminally victimized within some specified time frame, regardless of whether they reported the incident to police. Census Bureau personnel interview a national representative sample of people aged 12 or older on behalf of the Bureau of Justice Statistics (BJS) twice each year. This survey is known as the **National Crime Victimization Survey** (NCVS), and in 2005, 134,000 people from 77,200 households were interviewed (Catalano, 2006). The NCVS requests information on crimes committed against individuals and households, the circumstances of the offense, and personal information about victims (age, sex, race, income, and education level) and offenders (approximate age, sex, race, and victim/offender relationship). Figure 2.2 presents highlights from the 2006 NCVS report.

Victimization surveys have their own dark figures as well as other problems that make them almost as suspect as the UCR. These problems include the following:

- Crimes such as drug dealing and all "victimless" crimes such as prostitution and gambling are not revealed in the surveys, for obvious reasons. And because murder victims cannot be interviewed, this most serious of crimes is not included.

Figure 2.2 Highlights From the 2005 NCVS Survey Data

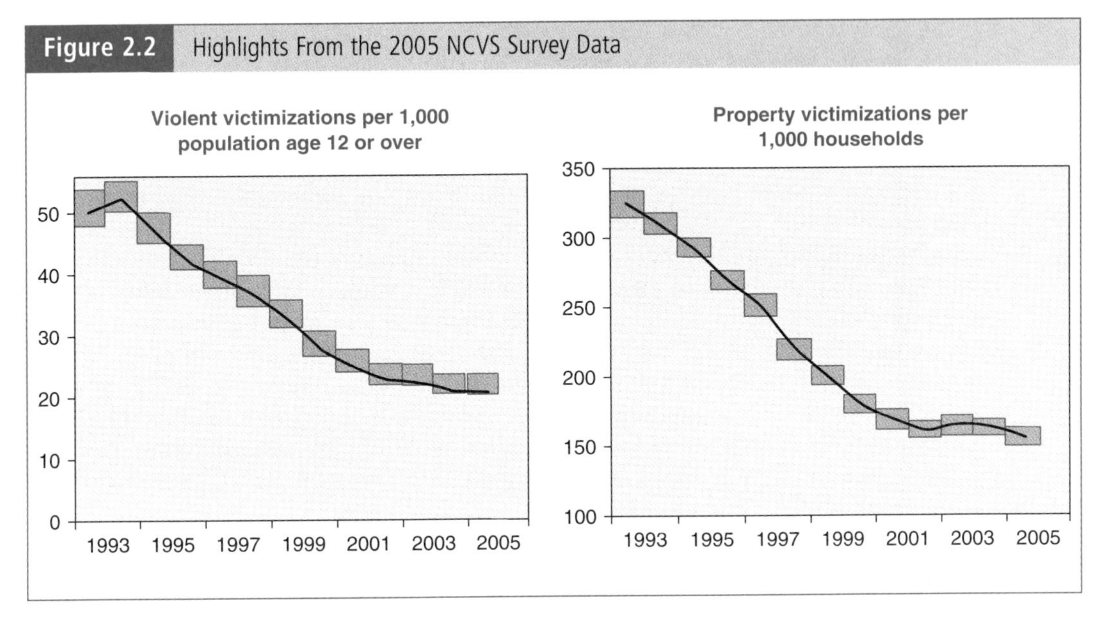

SOURCE: Catalano (2006).

- Because NCVS only surveys households, crimes committed against commercial establishments such as stores, bars, and factories are not included. This exclusion results in a huge underestimate of crimes such as burglaries, robberies, theft, and vandalism.

- Victimization data do not have to meet any stringent legal or evidentiary standards in order to be reported as an offense; if the respondent says he or she was robbed, a robbery will be recorded. UCR data, on the other hand, passes through the legal sieve to determine whether the reported incident was, indeed, a robbery.

- Other problems include memory lapses; providing answers the respondent thinks the interviewer wants to hear; forgetting an incident; embellishing an incident; and any number of other misunderstandings, ambiguities, and even downright lies, that occur when one person is asking another about his or her life experiences.

- Consistent with the above, there are suggestions that just as underreporting plagues UCR data, overreporting may plague NCVS data (O'Brien, 2001). Whatever the case may be, we find many anomalies when comparing the two sources of data. For instance, the 2006 UCR reports 93,934 cases of rape versus the 2006 NCVS's report of 191,670. The problem is that only 38.3% (73,410) of the NCVS rape victims said they reported it to the police, and that number is 20,524 *fewer* victims than were "known to the police" that year. The same situation exists for other crimes; that is, substantially more crimes appear in police records than NCVS victims claim to have reported to the police. The discrepancy is easily explained for burglary and motor vehicle theft because the NCVS does not include commercial establishments in its reports. It is more difficult to explain the violent crime discrepancy, however. One explanation for this is that the NCVS does not include victims less than 12 years of age, whereas the UCR does, although it is difficult to believe that children under 12 account for 15% to 20% of all rapes known to the police.

How is it that the NCVS can report 191,670 rapes when it only interviewed 134,000 people? As in any statistical analysis, sample results such as these are generalized to the population. If the sample reveals a rape rate of 0.66 per 1,000 individuals over age 12, for instance, the NCVS will report a rate of 66 per 100,000 for the United States. Assuming a population of 290 million, this will extrapolate to approximately the number of rapes reported by the NCVS. It is perfectly acceptable to make inferences from samples to populations when samples are truly representative of the population.

NCVS researchers are aware of the many problems that arise when asking people to recall victimization and have initiated many interview improvements in their methodology, one of which is the *bounding interview.* This technique involves comparing reported incidents from the same household in the current interview with those reported six months prior. When a report appears to be a duplicate, the respondent is reminded of the earlier report and asked whether the new report represents the incident previously mentioned or whether it is different. Other techniques used to minimize some of the reported problems mentioned above are available on the NCVS Web site at http://www.icpsr.umich.edu/NACJD/NCVS/.

Areas of Agreement Between the UCR and NCVS

The UCR and NCVS agree on the demographics of crime in that they both tell us that males, the young, the poor, and African Americans are more likely to be perpetrators and victims of crime than are females, older persons, wealthier persons, and persons of other races. Both sources also agree as to the geographic areas and times of the year and month when crimes are more likely to occur. Over a three year period, O'Brien (2001) found that NCVS victims reported that 91.5% of those who robbed them and 87.7% of their aggravated assault assailants were male, as were 91.2% and 84.3%, respectively of those arrested for those offenses. Likewise, NCVS victims reported that 64.1% of those who robbed them and 40% of their aggravated assault assailants were African American. These percentages fit the UCR arrest statistics for race almost exactly; 62.2% arrested for robbery were African American, as were 40% of those arrested for aggravated assault.

Comparisons of UCR and NCVS data have often proven very useful to resolve issues such as these. The article by Darrell Steffensmeir and his colleagues in this section uses a comparison of data trends reported in the UCR and the NCVS from 1980 to 2003 to explore gender ratios in violent crime. The issue for them is whether the gap between males and females is closing with regard to violent crime. They found that both data sources found little or no changes in the gender ratio for violent crimes such as murder and rape, but that the UCR reports indicated a sharp rise in assaults. Does this mean that women became more violent over the period examined, or does the increase reflect the behavior of the police more than the behavior of women? NCVS data do not bear out UCR trends, leaving Steffensmeir and his colleagues to conclude that the increase is due to police net-widening policies that mandate arrest for marginally serious offenses, and that there has been no actual increase in female violence.

Another area of broad agreement is recent crime trends. The 2005 NCVS (Catalano, 2006) reports that victimization rates for violent crimes declined 58% from 1993 to 2005, and that property crime victimization fell 50%. The UCR violent crime rate fell from 746 per 100,000 in 1993 to 469.2 per 100,000 in 2005 (FBI, 2006) for a decrease of 37.1%, and the UCR property crime rate fell from 4,737 to 3,429.8, a decrease of 27.6%.

Self-Reported Crime Surveys and Their Problems

Self-report surveys of offending provide a way for criminologists to collect data without having to rely on government sources. Questionnaires used in these surveys typically provide a list of offenses and ask subjects to check each offense they recall having committed and how often, and sometimes whether they have ever been arrested, and if so, how many times. Self-report surveys have relied primarily on college and high school students for subjects, although prison inmates and probationers/parolees have also been surveyed.

Several studies have addressed the issue of the accuracy and honesty of self-reported offenses in various ways, and the results have generally been encouraging, at least for uncovering the extent of minor offenses. On average, known delinquents and criminals disclose almost four times as many offenses as the non-delinquents. Had these differences not been found, the validity of the self-report procedure would have been in doubt.

The greatest strength of self-report research is that researchers can correlate admitted offenses with a variety of characteristics of respondents that go beyond the demographics of age, race, and gender. For instance, they can attempt to measure various constructs thought to be associated with offending, such as impulsiveness, lack of empathy, and sensation seeking, as well as their peer associations and their attitudes. The evidence indicates that self-report crime measures provide largely accurate information about some illegal acts sometime in their lives. However, there are a number of reasons why self-report crime surveys also provide a distorted picture of criminal involvement.

- The great majority of self-report studies survey "convenience" samples of high school and college students, populations in which we don't expect to find many seriously criminally involved individuals. Most self-report studies thus eliminate the very people we are most interested in gathering information about. One strength of the self-report method, however, is that it appears to capture the extent of illegal drug usage among high school and college students, something that neither the UCR nor the NCVS attempt to do.

- Self-report studies typically uncover only fairly trivial antisocial acts such as fighting, stealing items worth less than $5, smoking, and truancy. Almost everyone has committed one or more of these acts. These are hardly acts that help us to understand the nature of serious crime. A connected problem is that some researchers lump respondents who report one delinquent act together with adjudicated delinquents who break the law in many different ways many different times.

- Even though most people are forthright in revealing their peccadilloes, most people do not have a serious criminal history, and those who do have a distinct tendency to underreport their crimes (Hindelang, Hirschi, & Weis, 1981). As the number of crimes people commit increases, so does the proportion of offenses they withhold reporting, with those arrested for the most serious offenses having the greatest probability of denial (Farrington, 1982).

- Males tend to report their antisocial activities less honestly than females and African Americans less honestly than other racial groups (Cernkovich, Giordano, & Rudolph, 2000; Kim, Fendrich, & Wislar, 2000). This evidence renders suspect any statements about gender or racial differences in antisocial behavior that are based on self-report data. When it comes to relying on self-report data to assess the nature and extent of serious crime, it is well to remember the gambler's dictum: "Never trust an animal that talks."

The final article in this section illustrates the self-report method of data gathering. Jerome Cartier, David Farabee, and Michael Prendergast use the method to explore the link between methamphetamine use and self-reported crime and recidivism among 614 California parolees. They found that methamphetamine use was significantly related to self-reported violent crime and to recidivism. The authors claim a high level of agreement between self-reported crime and actual crimes committed.

The Dark Figure of Crime

The **dark (or hidden) figure of crime** is that portion of the total crimes committed each year that never comes to light. Figure 2.3 presents three diagrams that show the different dark figures for the three major measures of criminal behavior. (The dark figures are represented by the dark shading in each diagram.)

Each diagram shows the degree to which crimes of varying degrees of seriousness are most likely to be detected by each measure ("victimless" crimes excluded). In the top diagram displaying UCR data, you can see that very few trivial offenses are reported in official statistics, and most of those that are will be dismissed by the police as unfounded. For official statistics, then, the dark figures are highly concentrated at the non-serious end of the crime seriousness spectrum.

The middle diagram reveals that the dark figures for victimization data are primarily concentrated in the non-serious end of the spectrum also, although to a lesser degree than in the case of official data. The failure of victimization data to pick up these minor offenses is largely due to survey subjects not remembering all incidences of victimization.

In the bottom diagram, we see that most of the dark figures in the case of self-reports are concentrated in the upper end of the seriousness continuum rather than the lower end. This is partly due to (a) nearly all self-report surveys excluding most persistent serious offenders from their subject pools, and (b) many of the most serious offenders who remain in self-report subject pools not revealing the full extent of their criminal histories.

What Can We Conclude About the Three Main Measures of Crime in America?

All three main measures of crime in America are imperfect measures, and which one of them is "best" depends on what we want to know. UCR data is still probably the best single source of data for studying serious crimes and, indeed, the only one for studying murder rates and circumstances. For studying less serious but much more common crimes, either victimization or self-report survey data are best. If the interest is in drug offenses, self-reports are the preferable data source.

Because all three data sources converge on some very important points about crime, they enable us to proceed with at least some confidence in our endeavors to understand the whys of crime. The basic demographics of crime constitute the raw social facts that are the building blocks of our criminological theories. If street crime is concentrated among the lower socioeconomic classes and in the poorest neighborhoods, we can begin to ask such things as does poverty "cause" crime, or does some other variable cause both? Is social disorganization in a neighborhood independent of the people living in it, or completely dependent on the

Figure 2.3 Differing Proportions of Reported/Unreported Crimes for the Three Major Measures of Victimful Crimes

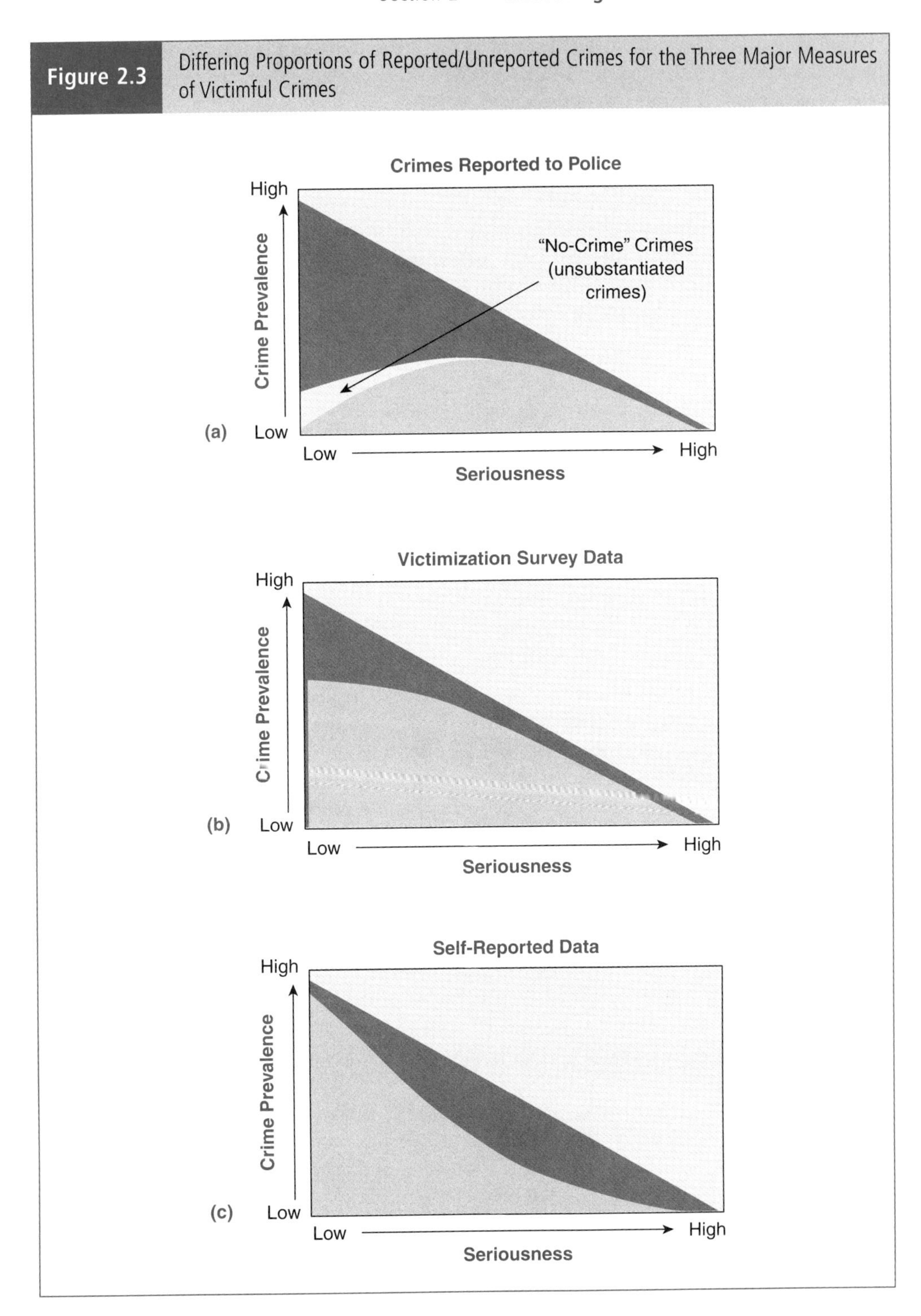

Note: Light shading = proportion of crimes reported. Dark shading = proportion not reported.

people living in it? Why do females always and everywhere commit far less crime (particularly the most serious crimes) than males? These and many dozens of other "why" questions can be asked once we have a firm grip on the raw facts.

Summary

- Crime and criminal behavior are measured in several ways in the United States. The oldest measure is the FBI's Uniform Crime Reports (UCR), which is a tabulation of all crimes reported to the police in most of the jurisdictions in the United States in the previous year. The UCR is divided into two parts: Part I records the eight Index crimes (homicide, rape, robbery, aggravated assault, burglary, larceny/theft, motor vehicle theft, and arson) and Part II records arrests made for all other crimes.

- UCR data seriously underestimates the extent of crime because it only records reported crimes, ignores drug offenses, and only reports the most serious crime in a multiple-crime event. The problems with the UCR led to the implementation of the National Incident-Based Reporting System (NIBRS).

- The second major source of crime statistics is the National Crime Victimization Survey (NCVS). This survey consists of many thousands of interviews of householders throughout the United States, asking them about their crime victimization (if any) during the previous six months. The NCVS also has problems because it leaves out crimes against commercial establishments and relies exclusively on the memory and the word of interviewees.

- The third source of crime data is self-report data collected by criminologists. The advantage of self-report data is that they are derived "from the horse's mouth," and typically the questionnaires used ask about "victimless" offenses not covered in either the UCR or NCVS. The major problems with self-report data are that it does not capture serious criminal behavior and is subject to dishonesty in the form of underreporting, especially underreporting by those most seriously involved in criminal activity.

- The UCR, NCVS, and self-report data come to different conclusions on a variety of points, but they agree about where, when, and among whom crime is most prevalent and the fact that crime has decreased dramatically in the United States over the past decade. Taken together, then, we have a fairly reliable picture of the correlates of crime from which to develop our theories about explanatory mechanisms.

EXERCISES AND DISCUSSION QUESTIONS

1. Consult your college library and browse one or more government documents (such as the *Source Book of Criminal Justice Statistics*, published annually by the U.S. Department of Justice) for information on some crime-related topic that interests you. Examples might be "United States crime trends" or "How age is related to crime rates." Then write a one- to two-page summary of what the document indicates about the topic.
2. Do you think it wise to make "authoritative" statements or formulate theories of criminal behavior, especially serious criminal behavior, based on self-report data?

3. Can you think of other problems possibly associated with asking people about their delinquent or criminal behavior or their victimization, other than those discussed in the chapter?
4. If you were the American "crime Czar," what would you do to get the various law enforcement agencies to fully implement NIBRS—no, you just can't order them to do so.

USEFUL WEB SITES

Bureau of Justice Statistics. http://www.ojp.usdoj.gov/bjs/.

National Archive of Criminal Justice Data. http://www.icpsr.umich.edu/NACJD/.

National Crime Victimization Survey Resource Guide. http://www.icpsr.umich.edu/NACJD/NCVS/.

National Incident-Based Reporting System Resource Guide. http://www.icpsr.umich.edu/NACJD/NIBRS/.

Uniform Crime Reports. http://www.fbi.gov/ucr/ucr.htm.

CHAPTER GLOSSARY

Cleared: A crime is cleared by the arrest of a suspect or by exceptional means (cases in which a suspect has been identified but he or she is not immediately available for arrest).

Crime rate: The rate of a given crime is the actual number of reported crimes standardized by some unit of the population.

Dark (or hidden) figure of crime: The dark (or hidden) figure of crime refers to all of the crimes committed that never come to official attention.

Hierarchy rule: A rule requiring the police to report only the most serious offense committed in a multiple-offense single incident to the FBI and to ignore the others.

National Incident-Based Reporting System (NIBRS): A comprehensive crime statistic collection system that is currently a component of the UCR program and is eventually expected to replace it entirely.

Part I offenses (or **Index Crimes**): The four violent offenses (homicide, aggravated assault, forcible rape, and robbery) and four property offenses (larceny/theft, burglary, motor vehicle theft, and arson) reported in the Uniform Crime Reports.

Part II offenses: The less serious offenses reported in the Uniform Crime Reports, recorded based on arrests made rather than cases reported to the police.

National Crime Victimization Survey (NCVS): A biannual survey of a large number of people and households requesting information on crimes committed against individuals and households (whether reported to the police or not) and the circumstances of the offense (time and place it occurred, perpetrator's use of a weapon, any injuries incurred, and financial loss).

Self-report surveys: The collecting of data by criminologists asking people to disclose their delinquent and criminal involvement on anonymous questionnaires.

Uniform Crime Reports (UCR): Annual report compiled by the Federal Bureau of Investigation (FBI) containing crimes known to the nation's police and sheriff's departments, the number of arrests made by these agencies, and other crime-related information.

READING

Gender Gap Trends for Violent Crimes, 1980 to 2003

A UCR-NCVS Comparison

Darrell Steffensmeier, Hua Zhong, Jeff Ackerman, Jennifer Schwartz, and Suzanne Agha

Darrell Steffensmeir and his colleagues assess whether or not female violent crime is rising relevant to male violent crime. Toward this end, they examine 1980 to 2003 trends in female-to-male interpersonal violence reported in Uniform Crime Reports (UCR) statistics and National Crime Victimization Survey (NCVS) data. Their analyses show much overlap yet differences in each source's portrayal of trends in female violence levels and the gender gap. Both sources show little or no change in the gender gap for homicide, whereas UCR police counts show a sharp rise in female-to-male arrests for criminal assault during the past one to two decades—but that rise is not borne out in NCVS counts. The authors conclude that net-widening policy shifts have apparently escalated the arrest proneness of females for "criminal assault" (e.g., policing physical attacks/threats of marginal seriousness that women, in relative terms, are more likely to commit); rather than women having become any more violent, official data increasingly mask differences in violent offending by men and women.

The past decade or so has witnessed a lively discussion in both the scholarly and popular literatures about whether female violence is rising and, in turn, causing the gap between female and male crime rates to close. The scholarly discussion centers mainly on the national arrest statistics from the FBI's (1979–2003) *Uniform Crime Reports (UCR)*, which show women making substantial gains on men in violent crime levels. For example, from 1980 to 2003, the female percentage (FP) of all arrests increased from one-fifth to one-third for simple or misdemeanor assault, from one-sixth to one-fourth for aggravated or felony assault, and from one-tenth to one-fifth for the Violent Crime Index (sum of homicide, forcible rape, robbery, and aggravated assault arrests). In the popular literature, media reports depict instances of high-profile violent crimes committed by women, which further serve to bolster the view that females appear to be "moving into the

---

SOURCE: Steffensmeier, D., Zhong, H., Ackerman, J., Schwartz, J. & Agha, S. (2006). Gender gap trends for violent crimes, 1980 to 2003: A UCR-NCVS comparison. *Feminist Criminology, 1*, 72–98. Reprinted with permission of Sage Publications, Inc.

world of violence that once belonged to males" (Ford, 1998, p. 13).

Increasing Female Arrest Rates: More Violence or More Enforcement?

We characterize the two competing explanations for the increased female arrest rate found in the *UCR* arrest statistics as the *behavior change* hypothesis and the *policy change* hypothesis. The behavior change explanation is supported to the degree that the NCVS victimization patterns are consistent with the *UCR* arrest patterns. In contrast, if the FP increase occurs only in the *UCR* but not the NCVS, the policy change explanation is more probable.

The Behavior Change Hypothesis

Although diverse in their speculations, an underlying theme of many recent writings on women's arrest trends is that the lives of girls and women have undergone major changes in ways that contribute to their greater involvement in physical aggression and violence. One popular view attributes women's arrest gains to changing gender role expectations that have allowed for greater female freedom and assertiveness. According to this view, these changes have subsequently "masculinized" female behavior and engendered in women an "imitative male machismo competitiveness," which has produced a greater penchant for physical aggression or attack (Adler, 1975, p. 12). Variations in this view also link the increases to today's entertainment media or to greater exposure to messages portraying or condoning women as violent.

A second view is that breakdowns in the effectiveness of the social control mechanisms found in the family, church, and community, which for years have been held responsible for violence among men, are finally catching up to women. Indeed, the recent trends toward higher divorce rates and the shifts in community social organization toward female-headed families can be seen as affecting female violence more than male violence because women's psychic and economic well-being is more dependent on the domestic sphere and because family and kin networks act as buffers against victimization and other conditions that lead to involvement in violence. Moreover, although these conditions affect women in general, their effects are likely to be felt most strongly by marginalized populations of females, especially minorities and low-income women living in depressed urban areas. Women may use violent offending in disadvantaged surroundings as coping strategies for dealing with abusive homes or for confronting interpersonal conflicts with partners, children, extended family, neighbors, or authority figures.

A third and overlapping perspective is that women experience greater role strain today, a combination of old stresses and new ones. They face greater struggles in maintaining a sense of self and confront a much more complex, multidimensional, and often-contradictory set of behavioral scripts that specify what is appropriate, acceptable, or possible for women to do. The stressful economic circumstances more common in poor and minority communities, brought about by recent changes in community social organization and family structure, may intensify these strains. The heightened role strain, along with greater freedom, subsequently increases women's propensities and opportunities for violence.

The Policy Change Hypothesis

The possibility that female arrest trends are by-products of policy changes that have led to greater visibility and to the increased reporting of women's violence rests on the interplay of three exigencies that surround the measurement of violence and its distribution by gender. These are (a) the elasticity of violence definitions, (b) the broadness of *UCR* violence categories such as aggravated or simple assault, and (c) the variability in the gender/violence relationship depending on behavioral or item content.

By *the elasticity of violence definitions*, we mean that citizens, police, and other officials have considerable discretion in defining violence and that the various ways in which they do so may mask considerable differences in its seriousness. Victims and those responsible for addressing victimization must decide on the lens they will use to determine the thresholds whereby a particular behavior is considered violent and the corresponding threshold between minor and serious violence.

The *broadness of the* UCR *offense categories* such as aggravated assault and simple assault produce a heterogeneous mix of behaviors and culpability levels. State laws and *UCR* offense definitions require police officials to distinguish one type of assault from another. Differences between assault categories frequently require inference about the offender's intent to commit harm and if harm was intended, the degree of injury the offender wished to inflict.

Along with considering the offender's perceived intent, other factors that may affect the decision on whether to file the most serious or the least serious charge may include the officer's perception of current injury seriousness and the officer's estimate of the ability of the offender to cause serious future injury if substantial sanctions are not imposed for the current transgression. Some of these factors may include (a) the size difference between offender and officer, (b) the level of physical exertion required by the officer(s) to make the arrest, and (c) beliefs about the judiciousness of devoting limited police resources to individuals not viewed as "real" or "serious" criminals. During times and in locations where criminal justice officials do not perceive women as serious violent offenders prone to repeat their behavior, the offense classifications of women in arrest situations similar to this example are more likely to fall in the lower region of the offense seriousness scale. Occasionally, police action related to perceived "minor" offenses may result in no criminal prosecution and no record of a *UCR*-classified arrest.

It is likely that relatively recent changes in criminal justice policy and practice have increased the variance of subjective violence thresholds longitudinally in addition to the cross-sectional variances already described. For example, the emerging practice of employing more female police officers, although innocuous or beneficial in many regards, may drastically alter officer/offender physical size and power differentials and, thus, may substantially change police perceptions about the level of physical exertion required to subdue the offender and the ability of the offender to cause future officer injury. Such changes may feed what several scholars have noted to be more inclusive, expansive definitions of what constitutes "violence" or "assault" (Steffensmeier, 1993).

By the *variability in the gender/violence relationship* we mean that the gender gap is smaller for violence involving (a) less seriousness, (b) less offender culpability, and (c) behaviors that occur in private settings and against intimates. Although serious, injury-producing violence is largely confined to men, women commit minor acts of violence nearly as often as do men. The more elastic or encompassing the definition of *violence*, the smaller the gender gap. Women's violence typically is perpetrated within or near the home and among family and other primary groups, whereas men are much more likely to commit serious or injury-producing acts within or near street or commercial settings and among acquaintances, strangers, or other secondary groups. Abundant research suggests that young women typically assume a less culpable role when engaged in aggression among mixed-sex peer groups. The role of females in these groups, for example, is often peripheral, such as being an accomplice or a bystander. In addition, the female role is often one that arguably might be considered self-defense (Miller, 2001). Thus, to the extent that measures of violence tap physical aggression within private settings and against intimates, women's violence will seem more frequent and the gender gap will appear to be narrower.

Gender-Specific Impact of Net-Widening Policy Shifts

There are at least three sets of developments that have escalated the arrest proneness of women today relative to women in prior decades and relative to men. The first set of developments involves recent policy changes toward criminalization and/or "charging up" of less serious or minor forms of violence, thus producing a net-widening effect that escalates female arrests because female violent offending is less serious and less chronic relative to male violent offending. Criminalization includes but goes beyond "zero-tolerance" policies to encompass quite broadly (a) the targeting of minor forms of physical or interpersonal aggression and (b) the charging up or converting of physical attacks or threats of marginal seriousness into offense classifications representing greater seriousness and harsher statutory penalties.

Recent enforcement practices have lowered the tolerance threshold for low-level crime, a shift that disproportionately will produce more arrests of less serious offenders. Analysts of crime trends point out that this net widening has been particularly robust in broad offense categories such as simple or aggravated assault. Today, it is more likely that (a) disorderly conducts, harassments, endangering, resisting arrest, and so forth will be categorized as simple assaults; and (b) former simple assaults will be charged up to aggravated assault. Regardless of whether these net-widening effects resulted from an overt bias toward women or from a more general trend toward zero tolerance by criminal justice officials, the effects are the same—a narrowing of the officially recorded gender gap in violence without any necessary changes in underlying behavior.

The second set of developments involves legal changes that have criminalized the violence occurring between intimates and in private or domestic settings. In these contexts, female violence levels more closely approximate male levels, whereas violence between strangers and in public or street settings is more typically male dominated.

After years of neglect and the questionable use of police discretion in domestic and/or intimate partner violence cases, state legislative bodies have implemented civil-legal protection and mandatory arrest policies as the appropriate response to such cases. These pro-arrest policies are based on the assumption "that the temporary removal of the perpetrators of domestic violence through arrest will immediately defuse the domestic violence situation and serve as a specific deterrent by reducing the individual's subsequent abusive behavior" (Mosher, Miethe, & Phillips, 2002, p. 177). Resulting changes in law, in policing, and in victim supports for battered women have greatly altered the response to domestic and relational violence from a private, family matter to a public, criminal one. Although the extent of pro-arrest policies varies considerably across state, county, city, and local departments, police no longer are discouraged from making arrests in domestic disputes or ignoring them.

Although established mainly for the protection of women against abusive partners, the reality of mandatory or pro-arrest policies always has been more complicated. It is ironic that both research and anecdotal evidence is accumulating that shows that pro-arrest policies disproportionately have affected female more so than male violence arrests. Many police departments have adopted the practice of arresting both parties in a family or partner violence incident (e.g., arrest the man and the woman, the parent and the son or daughter) if the "primary" aggressor is unclear. This practice partly reflects police departments' and officers' fears of liability if no arrest is made and a major incident subsequently occurs, as well as more altruistic motives on the part of some police to take immediate action to protect both parties or to prod the victim to seek help. Second, arrest increases have resulted in men becoming more savvy, knowing better the ins and outs of the criminal justice system, and being better able to manipulate it to their advantage. The new tactics used by men apparently include being the first one to call 911 to proactively define the situation ("get to the phone first"), self-inflicting wounds so that police would view the woman as assaultive and dangerous, and capitalizing on the outward calm they display once police arrive (his serenity contrasts with the hysterical, out-of-control woman). Some police have become more strict in following the rule of law, such that when a woman does commit (violent) crime or if allegations arise that she has, authorities believe she should be held accountable and arrested just as a man would. This new attitude persists in spite of the frequent scenario in which the woman is fighting back and "typically caused little damage to the man" (Miller, 2001, p. 1353).

The third set of developments includes the more punitive attitudes toward women and the more gender-neutral nature of law enforcement. Changes in police and public attitudes toward female suspects also may affect female violence rates. An increased emphasis on the legal equality of the sexes, the changing role of women in society, and the perception that they are becoming more violent may produce (a) an increased willingness on the part of the victims or the witnesses of female violence to report women suspects to the police and (b) an increased willingness for the police to proceed more bureaucratically and formally in processing female suspects after viewing them as having greater legal culpability.

Research Strategy

With these divergent positions as a backdrop, we turn now to an examination of female violence trends as reflected in arrest counts of the *UCR* (FBI, 1979–2003) vis-à-vis their trends as reflected in victims' reports of the NCVS (Bureau of Justice Statistics, 1979–2003, 1998, 2004). Aggravated assault typically involves bodily injury and/or the use of a weapon; simple assault (including attempts) does not involve a weapon or aggravated bodily injury. Similar to the *UCR*, the NCVS also gathers information for both types of assault (aggravated and simple). The NCVS provides no homicide counts (i.e., dead victims cannot report or identify the slayer). However, because homicide arrest figures are viewed as highly accurate, we can rely on them as robust markers of gender gap trends that require no contrast with alternative estimates.

UCR statistics can be criticized for being contaminated by changes in enforcement policy. The NCVS survey data, however, although derived from a sample intended to represent the noninstitutionalized U.S. population older than 12 years of age, are unlikely to include some groups most at risk for victimization (e.g., the transient or homeless), who are excluded from the NCVS sampling frame. The precision of NCVS rates is further confounded by the relatively small sample size and the low base rates of reported victimization for infrequent crimes such as aggravated assault.

We employed several methods to assess trends in female violence and the gender gap. First, we calculated sex-specific arrest rates for violent crime and the female-to-male percentage of arrests. The formula for female arrest rates per 100,000 is:

$$\frac{(\text{U.S. population/UCR covered population}) \times (\text{number of females arrested} \times 100{,}000)}{\text{Number of females in U.S. population (ages 12-64)}}$$

The rates adjust for the sex and age composition of the population and a correction factor is applied to account for variable coverage across jurisdictions in the *UCR* during the period from 1980 to 2003. Examining female and male rates yields evidence about (a) sex differences in violence both in general and by type of violent offense and (b) whether violence levels of either women or men are rising, falling, or holding steady. We used similar procedures to derive rates for NCVS data.

Second, we used the female-to-male percentage of violent offending to describe the gender gap. This measure indicates the female share of assault, or other measures of violence, after adjusting for the sex composition of the target population. Note that a narrowing gender gap does not necessarily imply that female rates of violence are rising. The female share of violence may increase because male rates are declining at a faster pace than female rates or female rates might be steady despite male declines.

Fourth, we conducted time-series tests on the data to assess statistically whether the trends in the gender gap have been converging, diverging, or essentially stable (trendless). The time-series method is well-suited for establishing statistically reliable patterns in the gender gap in violence during the 1980 to 2003 period, including (a) whether there are systematic year-to-year changes in the share of female offending after taking into account random fluctuations in the data, isolated "shocks" that cause rates to fluctuate and the aftermath of those shocks, and autocorrelated residuals; and (b) the direction of systematic trends in the gender gap (i.e., convergence or divergence).

UCR Results

(Only homicide and assault [simple and aggravated] are discussed here.) For homicide, the gender gap in arrests is essentially stable (i.e., year-to-year changes in female-to-male rates are not statistically significant). For assault, the gender gap in arrests has narrowed. Figure 2.4 A and B show these trends. In these figures, we can see that the gender gap in arrests is essentially stable or trendless for homicide. The female percentage (FP) for homicide fluctuates a bit between 10% and 13%, whereas the gender gap has narrowed considerably for assault. For the composite assault index, the FP rose from about 20% in 1980 to 32% in 2003.

Male and female rates rose during much of the past two decades, particularly during the 1986 to 1994 period, but leveled off or declined in the late 1990s—more so for males than females, whose rates had merely stabilized or continued to inch upward. Therefore, the narrowing gender gap for both types of assault is at least partly a function of the recent downward movement in male violence. It is not surprising that male interpersonal violence rates continue to be much higher than female rates.

So far, our analysis shows females making arrest gains on males for aggravated and simple assault but not for homicide. Thus, the arrest trends for the assault offenses support the widely publicized view that female violence is rising, but the trends for other violent crimes do not. The finding that female arrest gains are confined largely to assault, especially to simple assault, provides at least some inferential support for the view that police today tend to cast a wider net when making arrests for violent crimes—because criminal assault (physical attack or attempt or threat) is defined more ambiguously than homicide and, therefore, more subject to elastic definitions of violence.

NCVS Results

We turn next to a more discerning test of the policy change hypothesis by comparing female arrest trends for violence to their trends as reflected in victim's reports found in the NCVS. Unlike the *UCR*, victimization data are not limited to cases that come to the attention of the police or result in arrests. Recall that the policy change hypothesis implies disagreement across official and unofficial sources of data, with arrest data showing noticeably larger female gains in violence than do victimization data. In contrast, the behavioral change hypothesis implies general agreement across these data sources.

Figure 2.5 illustrates the 1980 to 2003 trends by displaying NCVS rates of male and female assault along with the relevant FPs. We see that the NCVS shows female assault levels as being much lower than male levels, and that male and female rates move largely in tandem; female rates of assault rose when male rates rose and declined when male rates declined, yielding a stable gender gap in overall violence. In combination with Figure 2.4 A, we see that both the NCVS and *UCR* trends show male and female assault rates rising during the late 1980s through the early 1990s and then tapering off. The rise, however, is smaller and the decline is greater in the NCVS series. The NCVS series in Figure 2.5 shows both female and male rates of assault dropping considerably in recent years as compared to the *UCR* series in Figure 2.4 A and B, where only male arrest rates for assault have been in sharp decline. This comparison supports a conclusion that recent policy shifts have affected women's more so than men's arrest proneness.

Figure 2.4a & b Trends in Female and Male Arrest Rates (per 100,000) and Female Percentage of Arrests for Interpersonal Violence: Uniform Crime Reports, 1980 to 2003

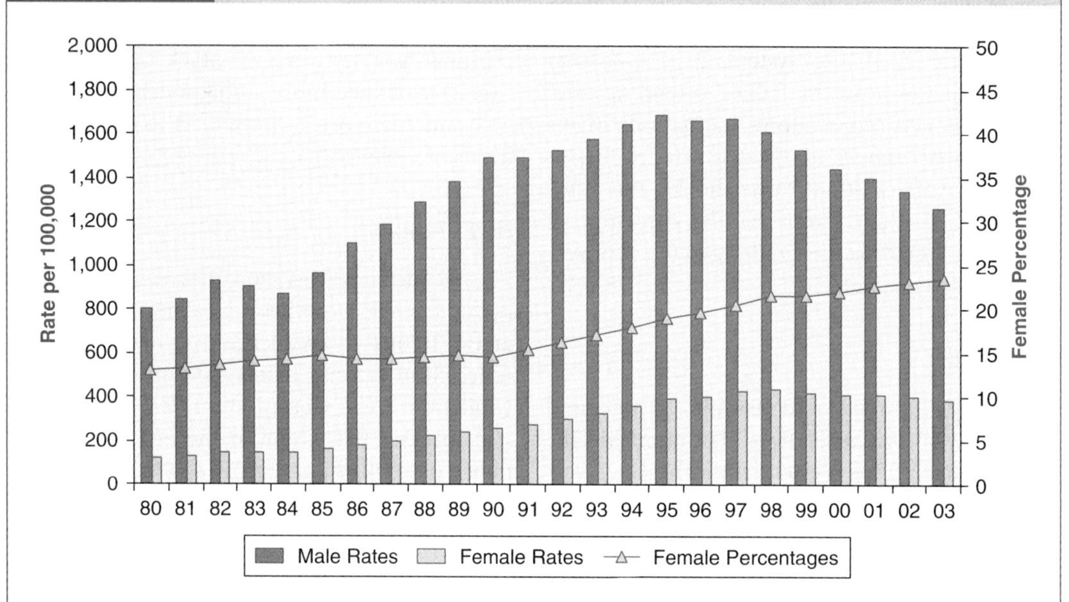

Note: The assault index includes aggravated and simple assaults.

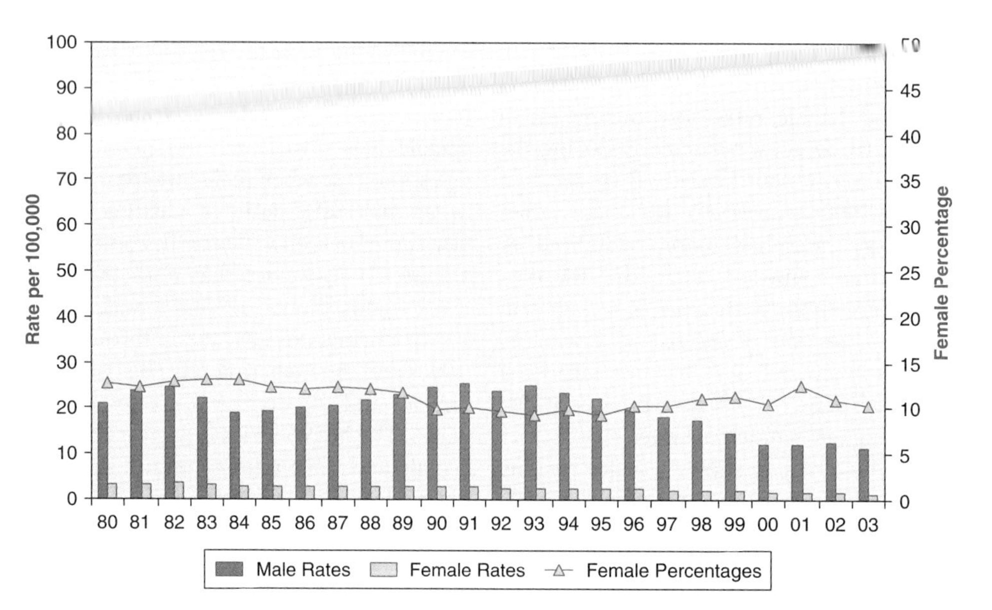

Note: Because the rates for homicide are much lower than the rates for assaults, the left scale for the histograms was reduced from 2,000 to 100.

Figure 2.5 Trends in Female and Male Assault Offending Rates (per 100,000) and the Female Percentage: National Crime Victimization Survey, 1980 to 2003

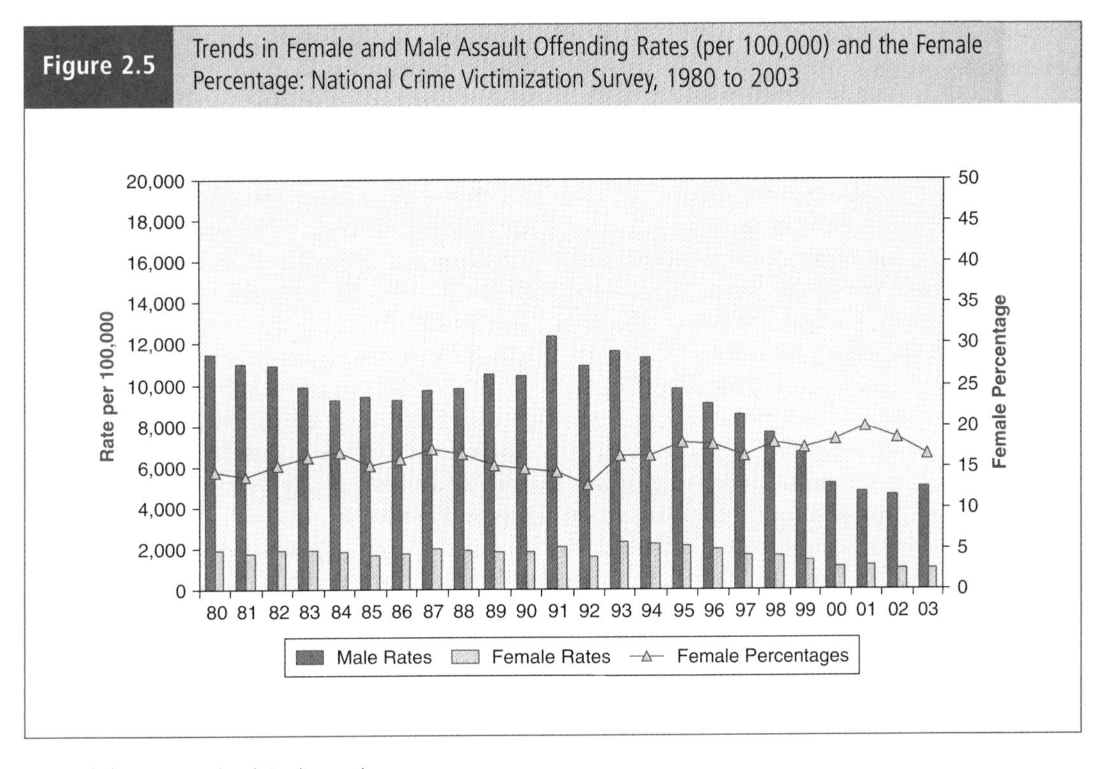

NOTE: Includes aggravated and simple assaults.

If we partition the NCVS findings into two decades and calculate an averaged FP for each decade, the gender gap for the assault index is roughly 17% in both the 1980s and the 1990s. These decade comparisons underscore the conclusion that female rates of violence typically rise when male rates rise and decline when male rates decline, yielding a stable or trendless gender gap in overall violence. In addition, the gender gap in assault is fairly comparable between NCVS and *UCR* figures in earlier years, but the two sources diverge in more recent years, as we would expect based on the policy change hypothesis. For example, the FP for aggravated assault in the 1980s was about 12% to 13% in both the NCVS and *UCR*, whereas by the early 2000s, the percentage in the NCVS had held at 12% but had jumped to about 20% in the *UCR*. Finally, there have been sizable declines in NCVS assault rates in recent years that considerably outpace the much smaller declines in *UCR* assault arrest rates, particularly for females.

The most important conclusion from the *UCR*-NCVS comparison is that the two sources differ sharply in their representation of gender gap trends in assault. In contrast to the *UCR*, the NCVS reveals very little change or a lack of convergence in the gender gap for assault crimes during the past one to two decades. Teased out, this comparison suggests that the sharp declines in assault crimes among both females and males since about the mid-1990s, as noted by the NCVS, have been partly offset by the greater proneness of police to arrest and charge persons with assault. The greater arrest proneness is salient particularly for females, as these trends show that female arrests have largely leveled, whereas victims' reports indicate sizable declines in female-perpetrated assaults since at least the mid-1990s.

Summary and Discussion of Gender Gap Trends

We began with two observations concerning the meaning of *UCR* arrest trends that show female violence rates rising and the gender gap closing. First, we noted the importance of examining alternative sources of data when generalizing about recent trends in female arrests for violence because these data are prone to criminal justice selection processes. At issue is whether the arrest trends reflect changes in underlying behaviors of females toward more violence or instead reflect policy changes that place females at risk of arrest without any change in underlying behavior. Second, we proposed that a comparison of *UCR* arrest counts with NCVS victimization reports is a particularly useful strategy for addressing the behavior change versus policy change hypotheses. Our analysis of these two sources reveals considerable overlap in their findings about female-to-male trends in violent crime but also important differences.

Both the overlap and the differences are summarized in Figure 2.6, which displays the female-to-male percentages for assault as based on *UCR* arrest figures with counts based on NCVS tabulations, together with the trend percentages in homicide arrests.

Figure 2.6 succinctly summarizes the main points of our analyses: (a) both arrest and victimization data show male levels of interpersonal violence as much higher than female levels both now and in the past; (b) both arrest and victimization data show male levels of interpersonal violence as much, much higher than female levels

Figure 2.6 Summary of Trends in the Gender Gap for Homicide and Assault Index: *Uniform Crime Reports* (*UCR*) Versus National Crime Victimization Survey (NCVS), 1980 to 2003

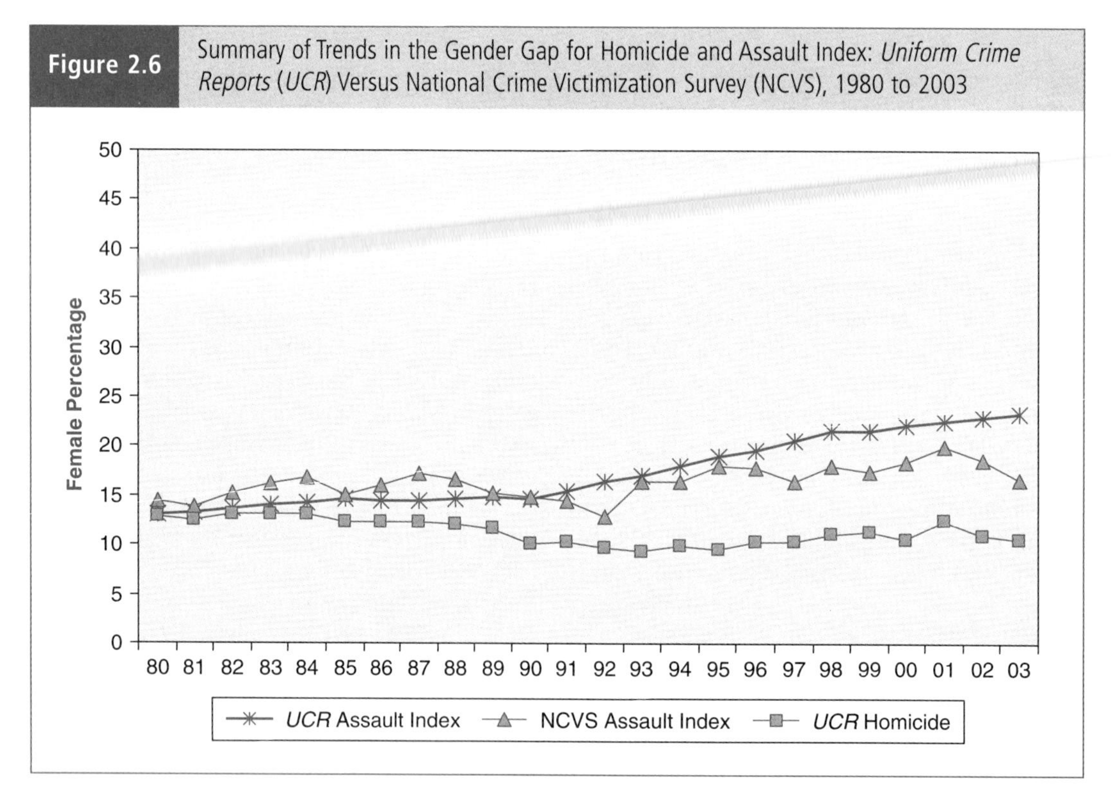

SOURCE: Bureau of Justice Statistics (1979–2003, 1998, 2004) and FBI (1979–2003).

for more serious forms of violence both now and in the past; (c) for the most serious kinds of interpersonal violent crime or "predatory violence," female rates have not been rising and the gender gap has not been closing; (d) only trends in assault differ by type of data—*UCR* arrest statistics show female rates rising and the gender gap closing (mainly since 1990), whereas victimization data show a stable gender gap; (e) arrest trends in assault are the driving force behind recent concerns about rising levels of female violence, but those trends are not borne out in victim's reports; and (f) the findings from both sources confirm what traditionally has been known about the gendered nature of interpersonal violence and its variation depending on behavioral item—the existence of a fairly small gender gap for minor kinds of physical attack or threat (e.g., misdemeanor assault) as compared to a very large gender gap for more serious forms.

These findings taken together are at odds with the behavior change hypothesis but supportive of the policy change hypothesis. In sum, there is no meaningful or systematic change in the violent-offending gender gap. First, the finding that female arrest gains are confined largely to aggravated assault and especially simple assault provides at least some inferential support for the view that police today tend to cast a wider net, particularly for female suspects in violent crime cases. Second, the reasoning of many criminologists that the NCVS series provides accurate estimates that are not confounded with changes in the behavior of criminal justice agents adds credibility to the NCVS finding of a lack of convergence in the gender gap for assault crimes during the past one to two decades. Third, the evidence from the victim surveys indicating no change in the female-to-male percentage of criminal violence is even more remarkable in light of caveats that a change in perceptions and expectations about women's violence in the society at large might itself have "self-fulfilling" effects leading to higher *reported* levels of women's assaults in survey responses. All else equal, just as police have become more prone to arrest females for violent misconduct, it also seems likely that victims have become more prone to identify female assailants in NCVS interviews.

Conclusion

A recurring theme in research and policy circles is the need for more accurate data on violence levels and trends both as a necessary foundation for informed opinion, theory, or policy and for correcting damaging myths and stereotypes that interfere with theory construction or effective policy approaches. Thus, the results of our analysis comparing arrest counts of the *UCR* with victims' reports of the NCVS (where victims identify the sex of the offender) are timely as well as cautionary in their portrayal of women's violence trends. Our key conclusion is that there has been no meaningful or systematic change in women's involvement in crimes of interpersonal violence and in the gender gap during the past couple of decades. This conclusion is based on (a) *UCR* trend data showing little or no change in the gender gap in arrests for homicide, a reliably reported crime; (b) *UCR* and NCVS trend data showing no change in the gender gap for rape/sexual assault; and (c) NCVS figures showing little or no change in female levels of assault and in the gender gap. We view the NCVS findings as particularly convincing because the data are independent of criminal justice selection biases and derived from nationally representative samples, because the NCVS is held in high regard within the social science research community, and because we observed these null results in the context of heightened perceptions about women today as being more violent and "male like." These perceptions might sway some victims and citizens to be less protective and more willing to identify women as violent offenders.

Our findings regarding the stability of the gender gap in interpersonal violence are hardly surprising in several respects. The first is that social change is seldom, if ever, so abrupt and robust as to bring about such a dramatic shift in behavior as characterizes the female-to-male trend in arrests for assault during the past decade or so (e.g., the FP of assault arrests jumped from roughly 15% to 24% since 1990). Second, it is doubtful that women's life experiences and the organization of gender have changed more drastically during the past decade or so than during the previous one or two decades. Third, even if women's lives are much more stressful today, this greater role strain may not translate into more violence on grounds that females tend to internalize stress and physical aggression, whereas males tend to externalize it. Fourth, it appears that women have not become more attracted to, or have much greater access to, violence-likely situations that historically overwhelmingly have involved males, such as violence occurring in the context of robbery, extortion, gangs, and drug or organized-crime networks.

Last, even if one accepts at face value the *UCR* arrest trends for criminal assault and also believes that women have become more prone for violence (e.g., because of greater stress or diminished femininity), it is also true that we have changed our laws, police practices, and policies in other ways toward enhanced identification and criminalization of violence in general and of women's violence in particular. The analysis here, based on the best data available, makes a strong case for the position that it is the cumulative effect of these policy shifts, rather than a change in women's behavior toward more violence, that accounts for their higher arrest rates and the narrowing gender gap in official counts of criminal assault. It is not so much that women have become any more violent as it is that the avenues to prevent or punish violence have grown so enormously and that the official data increasingly mask differences in violence among men and women.

References

Adler, F. (1975). *Sisters in crime.* New York: McGraw-Hill.

Bureau of Justice Statistics. (1979–2003). *National crime (victimization) survey: Criminal victimization in the United States.* Washington, DC: U.S. Department of Justice.

Bureau of Justice Statistics. (1998). *National crime surveys: National sample, 1973–1983* [Data file] (Conducted by U.S. Department of Commerce, Bureau of the Census, 6th ICPSR ed.). Ann Arbor, MI: Inter-University Consortium for Political and Social Research.

Bureau of Justice Statistics. (2004). *National crime victimization survey, 1992–2003* [Data file] (Conducted by U.S. Department of Commerce, Bureau of the Census, ICPSR ed.). Ann Arbor, MI: Inter-University Consortium for Political and Social Research.

Federal Bureau of Investigation. (1979–2003). *Uniform crime reports.* Washington, DC: U.S. Government Printing Office.

Ford, R. (1998, May 24). The razor's edge. *The Boston Globe Magazine*, pp. 13, 22–28.

Miller, S. L. (2001). The paradox of women arrested for domestic violence. *Violence Against Women, 7*, 1339–1376.

Mosher, C. J., Miethe, T. D., & Phillips, D. M. (2002). *The mismeasure of crime.* London: Sage.

Steffensmeier, D. (1993). National trends in female arrests, 1960–1990: Assessment and recommendations for research. *Journal of Quantitative Criminology, 9*, 413–441.

DISCUSSION QUESTIONS

1. What does this article tell us about the social construction of crime statistics?
2. What part may feminism have played in perpetrating the idea that women are as violent as men?
3. What part may the media have played in generating the idea that women are becoming as violent as men?

READING

Race and the Probability of Arrest

Stewart J. D'Alessio and Lisa Stolzenberg

The previous article showed how we can use the UCR and NCVS to bring light to bear on the issue of gender violence. This article by D'Alessio and Stolzenberg brings NIBRS data (which combines the best features of the UCR and NCVS) to bear on another vexing issue in criminology, the differential black/white arrest rate. Some criminologists maintain that the black/white difference is a function of discriminatory policing, others that it is a function of greater black participation in crime. D'Alessio and Stolzenberg use NIBRS data from 17 states and 335,619 arrests for rape, robbery, and aggravated and simple assault to assess the effect of offenders' race on the probability of arrest. Their results indicate the odds of arrest for robbery, aggravated assault, and simple assault were significantly greater for white offenders than for black offenders, but there was no racial difference in the probability of arrest for rape. D'Alessio and Stolzenberg conclude that the disproportionately high black arrest rate is most likely attributable to their disproportionately higher involvement in crime.

The relationship between race and arrest remains a topic of contentious debate. While blacks constitute about 12.8% of the population, they accounted for 38% of the arrests for violent crimes and 31% of the arrests for property crimes in 2000 (FBI 2001; U.S. Census Bureau 2001). Although it is readily acknowledged that blacks are arrested in numbers far out of proportion to their numbers in the population, considerable disagreement exists as to what this finding exactly means. Social scientists have proffered two major explanations. The first and most broadly solicited explanation employs normative theories to explain the overrepresentation of black citizens in official arrest statistics. Normative theories view the enforcing of criminal laws as unbiased, with little or no consideration being given to the offender's race or other demographic characteristics. These types of theories purport to attribute most of the disparity in arrest statistics between blacks and whites to differences in criminal involvement. It is argued that racial differences in arrest patterns occur primarily because blacks violate the law more frequently and commit more serious crimes than do whites.

Normative theories typically emphasize the nexus between social factors and crime to explicate differences in crime patterns between the races. The social factors most highlighted in the literature include poverty, economic inequality/deprivation, social disorganization, segregation, and family structure. Constitutional factors such as intelligence have also been adduced as engendering race differences in criminal behavior, but research on this topic is highly controversial. An alternative explanation for the differential arrest patterns of whites and blacks focuses on racially biased law enforcement practices. This perspective draws from conflict theory, which posits that the elevated arrest rate for black citizens is the consequence of discrimination by police. Conflict theorists view society as consisting of groups with differing and conflicting values and maintain that the state is organized to represent the interests of citizens who are wealthy and powerful. Criminal law is conceived as an instrument to protect the interests of the elite, and the severity of criminal sanction is based to a large degree on extralegal factors such as race and social class. Consequently, groups that challenge the status quo are more apt to be subjected to criminalization, arrest, and increased incarceration compared to groups that are perceived as less menacing. The conflict perspective thus suggests that blacks and other racial minorities will be more susceptible to biased law enforcement practices in order to ensure they are brought under state control.

It is also asserted that blacks, especially young black males, face a higher probability of arrest because the police have a negative perception of them. Disparaging labels such as "delinquents," "dope addicts," and "welfare pimps" are used frequently to depict black males (Gibbs 1988:2). Black males also epitomize an aggressive behavior style that is perceived by many whites to be threatening. The media often acts to render these stereotypes more negative. This general stereotype of blacks, especially young black males, as being dangerous and criminally inclined is thought to compel police to monitor and arrest black citizens more frequently than warranted based on their actual criminal behavior.

The Hindelang Study

Michael Hindelang compared race-specific arrest data derived from the UCR with reported offender data drawn from the National Crime Victimization Survey (NCVS) to ascertain the convergence of these two data sets in terms of the relative amount of crime committed by both blacks and whites. He used arrest data from the UCR because it is at this point in the criminal justice process that information about the race, sex, and age of offenders is first recorded. The NCVS contains data on the race of the offender as determined by the crime victim. Hindelang analyzed four crimes for which offenders and victims must come into contact and, thus, allow for the potential identification of the offender's race. These were rape, robbery, aggravated assault, and simple assault.

Hindelang (1978:99) theorized that, "if there are substantial biases in the UCR data for any reason, we would expect, to the extent that victimization survey reports are unbiased, to find large discrepancies between the UCR arrest data and victimization survey reports on the racial characteristics of offenders." His results showed that 62% of the robbery victims in the NCVS reported their assailants to be black, whereas 62% of the people arrested for robbery during the same year by police were also black. However, although Hindelang showed that data drawn from the UCR and NCVS did converge for the crime of robbery, he still evinced some evidence of racial bias in the arrest sanction for the crimes of rape and assault. Specifically, he found that blacks were overrepresented by about 10 percentage points in the UCR arrest data for the crimes of rape, aggravated assault, and simple assault.

Although these findings were consistent with the differential arrest hypothesis, Hindelang speculated that the observed dissimilarities between the UCR and the NCVS were in large part due to the fact that crimes involving black offenders were less apt to be reported to police than crimes involving white offenders. A supplemental analysis that only considered those crimes reported to the police by crime victims showed that these slight discrepancies in arrest patterns diminished substantially. Although the NCVS and UCR did not converge perfectly, Hindelang felt confident to conclude that blacks comprise a larger proportion of criminal offenders than their representation in the general population would warrant.

Because Hindelang analyzed only aggregate data, we cannot definitely say on the basis of his study whether blacks are more likely than whites to be arrested by police for similar types of crimes. All we are able to conclude from his work is that his findings are not consistent with the idea that blacks are more apt to be arrested by police than are whites. Taken in total, these problems compel us to seriously question whether the general absence of a race-arrest effect in Hindelang's research is not idiosyncratic to his reliance on the UCR and NCVS.

Our charge in this article is to resume where Hindelang halted. Using data from the NIBRS, we attempt to determine the extent that black overrepresentation in official arrest statistics is explained by differential offending or by differential selection into the criminal justice system via arrests by police. NIBRS represents the next generation of crime data and it is designed to replace the nearly 70-year-old UCR. The intent of NIBRS is "to enhance the quantity, quality, and timeliness of crime statistical data collected by the law enforcement community and to improve the methodology used for compiling, analyzing, auditing, and publishing the collected crime data" (FBI 2000:1). NIBRS is unique because rather than being restricted to a group of eight Index crimes that the summary-based program uses, it gathers information from individual crime reports recorded by police officers at the time of the crime incident for 57 different criminal offenses. The information collected by police typically includes victim and offender demographics, victim/offender relationship, time and place of occurrence, weapon use, and victim injuries. Because NIBRS is capable of producing more detailed and meaningful data than that generated by the traditional UCR, it is a valuable tool in the study of crime.

NIBRS data are well suited for our intentions because it is possible to link a reported crime incident to a subsequent arrest that was heretofore not feasible with the UCR. The ability to merge crime incident data with arrest data enables researchers to calculate the actual probability of arrest by race for crimes communicated to the police where the victim is able to identify the race of the offender. This is the most appropriate strategy for evaluating the discriminatory use of the arrest sanction because the police can only act upon illegal behaviors that come to their attention. These data also afford us the opportunity to examine how the arrest sanction is influenced by a number of salient factors about which Hindelang lacked data, such as whether the victim was injured, the race of the victim, the victim/offender relationship, and weapon use.

Data

The data used in this study were obtained from the NIBRS for 2,852 reporting jurisdictions in 17 states for 1999. Our sample comprised 9,551 forcible rapes, 12,315 robberies, 60,249 aggravated assaults, and 253,504 simple assaults where there was one offender and one victim. Of these crimes, approximately 25% of the forcible rapes, 16% of the robberies, 44% of the aggravated assaults, and 42% of the simple assaults resulted in an arrest. The offender was reported to be white in 63% of the forcible rapes, 21% of the robberies, 56% aggravated assaults, and 65% of the simple assaults.

Although our primary objective is to assess the influence of an offender's race on the likelihood of arrest, the multivariate model we develop allows us to discern the impact of other variables on the arrest sanction. If these additional variables are not controlled for, any observed relationship between an offender's race and the probability of arrest might be spurious. The control variables measure criminal offense characteristics, offender characteristics, and victim characteristics. Criminal offense characteristics account for whether the victim suffered a serious injury during the course of the crime, whether a deadly weapon was used in the crime, and whether the offender perpetrated less serious ancillary crimes during the commission of the primary offense. Criminal offense characteristics also include the relationship between the victim and offender and the location of the crime. The offender characteristic variables, in addition to the offender's race, include the age of the offender, the gender of the offender, and whether the offender was under the influence of drugs and/or alcohol during the commission of the crime. The variables measuring victim characteristics comprise the race, age, and gender of the crime victim. Table 2.2 provides the summary statistics and codings for the variables included in the study.

Bivariate Analyses

The differential arrest hypothesis predicts that controlling for crimes reported to the police, black citizens have a greater chance of being subjected to arrest. Looking simply at the two-way relationships presented in Table 2.3, we see that there is a consequential association between the race of the offender and the prospect of arrest for robbery, as 807/2,620 (31%) of the robberies with white offenders and 1,132/5,278 (21%) of the robberies with black offenders are cleared by arrest ($X^2 = 82.705$, $p < .001$).

Inspection of this table also reveals that aggravated assaults and simple assaults involving white offenders are significantly more likely to be cleared by arrest. Whites have about a 10% greater chance of being arrested for both aggravated and simple assault. Only for the crime of rape do blacks have an enhanced proclivity to be arrested by police. There is about a 28% chance of arrest for blacks, whereas whites have a 27% probability of arrest. This 1% difference, however, is not statistically significant ($X^2 = .694$, $p = .405$). These bivariate results are interesting because it appears that whites are more likely than blacks to be arrested by police. Such findings tend to cast doubt on the differential arrest thesis, which theorizes that black criminal offenders have a markedly higher prospect of arrest than do whites.

Table 2.4 shows the likelihood of arrest for white-on-white crimes, white-on-black crimes, black-on-black crimes, and black-on-white crimes. The results presented in this table indicate that the police are most disposed to effectuate an arrest for aggravated assaults and simple assaults involving white offenders and white victims. These findings run counter to much of the literature suggesting that blacks who victimize whites are more likely to be sanctioned severely by the state because of the elevated status of white victims in our society. In contrast, the odds of a black offender being arrested for raping a black is higher than for any other of the other victim/offender racial combinations. Table 2.4 also reveals that white-on-black robberies have the greatest likelihood of arrest.

Logistic Regression Analyses

A multivariate method is required to discern whether an offender's race influences the probability of arrest independently of other factors. We use logistic regression for this purpose because it is appropriate for analyzing a dichotomous dependent variable and allows use of both categorical and continuous independent variables. The regression coefficients from a logistic regression can also readily be translated into easily interpretable odds indicating the change in the likelihood of the dependent variable (probability of arrest) given a unit shift in an independent variable, holding other variables constant.

Table 2.5 presents the logistic regression models for the likelihood of arrest in forcible rape, robbery and assault cases. In addition to the offender's race and control variables, each model includes a logit-based "hazard rate" variable to account for the exclusion of crime incidents with multiple offenders and/or victims.

The first model in Table 2.5 estimates the effects of an offender's race and the control variables on the likelihood of arrest for the crime of forcible rape. The small and nonsignificant effect of an offender's race can be interpreted as evidence against the differential arrest hypothesis. It appears that black offenders are no more likely to be arrested for forcible rape, controlling for other factors, than are white offenders. One salient effect in this model is whether the offender perpetrated less serious ancillary crimes during the commission of the forcible rape. The presence of

Table 2.2 Percentage Distributions of Descriptive Characteristics of Crimes, Offenders and Victims by Type of Violent Crime, 1999

	Forcible Rape	Robbery	Aggravated Assault	Simple Assault
Offender arrested				
0 = No	75.4	84.0	55.5	58.5
1 = Yes	24.6	16.0	44.5	41.5
Offender white				
0 = No	24.1	42.9	31.8	27.9
1 = Yes	63.3	21.3	56.5	64.9
Missing data	12.6	35.9	11.7	72
Victim white				
0 = No	17.0	26.0	30.2	23.9
1 = Yes	78.8	53.4	66.1	73.1
Missing data	4.2	20.6	3.7	3.0
Offender male				
0 = No	1.5	3.9	19.7	21.0
1 = Yes	92.1	62.4	71.3	74.2
Missing data	6.3	33.7	9.0	4.8
Victim male				
0 = No	98.2	27.5	44.6	64.3
1 = Yes	1.8	54.9	54.7	35.2
Missing data	0.0	17.6	0.8	0.5
Offender stranger				
0 = No	78.7	54.0	78.1	85.8
1 = Yes	8.2	27.8	9.6	5.7
Missing data	13.1	17.3	12.3	8.4
Multiple offenses				
0 = No	95.5	97.1	96.3	97.0
1 = Yes	4.5	1.9	3.7	3.0
Residence				
0 = No	28.2	84.6	43.4	34.8
1 = Yes	71.8	15.4	56.6	65.2
Serious injury				
0 = No	94.7	78.0	75.2	100.0
1 = Yes	5.3	5.9	24.8	0.0
Missing data	0.0	16.1	0.0	0.0
Offender substance abuse				
0 = No	86.9	96.2	83.3	84.7
1 = Yes	13.1	3.8	16.7	15.3

(Continued)

Table 2.2 (Continued)

	Forcible Rape	Robbery	Aggravated Assault	Simple Assault
Deadly weapon				
0 = No	90.9	51.0	42.5	97.3
1 = Yes	3.7	42.0	54.6	0.0
Missing data	5.4	7.0	2.9	2.7
Offender's age				
Mean years	28.5	27.5	30.6	30.1
Missing data	14.7	43.9	13.7	8.4
Victim's age				
Mean years	20.8	34.6	30.2	29.1
Missing data	1.7	18.9	3.9	2.9
Total N	9,551	12,315	60,249	253,504

Table 2.3 Offenses Known and Clearances by Arrest by Race of the Offender, 1999

Offense	Crimes (N)	Arrests (N)	Cleared (%)
Forcible rape			
Offender white	6,043	1,642	27.2
Offender black	2,304	647	28.1
Robbery			
Offender white	2,620	807	30.8
Offender black	5,278	1,132	21.4
Aggravated assault			
Offender white	34,055	18,095	53.1
Offender black	19,137	8,141	42.5
Simple assault			
Offender white	164,543	76,966	46.8
Offender black	70,646	26,022	36.8

these ancillary offenses elevates the odds of arrest for forcible rape by approximately 210%. The coefficients for the age of the victim and criminal offender are also noteworthy in this equation. An arrest is more apt to occur in forcible rapes involving younger victims and older offenders. Injury to the victim also increases the likelihood of arrest by about 65%.

The second model reports a discernible relationship between an offender's race and the likelihood of arrest for robbery. Being white elevates the odds of arrest for robbery by 22%. This finding also fails to support the differential arrest hypothesis. Several other factors also directly impact the probability of arrest for robbery. One strong predictor is the relationship between the

Table 2.4 Offenses Known and Clearances by Arrest by Race of the Offender and Race of the Victim, 1999

Offense	Crimes (N)	Arrests (N)	Cleared (%)
Forcible rape			
White-on-white	5,733	1,538	26.8
White-on-black	128	25	19.5
Black-on-black	1,376	415	30.2
Black-on-white	884	217	24.5
Robbery			
White-on-white	1,733	486	28.0
White-on-black	154	47	30.5
Black-on-black	2,122	425	20.0
Black-on-white	2,377	467	19.6
Aggravated assault			
White-on-white	31,608	16,701	52.8
White-on-black	1,261	544	43.1
Black-on-black	15,248	6,356	41.7
Black-on-white	3,576	1,600	44.7
Simple assault			
White-on-white	156,261	73,117	46.8
White-on-black	4,310	1,441	33.4
Black-on-black	53,733	19,167	35.7
Black-on-white	15,798	6,296	39.8

NOTE: The total number of crimes reported within each offense category is less than the totals reported in Table 2.3 because missing data were excluded from this analysis.

victim and offender. When the victim and offender know each other, the probability of arrest is magnified. Whether the offender had been drinking or was under the influence of drugs also impacts the likelihood of arrest. The odds of arrest are 3.3 times as large for intoxicated offenders and/or drug-induced offenders as they are for offenders not under the influence of alcohol and/or drugs.

An examination of the third model reveals that the likelihood of arrest for aggravated assault varies directly with the offender's race, as reported by the victim. Being white heightens the odds of arrest for aggravated assault by 13%. Thus, relative to violation frequency as reported by crime victims, the likelihood of an arrest for aggravated assault is higher for whites than for blacks, net other factors. While the results for the offender's race variable is the most important substantively, the effects of some of the other variables are also worth noting. We again observe a rather pronounced effect of the victim/offender relationship variable on the odds of arrest. The effects of several other variables are also consequential. Net controls, the police are more likely to make an arrest in aggravated assaults with women victims, that include ancillary crimes, that result in victim injury, that include offenders under the influence of alcohol and/or drugs, and that involve older victims.

The final model also indicates an association between an offender's race and the probability of arrest for simple assault, as depicted previously in Table 2.3. Being white heightens the odds of arrest for simple assault by about 9%. The effect of the victim/offender variable remains stable.

Table 2.5 Logistic Regression Coefficients Predicting Probability of Arrest, 1999

	Forcible Rape (1)		Robbery (2)		Aggravated Assault (3)		Simple Assault (4)	
Offender white	−.088	(.082)	.202*	(.077)	.121**	(.032)	.089**	(.016)
Victim white	−.073	(.092)	.049	(.077)	−.085	(.034)	.191**	(.017)
Offender male	−1.983	(6.918)	.101	(.128)	−.044	(.023)	.096**	(.011)
Victim male	−1.886	(6.915)	−.114	(.067)	−.150**	(.020)	−.211**	(.010)
Offender stranger	−.235	(.109)	−.545**	(.077)	−.215**	(.033)	−.168**	(.020)
Multiple offenses	1.132*	(.385)	.277	(.208)	.613*	(.193)	−.755**	(.063)
Residence	.131	(.063)	.037	(.109)	−.074	(.098)	.916**	(.048)
Serious injury	.503*	(.174)	.311	(.223)	.128*	(.042)	—[a]	
Offender substance abuse	.497	(.271)	1.181**	(.271)	.774**	(.064)	.349**	(.022)
Deadly weapon	.243	(.255)	.543	(.508)	.146	(.059)	—[a]	
Offender's age	.008**	(.002)	.001	(.003)	.001	(.001)	−.005**	(.000)
Victim's age	−.020**	(.003)	.003	(.002)	.010**	(.001)	.013**	(.000)
Hazard rate	−9.205	(6.481)	−7.263	(4.555)	−5.492**	(1.518)	10.964**	(.826)
Constant	1.772		1.588		.499		−2.801	
−2 Log-likelihood	9,150.293		6,353.865		67,274.108		299,543.020	

NOTE: Standard errors are in parentheses.

a. The victim injury and weapon use variables were excluded from the simple assault equation because these offense characteristics are not applicable to simple assaults.

$*p < .01$; $**p < .001$ (two-tailed tests).

Controlling for other factors, simple assaults that involve nonstrangers are more likely to culminate in an arrest. Additionally, the police are more apt to make an arrest in simple assaults that involve white victims, male offenders, female victims, offenders under the influence of alcohol and/or drugs, younger offenders, older victims and that occur in a private residence. The multiple offenses variable also shows some predictive power in the simple assault equation, but in the negative direction. As the number of ancillary crimes rises, the likelihood of arrest decreases. Although this finding seems counterintuitive, it is most likely the result of police effectuating arrests in domestic violence cases. Approximately 70% of the arrests for simple assault without multiple offenses pertained to an offender and a victim involved in some type of domestic relationship. In contrast, the majority of simple assault incidents involving multiple offenses derived from disputes over property. It seems likely that police perceive these minor disagreements over property as less serious than domestic disputes. Additionally, many jurisdictions have mandatory arrest policies in domestic violence incidents, thereby increasing the likelihood of arrest even in relatively minor cases (Sherman 1992).

Overall, the bivariate and multivariate logistic regression results furnish little empirical evidence of systematic racial bias against blacks in the arrest decision for forcible rape, robbery, aggravated assault, and simple assault. These findings suggest some rethinking of traditional held notions about the underlying causes of the elevated arrest rate for blacks. Recall that a central aspect of the differential arrest thesis is that the police are racially biased, and that this bias is a major reason for blacks having a higher arrest rate. The differential arrest hypothesis is simply incapable of handling the disproportionate arresting of white offenders that we show to be associated with the crimes of robbery, aggravated assault, and simple assault. It is also incapable of explaining a lack of a race effect for the crime of forcible rape.

Conclusion

Debate persists as to whether the police perform their duties in a racially discriminatory fashion. The most frequently cited evidence for this assertion is the observation that blacks are arrested in numbers far out of proportion to their numbers in the general population. This observation, however, cannot in itself be taken as evidence of racial discrimination, since the elevated arrest rate for blacks may simply reflect their greater involvement in criminal activities. Because our legal system's claim to legitimacy is especially dependent on the public's perception of fairness and equity in the decision to arrest, we took a closer look at the evidence bearing on this issue.

Using data from the new NIBRS, we analyzed the effect of an offender's race, as perceived by the crime victim, on the probability of arrest for 335,619 incidents of forcible rape, robbery, aggravated assault, and simple assault in 17 states during 1999. Contrary to the theoretical arguments of the differential arrest hypothesis, and consistent with the tenets of the differential offending perspective, our analyses show that whites are considerably more likely than blacks to be arrested for robbery, aggravated assault, and simple assault. There are also no glaring differences in the data between white and black offenders regarding their chances of being arrested for forcible rape. This null finding also tends to refute the argument that racial bias in policing is affecting the arrest rate for blacks. Such findings beg the question: How can it be that whites and not blacks are more likely to be arrested for robbery and for assault, when many individuals who write about the criminal justice system assume precisely the opposite?

One likely explanation for our findings relates to black citizens' distrust of the police. In police work there are two basic ways that an individual is initially linked with the commission of a crime: (1) the police officer can observe the criminal offense and (2) a citizen can give testimony against the individual. In most cases, however, the police officer usually arrives too late to witness the criminal offense. Accordingly, the police are often forced to rely on the testimony of witnesses to gather the necessary evidence to effectuate an arrest. Our finding that whites are more likely to be arrested than blacks should be understood in this context. It is well known that blacks distrust the police more than whites (Sherman 2002). For example, a recent national Gallup Poll showed that 36% of black citizens, as compared to 13% of white citizens, have an unfavorable opinion of the local police (Gallup & Gallup 1999). Although speculative, this interpretation most likely explains our findings. Only future research designed to test this hypothesis can ascertain whether it is more than merely plausible and whether it actually produces the patterns we observe in this study.

Although our analyses present empirical evidence that whites generally have a higher expectation of arrest, our findings should be qualified by the fact that we analyze only certain types of crimes and cannot definitely say what the effects of an offender's race might be for other offenses. Our analyses are limited to rape, robbery, and assault because it is in these types of crimes that the victim is confronted by the offender and hence is able to infer his or her physical characteristics. Further insight into the nature and strength of the underlying structural relationship between race and the probability of arrest for other crimes such as drug or property offenses must await the development of richer data sets.

Contextual analyses are also needed because it is plausible that the impact of an offender's race on the likelihood of arrest varies across social contexts. It is often argued that the amount of social control experienced by blacks in society is greatest in areas where the size of the black population presents a serious challenge to the political and economic power of whites (Jacobs & Wood 1999). On the basis of this research, we believe that future investigations should concentrate on multilevel studies in which police actions are nested within differing social contexts.

Finally, our findings do not negate the possibility that some individual police officers discriminate against black citizens. What the present analysis does show is that regardless of whatever discrimination is present at the arrest stage, the outcome is generally a lower chance of arrest for blacks than for whites. If there is discrimination against blacks by some police officers, then, given the observed net result, it appears that there must also be some compensating effect. It is also important to consider that although we find no evidence of racial discrimination, such discrimination may manifest itself at later stages in the legal process.

A paramount concern about racial discrimination in the administration of justice relates to the unequal treatment of similarly situated individuals by law enforcement officials. Our findings have profound implications since they bear directly on the current debate as to whether the police perform their duties in a racially discriminatory fashion. The results of this study suggest that the disproportionately high arrest rate for black citizens is most likely ascribable to differential criminal participation in reported crime rather than to racially biased law enforcement practices. The new data presented here also suggest some caution in the pervasive practice of employing race-specific arrest rates as a surrogate measure of race-specific criminal offending, at least for the crimes of robbery, aggravated assault, and simple assault.

References

Federal Bureau of Investigation. (2000). *National incident-based reporting system, Volume 1: Data collection guidelines.* Washington, DC: U.S. Government Printing Office.

Federal Bureau of Investigation. (2001). *Crime in the United States, 2000.* Washington, DC: U.S. Government Printing Office.

Gallup, G., Jr., & Gallup, A. (1999). *The Gallup Poll monthly,* No 411.

Gibbs, J. (1988). *Young, black, and male in America.* Dover, MA: Auburn House.

Hindelang, M. J. (1978). Race and involvement in common law personal crimes. *American Sociological Review, 43,* 93–109.

Jacobs, D., & Wood, K. (1999). Interracial conflict and interracial homicide: Do political and economic rivalries explain white killings of blacks or black killings of whites? *American Journal of Sociology, 105,* 157–190.

Sherman, L. W. (1992). *Policing domestic violence.* New York: Free Press.

Sherman, L. (2002). Trust and confidence in criminal justice. *NIJ Journal, 248,* 23–31.

U.S. Census Bureau. (2001). *Resident population estimates of the United States by sex, race, and Hispanic origin: April 1, 1990 to July 1, 1999, with short-term projection to November 1, 2000.* Washington, DC: U.S. Government Printing Office.

DISCUSSION QUESTIONS

1. How is the NIBRS by itself better than a UCR/NCVS comparison in answering questions such as these?
2. Why do we continue to hear charges of discriminatory arrests of minorities when the variety of data source comparisons mentioned here do not support the contention?
3. What is your explanation for the greater probability of arrest for white offenders?

READING

Methamphetamine Use, Self-Reported Violent Crime, and Recidivism Among Offenders in California Who Abuse Substances

Jerome Cartier, David Farabee, and Michael L. Prendergast

The study by Jerome Cartier, David Farabee, and Michael L. Prendergast illustrates the use of self-report data by criminologists. Their study uses data from 641 state prison parolees in California to examine the associations between methamphetamine use and three measures of criminal behavior: (a) self-reported violent criminal behavior, (b) return to prison for a violent offense, and (c) return to prison for any reason during the first 12 months of parole. Methamphetamine use was significantly predictive of self-reported violent criminal behavior and general recidivism (i.e., a return to custody for any reason). However, methamphetamine use was not significantly predictive of being returned to custody for a violent offense. These trends remained even after controlling for involvement in the drug trade (i.e., sales, distribution, or manufacturing).

Methamphetamine (MA), also known as "speed," "meth," "ice," "crystal," and "crank," is a highly addictive (Schedule II) stimulant that acts on the central nervous system. Unlike other stimulants, such as cocaine, MA is metabolized at a slower rate, thus producing a sustained euphoric state for up to 8 hours (Anglin et al., 2000). Preliminary data from the Arrestee Drug Abuse Monitoring (ADAM) report for the year 2000 indicate that MA use among adult male arrestees is most prevalent in the western states, with the highest rates reported in three California cities: Sacramento (27%), San Diego (25%), and San Jose (22%; National Institute of Justice [NIJ], 2001).

Although researchers have hypothesized that causal relationships exist between MA use and violence, previous studies are ambiguous in demonstrating a significant association between the two. One of the barriers to measuring this association is the dichotomous separation of alcohol and "other" drugs when discussing links to violent behavior. A substantial literature demonstrates a significant correlation between alcohol and violence (e.g., Martin, 2001), and a growing body of literature suggests a strong correlation between drug misuse and violent crime (Grann & Fazel, 2004), as well as recidivism (Bonta, Law, & Hanson, 1998). Even so, these researchers do not distinguish between the types of other drugs that would make an assessment of the relationship between a specific drug and violence possible. However, because of the increased prevalence of MA use in the past decade many researchers are now beginning to focus on the specific relationship between MA use and subsequent human behavior.

Although the findings are inconclusive, clinical studies indicate that stimulants, including MA, may increase the likelihood of attack behaviors

Source: Cartier, J., Farabee, D., & Prendergast, M. L. (2006). Methamphetamine use, self-reported violent crime, and recidivism among offenders in California who abuse substances. *Journal of Interpersonal Violence, 21*(4), 435–445. Reprinted with permission of Sage Publications, Inc.

and aggression in humans (Pihl & Hoaken, 1997). Other, nonclinical researchers have warned that public safety may be threatened by high-level MA users whose irritability and paranoia may initiate a violent reaction when brought into contact with others, especially medical or law enforcement personnel. This is supported by a Japanese study (Yui et al., 2000) that demonstrated that even mild psychosocial stressors could initiate flashbacks to MA-induced psychosis in MA users even during periods of nonuse.

In a study of MA users admitted to treatment in Los Angeles, nearly two-thirds of the participants cited violent behavior as an outcome of their usage (von Mayrhauser, Brecht, & Anglin, 2002). Wright and Klee (2001) found that 47% of their participants who used MA reported being involved in violent crime, and 24% reporting that their involvement in violent crime was a direct result of their MA use. Another study, specific to MA use in five western cities, found that one third of arrestees using MA cited violent behavior as a consequence of their use. Moreover, arrestees using MA were more likely to have been arrested and incarcerated previously than their peers who did not use MA (Pennell et al., 1999).

Violent criminal behavior has also been associated with the manufacture, sales, and distribution of illicit drugs. Goldstein (1998) referred to this as the systemic model of drug-related violence and associated the high level of violence with the need to protect manufacturing sites, distribution operations, and trafficking territories in a black-market business environment. Methamphetamine production and trafficking have been demonstrated to be strongly associated with violent behavior and have forced local law enforcement agencies in those California counties with high levels of MA production to establish task forces specifically trained to interdict the production and distribution of the drug and cope with the associated violence (Blankstein & Haynes, 2001).

Because the prevalence of MA use remains substantially higher in California than in any other state—particularly among offenders—an opportunity exists to assess the specific relationships between MA use, violence, and recidivism among a sample of parolees who abused substances. The purpose of the current study was to test the hypothesis that, even after controlling for involvement in the drug trade (sales, distribution, or manufacture), MA use would be predictive of violent crime and recidivism in a population of adult male parolees during the 12-month period following release from prison.

Method Sample

The data for the current study were obtained from interviews conducted with 641 (321 treatment, 320 comparison) of the participants 12 months after their release to parole from baseline incarceration.

The interviews used for baseline and follow-up were modified versions of criminal justice treatment evaluations forms developed by researchers at Texas Christian University (Simpson & Knight, 1998). Current MA users were those who reported any MA use during the 30 days prior to the 12-month follow-up interview. This measure used a 9-point scale ranging from 0 (*no use*) to 9 (*use of methamphetamine 4 or more times daily*). The frequency of violent crime was based on the number of days a participant reported having committed robbery (e.g., armed robbery, mugging) or violence against other persons (e.g., homicide, aggravated assault, kidnapping, domestic violence, etc.) during the 30 days prior to the follow-up interview. Drug trade involvement was also determined by self-report and included sales, distribution, or manufacturing of drugs during the 30 days prior to the follow-up interview. Although not without controversy, the use of self-reported drug use and criminal behavior has substantial support in the literature (Nieves, Draine, & Solomon, 2000; Tran-Ha, Wiley, & Des Jarlais, 1998).

One-year recidivism outcomes in the current study were based on official records from the California Department of Corrections (CDC) records. *General recidivism* was defined as being returned to prison for any reason (including technical violations of parole conditions). Parolees were categorized as having been returned to custody for a violent crime if they had a conviction of any of 44 penal code offenses comprising four categories of violent crime: murder, manslaughter, robbery, or assault.

Results

Nearly 20% of the participants reported MA use in the 30 days prior to the interview. Those reporting MA use were significantly younger than the nonusers. Moreover, the majority (63.8%) of the sample that used MA was white. Although African Americans constituted almost 42% of the entire follow-up cohort, they were only 6% of those reporting any MA use. With regard to recidivism, those who used MA (81.6%) were significantly more likely than those who did not use MA (53.9%) to have been returned to custody for any reason or to report committing any violent acts in the 30 days prior to the follow-up interview (23.6% vs. 6.8%, respectively). However, the two groups did not differ in the likelihood of being returned to prison for a violent offense.

Self-reported frequency of violent activities during the past 30 days was significantly associated with younger age and Hispanic ethnicity (relative to white non-Hispanic). MA use over the past 30 days was significantly related to the frequency of self-reported violent activities. Likewise, involvement in the drug trade also has a significant, positive relationship to violence.

Table 2.6 shows the results of two logistic regressions predicting the likelihood of being returned to prison for any reason within 12 months after release. We see that after controlling for background covariates, MA use is a strong predictor of general 12-month recidivism. In Model 3, MA use is associated with a 30% increase in the likelihood of being returned to custody within 12 months of release; the drug trade variable in Model 4 appears to account for only a small amount of the maximum likelihood estimate attributed to MA use in Model 3.

Table 2.7 shows the results of two logistic regression models predicting 12-month return to custody for a violent offense. Recidivism for violence was associated with younger age. However, none of the other background variables was statistically significant. Moreover, MA use was not related to the likelihood of being returned for a violent offense. Although involvement in the drug trade is predicative of self-reported violent crime and return to custody for any reason, it is not predictive of return to custody for a violent offense.

Table 2.6 Odds Ratios Predicting 12-Month Return to Custody for Any Reason (N = 634)

	Model 3		Model 4	
Predictor	**OR**	**95% CI**	**OR**	**95% CI**
Age	.99	.97, 1.00	.97	.97,1.01
African American	1.72*	1.17, 2.52	1.66*	1.12, 2.45
Hispanic	1.08	.67, 1.73	1.06	.66, 1.73
High school degree	.84	.60, 1.18	.88	.62, 1.24
Full-time employment	.76	.54, 1.07	.79	.55, 1.11
Married	1.15	.71, 1.87	1.13	.69, 1.85
Methamphetamine use (Past 30 days)	1.30*	1.18, 1.43	1.27*	1.15, 1.40
Drug sales (Past 30 days)	—		1.04*	1.01, 1.017
Model χ^2	38.82**		45.95**	

NOTE: OR = odds ratio; CI = confidence interval.

*$p < .05$; **$p < .001$.

Table 2.7 Odds Ratios Predicting 12-Month Return to Custody for a Violent Offense (N = 445)

	Model 5		Model 6	
Predictor	**OR**	**95% CI**	**OR**	**95% CI**
Age	.94*	.91, .97	.94*	.91, .97
African American	1.11	.63, 1.93	1.21	.68, 2.14
Hispanic	.75	.37, 1.50	.76	.39, 1.55
High school degree	1.11	.68, 1.81	1.12	.68, 1.83
Full-time employment	.98	.60, 1.60	.97	.59, 1.58
Married	1.15	.58, 2.26	1.16	.59, 2.30
Methamphetamine use (Past 30 days)	.98	.86, 1.10	1.0	.90, 1.12
Drug sales (Past 30 days)	—		.98	.94, 1.01
Model χ^2	16.6**		18.4**	

NOTE: OR = odds ratio; CI = confidence interval.

*$p < .05$; **$p < .001$.

Discussion

The purpose of the current study was to test the hypothesis that, even after controlling for drug trade involvement (i.e., sales, distribution, manufacturing), MA use would be predictive of violent crime and recidivism among adult male parolees during their first 12 months of parole. Methamphetamine use was statistically significant in predicting self-reported violent crime and general recidivism, but not for a return to custody for a violent offense. These findings held even after controlling for background variables (age, ethnicity, education, marital status, and employment), though age and ethnicity were also predictive of outcomes.

While examining the relationship between MA use and involvement in the drug trade, we found that drug trade involvement was statistically significant for self-reported violent crime and general recidivism, but not for a return to custody for a violent offense. However, after controlling for drug trade involvement, MA use was still significantly predictive of self-reported violent crime and general recidivism.

Although MA use demonstrated reliable relationships with self-reported violent crime and general recidivism, it was not predictive of recidivism for a violent offense. This inconsistency may be a methodological artifact. Offenders are not arrested for every crime they commit, and a return to custody occurs when an offender is arrested and convicted for a crime or parole is revoked for technical reasons. In addition, the official charge or instant offense resulting in a return to custody may be the result of plea bargaining and should not be assumed to be a pristine measure of offender criminal behavior.

Although the use of self-report as a measure of violent crime may be a limitation in the current study, evidence exists that the concordance of self-report with actual crime committed is quite high. In a review of recidivism studies by Lagenbucher and Merrill (2001), the majority of these studies demonstrated a high level of agreement between self-report and official records in the measurement of lifetime arrest. Moreover, where discrepancies were evident, the number of self-reported arrests was more likely to be higher than those found in official records. Another limitation is the absence of arrest records. As previously mentioned, the findings in the current study relied on official return to custody records that contain only the offense for which the parolee was convicted, or pled guilty to, and not the full array of charges that may have been cited at the time of the parolee's arrest.

We lost 19% of the original cohort to follow-up, which raises the issue of bias in our outcomes. However, an analysis of baseline data and official records shows no significant differences between this subgroup and the larger study group in the basic demographic variables and recidivism rates. Although the lack of data on self-reported MA use and violent crime at follow-up precludes us from describing any correlation between MA use, violence, and recidivism for this subgroup, the fact that both groups were nearly identical suggests that the loss of follow-up data did not bias the results.

Notwithstanding these limitations, we believe these findings, in an environment of increasing levels of MA use and high levels of recidivism (Langan & Levin, 2002), provide impetus for further research into behaviors related to MA use. Furthermore, these findings demonstrate that offenders who use MA may differ significantly from their peers who do not use MA and may require more intensive treatment interventions and parole supervision than other types of offenders who use drugs.

References

Anglin, M. D., Burke, C., Perrochet, B., Stamper, E., & Dawud-Noursi, S. (2000). History of the methamphetamine problem. *Journal of Psychoactive Drugs, 32*(2), 137–141.

Blankstein, A., & Haynes, K. (2001, August 22). Los Angeles; 23 arrests made in series of drug task force raids; Crime: the long-running probe targeted alleged methamphetamine makers and traffickers. *Los Angeles Times*, p. B3.

Bonta, J., Law, M., & Hanson, R. K. (1998). The prediction of criminal and violent recidivism among mentally disordered offenders: A meta-analysis. *Psychological Bulletin, 123*(2), 123–142.

Goldstein, P. (1998). Drugs, violence and federal funding: A research odyssey. *Substance Use and Misuse, 33*(9), 1915–1936.

Grann, M., & Fazel, S. (2004). Substance misuse and violent crime: Swedish population study. *British Medical Journal, 328*, 1233–1234.

Lagenbucher, J., & Merrill, J. (2001). The validity of self-reported cost events by substance abusers. *Evaluation Review, 25*(2), 184–210.

Langan, P., & Levin, D. (2002). *Recidivism of prisoners released in 1994* [U.S. Department of Justice, Bureau of Justice Statistics]. Washington, DC: Government Printing Office.

Martin, S. (2001). The links between alcohol, crime and the criminal justice system: Explanations, evidence and interventions. *American Journal on Addictions, 10*, 136–158.

National Institute of Justice. (2001). *ADAM preliminary 2000 findings on drug use & drug markets.* Washington, DC: U.S. Department of Justice.

Nieves, K., Draine, J., & Solomon, P. (2000). The validity of self-reported criminal arrest history among clients of a psychiatric probation and parole service. *Journal of Offender Rehabilitation, 30*(3/4), 133–151.

Pennell, S., Ellett, J., Rienick, C., & Grimes, J. (1999). *Meth matters: Report on methamphetamine users in five western cities.* Washington, DC: U.S. Department of Justice.

Pihl, R. O., & Hoaken, P. (1997). Clinical correlates and predictors of violence in patients with substance use disorders. *Psychiatric Annals, 27*(11), 735–740.

Simpson, D. D., & Knight, K. (1998). *TCU data collection forms for correctional residential treatment.* Fort Worth, TX: Texas Christian University, Institute of Behavioral Research. Retrieved from www.ibr.tcu.edu/pubs/datacoll/Forms/cjlintake.pdf.

Tran-Ha, M. H., Wiley, D. E., & Des Jarlais, D. C. (1998). Validity of self-reported data, scientific methods and drug policy. *Drug and Alcohol Dependence, 51*(3), 265–266.

von Mayrhauser, C., Brecht, M., & Anglin, M. D. (2002). Use ecology and drug use motivation of methamphetamine users admitted to substance abuse treatment facilities in Los Angeles: An emerging profile. *Journal of Addictive Diseases, 21*(1), 45–60.

Wright, S., & Klee, H. (2001). Violent crime, aggression and amphetamine: What are the implications for drug treatment services? *Drugs: Education, Prevention, and Policy, 8*(1), 73–89.

Yui, K., Goto, K., Ikemoto, S., Ishuguro, T., & Kamata, Y. (2000). Increases sensitivity to stress in spontaneous recurrent of methamphetamine psychosis: Noradrenergic hyperactivity with contribution from dopaminergic hyperactivity. *Journal of Clinical Psychopharmacology, 20*(2), 165–174.

DISCUSSION QUESTIONS

1. Does this article increase or decrease your confidence in the validity of self-report data?
2. How is methamphetamine use related to violent behavior?
3. How could the authors have further validated their self-report measures?

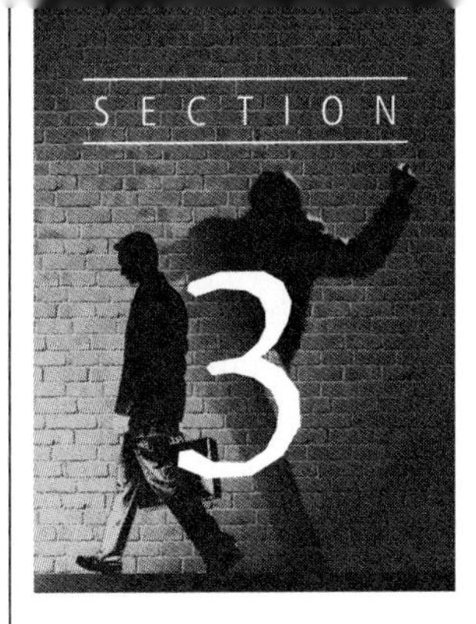

THE EARLY SCHOOLS OF CRIMINOLOGY AND MODERN COUNTERPARTS

"Lisa" is a 30-year-old mother of three children aged 8, 6, and 4. Her husband left her a year ago for another woman and his present whereabouts are unknown. Because Lisa only has a 10th-grade education and because she cannot afford child care costs, she was forced onto the welfare rolls. When Christmas rolled around she had no money to buy her children any presents, so she took a temporary Christmas job at the local Wal-Mart store where she earned $1,200 over a two-month period. Lisa did not report this income to the welfare authorities as required by law; a welfare audit uncovered her crime. The terrified and deeply ashamed Lisa pled guilty to grand theft, which carries a possible sentence of two years in prison, and she was referred to the probation department for a pre-sentence investigation report (PSI) and sentencing recommendation.

"Chris" is a 30-something male with a record of thefts and other crimes committed since he was 10 years old. Chris pled guilty before the same judge on the same day and was likewise referred for a PSI. Chris had stolen money and parts totaling $1,200 from an auto parts store during one of his very brief periods of employment.

These two cases point to a perennial debate among criminal justice scholars, with one side favoring the so-called **classical school** position and the other favoring the positivist position. Both positions are ultimately about the role of punishment in deterring crime, but the classical position maintains that punishment should fit the crime and nothing else; that is, all people convicted of similar crimes should receive the same punishment regardless of any differences they may have. Both Lisa and Chris freely chose to commit the crime, and the fact that Chris has a record and Lisa does not is irrelevant. The positivist position is that punishment should fit the offender and be appropriate to rehabilitation. Lisa's and Chris's crimes were motivated by very different considerations, they are very different people morally, and blindly applying similar punishments to similar crimes without considering the possible consequences is pure folly. Think about these two cases as you read about classical and positivists thought about human nature, punishment, and deterrence.

The Classical School

Modern criminology is the product of two main schools of thought: The classical school originating in the 18th century, and the positivist school originating in the 19th century. You may ask yourselves why a discussion of the "old masters" is necessary; after all, you don't see such discussions in physics, chemistry, or biology texts. The reason for this is that unlike those disciplines, modern criminology is still confronted by the same problems that confronted its pioneers, specifically the problem of explaining crime and criminality. Thus, their works are of more than passing interest to us.

The father of classical criminology is generally considered to be the Italian nobleman and professor of law, Cesare Bonesana, Marchese di Beccaria. In 1764, Beccaria published what was to become the manifesto for the reform of judicial and penal systems throughout Europe—*Dei Delitti e dellea Pene* (*On Crimes and Punishments*) (1764/1963). The book is an impassioned plea to humanize and rationalize the law and to make punishment more just and humane. The treatment of criminals at the time often included torture for all manner of offenses, and judges often levied vicious and arbitrary penalties, with the harshness of punishment often based on the social positions of offenders and victims and on a judge's penchant for mercy or cruelty. Public punishment was justified as social revenge and as a means of deterring others; the crueler the punishment, the greater its deterrent effect was assumed to be.

Beccaria believed that punishment should be identical for identical crimes, applied without reference to the social status of either the offender or the victim, and should be proportional to the level of damage done to society. Punishment must also be certain and swift to make a lasting impression on the criminal. He also argued for many of the due process rights, such as the right to confront accusers, to be informed of the charges, and to have the benefit of a public trial before an impartial judge as soon as possible after arrest and indictment.

Jeremy Bentham and Human Nature

Perhaps an even more prominent figure of the classical school was British lawyer and philosopher Jeremy Bentham. His major work, *Principles of Morals and Legislation* (1789/1948), is essentially a philosophy of social control based on the **principle of utility**, which prescribed "the greatest happiness for the greatest number." The principle posits that any human action at all should be judged moral or immoral by its effect on the happiness of the community. Thus, the proper function of the legislature is to promulgate laws aimed at maximizing the pleasure and minimizing the pain of the largest number in society—"the greatest good for the greatest number" (1789/1948, p. 151).

If legislators are to legislate according to the principle of utility, they must understand human motivation, which for Bentham was easily summed up: "Nature has placed mankind under the governance of two sovereign masters, pain and pleasure. It is for them alone to point out what we ought to do, as well as to determine what we shall do" (Bentham, 1789/1948, p. 125). Bentham more fully describes the principle of utility and his understanding of human nature in the piece in this section taken from his Principles.

▲ **Photo 3.1** Jeremy Bentham, often credited as the founder of University College London, insisted that his body be put upon display there after his death. You can visit a replica of it today.

The classical explanation of criminal behavior and how to prevent it can be derived from the Enlightenment assumption that human nature is hedonistic, rational, and endowed with free will. **Hedonism** is a doctrine whose central tenet is that the achievement of pleasure is the main goal of life. All other life goals are seen only as instrumentally desirable; that is, they are only desirable as means to the end of achieving pleasure or avoiding pain. Thus, hedonism is the greatest single motivator of human action.

Rational behavior is behavior that is consistent with logic. People are said to behave rationally when we observe a logical "fit" between the goals they strive for and the means they use to achieve them. The goal of human rationality is self-interest, and self-interest governs our behavior, whether in conforming or deviant directions.

Hedonism and rationality are combined in the concept of the **hedonistic calculus**, a method by which individuals are assumed to logically weigh the anticipated benefits of a given course of action against its possible costs. If the balance of consequences of a contemplated action is thought to enhance pleasure and/or minimize pain, then individuals will pursue it; if not, they will not.

Free will enables human beings to purposely and deliberately choose to follow a calculated course of action. If people seek to increase their pleasures illegally, they do so freely and with full knowledge of the wrongness of their acts, and thus society has a perfectly legitimate right to punish those who harm it.

It follows from these assumptions about human nature that if crime is to be deterred, punishment (pain) must exceed the pleasures gained from crime. Criminals will weigh the costs against the benefits of crime and desist if, on balance, the costs exceed the benefits. Estimations of the value of pleasures and pains are to be considered with reference to four circumstances: intensity (severity), duration, certainty, and propinquity (how soon after the crime pleasure or pain is forthcoming; Bentham, 1789/1948, p. 151).

The Rise of Positivism?

The explosion of technology and scientific knowledge in the 19th century led scholars to move away from classical assumptions and toward a more scientific view of human behavior. The increasingly popular view among criminologists of this period was that crime resulted from internal and/or external forces impinging on individuals, biasing or even completely determining their behavior. This position became known as *determinism*, and its adherents were known as *positivists.*

Positivism is simply the scientific method from which more *positive* knowledge can be obtained. Positivists insisted on divorcing science from metaphysics and morals and on looking only at what is, not what ought to be. The Enlightenment's flattering image of human beings gave way to the evolutionary view that we are different only in degree from other animal forms, and that science could explain human behavior just as it could explain events in the nonhuman world. Positivist criminologists were more concerned with discovering biological, psychological, or social determinants of criminal behavior than with the classical concerns of legal and penal reforms.

Cartographic Criminology

Some of the earliest positivist attempts to leave the armchair and collect facts about crime in order to understand it were cartographers: scholars who employ maps and other geographic information in their research. Rather than asking why individuals commit crimes, **cartographic criminologists** are more interested in where and when criminal behavior is most prevalent. The two most important cartographic criminologists were Andre-Michel Guerry and Lambert-Adolphe-Jacques Quetelet. Quetelet compared crime rates in France across ages, sexes, and seasons and saw the same reflections in his data that we see today in the American data: Young males living in poor neighborhoods commit the most crime. He thought socio-logically about crime before the discipline officially existed, writing that "society prepares the crime and the guilty is only the instrument by which it is accomplished" (in Vold & Bernard, 1986, p. 132).

This cartographic method crossed the English Channel to influence British researchers Henry Mayhew and Joseph Fletcher. Using British crime data from the 1830s to 1840s, both men independently mapped out the concentration of various kinds of criminal activity across England and Wales. They concluded that crime is concentrated in poor neighborhoods undergoing population changes. Many British cities experienced the same demographic changes in the early 1800s that American cities were to experience in the early 1900s. Rural people were flocking into the big cities to obtain work in the new factory system, and in the anonymity of these cities of strangers, social bonds were weakened, morals declined, and crime flourished (Levin & Lindesmith, 1971).

Biological Positivism: Cesare Lombroso and the Born Criminal

Italian army psychiatrist Cesare Lombroso published *Criminal Man* (1876), the first book devoted solely to the causes of criminality. His basic idea was that many criminals are evolutionary "throwbacks" to an earlier form of life. The term used to describe organisms resembling ancestral prehuman forms of life is **atavism.** Atavistic criminals could be identified by a number of measurable physical stigmata, included protruding jaws, drooping eyes, large ears, twisted and flattish noses, long arms relative to the lower limbs, sloping shoulders, and a coccyx that resembled "the stump of a tail," (Lombroso-Ferrero, 1911/1972, pp. 10–21). Lombroso was just one of many who sought to understand behavior with reference to the principles of evolution as they were understood at the time. If humans were just at one end of the continuum of animal life, it made sense to many people that criminals—who acted "beastly" and lacked reasoned conscience—were biologically inferior beings.

However, data did not support Lombroso's extreme views, so he modified his theory to include two other types in addition to his atavistic type: the **insane criminal**, and the **criminaloid.** Insane criminals bore some stigmata, but they were not born criminals; rather, they became criminal as a result of "an alteration of the brain, which completely upsets their moral nature" (Lombroso-Ferrero, 1911/1972, p. 74). Among the "insane" criminals were kleptomaniacs, nymphomaniacs, and child molesters. Criminaloids had none of the physical peculiarities of the born or insane criminal. They were further categorized as *habitual criminals,* who became so by contact with other criminals, the abuse of alcohol, or other "distressing circumstances"; *juridical criminals,* who fall afoul of the law by accident; and the *criminal by passion,* hot-headed and impulsive persons who commit violent acts when provoked.

▲ **Photo 3.2** Cesare Lombroso (1836–1909)

Raffael Garofalo: Natural Crime and Offender Peculiarities

Lombroso and two of his Italian contemporaries, Raffael Garofalo and Enrico Ferri, founded what became known as the Italian school of criminology. Garofalo (1885/1968) is perhaps best known for his efforts to formulate a "natural" definition of crime, wanting to anchor it in human nature. Garofalo believed that an act would be considered a crime if it was universally condemned, and it would be universally condemned if it offended the natural altruistic sentiments of probity (integrity, honesty) and pity (compassion, sympathy). Natural crimes are evil in themselves (mala in se), whereas other kinds of crimes (mala prohibita) are wrong only because they have been made wrong by the law.

Garofalo rejected the classical principle that punishment should fit the crime, arguing instead that it should fit the criminal, with the only question to be considered at sentencing being the danger the offender posed to society, as judged by his or her peculiarities. Peculiarities are characteristics that place offenders at risk for further criminal behavior. There were four such categories, each meriting different forms of punishment: extreme, impulsive, professional, and endemic. Society could only be defended from extreme criminals by swiftly executing them, regardless of the crime for which they were being punished. Impulsive criminals (alcoholics and the insane) were to be imprisoned. Professional criminals are normal individuals who choose to commit their crimes, and thus require "elimination," either by life imprisonment or transportation to a penal colony. Endemic crimes (crimes peculiar to a given region and mala prohibita crimes) could best be controlled by changes in the law.

Enrico Ferri and Social Defense

Enrico Ferri believed that moral insensibility underscored by low intelligence were the criminal's most marked characteristics: The criminal has "defective resistance to criminal tendencies and temptations, due to that ill-balanced impulsiveness which characterizes children and savages" (Ferri, 1897/1917, p. 11). Given this conception of criminals, his only rationale for punishment was **social defense.** This theory of punishment asserts that its purpose is not to deter or to rehabilitate but to defend society from criminal predation. Ferri reasoned that if criminals were not capable of basing their behavior on rational calculus, how could they be deterred? Rather, criminals must be locked up for as long as possible so that they no longer pose a threat to society.

Neoclassicism: Rational Choice Theory

A combination of high crime rates, the failure of existing criminological theories to adequately account for crime, and the emergence of more conservative attitudes in the country in the 1980s saw a swing away from the ideals of the positivist school back to the classical notion that offenders are free actors responsible for their own actions. Neoclassical criminologists are "soft determinists" because while they believe that criminal behavior is ultimately a choice, the choice is made in the context of personal and situational constraints and the availability of opportunities. In other words, rational choice theorists substitute the extremes of the classical free will concept (our actions are free of any causal chains, i.e., undetermined) for that of human agency. **Human agency** is a concept that maintains humans have the capacity to make choices and the moral responsibility to make moral choices regardless of the internal or external constraints on one's ability to do so. According to **rational choice theory**, rationality is the quality of thinking and behaving in accordance with logic and reason, such that one's reality is an ordered and intelligible system.

Rational choice theorists view rationality as a logical correspondence between the goals we have and the means we use to obtain them. This does not mean that they view people as walking calculating machines, or that they are even concerned about just how people actually go about their calculations. Rationality is both subjective and bounded, and unwanted outcomes can be produced by rational strategies (Boudon, 2003). We do not all make the same calculations or arrive at the same plan when pursuing the same goals, for we contemplate our anticipated actions with less than perfect knowledge, with different mind-sets, and with different reasoning abilities. Our emotions (guilt, shame, anxiety, etc.) also function to keep our temptations in check by "overriding" purely rational calculations of immediate gain (Mealey, 1995). We do the best we can to order our decisions relating to our self-interest with the knowledge and understanding we have about the possible outcomes of a particular course of action. All people have mental models of the world and behave rationally with respect to them, even if others might consider our behavior to be irrational. Criminals behave rationally from their private models of reality, but their rationality is constrained, as is everyone's, by ability, knowledge, emotional input, and time (Cornish & Clarke, 1986).

Rational choice theorists thus view crime in terms of Bentham's principle of maximizing pleasure and minimizing pain. People are conscious actors with the capacity to choose between alternative behaviors. They will choose crime if they perceive that its pleasures exceed the pains they might conceivably expect if discovered. The theory does not assume that we are all equally at risk to commit criminal acts, or that we do or do not commit crimes simply because we do or do not "want to." However, while the theory recognizes that factors such as temperament, intelligence, class, family structure, and neighborhood impact our choices (Clarke & Cornish, 1985, p. 168), it largely ignores these factors in favor of concentrating on the conscious thought processes involved in making decisions to offend.

Rather than focusing on the criminal, rational choice theory simply assumes a criminally motivated offender and focuses on the process of the choice to offend. This process is known as **choice structuring**, and is defined as "the constellation of opportunities, costs, and benefits attaching to particular kinds of crime" (Cornish & Clarke, 1987, p. 933). Thus, criminal events require motivated offenders meeting situations that they perceive as an opportunity to acquire something they want. Each event is the result of a series of choice-structuring decisions to initiate the event, continue, or desist, and each particular kind of crime is the result of a series of different decisions that can only be explained on their own terms: The decision to rape is arrived at quite differently than the decision to burglarize.

Routine Activities Theory

Lawrence Cohen and Marcus Felson (1979) have devised a neoclassical theory in the tradition of rational choice theory that may explain high crime rates in different societies and neighborhoods without invoking individual differences, but simply by pointing to the routine activities in that society or neighborhood. Routine activities are defined as "recurrent and prevalent activities which provide for basic population and individual needs" (Cohen & Felson, 1979, p. 593). In other words, they are the day-to-day activities characterizing a particular community. In disorganized communities, the routine activities are such that they practically invite crime.

According to Cohen and Felson (1979), crime is the result of (a) *motivated offenders* meeting (b) *suitable targets* that lack (c) *capable guardians.* If any one of these three elements is missing, crime is not likely to occur. Cohen and Felson take motivated offenders for granted and do not attempt to explain their existence. The theory is thus very much like rational choice theory in that it describes situations in which criminal victimization is likely to occur. In poor, disorganized communities, there is never a shortage of motivated offenders, and although the pickings are generally slim in such areas, victimization is more prevalent in them than in more affluent areas (Catalano, 2005). One of the obvious reasons for high victimization rates in poor, disorganized areas (besides the abundance of motivated offenders) is that they tend to lack capable guardians for either persons or property.

Routine activities theory looks at crime from the points of view of both the offender and crime prevention. A crime will only be committed when a motivated offender believes that he or she has found something worth stealing or someone to victimize who lacks a capable guardian. A capable guardian is a person or thing that discourages the motivated offender from committing the act. It can be the presence of a person, police patrols, strong security protection, neighborhood vigilance, or whatever. Because of disrupted families, transient neighbors, poverty, and all the other negative aspects of disorganized neighborhoods, except for police patrols, capable guardians are in short supply. Crime is a "situation," and crime rates can go up or down depending on how these situations (routine activities) change without any changes at all in offender motivation. Recurring situations conducive to acquiring resources with minimal effort may also tempt more individuals to take advantage of them.

Deterrence and Choice: Pain Versus Gain

That people respond to incentives and are deterred by the threat of punishment is the philosophical foundation behind all systems of criminal law. Rational choice theory evolved out of deterrence theory, which can be encapsulated by the principle of operant psychology that states that *behavior is governed by its consequences.* A positive consequence of crime for criminals is that it affords them something they want for little effort; a negative consequence is the possible punishment attached to their crimes.

Deterrence is the prevention of criminal acts by the use or threat of punishment and may be either specific or general. **Specific deterrence** refers to the effect of punishment on the future behavior of the person who experiences the punishment. For specific deterrence to work, a previously punished person must make a mental connection between an intended criminal act and the punitive consequences suffered as a result of similar acts committed in the past. Unfortunately, such connections, if made, rarely have the socially desired effect, either because memories of the previous consequences were insufficiently emotionally strong or the offender discounted them.

Committing another crime after previously being punished for one is called recidivism ("falling back" into criminal behavior). Recidivism is a lot more common among ex-convicts than repentance and rehabilitation. Nationwide, about 33% of released inmates recidivate within the first six months, 44% within the first year, 54% by the second year, and 67.5% by the third year (Robinson, 2005, p. 222). These are just the ones who are caught, so we can safely say that there is very little specific deterrent effect.

The effect of punishment on future behavior also depends on the **contrast effect**, which is the distinction between the circumstances of punishment and the usual life experience of the person being punished. The prospect of incarceration is a nightmarish contrast for those who enjoy a loving family and a valued career. The mere prospect of experiencing the embarrassment of public disgrace threatening families and careers is a strong deterrent. For those lacking these things, punishment has little effect because the contrast between the punishment and their normal lives is minimal. Specific deterrence thus works least for those who need deterrence the most.

General deterrence refers to the preventive effect of the threat of punishment on the general population; it is thus aimed at potential offenders. The existence of a system of punishment for law violators deters a large but unknown number of individuals who might commit crimes if no such system existed. Reviews of deterrence research indicate that legal sanctions do have "substantial deterrent effect" (Nagin, 1998, p. 16). Punishment has a greater deterrent effect for instrumental crimes (crimes that bring material rewards) than for expressive crimes (crimes that bring psychological rewards) the more certain it is, and the more swiftly it is applied, but as we have seen, there is little evidence that increasing the severity of sanction (in the form of sentence length) has any effect (McCarthy, 2002). These findings underscore the classical notions that individuals do (subconsciously at least) calculate the ratio of expected pleasures to possible pains when contemplating a course of action.

In their article in this section, Dutch criminologists Willem De Haan and Jaco Vos find the assumption of rational (reasoning) criminals to be a mistaken one. Based on their interviews with a number of street robbers, they conclude that the affective (emotional) aspects of criminal behavior are more important than the rational aspects in motivating offending behavior. They explore impulsiveness, self-expression, moral ambiguity, and shame in the accounts of the crimes and lives of their interviewees. They conclude that rather than simply arguing that robbers rob "to get money," we need to understand the meanings that criminals attribute to their behavior before and after they have committed their crimes.

In his article "The Economics of Crime," Nobel prize-winning economist Gary Becker, on the other hand, is more interested in the practical aspects of rationality rather than philosophizing about the relative merits of rationality and emotion in motivating behavior. Becker insists that people are rational insofar as reasoning is engaged to make choices about offending. Offenders and potential offenders are aware (more so that the average person, says Becker) of the relative probabilities attached to criminal activity. Policymakers cannot do anything about the varying ratios of rational and emotional factors motivating or inhibiting offenders, but they can increase the probability of punishment so that the deck can be stacked more favorably toward inhibiting the antisocial impulse. Becker knows that people are not calculating machines, but he also says that people are aware of the shifting probabilities of punishment and act accordingly.

Table 3.1 summarizes major differences between the classical and positivist schools.

Table 3.1 Summary and Comparisons of the Classical and Positivist Schools

	Classical	Positivist
Historical period	18th-century Enlightenment, early period of Industrial Revolution	19th-century Age of Reason, Mid-Industrial Revolution
Leading figures	Cesare Beccaria, Jeremy Bentham	Cesare Lombroso, Raffael Garofalo, Enrico Ferri
Purpose of school	To reform and humanize the legal and penal systems	To apply the scientific method to the study of crime and criminality
Image of human nature	Humans are hedonistic, rational, and have free will. Our behavior is motivated by maximizing pleasure and minimizing pain.	Human behavior is determined by psychological, biological, or social forces that constrain our rationality and free will.
Image of criminals	Criminals are essentially the same as noncriminals. They commit crimes after calculating costs and benefits.	Criminals are different from noncriminals. They commit crimes because they are inferior in some way.
Definition of crime	Strictly legal; crime is whatever the law says it is.	Based on universal human abhorrence; crime should be limited to inherently evil (mala in se) acts.
Purpose of punishment	To deter. Punishment is to be applied equally to all offenders committing the same crime. Judicial discretion to be imited.	Social defense. Punishment to be applied differently to different offenders based on relevant differences and should be rehabilitative.

Evaluation of Neoclassical Theories

Critics of neoclassical theories complain about the overemphasis on the rationality of human beings and criticize the theories for ignoring the social conditions that may make it rational for some to engage in crime (Curran & Renzetti, 2001, p. 21). We do need to understand what turns some people into "motivated offenders," that is, what it is that makes some of us willing to expend one resource (our potential loss of freedom) to attain another (the fruits of crime). Many of us don't spend our resources all that wisely because of a tendency to favor immediate gain over long-term consequences, and we would like to know why some of us more strongly favor immediate gain than others.

In response, neoclassical theorists might insist that they do not assume a model of "pure" rationality; rather, they assume a limited rationality constrained by ability, knowledge, and time (Cornish & Clarke, 1986, p. 1). These theories do not claim to explore the role of outside forces in producing criminals, but rather they explore criminal events with the purpose of trying to prevent them. They seek to deny the motivated offender the opportunity to commit a crime by target hardening. Additionally, the notion that individuals are responsible for their own actions meshes well with American values. If this assumption "grants society permission" (Williams & McShane, 2004, p. 242) to punish criminals who make purposeful decisions to flout the law, then so be it, for the act of punishment presupposes free human beings and thus dignifies them.

Policy and Prevention: Implications of Neoclassical Theories

If you were the kind of motivated rational criminal assumed by neoclassical theorists, what sorts of questions would you ask yourself at the potential crime site before you made your decision to commit the crime or not? We bet that among them would be "Is there a quick way out of the area after the job is done?" "How vulnerable are the targets (is the car unlocked, is the door open, is the girl alone)?" "What are my chances of being seen by people in the area?" "If people in this area do see me, do they look likely to do something about it?" The policy implications of neoclassical theories boil down to trying to arrange things to make criminals' choice structuring as difficult as possible, such that criminals will dissuade themselves from committing crimes.

Rational choice and routine activities theories thus shift the policy focus from large and costly social programs, such as antipoverty programs, to target hardening. They shift attention away from policies designed to change offenders' attitudes and behavior toward making it more difficult and more costly for them to offend. Examples of target hardening include antitheft devices on automobiles, the use of vandal-resistant materials on public property, improved city lighting, surveillance cameras in stores and at public gathering places, check guarantee cards, banning the sale of alcohol at sporting events, neighborhood watches, and curfews for teenagers.

Environmental design is primarily concerned with **defensible space**, defined as "a model for residential environments which inhibit crime by creating the physical expression of a social fabric that defends itself" (Newman, 1972, p. 3). It endeavors to bring people together into a tribe-like sense of community by designing the physical environment so as to awaken the human sense of territoriality. The best possible physical environment for the growth of crime is the large barracks-like blocks of apartments with few entrances, few private spaces, and few demarcation barriers that say, "This space is mine." Families must be given back a sense of ownership, for if everything is "owned" in common (elevators, walkways and staircases, balconies, grass and shrubberies), then no one takes care of it and it deteriorates rapidly. Streets must be blocked off, both to generate a sense of belonging to "my special little neighborhood" and so that criminals cannot easily access or escape them.

Summary

- The classical school of criminology began during the Enlightenment with the work of Cesare Beccaria, whose aim was to reform an arbitrary and cruel system of criminal justice.

- Jeremy Bentham, best known for his concept of the hedonistic calculus, was another leading figure. The hedonistic calculus summarized the classical notion of human nature as hedonistic, rational, and possessing free will.

- The positivist school aimed at substituting the methods of science for the armchair philosophizing of the classicists; i.e., they sought measurable causes of behavior.

- The cartographic criminologists such as Geurry, Quetelet, Mayhew, and Fletcher were among the first positivists. These scholars studied maps and statistics to pinpoint where and when crime was most likely to occur.

- Cesare Lombroso is widely considered the father of criminology. His work was much influenced by evolutionary thought as he understood it. Lombroso saw criminals as atavistic "throwbacks" to an earlier evolutionary period who could be identified by a number of bodily stigmata.

- Other early positivists included Raffael Garofalo and Enrico Ferri. Garofalo was interested in developing a "natural" definition of crime and in generating categories of criminals for the purpose of determining what should be done with them. Ferri was instrumental in formulating the concept of social defense as the only justification for punishment.

- Neoclassical theories reemerged in the form of rational choice and routine activities theories in the 1970s. These theories assume that humans are rational and self-seeking, although rationality is bounded by knowledge levels and thinking abilities. They downplay personal and background factors that influence choices in favor of analyzing the processes leading to offenders' choices to offend.

- Routine activities theory looks at a criminal event as a motivated offender meeting a suitable target lacking a capable guardian. These ideas show how crime rates can go up or down, without a change in the prevalence of motivated offenders, by increasing or decreasing suitable targets and capable guardians.

EXERCISES AND DISCUSSION QUESTIONS

1. If humans are primarily motivated by the hedonistic calculus, is simple deterrence the answer to the crime problem?

2. What advantages (or disadvantages) does positivism offer us over classicism?

3. Is Ferri's social defense rationale for punishment preferable to one emphasizing rehabilitation of offenders?
4. Use any search engine and type in "Beccaria preventing crime." How do Beccaria's ideas compare with those of the positivists for preventing crime? What is Beccaria's idea of "real crime," and how does it compare with Garofalo's?

USEFUL WEB SITES

Biological Positivism. http://seposition.redemption.co.nz/michaeltravis/writing/positivism.htm.

Classical and Positivist Schools of Criminology. http://faculty.ncwc.edu/TOCONNOR/301/301lect02.htm.

Enrico Ferri. http://enricoferri.com/.

Rational Choice Theory. http://privatewww.essex.ac.uk/~scottj/socscot7.htm.

Routine Activities Theory. http://www.crimereduction.gov.uk/learningzone/rat.htm.

CHAPTER GLOSSARY

Atavism: Cesare Lombroso's term for his "born criminals," meaning that they are evolutionary "throwbacks" to an earlier form of life.

Cartographic criminologists: Criminologists who employ maps and other geographic information in their research to study where and when crime is most prevalent.

Choice structuring: A concept in rational choice theory referring to how people decide to offend, and defined as "the constellation of opportunities, costs, and benefits attaching to particular kinds of crime."

Classical school: The classical school of criminology was a non-empirical mode of inquiry similar to the philosophy practiced by the classical Greek philosophers.

Contrast effect: The effect of punishment on future behavior depends on how much the punishment and the usual life experience of the person being punished differ or contrast.

Criminaloid: One of Lombroso's criminal types. They had none of the physical peculiarities of the born or insane criminal and were considered less dangerous.

Defensible space: A model for residential environments which inhibit crime by creating the physical expression of a social fabric that defends itself.

Deterrence: The prevention of criminal acts by the use or threat of punishment; deterrence may be either specific or general.

Free will: That which enables human beings to purposely and deliberately choose to follow a calculated course of action.

General deterrence: The assumed preventive effect of the threat of punishment on the general population; i.e., potential offenders.

Hedonism: A doctrine assuming that the achievement of pleasure or happiness is the main goal of life.

Hedonistic calculus: Combining hedonism and rationality to logically weigh the anticipated benefits of a given course of action against its possible costs.

Human agency: A concept that maintains humans have the capacity to make choices and the moral responsibility to make moral ones regardless of the internal or external constraints on one's ability to do so.

Insane criminal: One of Lombroso's criminal types. Insane criminals bore some stigmata but were not born criminals. Among their ranks were alcoholics, kleptomaniacs, nymphomaniacs, and child molesters.

Positivism: An extension of the scientific method—from which more *positive* knowledge can be obtained—to social life.

Principle of utility: A principle that posits that any human action at all should be judged moral or immoral by its effect on the happiness of the community.

Rational: Rational behavior is behavior consistent with logic, a logical "fit" between the goals people strive for and the means they use to achieve them.

Rational choice theory: A neoclassical theory asserting that offenders are free actors responsible for their own actions. Rational choice theorists view criminal acts as specific examples of the general principle that all human behavior reflects the rational pursuit of benefits and advantages. People are conscious social actors free to choose crime, and they will do so if they perceive that its utility exceeds the pains they might conceivably expect if discovered.

Routine activities theory: A neoclassical theory pointing to the routine activities in a society or neighborhood that invite or prevent crime. Routine activities are defined as "recurrent and prevalent activities which provide for basic population and individual needs." Crime is the result of (a) motivated offenders meeting (b) suitable targets that lack (c) capable guardians.

Social defense: A theory of punishment promulgated by the Italian School of criminology asserting that its purpose is not to deter or to rehabilitate but to defend society against criminals.

Specific deterrence: The effect of punishment on the future behavior of the person who experiences the punishment.

READING

An Introduction to the Principles of Morals and Legislation

Jeremy Bentham

This excerpt from Jeremy Bentham's 1789 book, *An Introduction to the Principles of Morals and Legislation*, lays out some of the classic assumptions of the classical school. Bentham was the most prominent member of this school, and it is to him that we owe the hedonistic (or "felicitous") calculus concept. People choose their activities according to the calculus of the pleasures they expect to receive and the pains that could be avoided. Note how Bentham qualifies the pleasure/pain motive by reference to such things as intensity, duration, and certainty.

Of the Principle of Utility

Nature has placed mankind under the governance of two sovereign masters, *pain* and *pleasure*. It is for them alone to point out what we ought to do, as well as to determine what we shall do. On the one hand the standard of right and wrong, on the other the chain of causes and effects, are fastened to their throne. They govern us in all we do, in all we say, in all we think: every effort we can make to throw off our subjection, will serve but to demonstrate and confirm it. In words a man may pretend to abjure their empire: but in reality he will remain subject to it all the while. The *principle of utility* recognizes this subjection and assumes it for the foundation of that system, the object of which is to rear the fabric of felicity by the hands of reason and of law. Systems which attempt to question it, deal in sounds instead of sense, in caprice instead of reason, in darkness instead of light.

But enough of metaphor and declamation: it is not by such means that moral science is to be improved.

The principle of utility is the foundation of the present work: it will be proper therefore at the outset to give an explicit and determinate account of what is meant by it. By the principle of utility is meant that principle which approves or disapproves of every action whatsoever, according to the tendency which it appears to have to augment or diminish the happiness of the party whose interest is in question or, what is the same thing in other words, to promote or to oppose that happiness. I say of every action whatsoever; and therefore not only of every action of a private individual, but of every measure of government.

By utility is meant that property in any object, whereby it tends to produce benefit, advantage, pleasure, good or happiness, (all this in the present case comes to the same thing) or (what comes again to the same thing) to prevent the happening of mischief, pain, evil, or unhappiness to the party whose interest is considered: if that party be the community in general, then the happiness of the community: if a particular individual, then the happiness of that individual. . . .

Value of a Lot of Pleasure or Pain, How to Be Measured

Pleasures then, and the avoidance of pains, are the *ends* which the legislator has in view: it behooves him therefore to understand their *value*. Pleasures and pains are the *instruments* he has to work with: it behooves him therefore to understand their force, which is again, in other words, their value. To a person considered by himself, the value of a pleasure or pain considered by *itself*, will be greater or less, according to the four following circumstances.

1. Its intensity.
2. Its duration.
3. Its certainty or uncertainty.
4. Its propinquity or remoteness.

These are the circumstances which are to be considered in estimating a pleasure or a pain considered each of them by itself. But when the value of any pleasure or pain is considered for the purpose of estimating the tendency of any act by which it is produced, there are two other circumstances to be taken into the account; these are,

5. Its *fecundity*, or the chance it has of being followed by sensations of the same kind: that is, pleasures, if it be a pleasure: pains, if it be a pain.
6. Its *purity*, or the chance it has of not being followed by sensations of the *opposite* kind: that is, pains, if it be a pleasure: pleasures, if it be a pain.

These two last, however, are in strictness scarcely to be deemed properties of the pleasure or the pain itself; they are not, therefore, in strictness to be taken into the account of the value of that pleasure or that pain. They are in strictness to be deemed properties only of the act, or other event, by which such pleasure or pain has been produced; and accordingly are only to be taken into the account of the tendency of such act or such event.

To a number of persons, with reference to each of whom the value of a pleasure or a pain is considered, it will be greater or less, according to seven circumstances: to wit, the six preceding ones; viz.

1. Its intensity.
2. Its duration.
3. Its certainty or uncertainty.
4. Its propinquity or remoteness.
5. Its fecundity.
6. Its purity.

And one other; to wit:

7. Its extent; that is, the number of persons to whom it extends; or (in other words) who are affected by it.

To take an exact account then of the general tendency of any act, by which the interests of a community are affected, proceed as follows. Begin with any one person of those whose interests seem most immediately to be affected by it: and take an account,

1. Of the value of each distinguishable *pleasure* which appears to be produced by it in the *first* instance.
2. Of the value of each pain which appears to be produced by it in the *first* instance.
3. Of the value of each pleasure which appears to be produced by it *after* the first. This constitutes the *fecundity* of the first pleasure and the *impurity* of the first *pain.*
4. Of the value of each pain which appears to be produced by it after the first. This constitutes the *fecundity* of the first pain, and the *impurity* of the first pleasure.
5. Sum up all the values of all the pleasures on the one side, and those of all the pains on the other. The balance, if it be on the side of pleasure, will give the good tendency of the act upon the whole, with respect to the interests of that *individual* person; if on the side of pain, the *bad* tendency of it upon the whole.

6. Take an account of the *number* of persons whose interests appear to be concerned; and repeat the above process with respect to each. *Sum up* the numbers expressive of the degrees of *good* tendency, which the act has, with respect to each individual, in regard to whom the tendency of it is *good* upon the whole: . . . do this again with respect to each individual, in regard to whom the tendency of it is *bad* upon the whole. Take the balance; which, if on the side of *pleasure*, will give the general *good tendency* of the act, with respect to the total number or community of individuals concerned; if on the side of pain, the general *evil tendency*, with respect to the same community.

It is not to be expected that this process should be strictly pursued previously to every moral judgment, or to every legislative or judicial operation. It may, however, be always kept in view: and as near as the process actually pursued on these occasions approaches to it, so near will such process approach to the character of an exact one.

The same process is alike applicable to pleasure and pain, in whatever shape they appear: and by whatever denomination they are distinguished: to pleasure, whether it be called *good* (which is properly the cause or instrument of pleasure) or *profit* (which is distant pleasure, or the cause or instrument of distant pleasure) or *convenience*, or *advantage*, *benefit*, *emolument*, *happiness*, and so forth: to pain, whether it be called *evil*, (which corresponds to *good*) or *mischief*, or *inconvenience*, or *disadvantage*, or *loss*, or *unhappiness*, and so forth.

Nor is this a novel and unwarranted, any more than it is a useless theory. In all this there is nothing but what the practice of mankind, wheresoever they have a clear view of their own interest, is perfectly conformable to. An article of property, an estate in land, for instance, is valuable, on what account? On account of the pleasures of all kinds which it enables a man to produce, and what comes to the same thing the pains of all kinds which it enables him to avert. But the value of such an article of property is universally understood to rise or fall according to the length or shortness of the time which a man has in it: the certainty or uncertainty of its coming into possession: and the nearness or remoteness of the time at which, if at all, it is to come into possession. As to the *intensity* of the pleasures which a man may derive from it, this is never thought of, because it depends upon the use which each particular person may come to make of it, which cannot be estimated till the particular pleasures he may come to derive from it, or the particular pains he may come to exclude by means of it, are brought to view. For the same reason, neither does he think of the *fecundity* or *purity* of those pleasures.

DISCUSSION QUESTIONS

1. If you were Bentham and were asked to devise a correctional system designed to minimize crime, what would you advise legislators to do?
2. Are we motivated at bottom by only pleasure and pain? If not, what else motivates us?
3. Describe what Bentham means by fecundity and purity of pleasure.

READING

A Crying Shame

The Over-Rationalized Conception of Man in the Rational Choice Perspective

Willem de Haan and Jaco Vos

In this article, de Haan and Vos take rational choice theory to task for its over-rational view of human nature. The rational choice perspective explains all forms of crime by viewing offenders as reasoning criminals, and the authors test its heuristic potential by looking at how well it works for one special type of crime, street robbery. On the basis of a detailed analysis of offender accounts, they argue that rational choice theory fails to adequately conceptualize some of the essential aspects of this form of criminal behavior, namely, impulsiveness, expressivity, moral ambiguity, and shame. They conclude that adequate explanations of criminality require taking more seriously the affective aspects of criminal behavior and the normative meanings that perpetrators attribute to their own behavior before, during, and after the crime.

Street robbery tends to be regarded as an "opportunistic" offence. It is viewed as an act of desperation, committed by "desperados" with a chaotic lifestyle, by "muggers" who act spontaneously out of boredom, or by "losers" who "accidentally" get caught in a theft or a break-in and, instantly, decide to rob the person. In the typical case of street robbery, there seems to be no obvious relationship between the means and the goal, like in the extreme example of the case of a woman who was beaten with an iron pipe one evening by two men who tried to rob her and, when they failed to get her handbag, threw both her and the bag into a canal. How can such excessively violent behaviour, which shows a large discrepancy between a small gain for the perpetrator and the serious violation of the personal integrity of the victim, be understood?

In this article, we take one of the most commonly used explanations for criminal behaviour, the rational choice perspective, and confront it with the accounts provided by the street robbers we interviewed and who participated in focus group discussions. In this way, we take the rational choice model to task by trying out its heuristic potential. It is not our intention to test empirically the rational choice theory (RCT) as this is impossible because RCT is not a theory but merely a heuristic model which, by definition, cannot be refuted but only evaluated in terms of its usefulness.

Rational Choice

Rational choice theory has its roots in utilitarian moral philosophy, political and legal theory, and

SOURCE: de Haan, W., & Vos, J. (2003). A crying shame: The over-rationalized conception of man in the rational choice perspective. *Theoretical Criminology, 7*, 29–54. Reprinted with permission of Sage Publications, Ltd.

economics. Classical sociological theorists like Marx, Durkheim, and Weber argued that the utilitarian view of the rational individual was a fundamentally flawed "fiction" (Marx). "Individuality" as a concept and as a way of existing is a product of the division of labour in society (Durkheim), and all social actions have meaning only within a context of norms and values (Weber). To these social theorists, the social construction of the individual and the problems of meaning and morality were essential to the explanation and understanding of social action. Rational choice theorists have considered this a fundamental straying and, therefore, claim that a focus on rational choice will offer a proper foundation for social theory and bring sociology back in line with the "hard" sciences of psychology and economics.

Clarke and Cornish (1985) were the first criminologists to offer a conception of crime as the outcome of rational choices and decisions, which was built on developments in the economic analysis of criminal behaviour. Reviving the cost-benefit analysis of criminal behaviour as one of the concerns of the utilitarian tradition in criminal law and criminology, these economic models assume that individuals, whether criminal or not, share in common the properties of being active, rational decision makers who respond to incentives and deterrents. Cornish and Clarke acknowledge that the purely economic models are too idealized and too abstract to be useful for empirical research. Therefore, they take as their own starting point "the assumption that offenders seek to benefit themselves by their criminal behaviour; that this involves the making of decisions and of choices, however rudimentary on occasion these processes might be; and that these processes exhibit a measure of rationality, albeit constrained by limits of time and ability and the availability of relevant information" (Cornish and Clarke, 1986a: 1).

The characteristic feature of their rational choice approach of reasoning criminals who make rational decisions based on "strategic thinking" is that it rejects deterministic and pathological explanations for criminality in favour of explanations for criminal behaviour, which give the goal-oriented, rational, and everyday aspects of human activity a central place. In this respect, the rational choice theory can be distinguished from traditional criminological theories that presuppose that criminals are different from "normal" people.

While critical criminologists reject rational choice theory mainly for its policy implications, which they see as reproducing conservative (neo)classical penal policies, they have only rarely taken the trouble of critically investigating the theoretical problems of the rational choice approach. However, a critical discussion of rational choice theory should not be limited to the realm of crime policy but should also, and more importantly, question the basic assumptions, limitations, and shortcomings of its theoretical model.

Sociologists critical of rational choice or rational action theory see it as a dubious form of "economic imperialism," misunderstanding or disregarding the importance of the problems of meaning and morality with which the classical sociological theorists were concerned, and unwarranted in its claim to be a general theory of social action. Donald Levine (1997: 7) calls on critics of rational choice theory "to mount a more vigorous offensive to engage its defenders in an exploration of the limits of that conceptual framework, insisting on a full accounting for the customary, habitual, emotional . . . and serendipitous dimensions of human action."

By disregarding the role of norms, values, and moral emotions like guilt and shame and leaving aside these normative and emotional elements of decision making, the rational choice perspective seems to misrepresent the nature of the action it explains in terms of rational choice. Moreover, the rational choice model "misconstrues rationality and choice, by neglecting that social actors often do not have or make choices and, if they do, these choices are not necessarily rational. As soon as there is no way of denying the prevalence of systematic irrational or non-rational action, rational action theory fails and clearly more explanatory work needs to be done.

[T]hese criticisms have not been very effective. Largely this has been the result of certain 'immunization strategies' that proponents of rational choice theory tend to apply. One important way in which rational action theorists counter these criticisms and try to defend their approach is by arguing that the question of what exactly counts as 'rational' is simply beside the point, if only because 'the very concept of rational action is one of "understandable" action that we need to ask no more questions about' (Coleman, 1986: 1)." In other words, "we need to know nothing more" (Boudon, 1998: 817), because "rational action is its own explanation" (Hollis, 1977: 21).

Boudon (1998) points out three major problems with the generality claim of rational choice. One problem is that rational choice assumes that individual action is instrumental and has to be explained by the actors' will to reach certain goals, whereas action is not always instrumental and, therefore, rational choice theory cannot be a general theory of action. The second problem is that RCT has never succeeded in explaining satisfactorily important classes of phenomena. Moreover, and this is a third problem, this version of rationality is not the only one representing the uniqueness of providing explanations without black boxes. As Boudon notes, classical sociologists like Max Weber have pointed out that the causes of action reside in their meaning, i.e., in the reasons the actor has of adopting this action, and these reasons can take the form of cost-benefit considerations, but they can also take other forms. Strategies currently used to make non-instrumental actions appear instrumental, e.g., by assuming that the causes of behaviour are unknowable in principle or by supposing that actions that seem to be non-instrumental are actually instrumental at a deeper level, appear to be unconvincing if only because they raise more problems than they solve.

Within criminology, Stanley Cohen addresses the same issue by accusing Cornish and Clarke of portraying their "reasoning criminal" as "someone who not only has more rationality than the determined creatures of sociological inquiry but also has nothing but choice and rationality. Disembodied from all social context—deprivation, racism, urban dislocation, unemployment. . . ." (Cohen, 1996: 5). Cornish and Clarke would [reply] that their model is "an idealized picture of decision making" (Clarke and Cornish, 1985:170), that their version of rational choice theory is merely "informal" and that accounts of criminal behaviour do not have to be "complete" explanations of criminal conduct. In their view, simple and parsimonious accounts of criminal offending can have considerable heuristic value and be "good enough" to accommodate existing research, suggest new directions for empirical inquiry, and provide new clues for preventing and combating criminality.

Left without any substantial criteria of rationality, this line of defence, ironically, creates several paradoxes. A first paradox is that in their eagerness to provide practical policy recommendations, Cornish and Clarke, prematurely, gave up the very normative concept of rationality that economists successfully applied to issues of crime, punishment, and social control. A second paradox is that, as a consequence of the fact that the number of subjective assumptions that need to be made as well as the possibly anticipated consequences of any course of action that the formal model requires are both seemingly endless, the basic idea of rational choice loses "its spartanic elegance" to become almost "baroque" (Karstedt and Greve, 1995: 187). And there is the paradox that, without a clearly defined concept of rationality, rational choice theorists need to differentiate between the decision-making processes of "reasoning criminals" and everybody else, which leads them right back to the deterministic and pathological explanations of a "types of people" criminology that they, initially, rejected (1995: 189).

If we accept that the rational choice perspective is not a theory but an idealized model of decision making, there is no point in empirically testing the rational choice approach. The value of a model should be in its heuristic usefulness and is, therefore, to be appreciated by applying it and seeing how much insight and understanding of criminal behaviour it provides. This is what we try to do in this article. As a hypothesis for this exercise, we take the conclusion that Cornish and Clarke's reduction of a rational choice theory to a heuristic rational choice model does not solve any of the theoretical problems that criticisms of RCT have revealed and that, rather, because of this reduction, the heuristic potential of the model cannot be fully developed.

Interviews and Focus Groups

As part of our investigation, we conducted several interviews and group discussions with perpetrators of street robberies. Some of the respondents were detained, while others underwent alternative punishment and still others were released. For this article, we have reassessed the accounts of the respondents and interpreted them from the rational choice perspective in order to get an answer to the following questions: What were their goals? What did they consider to be specific advantages in robbing passers-by? Were perpetrators aware of any disadvantages of street robbery? Why was street robbery preferred to other forms of violent criminality and offences against property?

Impulsivity

We came across examples of robbery that were committed without planned intentions or in which the perpetrator suddenly changed his mind, without being able to give a rational explanation for this in retrospect. The rational choice perspective assumes that the committing of a crime involves costs and benefits and that analysing them makes criminal behaviour more understandable. [M]ost street robbers hardly ever weigh the costs and benefits against each other and are in fact hardly capable of doing so. The majority of robberies, and most certainly street robberies, are "opportunistic," committed by impulsive, chaotic youngsters who seldom prepare their crimes and who are not capable of advance planning. Jacobs and Wright found that the "choice" to rob occurs in a context in which rationality not only is sharply bounded, but barely exists (1999: 167).

Cornish and Clarke recognize that many robberies are "impulsive and not planned" (1986a: 6). Nevertheless, they maintain that even in situations where an unexpected opportunity exists for committing a crime, "the offender still must decide to take advantage of the situation" (1986b: 6) and this can even be a "substantial degree of rationality" (Cornish and Clarke, 1986a: 14). This demand of substantial rationality seems to be easily fulfilled for "it seems likely that 'pattern planning' would be sufficient for offences that rely largely for their success on surprise, intimidation and a general ability to seize the initiative and think on one's feet" (Cornish and Clarke, 1986a: 14). Even if the choices made and decisions taken are far from optimal, if measured according to the results, "they may make sense to the offender and represent his best efforts at optimising outcomes" (Clarke and Cornish, 1985: 164).

Like Cornish and Clarke, Feeney (1986) believes that raiders' and street robbers' impulsively taken decisions may still be considered rational because committing robberies "clearly requires some thought" (1986: 66). In his view, the only exception is robbery committed under the influence of alcohol and/or drugs. From this point of view, criminal behaviour is rational if there is a conceivable framework in which this behaviour can be seen as functional, as some means to some end. However, as such a framework can (almost) always be conceived, the rational choice perspective, as a theory, is tautological and cannot be refuted. This leaves us the question, namely, to what extent does a rational choice perspective render this kind of criminal behaviour more plausible?

In our own research, we came across offenders who, in retrospect, found it difficult to explain or did not even "have a clue" why they had committed the crime. We would like to illustrate this with the following story about the robbery of a perfume shop where the perpetrators took money from the till while they walked past it. The answers to some questions that were asked during one of the focus group discussions indicate how difficult the offender found it to even imagine that committing the crime could be seen as involving a choice.

Question: At which point could you have chosen not to do it?

Andy: Have chosen not to do it? If we hadn't seen the price tag.

Question: Which price tag?

Andy: From where we were standing we saw bottles of seventeen hundred, eighteen hundred, sixteen hundred guilders. Yes, at that point we became a bit paranoid, you know

Ahmed: One bottle?

Andy: One bottle, man, of sixteen hundred guilders, man, "Giorgio Beverly Hills." . . . Yeah, man, we became all paranoid. . . .

Ahmed: What did you want to do with it anyway?

Andy: We didn't want to do anything. We had no money, man. We thought it would be nice to have all the money. . . .

Question: That was just an idea occurring to you? You had not decided in advance to rob the store?

Andy: No, not yet. . . . Yes, when we walked past and could see there was nobody in the store. . . .

Question: But that was a point at which you could have changed your mind.

Andy: No way, man. . . . We just had too much courage that day. We just went. . . . Yes, a mistake, yes.

Question: And you got the idea by looking through the display window?

Andy: Yes, you could put it that way. Then we got even more courage, when we saw that. . . . Sixteen hundred. . . . I hate it when I see a perfume shop. . . . I never look at perfume any more.

From a rational choice approach, street robbers' impulsivity can indeed be retrospectively reconstructed in terms of rational choices and decisions. Hence, the rational choice perspective enjoins us to assume that offenders make rational decisions in split seconds, even if they were not aware of doing so. We believe that we understand certain forms of behaviour better if the involved parties have told us they have weighed pros and cons we were perhaps not aware of or had imagined being different. However, forms of behaviour in which decisions are taken impulsively or intuitively, and in which other possibilities for action are not assessed in terms of pros and cons, do not become more understandable by applying the rational choice perspective.

When people try to make sense of their behaviour, their accounts usually go hand in hand with apologies and clarifications. For example, "impulsivity" not only provides the perpetrator with an excuse but also clarifies what s/he was experiencing before, during and after s/he committed the offence. When we discount such justifications as pure rhetoric, we neglect a useful opportunity for obtaining insight into some of the motives that played a role for the perpetrator. Instead of labelling perpetrators' accounts as unconvincing or discounting them as irrelevant, it may be more productive to acknowledge their reasons, analyse the context in which they are given, and interpret them from the standpoint of the perpetrators themselves.

In order to answer the question of why people commit offences, it is necessary to listen to what they have to say and to respect their moral feelings and emotions as authentic. Moral emotions that play a role during the commission of an offence are, for example, humiliation, self-righteousness, arrogance, ridiculousness, cynicism, horror, vengeance. Drawing upon the accounts of perpetrators, Katz made a case that perpetrators more or less consciously construct the "causes" themselves—"causes" that they feel compel them to commit the offence (1988: 216). For street robbers, this means that committing a robbery is more than an easy way to get money. Robbing people serves a "larger, more widely embraced fascination with the achievement of a morally competent existence" (1988: 272). The impulsivity of a street robber is ultimately inconceivable without a deep-seated conviction that he is a "real criminal."

Thus, there is more to say about how a criminal offence takes place than whether the perpetrator consciously or unconsciously chooses to do it. A rational choice approach is not only misleading because it emphasizes or even projects rational elements into this process. By putting human action in a cost benefit analysis it also overestimates the importance of rationality in these activities, and thereby underestimates, if not neglects, the relevance of the shadow side of these activities, namely, impulsivity, lack of self-control, and faulty awareness.

Moral Ambiguity

Many perpetrators try to justify the offence for themselves and for others. Sykes and Matza (1957) have described several fallacies that delinquents use to legitimate their crimes, namely, denying their own responsibility; denying the victim; denying injury, damage, or harm; and denying others the moral right to condemn their behaviour. The apparent need for self-justification suggests the presence of feelings of shame and guilt that are "neutralized" through the above-mentioned "neutralization techniques." Thus, these neutralization techniques can be used to shed more light on offenders' feelings of shame and guilt.

Feelings of guilt or shame can generally be only indirectly inferred from perpetrators' utterances. The most remarkable thing that emerged from our interviews and focus group discussions is that by far the majority of the respondents were not attracted to robberies and many even resented having to commit such offences. Most considered the crime as a last resort or desperate act or at least presented it as such.

Perpetrators' justification of street robbery as a desperate act is an example of a "neutralization technique." Strangely enough, the rational choice theory does not answer the question whether this neutralization technique is only used in retrospect as justification or whether it enables perpetrators to commit crimes without twinges of conscience. For the rational choice perspective, neutralization techniques or rationalizations are only interesting as part of a cost-benefit analysis. Truth and the function of these rationalizations are of no concern to the rational choice perspective. At best they can give insight into the boundaries of one's rationality. So Andy's saying he had too much courage the day he committed the robbery of the perfumery indicates his "bounded" rationality that day. By leaving the question unexplored whether these so-called "neutralization techniques" are used in prospect or in retrospect, their significance for perpetrators is misjudged, or, at best, underestimated. In contrast, we consider motivation to be both a rationalization in retrospect and a reference to motives that enables perpetrators to commit their crimes without fear or twinges of conscience. In other words, motivations offer both justifications and explanations. They refer to current as well as past feelings, thoughts, desires, and fantasies.

From a rational choice perspective, it is of no interest whatsoever how criminal perpetrators deal with their emotions or how they make sense of their feelings. Of interest is merely whether the result of their thoughts and feelings can be conceived as the outcome of a cost-benefit analysis. Employing a rational choice approach leads to evaluating thoughts and feelings in terms of their functionality, whereby feelings, in particular, are defined as primarily negative. In a rational choice model, emotions are subordinate to a mode of formal reasoning, and actors behave more rationally in a substantial, empirical way to the extent that feelings, which deter them from reaching their goals, are eliminated or suppressed. From a rational choice perspective, emotions are merely interesting as elements in a cost-benefit analysis and hardly as indicating different ways in which decisions can be made.

The notion that is sustained in RCT, that neutralization techniques serve to suppress or eliminate feelings, gives a rather limited perception of the role of emotionality and morality in the committing of a robbery. The habit of some street robbers, for example, of listening to loud, aggressive rap music and consuming drugs and alcohol before "having some fun" can hardly be understood as simply a way of suppressing emotions. Moral ambiguity is displayed in the fact that feelings are not only suppressed, but also evoked. Perpetrators must feel they are capable of doing anything, that they can control things, and that at the moment they are seduced by their surroundings they can "rise above" them. The behaviour of street robbers can be seen as rational to the extent that they act within the boundaries of their own limited rationality. From this point of view, drugs and alcohol can be seen as ways to manipulate the cost-benefit analysis in such a way that, subjectively, perceived costs decrease and benefits increase. It is our contention, however, that taking drugs and alcohol is a way of deliberately dismissing a cost-benefit analysis and of defying the normal injunction to think and act rationally.

Our research showed that street robbers adopt moral boundaries. A high moral barrier for the actual committing of a robbery must be overcome by "giving yourself some pep talk." There is always the danger of creating a situation that cannot be controlled and in which borders are crossed that the person would prefer not to cross. If s/he does not want to end up in a situation in which the means seem to justify the goal, it is better to anticipate this in advance than to place all hope in "common sense." For that reason, one of the participants in the focus group discussion never carried a knife with him because he feared that he would stab someone in blind anger: "You carry that knife for a reason."

Compared to the rational choice perspective, it is possible to approach criminal behaviour by stressing the often ambivalent emotional and moral aspects of a robbery. When one of the respondents hangs around the foreign exchange office, this should not only be understood as waiting for a suitable victim and a large profit. It is also a necessary "moral warm-up" to not only seeming insensitive but also being insensitive at the crucial moment (Katz, 1988: 173).

Expressivity

All human forms of behaviour consist of instrumental and expressive aspects. On the one hand, we can consider what is the function of a concrete form of behaviour, what purpose it is meant to serve. On the other hand, we can look for the meaning of a certain action—that is, what the perpetrator wants to make clear with his/her action. Blok suggests that expressive aspects of criminal violence are often misjudged. Violence is dismissed as "useless" because an easily recognizable goal is absent (2001: 189). But if forms of behaviour can be considered as instrumental, this does not necessarily mean that expressive aspects are missing altogether.

Criminal behaviour is rational if a framework can be conceived in which it can be seen as instrumental, even if such a framework can only be constructed afterwards. In this way, every robbery can be called instrumental and, therefore, rational, which leaves us wondering whether qualitative empirical research on the perpetrator's perspective makes any sense when everything will a priori be conceived as rational. In our view, this shows the poverty of the rational choice approach with its overly rationalized conception of man. It does not lead us into the field and into the lives of the actors (offenders) that we want to understand. Instead of trying to discover the objectives of criminal behaviour by inspecting it as closely as possible, a rational choice approach deflects researchers from ethnographic investigation, giving away the opportunity of fully developing its heuristic potential.

As an example, we would like to take a look at one of the group discussions, which involved what a street robber should do in the case that the victim resists.

Marciano: When you rob someone and you don't want to hurt them and he starts resisting, then you don't know what else to do, then you have to stab. I mean, you can run away, but well, then you're also stupid.

Question: Why would you be stupid?

Marciano: You're already committing a robbery, so then you should finish it. . . . Otherwise the victim will think, "Oh, that was easy." Running away while you have a weapon in your hand and while you're robbing someone

Question: Still, why would it be stupid to run away?

Marciano: You're robbing him. I mean why would one run away?

Robin: Either you do it, or you don't.

Marciano: You have different thoughts when you're robbing someone; then you don't remind yourself to run away. You remind yourself instead just to get him. . . . Not that you immediately want to stab him, but when he resists, you just see red. . . . And then you just start stabbing.

Marciano's violent behaviour could be considered functional. For example, he intends to make it clear to the victim that he is serious or to prevent the victim from running away. However, it remains to be seen whether this is really Marciano's intent. It is for a good reason that the question about the functionality of his behaviour was a surprise to Marciano. When he is committing a robbery Marciano does not realize what he is doing; he "just does it." Committing a robbery does not fulfill a particular function for him, but is an expression of his thoughts and feelings at that particular moment.

The rational choice theory does not do justice to the expressivity of committing robberies because the theory focuses on the functionality of the means that are used for achieving a certain goal. If one wishes to use the perpetrator's perspective to make the committing of robberies more understandable, then it is not sufficient to put all the elements of a robbery into a functional context. Committing a robbery has a meaning for the perpetrator and that meaning extends beyond the direct goal of his actions. As the rational choice theory asks the wrong questions based on wrong assumptions, the expressive meaning the perpetrator attaches to committing a robbery cannot be retrieved.

In an approach to criminal behaviour which leaves room for expressive aspects, robbery is not understood as a separate act that can be placed in a functional context and assessed in terms of rationality, but "as part of a larger ethnically or subculturally relevant project" (Katz, 1988: 272). From Marciano's perspective, functionality is not the essence. Far from it: his behaviour can only be understood as part of a larger subcultural "identity project."

In our research we came across different examples of such "projects," the core of which seems to lie in the norm of being "ruthless." In one of the group discussions Rico pointed out to Glenn what "ruthless" means.

Rico: Tight, no jokes, tough guys.

Glenn: More courage, dare to fight with the police.

Rico: For nothing, but if you're ruthless, then you should also look for other people. For instance, I hang out with them. He is a tight Antillean; I am a tight Antillean. We have a couple of friends; they are tight too.

For Rico and Glenn, committing robberies is assigned meaning partly by a subculturally valued attitude of being "tight" or "ruthless." By committing robberies they give shape to their lives. The risks that are attached to committing robberies are dealt with in a daring and non-rational way. They command respect, if not from others, then at least from each other, and this enables them to continue the business of committing robberies. They not only develop a personal style of robbing, but also a corresponding self-image. The committing of the robbery reflects who the perpetrator is. Thus, it becomes part of a lifestyle and the perpetrator considers committing robbery as an "identity project" (Giddens, 1991).

Living a life in which committing robberies goes hand in hand with an unrestrained consumption of sex, drugs, alcohol, and gambling entails creating circumstances which constantly pressure perpetrators into committing more robberies. Although many of our respondents indicated they preferred not to commit robberies and intended to stop doing it after a few times, most of them continued doing it. They committed robberies because, no matter how risky it was, it at least provided them with the opportunity to maintain a lifestyle that they had become used to and that suited them. Why they nevertheless labelled the robbing of passers-by as a desperate act can only be understood when we take their mixed feelings about doing this into account.

Conclusion

The rational choice perspective claims to shed light on all forms of criminality, including the impulsive or irrational ones, enabling such forms of criminal behaviour to become more plausible. At first sight, it appeared that street robbers chose to commit an offence only after they had weighed the relative advantages and disadvantages. However, after we examined to what extent impulsivity, moral ambiguity, and expressivity could make sense if considered as part of a rational choice process, we began to doubt whether the spontaneous and moral aspects of criminal behaviour can be understood if we assume that the crimes were committed as the result of a rational and deliberate choice.

Before, during and after an offence, perpetrators often experience contradictory feelings. In addition to relief and pride, they also experience feelings of fear, regret, shame, and guilt. In the rational choice perspective, these emotional aspects of criminal behaviour can be placed in a functional context without further ado as, from a rational choice point of view, goals like excitement, status, friendship or respect are, "of course," also rational. Although these emotions occur in different phases of the crime and are not necessarily inconsistent with a rational decision-making process of weighing costs and benefits, we would still argue that the three dimensions of impulsivity, moral ambiguity, and expressivity are essential in understanding street robberies and that a rational choice perspective is incapable of taking these into account without being abstract and artificial.

Within a rational choice perspective, it is not important whether the perpetrator has actually made a rational analysis of the costs and gains at the moment of his/her action, but rather whether his/her behaviour can be interpreted retrospectively as rational in the light of specific goals. Given that these goals can be reconstructed retrospectively, a tautology emerges in which the motives of perpetrators are irrelevant, simply because "the presumed causes of action are reconstructed in a circular fashion on the basis of what is actually chosen; they are "revealed" (Turner, 1992: 193). As any decision-making process can always be interpreted as rational by way of what Elster (1993) has called "backward induction," the rational choice perspective opens the door to an unrestrained "pseudo-rationalism" (Karstedt and Greve, 1995: 188).

In a rational choice perspective, even ostensibly senseless criminal behaviour is seen as "calculated" to meet more or less legitimate but unsatisfied needs of the perpetrator. By using a heuristic method of rational reconstruction, nearly all behaviour can be seen as rational. For the proponents of a rational choice approach, even an emotional outburst does not pose a problem: it has advantages and disadvantages and can, therefore, be interpreted as a choice. Thus, in this perspective "rationalizing everything is the solution, not the problem" (Turner, 1992: 193). It hinders rather than helps us to understand why offenders feel "they gotta do what they gotta do."

We believe that it is not possible to understand and explain criminal behaviour without paying attention to the way in which offenders themselves try to understand and explain the committing of crimes. Taking seriously the emotional aspects of criminal behaviour and the moral significance perpetrators assign to their actions calls for a theory that does justice to both the rational aspects of committing a crime and the moral feelings of a perpetrator before, during, and after the offence. A more adequate explanation of criminality needs to take account of the affective aspects of criminal behaviour, the normative meanings that perpetrators attribute to their own behaviour, and the social and cultural circumstances of the perpetrators. In order to explain criminal behaviour, a theory is required that does not treat perpetrators as rational actors, but rather assumes that they are moral subjects who are compelled to give meaning to their lives.

References

Blok, A. (2001). *Honour and violence.* Cambridge: Polity Press.

Boudon, R. (2003). Limitations of rational choice theory. *American Journal of Sociology, 104*(3), 817–828.

Clarke, R., & Cornish, D. (1985). Modelling offenders' decisions: A framework for research and policy. In T. Michael & N. Morris (Eds.), *Crime and justice. An annual review of research* (Vol. 6, pp. 147–185). Chicago: University of Chicago Press.

Cohen, S. (1996). Crime and politics: Spot the difference. *British Journal of Sociology, 47*(1), 1–22.

Coleman, J. (1986). *Individual interests and collective action.* Cambridge: Cambridge University Press.

Cornish, D., & Clarke, R. (Eds.). (1986a). *The reasoning criminal. Rational choice perspectives on offending.* New York: Springer Verlag.

Cornish, D., & Clarke, R. (1986b). Situational prevention, displacement of crime and rational choice theory. In K. Heal & G. Laycock (Eds.), *Situational crime prevention: From theory into practice* (pp. 1–16). London: HMSO.

Elster, J. (1993). Some unresolved problems in the theory of rational behaviour. *Acta Sociologica, 36*(3), 179–190.

Feeney, F. (1986). Robbers as decision-makers. In D. Cornish & R. Clarke (Eds.), *The reasoning criminal. Rational choice perspectives on offending* (pp. 53–71). New York: Springer Verlag.

Giddens, A. (1991). *Modernity and self-identity. Self and society in the late modern age.* Cambridge: Polity Press.

Hollis, M. (1977). *Models of man. Philosophical thoughts on social action.* Cambridge: Cambridge University Press.

Jacobs, B., & Wright, R. (1999). Stick-up, street culture, and offender motivation. *Criminology, 37,* 149–173.

Karstedt, S., & Greve, W. (1995). "Die vernunft des verbrechens." Rational, irrational oder banal? Der "rational-choice"-ansatz in der kriminologie. In K.-D. Bussman & R. Kreissl (Eds.), *Kritische kriminologie in der diskussion. Theorien, analyses, positionen* (pp. 171–210). Opladen: Westdeutscher Verlag.

Katz, J. (1988). *Seductions of crime. Moral and sensual attractions in doing evil.* New York: Basic Books.

Levine, D. (1997). What do we profess when we profess social theory? *The ASA Theory Section Newsletter, 19*(2), 2–8.

Sykes, G., & Matza, D. (1957). Techniques of neutralisation: A theory of delinquency. *American Sociological Review, 22*(6), 664–679.

Turner, S. (1992). Rationality today. *Sociological Theory, 9*(2), 191–194.

DISCUSSION QUESTIONS

1. If rationality is a fit between means and ends, can we call the kinds of behavior examined in the article rational or not?
2. Do you think rationality or emotion guides human behavior most of the time?
3. Since street muggers are hardly the brightest of individuals, is it fair to test rational choice theory using only them?

READING

The Economics of Crime

Gary S. Becker

In this article, Nobel Prize-winning economist Gary Becker offers the economist's "pure" rational explanations of why people commit crime, that is, logical cost/benefit analyses. Crime offers rewards, and if we are to prevent it, the costs associated with crime must outweigh the rewards. In contrast to the previous article by de Haan and Vos, Becker sees a much more rational criminal who will respond to reward and punishment contingencies just like everyone else. He shows how crime fell in the U.S. as punishments increased, and how crime increased in the U.K. as punishments decreased. He then offers a number of policy recommendations consistent with deterrence theory.

During the past 35 years, crime has grown enormously, not only in the United States, in Richmond, Chicago, and elsewhere in this country, but also in most other countries. The problems that we think are unique to the United States are found in most parts of Latin America, for example. Rio de Janeiro is at least as high a crime area as is Mexico City or Bogota. And crime is found increasingly in Great Britain and Europe. In Warsaw, Poland, people are more afraid to park their cars on the street without arranging for somebody to watch them than we would be in almost any part of the United States. The same is true in Prague. And while we all think of Switzerland as the epitome of a safe country, crime is growing. The crime rate there is perhaps modest compared to that in some parts of the United States, but it is not insignificant. In fact, property crimes are probably at about the same level in most parts of Europe as in the United States.

The question to ask then is whether a high crime rate is an inevitable part of life. Or, can we do something about crime, and if so, what? Or, can we at least understand a little better the causes of the growth in crime? Knowing the causes will help us learn how to combat crime in the future.

I believe that crime is not inevitable. It's not like death and taxes, which always will be with us. High crime rates have not prevailed throughout this country's history. The 1940s and 1950s were a period of relatively low crime rates. We should see if we can devise ways to go back to the levels of those times.

What is the best way of analyzing the crime problem? There may be no single way available at present that can touch on all aspects of crime and handle all the issues that we face. But there is an approach to crime that helps us understand a surprisingly large fraction of the regularities that we observe in crime. I call it the economic approach, although it certainly far transcends emphasis on monetary gains and benefits. Let me outline briefly what this approach is, then use it to explain some observations we have regarding crime, and finally use those explanations to discuss a few public policy issues and make some recommendations.

SOURCE: The economics of crime (1995), Becker, G, *Cross Sections*, Fall, 9-15. Reprinted with permission of the Federal Reserve Bank of Richmond.

The Economic Approach

The essence of the economic approach to crime is amazingly simple. It says that people decide whether to commit crime by comparing the benefits and costs of engaging in crime. True, the forces behind individuals' decisions to commit crimes differ. But I submit that some general principles apply in trying to understand the factors that determine whether people engage in crime.

It's pretty simple to assess the benefits from crime. For property crimes, the benefits are the car that is stolen, the money that is stolen, the goods that are taken in a burglary or a robbery, and the like. Forgeries, embezzlements, and many other white collar violations also yield monetary benefits. And there are psychic, even sick, thrills that criminals might get from assault and rape and other violent acts for which there are no monetary benefits. In understanding criminal behavior, we have to recognize that these benefits exist and are, for some people, important.

Turning to the costs of crime, we first can look at the simple monetary costs. If people are engaging in crime, they are not engaging in legal work, so the value of the time that they are forgoing working is a cost. In addition, there is the likelihood that they will be caught. The punishment, if caught, could take a variety of forms, ranging from simple probation to fines to imprisonment, and the like. So punishments can be monetary, but they also can take other forms.

The likelihood that criminals will be punished also affects the costs of criminal activity. Criminal behavior is risky: the returns are uncertain, and there is a good chance that you will get caught. I believe that criminals actually like risk—they're risk takers, not risk avoiders. What supports this belief? The economic approach implies that, for a risk taker engaged in crime, the certainty of punishment is more important than the magnitude of the punishment when or if you are convicted. The punishment may still be important, but certainty of punishment for a risk taker would be more of a deterrent that the magnitude of the punishment if convicted. It may still be important to punish, and not negligibly, but the certainty of conviction is crucial.

Unfortunately, the likelihood of being caught and convicted is low. For Great Britain, the data indicates that the probability of being caught, convicted, and sent to prison for committing a crime is under two percent. I think it is higher in the United States for most crimes, but still a conviction is far from certain for most crimes.

Crime also has psychic costs associated with it. Many people do not commit crimes because they believe doing so is ethically wrong. And the feelings we have about what is right and what is wrong are important. The decline in the attitude that crime is wrong has been one factor leading to the growth of crime.

If some events raise the benefits of crime—for example, the amount of money that can be stolen, the value of cars that can be stolen—then those events encourage crime. Similarly, if you lower the probability of being convicted, reduce the punishments if convicted, weaken the strength of the belief that it is wrong to commit crime, then you encourage crime. So changes in benefits and costs are the major tools for understanding why crime changes over time, and why certain individuals or groups are more likely to commit crimes than others. The economic approach means that people are acting rationally, driven in their behavior by the benefits and costs, taking account of all the ethical and psychic and other aspects that go into determining their behavior.

Understanding Crime

Let me consider, before getting into the public policy issues, a few observations about crime that are readily understandable with an economic approach to crime. Let's start out with a simple observation. In pretty much every society that we know about, the poor and less educated are more likely to commit more violent crimes. Contrariwise, the more educated are more likely to commit embezzlements and various white-collar crimes. Why should that be so? I think a good part of the answer is that the poor and less educated don't have as many opportunities to earn. So the gain to them from spending time stealing, rather than from working at some legal job, is greater than it is for the more educated. This story does not require assuming that the poor have low I.Q.s, an idea that has received some attention. It also does not require assumptions about genetics. It is simply that, being low educated and having fewer alternatives, you will be more likely to commit crime. That may be reinforced by the tendency of poor families to be less stable and thus less likely to instill the view that crime is bad.

We also observed that, among the poor, teenagers commit a disproportionate number of robberies, burglaries, thefts, and violent crimes. So in trying to understand crime, information about the age distribution of the population is important. There is some evidence that the young are committing an increasing share of these crimes and that the age at which the young are beginning to commit crimes is going down.

There are a few simple forces that explain the increase in crimes committed by teenagers. As I have already mentioned, low earnings are a factor behind crime, and teenagers have lower earnings and fewer opportunities. Some teenagers also may discount the future more heavily in assessing the costs and potential punishment. Punishment is something that will come in the future, so groups that more heavily discount the future will be more likely to engage in crime. Peer pressure and gangs may play a role, too.

Still another factor in teenage crime is the way most countries in the West, including the United States, have structured punishments. For teenagers especially, if their first crime is not too serious, it is free. By free I mean that there essentially is no punishment. Well, people respond to signals about punishment. If one can steal a bike or something else without expecting any serious punishment, combined with the other forces that I mentioned, then it is not at all surprising that teenagers are much more likely to engage in crime.

Let me now address an observation that often is cited as challenging the economic approach to crime: that recidivism is high. By recidivism I mean that people who are in prison and then released are likely to engage in further crime. This is said to be a puzzle because if they were caught, and if they are rational, it is claimed they would not want to engage in crime again. But that reasoning shows a misunderstanding of the way rationality operates. Let me give you an analogy. Suppose a construction worker falls and is seriously injured and out of work for a year. Does that mean he won't go back into construction work after recovering? Well, if the person knew the risks associated with construction work before the accident, and if the decision initially to go into construction work took into consideration these risks, then the decision to go into construction work before the accident was rational. And if the risks remain about the same after the accident, and if the person recovers fairly fully, then we should expect the same decision to be rational after the accident. So the fact that people go back to crime, or construction, or whatever, is a sign of rationality. If they did not go back, it would mean that they were not rationally weighing costs and benefits, which is inconsistent with the economic approach to decision making.

To understand why recidivism is high, you also have to consider that, when people go to prison, they may learn how to be better criminals. They also may have more difficulty getting good jobs when they come out because they have criminal records or their skills have rusted. Add in those factors, and you would surely expect high rates of recidivism. But that is true even if convicted criminals learn nothing in prison and even if the likelihood of their getting legal jobs when they come out is unaffected by their prison stays. So for me, the high rates of repeat crime, even for those who go to prison, support the economic approach to crime.

Another observation is that drug users commit crimes at high rates. The question is whether that observation reflects causation (whether drug use causes criminal behavior) or correlation (whether drug users tend also to be criminals). I do not think we know fully, but as I mentioned already, people are more likely to commit crimes if they discount the future heavily. That suggests that drug users would commit crimes more heavily or that people who commit crimes would more heavily use drugs because the cost of drug use comes later. But it also suggests—and this turns out to be true—that people who commit crimes are more likely to drink more heavily than others, to smoke more heavily than others, and to engage in other forms of addictive behavior for which the costs of the behavior are postponed until the future and the benefits come in the present. So a good part of the association between crime and drugs may be simply a common response to the role of discounting the future.

Policy Issues

Now let me apply this analysis of crime to address several quite controversial policy issues. First let me discuss gun control, a divisive issue. I agree that guns add to the likelihood of more serious crimes, as the evidence we have indicates. Nevertheless, the issue remains of what type of gun control policy we should have. On the one hand, there is considerable support for outlawing ownership of guns, not only assault weapons but also the weapons more typically used in the commission of crime. I don't support that policy. The question that I have asked is, "How effective will it be?" We have something like 70 million handguns in the United States, so about one out of every four have guns. With about three people per family, there is on average about one gun per family. Guns are not equally distributed among all families. I estimate that in the inner city of Chicago, for example, about 60 percent to 70 percent of families own guns.

Most of the guns in the hands of the public are obtained illegally, not by going to a registered gun dealer and through the waiting period. They are obtained in the significant underground that we have now in weapons. This, I believe, makes it difficult, if not impossible, to enforce effectively a law that prohibits people from having guns. Such a law would prevent the people you want to have guns from having guns, namely, shopkeepers, homeowners, and so on, who use guns in self-defense. There have been some studies by criminologists and economists showing that in many cases, guns are used in self-defense and to protect people against the commission of crime. There are far more of these cases than cases when guns accidentally go off and harm the owner of the gun.

So what should we do? I think what we're concerned about mainly is the use of guns to commit crime. The use of guns in criminal activity is worrisome because there is some chance that crime will be escalated. And gun use induces enormous fear in the minds of victims. For these reasons, the use of guns in crimes is something we want to discourage.

It seems to me that there are two ways to reduce gun use in crimes without attacking the ownership of guns. One way is to increase the sentence significantly for people who use guns to commit crimes. For example, if the normal punishment for a robbery or burglary is a year in jail in a particular jurisdiction, then the sentence might be doubled to two years if a gun was involved in the commission of a crime. The punishment for using guns could depend on the severity of the crime, whether the gun was fired, or whether the circumstances made it more likely that the gun would be fired. This would raise the costs—the magnitude of the punishment—so it would reduce the desire to use guns in committing crimes.

A second alternative is to give a little more freedom to police to frisk people whom they have a reasonable suspicion might be carrying weapons. Such searches once were common in the United States, but their use has declined in the last couple of decades. I think it would be wise to make them more commonplace so that, on the one hand, we make an effort to locate illegal guns directly, and on the other hand, we punish the use of guns in the commission of crimes.

Now let me turn to policies concerning conviction and imprisonment. There was a large increase in crime in the 1960s and 1970s. The explanation for it is not fully known; however, two important factors are known. One, and this has been documented, the likelihood of convicting somebody of a violent crime went down sharply in the 1960s and 1970s. This came about for a combination of reasons: more attention perhaps to criminals' rights and less to victims' rights; a general belief among many intellectuals, which spread to others, that deterrents were ineffective; the view that criminals are sick (the title of a book by a famous psychiatrist), and other views of that type. Whatever the reason, people have shown that the decline in the likelihood of apprehending and convicting criminals caused a significant part of the growth in crime over these two decades.

Families also began to deteriorate in the 1960s and 1970s, and we know that children of broken families, disruptive families, or parents on drugs, etc., are more likely to commit crime. So that was becoming more common at that time and clearly was also a factor. On this basis you might have expected crime would have continued to grow in the 1980s and 1990s. And in other nations, in many respects comparable to the United States, like Great Britain, that's precisely what happened. Property crime went up by more than 50 percent in Great Britain from 1980 to the early 1990s. Yet property crimes fell by more than 25 percent after 1979 in the United States, and violent crimes also fell by much less, maybe 10 percent. These statistics are based on household studies that ask people if they were victims of crimes.

The usual statistics you see quoted in the newspapers are not these but are based on FBI calculations of crimes reported to police departments. The trouble with the latter statistics is that many people don't report crimes. If you have a bicycle stolen, why report it? You know you'll never see your bike again.

When you compare the households survey data with the FBI data, the households survey shows a lot more crime than the FBI data, especially for the crimes that are less likely to be reported—the less serious crimes, or crimes like rape that people have been embarrassed to report. It's also true that the trends in these two sets of data are different, and the household data on victimization, which are the more reliable, actually show a rather significant decrease in crime since 1980, particularly in property crimes. And still families were dissolving and the rate of dissolution was growing in the 1980s. So why did crime decrease?

Nobody fully knows the answer yet, but an important part of the answer, although it has not been demonstrated, is that the probability of conviction and imprisonment rose substantially during this period. For a number of reasons, partly due to the Supreme Court's movement toward victims' rights and away from criminals' rights, partly because law and order became a major political issue, the states began to put more money into fighting crime. And there are other reasons. But we know that the number of persons in prisons grew significantly during the 1980s. And while the growth in federal prisoners was in good part due to growth in drug use, that was not true for the equally large growth in state prisoners, only a rather small fraction of whom were drug users. Meanwhile, in Great Britain the trend was toward less rather than more punishment, so it may not be surprising that crime in Britain went up while it went down in the United States.

This suggests, in my view, that imprisonment works and, barring any more effective methods, is useful. We should distinguish, however, between people committed for serious crimes and those committed for minor drug charges. I am not an enthusiast about giving people long prison sentences for drug activity. And for those who say we cannot afford to build more prisons, let me point out that about 6 percent of state and local budgets goes for incarceration and police. It's a rather small part of total state and local spending. It's even a small part of federal spending. Governments at all levels are taking over 40 percent of the U.S. gross national product (GNP), and of that, let's say 4 percent of their budgets is going into these activities. That means that about 1.5 percent of GNP is financing incarceration and police forces. We can well afford to put more money into police and punishment if we feel they are effective deterrents. The overwhelming fraction of our budgets goes for other activities that can be cut more easily, or we can increase total taxes in a minor way to have the significant impact on spending here.

Now let me discuss a third policy: the three-strikes-and-you're-out laws. A number of states have passed them, and some others are considering them. My own view on the laws is mixed. On the one hand, there is no question that, if you have committed one or two crimes and you face the risk of going to prison for life on a third crime, the laws will deter you by significantly raising the cost of committing another crime. This is the positive side. The negative side is that you also give criminals an incentive, if they do commit these additional crimes, to escalate the severity of the crimes. The reason is that they will go to prison for life if they're convicted a third time, so it won't matter to them if they get convicted for a more severe crime or a less severe crime. That's the risk of what's called "technically marginal punishment." The economic approach teaches that you always want to have punishments rising for more severe crimes because otherwise you won't discourage people from committing the more severe crimes. But the three-strikes-and-you're-out laws for people who already have been convicted of two crimes make the marginal punishment for raising the severity of crimes zero.

How do you balance the positive and negative aspects of these laws? It's not easy. My own conclusion is that you want to limit the three-strikes-and-you're-out laws only to quite serious crimes. The fact that this will escalate these crimes isn't so important because the laws are already limited to highly serious crime. But if you do restrict the laws to serious crimes, I think there's a lot to be said for them.

I have mentioned a few law-and-order-type policies that one can use—extra punishment for using guns, frisking people, convicting at a high rate, serious punishment for major crimes, building more prisons, and maybe three-strikes-and-you're-out for serious crimes. It might seem that all that comes out of this economic approach is a harsh law-and-order approach to crime. It's true that people respond to incentives; teenagers respond to the fact that they can get a free crime. A significant fraction of people respond to the fact that punishments might be weak or nonexistent, etc. And for those who will say, "Well, people don't have information about apprehension and conviction," let me point out to you that people who are thinking of engaging in crime, who live in areas with high crime rates, know far more about these probabilities than you and I do. Interviews of young people in high crime areas who do engage in crime show an amazing understanding of what punishments are, what the young people can get away with, how to behave when going before a judge, etc. So they do know, and punishments can deter.

But the economic approach to crime does not only suggest a focus on law and order to reduce crime. It also encompasses other more fundamental, or indirect, ways to attack crime. There's no question that we should devote resources to improving the opportunities in the legal sector for teenagers, the poor, and other groups who are more likely otherwise to turn to crime. One action that I think is important for improving opportunities is to improve the quality of schools, especially in inner cities. I've been a strong advocate of a school voucher system, tailored especially to inner-city families, that can be used for tuition to private and public schools.

Another action is to keep unemployment down overall and job opportunities up with sensible monetary fiscal policies. Further, reducing the unemployment of groups that are more likely to turn to crime can help, too. One policy that will help in this regard is to not increase the minimum wage. A minimum wage simply puts people out of work, which alone is bad for them, and, as a result, leads them to commit more crime.

The final and toughest issue is how to strengthen families. Many things have been suggested like welfare reform, changing divorce laws, taking children away more readily from parents who abuse them than we do currently, and still others. I think that if we can keep unemployment down and raise the quality of schooling and do some of these other things, then we will improve the quality of family life.

I've given a very brief glimpse of what I consider to be a powerful way of trying to understand crime, and a way that leads to suggestions for policies on how to reduce significantly the unacceptably high crime rate in this country. If there is a single message that I've been giving it is that crime is not inevitable. Rather, it's a result in part, at least, of public policies; policies not only about police and prisons, but about education and a number of other things. And we can improve those policies to have a significant effect on crime. Will we return this country to the way it was in the 1950s? Probably not, but we can make it a lot closer to the 1950s than it is today. If we can get crime to be at least half the way it was in the 1950s, living in major and medium-sized cities and almost everywhere else in the United States will be a lot more pleasant than it is now.

DISCUSSION QUESTIONS

1. Assume you are Gary Becker: How would you respond to the de Haan and Vos article?
2. Assume that you are either de Haan or Vos: How would you respond to Becker?
3. Can the two sides be reconciled?

SOCIAL STRUCTURAL THEORIES

On June 15, 1975, 12-year-old Kody Scott graduated from elementary school in Los Angeles. During the ceremony, his thoughts were on "the hood" and his one ambition in life, which was to join the Eight Tray Crips; become a "ghetto star"; and major in murder, robbery, and general mayhem. He went straight from the graduation to his initiation into the gang, which involved taking part in the gunning down of 15 members of a rival faction of L.A.'s other notorious gang, the Bloods. Two years later, during a robbery in which the victim tried to run, Kody beat and stomped the man into a coma. A police officer at the scene said that "whoever did this is a monster," a name Kody proudly took as his street moniker. Monster did time in juvenile detention and then served several prison terms. During one of these terms, he converted to Afrocentric Islam and changed his name to Sanyika Shakur. He also wrote *Monster: The Autobiography of an L.A. Gang Member,* which provides a frightening portrayal of the violence of ghetto life. Shakur was paroled in 1995; returned to prison on parole violations in 1996, 1997, and 1998; and again was incarcerated for a shooting in 2000. Paroled again sometime later, he was rearrested in 2004 for "battery with great bodily harm" and again sent to prison.

Shakur was allegedly the illegitimate son of an ex-football player named Dick Bass. His mother subsequently married another man and had four more children. She divorced their father when Shakur was 6 years old and had to raise the children alone. Shakur was mistreated by his stepfather and never included in family outings. He spent almost all his childhood in the wild and chaotic streets, which he says was the only thing that really interested him. As you read this chapter about disorganized neighborhoods, blocked opportunities to legitimate success, and lower-class values, try to imagine Shakur at the center of it all and how these things may have shaped his life.

The Social Structural Tradition

Almost all sociological theories of crime touch on structure to some degree. **Social structure** is the framework of social institutions—the family, educational, religious, economic, and political institutions that operate to structure the patterns of relationships members of a society have with one another. Structural theorists maintain that wholes (societies, institutions, groups) are greater than the sums of their parts, that these wholes are real and enjoy an existence of their own separate from their individual members. Although groups cannot exist apart from the individuals who comprise them, once formed they take on an existence independent of any one individual. Structural theorists work from assumptions made from general models of society and deduce everyday experiences of individuals from those models. Their philosophy is summed up in the proposition that society prepares the crime and individuals are only the instruments that give it life. They are thus interested in seeking the social structural causes of group crime rates, rather than why particular individuals commit crimes.

Structural criminologists tend to assume that human nature is socially constructed, an unconstrained vision position that avers human traits and characteristics are specific to local cultures. This is in opposition to the constrained vision that maintains that there are a broad range of human universals that are common to all cultures, but upon which cultures determine their expression (Mallon, 2007). Given the socially constructed assumption, the task of structural criminologists is to discover why social animals commit antisocial acts. If human nature is socially constructed, the presence of antisocial individuals reflects defective social practices such as competitiveness, poverty, racism, inequality, and discrimination rather than defective human materials.

Structural theorists follow one of two general models of society—the consensus perspective or the conflict perspective. This section examines the **consensus perspective** (sometimes known as the functionalist perspective), which views society as a system of mutually sustaining parts and characterized by broad normative consensus. When any one of these parts (the various institutions) malfunctions, all the other parts are affected.

The Chicago School of Ecology

The first criminological theory to be developed in the United States was the Chicago school of human ecology developed at the University of Chicago in the 1920s and 1930s primarily through the works of Clifford Shaw and Henry McKay (1972). Early social ecologists viewed the city as a kind of super organism with "natural areas" that were adaptive for different ethnic groups (little Italy's, Chinatowns, etc.). When natural areas are eroded by "alien" ethnic groups, large increases in deviant behavior ensue.

In their analysis of Cook County Juvenile Court records spanning the years from 1900 to 1933, Shaw and McKay (1972) noted that the majority of delinquents always came from the same neighborhoods. This suggested the existence of natural areas that facilitated crime and delinquency independent of other factors. It was not claimed that residential areas "caused" crime, but rather that it was heavily concentrated in certain neighborhoods regardless of the ethnic identities of their residents.

The mechanism that led to increases in antisocial behavior was identified by Shaw and MacKay (1972) as **social disorganization**, by which they meant the breakdown of the power of informal community rules to regulate conduct. Social disorganization is created by the continuous redistribution of neighborhood populations, bringing with them a wide variety of cultural traditions sometimes at odds with traditional American middle-class norms of behavior. A neighborhood invaded by members of alien racial or ethnic groups became rife with conflicting values and conduct norms and lost its sense of community.

Social disorganization impacts crime and delinquency in two ways. First, the lack of informal social controls within them facilitates crime by failing to inhibit it. Secondly, in the absence of prosocial values, a set of values supporting antisocial behavior is likely to develop to fill the vacuum. Slum youths thus have both negative and positive inducements to crime and delinquency, represented by the absence of social controls and the presence of delinquent values, respectively. These conditions are transmitted across generations until they become intrinsic properties of the neighborhood.

Shaw and McKay (1972) worked under the assumption that effective neighborhoods were characterized by warm emotional bonds based on shared ethnicity and values, and that social control was born from this shared intimacy. Robert Sampson's (2004) concept of collective efficacy, although rooted in the social disorganization approach, has updated the notion of neighborhood control of crime without reference to the narrow focus of traditional ethnicity-based emotional ties.

Collective efficacy is the shared power of a group of connected and engaged individuals to influence the maintenance of public order. Modern neighborhoods exercise social control based on shared rational goals, utilizing shared expectations that others can be counted on to take action to prevent crime (neighborhood watches, voluntary group associations, demand for police services, etc.). As we would expect from individuals bonded by shared interests rather than shared personal ties, neighborhood collective efficacy is dynamic and task specific rather than static and generalized. Sampson and his colleagues (1997) surveyed individuals for their study rather than looking only at census data as Shaw and McKay had, and they found that collective efficacy had a very positive impact on crime rates in Chicago. We should note the same things that predict the loss of collective efficacy—concentrated poverty, lack of home ownership, rundown buildings, family disruptions, and so on—are the same things that predict social disorganization. These factors accounted for 70% of the variance in collective efficacy among the 343 neighborhood clusters examined.

The article by D. Wayne Osgood and Jeff Chambers in this section finds that the principles of social disorganization apply to rural settings as well. In common with Shaw and McKay, Osgood and Chambers trace social disorganization to high population turnover and ethnic diversity, but identify high rates of female-headed households as the most important factor. This highlights the "people vs. places" argument that has been around as long as **social ecology** theory has existed. As Ruth Kornhauser (1978, p. 104) put it: "How do we know that area differences in delinquency rates result from the aggregated characteristics of communities rather than the characteristics of individuals selectively aggregated into communities?" Some theorists argue that ecological factors have no independent effect on crime once the human composition of areas (what kinds of people live there) is taken into consideration. Others argue the opposite, while still others argue that people and places are equally important in explaining crime.

The people vs. places issue has been well researched, with the overall conclusion being that neighborhood effects are small compared with the influences of individual differences (Oberwittler, 2004). However, neighborhoods do have effects independent of individual differences (Webster, MacDonald, & Simpson, 2006).

The Anomie/Strain Tradition

French sociologist Émile Durkheim provided criminology with one of its most revered and enduring concepts: anomie. **Anomie** is a term meaning "lacking in rules" or "normlessness." Durkheim was preeminently concerned with social solidarity and the threat posed to it by social change. He distinguished between mechanical and organic solidarity. **Mechanical solidarity** exists in small, isolated, prestate societies in which individuals share common experiences and circumstances, and thus common values and strong emotional ties to the collectivity. Under these circumstances, informal social controls are strong and antisocial behavior is minimal. **Organic solidarity** is characteristic of modern societies with high degrees of occupational specialization and diversity of experiences and circumstances. This diversity weakens common values and social bonds and antisocial behavior grows. Durkheim argued that because crime is found at all times and in all societies, it is a normal and inevitable social phenomenon, and even socially useful (functional) in that it serves to identify the limits of acceptable behavior. Too much repression of deviant behavior would lead to a pathological conformity that would stifle creativity, progress, and personal freedom. Crime is one of the prices we pay for personal freedom and for social progress. Durkheim (1982) even asserted that when crime drops significantly below average levels (such as in wartime), it is a sign that something is wrong (p. 102).

▲ Photo 4.1 Émile Durkheim (1858–1917)

Although Durkheim (1982) believed that crime was "a normal phenomenon of normal sociology," he did not imply that the criminal "is an individual normally constituted from the biological and psychological points of view" (pp. 106–107). He also asserted that in any society there are always individuals who "diverge to some extent from the collective type," and among them is the "criminal character" (p. 101). Durkheim thus left room for psychologists and biologists to get into the criminology game.

Durkheim (1951) also set the stage for later extensions of anomie theory by addressing "strains" when he wrote that although human beings are similar in their "essential qualities, . . . one sort of heredity will always exist, that of natural talent" (p. 251). The crux of the crime problem at the individual level is that "human activity naturally aspires beyond assignable limits and sets itself unattainable goals" (p. 247). When people get less than they expect, they are ripe for criminal behavior.

Robert Merton's Extension of Anomie Theory

Robert Merton (1938) expanded anomie theory to develop an explanation of crime that has come to be known as **strain theory**. The central feature of Merton's theory is that American culture defines monetary success (the "American dream") as the predominant cultural goal to which all its citizens should aspire, while at the same time, American social structure restricts access to legitimate means of attaining it to certain segments of the population.

The disjunction between cultural goals and the structural impediments to achieving them is the anomic gap in which crime is bred.

Rather than anomie being an occasional condition arising in periods of rapid social change, as Durkheim saw it, for Merton, anomie is a permanent condition of society caused by this disjunction. According to Merton, being unable to attain culturally defined wants legitimately invites frustration (strain) and may result in efforts to obtain them illegitimately. Merton disagreed with Durkheim's belief that acquisitiveness is intrinsic to human nature, viewing it instead as a culturally generated characteristic. Merton claims that American culture and social structure actually exert pressure on some people to engage in *nonconforming* (criminal) behavior rather than conforming behavior. Thus, society is the cause of anomie, not the victim of it as in Durkheim's view.

Merton identified five **modes of adaptation** that people adopt in response to this societal pressure, all of which, with the exception of **conformity**, are deviant.

Conformity is the most common mode of adaptation because most people have the means at their disposal to legally attain cultural goals. Conformists accept the success goals of American society and the prescribed means of attaining them (hard work, education, persistence, dedication).

Ritualism is the adaptation of the nine-to-five slugger who has given up on ever achieving material success but who, nevertheless, continues to work within legitimate boundaries because he or she accepts the legitimacy of the opportunity structure.

Innovation is the adaptation of the criminal who accepts the cultural goals of monetary success but rejects legitimate means of attaining them. Innovation is the mode of adaptation most associated with crime. For Merton, crime is an innovative avenue to success—a method by which deprived people (the not-so-deprived also) get what they have been taught by their culture to want.

Retreatism is adopted by those who reject both the cultural goals and the institutionalized means of attaining them. Retreatists drop out of society and often take refuge in drugs, alcohol, and transience and are frequently in trouble with the law.

Rebellion is the adaptation of those who reject both the goals and the means of American society but wish to substitute alternative legitimate goals and alternative legitimate means. Rebels may be committed to some form of sociopolitical ideal, such as socialism.

Merton never systematically explored why certain individuals took on one adaptation rather than another. In a sense, his theory of crime is about the envy and resentment that his innovators feel about being left out of the American Dream. The power of relative deprivation (the discontent felt when one compares what one has with what others have and discovers that one has less than one believes one should have) is often accompanied by negative feelings, which in turn may motivate adoption of deviant coping patterns (Stiles, Liu, & Kaplan, 2000).

As we will see below, Merton's anomie strain theory has generated a great deal of interest and theoretical extension over the decades. This being so, we believe that the inclusion of his original 1938 article in this section is warranted.

Institutional Anomie Theory

Institutional anomie theory (IAT) extends anomie theory and claims that "high crime rates are intrinsic to American society: in short, at all social levels, *America is organized for crime*" (Messner & Rosenfeld, 2001, p. 5). Messner and Rosenfeld show that inequality in the United States is not an aberration of the American Dream but an expression of it: "A competitive allocation of monetary rewards [that] requires both winners and losers, and winning and losing have meaning only when rewards are distributed unequally" (Messner & Rosenfeld, 2001, p. 9).

IAT posits that the root of the problem is the subjugation of all other social institutions to the economy in the United States. American culture devalues the non-economic function and roles of other social institutions and obliges them to accommodate themselves to economic requirements. Messner and Rosenfeld (2001) claim that a great deal of the focus of the family, religion, education, law, and government is brought to bear on instilling in Americans the beliefs and values of the marketplace to the detriment of the institution-specific beliefs and values they are supposed to inculcate. The dominance of the economy thus disrupts and devalues the prosocial functioning of the other institutions and substitutes an overweening concern for the pursuit of monetary rewards, which IAT sees as profoundly criminogenic.

Messner and Rosenfeld's (2001) plan to reduce crime is **decommodification**. Commodification means the transformation of social relationships formerly untainted by economic considerations into commodities; thus, decommodification refers to policies intended to free social relationships from economic considerations by freeing the other social institutions from the domination of the economy. For instance, few people complain about finding time for their jobs as opposed to time for their family, and how many students go to college "to get a job" rather than for the love of learning? Freeing people from economic domination would allow individuals to construct their lives unconstrained by market considerations. According to IAT, this would reduce cut-throat competition, which, in return, would reduce crime. IAT is further explored from a Marxist perspective in Barbara Sims's article in Section 6.

Robert Agnew's General Strain Theory

Robert Agnew (2002) has made several attempts to fine-tune strain theory, culminating in the laying of a foundation for a **general strain theory** (GSI). For Agnew, strain is primarily the result of negative emotions that arise from negative relationships with others, not from blocked opportunities to success, as Merton argued. Multiple strains are experiences throughout our lives, but their impact differs according to their *magnitude, recency, duration,* and *clustering* (miseries that cluster together produce a whole greater than the sum of its parts and may overwhelm coping resources). Agnew tells us that strain can result in crime and delinquency through the development of a generally negative attitude about other people, a lowering of social control, and a tendency to respond to it in an aggressive manner.

The presence of strain is less important than how one copes with it. According to Agnew (2002), the traits that differentiate people who cope poorly with strain from those who cope well include "temperament, intelligence, creativity, problem-solving skills, interpersonal skills, self-efficacy, and self-esteem. . . . These traits affect the selection of coping strategies by influencing the individual's sensitivity to objective strains and the ability to engage in cognitive, emotional, and behavioral coping" (p. 123). By adding psychological variables, Agnew has moved the anomie/strain tradition away from its "pure" sociological origins toward a more interdisciplinary future.

Subcultural Theories

Subcultures evolve when a significant number of people feel alienated from the host culture and forge a lifestyle that is distinguishable from the mainstream culture. Albert Cohen (1955) proposed a subcultural theory explaining how lower-class youths adapt to the limited avenues of success open to them. Cohen maintained that most criminal behavior in lower-class neighborhoods is not a rational method of acquiring assets, but is rather an expression of short-run hedonism. **Short-run hedonism** means that the actor is seeking immediate gratification of his or her desires without regard for any long-term consequences. Much delinquent behavior is also *non-utilitarian, malicious,* and *negative* in the sense that it turns middle-class norms of behavior upside down (e.g., destroying rather than creating).

Because lower-class boys cannot adjust to what Cohen (1955) calls **middle-class measuring rods**, they experience a status frustration and spawn an oppositional culture with behavioral norms that are consciously contrary to those of the middle class. Cohen saw criminal subcultures as a kind of mass reaction formation to the problem of blocked opportunities, although he saw **status frustration**, not blocked opportunity, as the real problem. Lower-class youth desire approval and status like everyone else, but seek it via alternative means. To gain status and respect, members of criminal subcultures establish "new norms, new criteria of status which defines as meritorious the characteristics they *do* possess, the kinds of conduct of which they *are* capable" (p. 66). These status criteria are most often physical in nature, such as being ready and able to respond violently to challenges to one's manhood, or gaining a reputation as a "stud."

Another influential extension of strain theory was Cloward and Ohlin's (1960) **opportunity structure theory**. Cloward and Ohlin accepted that delinquents and criminals want middle-class financial success, but have little interest in the usual indicators of middle-class economic success, preferring "big cars," "flashy clothes," and "swell dames" (p. 96). Their biggest contribution, however, was to point out that just as there are barriers to achieving legitimate success, there are barriers to achieving illegitimate (criminal) success—it takes more than talent and motivation to make it within either the legitimate or illegitimate opportunity structures. To obtain illegitimate opportunities, would-be crooks need a friend, relative, or acquaintance who can show them "the ropes." Youths born into an established and organized delinquent subculture—the illegitimate "opportunity structure"—have a career advantage over "wannabe" outsiders. Individuals within an illegitimate opportunity structure join criminal gangs. The best example of this type of gang is organized crime gangs such as the Mafia, which has a pool of aspiring "sponsored" recruits (see Section 13).

Cloward and Ohlin (1960) identified two other gang types that develop from the frustration in lower-class culture: *conflict gangs* and *retreatist gangs.* Conflict gangs are generated in slum areas with a high degree of transience and instability as opposed to stable areas with an established illegitimate opportunity structure. Members of these loose-knit gangs commit senseless acts of violence and vandalism, and their efforts to make a living from criminal activity tend to be "individualistic, unorganized, petty, poorly paid and unprotected" (p. 73). Retreatist gang members are more "escapist" in their attitudes than conflict gang members, in that almost all of them abuse drugs and/or alcohol. In both conflict and retreatist gangs, the concern is not with remote goals but rather with the immediate gratification of present wants.

Walter Miller's Focal Concerns

Walter Miller (1958) took issue with the idea that gangs are formed as a *reaction* to status deprivation. Criminals and delinquents may resent the middle class, but it is not a matter of "If you can't join 'em, lick 'em," because their resentment is born out of envy for what middle-class people have, not for what they are. Middle-class traits such as hard work, stable habits, and responsibility are not appealing to them. Miller asserted that lower-class values must be viewed on their own terms and not as simple negations of those of the middle class. He also identified six **focal concerns** that are part of a value system and lifestyle that has emerged from the realities of life on the bottom rung of society:

> *Trouble* is something to stay out of most of the time, but life is trouble and it is something that confers status if it is the right kind (being able to handle one's self).
>
> *Toughness* is very important to the status of lower-class males. Being strong, brave, macho, sexually aggressive, unsentimental, and "not taking any shit."
>
> *Smartness* refers to street smarts, the ability to survive on the streets using one's wits.
>
> *Excitement* is the search for fun, often defined in terms of fighting, sexual adventurism, gambling, getting drunk or stoned.
>
> *Fate* is a belief that the locus of control is external to oneself.
>
> *Autonomy* means personal freedom, being outside the control of authority figures such as teachers, employers, and the police, and thus being able to "do my own thing."

The hardcore lower-class lifestyle typified by these focal concerns catches those engaged in it in a web of situations that virtually guarantee delinquent and criminal activities. The search for *excitement* leads to sexual adventures in which little preventative care is taken (*fate*), and the desire for personal freedom (*autonomy*) is likely to preclude marriage if pregnancy results. Miller (1958) was concerned about the fact that many lower-class males thus grow up in homes lacking a father or any other significant male role model. This leaves them with little supervision and leads them to seek their male identities in what Miller called "one-sex peer units" (male gangs; p. 14).

Miller's (1958) ideas are given strong support by Elijah Anderson's (1999) ethnographic work in African American neighborhoods in Philadelphia. The concentration of disadvantages in such neighborhoods has spawned a hostile oppositional culture spurning most things valued by middle-class America, as in "rap music that encourages its young listeners to kill cops, to rape, and the like" (p. 107). Anderson points out that although there are many "decent" families in these neighborhoods, the cultural ambiance is set by "street" families, which often makes it necessary for decent people to "code-switch" (adopt street values) to survive. Striving for education and upward mobility is viewed as "dissing" the neighborhood, and street people do what they can to prevent their "decent counterparts from . . . acting white" (p. 65). The street code is primarily a campaign for respect ("juice") achieved by exaggerated displays of manhood, defined in terms of toughness and sexuality. These displays contribute greatly to the high rates of homicide and out-of-wedlock births in the kinds of neighborhoods Anderson describes.

Gangs Today

Figures from the National Youth Gang Survey estimated that there were 21,500 gangs and 731,500 gang members in 2002 and that approximately half of all homicides in Chicago and Los Angeles in that year were gang related (Egley & Major, 2004). Gang membership has increased dramatically since theorists such as Cohen were writing in the 1950s and 1960s. The increase has been largely attributed to the loss of millions of manufacturing jobs in the United States (Moore & Hagedorn, 2001, p. 2), a loss that has hit our most vulnerable citizens, the young and the uneducated, hardest. Deindustrialization has set in motion a chain of events that has created a large segment of the population that is economically marginalized and socially isolated from mainstream culture.

Marginalized and isolated people (mainly African Americans and Hispanics) have become known as the "underclass" and the "truly disadvantaged" (Wilson, 1987, p. 8), and their neighborhoods are fertile soil for the growth of gangs. A survey of the ethnic composition of gangs found that 47% were Hispanic, 31% African American, 13% white, and 7% Asian (Egley & Major, 2004). It is estimated that over 25% of black males between the ages of 15 and 24 in Los Angeles County are members of the nation's two most notorious youth gangs, the Crips and the Bloods (Shelden, Tracy, & Brown, 2001, p. 28).

Why Do Young People Join Gangs?

Irving Spergel (1995) writes that "Youths join gangs for many reasons: status, security, money, power, excitement, and/or new experiences" (pp. 108–109). Joining a gang has become a survival imperative in areas where unaffiliated youths are likely to be victimized. Having "homies" watching your back makes you feel safe and secure. Gang membership also provides a means of satisfying the need to belong—having a place in the world among people who care. Gang members often display their belonging through initiation rites, secret gang signals, special clothing, "colors," and tattoos, all of which shout out loud: "I belong!" "I'm valued!" A youth camp counselor describes this function of gangs well: "The gang serves emotional needs. You feel wanted. You feel welcome. You feel important. And there is discipline and there are rules" (Bing, 1991, p. 12).

Social institutions (especially the family and the economy) satisfy most needs for most of us, but in the virtual absence of the influence of these institutions in the lives of those most affected by the economic and demographic transitions of the past few decades, the gang offers an attractive substitute means of achieving their needs. Thus, for many of its members, the gang functions as (1) family, (2) friendship group, (3) play group, (4) protective agency, (5) educational institution, and (6) employer.

Martin Sánchez-Jankowski's article in this section offers an interesting history of gangs in the United States, as well as a summary of the literature on the subject. He wants us to examine the individualist assumptions of past gang researchers who tended to see gang members as people with weak egos from broken homes looking for an identity. Sánchez-Jankowski wants us to focus more on the relationship between the structure of society and the structure of gangs. His argument is rather like the argument of IAT, in that the individualistic and competitive nature of American society requires winners and losers, and gangs are composed of society's losers. Yet, he also sees gang members in Mertonian fashion, as accepting the economic principles of America and seeking economic success within the confines of the opportunities available to them.

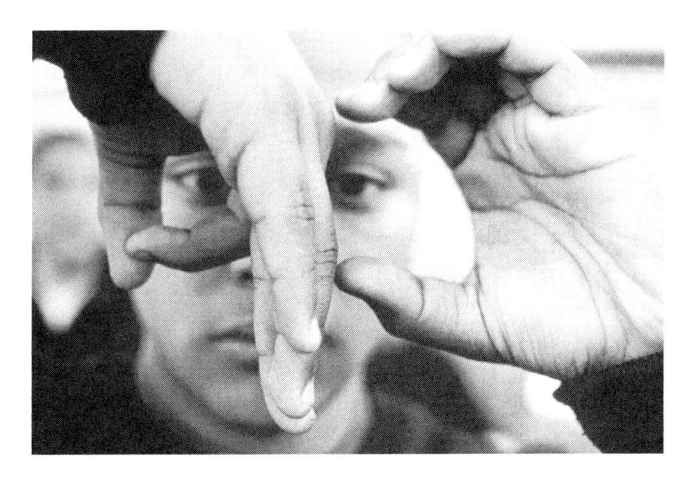

◀ **Photo 4.2** Young member of the Cypress gang in Los Angeles flashing a gang sign. Gangs claim to replace the family cohesiveness the youth may not have at home, while symbols such as colors and signs clearly demarcate members from outsiders or rival gangs.

Evaluation of Social Structural Theories

Shaw and McKay's (1972) social ecology points out that crime is concentrated in socially disorganized areas inhabited by economically deprived people. But causal direction has always been a problem; are neighborhoods run-down and criminogenic because people with personal characteristics conducive to both crime and poverty populate them, or do neighborhoods somehow "cause" crime independent of the characteristics of people living there? After all, when formerly blighted areas become "gentrified" and middle-class people move in, the neighborhood is no longer "criminogenic." Also, ecological theory cannot account for why the majority of people in disorganized neighborhoods (Anderson's "decent families") do not commit serious crimes, or why among those who do, a very small minority commits the majority of them.

Because of the emphasis on monetary success in the United States, anomie/strain theory should best explain rational crimes netting perpetrators monetary gain. Bartol and Bartol (1989) claim that anomie theory's strength lies in its "ability to explain why utilitarian crime rates are so high in one society (e.g., the United States) and so low in another (e.g., England)" (p. 110). Ever since the mid-1980s, however, England has had higher rates of utilitarian crimes (e.g., burglary, auto theft) than the United States (van Kesteren, Mayhew, & Nieuwbeerta, 2000). It is non-utilitarian crimes such as murder, rape, and assault that are more common in the United States than in England, which supports Cohen's (1955) contention that much of American lower-class crime is non-utilitarian and malicious rather than rationally instrumental.

The subcultural theories look at patterned ways of life in areas whose members set themselves apart and pride themselves in their distinctiveness. It is this patterned way of life that sustains delinquent values and goals. However, a number of theorists have cast doubt as to whether there are distinct lower-class subcultures in this sense. It is difficult to imagine that lower-class subculture arose by deliberately taking steps to take middle-class norms and turn them on their head, as Cohen's (1955) reaction formation hypothesis supposes. Nevertheless, there do seem to be areas in our cities in which middle-class values are disdained, not because they are defined as middle class, but rather because they demand self-control, delayed gratification, and the disciplined application of effort.

Miller's (1958) focal concerns are alive, well, and extensively documented. If Miller (and Anderson) is correct about the role of fate in lower-class life, then the whole anomie/strain argument about blocked opportunities may be well off base. If lower-class individuals perceive their opportunities in a fatalistic "live-for-the-moment" way, or spurn them as antithetical to the street code, they, or the visions of reality and the values imparted by their subculture, are blocking them, not the "system."

Table 4.1 summarizes the strengths and weaknesses of the social structural theories.

Policy and Prevention: Implications of Social Structural Theories

If socially disorganized slum neighborhoods are the "root cause" of crime, what feasible policy strategies might be recommended to public policymakers? One of the first things you might want to suggest would be the strengthening of community life, but how do we go about it? Clifford Shaw began by securing funds for the Chicago Area Project (CAP), which consisted of a number of programs aimed at generating or strengthening a sense of community in neighborhoods with the help and cooperation of schools, churches, recreational clubs, trade unions, and businesses. Athletic leagues, various kinds of clubs, summer camps, and many other activities were formed to busy the idle hands of the young. "Street corner" counselors were hired to offer advice to youths and to mediate with the police on their behalf when they got into trouble. Neighborhood residents were encouraged to form committees to resolve neighborhood problems.

Despite the money and energy investing in the CAP between 1932 and 1957, its effects were never evaluated in any systematic way. Similar programs in other cities had a number of positive outcomes, but their impact on crime and delinquency rates was negligible. Writing about the overall impact of CAP type programs, Rosenbaum, Lurigio, and Davis (1998) concluded that there were "few positive program effects. The local programs did not affect official crime rates and in some cases were associated with adverse change in survey-based victimization rates" (p. 214).

The ideas of strain theory had tremendous impact on public policy via President Lyndon Johnson's War on Poverty. If crime is caused by a disjunction between cultural values emphasizing success for all and social structure denying some access to legitimate means of achieving it, then the cure for crime is to increase opportunities or dampen aspirations. The latter option is not acceptable to policymakers of either the right or the left, so we are left with the task of trying to increase opportunities.

Following in the footsteps of CAP, Richard Cloward and Lloyd Ohlin (1960) developed a delinquency prevention project known as Mobilization for Youth (MFY), which concentrated on expanding legitimate opportunities for disadvantaged youths via a number of educational, training, and job placement programs. MFY programs received generous private, state, and federal funds and served as models for such federal programs as Head Start, the job corps, the Comprehensive Employment and Training Act (CETA), affirmative action, and many others (LaFree, Drass, & O'Day, 1992).

Table 4.1 Summarizing Social Structural Theories

Theory	Key Concepts	Strengths	Weaknesses
Social Disorganization	Poverty concentrates people of different cultural backgrounds and generates cultural conflict. The breakdown of informal social controls leads to social disorganization, and peer group gangs replace social institutions as socializers.	Explains high crime rates in certain areas. Accounts for intergenerational transmission of deviant values and predicts crime rates from neighborhood characteristics.	Cannot account for individuals and groups in the same neighborhood who are crime free or why a few individuals commit a highly disproportionate share of crime.
Anomie (Durkheim)	Rapid social change leads to social deregulation and the weakening of restraining social norms. This unleashes "insatiable appetites," which some seek to satisfy through criminal activity.	Emphasizes the power of norms and social solidarity to restrain crime and points to situations that weaken them.	Concentrates on whole societies and ignores differences in areas that are differentially affected by social deregulation.
Anomie/Strain (Merton)	All members of American society are socialized to want to attain monetary success, but some are denied access to legitimate means of attaining it. These people may then resort to crime to achieve what they have been taught to want.	Explains high crime rates among the disadvantaged and how cultural norms create conflict and crime. Explains various means of adapting to strain.	Does not explain why individuals similarly affected by strain to not react (adapt) similarly.
Institutional Anomie	America is literally organized for crime due to its overweening emphasis on the economy and material success. All other institutions are devalued and must accommodate themselves to the requirements of the economy.	Explains why crime rates are higher in America than in other capitalist societies. Points to decommodification as crime reduction strategy.	Concentrates on single cause of crime. Should predict high rates of property crime in America rather than violent crime, but the opposite is true.
General Strain	There are multiple sources of strain, and strain differs along numerous dimensions. Strain is the result of negative emotions that arise from negative relationships with other as well as from sociocultural forces. Individual characteristics help us to cope poorly or well with strain.	Reminds us that strain is multifaceted and that how we cope with it is more important than its existence. Adds individual characteristics to theory.	Criticized by structural theorists as reductionist because it fails to explore structural origins of strain.

(Continued)

Table 4.1 (Continued)

Theory	Key Concepts	Strengths	Weaknesses
Subcultural	Much delinquency is short-run hedonism rather than utilitarian. Lower-class youths cannot live up to middle-class measuring rods and thus develop status frustration. They seek status in ways peculiar to the subculture. Subcultural youths do not have equal illegitimate opportunities for attaining success. Those who do join criminal gangs; those who don't join retreatist and conflict gangs and engage in mindless violence and vandalism.	Extends the scope of anomie theory and integrates social disorganization theory. Focuses on processes by which lower-class youths adapt to their disadvantages and shows that illegitimate opportunities are also denied to some. Explains the patterned way of life that sustains delinquent values and goals.	Explains subcultural crime and delinquency only. There is some question as to whether a distinct lower-class culture exists in the sense that it is supported by proscriptive values that require antisocial behavior.

Some unknown number of people was diverted from a life of crime because of the opportunities presented to them by such programs, but unfortunately their heyday occurred at the same time that the United States was undergoing a huge jump in crime, from 1965 to 1980. This unfortunate convergence provided conservatives with arguments against the use of social welfare policies to combat crime.

The policy recommendation of institutional anomie theory would be to tame the power of the market via decommodification. For instance, the decision to have children could be freed from economic considerations by granting government-guaranteed maternity leave benefits and family allowances/income support, and higher education could be accessible to all people with talent without regard for the financial ability to pay. In other words, policies could ensure an adequate level of material well-being that is not so completely dependent on an individual's performance in the marketplace.

Any policy recommendations derived from subcultural theories would not differ from those derived from ecological or anomie/strain theories. Changing a subculture is extremely difficult. Insofar as a subculture is a patterned way of life, we cannot attack the problem in parts and expect to change the whole. One possible strategy would be to disperse "problem families" throughout a city rather than concentrating them in block-type projects as is typically done. But even if this was politically feasible, rather than breaking up the subculture and its values, it may result in its displaced carriers "infecting" areas previously insulated from deviant values.

The gang problem offers obvious policy recommendations in theory, such as increasing low-skill work opportunities by preventing American companies from moving overseas and strengthening the other social institutions for which gangs are a substitute. Gangs will always be a problem while legitimate social institutions in our poorest areas are too weak to provide young people with their basic needs.

Summary

- Social structural theories focus on social forces that influence people to commit criminal acts. Ecological theory emphasizes that "deviant places" can cause delinquent and criminal behavior regardless of the personal characteristics of individuals residing there. Such areas are characterized by social disorganization, which results from diverse cultural traditions within slum areas.

- One of the most interesting early findings of this perspective is that the same slum areas continued to have the highest crime rates in a city regardless of the ethnic or racial composition of its inhabitants. More recent ecological studies find that neighborhoods do have effects independent of the people who live in them, but most effects are mediated by individual differences.

- Collective efficacy is the opposite of social disorganization, but it does not focus on emotional bonds tied to ethnicity. The concept is about a neighborhood's ability to mobilize its residents as an effective force to fight problems, including crime.

- Anomie/strain theories focus on the strain generated by society's emphasis on success goals coupled with its denial to some of access to legitimate opportunities to achieve success. Merton's strain theory focuses on the ways people adapt to this situation via conformity, ritualism, retreatism, rebellion, and innovation (the modes of adaptation). Although the latter four modes are "deviant," they are not all criminal. The innovator and the retreatist modes are considered the most criminal.

- Institutional anomie theory argues that the United States is literally organized for crime because the institutional balance of power strongly favors the economy. All other American institutions are subordinate to our highly competitive economy, and the competition would be meaningless if there were not both winners and losers.

- General strain theory argues that there are many sources of strain other than Merton's disjunction between goals and means. These strains result in negative emotions that adversely affect relationships with others and may lead to crime. The important thing is not strain, however, but how people cope with it. Among the many attributes Agnew lists as coping resources are temperament, intelligence, and self-esteem.

- Subcultural strain theories have slightly different emphases. Albert Cohen noted that lower-class boys, knowing that they cannot live up to the middle-class measuring rod, form oppositional gangs that perpetuate an oppositional subculture. These gangs usually reject both the goals and the means of middle-class society, as gauged by the malicious and nonutilitarian nature of many of their crimes.

- Walter Miller augments Cohen's assertion that lower-class culture is oppositional to middle-class culture with his theory of focal concerns. Focal concerns—trouble, toughness, smartness, excitement, fate, and autonomy—are behavioral norms of lower-class culture that command strong emotional attention. Miller's thesis is supported by later work done by Elijah Anderson.

- Cloward and Ohlin emphasize that people have differential access to illegitimate, as well as legitimate, means to success and that sociological and psychological factors limit a person's access to both.

- Youth gangs have been noted throughout recorded history. The prevalence of gangs in the United States is greater than ever before and has been attributed to the deindustrialization of America. Deindustrialization has affected minorities the most and has tended to leave a sizable number of them marginalized from mainstream society and living in disorganized neighborhoods. The gang becomes an attractive option to many of these youths because it offers them many of the things that the ineffective social institutions in those neighborhoods do not.

EXERCISES AND DISCUSSION QUESTIONS

1. What is your position on the "kinds of people vs. kinds of places" argument in ecological theory? Do places matter independently of the people living in them?
2. Is the American stress on material success a good or bad one, overall? Is greed "good"? Does it drive the economy? Would we be psychologically better off with less?
3. Are lower-class delinquents reacting against middle-class values, as Cohen contends, or is there a lower-class culture with its own set of values and attitudes to which delinquents are conforming, as Miller contends?

USEFUL WEB SITES

Chicago Area Project. http://www.chicagoareaproject.org/.

Emile Durkheim. http://www.emile-durkheim.com/.

General Strain Theory. http://www.criminology.fsu.edu/crimtheory/agnew.htm.

The Chicago School of Ecology. http://www.research.umbc.edu/~lutters/pubs/1996_SWLNote96-1_Lutters,Ackerman.pdf.

Youth Gangs. http://www.ncjrs.gov/pdffiles/167249.pdf.

CHAPTER GLOSSARY

Anomie: A term meaning "lacking in rules" or "normlessness" used by Durkheim to describe a condition of normative deregulation in society.

Collective efficacy: The shared power of a group of connected and engaged individuals to influence the maintenance of public order.

Conformity: The most common of Merton's modes of adaptation; the acceptance of cultural goals and of the legitimate means of obtaining them.

Consensus perspective: A view of society as a system of mutually sustaining parts and characterized by broad normative consensus. Also known as the functionalist perspective.

Decommodification: The process of freeing social relationships from economic considerations.

Focal Concerns: Miller's description of the value system and lifestyle of the lowest classes: trouble, toughness, excitement, smartness, fate, and autonomy.

General strain theory: Agnew's extension of anomie theory into the realm of social psychology, stressing multiple sources of strain and how people cope with it.

Institutional anomie theory: Messner and Rosenfeld's extension of anomie theory, which avers that high crime rates are intrinsic to the structural and cultural arrangements of American society.

Mechanical solidarity: A form of social solidarity existing in small, isolated, prestate societies in which individuals sharing common experiences and circumstances share common values and develop strong emotional ties to the collectivity.

Middle-class measuring rods: According to Cohen, because low-class youths cannot measure up to middle-class standards, they experience status frustration and this frustration spawns an oppositional culture.

Modes of adaptation: Robert Merton's concept of how people adapt to the alleged disjunction between cultural goals and structural barriers to the means of obtaining them. These modes are conformity, ritualism, retreatism, innovation, and rebellion.

Opportunity structure theory: An extension of anomie theory claiming that lower-class youth join gangs as a path to monetary success.

Organic solidarity: A form of social solidarity characteristic of modern societies in which there is a high degree of occupational specialization and a weak normative consensus.

Short-run hedonism: The seeking of immediate gratification of desires without regard for any long-term consequences.

Social disorganization: The central concept of the Chicago School of Social Ecology. It refers to the breakdown or serious dilution of the power of informal community rules to regulate conduct in poor neighborhoods.

Social ecology: Terms used by the Chicago School to describe the interrelations of human beings and the communities in which they live.

Social structure: How society is organized by social institutions—the family, and educational, religious, economic, and political institutions—and stratified on the basis of various roles and statuses.

Status frustration: A form of frustration experienced by lower-class youth who desire approval and status but who cannot meet middle-class criteria and thus seek status via alternative means.

Strain theory: Merton's theory that American culture defines monetary success (the "American dream") as the predominant cultural goal to which all its citizens should aspire, while at the same time, American social structure restricts access to legitimate means of attaining it to certain segments of the population. The disjunction between cultural goals and the structural impediments to achieving them is the anomic gap in which crime is bred.

READING

Community Correlates of Rural Youth Violence

D. Wayne Osgood and Jeff M. Chambers

In this article, Osgood and Chambers test the concepts of social disorganization in a rural context to determine whether they hold up in non-urban areas. They examined data from 264 non-urban counties in Florida, Georgia, Nebraska, and South Carolina, focusing on violent crimes. Overall, Osgood and Chambers found that social disorganization had explanatory power in rural areas. Residential instability and ethnic diversity adversely impacted crime rates, but the most powerful predictive factor was the proportion of female-headed households. Poverty status, by itself, did not significantly increase the probability of delinquency. The authors conclude that family disruption, as indexed by female-headed households, was the critical element of social disorganization in non-metropolitan counties.

Introduction

Rates of crime and delinquency vary widely across communities, and research going back many decades provides a good understanding of the nature, correlates, and probable causes of these community differences. Social disorganization is the primary theory by which criminologists account for rates of crime in urban communities. If this theory also applies to rural settings, then what is known about crime in urban areas can provide a basis for developing programs that address the problem of delinquency in smaller communities. The research presented in this Bulletin indicates that the principles of social disorganization theory hold up quite well in rural settings. As in urban areas, rates of juvenile violence are considerably higher in rural communities that have a large percentage of children living in single-parent households, a high rate of population turnover, and significant ethnic diversity. These factors, it should be noted, are statistical correlates and not causes of such violence; nor are they the only correlates.

Several researchers on crime have called for more focus on rural settings, which have unique crime problems (e.g., the theft of agricultural equipment and commodities) (Weisheit, Wells, and Falcone, 1995). Equally important are the striking similarities that exist between urban and rural areas. For instance, there are comparable crime trends over time, and the relationship of crime to important factors such as age, sex, and race of the perpetrator and victim is nearly identical.

Social Disorganization and Rural Communities

Social disorganization is defined as an inability of community members to achieve shared values or to solve jointly experienced problems. In recent decades, the themes of social disorganization

Source: Osgood, D., & Chamber, J. (2003, May). Community correlates of rural youth violence. *Juvenile Justice Bulletin*, NCJ 193591. Retrieved January 11, 2008 from the U.S. Department of Justice Web site at http://www.ncjrs.gov/pdffiles1/ojjdp/193591.pdf

theory have been more clearly articulated and extended by Bursik and Grasmick (1993) and Sampson and Groves (1989). Shaw and McKay (1942) traced social disorganization to conditions endemic to the urban areas that were the only places the newly arriving poor could afford to live, in particular, a high rate of turnover in the population (residential instability) and mixes of people from different cultural backgrounds (ethnic diversity). Shaw and McKay's analyses relating delinquency rates to these structural characteristics established key facts about the community correlates of crime and delinquency, and their work remains useful today as a guide for efforts to address crime and delinquency at the community level.

Both theoretical development and empirical research in the study of community influences on crime and delinquency have focused on urban settings. For instance, studies of neighborhood differences in crime rates have been conducted in many of the largest cities in the United States (including Baltimore, MD; Boston, MA; Chicago, IL; New York, NY; and San Diego, CA), but only one such study has been conducted in a smaller city—Racine, WI. Nonmetropolitan areas have been included in some studies of communities and crime, but that research is of limited value for the purposes of this Bulletin. Some of those studies were based on national samples with both urban and rural respondents, but they did not separately examine patterns for nonmetropolitan communities.

Extending Social Disorganization Theory

Current versions of social disorganization theory assume that strong networks of social relationships prevent crime and delinquency. When most community or neighborhood members are acquainted and on good terms with one another, a substantial portion of the adult population has the potential to influence each child. The larger the network of acquaintances, the greater the community's capacity for informal surveillance (because residents are easily distinguished from outsiders), for supervision (because acquaintances are willing to intervene when children and juveniles behave unacceptably), and for shaping children's values and interests. According to the current theory, community characteristics such as poverty and ethnic diversity lead to higher delinquency rates because they interfere with community members' abilities to work together (see citations above).

Just as in urban areas, systems of relationships are relevant to crime and delinquency in small towns and rural communities. The only aspect of the theory specific to urban areas is the explanation of why social disorganization arises in some geographic locations and not in others.

Rural sociologists concerned with the disruptive effects of rapid population growth provide some evidence that the processes of social disorganization apply in rural settings. Freudenberg (1986), for example, argued that the "boomtown" phenomenon brings high rates of crime and other unacceptable behaviors but does not produce alienation or mental health difficulties. Furthermore, he explained these negative effects by the same logic as social disorganization theory: rapid growth greatly diminishes the proportion of people who know one another, which in turn interferes with surveillance and socialization of the young.

Community Correlates of Youth Violence Outside the City

Social disorganization theory specifies that several variables—residential instability, ethnic diversity, family disruption, economic status, population size or density, and proximity to urban areas—influence a community's capacity to develop and maintain strong systems of social relationships. To test the theory's applicability to nonmetropolitan settings, this Bulletin examines the relationships between these community variables and rates of offending because the same relationships provide the core empirical support for the theory in urban settings. This section discusses the relevance of each factor to delinquency rates in the social disorganization framework.

Residential Instability Based on research in urban settings, the authors expected that rates of juvenile violence in rural communities would increase as rates of residential instability increased. When the population of an area is constantly changing, the residents have fewer opportunities to develop strong, personal ties to one another and to participate in community organizations. This assumption has been central to research on social disorganization since its inception. Massive population change is also the essential independent variable underlying the boomtown research on rural settings.

Ethnic Diversity According to social disorganization theory, it could be expected that, as in urban areas, rates of juvenile violence would be higher in rural communities with greater ethnic diversity. According to Shaw and McKay (1942), ethnic diversity interferes with communication among adults. Effective communication is less likely in the face of ethnic diversity because differences in customs and a lack of shared experiences may breed fear and mistrust. It is important to distinguish this theoretically driven hypothesis about heterogeneity from simple ethnic differences in offense rates. In other words, this hypothesis sees crime as arising from relations between ethnic groups, not from some groups being more crime-prone than others.

Family Disruption Research in urban areas has found that delinquency rates are higher in communities with greater levels of family disruption, and the authors expected that this also would be true in rural areas. Sampson (1985) argued that unshared parenting strains parents' resources of time, money, and energy, which interferes with their ability to supervise their children and communicate with other adults in the neighborhood. Furthermore, the smaller the number of parents in a community relative to the number of children, the more limited the networks of adult supervision will be for all the children.

Economic Status Although rates of juvenile violence are higher in urban areas with lower economic status, it was not clear that this relationship should apply in rural settings. The role of economic status in social disorganization theory is based on patterns of growth in urban areas. In many major urban areas, growth leads to the physical, economic, and social decline of the residential areas closest to the central business district. These areas then become most readily available to the poor and to groups who migrate to the area. As a result, areas with the lowest average socioeconomic status will also have the greatest residential instability and ethnic diversity, which in turn will create social disorganization.

The processes that link poverty with population turnover are specific to urban settings. In nonmetropolitan settings, poor populations may be stable and ethnically homogeneous.

Population Density Population density is rather different from the other community factors for two reasons. First, evidence of a relationship between population density and urban crime and delinquency is inconsistent. Second, the meaning of density becomes quite different for non-urban communities, where, in the least dense areas, one must travel several miles to have significant contact with people outside of one's immediate family. The original reasoning for the urban context was that high population density creates problems by producing anonymity that interferes with accountability to neighbors. In the least dense rural areas, it may be social isolation, instead, that limits social support to monitor children and respond to problem behavior. On the other hand, Sampson (1983) suggested that density might be more important in terms of opportunities for offending than in terms of social disorganization. The relative isolation of living in a sparsely populated area may reduce opportunities for offending because of greater distance from targets and from potential companions in crime. Victimization rates are lowest in communities with the smallest populations, but only for populations of 25,000 or less. In larger communities, the rates were essentially unrelated to population size.

Proximity to Urban Areas This final community variable, which departs from the themes of current social disorganization theory, considers an issue specific to rural settings and to the linkages among communities. It is important to look beyond the internal dynamics of communities and consider ways in which rates of delinquency might be influenced by relationships between neighboring communities. Various rural and suburban communities have very different relationships with urban communities, and this is an important theme of research on rural settings. Heitgerd and Bursik suggested that "less delinquent groups of youths are being socialized into more sophisticated types of criminal behavior by youths in adjoining areas" (1987: 785). Because average crime rates are higher in communities with larger populations, this phenomenon would produce higher rates of delinquency in rural communities that are adjacent to metropolitan areas.

Methods, Sample, and Measures

The sample consists of the nonmetropolitan counties in Florida, Georgia, Nebraska, and South Carolina. The study sample included 264 counties with populations ranging from 560 to 98,000. Although these nonmetropolitan counties are much larger geographic units than areas analyzed in community-level research on crime in metropolitan settings, they are of equal or smaller size in terms of population. The average total population of these nonmetropolitan counties was roughly 10,000, which is comparable to the smallest units used in research on urban neighborhoods. This sample compares favorably with those in studies of urban areas in terms of the number of communities, the size of the populations, and the variety of communities included.

Delinquency

UCR data were used to measure each county's delinquency rate. No measure of crime or delinquency is perfect, and criminologists have long been concerned about potential biases in crime rates based on official records, especially that arrests might reflect the behavior of law enforcement officers more than the behavior of offenders. Fortunately, findings relating social disorganization to arrests have been replicated by more recent studies that measured offending through citizen calls for police assistance, self-reports of victims, and self-reports of offenders.

The measure of delinquency was the per capita arrest rate of juveniles ages 11–17 in each county, pooled over the 5-year period from1989 through 1993. The outcome measures were as follows: rates of arrest for homicide, forcible rape, aggravated assault, robbery, weapons offense, simple assault, and the UCR Violent Crime Index, which comprises the first four offenses. The study considered the full spectrum of violent offenses (capturing a large range of offense seriousness) for which recording is comparable across the four states. This approach provided a rich pool of information for establishing the consistency of the findings.

Explanatory Variables

Data from the 1990 census provide measures for most of the explanatory variables. The measure of residential instability was the proportion of households occupied by persons who had moved from another dwelling in the previous 5 years. Ethnic diversity was measured in terms of the proportion of households occupied by white versus nonwhite persons. Ethnic diversity was computed as the index of diversity, which reflects the probability that two randomly drawn individuals would differ in ethnicity. Family disruption was indexed by female-headed households, expressed as a proportion of all households with children. Low economic status was defined as the proportion of persons living below the poverty level.

Also included in the analysis was the number of youth ages 10–17, which is the population at risk for juvenile arrests. Population size serves as a proxy measure for population density because the two variables are so strongly correlated that they are effectively indistinguishable. Because states may differ in their statutes and in the organization, funding, and policies of their justice systems, it was important to make sure that differences among states were not confused with the contributions of the explanatory variables. Therefore, the analysis controls for differences among states in arrest rates for each offense. The outcome of interest in this study is the arrest rate, defined as the number of arrests in a county divided by the size of the juvenile population.

Results

Residential Instability, Ethnic Diversity, and Family Disruption

In research on social disorganization in urban settings, the three variables most strongly and consistently associated with rates of crime and delinquency are *residential instability, ethnic diversity,* and *family disruption.* In the four states in this study, a similar pattern was found in nonmetropolitan counties. Social disorganization theory holds that when turnover in the membership of a community is high, social relationships will weaken and juvenile violence will increase. Consistent with this theory, residential instability was significantly associated with higher rates of aggravated assault, simple assault, weapons violations (bivariate relationship only), and the overall Violent Crime Index. In the bivariate associations, a 10 percent increase in residential instability was associated with 29 to 65 percent higher rates of arrest for the various forms of juvenile violence, with the exception of homicide.

Ethnic diversity is also a key variable because cultural differences tend to interfere with adults' ability to work together in supervising and raising their children. The correlation between ethnic diversity and violent offenses was statistically significant in most instances. In the bivariate relationships, a 10 percent increase in ethnic diversity was associated with 20 to 35 percent higher rates of juvenile violence.

Higher levels of family disruption, as indexed by the proportion of female-headed households, also were strongly and consistently associated with higher rates of arrest for violent offenses other than homicide. According to social disorganization theory, this pattern arises from the burden of single parenting, which interferes with parents' abilities to work together and reduces the number of adults involved in the joint supervision of children. The relationship between family disruption and juvenile arrest rates was the strongest in the study's results. In the bivariate relationships, this relationship was significant for all offenses except homicide, and in the multivariate relationships, it was significant for all offenses except homicide and robbery. In the bivariate relationships, a 10 percent increase in female-headed households was associated with 73 to 100 percent higher rates of arrest for all offenses except homicide.

In combination, residential instability, ethnic diversity, and family disruption strongly differentiated counties with high rates of arrest from those with low rates. Compare, for example, a county with 35 percent residential instability, ethnic diversity of 0.23 (on a scale of 0 to 0.5), and 13 percent female-headed households, which would be a moderately low level of social disorganization, with one that has 45 percent residential instability, ethnic diversity of 0.33, and 23 percent female-headed households, which would be a moderately high level. The multivariate relationships shown (which control for all other explanatory variables) indicate that the arrest rate for the Violent Crime Index in the more disorganized county would be $2\frac{2}{3}$ times as great as that of the less disorganized county (217 per 100,000 versus 81 per 100,000).

Economic Status

The analysis did not find a meaningful relationship between rates of delinquency and rates of poverty. Instead of showing poverty to be associated with higher rates of delinquency, the relationships were either very slight or indicated an association between poverty and lower delinquency rates (significantly lower rates for simple assault and rape). To understand this finding, it is useful to examine the association between poverty rates and the other community correlates of juvenile violence. As research in urban areas has typically found, poverty rates in the study's nonmetropolitan counties were positively associated with both ethnic diversity (r = .48, controlling for state) and the rate of female-headed households (r = .55). In contrast to urban areas, however, the correlation between poverty and residential instability in these nonmetropolitan areas was negative rather than positive (r = –.39). This finding contradicts the classic pattern of relationships from Park and Burgess's (1924) theory of urban ecology, which was the basis for predicting that poverty would lead to social disorganization. Also in contrast to findings in urban areas, poverty rates were higher in nonmetropolitan counties with smaller populations than in those with larger populations (r = –.41). Poverty rates increase as ethnic diversity and the proportion of female-headed households increase, suggesting that delinquency rates will increase along with poverty rates. However, this source of positive correlation between the rates of poverty and delinquency is canceled out in nonmetropolitan areas, where rates of poverty are lower in areas with high residential instability and larger populations.

This pattern of relationships is consistent with research conducted by Fitchen (1994), who found that poorer residents do not make frequent moves in rural areas. Low-cost housing is often abundant, and residents have a support network of family and friends who can provide casual rent agreements and flexible payment schemes. It appears that—unlike in most urban areas—poverty does not disrupt the social fabric of small towns and rural communities. The reasons that a high rate of rural poverty does not increase the delinquency rate appear to be consistent with social disorganization theory.

Population Size and Density

Arrest rates for juvenile violence varied dramatically with differences in the sizes (and densities) of juvenile populations. [Figure 4.1] illustrates these findings with graphs for four of the studied offenses. For all violent offenses except homicide, differences in the size of county juvenile populations corresponded to differences of at least threefold in juvenile arrest rates. The figure shows that annual arrest rates for juvenile violence were uniformly lower in the rural counties with the smallest populations. Per capita arrest rates rose with increases in juvenile population, but only until the population size reached about 4,000. Beyond this level, increasing population had little impact on arrest rates for violent offenses other than robbery. These results are comparable to Laub's (1983) finding that victimization rates increased with population size for total populations (rather than juvenile populations) up to about 25,000, but did not increase further for larger populations. Arrest rates for the Violent Crime Index, rape, and aggravated assault appeared to decline somewhat in the upper range of juvenile population sizes, but it is unlikely that these decreases are statistically reliable because they are small.

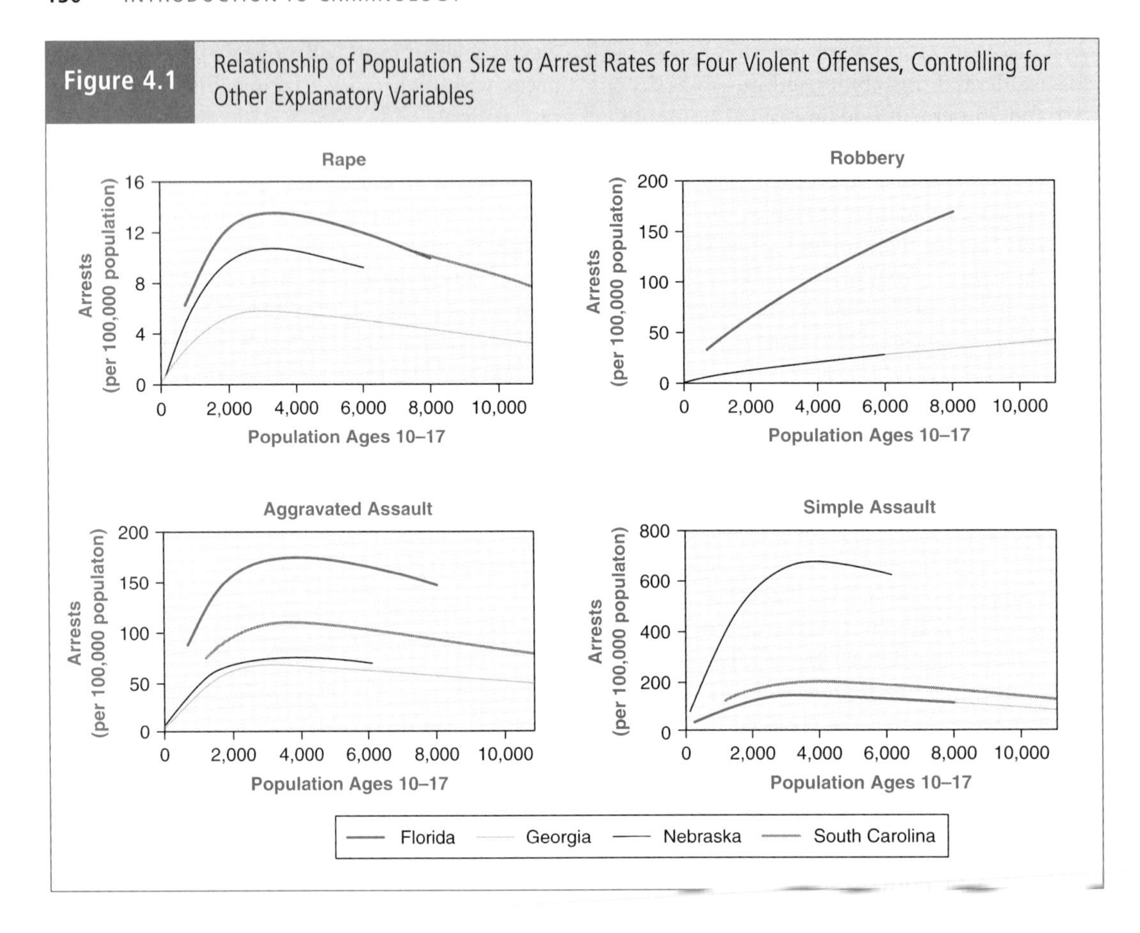

Figure 4.1 Relationship of Population Size to Arrest Rates for Four Violent Offenses, Controlling for Other Explanatory Variables

Proximity to Metropolitan Areas

Whether a rural county was adjacent to a metropolitan area had little bearing on its rate of juvenile arrests for violent offenses. None of the relationships for this explanatory variable approached statistical significance. If delinquency can spread from one community to another, the reason is not simple enough to be explained by the county's proximity to a metropolitan area.

Conclusion

The principles of social disorganization theory, developed in studies of urban neighborhoods, can be applied to rural communities. In the nonmetropolitan counties that made up the study sample, per capita rates of juvenile arrest for violent offenses were significantly and consistently associated with residential instability, ethnic diversity, and family disruption. Based on the strength and consistency of the findings, family disruption, in particular, appears to be a critical element of social disorganization in nonmetropolitan communities.

The study results diverged from the standard findings for urban areas in that they indicated no association between poverty and delinquency. When the correlates of poverty for this sample of nonmetropolitan communities are considered, however, this finding is consistent with the core logic of social disorganization theory. Shaw and McKay (1942) concluded that the relationship of poverty to delinquency in urban areas is produced by the connection of poverty with the combination of residential instability and ethnic diversity. This urban population dynamic does not exist in small towns and rural areas; outside the city, the populations of poorer communities are more stable than average, not less. Thus, these findings support Shaw and McKay's contention that it is not poverty per se but an association of poverty with other factors that weakens systems of social relationships in a community, thereby producing social disorganization.

The findings concerning the relationship of juvenile violence to the size and density of the juvenile population have interesting implications. Based on social disorganization theory, the authors hypothesized that high population density would interfere with social organization by creating anonymity and by increasing the difficulty of supervising children and adolescents. This reasoning implies that problems would intensify in areas with especially high population densities. The findings show the opposite: after reaching the modest density of about 4,000 juveniles in an entire county, population size makes little difference in the rate of juvenile violence. Clearly, another dynamic must be at work.

The relationship between population size and juvenile violence is more likely due to increased opportunities for offending in areas with larger populations. A small population reduces the chances that a potential robber will randomly encounter a likely victim or that two rivals will meet in an unguarded setting conducive to an assault. Furthermore, the company of peers provides support for engaging in delinquent behavior and a very low population density makes it more difficult for peers to get together.

The findings are consistent across the set of violent offenses. Many researchers limit their analyses to a few offenses that they presume to be most reliably recorded, such as homicide and robbery. Indeed, there can be little doubt that law enforcement officers have less discretion about whether to make arrests for these offenses or that victims and bystanders are more likely to report them. Nevertheless, the relationships of community characteristics to the rate of simple assaults are nearly identical to those for the other violent offense categories such as rape and aggravated assault. Thus, instead of finding inconsistent results for less serious offenses, the data provided additional confirmation for the overall pattern of results.

References

Bursik, R. J., Jr., & Grasmick, H. G. (1993). *Neighborhoods and crime: The dimensions of effective community control.* New York: Lexington Books.

Freudenberg, W. R. (1986). The density of acquaintanceship: An overlooked variable in community research. *American Journal of Sociology, 92*(1), 27–63.

Heitgerd, J. L., & Bursik, R. J., Jr. (1987). Extracommunity dynamics and the ecology of delinquency. *American Journal of Sociology, 92*(4), 775–787.

Laub, J. H. (1983). Urbanism, race, and crime. *Journal of Research in Crime and Delinquency, 20*(2), 183–198.

Park, R. E., & Burgess, E. W. (1924). *Introduction to the science of sociology* (2nd ed.). Chicago: University of Chicago Press.

Sampson, R. J. (1983). Structural density and criminal victimization. *Criminology, 21*(2), 276–293.

Sampson, R. J. (1985). Neighborhood and crime: The structural determinants of personal victimization. *Journal of Research in Crime and Delinquency, 22*(1), 7–40.

Sampson, R. J., & Groves, W. B. (1989). Community structure and crime: Testing social-disorganization theory. *American Journal of Sociology, 94*(4), 774–802.

Shaw, C. R., & McKay, H. D. (1942). *Juvenile delinquency and urban areas.* Chicago: University of Chicago Press.

Weisheit, R. A., Wells, L. E., & Falcone, D. N. (1995). *Crime and policing in rural and small-town America: An overview of the issues.* Washington, DC: U.S. Department of Justice, Office of Justice Programs, National Institute of Justice.

DISCUSSION QUESTIONS

1. What are the main differences related to crime and delinquency that would make the authors think that there may be a difference between urban and rural areas?
2. The most powerful finding relating to delinquency rates was the rate of female-headed households. Why would this be? Could this also explain poverty rate and residential mobility?
3. Why do we "celebrate" diversity in the U.S. when it is so consistently linked with crime and delinquency?

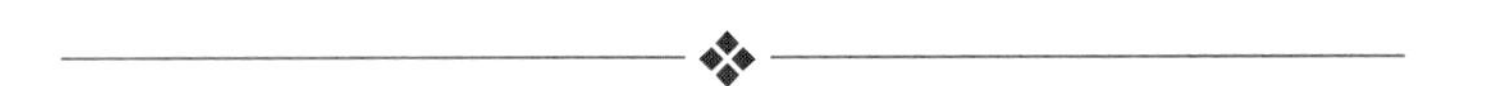

READING

Social Structure and Anomie

Robert K. Merton

Robert Merton's article expands Durkheim's anomie theory to add the social-psychological component of strain. He differs from Durkheim in denying that crime is the result of "natural appetites" that society has been unable to constrain. For Merton, these appetites are not "natural," but rather originate in capitalist, particularly American society. He also introduces here his concepts of how individuals adapt to a society in which everyone is supposed to strive for success goals, but in which some are denied access to legitimate means of achieving that goal.

There persists a notable tendency in sociological theory to attribute the malfunctioning of social structure primarily to those of man's imperious biological drives which are not adequately restrained by social control. In this view, the social order is solely a device for "impulse management" and the "social processing" of tensions. These impulses which break through social control, be it noted, are held to be biologically derived. Nonconformity is assumed to be rooted in original nature. Conformity is by implication the result of a utilitarian calculus or unreasoned conditioning. This point of view, whatever its other deficiencies, clearly begs one question. It provides no basis for determining the nonbiological conditions which induce deviations from prescribed patterns of conduct. In this paper, it will be suggested that certain phases of social structure generate the circumstances in which infringement of social codes constitutes a "normal" response.

The conceptual scheme to be outlined is designed to provide a coherent, systematic approach to the study of socio-cultural sources of deviate behavior. Our primary aim lies in discovering how some social structures exert a definite pressure upon certain persons in the society to engage in nonconformist rather than conformist conduct. The many ramifications of the scheme cannot all be discussed; the problems mentioned outnumber those explicitly treated.

Among the elements of social and cultural structure, two are important for our purposes. These are analytically separable although they merge imperceptibly in concrete situations. The first consists of culturally defined goals, purposes, and interests. It comprises a frame of aspirational reference. These goals are more or less integrated and involve varying degrees of prestige and sentiment. Some of these cultural aspirations are related to the original drives of man, but they are not determined by them. The second phase of the social structure defines, regulates, and controls the acceptable modes of achieving these goals. Every social group invariably couples its scale of desired ends with moral or institutional regulation of permissible and required procedures for attaining these ends. These regulatory norms and moral imperatives do not necessarily coincide with technical or efficiency norms. Many procedures which from the standpoint of particular individuals would be most efficient in securing desired values, e.g., illicit oil-stock schemes, theft, fraud, are ruled out of the institutional area of permitted conduct. The choice of expedients is limited by the institutional norms.

To say that these two elements, culture goals and institutional norms, operate jointly is not to say that the ranges of alternative behaviors and aims bear some constant relation to one another. The emphasis upon certain goals may vary independently of the degree of emphasis upon institutional means. There may develop a disproportionate, at times, a virtually exclusive, stress upon the value of specific goals, involving relatively slight concern with the institutionally appropriate modes of attaining these goals. The limiting case in this direction is reached when the range of alternative procedures is limited only by technical rather than institutional considerations. Any and all devices which promise attainment of the all important goal would be permitted in this hypothetical polar case. This constitutes one type of cultural malintegration. A second polar type is found in groups where activities originally conceived as instrumental are transmuted into ends in themselves. The original purposes are forgotten and ritualistic adherence to institutionally prescribed conduct becomes virtually obsessive. Stability is largely ensured while change is flouted. The range of alternative behaviors is severely limited. There develops a tradition-bound, sacred society characterized by neophobia. The occupational psychosis of the bureaucrat may be cited as a case in point. Finally, there are the intermediate types of groups where a balance between culture goals and institutional means is maintained. These are the significantly integrated and relatively stable, though changing, groups.

An effective equilibrium between the two phases of the social structure is maintained as long as satisfactions accrue to individuals who conform to both constraints, viz., satisfactions from the achievement of the goals and satisfactions emerging directly from the institutionally canalized modes of striving to attain these ends. Success, in such equilibrated cases, is twofold. Success is reckoned in terms of the product and in terms of the process, in terms of the outcome and in terms of activities. Continuing satisfactions must derive from sheer participation in a competitive order as well as from eclipsing one's competitors if the order itself is to be sustained. The occasional sacrifices involved in institutionalized conduct must be compensated by socialized rewards. The distribution of statuses and roles through competition must be so organized that positive incentives for conformity to roles and adherence to status obligations are provided for every position within the distributive order. Aberrant conduct, therefore, may be viewed as a symptom of dissociation between culturally defined aspirations and socially structured means.

Of the types of groups which result from the independent variation of the two phases of the social structure, we shall be primarily concerned with the first, namely, that involving a disproportionate accent on goals. This statement must be recast in a proper perspective. In no group is there an absence of regulatory codes governing conduct, yet groups do vary in the degree to which these folkways, mores, and institutional controls are effectively integrated with the more diffuse goals which are part of the culture matrix. Emotional convictions may cluster about the complex of socially acclaimed ends, meanwhile shifting their support from the culturally defined implementation of these ends. As we shall see, certain aspects of the social structure may generate countermores and antisocial behavior precisely because of differential emphases on goals and regulations. In the extreme case, the latter may be so vitiated by the goal-emphasis that the range of behavior is limited only by considerations of technical expediency. The sole significant question then becomes, which available means is most efficient in netting the socially approved value? The technically most feasible procedure, whether legitimate or not, is preferred to the institutionally prescribed conduct. As this process continues, the integration of the society becomes tenuous and anomie ensues.

Thus, in competitive athletics, when the aim of victory is shorn of its institutional trappings and success in contests becomes construed as "winning the game" rather than "winning through circumscribed modes of activity," a premium is implicitly set upon the use of illegitimate but technically efficient means. The star of the opposing football team is surreptitiously slugged; the wrestler furtively incapacitates his opponent through ingenious but illicit techniques; university alumni covertly subsidize "students" whose talents are largely confined to the athletic field. The emphasis on the goal has so attenuated the satisfactions deriving from sheer participation in the competitive activity that these satisfactions are virtually confined to a successful outcome. Through the same process, tension generated by the desire to win in a poker game is relieved by successfully dealing oneself four aces, or, when the cult of success has become completely dominant, by sagaciously shuffling the cards in a game of solitaire. The faint twinge of uneasiness in the last instance and the surreptitious nature of public delicts indicate clearly that the institutional rules of the game are known to those who evade them, but that the emotional supports of these rules are largely vitiated by cultural exaggeration of the success-goal. They are microcosmic images of the social macrocosm.

Of course, this process is not restricted to the realm of sport. The process whereby exaltation of the end generates a literal demoralization, i.e., a deinstitutionalization, of the means is one which characterizes many groups in which the two phases of the social structure are not highly integrated. The extreme emphasis upon the accumulation of wealth as a symbol of success in our own society militates against the completely effective control of institutionally regulated modes of acquiring a fortune. Fraud, corruption, vice, crime, in short, the entire catalogue of proscribed behavior, becomes increasingly common when the emphasis on the culturally induced success-goal becomes divorced from a coordinated institutional emphasis. This observation is of crucial theoretical importance in examining the doctrine that antisocial behavior most frequently derives from biological drives breaking through the restraints imposed by society. The difference is one between a strictly utilitarian interpretation which conceives man's ends as random and an analysis which finds these ends deriving from the basic values of the culture.

Our analysis can scarcely stop at this juncture. We must turn to other aspects of the social structure if we are to deal with the social genesis of the varying rates and types of deviate behavior characteristic of different societies. Thus far, we have sketched three ideal types of social orders constituted by distinctive patterns of relations between culture ends and means. Turning from these types of culture patterning, we find five logically possible, alternative modes of adjustment or adaptation by individuals within the culture-bearing society or group. These are schematically presented in Table 4.2, where (+) signifies "acceptance," (−) signifies "elimination" and (±) signifies "rejection and substitution of new goals and standards."

Our discussion of the relation between these alternative responses and other phases of the

Table 4.2 Alternative Modes of Adjustment or Adaptation by Individuals Within a Cultural Group

	Culture Goals	Institutionalized Means
1. Conformity	+	+
2. Innovation	+	−
3. Ritualism	−	+
4. Retreatism	−	−
5. Rebellion	+/−	+/−

social structure must be prefaced by the observation that persons may shift from one alternative to another as they engage in different social activities. These categories refer to role adjustments in specific situations, not to personality in toto. To treat the development of this process in various spheres of conduct would introduce a complexity unmanageable within the confines of this paper. For this reason, we shall be concerned primarily with economic activity in the broad sense, "the production, exchange, distribution and consumption of goods and services" in our competitive society, wherein wealth has taken on a highly symbolic cast. Our task is to search out some of the factors which exert pressure upon individuals to engage in certain of these logically possible alternative responses. This choice, as we shall see, is far from random.

In every society, Adaptation I (conformity to both culture goals and means) is the most common and widely diffused. Were this not so, the stability and continuity of the society could not be maintained. The mesh of expectancies which constitutes every social order is sustained by the modal behavior of its members falling within the first category. Conventional role behavior oriented toward the basic values of the group is the rule rather than the exception. It is this fact alone which permits us to speak of a human aggregate as comprising a group or society.

Conversely, Adaptation IV (rejection of goals and means) is the least common. Persons who "adjust" (or maladjust) in this fashion are, strictly speaking, in the society but not of it. Sociologically, these constitute the true "aliens." Not sharing the common frame of orientation, they can be included within the societal population merely in a fictional sense. In this category are some of the activities of psychotics, psychoneurotics, chronic autists, pariahs, outcasts, vagrants, vagabonds, tramps, chronic drunkards and drug addicts. These have relinquished, in certain spheres of activity, the culturally defined goals, involving complete aim-inhibition in the polar case, and their adjustments are not in accord with institutional norms. This is not to say that in some cases the source of their behavioral adjustments is not in part the very social structure which they have in effect repudiated nor that their very existence within a social area does not constitute a problem for the socialized population.

This mode of "adjustment" occurs, as far as structural sources are concerned, when both the culture goals and institutionalized procedures have been assimilated thoroughly by the individual and imbued with affect and high positive value, but where those institutionalized procedures which promise a measure of successful attainment of the goals are not available to the individual. In such instances, there results a twofold mental conflict insofar as the moral obligation for adopting institutional means conflicts with the pressure to resort to illegitimate means (which may attain the goal) and inasmuch as the individual is shut off from means which are both legitimate and effective. The competitive order is maintained, but the frustrated and handicapped individual who cannot cope with this order drops out.

Defeatism, quietism and resignation are manifested in escape mechanisms which ultimately lead the individual to "escape" from the requirements of the society. It is an expedient which arises from continued failure to attain the goal by legitimate measures and from an inability to adopt the illegitimate route because of internalized prohibitions and institutionalized compulsives, during which process the supreme value of the success-goal has as yet not been renounced. The conflict is resolved by eliminating both precipitating elements, the goals and means. The escape is complete, the conflict is eliminated and the individual is a socialized.

Be it noted that where frustration derives from the inaccessibility of effective institutional means for attaining economic or any other type of highly valued "success," that Adaptations II, III and V (innovation, ritualism and rebellion) are also possible. The result will be determined by the particular personality, and thus, the particular cultural background, involved. Inadequate socialization will result in the innovation response whereby the conflict and frustration are eliminated by relinquishing the institutional means and retaining the success-aspiration; an extreme assimilation of institutional demands will lead to ritualism wherein the goal is dropped as beyond one's reach but conformity to the mores persists; and rebellion occurs when emancipation from the reigning standards, due to frustration or to marginalist perspectives, leads to the attempt to introduce a "new social order."

Our major concern is with the illegitimacy adjustment. This involves the use of conventionally proscribed but frequently effective means of attaining at least the simulacrum of culturally defined success, wealth, power, and the like. As we have seen, this adjustment occurs when the individual has assimilated the cultural emphasis on success without equally internalizing the morally prescribed norms governing means for its attainment. The question arises, Which phases of our social structure predispose toward this mode of adjustment? We may examine a concrete instance, effectively analyzed by Lohman, which provides a clue to the answer. Lohman has shown that specialized areas of vice in the near north side of Chicago constitute a "normal" response to a situation where the cultural emphasis upon pecuniary success has been absorbed, but where there is little access to conventional and legitimate means for attaining such success. The conventional occupational opportunities of persons in this area are almost completely limited to manual labor. Given our cultural stigmatization of manual labor, and its correlate, the prestige of white collar work, it is clear that the result is a strain toward innovational practices. The limitation of opportunity to unskilled labor and the resultant low income cannot compete in terms of conventional standards of achievement with the high income from organized vice.

For our purposes, this situation involves two important features. First, such antisocial behavior is in a sense "called forth" by certain conventional values of the culture and by the class structure involving differential access to the approved opportunities for legitimate, prestige-bearing pursuit of the culture goals. The lack of high integration between the means-and-end elements of the cultural pattern and the particular class structure combine to favor a heightened frequency of antisocial conduct in such groups. The second consideration is of equal significance. Recourse to the first of the alternative responses, legitimate effort, is limited by the fact that actual advance toward desired success-symbols through conventional channels is, despite our persisting open-class ideology, relatively rare and difficult for those handicapped by little formal education and few economic resources. The dominant pressure of group standards of success is, therefore, on the gradual attenuation of legitimate, but by and large ineffective, strivings and the increasing use of illegitimate, but more or less effective, expedients of vice and crime. The cultural demands made on persons in this situation are incompatible. On the one hand, they are asked to orient their conduct toward the prospect of accumulating wealth and on the other, they are largely denied effective opportunities to do so institutionally. The consequences of such structural inconsistency are psycho-pathological personality, and/or antisocial conduct, and/or revolutionary activities. The equilibrium between culturally designated means and ends becomes highly unstable with the progressive emphasis on attaining the prestige-laden ends by any means whatsoever. Within this context, Capone represents the triumph of amoral intelligence over morally prescribed "failure," when the channels of vertical mobility are closed or narrowed in a society which places a high premium on economic affluence and social ascent for all its members.

This last qualification is of primary importance. It suggests that other phases of the social structure besides the extreme emphasis on pecuniary success must be considered if we are to understand the social sources of antisocial behavior. A high frequency of deviate behavior is not generated simply by "lack of opportunity" or by this exaggerated pecuniary emphasis. A comparatively rigidified class structure, a feudalistic or caste order, may limit such opportunities far beyond the point which obtains in our society today. It is only when a system of cultural values extols, virtually above all else, certain common symbols of success for the population at large while its social structure rigorously restricts or completely eliminates access to approved modes of acquiring these symbols for a considerable part of the same population, that antisocial behavior ensues on a considerable scale. In other words, our egalitarian ideology denies by implication the existence of noncompeting groups and individuals in the pursuit of pecuniary success. The same body of success-symbols is held to be desirable for all. These goals are held to transcend class lines, not to be bounded by them, yet the actual social organization is such that there exist class differentials in the accessibility of these common success-symbols. Frustration and thwarted aspiration lead to the search for avenues of escape from a culturally induced intolerable situation; or unrelieved ambition may eventuate in illicit attempts to acquire the dominant values. The American stress on pecuniary success and ambitiousness for all thus invites exaggerated anxieties, hostilities, neuroses and antisocial behavior.

This theoretical analysis may go far toward explaining the varying correlations between crime and poverty. Poverty is not an isolated variable. It is one in a complex of interdependent social and cultural variables. When viewed in such a context, it represents quite different states of affairs. Poverty as such, and consequent limitation of opportunity, are not sufficient to induce a conspicuously high rate of criminal behavior. Even the often mentioned "poverty in the midst of plenty" will not necessarily lead to this result. Only insofar as poverty and associated disadvantages in competition for the culture values approved for all members of the society is linked with the assimilation of a cultural emphasis on monetary accumulation as a symbol of success is antisocial conduct a "normal" outcome. Thus, poverty is less highly correlated with crime in southeastern Europe than in the United States. The possibilities of vertical mobility in these European areas would seem to be fewer than in this country, so that neither poverty per se nor its association with limited opportunity is sufficient to account for the varying correlations. It is only when the full configuration is considered, poverty, limited opportunity and a commonly shared system of success symbols, that we can explain the higher association between poverty and crime in our society than in others where rigidified class structure is coupled with differential class symbols of achievement.

In societies such as our own, then, the pressure of prestige-bearing success tends to eliminate the effective social constraint over means employed to this end. "The-end-justifies-the-means" doctrine becomes a guiding tenet for action when the cultural structure unduly exalts the end and the social organization unduly limits possible recourse to approved means. Otherwise put, this notion and associated behavior reflect a lack of cultural coordination. In international relations, the effects of this lack of integration are notoriously apparent. An emphasis upon national power is not readily coordinated with an inept organization of legitimate, i.e., internationally defined and accepted, means for attaining this goal. The result is a tendency toward the abrogation of international law, treaties become scraps of paper, "undeclared warfare" serves as a technical evasion, the bombing of civilian populations is rationalized, just as the same societal situation induces the same sway of illegitimacy among individuals.

The social order we have described necessarily produces this "strain toward dissolution." The pressure of such an order is upon outdoing one's competitors. The choice of means within the ambit of institutional control will persist as long as the sentiments supporting a competitive system, i.e., deriving from the possibility of outranking competitors and hence enjoying the favorable response of others, are distributed throughout the entire system of activities and are not confined merely to the final result. A stable social structure demands a balanced distribution of affect among its various segments. When there occurs a shift of emphasis from the satisfactions deriving from competition itself to almost exclusive concern with successful competition, the resultant stress leads to the breakdown of the regulatory structure. With the resulting attenuation of the institutional imperatives, there occurs an approximation of the situation erroneously held by utilitarians to be typical of society generally wherein calculations of advantage and fear of punishment are the sole regulating agencies. In such situations, as Hobbes observed, force and fraud come to constitute the sole virtues in view of their relative efficiency in attaining goals, which were for him, of course, not culturally derived.

It should be apparent that the foregoing discussion is not pitched on a moralistic plane. Whatever the sentiments of the writer or reader concerning the ethical desirability of coordinating the means-and-goals phases of the social structure, one must agree that lack of such coordination leads to anomie. Insofar as one of the most general functions of social organization is to provide a basis for calculability and regularity of behavior, it is increasingly limited in effectiveness as these elements of the structure become dissociated. At the extreme, predictability virtually disappears and what may be properly termed cultural chaos or anomie intervenes.

This statement, being brief, is also incomplete. It has not included an exhaustive treatment of the various structural elements which predispose toward one rather than another of the alternative responses open to individuals; it has neglected, but not denied the relevance of, the factors determining the specific incidence of these responses; it has not enumerated the various concrete responses which are constituted by combinations of specific values of the analytical variables; it has omitted, or included only by implication, any consideration of the social functions performed by illicit responses; it has not tested the full explanatory power of the analytical scheme by examining a large number of group variations in the frequency of deviate and conformist behavior; it has not adequately dealt with rebellious conduct which seeks to refashion the social framework radically; it has not examined the relevance of cultural conflict for an analysis of culture-goal and institutional-means malintegration. It is suggested that these and related problems may be profitably analyzed by this scheme.

DISCUSSION QUESTIONS

1. Do you believe that the disjunction between goals and means that Merton writes about still exists in American society?
2. Discuss what kind of society would be produced if cultural goals did not encourage competition, success, and achievement.
3. What are some reasons other than "denial by the system" why individuals cannot achieve American culturally prescribed goals.

READING

Gangs and Social Change

Martín Sánchez-Jankowski

In this article, Martin Sánchez-Jankowski complains that the extant literature on gangs has usually defined them as loose associations of individuals engaged in some type of delinquent or criminal activity. He want to explore gangs more in the context of social structure than in the context of individual differences, that is, sociologically. Yet researchers have failed to sociologically differentiate gangs from other types of collective behavior. In contrast, his article presents gangs as organizations influenced by the social structure of the urban areas in which they operate. Concentrating on gangs in the U.S. context, the article summarizes both common features and different forms gangs have assumed over five historical eras, arguing that gangs respond to, rather than create, significant social changes.

Introduction

Gangs have been the focus of so many studies in the United States that they have become a growth industry. Most recent research has treated gangs in one of two ways: either as a gathering of individuals with a specific negative set of personal attributes or as a group of individuals who act in a deviant and/or criminal manner. Troubling about both approaches, though, is their common underestimation of connections between the structural conditions of society at large and the form of collective behavior that is "the gang," and their similar (and sometimes unwitting) recycling of individualistic thought rampant in American culture. This has had the effect of

Source: Sánchez-Jankowski, M. (2003). Gangs and social change. *Theoretical Criminology, 7*(2), 191–216. Reprinted with permission of Sage Publications, Inc.

misrepresenting who joins gangs and confusing different forms of collective behavior under the same "gang" concept. For this reason the argument presented here purposely diverges from [these approaches]. Overall, my contention is that the role of structure in creating rational forms of human agency in the lower classes has been consistently underestimated in much of the sociological research on gangs. Moreover, this underappreciation has appeared in ways that are both obvious and subtle across a diverse and otherwise quite rich literature. For theories of gangs to be more precise, though, the importance of structure in creating and maintaining individuals' involvement in gangs has to be brought to the forefront—not relegated to the backdrop—of sociological approaches.

Understanding Gangs Differently

I use "structure" to refer to the configuration of material resources in a system of allocation that establishes various opportunity parameters for each social class. For people living in low-income communities, a scarcity of material resources organizes behavioral choices and influences people's efforts to become middle class. Consequently, many people who live in low-income communities have to fight their environment to find relief from the burdens it imposes. One of the products of this effort is the development of a "defiant individualist" personality. According to Fromm and Maccoby (1970), this personality characteristic combines dominant social values—i.e., a stress on being socio-economically mobile and on accumulating capital— with a paucity of resources available for people living in lower-income communities to achieve these objectives. Accordingly, "defiant individualism" leads people to become involved with money-producing economic activities whether legal or not; the trait carries along with it an edge that "defies" any and all attempts to thwart it.

While this "defiant individualist" personality is present among a number of residents from poor and working-class areas, nearly all gang members have it. This is because gangs comprise the very means and tools used to achieve dominant goals. For this reason, gangs themselves can be precisely defined as organized "defiant individualism"; but while gangs are organized "defiant individualism," this does not mean that they are simply loose associations. Rather, as organizational entities, gangs are capable of producing benefits for their members and other people in society, controlling the behavior of their rank and file, and regulating leadership changes in ways that ensure these entities' continuity.

Such organizational qualities suggest criteria by which gangs can be distinguished from other collective behaviors. Again, without such criteria, researchers are left unable to differentiate gang activities from those that are "pre-gang" (such as a "band" or parties aspiring to be a gang) or "post-gang" (such as a group that was once a gang, but has gone into a state of decline). For this reason, an organizational developmental continuum is needed that enables meaningful distinctions between groups to be made, and which can understand and predict divergent patterns of behavior in low-income communities. To concretize this call for a developmental organizational continuum, consider the following example. "Posse" is the name utilized by Jamaicans to identify a certain type of organization with which some people are involved. Although a "posse" assumes the same internal organizational structure as a gang, it is organized for the specific purpose of trafficking drugs. Since gangs assume multi-dimensional roles in their communities, they occupy a central institutional position. On the other hand, because "posses" assume a far more restricted role in their communities, they are not an organic part of the community. Instead the "posse" represents a new historical actor, with a role and behavior in lower-income communities that is both similar and dissimilar to that of a gang.

The case for gaining sociological precision by defining gangs along a developmental continuum can be illuminated by placing this organizational form in historical perspective. Gang organizations have been incredibly resilient over the more than 150 years they have been part of American society. While myriad social upheavals have affected gang dynamics over this period, missing from the sociological literature on gangs is an appreciation of how progressive social changes have produced concurrent transformations in the functional shape and behaviors of gangs. Gangs operate in society, and societies remain in a constant process of social change; both alter dialectically in relation to each other. Of course social changes occur incrementally, with the accumulation of these increments producing significant and unique changes in people, groups and institutions. Often we call these experiences "period effects" or "generation effects" but the meaning is the same: the social conditions of a particular time in history matters in societal form and development. Thus, to fully understand gangs in a particular era, one must consider broad social changes that have affected them at specific times.

Specifically, five critical periods have affected the social and organizational development of gangs over the last 150 years: the Great Wave of Immigration; the Expansion of Industrial Production; the Deregulation of the Illicit Drug Market; the Escalation of Mass Incarceration; and the Proliferation of Monopolistic Market Activity. Each period involved social structural changes that, in turn, affected the environment in which gangs at the time operated. Yet common to all five periods were two factors: first, poverty, or a very limited family income, was a key precondition generating gang formation and involvement; and, second, opportunities for socioeconomic mobility worsened during each period, making the gang increasingly attractive by contrast. Let me turn now to each period respectively, showing how they presented significant and sometimes unique material challenges for poor and low-income populations facing structural conditions that progressively worsened their life chances.

Gangs in Times of Immigration

Continuous waves of immigration to the United States has meant that, despite differences, varied ethnic groups have shared experiences of overcoming prejudice and discrimination from one generation to the next (Portes and Rumbaut, 1996). Beginning in the 18th century, gangs were associated with the lower classes of the various immigrant groups who found their way to the United States. Indeed, lower-class position was the main reason young people initiated gangs and became involved in delinquent behavior (Asbury, 1927). The combination of their parents' lower-class positions, the dominant ideology that the individual must make it on his/her own, and state support for this ideology offered little, if anything, in the way of a social safety net (Patterson, 2000). Therefore, individuals joined gangs because this form of association gave them camaraderie, entertainment and goods to consume—even if the latter were obtained through delinquent acts (Thrasher, 1927).

By the 1990s, though some of these conditions remain, the structure of the immigration experience has changed, affecting gangs in turn in two distinctive ways. First, as in the past, immigrants arriving in the USA have continued to establish their own communities (Waldinger, 2001); some of these groups, like the Chinese and Vietnamese, had a long history of gangs in their own societies prior to immigration (Chin, 1996). Different of late, though, is that the gangs in various sending countries have waited until their respective countrymen have established their own communities in the United States before sending elements of their organizations to set up enterprises thereafter (Chin, 1996). Resulting gangs have been primarily, though not exclusively, involved in drug trafficking and gambling establishments. Since the people living in communities are newcomers, language barriers and out-group prejudices they experience make them feel socially isolated. This provides these gangs with a fertile environment to develop their operations because such neighborhoods are a conveniently protected environment in which to sell illegal drug products to members of the more affluent sectors of the community. Residents' limited competency in English also ensures an economic niche to establish enterprises to satisfy the immigrant community's entertainment needs. This has been accomplished through the gang's installment of illegal gambling houses. Most significant here is that in addition to these gangs finding drug trafficking and gambling houses to be lucrative, the isolation of the immigrant community facilitates hiding their activities from the police.

Second, gangs have also developed in immigrant communities where the socio-economic mobility of the youth of these communities appears, in the 1990s, even more structurally blocked than in the past. Here a gang emerges when the youths in these areas, primarily those from the first and second generation, become frustrated and disillusioned upon realizing they are not likely to find jobs that can allow them to rise above the socioeconomic level attained by their parents. This leads some youths to form gangs to generate income they believe will provide them with a better life than their parents. These gangs have had two primary sources of money: extorted monies from small storeowners and restaurant workers living in their communities (Vigil and Yun, 1990), and heroin and cocaine bought with these extorted monies from the larger drug organizations (and then sold to the various retailers in the city). This has proved quite profitable for many of these gangs.

Thus, the character of the immigrant experiences has influenced the development of gangs past and present. This has included both the structural conditions existing in the sending communities, such as the presence of strong and sophisticated gangs, as well as the structural conditions in the host country including blocked mobility and sociogeographic concentration of poor non-white populations. In the contemporary period, the immigrant experience has produced gangs that have been primarily, although not exclusively, predatory on their community.

Gangs in Times of Blue-Collar Expansion

During the 1940s, 1950s, and 1960s, significant opportunities existed for working-class kids to secure working-class jobs. This has influenced the particular character of gangs in the following decades in communities where the opportunity to secure working-class jobs remains. In such communities, youth have grown up seeing and talking to family members, relatives or friends employed in blue-collar positions; they have come to know the social conditions that exist in the factories and the life that such work provides. On a personal level, many youths have found these jobs unattractive. Even when they talk about blue-collar jobs in the primary sector they believe that the work is boring and the hours long. When they discuss the blue-collar jobs in the secondary labor market, their views are even harsher concerning the actual working conditions and the chances of getting what they want out of life. The comments of Albert and Luis are typical in this regard. Albert is a 16-year-old African American whose father works in a factory that makes auto parts for General Motors: "I definitely don't want to do what my dad does. He is always complaining about how fast the production line is. He is always tired and even though he makes good money he never has anything to say about the job 'cause he does the same thing everyday. No wonder he drinks all the time." Luis is a 15-year-old Mexican whose father works in the garment industry: "My dad is like in a daze around the house. He comes home from work and he is dead tired. He works twelve hours a day, six days a week doing the same job. He has that dust from the machine all over him, and he coughs from not wearing a mask. I hope there is something more for me than a job like he's got."

Youths who see their parents' jobs negatively are likely to wish to prolong the time before they enter this particular job market and lifestyle. Even if high-paying production jobs are available, they are physically taxing because the firms that offer these jobs often pressure workers to work overtime whether the workers themselves want the extra money or not. Further, the work is usually repetitive and monotonous, making it also taxing psychologically. Under these conditions, gangs emerge as organizations that provide a social haven for young people to experience fun and pleasure before assuming jobs and a concomitant lifestyle they wished to avoid.

Thus, in this second structural situation—one that I am calling here "gangs in times of blue-collar expansion"—primary activities are oriented toward securing financial resources necessary to provide leisure for their members. This goal is pursued with resolve through members obtaining part-time jobs and paying dues to the gang organization, and/or by selling illegal drugs and stolen contraband. However, these gangs do not focus on accumulating profit to disperse to their members as they do under other structural conditions; rather their economic activities are concentrated on paying bills related to the entertainment they are providing. This is the reason that someone would get a temporary, or part-time job, because they could remain primarily involved in the gang and its leisure activities without having to commit themselves to a full-time work schedule that monopolizes their time and energy. Thus, the gang's primary character here is that it takes on the functions of a social organization (Schneider, 1999).

Gangs in Times of Drug Deregulation

In the past, the Italian Mafia monopolized the drug industry, including controls over both production and distribution. However, the Italian Mafia's total control of production and distribution evaporated for a variety of reasons, the most important of which involved ethnic conflict and market control. As ethnic antagonisms became more hostile between Italians and African Americans, Puerto Ricans and Mexicans, Italians in general, including the Mafia, found it nearly impossible to be physically safe in these groups' neighborhoods . . . . As a result the Mafia was forced to gradually withdraw their retail operations from most of these ethnic areas and simply wholesale drugs to local retailers (Robinson, 1993). Concomitantly, this opened up opportunities for other segments of the low-income community to become involved in the retail drug industry (Bourgois, 1995). As this happened through the 1970s and 1980s, gangs became involved in different capacities of the drug retail trade. Some gangs distributed drugs and also became involved in the production of crack cocaine and other drugs; some had drug mills that produced synthetic hallucinogens. Moreover, as market opportunities worsened for the poor and working class, increasing numbers of youth found gangs to be an attractive alternative (Wilson, 1996). It was the combination of contracting market opportunities in the production sector of the economy and the expanding market opportunities in both the production and retail illicit drug economy that stimulated youths from varied ethnic groups to become involved in gangs. For gangs could both recruit young people with the pitch that they could make substantial money, and convince them that they had the contacts necessary to produce a profitable business and the organizational capacity to protect them from other competitions. One of the most important by-products of this structural shift in the contemporary context in which gangs operate is that individuals have now increased the length of time during which they participate. In the past, gang participation would have been confined primarily to a young boy's teens whereas, at present and under the conditions just sketched, participation may extend to age 30 and beyond (Sánchez-Jankowski, 1991; Klein, 1998).

Gangs in Times of Mass Incarceration

Although there has been a 20-year "war on drugs," the industry has continued to grow and to provide a strong opportunity structure for individuals and groups like gangs and posses. However, the illicit drug trade presents individuals with very high risk, not only in the financial market, but vis-à-vis the law as well. As the drug economy has expanded and more people have become involved, the number of people who are incarcerated has correspondingly risen. In 2002, the United States, due to aggressive enforcement policies developed over the last several decades, holds the leading position among all nations on the numbers of people incarcerated. These policies have included enacting legislation to increase prison time for those incarcerated for gang-related crimes; mandatory sentencing laws; the trying of juveniles as adults when violent crimes have been committed; and increased prison building (Donziger, 1996). In turn, this has changed the demography of prisons, producing at least one interesting and unintended consequence.

More recently, as street gangs have become more involved in the drug industry, and as law enforcement policies have changed, the number of street gang members who are imprisoned has increased. In turn, as these happened, street gangs became more assimilated into prison gangs. For at least the last decade and a half, prison gangs, which are adult-organized crime syndicates, have been trying to organize street gangs under their authority within the prison in an effort to control greater portions of the drug market. Street gangs have consistently resisted these efforts. Since most street gang members are younger, one reason is that youthful gang members do not want "older boys" controlling them; adolescent rebelliousness has meant that authority imposed on them by older members of society, even if these members were part of the greater gang society, was resented. Yet, as more of these youths from street gangs went to prison, they were forced by the stark reality of the prison structure to affiliate with one of the prison gangs or else risk being vulnerable to the hostile predators within the prison population (Abbott, 1991).

Therefore, individual street gang members have entered an environment structured both by the state authorities and prison gangs; at the same time, many gang members perceive that it is likely they will do more than one stint, spending a considerable amount of their lives in prison. In California, over a period of time, this realization has influenced individuals from street gangs either to become members of prison gangs or to make formal alliances with them. For example, among California's Chicano gangs, prison gangs have divided the state in half: those who live south of Bakersfield are identified as *sureños* (symbolized by the color blue) and those living to the north are identified as *norteños* (symbolized by the color red). Consequently, whereas before the gang with which an inmate affiliated in prison was not the same as outside, now mergers take place: in this case, various street gangs of Mexican-origin inmates have unified with two dominant Mexican prison gangs, La Familia and the Mexican Mafia.

From this, one can surmise that state policy to incarcerate gang members for longer periods of time is producing unintended consequences. For one thing, an increasing tendency for street gangs and prison gangs to unify means that the resulting collective associations are even more organized and have greater resources to sustain themselves. [R]elatedly, an unintended consequence of the recent developments is that, by changing gang structures on the inside, an effect on gang structures outside prison has been to unify drug markets. None of this is likely to alter at the moment; indeed ongoing rises in incarceration suggest that the situation will continue to worsen at least into the first decade of the 21st century.

Gangs in Times of Monopoly Behavior

Yet another missing aspect of most academic analyses is insight into the structural conditions that influence gang violence. Before discussing the structural conditions that impact gang violence, though, it is necessary to clarify the concepts of "violence" and "gang violence." "Violence" may be defined as the use of force to achieve some desired end. It is a maximizing of physical force to achieve a desired end, and as such must be seen as a tool. "Gang violence" can be defined in relation to "gang-member violence": the former involves individuals committing violence as agents of the organization. On the other hand, "gang-member violence" involves individuals in gangs committing violence as independent agents. Clearly, distinguishing between these two types of violence is imperative for understanding particular violent acts in which gang members become involved.

The violence associated with gangs has been influenced by three conditions. The first is associated with "gang-member violence" and emerges from the material conditions in which gang members find themselves. In essence, "gang-member violence" is a product of both the structural conditions that permeate the scarcity of resources and the socialized manner that individuals learn to survive in such an environment. For example, gangs have consistently emerged from low-income communities where there has been a scarcity in resources. Thus, individuals brought up in such an environment learn that they must be aggressive in their efforts to compete and secure these resources because, if they are not, others will get them. This socialization process influences individuals from lower-income communities to be particularly cautious in their approach to others and to employ maximum power in their efforts to secure or maintain a possession or goal. Therefore, individuals who are in gangs, like other individuals from these environments, use violence to obtain their own individually oriented objectives. It is this individual-oriented type of violence that has been misrepresented by law enforcement, the media and some academics, since it would have occurred whether the individual was in a gang or not (Sánchez-Jankowski, 1991). In other words, "gang-member violence" does not involve the gang per se.

However, because gangs are organizational artifacts of their environments, they offer positive reinforcement to those who utilize violence to achieve their aims. Power is something that few people in lower-income communities have without resorting to physical force; thus, the more physical force one has as a resource, the more status one will enjoy in these communities. In this way, material conditions establish the foundations for a "culture of force" that can ultimately be labeled a "culture of violence." Through this culture, a view of the world develops with a rationale concerning how power is obtained; what this power can offer the individual or group in the way of material resources; and how status can be expected as a result. Force is the medium through which individuals try to monopolize the economic and social resources available to them.

The second way in which gang violence is structured has to do with the structure of the market in which gangs, as organizations, operate. As mentioned earlier, both individuals and gangs often use force in advancing their interests. However, in conjunction with the structure of the underground economy, levels of gang violence have increased in recent years. With the opening of new drug markets, gangs have behaved like any other capitalist-oriented organization. They have attempted to monopolize the various drug products and the markets wherein they are exchanged. This behavior has been, and will continue to be, particularly aggressive and violent because, contrary to other markets in the legal economy, there is no external party—e.g., a board, an agency, a court or a state—that can regulate the activity of competitors. Thus, in a market that has no outside party capable of intervening to regulate normal operations or disputes, this responsibility falls to the actors themselves and the power they bring to the exchange relationship. Those who possess the most physical power and are willing to use it are the most successful in their efforts to monopolize the various illegal product markets. When one of the competitors has considerably more physical power than the others, there tends to be less violence. However, when there is roughly equal power between competitors more violence will be present because of persistent initiatives to determine who is dominant. In all these situations, the structure of the market (type of products, amount of supplies and demand) and the structure of the organization (strength of its internal structure and power resources), along with the structure of the realm of competition (physical environment) determine the type and levels of ensuing violence.

In sum, gang violence, both individual member and organizational, results from a dynamic combination of three interacting conditions. These are material conditions of scarcity (encouraging competition over what little exists); a culture that sanctions physical force as the primary means to realize goals; and an available economy that has no formal state-authorized agency capable of regulating the monopolizing behavior of the individuals and organizations involved. Together these conditions form a peculiar structure that establishes parameters for risk and safety.

Conclusion

This article focused on the relationship between the social structure of American society and the social structure of gangs. The United States was founded on the image of being a revolutionary society. By this I mean it was founded on a political and social break with its historical origins. The new nation that emerged out of this break both established "the American" as a new identity and created new structures to help support this new identity. One of the factors that helped to shape the new identity of "the American" and the concomitant social structures that supported it, was the large and unsettled nature of the American geography. The U.S. was a "frontier nation" that taught people that there was unlimited opportunity, but that one must depend on oneself. This emphasis on the individual was also affected by the fact that the state was seen as a potential threat to individual liberty. Thus the state should not be involved in people's lives even if for the purpose of providing for the common good. What evolved was the belief that the state was incapable of being helpful because whenever it intervened it altered the very basis of what produced a productive society; the exact individual spirit necessary to overcome the hardship and defeat was destroyed. In essence one of the fundamental tenets of what became dominant American ideology was the principle that defeat was an important force that made for a great society. This was because both the quest for success, and the fear of defeat, made for a more productive citizen. In turn, it was believed that the fruit of people's labors was to either avoid defeat or to overcome it: this allegedly was what made the U.S. a great society.

These beliefs about inequality produced a political culture where the state is seen as undermining the very essence of what makes the society strong when providing social welfare to its citizens. Most citizens in the U.S. have learned these aspects of the political culture, especially members of the lower class. Individuals in the lower class know they must depend on themselves and that, if they are to improve their position in life, they must be creative and enterprising. As conditions among the lower class have declined in American inner cities, and the state has retrenched from intervening to improve them, young males (especially nonwhite males) have developed strategies to become more enterprising with the opportunities that they have. One, though not the only, strategy employed by some of the lower class has been to become involved in the underground economy, especially with illicit drugs; while individuals can enter this market, it is less personally risky in both economic and personal injury terms if one joins an organized group like a gang (Sánchez-Jankowski, 1991). Thus, instead of being deviants from the prevailing economic culture, gang members have accepted the principles of the dominant social ideology and economic culture, and have adapted their strategies to the opportunities and resources available to them. There have been losers in this effort, but, as in all capitalist markets, winners as well.

This situation has precipitated a structural response on the part of the state. In an effort to control the economic activity of gangs and other groups, the state has increased the number of law enforcement personnel responsible for gang activity. For example, starting in 1992–3, the Federal Bureau of Investigation decided, as a result of the ending of the cold war, to transfer those agents that had been assigned to counter-subversive units (anti-Communist units), into the newly organized anti-gang division. They have passed legislation that increased the amount of time individuals associated with a gang will serve in prison, and they have built more prisons where they dump ever larger quantities of lower-class men.

The state's response has caused gangs to react. Since more youths are incarcerated of late for longer periods of time, local street gangs have reacted by integrating themselves into the organized crime syndicates associated with prison. Thus, instead of weakening the organizational structure of gangs through the policy of increased incarceration of gang members, the state's policy worked to strengthened them. Ironically, despite the counter-productive results in affecting the gang phenomenon, the state's continued policy response is to build even more prisons and pass even more harsh rules.

Overall, then, the contemporary gang problem must be understood as a result of certain structural conditions that exist in the United States, or any society with similar structural and ideological conditions that have caused societal inequality to grow. Therefore, the gang must not be seen as a collection of deviants, or a deviant form of collective behavior. Rather, it must be seen as both an organization composed of people who have the values and goals of mainstream American society and as an organization that engages in a form of collective behavior that is particular to the socio-economic conditions its participants confront. Yet to understand the gang phenomenon in the United States, it is necessary to consider that socio-economic conditions of American society have continually changed, usually becoming worse for the lower class, and the structures associated with these changes have produced rational changes in the gang phenomenon. This article has attempted to sketch some of these changes and the effects they have had on gangs. In so doing, it has provided a theoretical outline that can be utilized to understand and forecast changes in the contours of the gang phenomenon in the future.

References

Abbott, J. H. (1991). *In the belly of the beast: Notes from prison.* New York: Vintage Books.

Asbury, H. (1927). *The gangs of New York: An informal history of the underworld.* Garden City, NY: Garden City.

Bourgois, P. (1995). *In search of respect: Selling crack in El Barrio.* New York: Cambridge University Press.

Chin, K.-L. (1996). *Chinatown gangs: Extortion, enterprise, and ethnicity.* New York: Oxford University Press.

Donziger, S. R. (1996). *The real war on crime: The report of the National Criminal Justice Commission.* New York: Harper Collins.

Fromm, E., & Maccoby, M. (1970). *Social character in a Mexican village: A socio-psychoanalytic study.* Englewood Cliffs, NJ: Prentice-Hall.

Klein, M. (1998). *The American street gang.* New York: Oxford University Press.

Patterson, J. T. (2000). *America's struggle against poverty in the twentieth century.* Cambridge: Harvard University Press.

Portes, A., & Rumbaut, R. (1996). *Immigrant America: A portrait.* Berkeley: University of California Press.

Robinson, C. D. (1993). Production of black violence. In D. F. Greenberg (Ed.), *Crime and capitalism: Readings in Marxist criminology* (pp. 279–333). Philadelphia, PA: Temple University Press.

Sánchez-Jankowski, M. (1991). *Islands in the street: Gangs and American urban society.* Berkeley: University of California Press.

Schneider, E. C. (1999). *Vampires, dragons, and Egyptian kings: Youth gangs in postwar New York.* Princeton, NJ: Princeton University Press.

Thrasher, F. M. (1927). *The gang: A study of 1313 gangs in Chicago.* Chicago: University of Chicago Press.

Vigil, J. D., & Yun, S. C. (1990). Vietnamese youth gangs in southern California. In C. R. Huff (Ed.), *Gangs in America* (pp. 146–162). Newbury Park, CA: Sage.

Waldinger, R. (2001). *Strangers at the gates: New immigrants in urban America.* Berkeley: University of California Press.

Wilson, W. J. (1996). *When work disappears: The world of the new urban poor.* New York: Knopf.

DISCUSSION QUESTIONS

1. Why does the author believe that the focus on individual differences to explain gang membership is misguided? Do you agree or disagree with him?
2. Why have the old Irish and Jewish gangs mostly gone while Hispanic and African American gangs survive?
3. What can be done to prevent the further integration of prison and street gangs?

❖

SOCIAL PROCESS THEORIES

The social structural theories discussed in the previous section only explain part of the possible reason that Kody Scott chose the path in life he did. Not all who experience the same conditions turn out the same way; indeed, only one of Kody's brothers ran afoul of the law. The social process theories discussed in this section take us a step further in understanding Kody's choices. Two of the theories in this section tell us that criminal behavior is learned in association with peers and that we choose to repeat behaviors that are rewarding to us. After he shot and killed the Blood gang members in his initiation, Kody tells us that he lay in bed that night feeling guilty and ashamed of his actions and that he knew they were wrong. Nevertheless, when the time came to do the same thing again, he chose to do what his peers told him to do because he valued their praise and approval more than anything else in life. His fellow Crips also provided him with rationales and justifications for his actions that neutralized his guilt.

Another theory in this section tells us that labels have the power to make us live up to them; we have seen how Kody proudly accepted the label of "Monster" and how he did his best to live up to it. Other theories stress the importance of attachment to social institutions, but he tells us that his "homeboys" were his only family and that his only commitment and involvement were to and with them and their activities.

On a personal level, he plainly lacked self-control; he was impulsive, hedonistic, and angry. Theorists in this chapter tell us that self-control is developed by consistent parental monitoring, supervision, and discipline, but his weary single mother lacked the time, resources, or incentive to provide Kody with proper parenting.

From the youngest age, he came and went as he pleased. His autobiography makes it plain that he was something of a "feral child," big enough, mean enough, and guiltless enough to be free to satisfy any and all urges as they arose. This section details many of the social processes by which Kody came to be the monster he claims himself to be.

Social process criminologists operate from a sociological perspective known as **symbolic interactionism**. Symbolic interactionists focus on how people interpret and define their social reality and the meanings they attach to it in the process of interacting with one another via language (symbols). Social process theorists believe that if we wish to understand social behavior we have to understand how individuals subjectively perceive their social reality and how they interact with others to create, sustain, and change it. The processes most stressed are socialization and cultural conflict; that is, they seek to describe criminal and delinquent socialization (how antisocial attitudes and behavior are learned) and how social conflict "pressures" individuals into committing antisocial acts. Some process theories focus on the reverse process of learning prosocial attitudes and behavior in the face of temptations to do otherwise. All social process theories represent the joining of sociology and psychology to varying extents, even if their authors explicitly deny that this is the case.

Differential Association Theory

Differential association theory (DAT) is the brainchild of Edwin Sutherland, whose ambition was to devise a theory that could explain both individual criminality and aggregate crime rates by identifying conditions that must be present for crime to occur and that are absent when crime is absent. His theory attempted to explain more precise mechanisms by which factors such as social disorganization led to crime. Like Walter Miller, Sutherland saw lower class culture has having its own integrity, and he disdained the phrase "social disorganization" as insulting, substituting "differential social organization." Although Sutherland explicitly denied the role of psychology in crime and delinquency, his theory is implicitly psychological in that it focuses on the process of becoming delinquent via subjective definitions of reality and attitude formation.

DAT takes the form of nine propositions outlining the process by which individuals come to acquire attitudes favorable to criminal behavior, which may be summarized as follows: Criminal behavior is learned (the motives, drives, and attitudes) in intimate social groups. By emphasizing social learning, DAT wants to guide criminologists away from the notion that criminal behavior is the result of biological or psychological abnormalities or invented anew by each criminal. Criminality is not the result of individual traits or learned from impersonal communication from movies or magazines and the like. The learning of criminal behavior involves the same mechanisms involved in any other learning and includes specific skills and techniques for committing crimes, as well as the motives, rationalizations, justifications, and attitudes of criminals.

The theory asserts that humans take on the hues and colors of their environments, blending in and conforming with natural ease. We view the world differently according to the attitudes, beliefs, and expectations of the groups around which our lives revolve. If you think about it, it could hardly be otherwise, particularly in our formative years.

▲ **Photo 5.1** Youthful racist skinheads in London give the fascist salute. Differential association theory would argue that if the people you spent most of your time with espouse deviant values, you are likely to adopt these as well.

The key proposition in DAT is "*A person becomes delinquent because of an excess of definitions favorable to violations of law over definitions unfavorable to violations of law*" (Sutherland & Cressey, 1974, p. 75). Learning criminal conduct is a process of modeling the self after and identifying with individuals we respect and value. **Definitions** refer to the meanings our experiences have for us; how we see things; our attitudes, values, and habitual ways of viewing the world.

Definitions become favorable to law violation according to the *frequency, duration, priority, and intensity* of exposure to them. That is, the earlier we are exposed to criminal definitions, the more often we are exposed to them, the longer they last, and the more strongly we are attached to those who supply us with them, the more likely we are to commit criminal acts when opportunities to do so arise. As we have already seen, antisocial definitions are more likely to be learned in lower-class neighborhoods. In such neighborhoods, children are surrounded by antisocial definitions (the code of the streets, focal concerns) and cannot help being influenced by them regardless of their individual characteristics.

Ronald Akers' Social Learning Theory

Ackers' **social learning theory** (SLT) goes beyond looking at frequency, priority, duration, and intensity of crime to identify mechanisms by which "definitions favorable" to crime are learned. Akers and his colleague, Robert Burgess (Burgess & Akers, 1966) applied the concepts of operant psychology to the vague "definitions favorable" of DAT. While psychological principles are central to SLT, Akers (2002) insists that it is in the same sociological tradition as DAT and that it retains all the processes found in Sutherland's theory, albeit modified and clarified.

Operant psychology is a theory of learning that asserts that behavior is governed by its consequences. When we behave we receive feedback from others that we interpret in terms of positive or negative consequences. Behavior has two general consequences; it is reinforced or it is punished. Behavior that has positive consequences for the actor is said to reinforce that behavior, making it more likely that the behavior will be repeated in similar situations. Behavior that is punished is less likely to be repeated.

Reinforcement is either positive or negative. The loot from a robbery or status achieved by facing down rivals are criminal examples of *positive reinforcement. Negative reinforcement* occurs when some aversive condition is avoided or removed, such as the removal of a street reputation as a "punk" following some act of bravado.

Punishment, which leads to the weakening or eliminating of the behavior preceding it, can also be positive or negative. *Positive punishment* is the application of something undesirable, such as a prison term. *Negative punishment* is the removal of a pleasant stimulus, such the loss of status in a street gang. The acquisition of Sutherland's "definitions favorable" to antisocial behavior (or prosocial behavior, for that matter) thus depends on each individual's history of reinforcement and punishment.

In any peer group, each member of the group has reciprocal effects on every other member via his or her participation in the reinforcement/punishment process. Of course, what is reinforcing for some may be punishing for others. For teens who value the approval of their parents and teachers, an arrest is punishment; for teens who value the approval of antisocial peers, such an outcome is a reinforcer because it marks them officially as a "bad ass." The social context is thus an extremely important component of SLT because most learning takes place in the presence of others who provide both the social context and the available reinforcers or punishments.

Discrimination is another important component of SLT. Whereas reinforcements or punishments *follow* behavior, discriminative stimuli are present *before* the behavior occurs. Discriminative stimuli are clues that signal whether a particular behavior is likely to be followed by reward or punishment. In other words, discrimination involves learning to distinguish stimuli that have been reinforced or punished in the past from similar stimuli you expect will result in the same response in the future. For instance, an unlocked car with the keys in it is a discriminative stimulus that signals "immediate reward" for the criminal, but for the average person who has never previously been rewarded for stealing a car, it probably signals nothing other than how foolish the owner is.

The excerpt from Akers presented in this chapter further explains SLT and how it is similar and different from DAT. Akers believes that his theory explains the link between social structure and individual behavior. For instance, differential social organization places individuals in contexts in which different types of behavior are reinforced or punished and in which social control systems provide different learning environments.

Social Control Theories

To ensure a peaceful and predictable social existence, all societies have created mechanisms that we may collectively call **social control,** designed to minimize nonconformity and deviance. In many senses, both Durkheim's anomie theory and Shaw and McKay's ecological theory are control theories because they point to situations or circumstances (anomie or social disorganization) that lessen control of individuals' behavior. Social control may be direct, formal, and coercive, as exemplified by uniformed symbols of the state. But indirect and informal social control is preferable because it produces prosocial behavior regardless of the presence or absence of external coercion. Obeying society's rules of proper conduct because we believe that the rules are right and just, not because we fear formal sanctions, means that we have our own internalized police officer and judge in the form of something called a conscience.

Travis Hirschi's Social Bond Theory

There is a variety of control theories, but most popular and enduring is Travis Hirschi's (1969) **social bond theory**. Previous theories we have examined assume that crime is something learned by good people living in bad environments, and ask, "What causes crime?" Control theorists believe that this question reveals a faulty understanding of human nature, and that the real question is not why some people behave badly, but why most of us behave well most of the time. After all, children who are not properly socialized hit, kick, bite, steal, and scream whenever the mood strikes them. They have to be taught not to do these things, which, in the absence of training, "come naturally." In this tradition, it is society that is "good" and human beings, in the absence of the proper training, who are "bad." Gwynn Nettler (1984) said it most colorfully: "If we grow up 'naturally,' without cultivation, like weeds, we grow up like weeds—rank" (p. 313).

Social control theory is thus about the role of social relationships that bind people to the social order and prevent antisocial behavior. Antisocial behavior will emerge automatically if controls are lacking; it needs no special motivating factors because human beings are assumed to be naturally self-centered. The classical assumption of self-interested persons anxious to experience pleasure and avoid pain is still there, but the theory tries to account for why some people pursue their self-interest in legitimate ways and others do not, with primary emphasis on socialization practices that do or do not produce individuals capable of reining in their natural instincts.

The Four Elements of the Social Bond

Hirschi finds that the "typical" criminal is a young male who grew up in a fatherless home in an urban slum, who has a history of difficulty in school, and who is unemployed. From this he deduces that those most likely to commit crimes lack the four elements of the social bond—*attachment, commitment, involvement,* and *belief*—that result in conformity with prosocial behavior.

Attachment is the emotional component of conformity and refers to the emotional bonds existing between the individual and key social institutions like the family and the school. Attachment to conventional others is the foundation for all other social bonds because it leads us to feel valued, respected, and admired and to value the favorable judgments of those to whom we are attached. Much of our behavior can be seen as attempts to gain favorable judgments from people and groups we care about. Lack of attachment to parents and lack of respect for their wishes easily spills over into a lack of attachment and respect for the broader social groupings of which the child is a part. Much of the controlling power of others outside the family lies in the threat of reporting misbehavior to parents. If the child has little respect for parental sanctions, the control exercised by others has little effect because parental control has little effect.

▲ **Photo 5.2** These students purchased similar jackets to display their solidarity and bonding to the group. Social control theory would predict that despite their quirky behavior as displayed here, they would not become involved in serious criminal behavior.

Commitment is the rational component of conformity and refers to a lifestyle in which one has invested considerable time and energy in the pursuit of a lawful career. People who invest heavily in a lawful career have a valuable stake in conformity and are not likely to risk it by engaging in crime. School dropouts and the unemployed do not have a strong investment in conventional behavior and therefore risk less by committing crime. Acquiring a stake in conformity requires disciplined application to tasks that children do not relish but which they complete in order to gain approval from parents. Attachment is thus the essential foundation for commitment to a prosocial lifestyle.

Involvement is a direct consequence of commitment; it is a part of an overall conventional pattern of existence. Involvement is a matter of time and energy constrictions placed on us by the demands of our lawful activities that reduce exposure to illegal opportunities. Conversely, noninvolvement in conventional activities increases the possibility of exposure to illegal activities.

Belief refers to the acceptance of the social norms regulating conduct. Persons lacking attachment, commitment, and involvement do not believe in conventional morality. A belief system empty of conventional morality is concerned only with narrow self-interest. It is important to realize that crime is not motivated by the absence of any of the social bonds;

their absence merely represents social deficiencies that result in a reduction of the potential costs of committing crime.

Leanne Alarid, Velmer Burton, and Francis Cullen's article in this section pits DAT and social bonding theory against one another to see which one better explains self-reported criminal behavior among a sample of incarcerated felons. They found support for both theories, but found that DAT had more consistent effects, especially for men. It was also apparent that the theories predicted differently by gender, with attachment being a stronger predictor of violent crime for females than for males. Note that "criminal friends" was the best overall predictor of the respondents' own criminal behavior. As we will see later, a major criticism of DAT is that it assumes that criminal friends exert a causal influence on a person's criminal behavior. Also note the types of questions often asked in self-report studies in the Appendix of this study. Ask yourselves (1) whether these kinds of questions adequately capture criminality among incarcerated felons, and (2) whether asking "Have you ever" seriously conflates individuals who have done these things once or twice and individuals who do these things frequently.

Gottfredson and Hirschi's Low Self-Control Theory

With colleague Michael Gottfredson, Hirschi has moved away from explaining crime and delinquency in terms of social control and toward explaining it in terms of self-control. *Self-control* is defined as the "extent to which [different people] are vulnerable to the temptations of the moment" (Gottfredson & Hirschi, 1990, p. 87). The theory accepts the classical idea that crimes are the result of unrestrained natural human impulses to enhance pleasure and avoid pain. Such a pathway to pleasure often leads to crimes, which Gottfredson and Hirschi define as "acts of force or fraud undertaken in pursuit of self-interest" (p. 15). Most crimes, they assert, are spontaneous acts requiring little skill and earn the criminal minimal, short-term satisfaction. People with low self-control possess the following personal traits that put them at risk for criminal offending:

- They are oriented to the present rather than to the future, and crime affords them immediate rather than delayed gratification.
- They are risk-taking and physical as opposed to cautious and cognitive, and crime provides them with exciting and risky adventures.
- They lack patience, persistence, and diligence, and crime provides them with quick and easy ways to obtain money, sex, revenge, and so forth.
- They are self-centered and insensitive, so they can commit crimes without experiencing pangs of guilt for causing the suffering of others. (pp. 89–90)

Low self-control is established early in childhood, tends to persist throughout life, and is the result of incompetent parenting. It is important to realize that children do not learn low self-control; low self-control is the default outcome that occurs in the absence of adequate socialization. Parental warmth, nurturance, vigilance, and the willingness to practice "tough love" are necessary to forge self-control in their offspring. Other factors that may result in low self-control include parental criminality (criminals are not very successful in socializing their children), family size (the larger the family the more difficult it is to monitor behavior), single-parent family (two parents are generally better than one), and working mothers, which negatively impacts the development of children's self-control if no substitute monitor is provided (Gottfredson & Hirschi, 1990, pp. 100–105).

Gottfredson and Hirschi (1990) argue that children acquire or fail to acquire self-control in the first decade of life, after which the attained level of control remains stable across the life course. Subsequent experiences, situations, and circumstances have little independent effect on the probability of offending. Because low self-control is a stable component of a criminal personality, most criminals typically fail in anything that requires long-term commitment, such as school, employment, and marriage, because such commitments get in the way of immediate satisfaction of their desires.

Low self-control is a necessary but not sufficient cause of criminal offending. What accounts for variation in criminal offending are the different opportunities criminals encounter that are conducive to committing crimes. A criminal *opportunity* is a situation that presents itself to someone with low self-control by which he or she can immediately satisfy needs with minimal mental or physical effort (Gottfredson & Hirschi, 1990, pp. 12–13). Crime is thus the result of people with low self-control meeting a criminal opportunity, and by virtue of differential placement in the social structure, some individuals are exposed to more criminal opportunities than others.

Gottfredson and Hirschi provide further insight into self-control theory in this section in the excerpt taken from their 1990 book introducing their theory.

Labeling Theory: The Irony of Social Reaction

Labeling or societal reaction theory (LT) takes seriously the power of bad labels to stigmatize, and by doing so labels evoke the very behavior the label signifies. The labeling perspective is interesting and provocative; unlike other theories, it does not ask why crime rates vary or why individuals differ in their propensity to commit crime. Rather, it asks why some behaviors are labeled criminal and not others, and thus shifts the locus from the actor (the criminal) to the reactor (the criminal justice system).

LT is often traced to Frank Tannenbaum's *Crime and the Community*, which emphasized that a major part in the making of a criminal is the process of identifying and labeling a person as such—the "dramatizing of evil" (1938, p. 20). Tannenbaum viewed the labeling of a person as a "criminal" as a self-fulfilling prophesy (a definition of something which becomes true if it is believed and acted upon), which means that processing law violators through the criminal justice system, rather than deterring them from future criminal behavior, may embed them further in a criminal lifestyle.

For labeling theorists, crime and other forms of deviance have no objective reality and are defined into existence rather than discovered. There is no crime independent of cultural values and norms, which are embodied in the judgments and reactions of others. To put it simply, no act is by its "nature" criminal, because acts do not have natures until they are witnessed, judged good or bad, and reacted to as such by others.

LT distinguishes between primary deviance and secondary deviance (Lemert, 1974). **Primary deviance** is the initial nonconforming act that comes to the attention of the authorities. Primary deviance can arise for a wide variety of reasons, but the reasons are of little interest because they have only marginal effects on the offender's self-concept as a criminal or non-criminal, and it is the individual's self-concept that is crucial in labeling theory. Primary deviance is of interest to labeling theorists only insofar as it is detected and reacted to by individuals with power to pin a stigmatizing label on the rule-breaker. Being caught in an act of primary deviance is either the result of police bias or sheer bad luck; the real criminogenic experience comes *after* a person is caught and labeled. The central concern of LT is thus to explain the consequences of being labeled.

Secondary deviance results from society's reaction to primary deviance. The stigma of a criminal label may result in people becoming more criminal than they would have been had they not been caught. This may occur in two ways. First, labeled persons may alter their self-concepts in conformity with the label ("Yes, I am a criminal, and will act more like one in future"). Second, the label may exclude them from conventional employment and lead to the loss of conventional friends. This may lead them to seek illegitimate opportunities to fulfill their financial needs and to seek other criminals to fulfill their friendship needs, which further strengthens their growing conception of themselves as "really" criminal. The criminal label becomes a self-fulfilling prophesy because it is a more powerful label than other social labels that offenders may claim.

Xiaoming Chen's article in this section offers an interesting look at the use of labeling and "reintegrative shaming" in China to maintain social control. Chen asserts that shaming has a much greater influence on Chinese behavior than on Western behavior because of China's culture of collectivism versus the Western culture of individualism. Chen sees shaming as a positive tool rather than a negative one for the Chinese because it serves (ideally) to integrate the miscreant back into "respectful" society. Chen also claims that the Chinese penal system claims an astoundingly low (by Western standards) recidivism rate of between 8% and 15% because of its methods.

Sykes and Matza's Neutralization Theory

Sykes and Matza's *neutralization theory* (NT) is a learning theory that attacks DAT's failure to explain why some people drift in and out of crime rather than being consistently criminal. It also runs counter to the assumption of DAT and the subcultural theories that give the impression that criminal behavior is endowed with positive value and condoned as morally right. It is difficult to believe that criminals do not know "deep down" that their behavior is wrong: "if there existed in fact a delinquent subculture such that the delinquent viewed his behavior as morally correct," he or she would show no shame when caught, but would instead show "indignation or a sense of martyrdom" (Sykes & Matza, 2002, p. 145). NT suggests that delinquents know their behavior is wrong, but they justify it on a number of grounds. In other words, they neutralize any sense of shame or guilt for having committed some wrongful act, which means that they are at least minimally attached to conventional norms. NT also runs counter to labeling theory because it shows how delinquents resist labeling rather than passively accepting it. Sykes and Matza's five **techniques of neutralization** are listed below.

1. *Denial of responsibility* shifts the blame for a deviant act away from the actor: "I know she's only six, but she seduced me."
2. *Denial of injury* is an offender's claim that no "real" offense occurred because no one was harmed: "He got his car back, and his insurance covers the damage, doesn't it?"

3. *Denial of victim* implies that the victim got what he or she deserves: "I guess I did beat her up, but she kept nagging; hell, she asked for it!"
4. *Condemnation of the condemners* involves attempts by the offender to share guilt with the condemners (parents, police, probation officers) by asserting that their behavior is just bad as his or hers is: "You drink booze, I smoke grass; what's the difference?"
5. *Appeal to higher loyalties* elevates the offender's moral integrity by claiming altruistic motives: "I have to cover my homies' backs, don't I?"

The motive behind the employment of these techniques is assumed to be the maintenance of a noncriminal self-image by individuals who have committed a criminal act and who have been asked to explain why. Such individuals "define the situation" in a way that mitigates the seriousness of their acts and simultaneously protects the image they have of themselves as non-criminals. A less benign interpretation of the use of these techniques is that rather than trying to protect their self-images, they are seeking to mitigate their punishment, or at least to "share" it with some convenient other. Conversely, intensive interviews with hardcore criminals indicate that they strive to maintain an image consistent with inner-city street codes, not with conventional ones; "they neutralize being good rather than being bad" (Topalli, 2005, p. 798).

Evaluation of Social Process Theories

DAT assumes that antisocial behavior is learned, not something that comes naturally in the absence of prosocial training. But as one early critic of DAT asked, "What is there to be learned about simple lying, taking things that belong to another, fighting and sex play?" (Glueck, 1956, p. 94). Individuals certainly learn to get better at doing these things in their associations with other like-minded individuals, but do they have to be taught them, or do they have to be taught how to curb them, what constitutes moral behavior, and how to consider the rights and feelings of others?

DAT is also criticized for ignoring individual differences in the propensity to associate with antisocial peers. Individual traits do sort people into different relationship patterns—as numerous studies of relationship patterns attest (Rodkin, 2000). Differential association may thus be a case of birds of a feather flocking together; their flocking may facilitate and accentuate their activities, but it does not "cause" them.

Mark Warr (2000) criticizes DAT for having a singular vision of peer influence. Warr makes a distinction between two approaches to the influence of delinquent peers—*private acceptance* and *compliance.* Compliance is "going through the motions" without privately accepting the appropriateness of what one is doing. Private acceptance refers to both the public and private acceptance of the attitudes, values, and behavior of the delinquent group. Warr says that DAT "was built squarely on the idea of private acceptance" (p. 7). This implies that pro-offending attitudes become an integral part of the person's psyche and lifestyle, and thus once a delinquent career is initiated, the person will continue offending across the lifespan. The problem with this is that we know that the great majority of delinquents limit their offending to adolescence (are temporary compliers) and do not become adult criminals (Moffitt & Walsh, 2003).

SLT adds some meat to DAT by specifying how definitions favorable to law violation are learned, although it neglects the role of individual differences in the ease or difficulty with which persons learn. Some people find general hell raising more exciting (and thus more reinforcing) than others. Some people are more susceptible to short-term rewards because they are especially impulsive, and some are better able to appreciate the long-term rewards of behaving well. Some are more ready to engage in aggressive behavior than others because of the nature of their temperaments and will find such behavior is reinforced in delinquent areas. As Gwynn Nettler (1984) put it, "Constitutions affect the impact of environment. What we learn and how well we learn it depends on constitution. . . . The fire that melts the butter hardens the egg" (p. 295).

All versions of control theory agree that the family is central to the control and developmental mechanisms that affect criminal behavior, and because of this they have been criticized for neglecting social structure (Grasmick, Tittle, Bursik, & Arneklev, 1993). The major criticism of self-control theory arises from the claim that it is a *general* theory meant to explain *all* crime. Although many crimes are impulsive spontaneous acts, many others are not.

The theory attributes variation in self-control solely to variation in parental behavior and ignores child effects, but the child development literature is unequivocal that socialization is a two-way street in which parental behavior is shaped by the evocative behavior of the child just as much as the child's behavior is shaped by its parents (Harris, 1998). Low self-control may be something children bring with them to the socialization process rather than a product of the failure of that process, given that a number of studies have found a strong genetic component to low self-control (Bernhardt, 1997; Goldman, Lappalainen, & Ozaki, 1996; Hur & Bouchard, 1997).

Labeling theory comes close to claiming that the original "causes" (primary deviance) do not matter. If the causes of primary deviance do not matter, then efforts to control crime via various structural changes would be abandoned in favor of reliance on labeling theory's "noninterventionism" ("Leave the kids alone, they'll grow out of it"). This advice may be prudent for teenage pot smokers or runaways, but hardly wise for teenage robbers and rapists. LT advises that such delinquents should be "treated" rather than "punished." But since they insist that there is nothing intrinsically bad about any action, what is the point of treatment? What is it that is to be treated?

One of the positive elements of NT is that it softens the overdetermined image of delinquents, many of whom are no more completely committed to antisocial values than they are to prosocial values. Neutralization techniques are not viewed as "causes" of antisocial behavior; rather, they are a set of justifications that loosen moral constraints and allow offenders to drift in and out of antisocial behavior because they are able to "neutralize" these constraints.

One of the major problems with the theory is that it says little about the origins of the antisocial behavior the actors seek to neutralize. To be a causal theory of criminal behavior rather than an explanation of a post hoc process of rationalization, it would have to show that individuals *first* neutralize their moral beliefs and *then* engage in antisocial acts. Some studies have found neutralization techniques were able to explain future deviance, but this should not surprise anyone, since persons in a position where they have to explain their offending behavior are more likely than those not in such a position to offend in the future—past is prologue regardless of our explanations of it.

Where we see the cause of crime is where we assume we will find the solution. However, although they deal with different units of analysis, very few policy recommendations not discussed in ecological and strain theories can be gleaned from DAT or SLT. The bottom line for all subcultural theories is that lower-class neighborhoods harbor values and attitudes conducive to criminal behavior. Thus, if learning crime and delinquency within a particular culture is the problem, then changing relative aspects of that culture is the answer. However, we have already seen that attempts to do that have met with only meager success at best.

Because DAT concerns itself with the influence of role models in intimate peer groups, the provision of prosocial role models to replace antisocial ones is an obvious thought. Probation and parole authorities have long recognized the importance of keeping convicted felons away from each other, making it a revocable offense to "associate with known felons." As every probation and parole officer knows, however, this is easier said than done. Programs that bring youths together for prosocial purposes, such as sports leagues and community projects, might be high on the agenda of any policymaker using differential association as a guide. But the lure of "the streets" and of the friends they have grown up with remains a powerful force retarding rehabilitation. The good news is that most delinquents will desist as they mature, and the breakup of the friendship group by the incarceration, migration, death, or marriage of some of its members will break the grip of antisocial behavior for many of the remaining members (Sampson & Laub, 1999).

The policy implications derivable from social control and self-control theories have to do with the family. Given the importance of nurturance and attachment, both versions of control theory support the idea of early family intervention designed to cultivate these things. In almost all advanced nations, families with children receive support via family allowances and receive paid maternal leave, but such programs do not exist in the United States, which shows that politics and ideology, more than criminological theory, dictate the direction of criminal justice policy.

Other attempts to increase bonding to social institutions would concentrate on increasing children's involvement in a variety of prosocial activities and programs centered in and around the school. These programs provide prosocial models, teach moral beliefs such as personal responsibility, and keep youths busy in meaningful and challenging ways. Social control theory might recommend more vocationally oriented classes to keep less academically inclined students bonded to school.

Neither version of control theory would advise increased employment opportunities as a way to control crime. The assumption of control theory is that people who are attached and who possess self-control will do fine in the job market as it is, and increasing job opportunities for those lacking attachment and self-control will have minimal effect. Because low self-control is the result of the absence of inhibiting forces typically experienced in early childhood, Gottfredson and Hirschi (1997) are pessimistic about the ability of less powerful inhibiting forces (such as the threat of punishment) present in later life to deter crime. They also see little use in satisfying the wants and needs other theories view as important in reducing crime (reducing poverty, improving neighborhoods, etc.) because crime's appeal is its provision of immediate gains and minimal cost. In short, "society" is neither the cause of nor the solution to crime.

Gottfredson and Hirschi advocate some of the same policies (e.g., target hardening) advocated by rational choice and routine activities theorists to reduce criminal opportunities. However, the most important policy recommendation is to strengthen families and improve parenting skills, especially skills relevant to teaching self-control. It is only by working with and through families that society can do anything about crime in the long run. Gottfredson and Hirschi's (1997) most important recommendation is stated as: "Delaying pregnancy among unmarried girls would probably do more to affect the long-term crime rates than all the criminal justice programs combined" (p. 33).

Labeling theory had an effect on criminal justice policy far in excess of what its empirical support warrants. If it is correct that official societal reaction to primary deviance amplifies and promotes more of the same, the logical policy recommendation is that we should ignore primary deviance for the sake of alleviating secondary deviance. Labeling theory recommends that we allow offenders to protect their self-images as noncriminals by not challenging their "techniques of neutralization." Juveniles must be particularly protected from labeling.

The only policy implication of neutralization theory is the exact opposite of that of labeling theory: criminal justice agents charged with managing offenders (probation/parole officers, etc.) should strongly challenge their excuse making. If offenders come to believe their own rationalizations, rehabilitative efforts will become more difficult. Thus, offenders must be shown that their thinking patterns have negative long-term consequences for them.

Summary

- Social process theories emphasize how people perceive their reality and how these perceptions structure their behavior. DAT is a learning theory that emphasizes the power of peer associations and the definitions favorable to law violation found within them to be the cause of crime and delinquency.

- SLT adds to DAT by stressing the mechanisms by which "definitions favorable" are learned. Behavior is either reinforced (rewarded) or punished. Behavior that is rewarded tends to be repeated; behavior that is punished tends not to be. Discriminative stimuli provide signals for the kinds of behaviors that are likely to be rewarding or punishing, and are based on what we have learned about those stimuli in the past.

- Control theories are in many ways the opposite of DAT and SLT because they don't ask why people commit crimes, but why most of us do not. Crime comes naturally to those who are not either socially controlled or self-controlled. Hirschi speaks of the social bonds (attachment, commitment, involvement, and belief) that keep us on the straight and narrow. These are not causes of crime; rather they are bonds, the absence of which allows our natural impulses to emerge.

- Gottfredson and Hirschi's self-control theory moves the focus from social to self-control, although our experiences within the family are still vital to learn self-control. Low self-control must be paired with a criminal opportunity for crime to occur.

- Labeling theory is not interested in why some people commit crimes (primary deviance), believing that the only thing that differentiates criminals from the rest of us is that they have been caught and labeled. The real problem for it is the affixing of a deviant label because it changes the person's self-concept, and he or she then engages in secondary deviance in conformity with the label.

- Sykes and Mazda's techniques of neutralization theory is contrary to labeling theory because it focuses on individuals' attempts to resist being labeled criminal by offering justifications or excuses for their behavior.

EXERCISES AND DISCUSSION QUESTIONS

1. Without indicating a particular theory, does the social structural or social process approach to explaining crime and criminality make most sense to you?
2. Compare differential association theory with control theory in terms of their respective assumptions about human nature. Which assumption makes more sense to you?
3. Is a delinquent or criminal label applied to someone sufficient in most cases to change a person's self-concept enough to lead him or her to continue offending?
4. Gottfredson and Hirschi claim that parents are to blame for an individual's lack of self-control. Are there some children who are simply more difficult to socialize than others, and are they, rather than their parents, at fault for their lack of self-control?
5. Why is attachment the most important of the four social bonds?

USEFUL WEB SITES

Control Theories of Crime. http://faculty.ncwc.edu/TOCONNOR/301/301lect11.htm.

Differential Association Theory. http://www.criminology.fsu.edu/crimtheory/sutherland.html.

Labeling Theories of Crime. http://faculty.ncwc.edu/toconnor/301/301lect12.htm.

Learning Theories of Crime. http://faculty.ncwc.edu/TOCONNOR/301/301lect10.htm.

Social Control Theories. http://www.indiana.edu/~theory/Kip/Control.htm

CHAPTER GLOSSARY

Attachment: One of the four elements of the social bond; the emotional component of conformity referring to one's attachment to others and to social institutions.

Belief: One of the four elements of the social bond; refers to the acceptance of the social norms regulating conduct.

Commitment: One of the four elements of the social bond; the rational component of conformity referring to a lifestyle in which one has invested considerable time and energy in the pursuit of a lawful career.

Definitions: Term used by Edwin Sutherland to refer to meanings our experiences have for us, our attitudes, values, and habitual ways of viewing the world.

Differential association theory: Criminological theory devised by Edwin Sutherland asserting that criminal behavior is behavior learned through association with others who communicate their values and attitudes.

Discrimination: A term applied to stimuli that provide clues that signal whether a particular behavior is likely to be followed by reward or punishment.

Involvement: One of the four elements of the social bond. A pattern of involvement with prosocial activities that prevents involvement with antisocial activities.

Operant psychology: A perspective on learning that asserts that behavior is governed and shaped by its consequences (reward or punishment).

Primary deviance: In labeling theory, the initial nonconforming act that comes to the attention of the authorities resulting in the application of a criminal label.

Punishment: A process that leads to the weakening or elimination of the behavior preceding it.

Reinforcement: A process that leads to the strengthening of behavior.

Secondary deviance: Deviance resulting from society's reaction to primary deviance.

Social bond theory: A social control theory focusing on a person's bonds to others.

Social control: Any action on the part of others, deliberate or not, that facilitates conformity to social rules.

Social learning theory: A theory about learning criminal behavior based on operant psychology's emphasis on reinforcement and punishment.

Symbolic interactionism: A perspective in sociology that focuses on how people interpret and define their social reality and the meanings they attach to it in the process of interacting with one another via language (symbols).

Techniques of neutralization: Techniques by which offenders justify their behavior (to themselves and others) as "acceptable" on a number of grounds.

READING

Social Learning Theory

Ronald L. Akers

As indicated in the text, Akers' social learning is an extension of Sutherland's differential association theory couched in the concepts of operant psychology. In the following portions of Akers' book, *A Social Learning Theory of Crime,* he outlines both how his theory reformulates Sutherland's theory and how he extends it. He agrees that criminals become criminal through harboring an excess of definitions favorable to crime, but goes on to more fully describe the nature of these definitions. Akers also argues that crime is learned by imitation of others and through reinforcements provided by others.

Concise Statement of the Theory

The basic assumption in social learning theory is that the same learning process, operating in a context of social structure, interaction, and situation, produces both conforming and deviant behavior. The difference lies in the direction of the process in which these mechanisms operate. In both, it is seldom an either-or, all-or-nothing process; what is involved, rather, is the balance of influences on behavior. That balance usually exhibits some stability over time, but it can become unstable and change with time or circumstances. Conforming and deviant behavior is learned by all of the mechanisms in this process, but the theory proposes that the principal mechanisms are in that part of the process in which differential reinforcement (instrumental learning through rewards and punishers) and imitation (observational learning) produce both overt behavior and cognitive definitions that function as discriminative (cue) stimuli for the behavior. Always implied, and sometimes made explicit when these concepts are called upon to account for deviant/conforming behavior, is the understanding that the behavioral processes in operant and classical conditioning are in operation (see below). However, social learning theory focuses on four major concepts—differential association, differential reinforcement, imitation, and definitions. The central proposition of the social learning theory of criminal and deviant behavior can be stated as a long sentence proposing that criminal and deviant behavior is more likely when, on balance, the combined effects of these four main sets of variables instigate and strengthen nonconforming over conforming acts.

The probability that persons will engage in criminal and deviant behavior is increased and the probability of their conforming to the norm is decreased when they differentially associate with others who commit criminal behavior and espouse definitions favorable to it, are relatively more exposed in-person or symbolically to salient criminal/deviant models, define it as desirable or justified in a situation discriminative for the behavior, and have received in the past and anticipate in the current or future situation relatively greater reward than punishment for the behavior.

The probability of conforming behavior is increased and the probability of deviant behavior is decreased when the balance of these variables moves in the reverse direction.

Each of the four main components of this statement can be presented as a separate testable hypothesis. The individual is more likely to commit violations when:

1. He or she differentially associates with others who commit, model, and support violations of social and legal norms.
2. The violative behavior is differentially reinforced over behavior in conformity to the norm.
3. He or she is more exposed to and observes more deviant than conforming models.
4. His or her own learned definitions are favorable toward committing the deviant acts.

General Principles of Social Learning Theory

Since it is a general explanation of crime and deviance of all kinds, social learning is not simply a theory about how novel criminal behavior is learned or a theory only of the positive causes of that behavior. It embraces variables that operate to both motivate and control delinquent and criminal behavior, to both promote and undermine conformity. The probability of criminal or conforming behavior occurring is a function of the variables operating in the underlying social learning process. The main concepts/variables and their respective empirical indicators have been identified and measured, but they can be viewed as indicators of a general latent construct, for which additional indicators can be devised.

Social learning accounts for the individual becoming prone to deviant or criminal behavior and for stability or change in that propensity. Therefore, the theory is capable of accounting for the development of stable individual differences, as well as changes in the individual's behavioral patterns or tendencies to commit deviant and criminal acts, over time and in different situations. . . . The social learning process operates in each individual's learning history and in the immediate situation in which the opportunity for a crime occurs.

Deviant and criminal behavior is learned and modified (acquired, performed, repeated, maintained, and changed) through all of the same cognitive and behavioral mechanisms as conforming behavior. They differ in the direction, content, and outcome of the behavior learned. Therefore, it is inaccurate to state, for instance, that peer influence does not explain adolescent deviant behavior since conforming behavior is also peer influenced in adolescence. The theory expects peer influences to be implicated in both; it is the content and direction of the influence that is the key.

The primary learning mechanisms are differential reinforcement (instrumental conditioning), in which behavior is a function of the frequency, amount, and probability of experienced and perceived contingent rewards and punishments, and imitation, in which the behavior of others and its consequences are observed and modeled. The process of stimulus discrimination/generalization is another important mechanism; here, overt and covert stimuli, verbal and cognitive, act as cues or signals for behavior to occur. As I point out below, there are other behavioral mechanisms in the learning process, but these are not as important and are usually left implied rather than explicated in the theory.

The content of the learning achieved by these mechanisms includes the simple and complex behavioral sequences and the definitions (beliefs, attitudes, justifications, orientations) that in turn become discriminative for engaging in deviant and criminal behavior. The probability that conforming or norm-violative behavior is learned and performed, and the frequency with which it is committed, are a function of the past, present, and anticipated differential reinforcement for the behavior and the deviant or nondeviant direction of the learned definitions and other discriminative stimuli present in a given situation.

These learning mechanisms operate in a process of differential association—direct and indirect, verbal and nonverbal communication, interaction, and identification with others. The relative frequency, intensity, duration, and priority of associations affect the relative amount, frequency, and probability of reinforcement of conforming or deviant behavior and exposure of individuals to deviant or conforming norms and behavioral models. To the extent that the individual can control with whom she or he associates, the frequency, intensity, and duration of those associations are themselves affected by how rewarding or aversive they are. The principal learning is through differential association with those persons and groups (primary secondary, reference, and symbolic) that comprise or control the individual's major sources of reinforcement, most salient behavioral models, and most effective definitions and other discriminative stimuli for committing and repeating behavior. The reinforcement and discriminative stimuli are mainly social (such as socially valued rewards and punishers contingent on the behavior), but they are also nonsocial (such as unconditioned physiological reactions to environmental stimuli and physical effects of ingested substances and the physical environment).

Sequence and Reciprocal Effects in the Social Learning Process

Behavioral feedback effects are built into the concept of differential reinforcement—actual or perceived changes in the environment produced by the behavior feedback on that behavior to affect its repetition or extinction, and both prior and anticipated rewards and punishments influence present behavior. Reciprocal effects between the individual's behavior and definitions or differential association are also reflected in the social learning process. This process is one in which the probability of both the initiation and the repetition of a deviant or criminal act (or the initiation and repetition of conforming acts) is a function of the learning history of the individual and the set of reinforcement contingencies and discriminative stimuli in a given situation. The typical process of initiation, continuation, progression, and desistance is hypothesized to be as follows:

1. The balance of past and current associations, definitions, and imitation of deviant models, and the anticipated balance of reinforcement in particular situations, produces or inhibits the initial delinquent or deviant acts.
2. The effects of these variables continue in the repetition of acts, although imitation becomes less important than it was in the first commission of the act.
3. After initiation, the actual social and nonsocial reinforcers and punishers affect the probability that the acts will be or will not be repeated and at what level of frequency.
4. Not only the overt behavior, but also the definitions favorable or unfavorable to it, are affected by the positive and negative consequences of the initial acts. To the extent that they are more rewarded than alternative behavior, the favorable definitions will be strengthened and the unfavorable definitions will be weakened, and it becomes more likely that the deviant behavior will be repeated under similar circumstances.
5. Progression into more frequent or sustained patterns, rather than cessation or reduction, of criminal and deviant behavior is promoted to the extent that reinforcement, exposure to deviant models, and norm-violating definitions are not offset by negative formal and informal sanctions and norm abiding definitions.

The theory does not hypothesize that definitions favorable to law violation always precede and are unaffected by the commission of criminal acts. Although the probability of a criminal act increases in the presence of favorable definitions, acts in violation of the law do occur (through imitation and reinforcement) in the absence of any thought given to whether the acts are right or wrong. Furthermore, the individual may apply neutralizing definitions retroactively to excuse or justify an act without having contemplated them beforehand. To the extent that such excuses become associated with successfully mitigating others' negative sanctions or one's self-punishment, however, they become cues for the repetition of deviant acts. Such definitions, therefore, precede committing the same acts again or committing similar acts in the future.

Differential association with conforming and nonconforming others typically precedes the individual's committing crimes and delinquent acts. This sequence of events is sometimes disputed in the literature because it is mistakenly believed to apply only to differential peer association in general or to participation in delinquent gangs in particular without reference to family and other group associations. It is true that the theory recognizes peer associations as very important in adolescent deviance and that differential association is most often measured in research by peer associations. But the theory also hypothesizes that the family is a very important primary group in the differential association process, and it plainly stipulates that other primary and secondary groups besides peers are involved (see Sutherland, 1947, pp. 164–65). Accordingly, it is a mistake to interpret differential association as referring only to peer associations. The theoretical stipulation that differential association is causally prior to the commission of delinquent and criminal acts is not confined to the balance of peer associations; rather, it is the balance (as determined by the modalities) of family, peer, and other associations. According to the priority principle, association, reinforcement, modeling, and exposure to conforming and deviant definitions occurring within the family during childhood, and such antisocial conduct as aggressiveness, lying, and cheating learned in early childhood, occur prior to and have both direct and selective effects on later delinquent and criminal behavior and associations. . . .

The socializing behavior of parents, guardians, or caretakers is certainly reciprocally influenced by the deviant and unacceptable behavior of the child. However, it can never be true that the onset of delinquency precedes and initiates interaction in a particular family (except in the unlikely case of the late-stage adoption of a child who is already delinquent or who is drawn to and chosen by deviant parents). Thus, interaction in the family or family surrogate always precedes delinquency.

But this is not true for adolescent peer associations. One may choose to associate with peers based on similarity in deviant behavior that already exists. Some major portion of this behavioral similarity results from previous association with other delinquent peers or from anticipatory socialization undertaken to make one's behavior match more closely that of the deviant associates to whom one is attracted. For some adolescents, gravitation toward delinquent peers occurs after and as a result of the individual's involvement in delinquent behavior. However, peer associations are most often formed initially around interests, friendships, and such circumstances as neighborhood proximity, family similarities, values, beliefs, age, school attended, grade in school, and mutually attractive behavioral patterns that have little to do directly with co-involvement or similarity in specifically law-violating or serious deviant behavior. Many of these factors in peer association are not under the adolescents' control, and some are simply happenstance. The theory does not, contrary to the Gluecks' distorted characterization, propose that "accidental differential association of non-delinquents with delinquents is the basic cause of crime (Glueck & Glueck, 1950, p. 164). Interaction and socialization in the family precedes and affects choices of both conforming and deviant peer associations.

Those peer associations will affect the nature of models, definitions, and rewards/punishers to which the person is exposed. After the associations have been established, their reinforcing or punishing consequences as well as direct and vicarious consequences of the deviant behavior will affect both the continuation of old and the seeking of new associations (those over which one has any choice). One may choose further interaction with others based on whether they too are involved in deviant or criminal behavior; in such cases, the outcomes of that interaction are more rewarding than aversive and it is anticipated that the associates will more likely approve or be permissive toward one's own deviant behavior. Further interaction with delinquent peers, over which the individual has no choice, may also result from being apprehended and confined by the juvenile or criminal-justice system.

These reciprocal effects would predict that one's own deviant or conforming behavioral patterns can have effects on choice of friends; these are weaker in the earlier years, but should become stronger as one moves through adolescence and gains more control over friendship choices. The typical sequence outlined above would predict that deviant associations precede the onset of delinquent behavior more frequently than the sequence of events in which the delinquent associations begin only after the peers involved have already separately and individually established similar patterns of delinquent behavior. Further, these behavioral tendencies that develop prior to peer association will themselves be the result of previous associations, models, and reinforcement, primarily in the family. Regardless of the sequence in which onset occurs, and whatever the level of the individual's delinquent involvement, its frequency and seriousness will increase after the deviant associations have begun and decrease as the associations are reduced. That is, whatever the temporal ordering, differential association with deviant peers will have a causal effect on one's own delinquent behavior (just as his actions will have an effect on his peers).

Therefore, both "selection," or "flocking" (tendency for persons to choose interaction with others with behavioral similarities), and "socialization," or "feathering" (tendency for persons who interact to have mutual influence on one another's behavior), are part of the same overall social learning process and are explained by the same variables. A peer "socialization" process and a peer "selection" process in deviant behavior are not mutually exclusive, but are simply the social learning process operating at different times. Arguments that social learning posits only the latter, that any evidence of selective mechanisms in deviant interaction run counter to social learning theory, or that social learning theory recognizes only a recursive, one-way causal effect of peers on delinquent behavior are wrong.

Behavioral and Cognitive Mechanisms in Social Learning

The first statement in Sutherland's theory was a simple declarative sentence maintaining that criminal behavior is learned, and the eighth statement declared that this involved all the mechanisms involved in any learning. What little Sutherland added in his (1947, p. 7) commentary downplayed imitation as a possible learning mechanism in criminal behavior. He mentioned "seduction" of a person into criminal behavior as something that is not covered by the concept of imitation. He defined neither imitation nor seduction and offered no further discussion of mechanisms of learning in any of his papers or publications. Recall that filling this major lacuna in Sutherland's theory was the principal goal of the 1966b Burgess-Akers reformulation. To this end we combined Sutherland's first and eighth statements into one: "Criminal behavior is learned according to the principles of operant conditioning." The phrase "principles of operant conditioning" was meant as a shorthand reference to all of the behavioral mechanisms of learning in operant theory that had been empirically validated.

Burgess and I delineated, as much as space allowed, what these specific learning mechanisms were: (1) operant conditioning, differential reinforcement of voluntary behavior through positive and negative reinforcement and punishment; (2) respondent (involuntary reflexes), or "classical," conditioning; (3) unconditioned (primary) and conditioned (secondary) reinforcers and punishers; (4) shaping and response differentiation; (5) stimulus discrimination and generalization, the environmental and internal stimuli that provide cues or signals indicating differences and similarities across situations that help elicit, but do not directly reinforce, behavior; (6) types of reinforcement schedules, the rate and ratio in which rewards and punishers follow behavior; (7) stimulus-response constellations; and (8) stimulus satiation and deprivation. We also reported research showing the applicability of these mechanisms of learning to both conforming and deviant behavior.

Burgess and I used the term "operant conditioning" to emphasize that differential reinforcement (the balance of reward and punishment contingent upon behavioral responses) is the basic mechanism around which the others revolve and by which learning most relevant to conformity or violation of social and legal norms is produced. This was reflected in other statements in the theory in which the only learning mechanisms listed were differential reinforcement and stimulus discrimination.

We also subsumed imitation, or modeling, under these principles and argued that imitation "may be analyzed quite parsimoniously with the principles of modern behavior theory," namely, that it is simply a sub-class of behavioral shaping through operant conditioning (Burgess and Akers, 1966b: 138). For this reason we made no specific mention of imitation in any of the seven statements. Later, I became persuaded that the operant principle of gradual shaping of responses through "successive approximations" only incompletely and awkwardly incorporated the processes of observational learning and vicarious reinforcement that Bandura and Walters (1963) had identified. Therefore, without dismissing successive approximation as a way in which some imitative behavior could be shaped,

I came to accept Bandura's conceptualization of imitation. That is, imitation is a separate learning mechanism characterized by modeling one's own actions on the observed behavior of others and on the consequences of that behavior (vicarious reinforcement) prior to performing the behavior and experiencing its consequences directly. Whether the observed acts will be performed and repeated depends less on the continuing presence of models and more on the actual or anticipated rewarding or aversive consequences of the behavior. I became satisfied that the principle of "observational learning" could account for the acquisition, and to some extent the performance, of behavior by a process that did not depend on operant conditioning or "instrumental learning." Therefore, in later discussions of the theory, while continuing to posit differential reinforcement as the core behavior-shaping mechanism, I included imitation as another primary mechanism in acquiring behavior. Where appropriate, discriminative stimuli were also specifically invoked as affecting behavior, while I made only general reference to other learning mechanisms.

Note that the term "operant conditioning" in the opening sentence of the Burgess-Akers revision reflected our great reliance on the orthodox behaviorism that assumed the empirical irrelevance of cognitive variables. Social behaviorism, on the other hand, recognizes "cognitive" as well as "behavioral." My social learning theory of criminal behavior retains a strong element of the symbolic interactionism found in Sutherland's theory. As a result, it is closer to cognitive learning theories, such as Albert Bandura's, than to the radical operant behaviorism of B. F. Skinner with which Burgess and I began. It is for this reason, and the reliance on such concepts as imitation, anticipated reinforcement, and self-reinforcement, that I have described social learning theory as "soft behaviorism."

The unmodified term "learning" implies to many that the theory only explains the acquisition of novel behavior by the individual, in contrast to behavior that is committed at a given time and place or the maintenance of behavior over time. It has also been interpreted to mean only "positive" learning of novel behavior, with no relevance for inhibition of behavior or of learning failures. As I have made clear above, neither of these interpretations is accurate. The phrase that Burgess and I used, "effective and available reinforcers and the existing reinforcement contingencies," and the discussion of reinforcement occurring under given situations make it obvious that we were not proposing a theory only of past reinforcement in the acquisition of a behavioral repertoire with no regard for the reward/cost balance obtaining at a given time and place. There is nothing in the learning principles that restrict them to prior socialization or past history of learning. Social learning encompasses both the acquisition and the performance of the behavior, both facilitation and inhibition of behavior, and both learning successes and learning failures. The learning mechanisms account not only for the initiation of behavior but also for repetition, maintenance and desistance of behavior. They rely not only on prior behavioral processes but also on those operating at a given time in a given situation.

Definitions and Discriminative Stimuli

In The Concept of Definitions, Sutherland asserted that learning criminal behavior includes "techniques of committing the crime which are sometimes very complicated, sometimes very simple" and the "specific direction of motives, drives, rationalizations and attitudes" (1947, p. 6). I have retained both definitions and techniques in social learning theory, with clarified and modified conceptual meanings and with hypothesized relationships to criminal behavior. The qualification that "techniques" may be simple or complex shows plainly that Sutherland did not mean to include only crime-specific skills learned in order to break the law successfully. Techniques also clearly include ordinary, everyday abilities. This same notion is retained in social learning theory.

By definition, a person must be capable of performing the necessary sequence of actions before he or she can carry out either criminal or conforming behavior—inability to perform the behavior precludes committing the crime. Since many of the behavioral techniques for both conforming and criminal acts are the same, not only the simple but even some of the complex skills involved in carrying out crime are not novel to most or many of us. The required component parts of the complete skill are acquired in essentially conforming or neutral contexts to which we have been exposed—driving a car, shooting a gun, fighting with fists, signing checks, using a computer, and so on. In most white-collar crime, the same skills needed to carry out a job legitimately are put to illegitimate use. Other skills are specific to given deviant acts—safe cracking, counterfeiting, pocket picking, jimmying doors and picking locks, bringing off a con game, and so on. Without tutelage in these crime-specific techniques, most people would not be able to perform them, or at least would be initially very inept.

Sutherland took the concept of "definitions" in his theory from W. I. Thomas's "definition of the situation" (Thomas and Thomas, 1928) and generalized it to orienting attitudes toward different behavior. It is true that "Sutherland did not identify what constitutes a definition 'favorable to' or 'unfavorable to' the violation of law." Nevertheless . . . there is little doubt that "rationalizations" and "attitudes" are subsumed under the general concept of definitions—normative attitudes or evaluative meanings attached to given behavior. Exposure to others' shared definitions is a key (but not the only) part of the process by which the individual acquires or internalizes his or her own definitions. They are orientations, rationalizations, definitions of the situation, and other attitudes that label the commission of an act as right or wrong, good or bad, desirable or undesirable, justified or unjustified.

In social learning theory, these definitions are both general and specific. General beliefs include religious, moral, and other conventional values and norms that are favorable to conforming behavior and unfavorable to committing any of a range of deviant or criminal acts. Specific definitions orient the person to particular acts or series of acts. Thus, there are people who believe that it is morally wrong to steal and that laws against theft should be obeyed, but at the same time see little wrong with smoking marijuana and rationalize that it is all right to violate laws against drug possession. The greater the extent to which one holds attitudes that disapprove of certain acts, the less likely one is to engage in them. Conventional beliefs are negative toward criminal behavior. The more strongly one has learned and personally believes in the ideals of honesty, integrity, civility, kindness, and other general standards of morality that condemn lying, cheating, stealing, and harming others, the less likely he or she is to commit acts that violate social and legal norms. Conversely, the more one's own attitudes approve of, or fail to condemn, a behavior, the greater the chances are that he or she will engage in it. For obvious reasons, the theory would predict that definitions in the form of general beliefs will have less effect than specific definitions on the commission of specific criminal acts.

Definitions that favor criminal or deviant behavior are basically positive or neutralizing. Positive definitions are beliefs or attitudes that make the behavior morally desirable or wholly permissible. They are most likely to be learned through positive reinforcement in a deviant group or subculture that carries values conflicting with those of conventional society. Some of these positive verbalizations may be part of a full-blown ideology of politically dissident, criminal, or deviant groups. Although such ideologies and groups can be identified, the theory does not rest only on this type of definition favorable to deviance; indeed, it proposes that such positive definitions occur less frequently than neutralizing ones.

Neutralizing definitions favor violating the law or other norms not because they take the acts to be positively desirable but because they justify or excuse them. Even those who commit deviant acts are aware that others condemn the behavior and may themselves define the behavior as bad. The neutralizing definitions view the act as something that is probably undesirable but, given the situation, is nonetheless justified, excusable, necessary all right, or not really bad after all. The process of acquiring neutralizing definitions is more likely to involve negative reinforcement; that is, they are verbalizations that accompany escape or avoidance of negative consequences like disapproval by one's self or by society.

While these definitions may become part of a deviant or criminal subculture, acquiring them does not require participation in such subcultures. They are learned from carriers of conventional culture, including many of those in social control and treatment agencies. The notions of techniques of neutralization and subterranean values (Sykes and Matza, 1957) come from the observation that for nearly every social norm there is a norm of evasion. That is, there are recognized exceptions or ways of getting around the moral imperatives in the norms and the reproach expected for violating them. Thus, the general prohibition "Thou shalt not kill" is accompanied by such implicit or explicit exceptions as "unless in time of war," "unless the victim is the enemy," "unless in self-defense" "unless in the line of duty," "unless to protect others"! The moral injunctions against physical violence are suspended if the victim can be defined as the initial aggressor or is guilty of some transgression and therefore deserves to be attacked.

The concept of neutralizing definitions in social learning theory incorporates not only notions of verbalizations and rationalizations and techniques of neutralization, but also conceptually similar if not equivalent notions of "accounts," "disclaimers," and "moral disengagement." Neutralizing attitudes include such beliefs as "Everybody has a racket"; "I can't help myself, I was born this way"; "It's not my fault"; "I am not responsible"; "I was drunk and didn't know what I was doing"; "I just blew my top"; "They can afford it"; "He deserved it." Some neutralizations (e.g., nonresponsibility) can be generalized to a wide range of disapproved and criminal behavior. These and other excuses and justifications for committing deviant acts and victimizing others are definitions favorable to criminal and deviant behavior.

Exposure to these rationalizations and excuses may be through after-the-fact justifications for one's own or others' norm violations that help to deflect or lessen punishment that would be expected to follow. The individual then learns the excuses either directly or through imitation and uses them to lessen self-reproach and social disapproval. Therefore, the definitions are themselves behavior that can be imitated and reinforced and then in turn serve as discriminative stimuli accompanying reinforcement of overt behavior. Deviant and criminal acts do occur without being accompanied by positive or neutralizing definitions, but the acts are more likely to occur and recur in situations the same as or similar to those in which the definitions have already been learned and applied. The extent to which one adheres to or rejects the definitions favorable to crime is itself affected by the rewarding or punishing consequences that follow the act.

DISCUSSION QUESTIONS

1. How does Akers' social learning theory differ from differential association theory in explaining why associating with delinquent peers usually results in delinquency?
2. Provide an example of learning that "sticks" resulting from positive and negative reinforcement.
3. How would Akers explain the large differences in crime and delinquency (especially the more serious kinds) between males and females living in the same neighborhood?

READING

The Nature of Criminality

Low Self-Control

Michael R. Gottfredson and Travis Hirschi

In this excerpt from Gottfredson and Hirschi's *A General Theory of Crime,* they emphasize the classical notion that we all would like to get everything we want with minimal effort. Underlying all criminal and other antisocial behaviors is the inability to do this; that is, the antisocial individual lacks self-control. Gottfredson and Hirschi focus on the family as the critical determinant of self-control. Individuals who fail to learn self-control when they are children ordinarily do not develop it later in life. Unless parents closely monitor their children, self-control will not be learned, and children retain their impulsive and insensitive natures considered to be the default in the absence of training. Gottfredson and Hirschi view crime as a consequence of individuals with low self-control coming into contact with opportunities to commit crime.

The Elements of Self-Control

Criminal acts provide immediate gratification of desires. A major characteristic of people with low self-control is therefore a tendency to respond to tangible stimuli in the immediate environment, to have a concrete "here and now" orientation. People with high self-control, in contrast, tend to defer gratification. Criminal acts provide easy or simple gratification of desires. They provide money without work, sex without courtship, revenge without court delays. People lacking self-control also tend to lack diligence, tenacity, or persistence in a course of action. Criminal acts are exciting, risky, or thrilling. They involve stealth, danger, speed, agility, deception, or power. People lacking self-control therefore tend to be adventuresome, active, and physical. Those with high levels of self-control tend to be cautious, cognitive, and verbal.

Crimes provide few or meager long-term benefits. They are not equivalent to a job or a career. On the contrary, crimes interfere with long-term commitments to jobs, marriages, family, or friends. People with low self-control thus tend to have unstable marriages, friendships, and job profiles. They tend to be little interested in and unprepared for long-term occupational pursuits.

Crimes require little skill or planning. The cognitive requirements for most crimes are minimal. It follows that people lacking self-control need not possess or value cognitive or academic skills. The manual skills required for most crimes are minimal. It follows that people lacking self-control need not possess manual skills that require training or apprenticeship. Crimes often result in pain or discomfort for the victim. Property is lost, bodies are injured, privacy is violated, trust is broken. It follows that people with low self-control tend to be self-centered, indifferent, or insensitive to the suffering and needs of others. It does not follow, however, that people with low self-control are routinely unkind or antisocial. On the contrary, they may discover the immediate and easy rewards of charm and generosity.

Recall that crime involves the pursuit of immediate pleasure. It follows that people lacking self-control will also tend to pursue immediate pleasures that are not criminal: they will tend to smoke, drink, use drugs, gamble, have children out of wedlock, and engage in illicit sex. Crimes require the interaction of an offender with people or their property. It does not follow that people lacking self-control will tend to be gregarious or social. However, it does follow that, other things being equal, gregarious or social people are more likely to be involved in criminal acts. The major benefit of many crimes is not pleasure but relief from momentary irritation. The irritation caused by a crying child is often the stimulus for physical abuse. That caused by a taunting stranger in a bar is often the stimulus for aggravated assault. It follows that people with low self-control tend to have minimal tolerance for frustration and little ability to respond to conflict through verbal rather than physical means.

Crimes involve the risk of violence and physical injury, of pain and suffering on the part of the offender. It does not follow that people with low self-control will tend to be tolerant of physical pain or to be indifferent to physical discomfort. It does follow that people tolerant of physical pain or indifferent to physical discomfort will be more likely to engage in criminal acts whatever their level of self-control. The risk of criminal penalty for any given criminal act is small, but this depends in part on the circumstances of the offense. Thus, for example, not all joyrides by teenagers are equally likely to result in arrest. A car stolen from a neighbor and returned unharmed before he notices its absence is less likely to result in official notice than is a car stolen from a shopping center parking lot and abandoned at the convenience of the offender. Drinking alcohol stolen from parents and consumed in the family garage is less likely to receive official notice than drinking in the parking lot outside a concert hall. It follows that offenses differ in their validity as measures of self-control: those offenses with large risk of public awareness are better measures than those with little risk.

In sum, people who lack self-control will tend to be impulsive, insensitive, physical (as opposed to mental), risk-taking, short-sighted, and nonverbal, and they will tend therefore to engage in criminal and analogous acts. Since these traits can be identified prior to the age of responsibility for crime, since there is considerable tendency for these traits to come together in the same people, and since the traits tend to persist through life, it seems reasonable to consider them as comprising a stable construct useful in the explanation of crime.

The Many Manifestations of Low Self-Control

Our image of the "offender" suggests that crime is not an automatic or necessary consequence of low self-control. It suggests that many noncriminal acts analogous to crime (such as accidents, smoking, and alcohol use) are also manifestations of low self-control. Our image therefore implies that no specific act, type of crime, or form of deviance is uniquely required by the absence of self-control. Because both crime and analogous behaviors stem from low self-control (that is, both are manifestations of low self-control), they will all be engaged in at a relatively high rate by people with low self-control. Within the domain of crime, then, there will be much versatility among offenders in the criminal acts in which they engage.

Research on the versatility of deviant acts supports these predictions in the strongest possible way. The variety of manifestations of low self-control is immense. In spite of years of tireless research motivated by a belief in specialization, no credible evidence of specialization has been reported. In fact, the evidence of offender versatility is overwhelming.

By versatility we mean that offenders commit a wide variety of criminal acts, with no strong inclination to pursue a specific criminal act or a pattern of criminal acts to the exclusion of others. Most theories suggest that offenders tend to specialize, whereby such terms as robber, burglar, drug dealer, rapist, and murderer have predictive or descriptive import. In fact, some theories create offender specialization as part of their explanation of crime. For example, Cloward and Ohlin (1960) create distinctive subcultures of delinquency around particular forms of criminal behavior, identifying subcultures specializing in theft, violence, or drugs. In a related way, books are written about white-collar crime as though it were a clearly distinct specialty requiring a unique explanation. Research projects are undertaken for the study of drug use, or vandalism, or teen pregnancy (as though every study of delinquency were not a study of drug use and vandalism and teenage sexual behavior). Entire schools of criminology emerge to pursue patterning, sequencing, progression, escalation, onset, persistence, and desistance in the career of offenses or offenders. These efforts survive largely because their proponents fail to consider or acknowledge the clear evidence to the contrary. Other reasons for survival of such ideas may be found in the interest of politicians and members of the law enforcement community who see policy potential in criminal careers or "career criminals."

Occasional reports of specialization seem to contradict this point, as do everyday observations of repetitive misbehavior by particular offenders. Some offenders rob the same store repeatedly over a period of years, or an offender commits several rapes over a (brief) period of time. Such offenders may be called "robbers" or "rapists." However, it should be noted that such labels are retrospective rather than predictive and that they typically ignore a large amount of delinquent or criminal behavior by the same offenders that is inconsistent with their alleged specialty. Thus, for example, the "rapist" will tend also to use drugs, to commit robberies and burglaries (often in concert with the rape), and to have a record for violent offenses other than rape. There is a perhaps natural tendency on the part of observers (and in official accounts) to focus on the most serious crimes in a series of events, but this tendency should not be confused with a tendency on the part of the offender to specialize in one kind of crime.

Recall that one of the defining features of crime is that it is simple and easy. Some apparent specialization will therefore occur because obvious opportunities for an easy score will tend to repeat themselves. An offender who lives next to a shopping area that is approached by pedestrians will have repeat opportunities for purse snatching, and this may show in his arrest record. But even here the specific "criminal career" will tend to quickly run its course and to be followed by offenses whose content and character is likewise determined by convenience and opportunity (which is the reason why some form of theft is always the best bet about what a person is likely to do next).

The evidence that offenders are likely to engage in noncriminal acts psychologically or theoretically equivalent to crime is, because of the relatively high rates of these "noncriminal" acts, even easier to document. Thieves are likely to smoke, drink, and skip school at considerably higher rates than nonthieves. Offenders are considerably more likely than nonoffenders to be involved in most types of accidents, including household fires, auto crashes, and unwanted pregnancies. They are also considerably more likely to die at an early age.

Good research on drug use and abuse routinely reveals that the correlates of delinquency and drug use are the same. As Akers (1984) has noted, "compared to the abstaining teenager, the drinking, smoking, and drug-taking teen is much more likely to be getting into fights, stealing, hurting other people, and committing other delinquencies." Akers goes on to say, "but the variation in the order in which they take up these things leaves little basis for proposing the causation of one by the other." In our view, the relation between drug use and delinquency is not a causal question. The correlates are the same because drug use and delinquency are both manifestations of an underlying tendency to pursue short-term, immediate pleasure. This underlying tendency (i.e., lack of self-control) has many manifestations, as listed by Harrison Gough (1948):

> unconcern over the rights and privileges of others when recognizing them would interfere with personal satisfaction in any way;

> impulsive behavior, or apparent incongruity between the strength of the stimulus and the magnitude of the behavioral response; inability to form deep or persistent attachments to other persons or to identify in interpersonal relationships; poor judgment and planning in attaining defined goals; apparent lack of anxiety and distress over social maladjustment and unwillingness or inability to consider maladjustment qua maladjustment; a tendency to project blame onto others and to take no responsibility for failures; meaningless prevarication, often about trivial matters in situations where detection is inevitable; almost complete lack of dependability . . . and willingness to assume responsibility; and, finally, emotional poverty. (p. 362)

This combination of characteristics has been revealed in the life histories of the subjects in the famous studies by Lee Robins. Robins is one of the few researchers to focus on the varieties of deviance and the way they tend to go together in the lives of those she designates as having "antisocial personalities." In her words:

> We refer to some one who fails to maintain close personal relationships with anyone else, [who] performs poorly on the job, who is involved in illegal behaviors (whether or not apprehended), who fails to support himself and his dependents without outside aid, and who is given to sudden changes of plan and loss of temper in response to what appear to others as minor frustrations. (1978, p. 255)

For 30 years Robins traced 524 children referred to a guidance clinic in St. Louis, Missouri, and she compared them to a control group matched on IQ, age, sex, and area of the city. She discovered that, in comparison to the control group, those people referred at an early age were more likely to be arrested as adults (for a wide variety of offenses), were less likely to get married, were more likely to be divorced, were more likely to marry a spouse with a behavior problem, were less likely to have children (but if they had children were likely to have more children), were more likely to have children with behavior problems, were more likely to be unemployed, had considerably more frequent job changes, were more likely to be on welfare, had fewer contacts with relatives, had fewer friends, were substantially less likely to attend church, were less likely to serve in the armed forces and more likely to be dishonorably discharged if they did serve, were more likely to exhibit physical evidence of excessive alcohol use, and were more likely to be hospitalized for psychiatric problems (1966, pp. 42–73).

Note that these outcomes are consistent with four general elements of our notion of low self-control: basic stability of individual differences over a long period of time; great variability in the kinds of criminal acts engaged in; conceptual or causal equivalence of criminal and noncriminal acts; and inability to predict the specific forms of deviance engaged in, whether criminal or noncriminal. In our view, the idea of an antisocial personality defined by certain behavioral consequences is too positivistic or deterministic, suggesting that the offender must do certain things given his antisocial personality. Thus we would say only that the subjects in question are more likely to commit criminal acts (as the data indicate they are). We do not make commission of criminal acts part of the definition of the individual with low self-control.

Be this as it may, Robins's retrospective research shows that predictions derived from a concept of antisocial personality are highly consistent with the results of prospective longitudinal and cross-sectional research: offenders do not specialize; they tend to be involved in accidents, illness, and death at higher rates than the general population; they tend to have difficulty persisting in a job regardless of the particular characteristics of the job (no job will turn out to be a good job); they have difficulty acquiring and retaining friends; and they have difficulty meeting the demands of long-term financial commitments (such as mortgages or car payments) and the demands of parenting. Seen in this light, the "costs" of low self-control for the individual may far exceed the costs of his criminal acts. In fact, it appears that crime is often among the least serious consequences of a lack of self-control in terms of the quality of life of those lacking it.

The Causes of Self-Control

We know better what deficiencies in self-control lead to than where they come from. One thing is, however, clear: low self-control is not produced by training, tutelage, or socialization. As a matter of fact, all of the characteristics associated with low self-control tend to show themselves in the absence of nurturance, discipline, or training. Given the classical appreciation of the causes of human behavior, the implications of this fact are straightforward: the causes of low self control are negative rather than positive; self-control is unlikely in the absence of effort, intended or unintended, to create it. (This assumption separates the present theory from most modern theories of crime, where the offender is automatically seen as a product of positive forces, a creature of learning, particular pressures, or specific defect. We will return to this comparison once our theory has been fully explicated.)

At this point it would be easy to construct a theory of crime causation, according to which characteristics of potential offenders lead them ineluctably to the commission of criminal acts. Our task at this point would simply be to identify the likely sources of impulsiveness, intelligence, risk-taking, and the like. But to do so would be to follow the path that has proven so unproductive in the past, the path according to which criminals commit crimes irrespective of the characteristics of the setting or situation.

We can avoid this pitfall by recalling the elements inherent in the decision to commit a criminal act. The object of the offense is clearly pleasurable, and universally so. Engaging in the act, however, entails some risk of social, legal, and/or natural sanctions. Whereas the pleasure attained by the act is direct, obvious, and immediate, the pains risked by it are not obvious, or direct, and are in any event at greater remove from it. It follows that, though there will be little variability among people in their ability to see the pleasures of crime, there will be considerable variability in their ability to calculate potential pains.

But the problem goes further than this: whereas the pleasures of crime are reasonably equally distributed over the population, this is not true for the pains. Everyone appreciates money; not everyone dreads parental anger or disappointment upon learning that the money was stolen. So, the dimensions of self-control are, in our view, factors affecting calculation of the consequences of one's acts. The impulsive or short-sighted person fails to consider the negative or painful consequences of his acts; the insensitive person has fewer negative consequences to consider; the less intelligent person also has fewer negative consequences to consider (has less to lose).

No known social group, whether criminal or noncriminal, actively or purposefully attempts to reduce the self-control of its members. Social life is not enhanced by low self-control and its consequences. On the contrary, the exhibition of these tendencies undermines harmonious group relations and the ability to achieve collective ends. These facts explicitly deny that a tendency to crime is a product of socialization, culture, or positive learning of any sort.

The traits composing low self-control are also not conducive to the achievement of long-term individual goals. On the contrary, they impede educational and occupational achievement, destroy interpersonal relations, and undermine physical health and economic well being. Such facts explicitly deny the notion that criminality is an alternative route to the goals otherwise obtainable through legitimate avenues. It follows that people who care about the interpersonal skill, educational and occupational achievement, and physical and economic well-being of those in their care will seek to rid them of these traits.

Two general sources of variation are immediately apparent in this scheme. The first is the variation among children in the degree to which they manifest such traits to begin with. The second is the variation among caretakers in the degree to which they recognize low self-control and its consequences and the degree to which they are willing and able to correct it. Obviously, therefore, even at this threshold level the sources of low self-control are complex.

There is good evidence that some of the traits predicting subsequent involvement in crime appear as early as they can be reliably measured, including low intelligence, high activity level, physical strength, and adventuresomeness. The evidence suggests that the connection between these traits and commission of criminal acts ranges from weak to moderate. Obviously, we do not suggest that people are born criminals, inherit a gene for criminality, or anything of the sort. In fact, we explicitly deny such notions. . . . What we do suggest is that individual differences may have an impact on the prospects for effective socialization (or adequate control). Effective socialization is, however, always possible whatever the configuration of individual traits.

Other traits affecting crime appear later and seem to be largely products of ineffective or incomplete socialization. For example, differences in impulsivity and insensitivity become noticeable later in childhood when they are no longer common to all children. The ability and willingness to delay immediate gratification for some larger purpose may therefore be assumed to be a consequence of training. Much parental action is in fact geared toward suppression of impulsive behavior, toward making the child consider the long-range consequences of acts. Consistent sensitivity to the needs and feelings of others may also be assumed to be a consequence of training. In deed, much parental behavior is directed toward teaching the child about the rights and feelings of others, and of how these rights and feelings ought to constrain the child's behavior. All of these points focus our attention on child-rearing.

Child-Rearing and Self-Control: The Family

The major cause of low self-control thus appears to be ineffective child-rearing. Put in positive terms, several conditions appear to produce a socialized child. Perhaps the place to begin looking for these conditions is the research literature on the relation between family conditions and delinquency. This research (e.g., Glueck & Glueck, 1950) has examined the connection between many family factors and delinquency. It reports that discipline, supervision, and affection tend to be missing in the homes of delinquents, that the behavior of the parents is often "poor" (e.g., excessive drinking and poor supervision (Glueck & Glueck 1950, pp. 10–11); and that the parents of delinquents are unusually likely to have criminal records themselves. Indeed, according to Michael Rutter and Henri Giller, "of the parental characteristics associated with delinquency, criminality is the most striking and most consistent" (1984, p. 182).

Such information undermines the many explanations of crime that ignore the family, but in this form it does not represent much of an advance over the belief of the general public (and those who deal with offenders in the criminal justice system) that "defective upbringing," or "neglect" in the home is the primary cause of crime.

To put these standard research findings in perspective, we think it necessary to define the conditions necessary for adequate child-rearing to occur. The minimum conditions seem to be these: in order to teach the child self-control, someone must (1) monitor the child's behavior; (2) recognize deviant behavior when it occurs; and (3) punish such behavior. This seems simple and obvious enough. All that is required to activate the system is affection for or investment in the child. The person who cares for the child will watch his behavior, see him doing things he should not do, and correct him. The result may be a child more capable of delaying gratification, more sensitive to the interests and desires of others, more independent, more willing to accept restraints on his activity, and more unlikely to use force or violence to attain his ends.

When we seek the causes of low self-control, we ask where this system can go wrong. Obviously, parents do not prefer their children to be unsocialized in the terms described. We can therefore rule out in advance the possibility of positive socialization to unsocialized behavior (as cultural or subcultural deviance theories suggest). Still, the system can go wrong at any one of four places. First, the parents may not care for the child (in which case none of the other conditions would be met); second, the parents, even if they care, may not have the time or energy to monitor the child's behavior; third, the parents, even if they care and monitor, may not see anything wrong with the child's behavior; finally, even if everything else is in place, the parents may not have the inclination or the means to punish the child.

So, what may appear at first glance to be nonproblematic turns out to be problematic indeed. Many things can go wrong. According to much research in crime and delinquency, in the homes of problem children many things have gone wrong: "Parents of stealers do not track ([they] do not interpret stealing . . . as 'deviant'); they do not punish; and they do not care" (Patterson 1980, pp. 88–89; see also Glueck & Glueck, 1950; McCord & McCord, 1959; West & Farrington, 1977). Let us apply this scheme to some of the facts about the connection between child socialization and crime, beginning with the elements of the child-rearing model.

The Attachment of the Parent to the Child

Our model states that parental concern for the welfare or behavior of the child is a necessary condition for successful child rearing. Because it is too often assumed that all parents are alike in their love for their children, the evidence directly on this point is not as good or extensive as it could be. However, what exists is clearly consistent with the model. Glueck and Glueck (1950, pp. 125–128) report that, compared to the fathers of delinquents, fathers of nondelinquents were twice as likely to be warmly disposed toward their sons and one-fifth as likely to be hostile towards them. In the same sample, 28 percent of the mothers of "delinquents "were characterized as "indifferent or hostile" toward the child as compared to 4 percent of the mothers of nondelinquents. The evidence suggests that stepparents are especially unlikely to have feelings of affection toward their stepchildren (Burgess, 1980), adding in contemporary children will be "reared" by people who do not especially care for them.

Parental Supervision

The connection between social control and self-control could not be more direct than in the case of parental supervision of the child. Such supervision presumably prevents criminal or analogous acts and at the same time trains the child to avoid them on his own. Consistent with this assumption, supervision tends to be a major predictor of delinquency, however supervision or delinquency is measured.

Our general theory in principle provides a method of separating supervision as external control from supervision as internal control. For one thing, offenses differ in the degree to which they can be prevented through monitoring; children at one age are monitored much more closely than children at other ages; girls are supervised more closely than boys. In some situations monitoring for some offenses is virtually absent. In the present context, however, the concern is with the connection between supervision and self-control, a connection established by the stronger tendency of those poorly supervised when young to commit crimes as adults.

Recognition of Deviant Behavior

In order for supervision to have an impact on self-control, the supervisor must perceive deviant behavior when it occurs. Remarkably, not all parents are adept at recognizing lack of self-control. Some parents allow the child to do pretty much as he pleases without interference. Extensive television viewing is one modern example, as is the failure to require completion of homework, to prohibit smoking, to curtail the use of physical force, or to see to it that the child actually attends school.

DISCUSSION QUESTIONS

1. Explain how Gottfredson and Hirschi's low self-control theory fits in with Sowell's constrained vision we examined in Section 1.
2. Gottfredson and Hirschi appear to believe that all children are equally ready and able to learn self-control. Do you believe that this is so? Are different children different in their readiness to be socialized?
3. After you have read Xiaoming Chen's article below, discuss possible reasons for the differences in American and Chinese crime rates from the perspective of Gottfredson and Hirschi's low self-control theory.

❖

READING

Social Control in China

Applications of the Labeling Theory and the Reintegrative Shaming Theory

Xiaoming Chen

Xiamong Chen's article provides us with a fascinating look at Chinese culture as it relates to crime control via shaming and reintegration of criminals. Chen delineates the underlying philosophy and functions of social control in the Chinese society. This topic is particularly interesting because specific control functions are grounded in a unique macro-control system, which is totally different from that typical of Western countries. He also scrutinizes the implications of labeling theory and reintegrative shaming theory, as they are elaborated in the West, and tests their sensitivity to cross-cultural differences. Although some caveats are in order, the evidence presented within the Chinese context tends to support reintegrative shaming theory rather than labeling theory.

Social Control: Chinese Philosophy and Perspectives

In every society, a normative order must be accompanied by a formal and informal structure for maintaining this normative order. Although crime and social control are universal facets of social life, every society exercises a cultural option when it develops a specific way of looking at crime and delinquency and a characteristic mechanism for controlling them. China offers a valuable context for

Source: Chen, X. (2002). Social control in China: Applications of the labeling and the reintegrative shaming theory. *International Journal of Offender Therapy and Comparative Criminology, 46*(1), 45–63. Reprinted with permission of Sage Publications, Inc.

extending previous—mainly American and European—research because of the profound social and cultural differences between Chinese society and Western societies.

Social control is found wherever and whenever people hold each other to certain standards. The concept of social control has appeared in various radical, critical, or revisionist writings about crime, delinquency, and law over the past three decades. Social control and crime cannot be studied in the abstract, but certain abstract concepts do have general applicability. Although social control takes many forms and appears in many places, it always is a manifestation of the same essence. "Social control" should be considered as a generic term for response to nonconformity, including both perception of and reaction to rule breaking. In other words, it includes the formal and informal ways society has developed to help ensure conformity to social norms. Social control takes place when a person is induced or forced to act according to the values of the given society, whether or not it is in accordance with his or her own interests. The aims of social control are "to bring about conformity, solidarity, and continuity of a particular group or society" (Young, 1942, p. 898). Due to different cultures and traditions, there exist different models of social control in different societies. Compared with the West, Chinese society has developed a very different model of social control. As Xin Ren (1997) stated,

> The most important distinction, perhaps, is the efforts of the Chinese state to control both the behavior and the minds of the people. Social conformity in the Chinese vocabulary is not limited to behavioral conformity with the rule of law but always moralistically identifies with the officially endorsed beliefs of social standards and behavioral norms. Perhaps it goes against the free-will notion of the classical theory of criminology or perhaps it is socialist totalitarianism. Whatever it is, the Chinese tradition of so-called "greatest unity" has always attempted to achieve ultimate uniformity of both mind and act within Chinese society. (p. 6)

Brief Reviews of Labeling and Reintegrative Shaming Theories

Labeling theory holds the view that formal and informal societal reactions to delinquency can influence delinquents' subsequent attitudes and behavior. This was recognized early in the 1900s. Frederick Thrasher's work on juvenile gangs in Chicago was one of the first instances in which the consequences of official labels of delinquency were recognized as potentially negative. Thereafter, Frank Tannenbaum introduced the term "dramatization of evil," in which he argued that officially labeling someone as a delinquent can result in the person becoming the very thing he is described as being. Edwin Lemert (1951) developed the concepts of primary and secondary deviance, which became the central elements of the first systematic development of what has come to be known as labeling theory. These concepts all stress the importance of social interactions in the development of self-images and social identities.

Labeling theory attempts to account for the mutual effects of the actor and his audience. The theory is concerned not only with what the actor and reactor do, but also with how each one's actions affect the behavior of the other. It is thus clear that one of the most central issues with labeling theory is the connection between behavior and the societal reaction to it.

An issue with labeling theory is the matter of how the label is handled by the labelee. The importance of groups and associations in the actor's reception of societal reactions to his behavior is emphasized. For example, Howard Becker (1973, p. 37) thought that a final step in the career of a deviant is the identification with an "organized deviant group." It suggests that group support of a labeled deviant may either push him further into an identity as a deviant or serve as a catalyst for a transformation from a socially shunned role to a more positive social status.

The origins of the reintegrative shaming theory may be traced to John Braithwaite's (1989) publication of *Crime, Shame and Reintegration.* Here, he wrote,

> The first step to productive theorizing about crime is to think about the contention that labeling offenders makes things worse. The contention is both right and wrong. The theory of reintegrative shaming is an attempt to specify when it is right and when wrong. The distinction is between shaming that leads to stigmatization—to outcasting, to confirmation of a deviant master status—versus shaming that is reintegrative, that shames while maintaining bonds of respect or love, that sharply terminates disapproval with forgiveness, instead of amplifying deviance by progressively casting the deviant out. Reintegrative shaming controls crime; stigmatization pushes offenders toward criminal subcultures. (pp. 12–13)

Reintegrative shaming theory is about the positive effectiveness of reintegrative shaming and the counterproductiveness of stigmatization in controlling crime. Individuals should be confronted in the communities to which they belong and reintegrated back into them rather than cast out of the community. Otherwise, they may be driven into one of the variety of criminal subcultures that are available. Generally, the reintegrative shaming theory and practice aim at maximizing shame while maintaining community links and providing opportunities for the offender to make amends to those injured. According to John Braithwaite (1989), crime is best controlled when members of the community are the primary controllers through active participation in both shaming offenders and concerted efforts to reintegrate them back into the community of law-abiding citizens.

Social Control in China and the Postulates of the Labeling Theory and the Reintegrative Shaming Theory

Interestingly, the Chinese system seems to rely greatly on benefits of labeling in preventing and controlling juvenile delinquency. Evidence available from Chinese society will be used to assess the main tenets of these theories and to probe both their cultural specificity and possible universality. A closer, cross-cultural look at these theories may prove illuminating for both Western and Chinese scholars. Given, however, the cultural disparity between the Western and Chinese ways of thinking, theoretical concepts have to be contextualized and explored differently in each culture.

> LT: The primary factor in the repetition of delinquency is the fact of having been formally labeled as a delinquent.
>
> RS: Shaming carries a risk of alienating the first-time offender if it is not combined with positive reintegrative efforts. These efforts are more likely to succeed in a society tied by strong informal bonds.

The Chinese approach to delinquency assumes that labeling a delinquent does not necessarily produce his or her secondary deviance in spite of the existence of an effect of official labels on delinquent identities and behavior. In Chinese society, it is recognized that the social reaction to delinquency (labeling) may have both positive and negative effects, but a greater emphasis is placed on its potential for positive effects. In China, stigma is regarded as a great deterrent; it is shameful and therefore people try to avoid it. It is also believed that labeling contributes to rehabilitation. In Chinese society, on being labeled deviant, a person is subject to negative reactions and moral condemnation from society and is initially removed from his or her normal position in society and assigned a special role—a distinctive deviant role. This strategy is expected to force the individual to feel "pain" and recognize that he or she indeed has some problems and therefore become motivated to rehabilitate himself or herself.

This process of labeling takes place in a distinct cultural context, however. Chinese society is a highly interdependent and relatively communitarian society in which individual survival depends on the survival of groups, and the functioning of groups depends on the capacity of their members to cooperate (Wrong, 1994). In such a society, social comparison becomes a major concern of members of society. Individuals care profoundly about what others think of them and their positions in the social world.

Because individuals strongly depend on others for fulfillment of basic needs, pride and shame serve as indicators of the strength of the individual's social connection to others in Chinese society. Because of this sense of interdependence, the Chinese have the greatest esteem for those who make progress in accepting conventional norms. The role of a (positively labeled) model is always emphasized. To the Chinese people, to be praised as a model citizen—as the chaste widow or the dutiful son were praised by Confucians—is regarded as the most legitimate and desirable form of reward because it moves others to imitate and respect them. This thought continues to exert a strong influence in the present China. Many officials of correctional institutions are convinced that the use of models provides the most desirable mode of learning and is especially effective because peer respect is a powerful human need. Therefore, the inmates are urged to learn from good examples. Emulated models may be living or dead, national or local. A model may exhibit general virtues, such as "selflessly serving the people," or particular ones, such as "working or study hard." The other side of this emphasis on the model is to encourage people to become models themselves. Inmates thus are exhorted to act as models for other inmates. Those chosen as models of neatness, productivity, helpfulness, or some other virtue are often lauded at many different occasions, and others are encouraged to follow their example. For example, there are blackboards in almost every correctional institution that display inmates' good deeds. Many officials believe that the model's example helps to develop an entrenched attitude, which is far more stable than a fear of punishment.

For social control to be effective, however, the effect of negative labeling and shame is also emphasized. The effect of shaming lies not only in deterrence but also reintegration. Although at first, shaming can stigmatize and alienate the juvenile offender when the community expresses disapproval not only of the act but also of the person, need to regain or protect one's social status is a strong motivation for conforming behavior. This is facilitated by the subsequent reintegrative efforts of the collective.

Shaming is particularly meaningful in societies with strong social bonds. In China, people, especially juveniles, are closely attached to their family, school, and neighborhood. Strong social bonds make shame have a greater effect on offenders because shame is the emotional cognate to the social bond and shame is felt when the bond is threatened. Once a person is shamed, the shame is often also borne by his or her family and, although to a lesser degree, by neighborhood and school. Under these circumstances, juvenile offenders as well as their families, schools, and neighborhoods endure painful humiliation. This situation can force shamed persons to regret their behavior because they accrue greater interpersonal costs from shame.

Shaming is powerful and where there is power, there is also a risk. John Braithwaite and Stephen Mugford (1994) argued that stigmatization and exclusion are the most significant risks of shaming. Braithwaite (1989) pointed out that shame can take dramatically different forms. They range from stigmatizing or disintegrative shaming that brings a degradation in social status to RS. He further explained,

> Potent shaming directed at offenders is the essential necessary condition for low crime

> rates. Yet shaming can be counterproductive if it is disintegrative rather than reintegrative. Shaming is counterproductive when it pushes offenders into the clutches of criminal subcultures; shaming controls crime when it is at the same time powerful and bounded by ceremonies to reintegrate the offender back into the community of responsible citizens. (p. 4)

Indeed, a shame-induced negative, or delinquent, self-image has a detrimental effect on some juveniles. In China, some informal or formal control measures, such as early social-educational intervention and confinement in a work-study school or juvenile reformatory, result in considerable stigma. This may persist even though these measures do not attempt to demean and humiliate offenders and emphasize reintegration into the community following treatment at an institution. For example, trials of those who are 18 years of age and younger are not open to the public. Clearly, this provision of the Law on Criminal Procedure is intended to protect juveniles from stigmatization and increase the possibility that a rehabilitated offender becomes again part of the community. In addition, social-educational teams, which are composed of not only their parents, relatives, teachers, colleagues, and neighbors, but also the police, may be assigned to monitor and help reintegrate the returning delinquent. Concrete measures are supposed to be selected according to the nature of the offender and the offense to "suit the remedy to the case" (a Chinese proverb). For different cases, there are different educators and methods, ranging from heart-to-heart talks to group study of the Party and governmental documents, laws and regulations; to public discussion of the deviant's problems and offering of suggestions and criticisms; to visiting factories, farms, and historical and cultural sites and solving young people's practical problems, such as housing, schooling, employment, and so on.

Although many measures have been taken, it is impossible to erase stigma completely. Once stigmatized with a deviant label, some juveniles, especially those from broken families, may lose love, dignity, and respect. In some areas, many schools, factories, and other institutions are reluctant to accept youngsters with a criminal or delinquent past. These young people usually have fewer opportunities to enter school, to find jobs, to join the army, even to establish a family. Thus, they may experience a sense of injustice at the way they are victimized by agents of social control. This situation may indeed influence some juvenile offenders to have more negative attitudes toward society, become more involved with delinquent peers, and regard themselves as more delinquent. Under these circumstances, deviance becomes and is rationalized as a defensible lifestyle, which is difficult to change. This partly explains the phenomenon of recidivism in China.

> LT: Formal labels eventually alter a person's self-image to the point at which the person begins to identify himself or herself as a delinquent and acts accordingly.
>
> RS: Shaming may damage a person's self-esteem if it is directed at the individual identity not the act and is not followed by positive reintegrative ceremonies.

According to the Chinese experience, whether one resists the negative effects of labeling and becomes positively influenced by the label depends not only on the labelees themselves but also on the efforts of the whole society. On one hand, stigmatization by certain agents of social control can indeed increase the attraction of outcasts to subcultural groups that provide social support for crime. On the other hand, concerted efforts of the community can change this situation. In China, the belief that society is both a cause and a victim of crime and delinquency confers on society the right and duty to be involved in such matters. Intrusion into other people's lives is taken for granted and viewed positively. It is perceived as a sign of caring rather than meddling. To the Chinese mind, offenders come from society and will eventually go back to the society. Social education is an indispensable factor in the process of the reform. People believe that participation of the whole society in reforming offenders will greatly accelerate the process of socialization of offenders. Therefore, an unusually broad range of citizens and groups are drawn into the process. The deep-seated interdependency and communitarianism in Chinese society and culture are also conducive to the public support and enthusiasm for dealing with any problems affecting the community. Therefore, Chinese crime control does not just work from the top down through the formal criminal justice system but, and above all, from the bottom up.

Families, friends, neighbors, schools, work unit, and the police intervene at the first sign of possible trouble. Chinese people do not mind only their own business; they prefer to handle crime and juvenile problems in their neighborhood rather than hand them over to professionals. As a result, almost all members of the society seem to have become active controllers of crime and delinquency. Families, neighborhood communities, work units, and schools act as positive solvers of problems instead of silent observers. Their active participation is believed to have great effect on improving the offenders' legal and moral conscience, heightening their immunity to delinquent influences, decreasing attractive opportunities for participation in subcultures, and reintegrating them back into the community of law-abiding citizens.

As the first line of defense against criminal activity, the family and the grassroots community organizations play the most important role. In the vast majority of cases, delinquents are put under the supervision of their families and the relevant local authority—neighborhood committee, school, or work unit (depending on the delinquent's status). According to Article 17 of the Criminal Code of China, parents or guardians of offenders under the age of 16 have the obligation to discipline their children. Under the watchful eyes of those with whom the delinquents have the most frequent daily contact, they are encouraged to reform and develop a sense of social responsibility, which will lead to their reintegration into the society.

Various specialized organizations, through their local branches, also take a hand in urging delinquents to rehabilitate themselves. Many state or regional leaders and other high-ranking officials often visit the institutions and show interest in inmates' study, labor, and life. Different levels of the People's Congress, government, and other organizations, such as labor unions, the Young League, women's associations, even military organs, all are assisting reformatories with reforming delinquents. For example, one naval school of the People's Liberation Army in Shanghai established a link with a local juvenile reformatory. The school regularly sent its staff members and students to the reformatory. Not only did they give lectures on Chinese history, tradition, and moral and spiritual civilization but also conducted many kinds of cultural and entertainment programs for inmates. Furthermore, they organized inmate visits to their school and enabled the offenders to watch their military training. Perhaps, mass involvement in reforming delinquents is one of the most striking aspects of the Chinese justice system that differentiates it from Western models.

Unlike in the RS approach, however, the deliberate focus of labeling in China seems to be both on the offense and the offender. Offenders are first labeled deviants and educated to understand that what they have done has detrimental psychological, social, and economic consequences for victims and also for themselves and other parties and to assume responsibility for what they have done. Then they are shown concern and love, accompanied by attempts to solve their practical problems, such as housing, schooling, and employment. This two-stage approach both expresses community disapproval and symbolizes reacceptance of offenders while offering a practical basis for reintegrating offenders. Thus, community not only reaffirms the normative order by shaming offenders but also provides them with opportunities for conventional reintegration.

In China, educating and helping offenders, especially juvenile delinquents, has become one of the main tasks of the police who are working in neighborhood police stations. They always work closely with local government, factories, enterprises, schools, neighborhoods, and other institutions to maintain public order, "prevent, reduce and forestall crimes through ideological, political, economic, educational, administrative and legal work," and create a better social environment (Gao, 1983, p. 4). They not only provide community surveillance but also conduct many education programs at the neighborhood level, such as distributing legal materials and educating the residents in the neighborhood about laws and rights; holding community meetings to discuss justice and social problems; preparing messages involving legal and crime prevention matters and persuading residents of the necessity of rules for public security and individual protection; and visiting offenders and their families, friends, and relatives to determine what the problem is, seek solutions outside of the criminal justice system, and help them solve such problems as unemployment. In some cases, according to Leng (1977), the police station in a neighborhood community may draw up a contract between the juvenile and his or her parents, the neighborhood, and the school. If the juvenile meets the terms of the contract within a specified time, then a certificate of good behavior is issued to the youngster, and the case is considered closed. If the adolescent does not meet the terms of the contract or commits a new violation, then the contract is extended or revised. Although the emphasis is on the collective nature of the treatment program—it is more than just an agreement between the juvenile and the police—direct involvement by the police is also intended as a reminder that an arrest can still be effected if the juvenile offender strays further from the beaten path. The aim is to put greater pressure on delinquents. In reality, although the police participate directly in the early intervention, they are not concerned with law and punishment as such but rather with helping offenders become law-abiding and useful citizens. The mere presence of the police is considered to be able to strengthen intervention and has in fact become an integral part of it. The extent and the diversity of activity performed by the Chinese police indicate that policing not only is a prime force in enforcing the criminal law but also plays an important role in the elaborate system of informal social control.

Although the judicial shame penalties, such as public exposure and debasement penalties (generally achieved by associating the offender with a noxious activity) that exist in the United States do not exist currently in China, China does have some judicial options, such as mass trials and sentence-pronouncing rallies, which can reach the same effect as shame penalties. They are designed to convey moral condemnation and inform the public about the offense and the offender to elicit public shaming of the offender. In the Chinese social environment, these legal measures can work as both a specific and general deterrent of possible future transgressions. With respect to young people, the educational sessions conducted by juvenile courts before disposition of cases may have a similar effect.

The Chinese experience suggests that informal and quasi-informal social education, including various education and training programs operated by parents, relatives, friends, or a relevant collectivity, has a greater effect on the juvenile's future behavior and self-identification than sanctions imposed by a remote legal authority. Social education can induce juveniles to recognize the harmfulness of their conduct and eventually to rehabilitate themselves. This is especially true in Chinese society because of its collectivist nature.

Through many kinds of education and help, and through positive inducement rather than merely criticism or negative incentives, the vast majority of the young people initially labeled as criminals or delinquents eventually leave these deviant statuses well and truly behind them. This is indicated by the low rate of recidivism in China. A number of different sources suggest that recidivism rates among those released from reformatories are between 8% and 15% (Li, 1992, p. 55). In spite of different estimates, it is clear that the recidivism rate is quite low. From the limited data available, it is difficult to judge what causes low recidivism rates for the released inmates; however, community involvement, tight social control, or postrelease surveillance and rehabilitation all should be regarded as contributing factors.

Conclusion

The limitations of LT are quite clear in the light of Chinese experience. LT ignores the cultural characteristics of different countries. For example, the concept of shaming is consistent with the cultural ethos of Chinese collectivism rather than Western individualism. Therefore, LT assertions regarding the relationship of a formal delinquency label and secondary deviance do not seem particularly relevant to the Chinese experience. Rather, the Chinese experience tends to support the RS theory developed by John Braithwaite—the collectivist nature, strong social bonds, effective informal social control, and emphasis on social education all make shaming a positive rather than negative tool of social control in Chinese society.

References

Becker, H. (1973). *Outsiders.* New York: Free Press.

Braithwaite, J. (1989). *Crime, shame and reintegration.* Cambridge, UK: Cambridge University Press.

Braithwaite, J., & Mugford, S. (1994). Conditions of successful reintegration ceremonies. *British Journal of Criminology, 34*(2), 139–170.

Gao, X. (1983). Combating criminal offenders by People's Republic of China. *Police Studies, 6,* 3–5.

Lemert, E. (1951). *Social pathology.* New York: McGraw-Hill.

Leng, S. (1977). The role of law in the People's Republic of China as reflecting Mao Zedong's influence. *Journal of Criminal Law and Criminology, 68*(3), 356–373.

Li, J. (1992). *Zhongguo chongxin fanzui yanjiu* [Research on recidivism in China]. Beijing, China: Falu Chubanshe.

Ren, X. (1997). *Tradition of law and law of the tradition.* Westport, CT: Greenwood.

Wrong, D. (1994). *The problem of order.* Cambridge, MA: Harvard University Press.

Young, K. (1942). *Sociology.* Cincinnati, OH: American Book.

DISCUSSION QUESTIONS

1. Would the methods described in this article work in America? Why or why not?

2. Given that China executes proportionately far more criminals than any other nation, how do you think Chen would reconcile this with his claim that rehabilitation is the goal of the Chinese criminal justice system?

3. What is the influence of collectivism vs. individualism in relation to crime control?

READING

Gender and Crime Among Felony Offenders

Assessing the Generality of Social Control and Differential Association Theories

Leanne Fiftal Alarid, Velmer S. Burton, Jr., and Francis T. Cullen

This article by Alarid, Burton, and Cullen assesses the general applicability of two of the theories presented in this chapter—differential association and social control theories. Their sample consists of 1,153 newly incarcerated male and female felons, a big improvement over sampling high school and college subjects. Their findings support both theories' claims that they are "general" theories of crime. They also found that differential association theory had more consistent effects, especially for men. For females, parental attachment was more important for explaining violent crime than it was for men.

Since the 1969 publication of Hirschi's *Causes of Delinquency*, social control or control theory and differential association or "cultural deviance" theory have competed vigorously against one another for the status of the preeminent microlevel sociological theory of crime.

A key concern is the generality of social control and differential association theories. These perspectives typically have been seen as having the ability to explain various forms of crime for all people (see, e.g., Akers 1998). Clearly, the theoretical power of each of these perspectives hinges on their ability to achieve the status of a general theory. Accordingly, a central empirical and thus theoretical issue is whether social control and differential association theories can explain criminal behavior that extends beyond the delinquency found among community samples of youth.

Research on social control theory with adult samples also suggests that adult social bonds may reduce criminal behavior (Sampson and Laub 1993). Even so, the empirical literature assessing whether social control and differential association theories can explain adult criminal behavior remains limited, especially when one searches for studies that include in their analyses measures of both theoretical perspectives. Among the few adult studies testing social control versus differential association theory, the results show mixed support for both perspectives. Thus, two studies using community samples of adults found stronger support for differential association theory (Burton 1991; Macdonald 1989), while another study revealed that social control theory was better able to account for variation in arrests for men and women drug offenders (Covington 1985).

Even with this research, however, there are typically two limitations. First, especially with adult samples, studies often do not explore whether the effects of variables measuring traditional criminological theories vary by gender. Traditional theories have been criticized for their perceived inability to explain female criminality and for ignoring how gender-related factors—such as patriarchal power relations—differentially shape the involvement of gender groups in crime

SOURCE: Alarid, L., Burtin, V., & Cullen, F. (2000). Gender and crime among felony offenders: Assessing the generality of social control and differential association theories. *Journal of Research in Crime and Delinquency* 37(2):171-199. Reprinted with the permission of Sage Publications, Inc.

victimization (Chesney-Lind and Shelden 1992). In turn, it is argued that scholars should formulate separate or different theories of crime for women (Leonard 1982).

In light of the status of the existing research, the current study attempts to test the generality of social control and differential association theories to a population that has been infrequently studied: young adult felons drawn from the inner city. Social control and differential association theories were selected because they have been found, among existing sociological theories, to have the most consistent empirical support in studies of juveniles. Social control theory is also used to explain adult crime in recent important studies (Horney et al. 1995; Sampson and Laub 1993). These theories both predict that their effects will be general—that is, they will account for variation in offending and that this effect will hold for males and females.

Thus, the purpose of this article is to test the generality of social control and differential association theories with a sample of young adult felons drawn from the inner city. The study also has the advantage of including males and females. Using a self-report survey, the study examines both overall offending and involvement in specific forms of criminal conduct. By testing the two theories against each other, the central focus will be to discern which theory—if not both theories—can explain variation in offending among young adult men and women felony offenders from an urban area. That is, are social control and differential association theories mainly explanations of juvenile delinquency, or can they help account for participation in crime among young adult male and female felons?

Method

The procedures were used to test differential association and social control theory with 1,153 felony women and men in a community corrections program. Data for this study were gathered from felons sentenced to a residential court-ordered boot camp program in Harris County, Texas. Of the 1,153 felons, 1,031 men and 122 women indicated that the vast majority lived in the inner city or surrounding urban area at the time of arrest for their most recent conviction. Men and women were both between the ages of 17 and 28 years, with the average age calculated at 19.5 for men and 20.6 years for women. The African American composition of both samples was identical (44 percent).

Dependent Variables

Dependent variable measures for the analysis were originally derived from Elliott and Ageton's (1980) scale of delinquent behaviors offenses found in the National Youth Survey (NYS). We adjusted certain measures to apply to our young adult sample. Respondents were asked whether they had ever committed 35 different deviant and criminal acts. Our prevalence dimension of crime measures offense participation or the number of crime types committed. Other researchers have preferred the use of the prevalence measure of variety (see Table 5.2).

In the current study, one point was assigned if a respondent admitted involvement in a specific act; if respondents did not admit involvement, no points were assigned. The general crime scale was summed to obtain a composite score, which ranged from 0 to 35. Items in the general crime scale have been employed in previous tests of criminological theories using samples of adults (Burton et al., 1993). In addition, we distinguished type of criminal behavior by dividing the offenses into three subscales: violent, property, and drug. Table 5.3 shows the means, standard deviations, and reliability for all dependent and independent variables used in the analysis.

Independent Variables

The items used in this study to measure social control and differential association variables were all drawn from previous theoretical tests (e.g., Burton et al. 1998). Participants in our sample responded to each item by using a 6-point Likert scale that ranged from **strongly agree** (6) **to strongly disagree** (1).

Table 5.1 Prevalence of Self-Reported Criminal Behavior: "Have You Ever . . . ?"

	Women		**Men**	
	n	(%)	*n*	(%)
Property crime subscale (15 items)				
1. Avoided paying at restaurants, the movie theater, etc.?	29	(23.8)	346	(33.7)*
2. Knowingly bought, held, sold stolen property?	50	(41)	531	(51.8)*
3. Taken someone else's vehicle without their permission?	33	(27)	379	(36.9)*
4. Taken anything ($5 or less) from your job?	11	(9)	152	(14.8)
5. Taken anything (between $5 and $50) from your job?	11	(9)	118	(11.5)
6. Taken anything (over $50) from your job?	3	(2.5)	102	(10.0)**
7. Taken anything ($5 or less) from someone (other than work)?	46	(37.7)	317	(31.0)
8. Taken anything (between $5 and $50) from someone (other than work)?	25	(20.5)	306	(30.0)*
9. Taken anything (over $50) from someone else (other than work)?	31	(25.4)	346	(33.7)
10. Damaged/destroyed property belonging to relative/family of origin?	15	(12.3)	166	(16.2)
11. Purposely damaged/destroyed property belonging to an employer?	2	(1.6)	59	(5.7)*
12. Purposely damaged/destroyed property belonging to spouse/partner/friend (someone other than family of origin or an employer)?	22	(18.0)	327	(31.9)**
13. Broken into a building/vehicle?	25	(20.5)	423	(41.2)**
14. Stolen or attempted to steal a motor vehicle?	19	(15.6)	314	(30.6)**
15. Thrown objects at cars/property?	41	(33.6)	419	(40.9)
Drug crime subscale (11 items)				
1. Drank alcoholic beverages before age 21?	92	(75.4)	788	(76.8)
2. Drove a car while drunk?	51	(41.8)	471	(46.0)
3. Bought/provided beer/liquor for someone under 21?	51	(41.8)*	340	(33.2)
4. Had marijuana/hashish?	86	(70.5)**	572	(55.8)
5. Used hallucinogens/LSD?	26	(21.3)	226	(22.0)
6. Had amphetamines?	17	(13.9)	129	(12.6)
7. Had barbiturates?	14	(11.5)	91	(8.9)
8. Had heroin?	6	(4.9)	32	(3.1)
9. Had cocaine?	60	(49.2)**	278	(27.1)
10. Sold marijuana?	34	(27.9)	359	(35.2)
11. Sold hard drugs (cocaine, crack, heroin)?	46	(37.7)	435	(42.5)

(Continued)

Table 5.1 (Continued)

	Women		Men	
	n	(%)	*n*	(%)
Violent crime subscale (9 items)				
1. Been involved in a gang fight?	25	(20.5)	232	(22.6)
2. Used physical force to get money from a relative/family of origin?	4	(3.3)	48	(4.7)
3. Used physical force to get money from someone at work?	1	(0.8)	35	(3.4)
4. Used physical force to get money from someone, a family member, or someone at work?	12	(9.8)	151	(14.7)
5. Hit or threatened to hit a relative/member of family of origin?	42	(34.4)	337	(32.9)
6. Hit or threatened to hit someone at work?	5	(4.1)	144	(14.0)**
7. Hit or threatened to hit friend/partner/spouse?	80	(65.6)**	528	(51.7)
8. Had or tried to have sex with someone against their will?	1	(0.8)	31	(3.0)
9. Attacked someone with the idea of seriously hurting/killing them?	30	(24.6)	264	(25.7)

NOTE: Women, $n = 122$; men, $n = 1{,}031$.

$^{*}p < .05$; $^{**}p < .01$.

Social Control Theory

Hirschi's (1969) social control theory was originally formulated to rival differential association or "cultural deviance" theory as an explanation of why some juveniles conform and others engage in delinquency. More recent research has applied this perspective to adult criminality. Social control theory asserts that individuals with strong ties to family and friends will be protected from criminal involvement. In this study, we measure the adult social bond by marital attachment, attachment to parents, attachment to friends, involvement, and belief.

Marital/partner attachment. Attachment to a significant other of the opposite sex by cohabiting or marriage is perceived as providing informal social control by protecting individuals from participation in criminal and other antisocial activities.

Attachment to parents. We include three items to measure attachment to parents to determine the effect of longstanding connections to parents by gender:

Attachment to friends. The bond to friends is an important part of conventional attachment.

Involvement. Involvement in conventional activities is measured by how much free time individuals have.

Belief. Finally, the conventional bond is measured by moral belief in the law, specifically in the police.

Differential Association Theory

According to Sutherland's (1947) theory of differential association, individuals develop internalized definitions that are favorable or nonfavorable toward violating the law. As individual

Table 5.2 Sample Characteristics

Characteristic	Women (n = 122)	Men (n = 1,031)
Age (17–28 years)	20.6	19.5
Race/ethnicity (percentage)		
African/American	44.3	44.4
Caucasian	45.1	31.4
Hispanic	10.7	23.8
Formal education (years)	10.5	10.3
Marital/partner attachment		
(percentage attached)	32.0	35.1
Number of children (percentage)		
None	46.7	69.8
One	30.3	21.3
Two	9.8	6.1
Three or more	13.1	2.9
Median household income (in dollars)	16,800	27,500
Current conviction (percentage)		
Drug	41.8	28.9
Person	19.7	19.5
Property	38.5	51.6

exposure to procriminal values, patterns, and associates increases, the likelihood of criminal involvement also increases. As in previous studies, we use three differential association variables to measure the likelihood of criminal exposure: individual definitions toward the law, others' definitions toward the law, and number of criminal friends.

> *Individual definitions toward crime.* The five-item scale measuring individual definitions toward crime examined an individual's degree of tolerance for criminal behavior.
>
> *Others' definitions toward crime.* According to Sutherland (1947), criminal behavior is influenced primarily through exposure to other individuals holding definitions favorable toward violating the law.
>
> *Criminal friends.* Many empirical tests of differential association theory have relied on the number of criminal associates as evidence of interaction with criminal members within an individual's primary group.

Findings

Social Control

Table 5.4 shows that three out of five social control variables were significantly related to the overall measure of criminal behavior for the sample. Attachment to peers was significantly and positively related to criminal behavior. Attachment to parents and involvement in conventional activities were significantly and inversely related to involvement in criminal behavior. Those who got along well with their

Table 5.3 Individual Dependent and Independent Variables by Gender

Variable	Number of Items	Women		Men		Sample *F*-Ratio
		Reliability	Mean (*SD*)	Reliability	Mean (*SD*)	
Dependent						
General crime	35	.82	8.57 (5.12) (range: 0-26)	.90	9.49 (6.85) (range: 0-31)	2.06
Violent crime	9	.61	1.64 (1.25) (range: 0-6)	.66	1.72 (1.70) (range: 0-9)	0.27
Property crime	15	.72	2.98 (2.60) (range: 0-11)	.85	4.19 (3.76) (range: 0-17)	11.99**
Drug crime	11	.76	3.96 (2.55) (range: 0-11)	.81	3.62 (2.72) (range: 0-11)	1.68
Independent						
Social control						
Marital/partner attachment	1	—	0.32 (0.47)	—	0.35 (0.48)	0.48
Parental attachment	3	.75	15.25 (2.98)	.64	14.87 (3.02)	1.77
Peer attachment	2	.81	6.84 (2.91)	.73	7.68 (2.79)	9.58**
Involvement	2	.69	6.44 (2.89)	.56	6.09 (2.69)	1.82
Belief	2	.63	8.88 (2.41)	.51	8.11 (2.40)	11.23**
Differential association						
Individual definitions	5	.77	10.97 (4.95)	.62	11.98 (4.52)	5.30*
Others' definitions	3	.71	9.50 (3.79)	.66	10.11 (3.61)	3.11
Criminal friends	1	—	2.87 (1.66)	—	2.77 (1.85)	0.30

$^*p < .05$; $^{**}p < .01$.

parents and considered them an important part of their lives were less likely to engage in criminal behavior. Furthermore, respondents who were immersed in time-consuming activities (i.e., work, family, and community activities) were also less likely to commit crimes.

One of the strongest predictors for a woman's involvement in crime was whether she lived with a mate or was married. The significance of this relationship was not in the predicted direction of social control theory, which would hypothesize that family relationships, including marriage, insulate against criminal involvement. On the other hand, marital attachment was not significant for predicting men's criminal behavior.

One factor that was significantly related to men's participation in crime was peer attachment. Young men who were attached to their friends were more likely to engage in criminal behavior—a relationship in the direction opposite to that predicted by social control theory.

Table 5.4 Effects of Social Control and Differential Variables on General Crime: Standardized Betas (unstandardized B coefficients) Reported

Variables	Women (n = 122)		Men (n = 1,031)		Women and Men (N = 1,153)	
Social control						
Marital/partner attachment (0 = not attached)	.21	(2.27)**	.01	(.08)	.02	(.33)
Parental attachment	–.31	(–.53)**	–.09	(–.21)**	–.12	(–.32)**
Peer attachment	.07	(.12)	.06	(.16)*	.07	(.16)*
Involvement	–.24	(–.43)**	–.06	(–.16)*	–.07	(–.18)*
Belief	.03	(.05)	–.06	(–.17)	–.05	(–.15)
Differential association						
Individual definitions	.11	(.11)	.12	(.19)**	.12	(.17)**
Others' definitions	.18	(.31)*	.16	(.31)**	.16	(.30)**
Criminal friends	.16	(.50)*	.26	(.98)**	.26	(.94)**
Control variables						
Age	–.08	(–.15)	.03	(.08)	.01	(.04)
Gender (0 = male)	—	—	—	—	–.05	(–.93)
Race (0 = non-White)	.22	(2.23)**	.19	(2.75)**	.19	(2.64)**
R^2		.51		.29		.30

$^{*}p < .05$; $^{**}p < .01$

Other studies have interpreted peer attachment and increased criminal involvement as indirect support for differential association theory, depending on whether the friends were delinquent (Conger 1976).

Differential Association

In Table 5.4, others' definitions and criminal friends significantly predicted both male and female involvement in crime. A difference of slopes test on criminal friends indicated there were no significant differences between women and men; the effect of criminal friends on participation in crime is similar for men and women.

Individual definitions, although in the expected direction for women, were significant only for men. Furthermore, for men, differential association variables had stronger effects than social control theory variables in predicting their participation in crime. For women, there was some tendency for the social control variables to have stronger effects. The overall explanatory power for both theories in the model for women was $R^2 = .51$ and for men was .29.

Control Variables

Race/ethnicity significantly predicted criminal involvement for women and men. Specifically, Anglo men and women were more likely to have been involved in crime than non-White individuals. Age was not a significant predictor of overall rates of criminal behavior. To more precisely measure the nature of gender differences and similarities for overall crime, a separate *t*-test for independent samples was conducted using unstandardized partial regression coefficients while using gender as the grouping variable. Results indicated that the mean for the men (9.88) was significantly higher than for women (8.56) for the general crime scale.

Discussion

Both social control theory and differential association theory appear to have general effects. The support for differential association is especially strong and consistent: The three differential association variables are significantly related to the general crime scale drug, property, and violent crime subscales. Note that many studies that assess differential association use only the single measure of "number of delinquent peers." Some additional support for differential association theory can be drawn from the positive relationships of peer attachment to all the crime measures for the sample. Furthermore, to the extent that the members of the sample are likely to have delinquent friends, then such attachment might be viewed as an indicator of differential association.

The findings for the social control measures are less impressive but nonetheless offer support for the perspective. Parental attachment had a consistent negative relationship to all of the crime measures for the sample. It appears, therefore, that bonds to parents have continuing effects on serious felons into early adulthood.

The analyses suggest that the "traditional" criminological perspectives can account for offending across the genders. This is not to say, however, that gender-specific theories could not expand our understanding not only of female crime but also of male criminal participation (see, e.g., Messerschmidt 1993). Still, the data reveal that differential association and social control variables—have similar effects across male and female felons. In fact, for every crime measure, the amount of explained variation is higher for the female sample than it is for the male sample.

The effects of social control variables were stronger for the women than for the men. We found that the effects of parental attachment significantly and inversely affected males' offending, but the impact of parental attachment on females was significantly stronger for participation in violent offenses. This difference between men and women was not found for property crime, and parental attachment was not significant for male drug crime participation. The effect of lack of parental attachment as a significant predictor of female violent crime is especially noteworthy for furthering our understanding of female offending.

In closing, the analyses reported here lend continuing support to the traditional theories of social control and, especially, differential association/social learning. Importantly, the data support the claim that these are "general" theories, explaining variations in self-report criminality among felony offenders and across men and women.

References

Akers, R. L. (1998). *Social learning and social structure: A general theory of crime and deviance.* Boston: Northeastern University Press.

Burton, V. S. (1991). *Explaining adult criminality: Testing strain, differential association, and control theories.* Ph.D. dissertation, University of Cincinnati.

Burton, V. S., Marquart, J. W., Cuvelier, S. J., Hunter, R. J., & Fiftal, L. (1993). The Harris County CRIPP program: An outline for evaluation, Part 1. *Texas Probation, 8,* 1–8.

Chesney-Lind, M., & Shelden, R. G. (1992). *Girls, delinquency, and juvenile justice.* Pacific Grove, CA: Brooks/Cole.

Conger, R. (1976). Social control and social learning models of delinquency: A synthesis. *Criminology, 14,* 17–40.

Covington, J. (1985). Gender differences in criminality among heroin users. *Journal of Research in Crime and Delinquency, 22,* 329–353.

Elliott, D. S., & Ageton, S. S. (1980). Reconciling race and class differences in self-reported and official estimates of delinquency. *American Sociological Review, 45,* 95–110.

Hirschi, T. (1969). *Causes of delinquency.* Berkeley: University of California Press.

Horney, J. D., Osgood, W., & Marshall, I. H. (1995). Criminal careers in the short-term: Intra-individual variability in crime and its relation to local life circumstances. *American Sociological Review, 60,* 655–673.

Leonard, E. B. (1982). *Women, crime, and society: A critique of theoretical criminology.* New York: Longman.

Macdonald, P. T. (1989). Competing theoretical explanations of cocaine use: Differential association versus control theory. *Journal of Contemporary Criminal Justice, 5,* 73–88.

Messerschmidt, J. W. (1993). *Masculinities and crime.* Lanham, MD: Rowman & Littlefield.

Sampson, R. J., & Laub, J. H. (1993). *Crime in the making.* Cambridge, MA: Harvard University Press.

Sutherland, E. H. (1947). *Principles of criminology* (4th ed.). Philadelphia, PA: Lippincott.

DISCUSSION QUESTIONS

1. What do the authors mean by a "general theory" of crime? Do you think either of the theories assessed really fits the bill?
2. The Appendix is presented to show you the kinds of items often used in self-report studies. How honest do you think the newly incarcerated felons were in answering?
3. Why do different variables affect males and females differently?

CRITICAL THEORIES

Marxist, Conflict, and Feminist

At the heart of the theories in this section is social stratification by class and power and how it generates conflict. The theories address how those at the top of the social heap pass laws to maintain their privileged position and how those at the bottom often violate those laws to improve their position. These theories are thus the most "politicized" of all criminological theories. Sanyika Shakur, a.k.a. Kody Scott, came to embrace this politicized view of society as he grew older and was converted to Afrocentric Islam. Shakur was very much a member of the class Karl Marx called the "**lumpenproletariat**," which is the very bottom of the class hierarchy. Many critical theorists would view Shakur's criminality as justifiable rebellion against class and racial exploitation. Shakur wanted all the material rewards of American capitalism, but he perceived that the only way he could get them was through crime. He was a thoroughgoing egoist, but many Marxists would excuse this as a trait in him nourished by capitalism, the "root cause" of crime. From his earliest days, he was on the fringes of a society he plainly disdained. He frequently referred to whites as "Americans" to emphasize his distance from them, and he referred to black cops as "Negroes" to distinguish them from the "New African Man." He called himself a "student of revolutionary science," referred to the 1992 L.A. riots as "rebellion," and advocated a separate black nation in America.

Even at a less politicized level, conflict concepts dominated Shakur's life as he battled the Bloods and other Crip "sets" who had interests at odds with his set. It is easy to imagine his violent act as the outlets of a desperate man struggling against feelings of class and race inferiority. Perhaps he was only able to achieve a sense of power when he held the fate of another human being in his hands. His fragile narcissism often exploded into violent fury whenever he felt himself being "dissed." How much of Shakur's behavior and the behavior of youth gangs in general is explained by the concepts of critical theories? Is violent conflict a justifiable response to class and race inequality in a democratic society, or are there more productive ways to resolve such conflicts?

The Conflict Perspective of Society

Although all sociological theories of crime contain elements of social conflict, consensus theories tend to judge alternative normative systems from the point of view of mainstream values, and they do not call for major restructuring of society. However, theories presented in this section do just that and concentrate on power relationships as explanatory variables, to the exclusion of almost everything else. They view criminal behavior, the law, and the penalties imposed for breaking it as originating in the deep inequalities of power and resources existing in society.

One does not have to be a radical or even a liberal to acknowledge that great inequalities exist and that the wealthy classes have the upper hand in all things. History is replete with examples: Plutarch wrote of the conflicts generated by disparity in wealth in Athens in 594 B.C. (Durrant & Durrant, 1968, p. 55), and U.S. President John Adams wrote that American society in the late 18th century was divided into a small group of rich men and a great mass of poor engaged in a constant class struggle (Adams, 1778/1971, p. 221).

Karl Marx and Revolution

Karl Marx, philosopher, journalist, and revolutionary, is the father of critical sociology. The core of Marxist philosophy is the concept of **class struggle:** "Freeman and slave, patrician and plebian, lord and serf, guildmaster and journeyman, in a word, oppressor and oppressed, stood in constant opposition to one another" (Marx & Engels, 1948, p. 9). The oppressors in Marx's time were the owners of the means of production (the **bourgeoisie**) and the oppressed were the workers (the **proletariat**). The bourgeoisie strive to keep the cost of labor at a minimum, and the proletariat strives to sell its labor at the highest possible price. These opposing goals are the major source of conflict in a capitalist society. The bourgeoisie enjoy the upper hand because capitalist societies typically have large armies of unemployed workers anxious to secure work at any price, thus driving down the cost of labor. According to Marx, these economic and social arrangements—the material conditions of people's lives—determine what they will know, believe, and value, and how they will behave.

Marx and his collaborator, Freidrich Engels, saw crime as a social cancer and made plain their disdain for criminals, calling them "The dangerous class, the social scum, that rotting mass thrown off by the lowest layers of the old society" (1948, p. 22). This "social scum" came from a third class in society—the lumpenproletariat—who would play no decisive role in the expected revolution. It is probably for this reason that Marx and Engels only wrote about crime to illustrate the bitter fruits of capitalism and produced no coherent theory of crime. For Marx and Engels (1965, p. 367) crime was simply the product of unjust and alienating social conditions—"the struggle of the isolated individual against the prevailing conditions."

This became known as the **primitive rebellion hypothesis**, one of the best modern statements of which is Bohm's (2001): "Crime in capitalist societies is often a rational response to the circumstances in which people find themselves" (p. 115).

Willem Bonger: The First Marxist Criminologist

Dutch criminologist Willem Bonger's *Criminality and Economic Conditions* (1905/1969) is the first work devoted to a Marxist analysis of crime. For Bonger, the roots of crime lay in the exploitive and alienating conditions of capitalism, although some individuals are at greater risk for crime than others because people vary in their "innate **social sentiments**"—*altruism* (an active concern for the well-being of others) and its opposite, *egoism* (a concern only for one's own selfish interests). Bonger believed that capitalism generates egoism and blunts altruism because it relies on competition for wealth, profits, status, and jobs, setting person against person and group against group, leaving the losers to their miserable fates. Such a moral climate generates alienation, greed, and crime. Bonger believed that poverty was the major cause of crime, but traced poverty's effects to family structure (broken homes, illegitimacy), poor parental supervision of their children, and "the lack of civilization and education among the poorer classes' (1905/1969, p. 195).

The excerpt from Ian Taylor, Paul Walton, and Jock Young's (1973) important Marxist work on criminology in this section provides us with a brief history of Marxist criminology, concentrating on Willem Bonger. As orthodox Marxists, Taylor, Walton, and Young severely criticize Bonger's emphasis on family structure and the moral deficits of the poor as being anti-Marxist because it takes blame away from the capitalist mode of production as uniquely responsible for crime. Their writings well illustrate the tendency of Marxists to excuse and even romanticize criminal activity as an understandable rebellion against the socioeconomic conditions of capitalism.

Modern Marxist Criminology

Because Marx wrote so little about crime, it is better to characterize modern Marxist criminologists as radicals for whom Marxism serves as a philosophical underpinning. Contrary to Marx, Marxist criminologists have a propensity to excuse criminals. William Chambliss (1976, p. 6), for instance, views some criminal behavior to be "no more than the 'rightful' behavior of persons exploited by the extant economic relationships," and Ian Taylor (1999, p. 151) sees the convict as "an additional victim of the routine operations of a capitalist system—a victim, that is of 'processes of reproduction' of social and racial inequality." David Greenberg (1981, p. 28) even elevated Marx's despised lumpenproletariat to the status of revolutionary leaders: "criminals, rather than the working class, might be the vanguard of the revolution." Many Marxist criminologists also appear to view the class struggle as the *only* source of *all* crime and to view "real" crime as violations of human rights, such as racism, sexism, imperialism, and capitalism, and accuse other criminologists of being parties to class oppression. Tony Platt, for instance, wrote that "it is not too far-fetched to characterize many criminologists as domestic war criminals" (in Siegel, 1986, p. 276).

A second example of modern Marxist criminology is presented in the paper by Barbara Sims in this section. What Sims is attempting to do is to recast institutional anomie theory

(IAT) in more thoroughly Marxist terms. Having shown where IAT "went wrong" in relying on theories other than Marxism, Sims goes on to show how America's "imprisonment binge" of the last two decades is a reflection of a society whose institutions are in disarray. She concludes by saying that we cannot wait for the overthrow of the capitalist system (which she believes will occur in its own good time), but we should agitate for change now within the context of the current system.

Left Realism

Sims's recommendation puts her among Marxists calling themselves **left realists.** Left realists acknowledge that predatory street crime is a *real* source of concern among the working class, who are the primary victims of it, and they have to translate their concern for the poor into practical, *realistic* social policies. This theoretical shift signals a move away from the former singular emphasis on the political economy to embrace the interrelatedness of the offender, the victim, the community, and the state in the causes of crime. It also signals a return to a more orthodox Marxist view of criminals as people whose activities are against the interests of the working class, as well as against the interests of the ruling class. Although unashamedly socialist in orientation, left realists have been criticized by more traditional Marxists, who see their advocacy of solutions to the crime problem within the context of capitalism as a sell out (Bohm, 2001).

Conflict Theory: Max Weber, Power, and Conflict

In common with Marx, German lawyer and sociologist Max Weber (1864–1920) saw society as best characterized by conflict. They differed on three key points, however: (1) while Marx saw cultural ideas as molded by its economic system, Weber saw a culture's economic system being molded by its ideas; (2) whereas Marx emphasized economic conflict between only two social classes, Weber saw conflict arising from multiple sources; (3) Marx envisioned the end of conflict with the destruction of capitalism, while Weber contended that it will always exist, regardless of the social, economic, or political nature of society.

Even though individuals and groups enjoying wealth, prestige, and power have the resources necessary to impose their values and vision for society on others with fewer resources, Weber viewed the various class divisions in society as normal, inevitable, and acceptable, as do many contemporary conflict theorists (Curran & Renzetti, 2001). Weber saw the law as a resource by which the powerful are able to impose their will on others by criminalizing acts that are contrary to their class interests. Because of this, wrote Weber, "criminality exists in all societies and is the result of the political struggle among different groups attempting to promote or enhance their life chances" (in Bartollas, 2005, p. 179).

George Vold produced a version of conflict theory that moved conflict away from an emphasis on value and normative conflicts (as in the Chicago ecological tradition) to include *conflicts of interest.* Vold saw social life as a continual struggle to maintain or improve one's own group's interests—workers against management, race against race, ecologists against land developers, and the young against adult authority—with new interest groups continually forming and disbanded as conflicts arise and are resolved.

Conflicts between youth gangs and adult authorities were of particular concern to Vold, who saw gangs in conflict with the values and interests of just about every other interest group, including those of other gangs. Gangs are examples of *minority power groups*, or groups whose interests are sufficiently on the margins of mainstream society that just about all their activities are criminalized. Vold's theory concentrates entirely on the clash of individuals loyally upholding their differing group interests and is not concerned with crimes unrelated to group conflict (Vold & Bernard, 1986, p. 276).

Vold's thinking is in the Weberian tradition in that he viewed conflict as normal and socially desirable. Conflict is a way of assuring social change, and in the long run, a way of assuring social stability. A society that stifles conflict in the name of order stagnates and has no mechanisms for change short of revolution. Since social change is inevitable, it is preferable that it occur peacefully and incrementally (evolutionary) rather than violently (revolutionary). Even the 19th-century arch-conservative British philosopher Edmund Burk saw that conflict is functional in this regard, writing that "A state without the means of some change is without means of its conservation" (in Walsh & Hemmens, 2000, p. 214).

Situating Conflict Theory in Relation to Marxist and Labeling Theory

All versions of conflict theory share with labeling and Marxist theories the characteristic of being critical of the status quo, although there are differences. Conflict criminology differs from Marxist criminology in that it concentrates on the *processes* of value conflict and lawmaking rather than on the social structural elements underlying those things. It is also relatively silent about how the powerful got to be powerful and makes no value judgments about crime (is it the activities of "social scum" or of "revolutionaries"?); conflict theorists simply analyze the power relationships underlying the act of criminalization.

Conflict theory shares with labeling theory the idea that crime is a social construct with no intrinsic meaning. "Criminal" behavior is normal behavior subject to criminalization and decriminalization depending on the power relationship existing between those who "do it" and those who don't want them to. Conflict and labeling theories differ in one important regard, however. Labeling theory concerns itself with the application of a deviant label to the powerless and the consequences that follow, but is not concerned with the process of how particular labels come to be stigmatized, while that process is of central importance to conflict theorists. There is quite a difference between tagging an *individual* with a criminal label that is already available for use and labeling a previously permissible *act* as criminal (Triplett, 1993, p. 546).

Because Marxist and conflict theories are frequently confused with one another, Table 6.1 summarizes the differences between them on key concepts.

Peacemaking Criminology

Peacemaking criminology is a fairly recent addition to the growing number of theories in our discipline. It is situated squarely in the postmodernist tradition (a tradition that rejects the notion that the scientific view is any better than any other view, and which disparages the claim that any method of understanding can be objective) and has drawn a number of disillusioned former Marxists into its fold. In its peacemaking endeavors it relies heavily on

Table 6.1 Comparing Marxist and Conflict Theory on Major Concepts

Concept	Marxist	Conflict
Origin of conflict	The powerful oppressing the powerless (e.g., the bourgeoisie oppressing the proletariat under capitalism).	It is generated by many factors regardless of the political and economic system.
Nature of conflict	It is socially bad and must and will be eliminated in a socialist system.	It is socially useful and necessary and cannot be eliminated.
Major participants in conflict	The owners of the means of production and the workers are engaged in the only conflict that matters.	Conflict takes place everywhere between all sorts of interest groups.
Social class	Only two classes defined by their relationship to the means of production, the bourgeoisie and proletariat. The aristocracy and the lumpenproletariat are parasite classes that will be eliminated.	There are number of different classes in society defined by their relative wealth, status, and power.
Concept of the law	It is the tool of the ruling class that criminalizes the activities of the workers harmful to its interests and ignores its own socially harmful behavior.	The law favors the powerful, but not any one particular group. The greater the wealth, power, and prestige a group has, the more likely the law will favor it.
Concept of crime	Some view crime as the revolutionary actions of the downtrodden, others view it as the socially harmful acts of "class traitors," and others see it as violations of human rights.	Conflict theorists refuse to pass moral judgment because they view criminal conduct as morally neutral with no intrinsic properties that distinguish it from conforming behavior. Crime doesn't exist until a powerful interest group is able to criminalize the activities of another less powerful group.
Cause of crime	The dehumanizing conditions of capitalism. Capitalism generates egoism and alienates people from themselves and from others.	The distribution of political power that leads to some interest groups being able to criminalize the acts of other interest groups.
Cure for crime	With the overthrow of the capitalist mode of production, the natural goodness of humanity will emerge, and there will be no more criminal behavior.	As long as people have different interests and as long as some groups have more power than others, crime will exist. Since interest and power differentials are part of the human condition, crime will always be with us.

"appreciative relativism," a position that holds that all points of view, including that of criminals, are relative, and all should be appreciated. It is a compassionate and spiritual criminology that has much of its philosophical roots in humanistic religion.

Peacemaking criminology's basic philosophy is similar to the 1960s Hippie adage, "Make love, not war," without the sexual overtones. It shudders at the current "war on crime" metaphor and wants to substitute "peace on crime." The idea of making peace on crime is perhaps best captured by Kay Harris in writing that we "need to reject the idea that those who cause injury or harm to others should suffer severance of the common bonds of respect and concern that binds members of a community. We should relinquish the notion that it is acceptable to try to 'get rid of' another person whether through execution, banishment, or caging away people about whom we do not care" (1991, p. 93). While recognizing that many criminals should be incarcerated, peacemaking criminologists aver that an overemphasis on punishing criminals escalates violence. Marxist-cum-peacemaker Richard Quinney has called the American criminal justice system the moral equivalent of war and notes that war naturally invites resistance by those it is waged against. He further adds that when society resists criminal victimization, it "must be in compassion and love, not in terms of the violence that is being resisted" (Vold, Bernard, & Snipes, 1998, p. 274).

In place of imprisoning offenders, peacemaking criminologists advocate **restorative justice,** which is basically a system of mediation and conflict resolution. Restorative justice is "every action that is primarily oriented toward justice by repairing the harm that has been caused by the crime" and "usually means a face-to-face confrontation between victim and perpetrator, where a mutually agreeable restorative solution is proposed and agreed upon" (Champion, 2005, p. 154). Restorative justice has been applauded because it humanizes justice by bringing victim and offender together to negotiate a mutually satisfying way to correct the wrong done. Although developed for juveniles and primarily confined to them, restorative justice has also been applied to nonviolent adult offenders in a number of countries as well as the United States. The belief behind restorative justice is that, to the extent that both victim and victimizer come to see that justice is attained when a violation of one person by another is made right by the violator, the violator will have taken a step to reformation and the community will be a safer place in which to live.

▲ **Photo 6.1** The friendly presence of police at a large ethnic festival demonstrates the peacekeeping approach to crime prevention.

Feminist Criminology

Feminist criminology is firmly in the critical/conflict camp of criminology. Feminists see women as being doubly oppressed by gender inequality (their social position in a sexist culture) and by class inequality (their economic position in a capitalist society). Some feminists view the answer to women's oppression to be the overthrow of the two-headed monster—capitalism and patriarchy. In the meantime, they want to be able to interpret female crime from a feminist perspective. With this in mind, feminist criminology wrestles with two major concerns: "Do traditional male-centered theories of crime apply to women?" This is known as the **generalizability problem**. The second concern is "What explains the universal fact that women are far less likely than men to involve themselves in criminal activity?" This is known as the **gender-ratio problem** (Daly & Chesney-Lind, 2002, pp. 269–270).

With regard to the generalizability problem, many feminist criminologists have concluded that male-centered theories have limited applicability to females (Leonard, 1995), and that despite the best efforts of many there is still no female-specific theory of crime. Some feminist scholars believe that no such theory is possible and that they must be content to focus on crime-specific "mini-theories" (Daly & Chesney-Lind, 2002, p. 268). Nevertheless, most female offenders are found in the same places as their male counterparts; that is, among single-parent families located in poor, socially disorganized neighborhoods. Male and female crime rates march in lockstep across different nations and across communities in the same nation (as male rates increase so do female rates, and vice versa), indicating that females are broadly responsive to the same environmental conditions as males (Campbell, 1999). Given this evidence, Daly and Chesney-Lind ask, "Why do similar processes produce a distinctive, gender-based [male] structure to crime and delinquency?" (1996, p. 349).

This question leads us to the gender-ratio problem. Two early attempts to answer the question were Freda Adler's (1975) **masculinization hypothesis** and Rita Simon's (1975) **emancipation hypothesis**. Both hypotheses accepted the traditional sociological notion that gender differences are mostly products of differential socialization; that is, men are socialized to be assertive, ambitious, and dominant, and women are socialized to be nurturing, passive, and home and family oriented. In Adler's view, as females increasingly adopt "male" roles they will increasingly masculinize their attitudes and behavior, and will thus become as crime-prone as men. Simon's view was that increased participation in the workforce affords women greater opportunities to commit job-related crime, and that there was no reason for them to first undergo Adler's masculinization. Neither hypothesis proved useful in explaining the gender crime ratio. Female crime rates have increased over the past 30 years, but as a proportion of total arrests they have not varied by more than 5 percentage points, and the male/female gap has remained essentially unchanged (Campbell, 1999).

It has been proposed that the gender ratio exists because the genders differ in exposure to delinquent peers and males are more influenced by delinquent peers than females, and because of females' greater inhibitory morality (Mears, Ploeger, & Warr, 1998). This has been called nothing more than claiming that "boys will be boys," and "girls will be girls," because it begs the questions of why males are more "exposed" and more "influenced" than females and why females have a stronger sense of morality (Walsh, 2002, p. 207). One of the standard answers to these questions is that girls are more closely supervised than boys, yet controlling for supervision level results in the same large gender gap in offending (Gottfredson & Hirschi, 1990), and a meta-analysis of 172 studies found a slight tendency for girls to be *less* strictly supervised than boys (Lytton & Romney, 1991). Many others studies have shown that large sex differences in antisocial behavior exist regardless of what control variables are introduced (Chesney-Lind & Shelden, 1992). As Dianna Fishbein (1992) has summed up the issue: "Cross cultural studies do not support the prominent role of structural and cultural influences of gender-specific crime rates as the type and extent of male versus female crime remains consistent across cultures" (p. 100).

Others calling themselves "radical feminists" have argued that because the magnitude of the gender gap varies across time and space and yet still remains constantly wide at all times and in all places, biological factors *must* play a large part. If social factors accounted for gender differences, there should be a set of cultural conditions under which crime rates would be equal for both sexes (or even higher for females), but no such conditions have ever been found. Sex differences in dominance and aggression are seen in all human cultures from the earliest days of life and are underscored during the teen years. Furthermore, these differences are observed in all primate and most mammalian species, and no one would evoke socialization as an explanation in these instances (Archer, 1996; Geary, 1998).

▲ **Photo 6.2** Did serial killer nurse Kristen Gilbert poison her patients as acts of mercy killing or from other motivations? What might the radical feminist perspective have to say on this?

Biologically informed feminists embrace evidence from the neurosciences indicating that hormones organize the brain in male or female directions during sensitive prenatal periods (Amateau & McCarthy, 2004), and that this process organizes male brains in such a way that males become more vulnerable to the various traits associated with antisocial behavior via the regulation of brain chemistry (Ellis, 2003). According to Doreen Kimura (1992), males and females come into this world with "differently wired brains," and these brain differences "make it almost impossible to evaluate the effects of experience [the socialization process] independent of physiological predisposition" (p. 119). The major biological factor said to underlie gender differences in aggression, violence, and general antisocial behavior is testosterone (Kanazawa, 2003). Note that these theorists are not saying that testosterone is a major or even minor cause of crime and general mayhem, only that it is the major factor that underlies gender differences in crime and general mayhem.

Anne Campbell's Staying Alive Hypothesis

Why do "differently wired brains" exist in the first place? Sex differences do not arise without there being an adaptive evolutionary reason behind them. Biologists note that sex differences in aggression and dominance seeking are related to parental investment (time and resources devoted to parental care), not biological sex per se. It is parental investment that provokes evolutionary pressures for the selection of the mechanisms that underlie these behaviors. In some bird and fish species, males contribute greater parental investment (e.g., incubating the eggs and feeding the young) than females, and females take more risks, are more promiscuous and aggressive in courtship, have higher testosterone levels, and engage in violent competition for mates (Barash & Lipton, 2001; Betzig, 1999). In these species, sex-related characteristics are the opposite of those found in species in which females assume all or most of the burden of parenting (the vast majority of species).

Anne Campbell (1999) has attempted to account for the gender-ratio problem using the logic of evolutionary theory in her **staying alive hypothesis**. Campbell argues that because the *obligatory* parental investment of females is greater than that of males, and because of the infant's greater dependence on the mother, a mother's presence is more critical to offspring survival than is a father's. She notes that offspring survival is more critical to female reproductive success (the passing of one's genes to subsequent generations—the ultimate "goal" of all life forms) than to male reproductive success. Because of the limits placed on female reproductive success by long periods of gestation and lactation, females have more investment tied up in children they already have than do males (male reproductive success is only limited by access to willing females).

Campbell (1999) argues that because offspring survival is so enormously important to their reproductive success, females have evolved a propensity to avoid engaging in behaviors that pose survival risks. The practice of keeping nursing children in close proximity in ancestral environments posed an elevated risk of injuring the child as well as herself if the mother placed herself in risky situations. Thus, it became adaptive for females to experience many different situations as fearful. There are no sex differences in fearfulness *unless* a situation contains a significant risk of physical injury, and it is this fear that accounts for the greater tendency of females to avoid or remove themselves from potentially violent situations and to employ low-risk strategies in competition and dispute resolution relative to males. Females do engage in competition with one another for resources and mates, of course, but it is rarely violent competition. Most of it is decidedly low key, low risk, and chronic as opposed to high key, high risk, and acute male competition.

Campbell (1999) shows that when females engage in crime, they almost always do so for instrumental reasons, and their crimes rarely involve risk of physical injury. There is no evidence, for instance, that female robbers crave the additional payoffs of dominance that male robbers do, or seek reputations as "hard asses." Any woman with a reputation as a "hard ass" would not be very desirable as a mate. Thus, Campbell notes that while women do aggress and do steal, "they rarely do both at the same time because the equation of resources and status reflects a particularly masculine logic" (p. 210).

Meda Chesney-Lind's article in this section is typical of all critical theories in making a call for more activism among feminist criminologists. She asserts that certain feminist successes have created a backlash against females in general that is manifested in anti-feminist and racist agendas in the criminal justice system. Women have been demonized in the media with lurid stories of "bad girl" violence, which in turn has resulted in ever-larger numbers of women, especially women of color, being incarcerated. Chesney-Lind addresses all the major themes of this section (patriarchy and the generalizability and gender-ratio problems). She maintains that much of the so-called increase in female violence is the result of such things as zero tolerance in schools and of policies that mandate arrest for the most minor of domestic assaults (see the Steffensmeier et al. article in Section 2 for a similar conclusion).

Evaluation of Critical Theories

It is often said that Marxist theory has very little that is unique to add to criminology theory: "When Marxist theorists offer explanations of crime that go beyond simply attributing the causes of all crime to capitalism, they rely on concepts taken from the same 'traditional' criminological theories of which they have been so critical" (Akers, 1994, p. 167).

Marxists also tend to be hostile to empiricism, preferring historical, descriptive, and illustrative research. The tendency to romanticize criminals as revolutionaries has long been a major criticism, although Marxist criminologists are less likely to do this today.

Can Marxists claim empirical support for their contention that capitalism causes crime and socialism "cures" it? That capitalism is *associated* with higher crime rates than socialism is uncontested, but the question is whether the Marxist *interpretation* of this fact is correct. Analyses of previously secret crime figures from the former Soviet Union reveal that crime rates there fluctuated over the years almost as much as they did in capitalist societies, and crime started to increase significantly there even before the implementation of the liberalization policies of the late 1980s (Butler, 1992). The lower crime rates in socialist societies probably have more to do with repressive law enforcement practices than with any altruistic qualities intrinsic to socialism.

Marxist criminology also appears to be in a time warp in its implicit assumption that the conditions prevailing in Marx's time still exist today in advanced capitalist societies. People from all over the world have risked life and limb to get into capitalist countries because those countries are where human rights are most respected and human needs most readily accessible. Even before the collapse of the Soviet system, left realists realized that utopia would be a long time coming and that it would be more realistic to work within the system to achieve reforms. Left realism is thus more the reform-minded "practical" wing of Marxism than a theory of crime that has anything special to offer criminology.

Conflict theory is challenging and refreshing because its efforts to identify power relationships in society have applications that go beyond criminology. But there are problems with it as a theory of criminal behavior. It has even been said that "Conflict theory does not attempt to explain crime; it simply identifies social conflict as a basic fact of life and a source of discriminatory treatment" (Adler, Mueller, & Laufer, 2001, p. 223).

Conflict theory's assumption that crime is just a "social construct" without any intrinsic properties minimizes the suffering of those who have been assaulted, raped, robbed, and otherwise victimized. These acts *are* intrinsically bad (mala-in-se) and are not arbitrarily criminalized because they threaten the privileged world of the powerful few. The is wide agreement among people of various classes in the United States about what crimes are—laws exist to protect everyone, not just "the elite" (Walsh & Ellis, 2007).

Peacemaking criminology urges us to make peace on crime, but what does such advice actually mean? As a number of commentators have pointed out, "being nice" is not enough to stop others from hurting us (Lanier & Henry, 1998). It is undoubtedly true that the reduction of human suffering and achieving a truly just world will reduce crime, as advocates of this position contend, but they offer us no notion of how this can be achieved beyond counseling that we should appreciate criminals' points of view and not be so punitive.

Despite the best efforts of many feminist criminologists, there is still no gender-specific theory of crime, and some have even pointed out that no such theory is possible (Daly & Chesney-Lind, 2002, p. 268). Feminist theorists have thus been content to focus on descriptive studies or on crime-specific "mini-theories." When all is said and done, maleness is without doubt the best single predictor of criminal behavior, which leaves feminist theorists without much left to explain in *specific* female terms about female offending.

Campbell's staying alive/high-fear hypothesis is about why females commit so little crime, not why some females commit it. Because of its biological underpinnings, it may not be acceptable to many traditional feminists, although only 4 of the 27 commentators on her target article argued that strictly social theories better accounted for gender differences in crime. Campbell's hypothesis must be augmented with cultural factors, though, because we sometimes do see females committing more serious crimes than males. For many years, African American females have had higher homicide rates than white males. This does not negate the basic gender-ratio argument because *within* the African American community the gender ratio is generally higher than it is in the white community; that is, there is a bigger gap between the homicide rates of black males and females than there is between the homicide rates of white males and females (Laub & McDermott, 1985).

Policy and Prevention: Implications of Critical Theories

The policy implications of Marxist theory are straightforward: Overthrow capitalism and crime will be reduced. Marxist criminologists realize today that the abolition of capitalism is unrealistic, a fact underlined for them by the collapse of Marxism across Eastern Europe. They also realize that emphasis on a single cause of crime (the class struggle) and romanticizing criminals is equally unrealistic. Rather than throw out their entire ideological agenda, left realists now temper their views while at the same maintaining their critical stance toward the "system." Policy recommendations made by left realists have many things in common with those made by ecology, anomie, and routine activities theorists. Community activities, neighborhood watches, community policing, dispute resolution centers, and target hardening are among the policies suggested.

Because crime is viewed as the result of conflict between interest groups with power and wealth differences, and since conflict theorists view conflict and the existence of social classes as normal, it is difficult to recommend policies *specifically* derived from conflict theory. We might logically conclude from this view of class and conflict that if these things are normal and perhaps beneficial, then so is crime, in some sense. If we want to reduce crime, we should equalize the distribution of power, wealth, and status, thus reducing the ability of any one group to dictate what is criminalized. Generally speaking, conflict theorists favor programs such as minimum wage laws, sharply progressive taxation, a government-controlled comprehensive health care system, maternal leave, and a national policy of family support as a way of reducing crime (Currie, 1989).

The policy recommendations of feminist theory range from the liberal's affirmative action programs to the Marxist's revolutionary overthrow of capitalism. The former has been relatively successful in moving women into what had formerly been "male" occupations, and the latter is hardly likely to occur. There are all sorts of other recommendations in between, the major one being the reform of our patriarchal society. Other recommendations include the more equal (less paternalistic) treatment of girls and boys by juvenile authorities, increased educational and occupational choices for women so that those in abusive relationships can leave them, more day care centers, and so forth. Feminist theory suggests that gender sensitivity education in the schools and workplaces may lead men to abandon many of their embedded sexist ideas pertaining to the relationship between the sexes.

Summary

- Critical criminology is a generic term encompassing many different theoretical positions united by the common view that society is best characterized by conflict and power relations rather than by value consensus.

- Marxist criminologists follow the theoretical trail of Karl Marx, who posited the existence of two conflicting classes in society, the bourgeoisie and the proletariat. While some modern Marxists tend to romanticize criminals as heroic revolutionaries, Marx considered them "social scum" who preyed upon the working class.

- Willem Bonger is credited with being the first Marxist criminologist. He was concerned with two opposite "social sentiments": altruism and egoism. The sentiment of altruism is killed in a capitalist social system because it generates competition for wealth, status, and jobs. Thus, capitalism produces egoism, which leads to criminal behavior on the part of both the poor and the rich.

- Marxists tend to view capitalism as the only cause of crime, and they insist that class and class values are generated by the material conditions of social life. Because only the material conditions of life really matter, the only way to make any serious impact on crime is to eliminate the capitalist mode of production and institute a Marxist social order. Left realists realize that such a radical transformation is highly unlikely in modern times, and although they maintain a critical stance toward the system, they work within it in an effort to influence social policy.

- Conflict theorists share some sentiments with Marxists, but view conflict in pluralistic terms and as intrinsic to society, not something that can be eliminated. Crime is the result of the ability of powerful interest groups to criminalize the behavior of other less powerful interest groups when that behavior is contrary to their interests.

- Conflict criminological research tends to focus on the differential treatment by the criminal justice system of individuals who are members of less powerful groups such as minorities, women, and working-class whites.

- Peacemaking criminology is based on religious principles more than empirical science. It wants to make peace on crime, counsels us that we should appreciate the criminals' point of view, and wants us to be less punitive.

- Feminist criminology focuses on trying to understand female offending from the feminist perspective, which contends that women are faced with special disabilities living in an oppressive sexist society.

- The two big issues in feminist criminology are the *generalizability* problem (do traditional theories of crime explain female as well as male offending?) and the *gender-ratio* problem (what accounts for the huge gap in offending between males and females?).

- Early attempts to explain female crime from the feminist tradition emphasized the masculinization of female attitudes as they increasingly adopted "male" roles, or simply that as women move into the workforce in greater numbers they found greater opportunities to commit job-related crimes. Many feminists rejected both positions, pointing out that such theorizing provided ammunition for those who opposed the women's movement and that regardless of any increase in female offending, the male/female gap remains as wide as ever.

- The size and universality of the gender gap suggests to some that the most logical explanation for it must lie in some fundamental differences between the sexes rather than socialization, such as neurological and hormonal differences.

- Anne Campbell's staying alive hypothesis attempts to explain the gender-ratio problem in terms of differential evolutionary selection pressures between the sexes. Female survival was more crucial to their reproductive success than male survival was to theirs. Natural selection exerted pressure for females to be more fearful of dangerous situations, whereas for males the seeking of dominance and status, which aided their reproductive success, often placed them in such situations.

EXERCISES AND DISCUSSION QUESTIONS

1. Do you think that the "material conditions of life" largely determine what we will know, believe, and value and how we will behave?
2. Do you believe that social conflict is inevitable? In what ways is conflict a good thing?
3. Do we really need a feminist criminology, or do the traditional theories suffice to explain both male and female criminality?

USEFUL WEB SITES

Conflict Theory. http://web.grinnell.edu/courses/soc/s00/soc111-01/IntroTheories/Conflict.html.

Feminist Criminology. http://faculty.ncwc.edu/toconnor/301/301lect14.htm.

Karl Marx. http://www.historyguide.org/intellect/marx.html.

Peacemaking Criminology. http://www.greggbarak.com/whats_new_2.html.

Postmodern Criminology. http://uwacadweb.uwyo.edu/RED_FEATHER/lectures/051techcrm7.htm.

CHAPTER GLOSSARY

Bourgeoisie: The wealthy owners of the means of production.

Class struggle: The core concept of Marxist philosophy. It asserts that all history is one of class struggles between the "haves" (the oppressors) and "have not's" (the oppressed).

Emancipation hypothesis: The assumption that as women become freer to move into male occupations they will find and take advantage of more criminal opportunities.

Generalizability problem: In feminist criminology, the question about whether traditional male-centered theories of crime apply to women.

Gender-ratio problem: In feminist criminology, the question of what explains the universal fact that women are far less likely than men to involve themselves in criminal activity.

Left realists: An approach to crime that maintains that although inequality is a cause of crime, the best solution is to work within the system to prevent and control crime.

Lumpenproletariat: The lower classes; the criminal class.

Masculinization hypothesis: The assumption that as females adopt "male" roles and masculinize their attitudes and behavior they will commit as much crime as men.

Peacemaking criminology: A postmodernist theory that relies heavily on "appreciative relativism," a position that holds that all points of view, including that of criminals, are relative, and all should be appreciated.

Primitive rebellion hypothesis: The Marxist idea that crime is the product of unjust and alienating social conditions that deny productive labor to masses of unemployed.

Proletariat: The working class.

Restorative justice: A system of mediation and conflict resolution that brings victim and offender together to negotiate a mutually satisfying way to correct the wrong done.

Social sentiments: In Willem Bonger's theory, they are *altruism* (active concern for the well-being of others) and *egoism* (concern only for one's own selfish interests).

Staying alive hypothesis: The idea that women are less criminal than men because they have evolved a propensity to experience more situations as fearful than men do. This fear keeps women and their children away from danger and thus aids their reproductive success.

READING

Marx, Engels, and Bonger on Crime and Social Control

Ian Taylor, Paul Walton, and Jock Young

In this excerpt from their book *The New Criminology: For a Social Theory of Deviance*, Taylor, Walton, and Young make the typical Marxist argument that crime must be placed in the context of a capitalist political economy. They explicitly deny that crime, or the laws forbidding it, can be understood without an examination of power relationships within a capitalist society. Taylor, Walton, and Young provide us with historical background of Marxist theory through the work of Willem Bonger, and also a brief overview of the more contemporary contributions of conflict theorists. Although Bonger viewed capitalism as criminogenic, note how Taylor, Walton, and Young take him to task for deviating from the "official" Marxist line by departing from purely structural explanations for crime and delving into individual differences.

In part, Marxism stands or falls on the basis of certain assumptions it makes about the nature of man. Where other social theories operate with implicit assumptions about man's nature, Marx made his starting point a quite explicit philosophical anthropology of man. In *The Economic and Philosophical Manuscripts* of 1844, Marx is concerned to show that man is distinct in a crucial and precise way from the members of the animal world.

Man is a *species-being* not only in the sense that he makes the community (his own as well as those of other things) his object both practically and theoretically, but also (and this is simply another expression of the same thing) in the sense that he treats himself as the present living species, as a *universal* and consequently free being.

The bulk of Marx's later work is concerned with the demonstration of the ways in which man's social nature and consciousness have been distorted, imprisoned or diverted by the social arrangements developed over time. These social arrangements are the product of man's struggle to master the conditions of scarcity and material underdevelopment. These social arrangements, developed as a response to man's domination by poverty, imprison man tightly in social relationships of an exploitative nature and alienate men from men, and thus from the objects of their labour. Man is struggling to be free, but cannot realize freedom (or himself as a fully-conscious, "sensuous" species-being) until such time as he is free of the exploitative relationships which are outmoded and unnecessary.

The continuing debates over Marxism in sociology and philosophy, (as well as within socialist movements) in the twentieth century, therefore, have had to do with problems of consciousness, contradictions and social change. That is, the image of society offered out by classical Marxism is one of competing social groups, each with a distinct set of interests and cultural world views, caught within a network of essentially temporary (or historically specific) social arrangements, which in their turn are more or less likely to be revolutionized in periods of crisis. Capitalism, as a set of social relationships, is conceptualized as the most highly developed form of social exploitation, within which are sown the seeds of man's leap to a liberating consciousness. Capitalism "contains the seeds of its own destruction" not only in the sense that it creates the technology whereby physical and material need may be satisfied but also because it prevents a more sophisticated set of social relationships developing alongside such productive forces.

A full-blown Marxist theory of deviance, or at least a theory of deviance deriving from a Marxism so described, would be concerned to develop explanations of the ways in which particular historical periods, characterized by particular sets of social relationships and means of production, give rise to attempts by the economically and politically powerful to order society in particular ways. It would ask with greater emphasis the question that Howard Becker poses (and does not face), namely, who makes the rules, and why? A Marxist theory would attempt, that is, to locate the defining agencies, not only in some general market structure, but quite specifically in their relationship to the overweening structure of material production and the division of labour. Moreover, to be a satisfactory explanation, a Marxist theory would proceed with a notion of man which would distinguish it quite clearly from, classical, positivist, or interactionist "images" of man. It would assume, that is, a degree of consciousness, bound up with men's location in a social structure of production, exchange and domination, which of itself would influence the ways in which men defined as criminal or deviant would attempt to live with their outsider's status. That is, men's reaction to labeling by the powerful would not be seen to be simply a cultural problem—a problem of reacting to a legal status or a social stigma: it would necessarily be seen to be bound up with men's degree of consciousness of domination and subordination in a wider structure of power relationships operating in particular types of economic context. One consequence of such an approach—which, it must be stated, has been conspicuous for its absence in deviancy theory—would be the possibility of building links between the insights of interactionist theory, and other approaches sensitive to men's subjective world, and the theories of social structure implicit in orthodox Marxism. More crucially, such a linkage would enable us to escape from the strait-jacket of an economic determinism and the relativism of some subjectivist approaches to a theory of contradiction in a social structure which recognizes in "deviance" the acts of men in the process of actively making, rather than passively taking, the external world. It might enable us to sustain what has until now been a polemical assertion, made (in the main) by anarchists and deviants themselves, that much deviance is in itself a political act, and that, in this sense, deviance is a property of the act rather than a spurious label applied to the amoral or the careless by agencies of political and social control.

Willem Bonger and Formal Marxism

In the study of crime and deviance, the work of Willem Bonger (1876–1940) . . . has assumed the mantle of the Marxist orthodoxy—if only because (with the exception of untranslated writers inside the Soviet bloc) no other self-proclaimed Marxist has devoted time to a full-scale study of the area. Bonger's criminology is an attempt to utilize some of the formal concepts of Marxism in the understanding of the crime-rates of European capitalism in the late nineteenth and early twentieth centuries. Importantly, however, Bonger's efforts appear, for us, not so much the application of a fully-fledged Marxist theory as they are a recitation of a "Marxist catechism" in an area which Marx had left largely untouched—a recitation prompted by the growth not of the theory itself, but by the growth of a sociological pragmatism. Bonger must, therefore, be evaluated in his own terms, in terms of the competence of his extension of the formal concepts of Marxism to the subject-matter, rather than in terms of any claim that might be made for him as *the* Marxist criminologist.

In at least two respects, Bonger's analysis of crime differs in substance from that of Marx. On the one hand, Bonger is clearly very much more seriously concerned than Marx with the causal chain linking crime with the precipitating economic and, social conditions. On the other, he does not confine his explanations to working-class crime, extending his discussions to the criminal activity of the industrial bourgeoisie as defined by the criminal laws of his time. Whilst differing from Marx in these respects, however, Bonger is at one with his mentor in attributing the activity itself to demoralized individuals, products of a dominant capitalism.

Indeed, in both Marx and Bonger, one is aware of a curious contradiction between the "image of man" advanced as the anthropological underpinning of "orthodox" Marxism and the questions asked about men who deviate. . . . The criminal thought, which runs through the bulk of Bonger's analysis of crime, is seen as the product of the tendency in industrial capitalism to create "egoism" rather than "altruism" in the structure of social life. It is apparent that the notion performs two different notions for Bonger, in that he is able to argue, at different points, that, first, "the criminal thought" is engendered by the conditions of misery forced on sections of the working class under capitalism and that, second, it is also the product of the greed encouraged when capitalism thrives. In other words, as an intermediary notion, it enabled Bonger to circumvent the knotty problem of the relationship between general economic conditions and the propensity to economic crime.

Now, whilst the ambiguity in the notion may help Bonger's analysis, it does not stem directly from his awareness of dual problems. For Bonger, it does appear as an autonomous psychic and behavioural quality which is to be deplored and feared; "the criminal thought," and its associated "egoism" are products of the brutishness of capitalism, but at the same time they do appear to "take over" individuals and independently direct their actions.

The Marxist perspective, of course, has always emphasized the impact that the dominant mode of production has had on social relationships in the wider society, and, in particular, has spelt out the ways in which a capitalist means of production will tend to "individuate" the nature of social life. But to understand that "egoism" and "individuation" are products of particular sets of social arrangements is to understand that egoism and individuation have no force or influence independently of their social context. For Bonger, the "criminal thought"—albeit a product of the egoistic structure of capitalism—assumes an independent status as an intrinsic and behavioural quality of certain (criminal) individuals. It is enormously paradoxical that a writer who lays claim to be writing as a sociologist and a Marxist should begin his analysis with an assumed individual quality (which he deplores) and proceed only later to the social conditions and relationships sustaining and obstructing the acting-out quality.

In the first place, the emphasis in Bonger on "the criminal thought" as an independent factor for analysis is equivalent to the biological, physiological, and sociological (or environmental) factors accorded an independent and causative place in the writings of the positivist theorists of crime. The limitations of this approach have been pointed out, amongst others, by Austin T. Turk:

> Students of crime have been preoccupied with the search for an explanation of distinguishing characteristics of criminality, almost universally assuming that the implied task is to develop scientific explanations of the behaviour of persons who deviate from "legal norms." The quest has not been very successful . . . the cumulative impact of efforts to specify and explain differences between "criminal" and "non-criminal" cultural and behaviour patterns is to force serious consideration of the possibility that there may be *no* significant differences between the overwhelming majority of legally identified criminals and the *relevant* general population, i.e., that population whose concerns and expectations impinge directly and routinely upon the individuals so identified.

More succinctly: "the working assumption has been that *crime* and *non-crime* are classes of behaviour instead of simply labels associated with the processes by which individuals come to occupy ascribed . . . *statuses* of criminal and non-criminal."

It is a comment on the nature of Bonger's Marxism that the actor is accorded such an idealistic independence; when to have started with a model of a society within which there are conflicting interests and a differential distribution of power would have revealed the utility of the criminal law and the "criminal" label (with a legitimating ideology derived from academia) to the powerful elites of capitalist society. In fact, of course, a criminology which proceeds in recognition of competing social interests has two interrelated tasks of explanation. Certainly it has the task of explaining the causes for an individual's involvement in "criminal" behaviour: but, prior to that, it has the task of explaining the derivation of the "criminal" label (whose content, function and applicability we have argued will vary across time, across cultures, and internally within a social structure).

One cannot entirely avoid the conclusion that Bonger's analysis, irrespective of the extent to which it is guided by a reading and acceptance of Marxist precepts, is motivated (and confused) by a fear of those with "criminal thoughts." For Bonger, "criminal thought" is by and large a product of the lack of moral training in the population. Moral training has been denied the proletariat, in particular, because it is not the essential training for work in an industrializing society. The spread of "moral training" is the antidote to "criminal thoughts," but, since such an education is unlikely under the brutish capitalism of the imperialist period, capitalism—or more precisely, the economic conditions (of inequality and accumulation)—are indeed a cause of crime.

Insofar as Bonger displays any concern for the determinant nature of social relationships of production, he does so in order to illustrate the tendencies of different social arrangements to encourage "criminal thoughts" in the population at large. As against the ameliorist school, which saw an inevitable advance of man from conditions of primitive and brutish living to societies in which altruistic relationships would predominate, Bonger, in fundamental agreement with the value placed on altruism and liberalism, identified the advent of capitalism with the break in the process of civilizing social relationships. . . . Bonger comments: "The fact that the duty of altruism is so much insisted upon is the most convincing proof that it is not generally practised."

The demise of egoism, and the creation of social conditions favourable to the "criminal thought" parallels, for Bonger, the development of social arrangements of production as described by Marx. . . . Under capitalism, the transformation of work from its value for use to its value for exchange (as fully described by Marx) is responsible for the "cupidity and ambition," the lack of sensitivity between men, and the declining influence of men's ambitions on the actions of their fellows. . . . Capitalism, in short, "has developed egoism at the expense of altruism."

"Egoism" constitutes a favourable climate for the commission of criminal acts, and this, for Bonger, is an indication that an environment in which men's social instincts are encouraged has been replaced by one which confers legitimacy on asocial or "immoral" acts of deviance. The commission of these acts, as Bonger explicitly states in *Introduction to Criminology,* has a demoralizing effect on the whole of the body politic.

Bonger's substantive analysis of types of crime, covering a range of "economic crimes," "sexual crimes," "crimes from vengeance and other motives," "political crimes" and "pathological crimes," is taken up with a demonstration of the ways in which these crimes are causatively linked with an environment encouraging egoistic action. Even involvement of persons born with "psychic defects" in criminal activity can be explained in terms of these enabling conditions:

> These persons adapt themselves to their environment only with difficulty . . . have a smaller chance than others to succeed in our present society, where the fundamental principle is the warfare of all against all. Hence they are more likely to seek for means that others do not employ (prostitution, for example).

The whole of Bonger's analysis, however much it is altered or qualified at particular points in his discussion, rests on the environmental determinism of his "general considerations." In a social structure encouraging of egoism, the obstacles and deterrents to the emergence of the presumably ever-present "criminal thought" are weakened and/or removed; whereas, for example, under primitive communism, the communality was constructed around, and dependent upon, an interpersonal altruism. Capitalism is responsible for the free play granted to the pathological will, the "criminal thought" possessed by certain individuals.

The bulk of Bonger's work, indeed, so far from being an example of dialectical procedure, is a kind of positivism in itself, or at least an eclecticism reminiscent of "inter-disciplinary" positivism. Where the general theory appears not to encompass all the facts (facts produced by positivist endeavor), mediations of various kinds are introduced. In Bonger, it is possible to find examples of the elements of anomie theory, differential opportunity theory and, at times, the frameworks of structural-functionalism (much of it well in advance of its time). In his discussion of economic crime, for example, Bonger approached a Mertonian stance on larceny:

> Modern industry manufactures enormous quantities of goods without the outlet for them being known. The desire to buy must, then, be excited in the public. Beautiful displays, dazzling illuminations, and many other means are used to attain the desired end. The perfection of this system is reached in the great modern retail store, where persons may enter freely, and see and handle everything—where, in short, the public is drawn as a moth to a flame. The result of these tactics is that the cupidity of the crowd is highly excited.

And Bonger is not unaware of the general, or the more limited, theories of criminality and deviance produced by the classical thinkers of his time and earlier. Where appropriate, Bonger attempts to incorporate elements of these competing theorists, though always in a way which subordinates their positions to his own "general considerations." On Gabriel Tarde's "law of imitation," for example, which purports to explain criminality as a function of association with "criminal types," Bonger writes:

> In our present society, with its pronounced egoistic tendencies, imitation strengthens these, as it would strengthen the altruistic tendencies produced by another form of society. . . . It is only as a consequence of the predominance of egoism in our present society that the error is made of supposing the effect of imitation to be necessarily evil.

Our concern here is not to dispute particular arguments in Bonger for their own sakes, but rather to point to the way in which a single-factor environmentalism is given predominance, with secondary considerations derived from the body of existing literature being introduced eclectically. That is, Bonger's method, though resting on an environmentalism explicitly derived from Marx, appears in the final analysis as a method reminiscent of the eclectism practised by positivist sociologists operating with formal concepts lacking a grounding in history and structure.

This eclectic approach is accompanied by a crudely statistical technique of verification and elaboration. We are presented, amongst other things, with statistical demonstrations of the relationship between levels of educational attainment and violent crime, declines in business and "bourgeois" crime (fraud, etc.), degrees of poverty and involvement in sexual crime (especially prostitution), crimes of "vengeance" and the season of the year and many more. Consistently, the objective is to demonstrate the underlying motivation as being bound up with an egoism induced and sustained by the environment of capitalism. . . . And, lest we should think that egoism is directly a product of poverty and subordination, as opposed to being a central element of a general moral climate, Bonger is able to offer explanations of crime among the bourgeoisie. These crimes he sees to be motivated by need, in cases of business decline and collapse, or by cupidity. In the latter case, "what [men] get by honest business is not enough for them, they wish to become richer." In either case, Bonger's case is contingent on the moral climate engendered by the economic system:

> It is only under special circumstances that this desire for wealth arises, and . . . it is unknown under others. It will be necessary only to point out that although cupidity is a strong motive with all classes of our present society, it is especially so among the bourgeoisie, as a consequence of their position in the economic life.

Now, Bonger's formal Marxism does enable him to make an insightful series of comments about the nature of the deprivations experienced under capitalism. Judged in Bonger's own terms—that is, in terms of the social positivism of his time—his work surpasses much that was, and is, available. Notably, Bonger's discussion of the effects of the subordination of women (and its contribution to the aetiology of female criminality) and of "militarism" (in sustaining an egoistic and competitive moral climate) seem far ahead of their time.

Writing of the criminality of women, for example, Bonger asserts that:

> The great power of a man over his wife, as a consequence of his economic preponderance, may equally be a demoralizing cause. It is certain that there will always be abuse of power on the part of a number of those whom social circumstances have clothed with a certain authority. How many women there are now who have to endure the coarseness and bad treatment of their husbands, but would not hesitate to leave them if their economic dependence and the law did not prevent.

The contemporary ring of these comments is paralleled in Bonger's comments, made, it should be remembered, at the time when the "Marxist" parties of Europe found their members rushing to the "national defence" in the "Great War". . . . Thus, whilst much of Bonger's formal Marxism appears as a form of abstracted and eclectic positivism when viewed across its canvas, he still derives a considerable benefit and understanding from the Marxist perspective in his sensitivity to the demoralizing and destructive consequences of the forms of domination characteristic of a capitalist society. Paradoxically, however, this sensitivity does not extend to an understanding of the nature of domination and social control in defining and delineating the field of interest itself, namely what passes for crime and deviance in societies where "law is the law determined by powerful interests and classes in the population at large. . . ."

Bonger asserts that "there are instances where an action stamped as criminal is not felt to be immoral by anybody." But these statements, and others like them, are made in passing and do not constitute the basis for the thoroughgoing analysis of the structure of laws and interests. And Bonger is ambivalent throughout on the role of social control in the creation of crime. He seems aware only in certain cases, of "societal reaction" in determining degrees of apprehension. So, for example: "the offences of which women are most often guilty are also those which it is most difficult to discover, namely those committed without violence. Then, those who have been injured are less likely to bring a complaint against a woman than against a man."

But later, in dealing with sexual crimes in general, Bonger uncritically accepts the official statistics of apprehension as an indication of "the class of the population that commits these crimes." In fact, Bonger's position is that the law (and its enforcement)—whilst certainly the creation of a dominant class—is a genuine reflexion of some universal social and moral sentiment. . . ."

The manifest explanation for the inclusion within the criminal law of sanctions controlling behaviour which is not directly harmful to the class interests of the powerful is that the working classes themselves are not without power. That is, one supposes, it is in the interests of the powerful to operate a system of general social control in the interests of order (within which individual and corporate enterprise can proceed unimpeded). However, there is more than a suspicion that Bonger's equation of social control with a universal moral sentiment is based on a belief he shares with the bourgeoisie in order for its own sake. Socialism is preferable to capitalism because it is more orderly. . . .

Bonger's formal Marxism, therefore, tells us that the solution to the problems of criminality is not so much to challenge the labels and the processing of capitalist law as it is to wage a responsible and orderly political battle for the reform of a divisive social structure. Even in the case of political opposition, a crucial distinction is to be drawn between responsible activity (the acts of a noble man) and the irresponsible and pathological activity—especially that of the anarchist movement (characterized, argues Bonger, by "extreme individualism," "great vanity," "pronounced altruistic tendencies" "coupled with a lack of intellectual development"). . . .

. . . . For us, the outstanding feature of Bonger's essentially correctional perspective is that, quite aside from the premises on which it operates (the contingency of criminality on an egoistic moral climate), it does not reveal a consistent social psychology, or, by the same token, a systematic social theory. At one moment, the actor under consideration is seen to be inextricably caught up in a determined and identifiable set of circumstances (or, more properly, a set of economic relationships); at another, he appears as the victim of an assumed personal quality (the criminal thought) sustained and (often) apparently developed by the moral climate of industrial capitalism.

Insofar as a social theory reveals itself in Bonger, the central assumptions on which it is built appear to be Durkheimian in nature rather than to derive from the avowedly Marxist theory of its author. Criminal man is consistently depicted not so much as a man produced by a matrix of unequal social relationships, nor indeed as a man attempting to resolve those inequalities of wealth, power and life chances; rather, criminal man is viewed as being in need of social control. "Socialism," in this perspective, is an alternative and desirable set of social institutions, which carry with them a set of Durkheimian norms and controls. "Socialism" thus expressed is the resort of an idealist, wishing for the substitution of a competitive and egoistic moral climate by a context in which the cooperativeness of men is encouraged. Socialism is preferable to capitalism, most of all, because it will control the baser instincts of man, Bonger does not assert that the "egoistic" man will "wither away" under socialism: it is only that the social relationships of socialism will not reward the endeavours of an egoist. . . .

DISCUSSION QUESTIONS

1. How does Bonger's approach differ from Marx's; does it add to or subtract from Marx?
2. Discuss Taylor, Walton, and Young's "cure" for crime.
3. Why and how does capitalism encourage egoism? Do you think that people in socialist countries are more altruistic than people in capitalist countries? How would you find this out?

READING

Crime, Punishment, and the American Dream

Toward a Marxist Integration

Barbara A. Sims

Barbara Sims's article serves to illustrate differences between two theories that are highly critical of American capitalism, with one advocating reform (institutional anomie) and the other advocating revolution (Marxism). In Section 4 it was noted that institutional anomie theory claims that all other institutions in America are subordinate to the economy. Sims agrees with this and says that it is in accordance with Marxist thought. However, she argues that the theory does not take the contribution of Marxist criminology into consideration. Sims seeks to rectify this by showing how economic and other inequalities are naturally occurring events in capitalist America. After doing this, she attempts to show how her analysis can be applied to address what can be done about America's imprisonment binge.

According to a 1994 Gallup poll, 52% of Americans believe that crime is the most important problem facing the United States today. In spite of research that has shown a great disparity between individuals' fear of crime and the chance of them actually becoming victims, the media, legislators, and U.S. presidents have continued to paint a picture of "soaring crime rates' (Walker 1994, p. 4). In his State of the Union Address in January 1994, President Bill Clinton placed a great emphasis on crime and announced that he was introducing a new crime bill. During the months that followed, as the U.S. Congress began considering the bill, extensive media coverage kept the issue of crime alive in the minds of Americans.

Although the picture of crime that Americans are left with may be distorted and used to feed a politically fruitful "get tough" approach to crime, the fact is that crime is a major problem in the United States. Whether or not crime is increasing in such drastic measures, as has been portrayed by the press and politicians, is a question that is not answered in this article. Instead, an argument is made that the resulting "lock 'em up" policy—fueled by the media, politicians, and the public's increased punitiveness toward offenders—has diverted attention away from a close examination of a social structure and culture that produces criminal activity in the first place. This punitive approach to crime has had a tremendous impact on the criminal justice system (police, courts, and corrections)—an impact that has been felt throughout other social institutions across U.S. communities. To lay the theoretical foundation for this argument, I apply Messner and Rosenfeld's (1994) "sociological paradigm" in *Crime and the American Dream* to Irwin and Austin's (1994) "imprisonment binge" in *It's About Time: America's Imprisonment Binge.*

AUTHOR'S NOTE: This article was originally published in the *Journal of Research in Crime and Delinquency*, Vol. 34, No. 1, February 1997 5–24. Reprinted with the permission of Sage Publications, Inc.

The Missing Link in Messner and Rosenfeld's "Sociological Paradigm"

Messner and Rosenfeld (1994, p. 8) argue that there is a "dark side" to the American Dream. The thesis of their book is that the American Dream itself encourages an exaggerated emphasis on monetary achievements while devaluing alternative criteria of success; it promotes a preoccupation with the realization of goals while deemphasizing the importance of the ways in which these goals are pursued; and it helps create and sustain social structures with limited capacities to restrain the cultural pressures to disregard legal restraints. Out of their thesis, the authors form the hypothesis that high crime rates in America are intrinsic to the basic cultural commitments and institutional arrangements of American society. Both the cultural and structural underpinnings of U.S. communities are, for Messner and Rosenfeld, "organized for crime" (p. 6).

The intellectual roots of *Crime and the American Dream* are found in Durkheim and his examination of the critical role that social forces play in explaining human behavior. For Durkheim (1933), any explanation of human behavior must take into account the various social forces surrounding the individual. His key concept, anomie (a sense of normlessness brought about by the breakdown in social institutions), is a natural result, a state of *confusion,* when societies are transformed from the mechanical (traditional/rural) to the organic (modern/urban).

Messner and Rosenfeld rely on the later works by Merton (1938), however, and his expansion of Durkheim's anomie, to provide the underlying premise for their own work. They accept Merton's notion that motivations for crime do not result simply from the flaws, failures, and/or free choices of individuals and that a complete examination of crime ultimately must consider the sociocultural environments in which people are located. They suggest, however, that Merton's argument, by itself, is not enough because it does not provide a "fully comprehensive, sociological explanation of crime in America" (Messner and Rosenfeld 1994, p. 15).

To achieve this comprehensive, sociological explanation (their sociological paradigm), Messner and Rosenfeld look to the "levels of explanation" in social research, namely, micro and macro. Primarily, micro explanations of crime focus on individual behavior, whereas macro explanations shift attention from individuals to social collectivities. Messner and Rosenfeld integrate six theories across these levels of explanations to form the theoretical argument for *Crime and the American Dream.* Their assumptions are detailed in the following.

1. *Social learning* (micro) theories are associated with *cultural deviance* (macro) theories that explain crime as the product of cultural (or subcultural) values and norms.
2. *Social control* (micro) theory is most closely connected with *social disorganization* (macro) theory as it refers to the inability of groups or communities to realize collective goals, including the goal of crime control.
3. *Anomie* (macro) is a result of *strain* (micro) experienced as a result of a differential distribution of opportunities to achieve highly valued goals.

Thus, for Messner and Rosenfeld, as communities become socially disorganized, they lose their ability to maintain sufficient control (both formal and informal) over their members such as to deter them from adopting delinquent lifestyles. As a result, subcultures can arise that create a new set of values and norms that can be learned in the same manner that values and norms in the mainstream culture are learned. In a situation of this sort, with institutions in decay, the problem is further exacerbated by an unequal opportunity structure in which not only the institutions but also the individuals who compose them are likely to suffer strain, the result of which is Durkheim's state of *confusion* (anomie).

According to these assumptions, Messner and Rosenfeld suggest that the answer to America's crime problem is to be found in a strengthening of its institutions (family, school, and polity) through social reorganization. Part of this reorganization process will entail the reassessing of the cultural values found in American society with its "exaggerated emphasis on monetary success" (p. 76). For too long, say the authors, the economy has maintained such a grip on American life in general as to short-change families, schools, local communities, and even the one institution that is supposed to give a voice to Americans themselves—the political arena.

In their discussion on social structure and culture, Messner and Rosenfeld refer briefly to Marx. They draw from Marx's "insight that the distribution of the means of consumption is ultimately dependent on the conditions of production themselves" (p. 108).

Marx's Contribution to Messner and Rosenfeld

Messner and Rosenfeld do not fail to give some credit to Marx. Yet, their brief mention of the contribution Marxist criminology makes to their own work is more *implicit* than *explicit* and, therefore, easy for the reader to miss. To synthesize, this contribution stems from an argument that the economic mode of production in America (i.e., capitalism) sets up a society for conflict and crime. It forms the foundation of a society from which other social institutions arise—the institutions that, for Messner and Rosenfeld, become so important.

Marxist criminologists align with strain, social disorganization, and cultural theories at this juncture and with the assumptions in *Crime and the American Dream.* When Messner and Rosenfeld talk about value patterns such as achievement, individualism, universalism, and the fetishism of money, Marxists point out that these values are derivatives of a capitalist mode of production. Its members are socialized to overemphasize materialism, which quite often leads to greed. Yet, for Marxists, it is not the consciousness of men that determines their existence; rather, it is their social existence that determines their consciousness (Marx 1978). The cause of crime, then, is determined by social forces outside the control of individuals. Those who engage in criminal activity may be "freely acting," but, for Marxists and for Messner and Rosenfeld, they are acting within a determined social, political, and economical setting.

Figure 6.1 argues conceptually that Marxist criminology can provide an explanation of how social relationships in a capitalist economic system (the *social formation* for Marx) can produce structural and economic inequalities. The economic and structural inequalities that arise in the social environment can produce a state of confusion (anomie). Within this anomic cultural environment, individuals are likely to suffer strain, and institutions (schools, communities, families, and the criminal justice system) lose their ability to control their members. As a result, a new culture may form whose members reject the norms and values of the mainstream culture. Within this subculture and, in particular, a *delinquent* subculture, a new set of norms and values are adopted that could provide fertile ground for the learning of criminal behavior.

The basic premise for this article is that Messner and Rosenfeld fail to adequately address the system under which social and economic inequalities arise, namely, a capitalist economy. This effort to call attention to the missing link (Marxist criminology) in their sociological paradigm is not completely at odds with the not-too-distant literature.

Examining further the similarities between traditional and radical criminological theories, Lynch and Groves (1989) argue that Merton failed to address exactly where the goals—which he argues are not capable of being achieved by all

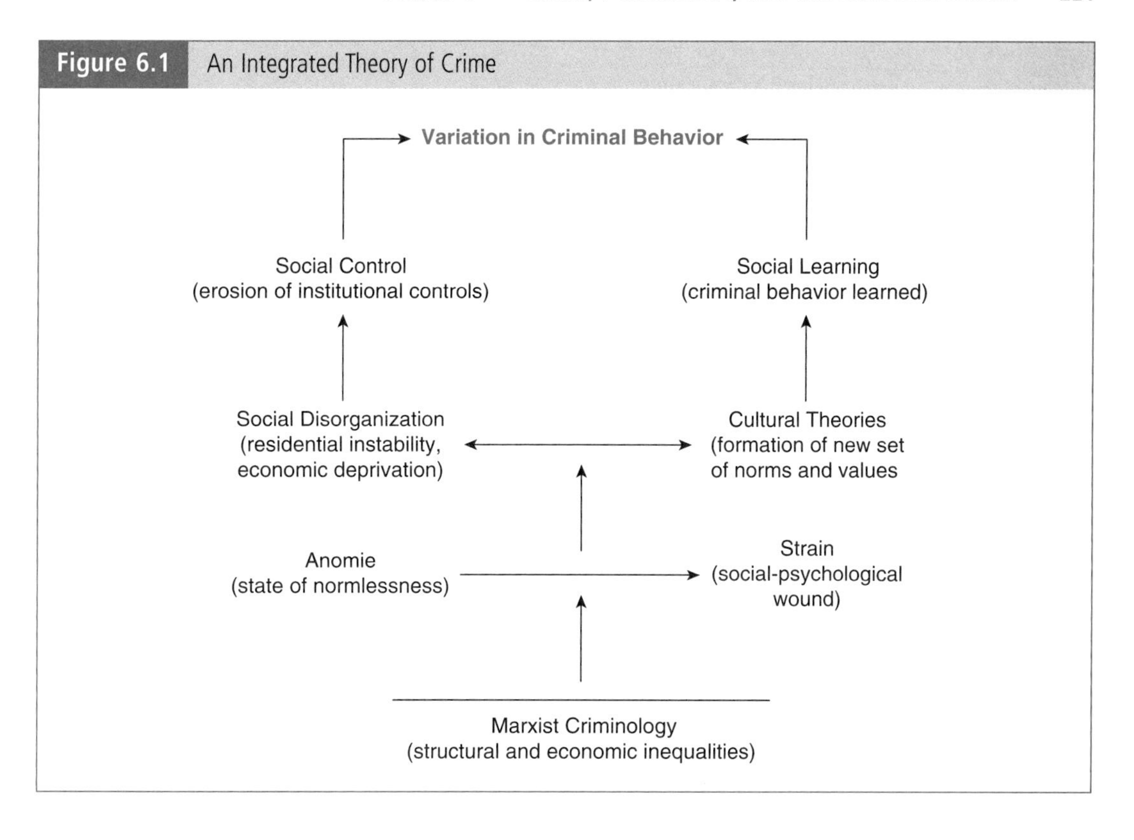

Figure 6.1 An Integrated Theory of Crime

in society—originate. Nor did Merton explain how various social classes are formed. Goals, according to Marxists, are materialistic in nature and, as such, can be traced to the "economic requirements of capitalism" (Lynch and Groves 1989, p. 74). Opportunity structures are, in Marxist terms, distributed along class lines created by stratifications that are dictated by society's economic structure. Classes are not created equally and, as pointed out by Lynch and Groves, do not "pop out of thin air" (p. 75). Thus they amend strain theory by supplying an explanation of how they are formed and further suggest the following.

Consistent with radical expectations, strain theorists suggest that crimes are not sporadic occurrences committed by isolated and abnormal individuals, but are regular and institutionalized features of a social system characterized by intense stratification and pervasive class conflict.

Lynch and Groves (1989) also bridge social disorganization and control theories with Marxist criminology. This bridge, for Lynch and Groves, is based on Durkheim's argument that when there is a breakdown in institutional integration, an imbalance occurs and social disorganization is the result. Marxists criminologists view *imbalance* as a logical result of a capitalist economic mode of production. Messner and Rosenfeld do an excellent job of picking up the ball from this point and arguing how, under a system in which all other institutions take a backseat to the economic system, such an imbalance can occur. Social disorganization and control theories are solid arguments for a breakdown in community characteristics and the resulting loss of control over its members.

In the context of a community in which poverty and inequality are a way of life, a new set of values may form that seem to be in direct opposition to those of the larger society (criminogenic in nature). Sutherland (1947) suggested that persons engage in criminal behavior because it is demanded by their culture; they commit crimes because they have learned that it is the correct thing to do. Marxist's criminologists ask the question: Where do these criminogenic values come from? Like cultural goals, class strata, and socially disorganized institutions, they originate somewhere. Miller (1979) attempts to answer the question by arguing that "lower class structure is a distinctive tradition many centuries old with an integrity all its own" (p. 167).

However, Miller and other cultural theorists would have us believe that lower class culture is "something that floats through history and just happens to be adhered to by those at the bottom of the class hierarchy." Argued from a Marxist perspective, persons are in fact motivated by their values, beliefs, and ideas, which in turn are determined by structurally defined conditions of life. Cultural theorists ignore what produces the idea of an act, or of a particular value system, and draw a line straight from the idea of committing a crime to the criminal act itself.

In Messner and Rosenfeld's sociological paradigm, they include social learning theory as the micro analog of cultural theories, that is, criminal behavior is learned just like any other behavior is learned. They also include social control theory as the micro analog of social disorganization theory; that is, when a person's bond to society is weak or broken, he or she is more likely to engage in deviant behavior. At the individual level, both social learning and social control variables are able to explain much of the variance in deviant behavior. This is to be expected given that those variables are the most proximate to the act itself. What researchers fail to recognize, however, is the fact that the more distal variables (i.e., the structural or macro-level variables) are contributing much to any model of crime. What is it about the way in which society is structured that allows—or disallows—social bonding to take place? What is it about the way in which society is structured that produces variation in value systems where social learning takes place?

The Interaction of Culture With Economic and Structural Components

In *Crime and the American Dream*, Messner and Rosenfeld (1994) state that the "American Dream is a broad cultural ethos that entails a commitment to the goal of material success to be pursued by everyone in society under conditions of open, individual competition" (p. 6). This cultural ethos sets up all members of American society to want the same things, to view success in terms of material items, with other success criteria (educational achievement, artistic talent, etc.) taking a backseat to monetary success. Messner and Rosenfeld call this phenomenon the "fetishism of money" and argue that the American Dream, then, can never actually be realized by anyone given that members of society always are working to accumulate more material goods.

Marxist criminologists ask the question: Just what is it about American society that produces this fetishism of money? Messner and Rosenfeld argue that in American society, individuals are encouraged to succeed "by any means necessary" (p. 9). They talk about *universalism* in which members of American society are socialized to "embrace the tenets of the dominant cultural ethos" (p. 9), an ethos that says that, regardless of your position in society, you can, and in fact should, achieve the American Dream.

Just how is it that success, in American culture, became so entangled with material goods? From a Marxist perspective, that question is not difficult to answer. During the Industrial Revolution, as American society moved from an agrarian to a market economy, individual workers were transformed from producers of goods for their own consumption to producers of goods that would produce profits for the factory owners. During this process, the early proletariats were twice "duped" by the capitalists. First, they were transformed into a "disposal industrial army" (Marx 1967, p. 632) to be used by the industrialists in the *production* of various commodities. Second, and here is the rub, the workers then again were transformed, through a massive advertising campaign that took root during the 1920s, into the *consumers* of the very products they were producing. Workers were paid wages that barely kept their families housed, fed, and clothed and then were taught that no longer was it enough to have what they needed in life to survive; rather, somehow they could be more "civilized," and not "social failures," once a certain amount of the produced commodities were accumulated (Ewen 1976, pp. 42–44).

For industry to succeed, it had to create in workers a desire to consume. What better way to mobilize the desires of individuals than through massive advertising. In 1926, U.S. President Calvin Coolidge remarked that advertising, if applied correctly, could be the "method by which the desire is created for better things" (in Ewen 1976, p. 37). The new advertising age was up and running, and with it came a host of ads that were meant to keep the workers dissatisfied with their way of life; dissatisfied customers, after all, "are much more profitable than satisfied ones" (p. 39). A new culture was created with industry, and thus the economy, becoming the "captain of consciousness" for the new army of workers (Ewen 1976). A sense of "excessiveness" replaced "thrift" as the *old ways* were usurped by this new culture of mass consumption, a social phenomenon that many would argue is very much alive and well in American society today. Messner and Rosenfeld argue that this drive to accumulate material goods contributes greatly to the dark side of the American Dream, and suggest that all are equally encouraged to consume but that there is a "relatively weak emphasis on the importance of using the legitimate means to do so" (p. 69).

Given an overemphasis on monetary success coupled with an unequal means to achieve it, it is more likely than not that individuals could, under these circumstances, become frustrated over the inability to accumulate material goods. The concept of "relative deprivation" has a rich history in much of the literature and describes just such a situation. Gurr (1970) defines relative deprivation as "actors' perception of discrepancy between their value *expectations* and their value *capabilities*" (p. 24). Value is a term used to describe the goods, and the conditions of life in general, to which individuals believe they are entitled. Gurr uses relative deprivation to argue that the potential for collective violence could vary strongly with the "intensity and scope" (p. 24) of relative deprivation among members of a group.

If, as has been argued, there is as much a cultural component that produces crime in one capitalist country—namely, in America—as there is a structural component, then one question comes quickly to mind: How is it that in other capitalist nations, the crime rates have not soared to the same heights as the crime rate seemingly has soared in America? Is it because the institutions themselves vary greatly from one country to the other, or could it be that not so heavy an emphasis is placed on mass consumerism and the accumulation of material goods in other countries relative to that found in the United States?

Initial reactions to this question might produce, and not unrightfully so, answers coming from perspectives based on the easy accessibility of guns in American society or the fact that minorities are disproportionately represented in arrest reports in the United States. Messner and Rosenfeld (1994) address both of these issues in great detail. They argue that although rates of gun ownership are much higher in America than they are in other industrial nations, if gun-related homicides were left out of the U.S. homicide rate, then the United States still would have a non-gun homicide rate higher than the total homicide rates of other industrial nations.

Messner and Rosenfeld argue further that although lethal violence among young African American men is extremely high, this, in and of itself, does not explain the differences in homicide rates between the United States and other industrial societies. Excluding African Americans from the equation, Messner and Rosenfeld state that the "homicide rate for *Whites* is more than four times the average rate of homicide" among other industrial nations (p. 29). Thus the answer to the question of what explains the high rate of violent crime in the United States cannot be reduced to guns and ethnicity.

When opponents of a Marxist perspective attempt to show how the argument that "capitalism produces crime" fails, they usually point to the Japanese society. In spite of low levels of crime in Japanese society, however, the fact remains that the country is not crime free. According to Fenwick (1985), Japan does have concerns in areas such as juvenile delinquency, drug abuse, and organized crime. Although much lower than other industrial nations, the Japanese crime rate has been on the increase since the late 1970s. Looking specifically at increases in juvenile crime, Chang (1988) offers the following reason for this increase. In Tokyo, the Japanese city with the highest mobility rate, the value system of youth has begun a transformation of sorts. No longer is personal interdependence the key to social status and satisfaction; *money* quickly has become the ticket to personal success. Youths commute long distances to schools and neighborhoods where their families are not known. Instead of living in houses owned by their forefathers for generations, they live in crowded multiple-family housing units—a situation not conducive to pride and healthy self-images. Under these circumstances, the notion of honor, so important in the Japanese culture, begins to break down, and the responsibility to the community and the family is weakened.

Punitive Crime Control: America's Imprisonment Binge

In a society with its institutions in disarray and thus unable to maintain control over its members, and in a society where the cultural expectations are geared toward the accumulation of material goods with unequal means to achieve those goods, crime often is an unintended consequence. As crime increases, so too does the method of social control over members of society. It follows, then, that America's "war on crime" would result in higher numbers of incarceration for members of its lower classes given that this war has been waged on "street crime" as opposed to white-collar crime. Marxist criminologists view laws in a capitalist society as tilted in favor of the ruling class and more punitive toward the lower classes.

Inner-city youths have been, and often unrightfully so, connected with the use and sale of illegal drugs. The war on drugs, with its more punitive sentencing policies for drug offenders, has focused heavily on crack cocaine. Because this drug is mainly sold and used in inner-city communities, Hispanics and African Americans are the ones who have felt the strong arm of the law. As of 1989, the African American prison population, for example, had increased to 46%—from 21% of all prison admissions in 1926, when the race of the prison population began to be recorded (Irwin and Austin 1994).

In *It's About Time: America's Imprisonment Binge,* Irwin and Austin (1994) argue that punitive approaches to crime have left America's correctional institutions overburdened and unable to meet the demands of a coerced public. This coercion stems from a public misperception about crime brought about primarily by the attention paid to such by politicians and the media. Marxist criminologists would relate this misperception to the concept of "false consciousness." The attention politicians have paid to crime has acted as a smokescreen—a diversion tactic of sorts. With the public's attention focused on crime, and with a public willing to pay whatever cost it must to get the crime rate down, issues such as the threat of nuclear war, unemployment, high living costs, and the economy are placed on the back burner. Irwin and Austin (1994) point out that any attempts made to solve these problems, along with a host of other social problems, would be met by the powerful forces of political, legal, and economic interest groups—an argument with which Marxist criminologists would readily agree.

As Messner and Rosenfeld point out, white-collar crime costs society approximately $200 billion a year—roughly 20 times the annual monetary loss or damage due to street crimes in the

United States. Irwin and Austin (1994) cite a 1990 report by the U.S. Department of Justice's National Crime Victim Survey that puts the cost of crimes to victims at about $19.2 billion. Still, it is not chief executive officers or the owners (including stockholders) of major corporations who form the bulk of the U.S. inmate population.

Messner and Rosenfeld, as well as Irwin and Austin, move down the crime continuum and look closely at the inequalities in America's social structure that lead individuals to the threshold of the criminal justice system. The former spend a great amount of energy arguing that America's social system is tilted in favor of the economy—an argument with which Marxist criminologists would not disagree. The latter argue that America must begin to rethink its punitive approach to crime because it cannot afford the price tag of exponential incarceration—an argument that would get no disagreement from Marxist criminologists. To "catch criminals and lock them up; if they hit you, hit them back" may seem a *commonsense* approach to crime (Menninger 1969, p. 5). As expressed by Menninger (1969), it is also a commonsense approach that if it gets dark, you go to bed. Yet, the *uncommon sense* of science has made it no longer necessary to go to bed when it gets dark and has taught us that simply to lock up individuals who commit crimes does little to curtail them (p. 5).

Theory and Its Application in the Reduction of Crime in American Society

Before moving to a discussion of solutions for America's crime problem, there is one clarification with regard to Marxist criminology that should be made. Some factions of Marxist criminology believe that no solution outside of an overthrow of the current economic mode of production in America (i.e., capitalism) can expect to turn around the crime tide—be it street crime or suite crime. Neo-Marxists, however, take a more realistic approach through the realization that, as Marx himself proclaimed, history will usher in a new economic order in its own good time. In the meantime, while we wait and agitate for large-scale transformation, there is a great deal to be accomplished in terms of middle-range policy alternatives which do not compromise any overall design for fundamental social change.

What follows is a list of these policy alternatives offered up by Marxist criminologists.

1. Crime should be defined according to the amount of harm inflicted on society. The definition of crime and the practice of crime control should no longer be organized along class lines.
2. Reduce the capacity of capital to displace labor. Develop tax initiatives to establish a surcharge on any industry attempting to close plants or permanently reduce the workforce in a given community, introduce legislation requiring the retraining and placement of displaced workers, and increase minimum wages to a sufficient amount so as to decrease the numbers of working poor.
3. Reduce inequality in the existing social structure, because it is a strong predictor of U.S. homicide rates.
4. Abolish mandatory sentences that discriminate against lower class America and greatly overburden its correctional institutions.
5. Curb white-collar crime by improving enforcement through the allocation of more resources to regulatory agencies, implementing structural reforms, and enacting political reforms designed to minimize conflict of interest.
6. Take a closer look at the enactment of laws that require punitive governmental

intervention such as those that create new classes of criminals, for example, laws that criminalize the homeless.

In their passion for seeking social change, critical theorists quite often make the same mistake as their more conservative counterparts. They wrap themselves in a rhetoric that is overbearing and quite often misunderstood by fellow theorists as well as the public and decision makers. The result often is deadlock; nothing is accomplished, and society continues down a spiraling path of institutional decay and loss of control over its members.

The future of theoretical development should take a two-prong approach. One is to be found in a coming together of consensus and conflict theorists. As shown here, both camps are closer to agreement than much of the literature would dictate. The other approach is to move the research findings out of the secluded halls of academe and into the halls of the body politic. With the increase in social science techniques, criminologists, sociologists, psychologists, and biologists all have produced a sufficient amount of evidence supporting possible solutions to the problem of crime in America. Until the message reaches the public, however, these findings will forever be concealed in the dustbins of academe.

References

Chang, D. H. (1988, Winter). Crime and delinquency control strategy in Japan: A comparative note. *International Journal of Comparative and Applied Criminal Justice, 12,* 139–149.

Durkheim, E. (1933). *The division of labor in society.* New York: Free Press.

Ewen, S. (1976). *Captains of consciousness: Advertising and the social roots of the consumer culture.* New York: McGraw-Hill.

Fenwick, C. R. (1985, Spring). Culture, philosophy and crime: The Japanese experience. *International Journal of Comparative and Applied Criminal Justice, 9,* 67–81.

Gurr, T. R. (1970). *Why men rebel.* Princeton, NJ: Princeton University Press.

Irwin, J., & Austin, J. (1994). *It's about time: America's imprisonment binge.* Belmont, CA: Wadsworth.

Lynch, M. J., & Groves, W. B. (1989). *A primer in radical criminology* (2nd ed.). Albany, NY: Harrow & Heston.

Marx, K. (1967). *Capital* (Vol. I). New York: International.

Marx, K. (1978). *The Marx Engels reader* (2nd ed., R. C. Tucker, Ed.). New York: Norton.

Menninger, K. (1969). *The crime of punishment.* New York: Viking.

Merton, R. K. (1938). Social structure and anomie. *American Sociological Review, 3,* 672–682.

Messner, S. F., & Rosenfeld, M. (1994). *Crime and the American dream.* Belmont, CA: Wadsworth.

Miller, W. (1979). Lower class culture as a generating milieu of gang delinquency. In J. E. Jacoby (Ed.), *Classics of criminology* (pp. 155–168). Prospect Heights, IL: Waveland.

Sutherland, E. H. (1947). *Principles of criminology.* Philadelphia: Lippincott.

Walker, S. (1994). *Sense and nonsense about crime and drugs.* Belmont, CA: Wadsworth.

DISCUSSION QUESTIONS

1. Someone once said about Marxism: "Beautiful theory; wrong species." How would Sims respond to this one-liner criticism of her criminological orientation?

2. Sims complains about the severe penalties applied to the sale of crack cocaine. Has the black community been harmed by or benefited from these penalties?

3. Sims writes about laws that criminalize the homeless. Examine your state's criminal codes to see whether being homeless has been criminalized.

READING

Patriarchy, Crime, and Justice

Feminist Criminology in an Era of Backlash

Meda Chesney-Lind

In this article, Meda Chesney-Lind argues that feminist criminology, as an outgrowth of the second wave of feminism, came of age during a period of considerable change and political optimism. She avers that it now inhabits a social and political landscape radically altered and increasingly characterized by the politics of backlash. Given feminist criminology's dual focus on gender and crime, it is uniquely positioned to respond to two core aspects of the current backlash political agenda: racism and sexism. To do this effectively, feminist criminology must prioritize research on the race/gender/punishment nexus. This article provides three examples of how such a focus exposes the crucial roles played by constructions of the crime problem as well as current crime control strategies in the ratification and enforcement of antifeminist and racist agendas. She concludes by asserting that the field must seek creative ways to blend scholarship with activism while simultaneously providing support and encouragement to emerging feminist criminologists willing to take such risks.

Feminist Criminology in the 20th Century: Looking Backward, Looking Forward

The feminist criminology of the 20th century clearly challenged the overall masculinist nature of theories of crime, deviance, and social control by calling attention to the repeated omission and misrepresentation of women in criminological theory and research. Turning back the clock, one can recall that prior to path-breaking feminist works on sexual assault, sexual harassment, and wife abuse, these forms of gender violence were ignored, minimized, and trivialized. Likewise, girls and women in conflict with the law were overlooked or excluded in mainstream works while demonized, masculinized, and sexualized in the marginalized literature that brooded on their venality.

The enormity of girls' and women's victimization meant that the silence on the role of violence in women's lives was the first to attract the attention of feminist activists and scholars. Compared to the wealth of literature on women's victimization, interest in girls and women who are labeled, tried, and jailed as "delinquent" or "criminal" was slower to fully develop in part because scholars of "criminalized" women and girls had to contend early on with the masculinization (or "emancipation") hypothesis of women's crime, which argues in part that "in the same way that women are demanding equal opportunity in the fields of legitimate endeavor, a similar number of determined women are forcing their way into the world of major crimes" (Adler, 1975, p. 3). Feminist criminologists, as well as mainstream criminologists, debated the

Source: Chesney-Lind, M. (2006). Patriarchy, crime, and justice: Feminist criminology in an era of backlash. *Feminist Criminology, 1*, 6-26.

nature of that relationship for the next decade and ultimately concluded it was not correct (Chesney-Lind, 1989), but this was a costly intellectual detour.

Instead of the "add women and stir" (Chesney-Lind, 1988) approach to crime theorizing of the past century (which often introduces gender solely as a "variable" if at all), new important work on the gender/crime nexus *theorizes gender.* This means, for example, drawing extensively on sociological notions of "doing gender" (West & Zimmerman, 1987) and examining the role of "gender regimes" (Williams, 2002) in the production of girls' and women's behavior. Contemporary approaches to gender and crime (see Messerschmidt, 2000) tend to avoid the problems of reductionism and determinism that characterize early discussions of gender and gender relations, stressing instead the complexity, tentativeness, and variability with which individuals, particularly youth, negotiate (and resist) gender identity.

Feminist Criminology and the Backlash

Feminist criminology in the 21st century, particularly in the United States, finds itself in a political and social milieu that is heavily affected by the backlash politics of a sophisticated and energized right wing—a context quite different from the field's early years when the initial intellectual agenda of the field evolved. Political backlash eras have long been a fixture of American public life. The centrality of both crime and gender in the current backlash politics means that feminist criminology is uniquely positioned to challenge right-wing initiatives. To do this effectively, however, the field must put an even greater priority on *theorizing patriarchy and crime,* which means focusing on the ways in which the definition of the crime problem and criminal justice practices support patriarchal practices and worldviews.

To briefly review, patriarchy is a sex/gender system in which men dominate women and what is considered masculine is more highly valued than what is considered feminine. Patriarchy is a system of social stratification, which means that it uses a wide array of social control policies and practices to ratify male power and to keep girls and women subordinate to men. Often, the systems of control that women experience are explicitly or implicitly focused on controlling female sexuality. Not infrequently, patriarchal interests overlap with systems that also reinforce class and race privilege, hence, the unique need for feminist criminology to maintain the focus on intersectionality that characterizes recent research and theorizing on gender and race in particular.

Again, in this era of backlash, the formal system of social control (the law and criminal justice policies) play key roles in eroding the rights of both women and people of color, particularly African Americans but increasingly, other ethnic groups as well. Feminist criminology is, again, uniquely positioned to both document and respond to these efforts. To theorize patriarchy effectively means that we have done cutting-edge research on the interface between patriarchal and criminal justice systems of control and that we are strategic about how to get our findings out to the widest audience possible, issues to which this article now turns.

Race, Gender, and Crime

America's long and sordid history of racism and its equally disturbing enthusiasm for imprisonment must be understood as intertwined, and both of these have had a dramatic effect on African American women in particular. More than a century ago, W. E. B. Du Bois saw the linkage between the criminal justice system and race-based systems of social control very clearly. Commenting on the dismal failure of "reconstruction," he concluded: "Despite compromise, war, and struggle, the Negro is not free. In well-nigh the whole rural South the black farmers are peons, bound by law and custom to an economic slavery from which the only escape is death or the penitentiary" (as quoted in Johnson, 2003, p. 284).

Although the role of race and penal policy has received increased attention in recent years, virtually all of the public discussion of the issues has focused on African American males. More recent, the significant impact of mass incarceration on African American and Hispanic women has received the attention it deserves. Current data show that African American women account for "almost half (48 percent)" of all the women we incarcerate (Johnson, 2003, p. 34). Mauer and Huling's (1995) earlier research adds an important perspective here; they noted that the imprisonment of African American women grew by more than 828% between 1986 and 1991, whereas that of White women grew by 241% and of Black men by 429%. Something is going on, and it is not just about race or gender; it is about both—a sinister synergy that clearly needs to be carefully documented and challenged.

An examination of Black women's history from slavery through the Civil War and the postwar period certainly justifies a clear focus on the role that the criminal justice system played in the oppression of African American women and the role of prison in that system. And the focus is certainly still relevant because although women sometimes appear to be the unintended victims of the war on drugs, this "war" is so heavily racialized that the result can hardly be viewed as accidental. African American women have always been seen through the "distorted lens of otherness," constructed as "subservient, inept, oversexed and undeserving" (Johnson, 2003, pp. 9–10), in short, just the "sort" of women that belong in jail and prison.

Media Demonization and the Masculinization of Female Offenders

In her book *Backlash: The Undeclared War Against American Women*, Susan Faludi (1991) was quick to see that the media in particular were central, not peripheral, to the process of discrediting and dismissing feminism and feminist gains. She focused specific attention on mainstream journalism's efforts to locate and publicize those "female trends" of the 1980s that would undermine and indict the feminist agenda. Stories about "the failure to get husbands, get pregnant, or properly bond with their children" were suddenly everywhere, as were the very first stories on "bad girls."

Faludi's (1991) recognition of the media fascination with bad girls was prescient. The 1990s would produce a steady stream of media stories about violent and bad girls that continues unabated in the new millennium. Although the focus would shift from the "gangsta girl," to the "violent girl," to the "mean girl" (Chesney-Lind & Irwin, 2004), the message is the same: Girls are bad in ways that they never used to be.

Media-driven constructions such as these generally rely on commonsense notions that girls are becoming more like boys on both the soccer field and the killing fields. Implicit in what might be called the "masculinization" theory of women's violence is the idea that contemporary theories of violence (and crime more broadly) need not attend to gender but can, again, simply add women and stir. The theory assumes that the same forces that propel men into violence will increasingly produce violence in girls and women once they are freed from the constraints of their gender. The masculinization framework also lays the foundation for simplistic notions of "good" and "bad" femininity, standards that will permit the demonization of some girls and women if they stray from the path of "true" (passive, controlled, and constrained) womanhood.

Ever since the first wave of feminism, there has been no shortage of scholars and political commentators issuing dire warnings that women's demand for equality would result in a dramatic change in the character and frequency of women's crime. Again, although this perspective was definitely refuted by the feminist criminology of the era, media enthusiasm about the idea that feminism encourages women to become more like men and, hence, their "equals" in crime, remains undiminished.

In virtually all the stories on this topic, the issue is framed as follows. A specific and egregious example of female violence is described, usually with considerable, graphic detail about the injury suffered by the victim—a pattern that has been dubbed "forensic journalism" (Websdale & Alvarez, 1997, p. 123). In the *Mercury News* article, for example, the reader hears how a 17-year-old girl, Linna Adams, "lured" the victim into a car where her boyfriend "pointed a .357 magnum revolver at him, and the gun went off. Rodrigues was shot in the cheek, and according to coroner's reports, the bullet exited the back of his head" (Guido, 1998, p. 1B).

These forensic details are then followed by a quick review of the Federal Bureau of Investigation's arrest statistics showing what appear to be large increases in the number of girls arrested for violent offenses. Finally, there are quotes from "experts," usually police officers, teachers, or other social service workers, but occasionally criminologists, interpreting the narrative in ways consistent with the desired outcome: to stress that girls, particularly African American and Hispanic girls whose pictures often illustrate these stories, are getting more and more like their already demonized male counterparts and, hence, becoming more violent.

Although arrest data consistently show dramatic increases in girls' arrests for "violent" crimes (e.g., arrests of girls for assault climbed an astonishing 40.9%, whereas boys' arrests climbed by only 4.3% in the past decade; Federal Bureau of Investigation, 2004), other data sets, particularly those relying on self-reported delinquency, show no such trend (indeed they show a decline; Chesney-Lind, 2004). It seems increasingly clear that forces other than changes in girls' behavior have caused shifts in girls' arrests (including such forces as zero-tolerance policies in schools and mandatory arrests for domestic violence). There are also indications that although the hype about bad girls seems to encompass all girls, the effects of enforcement policies aimed at reducing "youth violence" weigh heaviest on girls of color whose families lack the resources to challenge such policies.

Between 1989 and 1998, girls' detentions increased by 56% compared to a 20% increase seen in boy's detentions, and the "large increase was tied to the growth in the number of delinquency cases involving females charged with person offenses (157%)" (Harms, 2002, p. 1). At least one study of girls in detention suggests that "nearly half" the girls in detention are African American girls, and Latinas constitute 13%; Caucasian girls, who constitute 65% of the girl population, account for only 35% of those in detention.

It is clear that two decades of the media demonization of girls, complete with often racialized images of girls seemingly embracing the violent street culture of their male counterparts, coupled with increased concerns about youth violence and images of "girls gone wild," have entered the self-fulfilling prophecy stage. It is essential that feminist criminology understand that in a world governed by those who self-consciously manipulate corporate media for their own purposes, newspapers and television may have moved from simply covering the police beat to constructing crime "stories' that serve as a "nonconspiratorial source of dominant ideology" (Websdale & Alvarez, 1997, p. 125). Feminist criminology's agenda must consciously challenge these backlash media narratives, as well as engage in "newsmaking criminology" (Barak, 1988), particularly with regard to constructions of girl and women offenders.

Criminalizing Victimization

Many feminist criminologists have approached the issue of mandatory arrest in incidents of domestic assault with considerable ambivalence. On one hand, the criminalization of sexual assault and domestic violence was in one sense a huge symbolic victory for feminist activists and criminologists alike. After centuries of ignoring the private victimizations of women, police and courts were called to account by those who founded rape crisis centers and shelters for battered women and those whose path-breaking research laid the foundation for major policy and legal changes in the area of violence against women.

On the other hand, the insistence that violence against women be handled as a criminal matter threw victim advocates into an uneasy alliance with police and prosecutors—professions that feminists had long distrusted and with good reason. Ultimately, the combined effects of the early scientific evidence; political pressure from the attorney general of the United States, the American Bar Association, and others; and the threat of lawsuits against departments who failed to protect women from batterers "produced nearly unanimous agreement that arrest was the best policy for domestic violence" (Ferraro, 2001, p. 146).

As the academic debate about the effectiveness of arrest in domestic violence situations continued unabated, the policy of "mandatory arrest" became routinized into normal policing and quite quickly, other unanticipated effects began to emerge. When arrests of adult women for assault increased by 30.8% in the past decade (1994 to 2003), whereas male arrests for this offense fell by about 5.8%, just about everybody from the research community to the general public began to wonder what was happening. Although some, such as criminologist Kenneth Land, quoted in a story titled "Women Making Gains in Dubious Profession: Crime," attributed the increase to "role change over the past decades" that presumably created more females as "motivated offenders" (Anderson, 2003, p. 1), others were not so sure. Even the Bureau of Justice Statistics looked at a similar trend (increasing numbers of women convicted in state courts for "aggravated assault") and suggested the numbers might be "reflecting increased prosecution of women for domestic violence" (Greenfeld & Snell, 1999, pp. 5–6).

Despite the power of the stereotypical scenario of the violent husband and the victimized wife, the reality of mandatory arrest practices has always been more complicated. Early on, the problem of "mutual" arrests—the practice of arresting both the man and the woman in a domestic violence incident if it is not clear who is the "primary" aggressor—surfaced as a concern. Nor has the problem gone away, despite efforts to clarify procedures; indeed, many jurisdictions report similar figures. In Wichita, Kansas, for example, women were 27% of those arrested for domestic violence in 2001. Prince William County, Maryland, saw the number of women arrested for domestic violence triple in a 3-year period, with women going from 12.9% of those arrested in 1992 to 21% in 1996. In Sacramento, California, even greater increases were observed; there the number of women arrested for domestic violence rose by 91% between 1991 and 1996, whereas arrests of men fell 7% (Brown, 1997).

Susan Miller's (2005) study of mandatory arrest practices in the state of Delaware adds an important dimension to this discussion. Based on data from police ride alongs, interviews with criminal justice practitioners, and observations of groups run for women who were arrested as offenders in domestic violence situations, Miller's study comes to some important conclusions about the effects of mandatory arrest on women. According to beat officers S. Miller (2005) and her students rode with, in Delaware, they do not have a "pro-arrest policy, we have a pro-paper policy" (p. 100) developed in large part to avoid lawsuits. As a consequence, although the officers "did not believe there was an increase in women's use of violence" (S. Miller, 2005, p. 105), "her fighting back now gets attention too" (S. Miller, 2005, p. 107) because of this sort of broad interpretation of what constitutes domestic violence. None of the social service providers and criminal justice professionals S. Miller (2005) spoke with felt women had become more aggressively violent; instead, they routinely called the women "victims." They noted that at least in Delaware, as the "legislation aged," the name of the game began to be "get to the phone first" (S. Miller, 2005, p. 127).

Essentially, it appears that many mandatory arrest policies have been interpreted on the ground to make an arrest if any violent "incident" occurs, rather than considering the context within which the incident occurs (Bible, 1998). Like problematic measures of violence that simply count violent events without providing information on the meaning and motivation, this definition of *domestic violence* fails to distinguish between aggressive and instigating violence from self-defensive and retaliatory violence. According to S. Miller (2005) and other critics of this approach, these methods tend to produce results showing "intimate violence is committed by women at an equal or higher rate than by men" (p. 35). Although these findings ignited a firestorm of media attention about the "problem" of "battered men" in the United States (Ferraro, 2001, p. 137), the larger question of how to define *domestic violence* in the context of patriarchy is vital. Specifically, much feminist research of the sort showcased here is needed on routine police and justice practices concerning girls' and women's "violence."

Women's Imprisonment and the Emergence of Vengeful Equity

When the United States embarked on a policy that might well be described as mass incarceration, few considered the impact that this correctional course change would have on women. Yet the number of women in jail and prison continues to soar (outstripping male increases for most of the past decade), completely untethered from women's crime rate, which has not increased by nearly the same amount. The dimensions of this shift are staggering: For most of the 20th century, we imprisoned about 5,000 to 10,000 women. At the turn of the new century, we now have more than 100,000 women doing time in U.S. prisons (Harrison & Beck, 2004, p. 1). Women's incarceration in the United States not only grew during the past century but also increased tenfold; and virtually all of that increase occurred in the final two decades of the century.

The number of women sentenced to jail and prison began to soar at precisely the same time that prison systems in the United States moved into an era that abandoned any pretense of rehabilitation in favor of punishment. As noted earlier, decades of efforts by conservative politicians to fashion a crime policy that would challenge the gains of the civil rights movement as well as other progressive movements in the 1960s and 1970s had, by the 1980s, born fruit (Chambliss, 1999). Exploiting the public fear of crime, particularly crime committed by "the poor, mostly nonwhite, young, male inner-city dwellers" (Irwin, 2005, p. 8), all manner of mean-spirited crime policies were adopted. The end of the past century saw the war on drugs and a host of other "get tough" sentencing policies, all of which fueled mass imprisonment (see Mauer, 1999).

Although feminist legal scholars can and do debate whether equality under the law is necessarily good for women, a careful look at what has happened to women in U.S. prisons might serve as a disturbing case study of how correctional equity is implemented in practice. Such a critical review is particularly vital in an era where decontextualized notions of gender and race "discrimination" are increasingly and successfully deployed against the achievements of both the civil rights and women's movements. Consider the account of Martha Sierra's experience of childbirth. As she

> writhed in pain at a Riverside hospital, laboring to push her baby into the world, Sierra faced a challenge not covered in the childbirth books: her wrists were shackled to the bed. Unable to roll on her side or even sit straight up, Sierra managed as best she could. The reward was fleeting . . . she watched as her daughter, hollering and flapping her arms, was taken from the room (Warren, 2005, p. A1).

Sierra's story is unfortunately all too familiar to anyone who examines gender themes in modern correctional responses to women inmates. In fact, her experience is less horrific than the case of Michelle T., a former prisoner from Michigan who told Human Rights Watch (1996) that she was accompanied by two male correctional officers into the delivery room: "According to

Michelle T., the officers handcuffed her to the bed while she was in labor and positioned themselves where they could view her genital area while giving birth. She told [Human Rights Watch] they made derogatory comments about her throughout the delivery" (p. 249).

Basically, male prisoners have long used visits to hospitals as opportunities to escape, so correctional regimes have generated extensive security precautions to assure that escapes do not occur, including shackling prisoners to hospital beds (Amnesty International, 1999, p. 63). This is the dark side of the equity or parity model of justice—one which emphasizes treating women offenders as though they were men, particularly when the outcome is punitive, in the name of equal justice—a pattern that could be called vengeful equity.

Other examples of vengeful equity can be found in the creation of women's boot camps, often modeled on the gender regimes found in military basic training. These regimes, complete with uniforms, shorn hair, humiliation, exhausting physical training, and harsh punishment for even minor misconduct have been traditionally devised to "make men out of boys." As such, feminist researchers who have examined them contended, they "have more to do with the rites of manhood" than the needs of the typical woman in prison (Morash & Rucker, 1990). Reviewing the situation of women incarcerated in five states (California, Georgia, Michigan, Illinois, and New York) and the District of Columbia, Human Rights Watch (1996) concluded,

> Our findings indicate that being a woman prisoner in U.S. state prisons can be a terrifying experience. If you are sexually abused, you cannot escape from your abuser. Grievance or investigatory procedures where they exist, are often ineffectual, and correctional employees continue to engage in abuse because they believe they will rarely be held accountable, administratively or criminally. Few people outside the prison walls know what is going on or care if they do know. Fewer still do anything to address the problem. (p. 1)

Finally, it appears that women in prison today are also recipients of some of the worst of the more traditional, separate-spheres approach to women offenders. Correctional officers often count on the fact that women prisoners will complain, not riot, and as a result, often punish women inmates for offenses that would be ignored in male prisons. McClellan (1994) found this pattern quite clearly in her examined disciplinary practices in Texas prisons. Following up two cohorts of prisoners (one male and one female), she found most men in her sample (63.5%) but only a handful of women (17.1%) had no citation or only one citation for a rule violation. McClellan found that women prisoners not only received numerous citations but also were charged with different infractions than men. Most frequent, women were cited for "violating posted rules," whereas males were cited most often for "refusing to work" (McClellan, 1994, p. 77). Women were more likely than men to receive the most severe sanctions.

Much good, early feminist criminology focuses on the conditions of girls and women in training schools, jails, and prisons. Unfortunately, that work is now made much harder by a savvy correctional system that is extremely reluctant to admit researchers, unless the focus of the research is clearly the woman prisoner and not the institution. That said, there is much more need for this sort of criminology in the era of mass punishment, and the work that is being done in this vein points to the need for much of the same. Huge numbers of imprisoned girls and women are targeted by male-based systems of "risk" and "classification" and then subjected to male-based interventions such as "cognitive behaviorism" to address their "criminal" thinking as though they were men. Good work has also been done on the overuse of "chemical restraints' with women offenders (Leonard, 2002). In short, as difficult as it might be to do, in this era of mass imprisonment, feminist criminology needs to find creative ways to continue to engage core issues in girls' and women's carceral control as a central part of our intellectual and activist agenda.

Theorizing Patriarchy: Concluding Thoughts

Given the focus of the backlash, this article argues that feminist criminology is uniquely positioned to do important work to challenge the current political backlash. To do so effectively, however, it is vital that in addition to documenting that gender matters in the lives of criminalized women, we engage in exploration of the interface between systems of oppression based on gender, race, and class. This work will allow us to make sense of current crime-control practices, particularly in an era of mass incarceration, so that we can explain the consequences to a society that might well be ready to hear other perspectives on crime control if given them. Researching as well as theorizing both patriarchy and gender is crucial to feminist criminology so that we can craft work, as the right wing does so effectively, that speaks to backlash initiatives in smart, media-savvy ways. To do this well means foregrounding the role of race and class in our work on gender and crime, as the work showcased here makes clear. There is simply no other way to make sense of key trends in both the media construction of women offenders and the criminal justice response that increasingly awaits them, particularly once they arrive in prison.

Finally, we must also do work that will document and challenge the policy and research backlash aimed at the hard fought and vitally important feminist and civil rights victories of the past century. To do any less would be unthinkable to those who fought so long to get us where we are today, and so it must be for us.

References

Adler, F. (1975). *Sisters in crime.* New York: McGraw-Hill.

Amnesty International. (1999). *Not part of my sentence: Violations of the human rights of women in custody.* Washington, DC: Author.

Anderson, C. (2003, October 28). Women making gains in dubious profession: Crime. *Arizona Star*, p. A1.

Barak, G. (1988). Newsmaking criminology: Reflections on the media, intellectuals, and crime. *Justice Quarterly, 5*(4), 565–587.

Bible, A. (1998). When battered women are charged with assault. *Double-Time, 6*(1/2), 8–10.

Brown, M. (1997, December 7). Arrests of women soar in domestic assault cases. *Sacramento Bee.* Retrieved July 31, 2005, from http://www.sacbee.com/static/archive/news/projects/violence/part12.html

Chambliss, W. (1999). *Power, politics and crime.* Boulder, CO: Westview.

Chesney-Lind, M. (1988, July-August). Doing feminist criminology. *The Criminologist, 13*, 16–17.

Chesney-Lind, M. (1989). Girls' crime and woman's place: Toward a feminist model of female delinquency. *Crime & Delinquency, 35*(10), 5–29.

Chesney-Lind, M., & Irwin, K. (2004). From badness to meanness: Popular constructions of contemporary girlhood. In A. Harris (Ed.), *All about the girl: Culture, power, and identity* (pp. 45–56). New York: Routledge.

Faludi, S. (1991). *Backlash: The undeclared war against American women.* New York: Anchor Doubleday.

Federal Bureau of Investigation. (2004). *Crime in the United States, 2003.* Washington, DC: Government Printing Office.

Ferraro, K. (2001). Women battering: More than a family problem. In C. Renzetti & L. Goodstein (Eds.), *Women, crime and criminal justice* (pp. 135–153). Los Angeles: Roxbury.

Greenfeld, A., & Snell, T. (1999). *Women offenders: Bureau of Justice Statistics, special report.* Washington, DC: U.S. Department of Justice.

Guido, M. (1998, June 4). In a new twist on equality, girls' crimes resemble boys. *San Jose Mercury News*, p. 1B–4B.

Harms, P. (2002, January). *Detention in delinquency cases, 1989–1998* (OJJDP Fact Sheet No. 1). Washington, DC: U.S. Department of Justice.

Harrison, P. M., & Beck, A. J. (2004). *Prisoners in 2003.* Washington, DC: U.S. Department of Justice, Bureau of Justice Statistics.

Human Rights Watch. (1996). *All too familiar: Sexual abuse of women in U.S. state prisons.* New York: Human Rights Watch.

Irwin, J. (2005). *The warehouse prison.* Los Angeles: Roxbury.

Johnson, P. (2003). *Inner lives: Voices of African American women in prison.* New York: New York University Press.
Leonard, E. (2002). *Convicted survivors: The imprisonment of battered women.* New York: New York University Press.
Mauer, M. (1999). *Race to incarcerate.* New York: New Press.
Mauer, M., & Huling, T. (1995). *Young black Americans and the criminal justice system: Five years later.* Washington, DC: The Sentencing Project.
McClellan, D. S. (1994). Disparity in the discipline of male and female inmates in Texas prisons. *Women & Criminal Justice, 5*(20), 71–97.
Messerschmidt, J. W. (2000). *Nine lives: Adolescent masculinities, the body, and violence.* Boulder, CO: Westview.
Miller, S. (2005). *Victims as offenders: Women's use of violence in relationships.* New Brunswick, NJ: Rutgers University Press.
Morash, M., & Rucker, L. (1990). A critical look at the idea of boot camp as a correctional reform. *Crime & Delinquency, 36*(2), 204–222.
Warren, J. (2005, June 19). Rethinking treatment of female prisoners. *Los Angeles Times,* p. A1.
Websdale, N., & Alvarez, A. (1997). Forensic journalism as patriarchal ideology: The newspaper construction of homicide-suicide. In D. Hale & F. Bailey (Eds.), *Popular culture, crime and justice* (pp. 123–141). Belmont, CA: Wadsworth.
West, C., & Zimmerman, D. H. (1987). Doing gender. *Gender & Society, 1,* 125–151.
Williams, L. S. (2002). Trying on gender, gender regimes, and the process of becoming women. *Gender & Society, 16,* 29–52.

DISCUSSION QUESTIONS

1. What does Chesney-Lind mean by "backlash"?
2. Give an example of an anti-feminist agenda mentioned and decide whether you agree.
3. Give an example of a racist agenda mentioned in the article. Do you agree with the characterization?

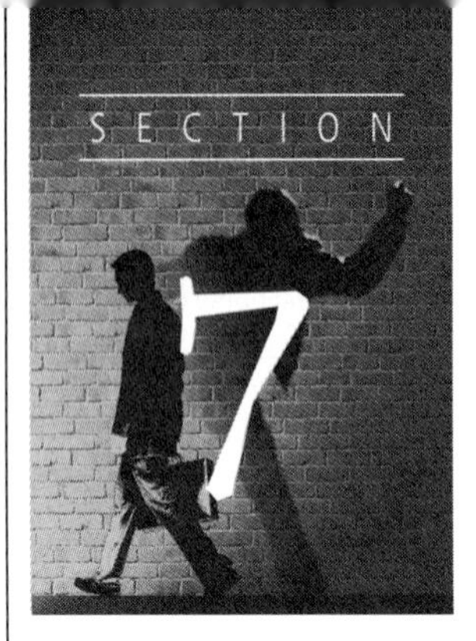

PSYCHOSOCIAL THEORIES

Individual Traits and Criminal Behavior

Little Jimmy Caine, a pug-nosed Irish American, is an emotionless, guiltless, walking id, all 5′5″ and 130 pounds of him. By the time he was 26, Jimmy had accumulated one of the worst criminal records the police in Toledo, Ohio, had ever seen. Burglary, robbery, aggravated assault, rape—name it, Jimmy had done it. This little tearaway had been arrested for the brutal rape of a 45-year-old barmaid. Jimmy entered an unlocked bar after closing time to find the lone barmaid attending to some cleaning chores. Putting a knife to the terrified woman's throat, he forced her to strip and proceeded to rape her. Because she was not sexually responsive, Jimmy became angry and placed her head over the kitchen sink and tried to decapitate her. His knife was a dull as his conscience, which only increased his anger, so he picked up a bottle of liquor and smashed it over her head. While the woman lay moaning at his feet, he poured more liquor over her, screaming: "I'm going to burn you up, bitch!" The noisy approach of the bar's owner sent Jimmy scurrying away. He was arrested 45 minutes later casually eating a hamburger at a fast food restaurant.

Jimmy didn't fit the demographic profile of individuals who engage in this type of crime. Although he had a slightly below average IQ, he came from a fairly normal intact middle-class home. However, Jimmy had been in trouble since his earliest days and had been examined by a variety of psychiatrists and psychologists. When he was 8 years old, psychiatrists diagnosed him with something called conduct disorder and at 18 he was diagnosed as having antisocial personality disorder.

Jimmy's case reminds us that we have to go beyond factors such as age, race, gender, and socioeconomic status to explain why individuals commit criminal acts. In this chapter, we look at many of the traits that psychologists and psychiatrists have examined to explain individual criminality. These explanations do not compete with sociological explanations; rather, they complete them.

We have called this section *psychosocial* rather than *psychological* because we consider it artificial to strictly separate social and psychological approaches. However, the emphasis is placed much more on the psychology than on the sociology of behavior. Psychosocial theories of criminal behavior are more interested in individual differences in the propensity to commit crimes than in environmental conditions that facilitate it. Early theories strongly emphasized **intelligence** and temperament, assuming that low intelligence hampers the ability to calculate the pleasures and pains involved in undertaking criminal activity, and that certain types of temperament make those possessing them impulsive and difficult to socialize.

The IQ/Crime Connection

David Wechsler (who devised many of the IQ tests in use today) defined intelligence as "the aggregate or global capacity of the individual to act purposefully, to think rationally, and to deal effectively with his [or her] environment" (in Matarazzo, 1976, p. 79). Although some social scientists still claim that IQ tests are biased, according to the National Academy of Sciences (Seligman, 1992) and the American Psychological Association's (APA) Task Force on Intelligence (Neisser et al., 1995), no study designed to detect such bias has ever done so.

A number of reviews of the IQ/crime relationship find it to be robust (Ellis & Walsh, 2003; Lynham, Moffitt, & Stouthamer-Loeber, 1993), although many studies lump together boys who commit only minor delinquent acts during their teenage years with boys who will continue to seriously and frequently offend into adulthood. Simple arithmetic tells us that pooling these two groups hides the magnitude of IQ differences between non-offenders and serious offenders if the latter have lower IQs than the former. Casual and less serious offenders differ from non-offenders by about one point while serious persistent offenders differ from non-offenders by about 17 points (Moffitt, 1993).

Most IQ studies look at *full-scale* IQ (FSIQ), which is obtained by averaging the scores on *verbal* (VIQ) and *performance* (PIQ) IQ sub-scales. While most people have VIQ and PIQ scores that closely match, offender populations are almost always found to have significantly lower than average VIQ scores, but not lower PIQ scores, than non-offenders. This VIQ/IQ discrepancy is called **intellectual imbalance.** As Miller (1987) remarks, "This PIQ>VIQ relationship was found across studies, despite variations in age, sex, race, setting, and form of the Wechsler [IQ] scale administered, as well in differences in criteria for delinquency" (p. 120). A literature review found that overall VIQ>PIQ boys are underrepresented in delinquent populations by a factor of about 2.6, and PIQ>VIQ boys are overrepresented by a factor of about 2.2 (Walsh, 2003).

The most usual explanation for the IQ/antisocial behavior link is that it works via poor school performance, which leads to dropping out of school and then associating with delinquent peers (Ward & Tittle, 1994). The idea that IQ influences offending via its influences on school performance is supported by a review of 158 IQ/criminal behavior studies that found it supported by 89% based on official statistics and 77.7% based on self-reports (Ellis & Walsh, 2000). On the other hand, all 46 studies exploring the link between grade point average (GPA) and antisocial behavior did so. Actual performance measures of academic achievement such as GPA are thus probably better predictors of antisocial behavior than is IQ. Academic achievement is a measure of the kind of intelligence IQ measures plus many other personal and situational characteristics, such as conscientious study habits and supportive parents.

IQ is related to a wide range of life outcomes that are themselves related to criminal and antisocial behavior such as poverty, lack of education, and unemployment. The data presented in Table 7.1 come from 12,686 white males and females in the National Longitudinal Study of Youth (NLSY). This study began in 1979 when subjects were 14 to 17 years old; the data were collected in 1989 when the subjects were 24 to 27 years old. The bottom 20% on IQ had scores of 87 and below; the top 20% had scores 113 and above. Note the large ratios between the two groups on all outcomes. For instance, for every 1 high-IQ subject ever interviewed in jail or prison there were 31 low-IQ subjects ever interviewed in jail or prison. Low IQ thus impacts many areas of life that increase the probability of offending.

Table 7.1 The Impact of High and Low IQ on Selected Life Outcomes

	IQ Level		
Social Behavior	**Bottom 20%**	**Top 20%**	**Ratio**
Dropped out of high school	66%	2%	33.0:1
Living below poverty level	48%	5%	9.6:1
Unemployed entire previous year*	64%	4%	16.0:1
Ever interviewed in jail or prison	62%	2%	31.0:1
Chronic welfare recipient	57%	2%	28.5:1
Had child out of wedlock**	52%	3%	17.3:1

SOURCE: NLSY data taken from various chapters in Herrnstein and Murray (1994).

*Males only; **females only.

The excerpt from Henry Goddard's 1914 book *Feeblemindedness* shows the long history in psychology of marrying intellectual ability to temperament to explain criminal behavior. Many of the terms Goddard used to describe people (imbeciles, morons, etc.) were once acceptable classification terms, but are now considered politically incorrect. It is clear that Goddard exaggerated the degree to which the typical criminal was feebleminded and the degree to which such a condition is inherited. Because the idea of feeblemindedness led to eugenics (the selective breeding of individuals of "good stock" and sterilization of "bad stock"), Goddard eventually withdrew his beliefs on the hereditary nature of feeblemindedness and on the morality of eugenic solutions to the problem of mental deficiency.

Temperament and Personality

Temperament is an individual characteristic that constitutes a person's habitual mode of emotionally responding to stimuli. Temperamental components include mood (happy/sad), sociability (introverted/extraverted), activity level (high/low), reactivity (calm/excitable), and affect (warm/cold), which make it easy or difficult for others to like us and to get along with us. Temperamental differences are largely a function of genes governing physiological arousal patterns, although arousal systems are fine-tuned by experience (Rothbart, Ahadi, & Evans, 2000). Temperamental differences in children make some easy to socialize and others difficult. Children who throw temper tantrums and reject warm overtures from others may adversely affect the quality of parent-infant interactions regardless of their parents' temperaments, and thus lead to poor parent-child attachment and all the negative consequences that result. Numerous studies have shown that parents, teachers, and peers respond to children with bad temperaments negatively, and that such children find acceptance only in association with others with similar dispositions (Caspi, 2000).

Personality is an individual's set of relatively enduring and functionally integrated psychological characteristics that result from his or her temperament interacting with cultural and developmental experiences. There are many different components of personality that psychologists call **traits,** some of which are associated with the probability of committing antisocial acts. People differ only on the strength of these traits; they are not characteristics that some people possess and others do not. We briefly address the major traits associated with offending, which is always the result of a constellation of risk factors rather than of any one factor.

Perhaps the trait most often linked to criminal behavior is **impulsiveness**, the tendency to act without giving much thought to the consequences. A review of 80 studies examining the relationship between impulsivity and criminal behavior found that 78 were positive and the remaining two were non-significant (Ellis & Walsh, 2000). Although impulsiveness is a potent risk factor for criminality in its own right, it becomes more potent if negative emotionality is added to the mix.

Negative emotionality refers to the tendency to experience many situations as aversive and to react to them with irritation and anger more readily than with positive affective states (McGue, Bacon, & Lykken, 1993). It is strongly related to self-reported and officially recorded criminality "across countries, genders, races, and methods" (Caspi, Moffitt, Silva, Stouthamer-Loeber, Krueger, & Schmutte, 1994). People who are low on constraint (they are impulsive) tend to be high on negative emotionality. Low levels of a brain chemical called serotonin underlie both high levels of negative emotionality and impulsivity, and Caspi and his colleagues (1994) claim that low serotonin may represent a constitutional predisposition for these traits, and thus a general vulnerability to criminality.

Sensation seeking refers to the active desire for novel, varied, and risky sensations (Zuckerman, 1990). Socialized sensation seekers will want to work as fire fighters, police officers, or any other job that provides physical activity, variety, and excitement; unsocialized sensation seekers, on the other hand, may well find their kicks in carjacking and burglary. A review of the literature found that 98.4% of the studies reported a statistically significant relationship between sensation seeking and antisocial behavior (Ellis & Walsh, 2000).

Empathy is the emotional and cognitive ability to understand the feelings and distress of others as if they were your own. The emotional component allows you to "feel" the other person's pain, and the cognitive component allows you to understand that person's pain and why he or she is feeling it. Some people carry the pains of the world on their shoulders while others couldn't care less about anyone. Most criminals fall into the "care less" category, for obvious reasons—you are less likely to victimize someone if you feel and understand what the consequences may be for him or her (Covell & Scalora, 2002).

Altruism can be thought of as the action component of empathy; if you feel empathy for someone, you will probably feel motivated to take some sort of action to alleviate his or her distress, if you are able. As with empathy, altruism lies on a continuum, with criminals (again for obvious reasons) on the wrong end of it. Lack of empathy and altruism is considered one of the most salient characteristics of psychopaths, the worst of the worst among criminals (Fishbein, 2001). A review of 24 studies of those traits found that 23 of them were statistically significant in the predicted direction; that is, the lower the level of empathy/altruism, the greater the antisocial behavior (Ellis & Walsh, 2000).

The final trait of concern to us is **conscientiousness**, which is a primary trait composed of several secondary traits: well-organized, disciplined, scrupulous, responsible, and reliable at one pole, and disorganized, careless, unreliable, irresponsible, and unscrupulous at the other. It is easy to see how conscientiousness could be directly related to crime through the inability of people who lack it to follow a legitimate path to the American Dream: "It is not merely a matter of talented individuals confronted with inferior schools and discriminatory hiring practices. Rather, a good deal of research indicates that many delinquents and criminals are untalented individuals who cannot compete effectively in complex industrial societies" (Vold, Bernard, & Snipes, 1998, p. 177). In other words, persons with certain kinds of temperament do not develop the qualities needed to apply themselves to the long and arduous task of achieving financial success legitimately and, as a consequence, may attempt to obtain it through crime.

Conscience and Arousal

One of the basic ideas of psychosocial criminology is that different levels of physiological arousal correlate with different personality and behavioral patterns because arousal levels determine what we pay attention to and the ease or difficulty of acquiring a conscience. **Conscience** is a complex mix of emotional and cognitive mechanisms that we acquire by internalizing the moral rules of our social group. People with strong consciences will feel guilt, shame, stress, and anxiety when they violate, or even contemplate violating, these rules. Differences in the emotional component of conscience reflect variation in **autonomic nervous system** arousal patterns (Kochanska, 1991). The autonomic nervous system (ANS) carries out the basic housekeeping functions of the body by funneling messages, which never reach our conscious awareness, from the environment to the various internal organs so that they may keep the organism in a state of biological balance (e.g., adjusting pupil size, shivering or sweating in response to temperature, etc.). Messages that influence ANS functioning and do reach awareness are important for the acquisition of conscience via a process called conditioning.

Classical conditioning is a form of learning that depends more on ANS arousal than on cognition, is visceral in nature, and forms an association between two paired stimuli. It is by way of these associations that we develop the "gut level" emotions of shame, guilt, and embarrassment that make up the emotional ("feeling") scaffolding of our consciences. Children must learn which behaviors are acceptable and which are not (the "knowledge" part of our conscience), most of which comes via parental teaching. Once children know the behavior expected of them, the degree to which emotions influence future behavior depends on the severity of the reprimand interacting with the responsiveness of their ANS (Pinel, 2000). Assuming an ANS that is adequately responsive to discipline, refraining from unacceptable behavior in the future it is not simply a rational calculation of cost and benefits, but rather a function of the emotional component of conscience strongly discouraging the behavior by generating unpleasant feelings.

Individuals with a readily aroused ANS are easily socialized; they learn their moral lessons well because ANS arousal ("butterflies in the stomach") is subjectively experienced as fear and anxiety. A hyper-responsive ANS generates high levels of fear and anxiety and is a protective factor against antisocial behavior. Studies have shown that males with *hyper*-arousable ANS's living in environments that put them at high risk for antisocial behavior were less involved with antisocial behavior than males living in low-risk communities with *hypo*-arousable (slow to arouse) ANS's (Brennan et al., 1997).

Individuals with relatively unresponsive ANS's are difficult to socialize because they experience little fear, shame, or guilt when they transgress, even when discovered and punished. Measures of ANS underarousal in childhood have enabled researchers to accurately predict which of their subjects would be criminal at age 24, with 75% accuracy (Raine, 1997). Across a wide variety of subjects and settings, it is consistently found that antisocial individuals evidence relatively unresponsive ANS's, and the reason for this is hypo-arousable ANS's do not allow for adequate conditioning of the social emotions. Having knowledge of what is right or wrong without that knowledge being paired with emotional arousal is rather like knowing the words to a song but not the music.

▲ **Photo 7.1** Arousal theory states that human beings have varying internal "thermostats," which explains why people differ on the levels of arousal or stimulation they need to feel comfortable. Those already set on "high" may attempt to avoid noise, activity, or crowds. On the other hand, those who crave stimulation might climb mountains, go to loud parties, or watch slasher movies.

Another form of arousal of interest to psychologists is neurological, the regulator of which is the brain's reticular activating system (RAS). Some individuals possess a RAS that is highly sensitive to incoming stimuli (augmenters), and others possess one that is unusually insensitive (reducers). Thus, in identical environmental situations, some people are under-aroused and other people are over-aroused, and both levels are psychologically uncomfortable. There is no conscious attempt to augment or reduce incoming stimuli; as is the reactivity of the ANS, augmentation or reduction is solely a function of differential physiology. Augmenters tend to be the people with hyperactive ANS's, and reducers tend to be people with hypoactive ANS's. Under-arousal of the ANS is associated with fearlessness and under-arousal of the RAS with sensation seeking. We can readily appreciate that sensation seeking and fearlessness are correlated because sensation seeking is aided by fearlessness (Raine, 1997).

Reducers are easily bored with levels of stimulation that are "just right" for most of us and continually seek to boost stimuli to what are, for them, more comfortable levels. They also require a high level of punishing stimuli before learning to avoid the behavior that provokes punishment, and are thus unusually prone to criminal behavior. A number of studies have shown that relative to the general population, criminals, especially those with the most serious records, are chronically under-aroused, as determined by EEG brain wave patterns, resting heart rate, and skin conductance (Ellis, 2003).

The article by Rene Veenstra and colleagues in this section discusses several of the traits we have discussed and their relationship to antisocial behavior. These Dutch researchers focus mainly on what they call negative affect and effortful control, which are analogous to what American researchers call negative emotionality and self-control (or constraint). The article addresses the interaction of these temperamental variables with socioeconomic status (SES) and type of parenting (warm, rejecting, overprotective). Parental rejection was positively linked to antisocial behavior, and parental warmth was negatively linked. The temperamental variables were both associated with antisocial behavior in predicted directions (the more the negative affect, the more the antisocial behavior; the more the effortful control, the less the antisocial behavior). SES was only related to antisocial behavior among boys high on negative affect and low on effortful control. The article highlights the importance of examining both environmental and individual-level variables in order to gain an adequate understanding of causality.

Glen Walters's Lifestyle Theory

Perhaps the best-known modern psychosocial criminological theory is Glen Walters's (1990) **lifestyle theory**. As the term *lifestyle* implies, Walters believes that criminal behavior is part of a general pattern of life characterized by irresponsibility, impulsiveness, self-indulgence, negative interpersonal relationships, and the chronic willingness to violate society's rules. Lifestyle theory has three key concepts: *conditions, choice,* and *cognition.* A criminal lifestyle is the result of *choices* criminals make "within the limits established by our early and current biological/environmental *conditions*" (Walters & White, 1989, p. 3). Thus, various biological and environmental conditions lay the foundation of future choices. Walters stresses impulsiveness and low IQ as the most important choice-biasing conditions at the individual level and attachment to significant others as the most important environmental condition.

The third concept, *cognition,* refers to cognitive styles people develop as a consequence of their biological/environmental conditions and the pattern of choices they have made in response to them. According to this theory, criminals display eight major cognitive features or **thinking errors** that make them what they are (Walters, 1990). Examples of criminal thinking errors are *cutoff* (the ability to discount the suffering of their victims), *entitlement* (the world owes them a living), *power orientation* (viewing the world in terms of weakness and strength), *cognitive indolence* (orientation to the present; concrete in thinking), and *discontinuity* (the inability to integrate thinking patterns). Little can be done to change criminal behavior until criminals change their pattern of thinking.

These thinking errors lead to four interrelated behavioral patterns or styles that almost guarantee criminality: *rule breaking, interpersonal intrusiveness* (intruding into the lives of others when not wanted), *self-indulgence,* and *irresponsibility.* These behavioral patterns are the result of faulty thinking patterns, which arise from the consequences (reward and punishment) of choices in early life, which are themselves influenced by biological and early environmental conditions.

The Antisocial Personalities

Depending on whom you ask, **antisocial personality disorder**, psychopathy, and sociopathy are terms describing the same constellation of traits or separate concepts with fuzzy boundaries. We use the generic term **psychopathy**, a syndrome we can define psychologically as characterized by egocentricity; deceitfulness; manipulativeness; selfishness; and a lack of empathy, guilt, or remorse, or physiologically as a syndrome characterized by the inability to tie the social emotions with cognition. Some researchers believe that there is a subset of psychopaths (so-called primary psychopaths) whose behavior is biological in origin, and a more numerous group (secondary psychopaths) whose behavior is the result of genetics and adverse environments (Mealey, 1995). Other researchers view psychopathy not as something one is or isn't, but rather as a name we have applied to the most serious and chronic criminal offenders. Psychiatrists apply the label antisocial personality disorder (APD) to such criminals. APD is defined by the American Psychiatric Association (APA) as "a pervasive pattern of disregard for, and violation of, the rights of others that begins in childhood or early adolescence and continues into adulthood" (APA, 1994, p. 645). The criteria for diagnosing APD are both clinical and legal, but rest primarily on behavior.

Criminologists, however, generally want to define individuals according to criteria that are independent of their behavior and then determine in what ways those so defined differ from individuals not so defined. The most widely used measure of psychopathy is the **Psychopathy Checklist-Revised** (PCL-R), which was devised by Robert Hare, the leading expert in psychopathy in the world today (Bartol, 2002). Hare (1993) is a leading proponent of the idea that psychopathy (which he considers to be a different construct from APD) is primarily biological in origin: "I can find no convincing evidence that psychopathy is the direct result of early social or environmental factors" (p. 170). Hare is among a subset of researchers that echo Cesare Lombroso, who probably had psychopaths in mind with his "morally insane" born criminals; namely, those "who appear normal in physique and intelligence but cannot distinguish good from evil" (in Gibson, 2002, p. 25). Modern researchers no longer talk of criminals as evolutionary throwbacks whose behavior is "unnatural"; rather, they view psychopaths as behaving exactly as they were designed by natural selection to behave (Harris, Skilling, & Rice, 2001; Quinsey, 2002). This does not mean that their behavior is acceptable or that we cannot consider it *morally* pathological and punish it accordingly.

▲ **Photo 7.2** Although never formally diagnosed as a psychopath, John Wayne Gacy's behavior was psychopathic to the core. Gacy raped, tortured, and killed at least 30 young males and buried them in his crawlspace. He was executed in 1994.

If psychopathy is a behavioral strategy forged by natural selection, there must be a number of markers that distinguish psychopaths from the rest of us. One of the most consistent physiological findings about psychopaths is their greatly reduced ability to experience the social emotions of shame, embarrassment, guilt, and empathy (Scarpa & Raine, 2003; Weibe, 2004). Social emotions modify brain activity in ways that lead us to choose certain responses over others. Feelings of guilt, shame, and empathy prevent us from doing things that might be to our immediate advantage (steal, lie, cheat) but would cost us in reputation and future positive relationships if discovered. Hundreds of studies using many different methods have revealed over and over that the defining characteristic of psychopaths is their inability to "tie" the brain's cognitive and emotional networks together, and thus to form a conscience (Scarpa & Raine, 2003).

Some theorists believe that sociopaths are different from psychopaths. Lykken (1995) colorfully describes sociopaths as "feral creatures, undomesticated predators, stowaways on our communal voyage who have never signed the Social Contract," and states that their behavior is "traceable to deviant learning histories interacting, perhaps, with deviant genetic predilections" (p. 22). One of the biggest factors contributing to sociopathy is poor parenting, which is itself a function of the increase in the number of children being born out of wedlock (Rowe, 2002).

According to a study taken from the National Longitudinal Survey of Youth of 1,524 sibling pairs from different family structures, Cleveland and his colleagues (2000) found that, on average, unmarried mothers have a tendency to follow an impulsive and risky lifestyle and to have a number of antisocial personality traits, be more promiscuous, and to have a below-average IQ. Families headed by single mothers with children fathered by different men were found to be the family type that put offspring most at risk for antisocial behavior. Two-parent families with full siblings placed offspring at lowest risk. Similar findings and conclusions from a large-scale British study have been reported (Moffitt & The E-Risk Study Team, 2002). Finally, Barber (2004) found that illegitimacy was by far the strongest predictor of a composite measure of violent crime (murder, rape, assault) in his sample of 39 countries from Argentina to Zimbabwe.

Given the many and severe deficits faced by many children born out of wedlock, it is no surprise that Gottfredson and Hirschi (1997) concluded that "delaying pregnancy among unmarried girls would probably do more to affect the long-term crime rates than all the criminal justice programs combined" (p. 33). This view is shared by the Office of Juvenile Justice and Delinquency Prevention (OJJDP), which claims that delaying pregnancy until 20–21 years of age would lead to a 30% to 40% reduction in child abuse and neglect and could potentially save $4 billion in law enforcement and corrections costs because offspring of teenage mothers are 2.7 times more likely than offspring of adult mothers to be incarcerated (Maynard & Garry, 1997).

British researchers Ahn Vien and Anthony Beech offer an excellent overview of the theories of psychopathy, its measurement, and its treatment in their article included in this section. Vien and Beech conclude that psychopathy is the result of the interplay of neurobiological, psychological, and social factors, but that the only agreement among all researchers in this area is that it involves the amygdalae (part of the brain that plays a role in emotional memories and contains areas implicated in rage, fear, and sexual feelings) and the prefrontal cortex. This goes a long way in explaining the difficulties psychopaths have in processing emotions. The section on treatment is particularly interesting, given that most psychopathy researchers believe that treatment only makes psychopaths worse.

Table 7.2 summarizes the strengths and weaknesses of the psychosocial theories described above.

Evaluation of the Psychosocial Perspective

Psychologists are always happy to point out that whatever social conditions may contribute to criminal behavior, they must influence individuals before they affect crime. Social factors matter and may well "prepare the scene" for crimes, but real flesh and blood people commit them.

Table 7.2 Summarizing Psychosocial Theories

Theory	Key Concepts	Strengths	Weaknesses
Arousal	Because of differing ANS and RAS physiology, people differ in arousal levels they consider optimal. Under-arousal under normal conditions poses an elevated risk of criminal behavior because it signals fearlessness, boredom, and poor prospects for socialization.	Allows researchers to use "harder" assessment tools such as EEG's to measure traits. Ties behavior to physiology. Explains why individuals in "good" environments commit crimes and why individuals in "bad" ones do not.	May be too individualistic for some criminologists. Puts all the "blame" on the individual's physiology. Ignores environmental effects.
Lifestyle	Crime is a patterned way of life (a lifestyle) rather than simply a behavior. Crime is caused by errors in thinking, which results from choices previously made, which are the results of early negative biological and environmental conditions.	Primarily a theory useful for correctional counselors dealing with their clients. Shows how criminals think and how these errors in thinking lead them into criminal behavior.	Concentrates only on thinking errors. Does talk about why they exist but pays scant attention to these reasons.
Antisocial personality	There is a small stable group of individuals who may be biologically obligated to behave antisocially (psychopaths) and a larger group who behave similarly but whose numbers grow or subside with changing environmental conditions (sociopaths).	Concentrates on the scariest and most persistent criminals in our midst. Uses theories from evolutionary biology and "hard" brain imaging and physiological measures to identify psychopaths.	Some doubt the division of psychopath and sociopath as separate entities. While they are the scariest criminals, they constitute only a small proportion of all criminals.

Individuals are differentially vulnerable to the criminogenic forces existing in the environment because they have different personalities: "The heat that melts the butter hardens the egg." Psychologists largely take the heat (the environment) for granted and look for how the butter and eggs of our differing constitutions relate to the heat.

The relationship between IQ and criminal behavior has always been contentious. Adler, Mueller, and Laufer (2001, p. 109) voice the familiar criticism that IQ tests are culturally biased, despite the findings of the National Academy of Sciences and the APA's Task Force on Intelligence cited earlier. They also cite the "debate" over whether genetics or the environment "determines" intelligence. This implies that an either/or answer is possible, but since scientists involved in the study of intelligence are unanimous that all traits are *necessarily* the result of both genes and environment, it is a monumental non-debate (Carey, 2003, p. 4).

One of the most pervasive criticisms of psychological theories is that they focus on "defective" or "abnormal" personalities (Akers, 1994, p. 86). If by "abnormal," critics mean *statistical* abnormality (below or above the average on a variety of traits), however, then by definition all theories of criminality focus on abnormality. Our personalities consist of normal variation in traits we *all* possess, and these are products of the interaction of our temperaments and our developmental experiences. Lifestyle theory emphasizes that criminals are at the tails of normal distributions of many traits, but focuses mainly on how they think as a result.

Policy and Prevention: Implications of Psychosocial Theories

The best anticrime policies are environmental because they are aimed at reducing the prevalence of crime in the population. But because such policies have had little impact on the crime problem in the past (Rosenbaum, Lurigio, & Davis, 1998), perhaps it is wise at present to focus efforts on those who are already committing crimes, rather than on conditions external to them. A variety of such programs aimed at rehabilitating offenders operate under the assumption that they are rational beings who are, however, plagued by ignorance of the long-term negative consequences of their offending behavior.

There is wide disagreement on how well rehabilitative programs work and even about the criteria for success. Reviews of studies with strict criteria for determining success find recidivism rates lowered by between 8% and 10% (Andrews & Bonta, 1998). Effective programs use multiple treatment components; are structured and focus on developing social, academic, and employment skills; use directive cognitive-behavioral counseling methods; and provide substantial and meaningful contact between treatment personnel and offenders (Sherman, Gottfredson, McKenzie, Eck, Reuter, & Bushway, 1997).

Glen Walters's theory deals with what correctional counselors call "stinkin' thinkin'." Counselors see their task as guiding offenders to realize how destructive their thinking has been in their lives. The counselor sees offenders' problems as resulting from illogical and negative thinking about experiences that they reiterate in self-defeating monologues. The counselor's task is to strip away self-damaging ideas (such as techniques of neutralization) and beliefs by attacking them directly and challenging offenders to reinterpret their experiences in a growth-enhancing fashion. The cognitive-behavioral counselor operates from the assumption that no matter how well offenders come to understand the remote origins of their behavior, if they are unable to make the vital link between those origins and current behavioral problems, it is of no avail.

Psychopaths are poor candidates for treatment. Robert Hare (1993) states that because they are largely incapable of the empathy, warmth, and sincerity needed to develop an effective treatment relationship, treatment often makes them worse because they learn how to better push other people's buttons. Old age seems to be the only "cure" for the behaviors associated with this syndrome.

Summary

- Psychosocial criminology focuses largely on intelligence and temperament as the most important correlates of criminal behavior. Low intelligence, as measured by IQ tests, is thought to be linked to crime because people with low IQ are said to lack the ability to correctly calculate the costs and benefits of committing crimes, and temperament is linked to crime largely in terms of impulsiveness. Intelligence is the product of both genes and environment. We concluded that IQ is probably related to crime and delinquency through its effect on poor school performance.

- Temperament constitutes a person's habitual way of emotionally responding to stimuli. The kind of temperament we inherit makes us variably responsive to socialization, although patient and caring parents can modify a difficult temperament.

- Our personalities are formed from the joint raw material of temperament and developmental processes. A number of personality traits are associated with the probability of engaging in antisocial behavior, particularly being high on impulsiveness, negative emotionality, and sensation seeking, and being low on conscientiousness, empathy, altruism, and moral reasoning.

- Classical conditioning via the ANS constitutes the emotional component of conscience and precedes the cognitive component. People differ in the responsiveness of their ANS, with people having a sluggish ANS being difficult to socialize. RAS arousal is also important to understanding criminal behavior because RAS reducers are chronically bored and seek to increase stimuli to alleviate that boredom. This may result in criminal behavior. Although people will differ greatly in their behavior depending on their innate temperaments (a function of arousal levels), their developmental and other environmental experiences also play huge parts.

- Lifestyle theory views criminal behavior as a lifestyle rather than just another form of behavior. The lifestyle begins with biological and environmental conditions that lead criminals to make certain choices, which in turn leads to criminal cognitions. The theory focuses on these cognitions, or "thinking errors." Thinking errors lead criminals into behavioral patterns that virtually guarantee criminality. The theory was devised primarily to assist correctional counselors to change criminal thinking patterns.

- Psychopaths are at the extreme end of the antisocial personality continuum. Most researchers regard the psychopathy syndrome as biological in origin, whereas some view sociopaths as formed both by genetics and the environment, with the environment playing the larger role. Many hundreds of studies have shown psychopaths have limited ability to tie the rational and emotional components of thinking together.

◆ Some researchers assert that the primary cause of psychopathy is inept parenting by single mothers. Other theorists point to the fact that children born to such mothers also receive genes advantageous to antisocial behavior from both parents, in addition to an environment conducive to its expression.

EXERCISES AND DISCUSSION QUESTIONS

1. Since psychologists have long identified different temperaments as something that makes it easy or difficult to socialize children, why do you think Gottfredson and Hirschi ignored it in their self-control theory?
2. Honestly rate yourself from 1 to 10 on the traits positively associated with antisocial behavior (impulsiveness, negative emotionality, and sensation seeking), then on the traits negatively related with antisocial behavior (conscientiousness, empathy, and altruism). Subtract the latter from the former. If the difference is a positive number greater than 10, or a negative number less than –10, does this little exercise correspond to your actual behavior?
3. Explain what role low arousal of the autonomic nervous system and the reticular activating system play in psychopathy.

USEFUL WEB SITES

IQ. http://www.psyonline.nl/en-iq.htm.

IQ/aggression Connection. http://www.crimetimes.org/95c/w95cp10.htm.

Mental Deficiency and Crime. http://faculty.ncwc.edu/TOConnor/301/301lect04.htm.

Mental Health. http://www.mentalhealth.com/.

Personality Disorders. http://www.focusas.com/PersonalityDisorders.html.

CHAPTER GLOSSARY

Altruism: An *active* concern for the well-being of others.

Antisocial personality disorder: A pervasive pattern of antisocial behavior that continues across the lifespan.

Autonomic nervous system: Part of the body that carries out the basic housekeeping functions; also the physiological basis of the conscience.

Classical conditioning: A form of learning depending on ANS arousal that forms an association between two paired stimuli.

Conscience: A mix of emotional and cognitive mechanisms acquired by socialization.

Conscientiousness: A primary trait composed of several secondary traits such as being well-organized, disciplined, scrupulous, responsible, and reliable.

Empathy: The emotional and cognitive ability to understand the feelings and distress of others as if they were your own.

Impulsiveness: The tendency to act on the spur of the moment without thinking.

Intellectual imbalance: A significant difference between a person's VIQ and PIQ.

Intelligence: The capacity to act purposefully, to think rationally, and to deal effectively within one's environment.

Lifestyle theory: A theory stressing that crime is part of a general pattern of life.

Negative emotionality: The tendency to experience many situations as aversive and to react to them with irritation and anger.

Personality: The relatively enduring set of psychological characteristics that results from people's temperaments interacting with their cultural and developmental experiences.

Psychopathy: A syndrome characterized psychologically by egocentricity; deceitfulness; manipulativeness; selfishness; and a lack of empathy, guilt, or remorse or physiologically by the inability to tie brain areas controlling social emotions with areas controlling cognition.

Psychopathy Checklist-Revised: The most widely used measure of psychopathy, devised by Robert Hare.

Sensation seeking: The desire for novel, varied, and extreme sensations and experiences.

Temperament: A person's habitual mode of emotionally responding to stimuli; the biological basis of personality.

Thinking errors: Criminals' typical patterns of faulty thoughts and beliefs.

Traits: The many different components of personality, some of which are associated with the probability of committing antisocial acts.

READING

Feeble-Mindedness

H. H. Goddard

This excerpt from Henry Goddard's 1914 book *Feeblemindedness: Its Causes and Consequences* illustrates the thinking about criminals and the mentally deficient at the beginning of the 20th century. Terms such as "moron," "idiot," and "imbecile" have long been discarded by psychologists. Goddard pitied those he called feeble-minded and considered them mistreated, misunderstood, and driven into criminality, although he considered them to be "well fitted by nature" to it. Note that Goddard's thinking is consistent with modern psychosocial criminology; he considered criminality to be related to low intellectual functioning when combined with the right kind of temperament and the right kind of environment.

Society's attitude toward the criminal has gone through a decided evolution, but that evolution has been in the line of its treatment rather than of its understanding of him and of his responsibility. Almost up to the present time there has been a practically universal assumption of the responsibility of all except the very youngest children and those recognized as idiots, imbeciles or insane. The oldest method of treatment was in accordance with the idea of vengeance, an eye for an eye. The god Justice was satisfied if the offender suffered an equal amount with those whom he had made suffer. Later came the idea of punishing an offender for the sake of deterring others from similar crimes. This is the basis of much of our present penal legislation. But students of humanity have gone farther and now realize that the great function of punishment is to reform the offender.

We have had careful studies of the offender from this standpoint. Studies have been made of his environment and of those things which have led him into crime. Attempts have been made to remove these conditions, so that criminals shall not be made, or having reformed, they shall not again be led into a criminal life. A great deal has been accomplished along these lines. But we shall soon realize, if we have not already, that on this track there is a barrier which we cannot cross. Environment will not, of itself, enable all people to escape criminality. The problem goes much deeper than environment. It is the question of responsibility. Those who are born without sufficient intelligence either to know right from wrong, or those, who if they know it, have not sufficient will-power and judgment to make themselves do the right and flee the wrong, will ever be a fertile source of criminality. This is being recognized more and more by those who have to do with criminals. We have no thought of maintaining that all criminals are irresponsible. Although we cannot determine at present just what the proportion is, probably from 25 percent to 50 percent of the people in our prisons are mentally defective and incapable of managing their affairs with ordinary prudence. A great deal has been written about the criminal type and its various characteristics. It is interesting to see in the light of modern knowledge of the defective that these descriptions are almost without exception accurate descriptions of the feeble-minded.

The hereditary criminal passes out with the advent of feeble-mindedness into the problem. The criminal is not born; he is made. The so-called criminal type is merely a type of feeble-mindedness, a type misunderstood and mistreated, driven into criminality for which he is well fitted by nature. It is hereditary feeble-mindedness not hereditary criminality that accounts for the conditions. We have seen only the end product and failed to recognize the character of the raw material.

Perhaps the best data on this problem come from the prisons and the reformatories. It is quite surprising to see how many persons who have to do with criminals are coming forward with the statement that a greater or less percentage of the persons under their care are feeble-minded. They had always known that a certain proportion were thus affected, but since the recognition of the moron and of his characteristics, the percentage is found ever higher and higher. The highest of all come from the institutions for juveniles, partly because it is difficult to believe that an adult man or woman who makes a fair appearance but who lacks in certain lines, is not simply ignorant. We are more willing to admit the defect of children. The discrepancy is also due to the fact that the mental defectives are more apt to die young leaving among the older prisoners those who are really intelligent. . . .

From [various] studies, we might conclude that at least 50 percent of all criminals are mentally defective. Even if a much smaller percentage is defective, it is sufficient for our argument that without question one point of attack for the solution of the problem of crime is the problem of feeble-mindedness. It is easier for us to realize this if we remember how many of the crimes that are committed seem foolish and silly. One steals something that he cannot use and cannot dispose of without getting caught.

Sometimes the crime itself is not so stupid but the perpetrator acts stupidly afterwards and is caught, where an intelligent person would have escaped. Many of the "unaccountable" crimes, both large and small, are accounted for once it is recognized that the criminal may be mentally defective. Judge and jury are frequently amazed at the folly of the defendant—the lack of common sense that he displayed in his act. It has not occurred to us that the folly, the crudity, the dullness, was an indication of an intellectual trait that rendered the victim to a large extent irresponsible. . . .

As a rule, our workers have easily been able to decide the mentality of the persons they saw. In some cases, indeed, this was not so easy and only after much observation and questions of neighbors and friends as to the conduct and life of these persons was it possible to come to a reasonably satisfactory conclusion. In many cases it has been impossible to decide even after all our care; and these cases are therefore left undetermined.

In regard to the persons not seen, and especially those of earlier generations who are no longer living, the task at first sight seems more difficult. Some even assume that it is impossible to determine the mentality of such cases unless they were commonly recognized idiots or imbeciles. That such is not the fact however will become evident from a little thoughtful consideration. It must be remembered that the field worker goes out with a background of knowledge of four hundred feeble-minded boys and girls, men and women, of all grades of intelligence, and a great variety of temperaments and hereditary influences. With this background, it is possible to project any individual into a known group and decide that he is or is not like someone in the group. This of course must not be done, and is not done, by any superficial resemblance but the basis of many fundamental characteristics.

The idea that it is impossible to determine the mentality of a person who is three or four generations back of the present is partly an ill-considered one and partly the result of erroneous logic. One says—"I don't know my own grandparents; I do not even know their names." And the implied argument is "If a person as intelligent as I am does not know his grandparents, how can these ignorant defectives know theirs?"

Again, the fact that I do not know my grandparents does not prove that no one now living knew them. As a matter of fact, there are numerous people now living who knew them well. It is a surprise to us to be told that there are persons now living who remember heroes of the American Revolution! John Doe enlisted in the Continental army on 1775 at the age of twenty. He died in 1845 at the age of ninety. Richard Roe was twelve years old at that time and vividly remembers hearing the old man Doe tell of the exciting experiences of '76. Richard Roe is eighty-one years old now. That is a rare occurrence? Certainly. And we have been able to determine that a person in the sixth generation back was feeble-minded in only one family out of 327—the Kallikak family. For the fifth generation we have made determinations in only four cases and even these are not involved in our conclusions.

The ease with which it is sometimes possible to get satisfactory evidence on the fifth generation is illustrated in *The Kallikak Family*. The field worker accosts an old farmer—"Do you remember an old man, Martin Kallikak (Jr.), who lived on the mountain edge yonder?" "Do I? Well, I guess! Nobody'd forget him. Simple, not quite right here (tapping his head) but inoffensive and kind. All the family was that. Old Moll, simple as she was, would do anything for a neighbor. She finally died—burned to death in the chimney corner. She had come in drunk and sat down there. Whether she fell over in a fit or her clothes caught fire, nobody knows. She was burned to a crisp when they found her. That was the worst of them, they would drink. Poverty was their best friend in this respect, or they would have been drunk all the time. Old Martin could never stop as long as he had a drop. Many's the time he's rolled off of Billy Parson's porch. Billy always had a barrel of cider handy. He'd just chuckle to see old Martin drink and drink until finally he'd lose his balance and over he'd go!" Is there any doubt that Martin was feeble-minded?

Classification

Our 327 families naturally fall into six fundamental groups as follows:

1. Where feeble-mindedness is *certainly* hereditary—designated hereafter for brevity's sake as the Hereditary Group or Hereditary (H). *164 families.*
2. A group which, while not so *certainly* hereditary, yet shows high degrees of probability that the feeble-mindedness is hereditary—designated as Probably Hereditary (PH.). *34 families.*
3. A group in which there is no evidence of hereditary feeble-mindedness, but in which the families show marked neuropathic conditions—designated as the Neuropathic Group or Neuropathic (Neu.). *37 families.*
4. A group where it is clear that some accident either to mother or child, including disease, injury at birth, etc., is the cause of the feeble-mindedness—designated the Accident Group. *57 families.*
5. A small group where it has been impossible to assign a cause. The family history is known and is good; there are no accidents. We have designated this No Cause Discovered, or briefly, No Cause (NC.). *8 families.*
6. A group where so little of a definite character could be learned that it was impossible to classify them—designated as Unclassifiable (Uncl.). *27 families.* This group is not counted at all in making up the percentages. One case in this group was thrown out because it proved to be a case of insanity and not of feeble-mindedness.

Each of these fundamental groups of charts is subdivided and arranged according to mental age as determined by the Binet-Simon Measuring Scale of Intelligence. This gives the child's mentality in terms of a normal child, e.g., mentality 7 means like a normal or average child of 7 years in intelligence. We may speak of a man 40 years old as having a mentality of any age from 1 to 12. We say he tests 7, or his mental age is 7.

Each chart is accompanied by a condensed description of the child. The information comes from parents, physicians and the Institution records, including the school department and the department of research. The latter are incomplete on the physical and the physiological (bio-chemical) side because we have not yet completed systematic studies of these cases. That must wait for a later report. . . .

Criminality and Feeble-Mindedness

Every feeble-minded person is a potential criminal. This is necessarily true since the feeble-minded lacks one or the other of the factors essential to a moral life—an understanding of right and wrong, and the power of control. If he does not know right and wrong does not really appreciate this question, then of course he is as likely to do the wrong thing as the right. Even if he is of sufficient intelligence and has had the necessary training so that he does know, since he lacks the power of control he is unable to resist his natural impulses.

Whether the feeble-minded person actually becomes a criminal depends upon two factors, his temperament—and his environment. If he is of a quiet, phlegmatic temperament with thoroughly weakened impulses, he may never be impelled to do anything seriously wrong. In this case when he cannot earn a living he will starve to death unless philanthropic people provide for him. On the other hand, if he is a nervous, excitable, impulsive person he is almost sure to turn in the direction of criminality. Fortunately for the welfare of society the feeble-minded person as a rule lacks energy. But whatever his temperament, in a bad environment he still becomes a criminal, the phlegmatic temperament becoming simply the dupe of more intelligent criminals, while the excitable, nervous, impulsive feeble-minded person may escape criminality if his necessities are provided for, and his impulses and energies are turned in a wholesome direction.

It is not easy to decide beforehand which of these conditions is fulfilled in a particular group. In the data that we are studying, criminality seems at first sight to be surprisingly small. This is partly explained by the fact that our cases include only those who have been under arrest, Thirty-two charts with a total of forty-five individuals show criminality. That is, criminality appears on 10 percent of the charts, but only one-third of 1 percent of the individuals are criminalistic. It is perhaps significant that the greater proportion of these are in the Hereditary Group. Thirty of the charts in the Hereditary Groups, or 15.1 percent, have criminals on them; in the Neuropathic Group two charts or 5.4 percent, in the Accidents none. The criminal individuals are 0.52 percent of the persons in the Hereditary Groups; 0.24 percent of those in the Neuropathic Group and none of the accidents. Of the 45 criminals 41 are men, 4 are women, while 24 are known to be feeble-minded. 1 is normal and 20 unknown.

It is probable that in these cases two factors account for the small proportion of criminals. These people are very largely from rural districts, and their temptations perhaps have not been so great. But more significant is the fact in such communities minor kinds of crime are not taken account of, so that they do not get marked "criminal" because they were never arrested. In the city cases, our data are always much less complete. There are individuals of whom we have learned enough to determine their mentality while not being able to follow their careers. They have left home or have been lost sight of and may be today in prison without their family and relatives knowing anything about it. Undoubtedly, there are cases that escape in this way, but on the whole it seems probable that the fact of a criminal life would be one that we would be likely to discover if it existed. Such facts are hard to conceal.

There are nine criminalistic individuals on the charts that do not belong to the family, that is to say, they have married in, and they are only significant as showing the kind of company these people keep. . . .

DISCUSSION QUESTIONS

1. What does Goddard say about the relationship between "feeblemindedness" and criminal behavior? How different is it from what others say today?
2. What else other than feeblemindedness is required to increase the probability of offending?
3. What would Goddard have to say about low self-control theory? Would he support it or would he say that it is a function of something else?

READING

Temperament, Environment, and Antisocial Behavior in a Population Sample of Preadolescent Boys and Girls

René Veenstra, Siegwart Lindenberg, Albertine J. Oldehinkel, Andrea F. De Winter, and Johan Ormel

This article by René Veenstra and colleagues examines the risk-producing and risk-buffering interactions between temperament, perceived parenting, socioeconomic status, and sex in terms of antisocial behavior. Their sample is composed of preadolescent Dutch children (N = 2230). The researchers found that all parenting and temperament factors were significantly associated with antisocial behavior. The strongest risk-buffering interactions were found for SES, which was only related to antisocial behavior among children with a low level of effortful control or a high level of frustration. Furthermore, the associations of SES with antisocial behavior were more negative for boys than for girls. Thus, the effects of SES on antisocial behavior depend on both temperament and child's sex.

Introduction

Both temperament and parenting are important in understanding antisocial behavior. Environmental factors may vary in their developmental influence as a function of attributes of the child. Research has also shown that how parents rear their children is partially shaped by the parents' own characteristics and partially by the characteristics of the children that they bring up. Thus, a difficult temperament does not necessarily lead to antisocial behavior by itself; it does so in conjunction with particular environments. Thomas and Chess (1977) called this a goodness of fit between an individual's temperament and the expectations and resources of specific contexts. Others talked about "risk-buffering" effects with regard to temperament-by-environment interactions (e.g., Belsky, Hsieh, & Crnic, 1998). The current study deals with such risk buffering effects for preadolescents on antisocial behavior.

The home environment of children, with as important indicators the parenting style and the socio-economic status, has a profound impact on the development of children. It has been suggested that research that focuses on the interaction effects of parenting and child temperament might do more justice to the complexity of child development (Magnusson & Stattin, 1998). For example, Kochanska (1997) studied the development of conscience in young children, and discovered that for shy, temperamentally-fearful children, parental power-assertion does not appear to promote conscience development.

Source: Veenstra, R. S., Lindenberg, A., Oldehinkel, A., De Winter, A., & Ormel, J. (2006). Temperament, environment, and antisocial behavior in a population of pre-adolescent boys and girls. *International Journal of Behavioral Development, 30*(5), 422–432. Reprinted with permission of Sage Publications, Ltd.

Gentler techniques are called for with such slow-to-warm-up children. But with bold assertive children, effective parenting involves firmness, along with maternal responsiveness and the formation of a close emotional bond with the child. Belsky et al. (1998) concluded that children with a difficult temperament are most susceptible to parenting practices. Bates et al. (1998) investigated the interaction effect of the child's temperamental resistance to control and parents' restrictive control at an early age on externalizing behavior at ages 7 to 11 years. A robust finding was that early resistance to control predicted later externalizing behavior better when the mother had been relatively low in control actions, which fits with Kochanska's (1997) findings for toddlers. Given high resistance to control, the risk of later externalizing behavior by the child seems to be buffered by high control actions by the mother. Most research on such temperament-by-environment interactions has been done with toddlers and young children, and the question is whether and to what extent we can generalize the results of such studies to late childhood or adolescence.

There are a number of studies that have examined temperament-by-environment interactions in late childhood or adolescence. For example, Stice and Gonzales (1998) found that temperament interacted with perceived parenting in their effects on antisocial behavior. Effective parenting (i.e., maternal control) was most important for youths that were temperamentally at risk (i.e., high on behavioral undercontrol). Stice and Gonzales (1998) reasoned that adolescents who are behaviorally controlled are unlikely to evidence problem behaviors, regardless of the parenting they experience.

Consistently, other studies found that ineffective parenting was least harmful for youths that were not temperamentally at risk. For example, Wills, Sandy, Yaeger, and Shinar (2001) found that the impact of parental risk factors, i.e., parent-child conflict and parental tobacco and alcohol use, on adolescent substance abuse decreased with higher task attention (focusing on tasks and persisting until finished) and positive emotionality (generally being in a cheerful mood and smiling frequently) of the preadolescent. Wills et al. (2001) argue that these temperamental factors promote adaptation through reducing reactivity to aversive stimuli, a resilience effect. Wills and Dishion (2004) say that, for example, the emergence of good self-control can serve as a resilience factor. Maziade et al. (1990) found that only the combination of difficult temperament and dysfunctional parenting (in particular inadequate behavioral control) in childhood was associated with an increased risk of developing psychiatric disorders in adolescence. Van Leeuwen, Mervielde, Braet, and Bosmans (2004) found that negative parental control was more related to externalizing behavior for undercontrollers (i.e., low on conscientiousness and benevolence and around the mean on imagination, extraversion, and emotional stability) than for resilients (i.e., high on imagination, conscientiousness, extraversion, benevolence, and emotional stability). Sanson, Hemphill, and Smart (2004) concluded that the combination of irritable, difficult child temperament with poor, especially punitive, parenting adds to the prediction of antisocial behavior beyond their independent effects.

The Study

To do more justice to the complexity of child development, we focus not only on main effects of sex, temperament, and environment but also on risk-buffering interactions between temperament and environment in relation to antisocial behavior. We define antisocial behavior as behavior that inflicts physical or mental harm or property loss or damage on others. It is behavior that intends to lower the well-being of other persons that may or may not constitute the breaking of criminal laws.

Environment

We will make use of perceived parenting rather than observed or parent-reported parenting. The choice of perceived parenting styles included in our study has been inspired by the fact that several studies have found them to be strongly linked to antisocial behavior. It has been found that *perceived rejection* (characterized by hostility, punishment, derogation, and blaming of subject), and *perceived overprotection* (characterized by fearfulness and anxiety for the child's safety, guilt engendering, and intrusiveness) are both positively linked, whereas *perceived emotional warmth* (characterized by affection, attention, and support) is negatively linked to antisocial behavior. In addition to the three perceived parental environments, we also consider SES of the parents as an (objective) environment of the child. SES is a proxy for a number of important aspects of parenting. SES has proven to be an important proxy for effects of social, cultural, and financial capital on child development that cannot easily be unpacked into factors such as parenting styles made clear that socioeconomic conditions in childhood have a big impact on the life chances of children. For example, the higher the occupational level of parents the higher their autonomy. This autonomy is related to other characteristics of the family, such as the lifestyle, the parenting style, and the aspirations of themselves and their children. Parents with a high occupational level educate their children more authoritatively, whereas parents with a low occupational level educate their children more restrictively. Thus, SES can be seen as a proxy for the human, cultural, and financial capital of a family that will not quite be made superfluous by parenting style.

Temperament

We focus on two temperamental aspects—effortful control and frustration. Effortful control, denoting the ability to regulate attention and behavior, is believed to make major contributions to social development. Children with low effortful control are less likely to consider the possible consequences of their actions, especially consequences that are likely to be long-delayed. Frustration is a temperament feature characterized by negative affect related to goal blocking or an interruption of ongoing tasks. In other words, children with a high level of frustration react strongly and aversively to obstacles that prevent them from doing what they want.

Potential Confounders

Interpretation of associations between family circumstances and (risk factors for) antisocial behavior is hampered by potential confounders, including sex and genetic disposition.

Genetic risk factors may have a (direct) effect on both temperament and antisocial behavior, which could mean that observed associations between the two are spurious. In addition, the effect of genetic risk can be indirect, through gene-environment correlations. In other words, what seems to be the effect of poor parenting behavior may actually be the effect of susceptibility genes, or vice versa. To assess possible confounding, we included a proxy of genetic risk, that is, an index of familial externalizing psychopathology, in the analyses.

Hypotheses

[O]ur *environment hypotheses* are that rejection and overprotection will be positively associated with antisocial behavior and that emotional warmth and SES will be negatively associated with antisocial behavior. [O]ur *temperament hypotheses* are that effortful control will be negatively and frustration positively associated with antisocial behavior. Our *sex hypothesis* states that the risk of antisocial behavior will be higher for boys than for girls.

With regard to the temperament-by-environment interaction, we base our hypotheses on a *risk-buffering model*, which implies that we mix the perspectives of the environment effects being moderated by temperament and the effect of temperament being moderated by the environment. Our *protective environment hypotheses* are that the environment protective factors (emotional warmth and SES) will help reduce the more antisocial behavior where the child is temperamentally more at risk (low effortful control and high frustration). The same should hold for the higher risk due to sex. Boys can be expected to profit more than girls from environmental protection against undesirable, hedonic urges that result in antisocial behavior. Our *protective temperament hypotheses* are that the temperamental protective factors (high effortful control and low frustration) will help reduce antisocial behavior more where the environment (overprotection, rejection) puts the child more at risk.

Method

This study is part of the TRacking Adolescents' Individual Lives Survey (TRAILS), a new prospective cohort study of Dutch preadolescents. The present study involves the first assessment wave of TRAILS. [The size of the present sample is] 2,230 [with a] mean age of 11.09 years (*SD* = 0.55).

Measures

Antisocial Behavior Antisocial behavior was assessed by the Child Behavior Checklist (CBCL), one of the most commonly-used questionnaires in current child and adolescent psychiatric research. It contains a list of 112 behavioral and emotional problems, which parents can rate as 0 = not true, 1 = somewhat or sometimes true, or 2 = very or often true in the past six months. In addition to the CBCL, we administered the self-report version of this questionnaire, the Youth Self-Report. Antisocial behavior that is rated as present by both parent and child is assumed to be more severe (more generalized) than problems rated by only one informant. Based on this assumption, we used the mean of the parent and child scores as a measure of antisocial behavior in this study.

Perceived Parenting The My Memories of Upbringing for Children [EMBU-C] has been developed to assess the perception of actual parental rearing by children and early adolescents. Children could rate the EMBU-C as 1 = no, never, 2 = yes, sometimes, 3 = yes, often, 4 = yes, almost always. Each item was asked for both the father and the mother. The EMBU-C contains the factors Emotional Warmth, Rejection, and Overprotection. The main concepts of Emotional Warmth are giving special attention, praising for approved behavior, unconditional love, and being supportive and affectionately demonstrative.

Temperament Temperament was assessed by the parent and the child version of the Early Adolescent Temperament Questionnaire-Revised [EATQ-R], a 62-item questionnaire. Frustration is the negative affect related to goal blocking or an interruption of ongoing tasks.

Sex and SES The sample consisted of 50.8% girls and 49.2 % boys. The TRAILS database contains various variables for SES: income level, educational level of both the father and the mother, and occupational level of each parent.

Familial Vulnerability to Externalizing Psychopathology Parental psychopathology with respect to depression, anxiety, substance abuse, antisocial behavior, and psychoses was measured by means of the Brief TRAILS Family History Interview.

Results

Except for SES and familial externalizing psychopathology, all variables showed significant sex differences. Girls perceived less Overprotection and Rejection, and more Emotional Warmth than boys. Furthermore, they scored higher on Effortful Control and lower on Frustration and antisocial behavior. All parenting and temperament variables were moderately associated with antisocial behavior and with each other. Familial externalizing psychopathology was positively associated with antisocial behavior and negatively associated with Effortful Control and SES but not related to perceived parenting behaviors. The correlation between Overprotection and Emotional Warmth was higher for boys (.25) than for girls (.13). Rejection and SES were significantly related for girls, but not for boys.

Testing the Hypotheses

Direct Effects Table 7.3 shows the results of the analyses with respect to the interaction of temperament (Effortful Control and Frustration) and environment (Overprotection, Rejection, Emotional Warmth, and SES). In order to control for possible confounding effects of genetic risk which may affect both temperament and antisocial behavior, we controlled for familial vulnerability to externalizing psychopathology (as a proxy for genetic risk) in the regression. Our environment hypotheses stated that Rejection and Overprotection will be positively associated with antisocial behavior and that Emotional Warmth and SES will be negatively associated with antisocial behavior. We see from Table 7.3 that Overprotection and Rejection are significantly positively associated and Emotional Warmth and SES significantly negatively associated with antisocial behavior. Our temperament hypotheses stated that Effortful Control will be negatively and Frustration positively associated with antisocial behavior. These hypotheses are also supported by our results. Our sex hypothesis stated that being a boy will be more positively related to antisocial behavior than being a girl. Our results are in line with this hypothesis.

Interaction Effects Our *protective environment* hypotheses stated that the environment protective factors (Emotional Warmth and SES) will help reduce the more antisocial behavior where the preadolescent is temperamentally more at risk (low Effortful Control and high Frustration). From Table 7.3, we see that there are significant interactions of Emotional Warmth with Frustration and SES with Effortful Control and Frustration, consistent with the hypothesis. Observe though, that in the simultaneous model, the interaction of Emotional Warmth and Frustration on antisocial behavior is only significant at the .10 level.

With regard to sex, we see in the last model in Table 7.3 an interaction effect with SES, which indicates that SES relates negatively to antisocial behavior for boys ($b = -.10$, $t = -3.99$, $p < .01$) and not for girls ($b = -.03$, $t = -1.23$, $p = .22$). This is in line with our expectation. Against our prediction, we find no extra protective effect of emotional warmth for boys.

Our *protective temperament hypotheses* stated that the temperamental protective factors (high Effortful Control and low Frustration) will help reduce antisocial behavior the more the environment (Overprotection, Rejection) puts the child at risk. From Table 7.3, we see that there is a significant interaction effect of Overprotection as well as Rejection with Frustration. The protective temperament hypotheses with Effortful Control were disconfirmed. Observe though, that in the simultaneous model, only the interaction between Rejection and Frustration remains (marginally) significant. Simple slope analyses revealed that Rejection was indeed a somewhat weaker predictor of antisocial behavior at one standard deviation below the mean of Frustration ($b = .21$, $t = 6.71$, $p < .01$) than at one standard deviation above the mean of Frustration ($b = .29$, $t = 9.42$, $p < .01$).

Discussion

The results support our *environment hypotheses* and reaffirm similar findings in other studies. All parenting characteristics examined in our study (emotional warmth, overprotection, and rejection) appeared to be related to antisocial behavior. Because they were adjusted for familial externalizing psychopathology, the associations are unlikely to be spurious on account of genetic

Table 7.3 Main Effect and Interactions (Standardized Coefficients) of Sex, Temperament, Environment, and Their Interactions on Antisocial Behavior: Separately and Simultaneously for Overprotection, Rejection, Emotional Warmth, and SES

Variable	Model 1: Overprotection ($R^2 = .34$)		Model 2: Rejection ($R^2 = .41$)		Model 3: Em.Warmth ($R^2 = .34$)		Model 4: SES ($R^2 = .33$)		Simultaneous Model ($R^2 = .43$)	
	B	*SE*[a]	B	*SE*[a]	B	*SE*[a]	B	*SE*[a]	B	*SE*[a]
Main Effects										
Sex (1 = boys)	.27	(.04)**	.24	(.03)**	.26	(.04)**	.28	(.04)**	.22	(.03)**
Overprotection	.17	(.02)**	—		—		—		.07	(.02)**
Rejection	—		.31	(.02)**	—		—		.25	(.02)**
Emotional Warmth	—		—		–.15	(.02)**	—		–.06	(.03)*
SES	—		—		—		–.09	(.02)**	–.03	(.02)
Effortful Control	–.20	(.02)**	–.17	(.02)**	–.18	(.02)**	–.20	(.02)**	–.15	(.02)**
Frustration	.37	(.02)**	.35	(.02)**	.38	(.02)**	.38	(.02)**	.34	(.02)**
Familial Vulnerability	.06	(.02)**	.03	(.02)**	.07	(.02)**	.05	(.02)*	.06	(.02)**
Temperament–by–environment										
Overprotection × Effortful Control	–.02	(.02)	—		—		—		—	
Overprotection × Frustration	.04	(.02)*	—		—		—		.01	(.02)
Rejection × Effortful Control	—		01	(.02)	—		—		—	
Rejection × Frustration	—		.07	(.02)**	—		—		.04	(.02)~
Em. Warmth × Effortful Control	—		—		.01	(.02)	—		—	
Em. Warmth × Frustration	—		—		–.08	(.02)**	—		–.04	(.02)~
SES × Effortful Control	—		—		—		.06	(.02)**	.05	(.02)**
SES × Frustration	—		—		—		–.08	(.02)**	–.06	(.02)**
Sex Interactions										
Sex × Emotional Warmth	—		—		—		—		–.04	(.04)
Sex × SES	—		—		—		—		–.07	(.03)*

NOTE: Em. Warmth = emotional warmth; SES = socioeconomic status.

a. Tests were two-tailed. ~$p < .10$; *$p < .05$; **$p < .01$.

risk. Consistent with previous studies, we found that rejection was positively linked and that emotional warmth was negatively linked to antisocial behavior. These results are in line with Farrington (1997) who argued that children who are exposed to poor parenting practices may be more likely to offend because they do not build up internal inhibitions against socially disapproved behavior.

Our *temperament hypotheses* that effortful control will be negatively and frustration positively associated with antisocial behavior were also supported. Again, spurious associations due to familial vulnerability are unlikely. Consistent with earlier studies, we found effortful control to be negatively associated with antisocial behavior. Children with low effortful control, that is, with a limited ability to regulate attention and behavior, are less likely to consider the possible consequences of their actions, especially consequences that are likely to be long-delayed. The inability to restrain undesirable, hedonic urges by considering their consequences may result in antisocial behavior. Frustration reflects the tendency to experience negative feelings if things do not run according to plan. It was positively associated with antisocial behavior. If the efforts to reach a goal do not succeed, the situation involves loss, and the irritation and anger associated with blocked goals renders highly frustrated children prone to externalizing. We found that boys were more at risk of developing antisocial behavior than girls. This replicates previous findings and is consistent with our *sex hypothesis.*

The hypotheses on *temperament-by-environment* . . . predicted that environmental factors that protect against the risk of antisocial behavior (emotional warmth, SES) are assumed to be more helpful for children who are temperamentally (or because they are boys) more at risk of committing antisocial behavior. Conversely, temperamental factors that protect against the risk of antisocial behavior (effortful control and low frustration) are assumed to work better for children who are environmentally more at risk (because of overprotection and rejection). Seemingly, SES has extra protective effects for preadolescents who are temperamentally at risk of committing antisocial behavior. This is in line with other research showing that relations of parenting to self-regulation have been found to be stronger in more disadvantaged, i.e., low SES, populations.

It turned out that SES is almost exclusively protective for preadolescents who are at risk either because of a difficult temperament (low effortful control or high frustration) or because of sex (being a boy means a higher risk of antisocial behavior). This makes it extra important to consider interaction effects when studying the impact of SES on antisocial behavior. Lynam et al. (2000) found a similar interaction between temperament (high impulsivity) and environment (neighborhood). A poor neighborhood, defined by the census-SES, had only an effect on juvenile offending for impulsive boys and not for non-impulsive boys.

We have no ready explanation as to why the interactions with emotional warmth are weaker (with frustration) or absent (with effortful control). We can only speculate that parental emotional warmth bleeds into overprotection for children who are at risk of committing antisocial behavior. [E]motional warmth correlates .25 with overprotection for boys and .13 for girls. This suggests that emotional warmth does indeed bleed into overprotection and that it does so more for boys than for girls. The extra protective effect of emotional warmth may thus be counteracted by its closeness to overprotection exactly for those preadolescents who are most at risk of antisocial behavior.

With regard to the fact that we found no strong indications that favorable temperaments are extra protective when environmental risk is high, we also have no ready explanation (we only found interaction effects of overprotection and rejection with frustration, indicating that low frustration buffers the effect of environmental risk, but this effect is considerably weakened when the interactions with SES are added). Here too we can only speculate. It is possible that with regard to rejection, there is a confounding effect with negative aspects of temperament. In part, parents may reject the child because of the negative temperamental aspects. This confound may mask the risk-buffering effect of favorable temperament. For overprotection, we mentioned already that preadolescents might commit antisocial behavior as an act of protest against excessive parental interference. In this case high effortful control and low frustration would not have much mitigating influence. This interpretation is supported by the fact that at least in Dutch society, individual autonomy is considered very important, especially by young generations. In older age groups and in cultures that value maintenance of affective bonds among family members more highly, parental over-interference is less likely to cause protest behavior.

References

Bates, J. E., Dodge, K. A., Pettit, G. S., & Ridge, B. (1998). Interaction of temperamental resistance to control and restrictive parenting in the development of externalizing behavior. *Developmental Psychology, 34,* 982–995.

Belsky, J., Hsieh, K.-H., & Crnic, K. (1998). Mothering, fathering, and infant negativity as antecedents of boys' externalizing problems and inhibition at age 3 years: Differential susceptibility to rearing experience? *Development and Psychopathology, 10,* 301–319.

Farrington, D. P. (1997). Human development and criminal careers. In M. Maguire, R. Morgan, & R. Reiner (Eds.), *The Oxford handbook of criminology* (2nd ed., pp. 361–408). Oxford: Clarendon.

Kochanska, G. (1997). Multiple pathways to conscience for children with different temperaments: From toddlerhood to age 5. *Developmental Psychology, 33,* 228–240.

Lynam, D. R., Caspi, A., Moffitt, T. E., Wikström, P.-O. H., Loeber, R., & Novak, S. (2000). The interaction between impulsivity and neighborhood context on offending: The effects of impulsivity are stronger in poorer neighborhoods. *Journal of Abnormal Psychology, 109,* 563–574.

Magnusson, D., & Stattin, H. (1998). Person-context interaction theories. In W. Damon & R. M. Lerner (Eds.), *Handbook of child psychology. Volume 1. Theoretical models of human development* (5th ed., pp. 685–759). New York: John Wiley & Sons.

Maziade, M., Caron, C., Cote, R., Merette, C., Bernier, H., Laplante, B., et al. (1990). Psychiatric status of adolescents who had extreme temperaments at age seven. *American Journal of Psychiatry, 147,* 1531–1536.

Sanson, A., Hemphill, S. A., & Smart, D. (2004). Connections between temperament and social development: A review. *Social Development, 13,* 142–170.

Stice, E., & Gonzales, N. (1998). Adolescent temperament moderates the relation of parenting to antisocial behavior and substance use. *Journal of Adolescent Research, 13,* 5–31.

Thomas, A., & Chess, S. (1977). *Temperament and development.* New York: Brunner and Mazel.

Van Leeuwen, K. G., Mervielde, I., Braet, C., & Bosmans, G. (2004). Child personality and parental behavior as moderators of problem behavior. Variable- and person-centered approaches. *Developmental Psychology, 40,* 1028–1046.

Wills, T. A., & Dishion, T. J. (2004). Temperament and adolescent substance use: A transactional analysis of emerging self-control. *Journal of Clinical Child and Adolescent Psychology, 33,* 69–81.

Wills, T. A., Sandy, J. M., Yaeger, A., & Shinar, O. (2001). Family risk factors and adolescent substance use: Moderation effects for temperament dimensions. *Developmental Psychology, 37,* 283–297.

DISCUSSION QUESTIONS

1. Parenting is seen in this study as playing an important role in preventing or contributing to antisocial behavior. To what extent do you think that parenting style is a response to children's temperaments, as opposed to an independent factor?
2. Why is the association between SES and antisocial behavior stronger for boys than for girls?
3. The authors state that warm and close-monitoring parental behavior is most needed for children temperamentally at risk for antisocial behavior; why are these children not likely to receive such parenting?

READING

Psychopathy

Theory, Measurement, and Treatment

Anh Vien and Anthony R. Beech

In this article, Ahn Vien and Anthony Beech provide us with an excellent overview of the concept, measurement, and treatment of psychopathy. They first link empirical literature to the theoretical background of the concept of psychopathy and the impact that this has had on the development of treatment and intervention procedures for psychopathic offenders. They discuss the different theories of psychopathy, which leads into considerations of different developmental pathways of psychopathy. They then discuss the psychometrics and measurement tools used to assess psychopathy, focusing on the Psychopathy Checklist–Revised (PCL-R), the most frequently used and validated measure of psychopathy. The relationship between psychopathy and different types of crime is also discussed. The final section of the article considers the treatment and interventions that are available to psychopathic offenders.

Introduction

Hare (1995) estimated that 15% to 20% of all prisoners are psychopaths and that these individuals account for a large proportion of serious crimes involving violence in relation to treatment and the implications of this for future research. Researchers theorizing about *psychopathy* have attempted to provide a comprehensive and integrated general definition of the concept; however, there is little agreement of exactly what it is. The only aspect that has been agreed on is that psychopathy is an integration of motivational dispositions, developmental factors, individual differences, and mental health. Although the evolution of psychopathy as a formal clinical disorder began more than a century ago, it is only recently that scientifically sound psychometric procedures for its assessment have become available, making it possible to measure *psychopathy* as a construct.

Cleckley in *The Mask of Sanity* (1941) was the first to look extensively at the personality traits of the psychopath. Cleckley considered *psychopathy* as concealed psychosis and is only primarily revealed through the disintegrating effects of strong emotion, as well as inherent disjointing of the relationship between word and deed in which he terms "semantic dementia." Cleckley proposed 16 core personality traits in psychopaths (see Table 7.4). Hare has extended Cleckley's ideas to develop the Hare Psychopathy Checklist and the revised version (PCL-R; see Hare, 2003) as a standardized and clinical measure of psychopathy. Table 7.4 shows Cleckley's and Hare's conceptualization of psychopathic traits.

SOURCE: Vien, A., & Beech, A. R. (2006). Psychopathy: Theory, measurement, and treatment. *Trauma, Violence, and Abuse, 7*(3), 155–174. Reprinted with permission of Sage Publications, Inc.

Table 7.4 Cleckley's Core Traits of Psychopathic Personality and Items From the PCL-R

Cleckley's Core Traits	Hare's PCL-R Items
1. Superficial charm and good intelligence	1. Glibness/superficial charm[a]
2. An absence of delusions & other signs of irrational thinking	2. Grandiose sense of self-worth[a]
3. An absence of "nervousness" or psychopathic manifestations	3. Need for stimulation and/or proneness to boredom
4. Unreliability	4. Pathological lying[a]
5. Untruthfulness and insincerity	5. Conning and/or manipulative
6. A lack of remorse or shame for their behavior	6. Lack of remorse or guilt[a]
7. Inadequately motivated antisocial behavior	7. Shallow affect
8. Poor judgment and failure to learn from previous experiences	8. Callous and/or lack of empathy[a]
9. Pathologic egocentricity and incapacity for love	9. Parasitic lifestyle
10. General poverty in any major affective reactions or emotions	10. Poor behavioral controls[a]
11. A specific loss of insight	11. Promiscuous sexual behavior
12. A general unresponsiveness to interpersonal relationships	12. Early behavioral problems
13. Fantastic and uninviting behavior with or without alcohol	13. Lack of realistic and long-term goals[a]
14. Suicide is rarely carried out because of love of the self	14. Impulsivity[a]
15. Sex life will be impersonal, trivial, and poorly integrated	15. Irresponsibility[a]
16. A failure to follow any kind of life plan	16. Failure to accept responsibility for own actions[a]
	17. Many short-term marital relationships
	18. Juvenile delinquency[a]
	19. Revocation of conditional release
	20. Criminal versatility

a. These items are those included in the Psychopathy Checklist–Screening Version, plus one additional item of adult antisocial behavior.

Theories of Psychopathy

The psychopath is not a person that can be readily recognized by physical symptoms that are seen within other types of mental illnesses or disorders. Although psychopaths cannot be readily recognized by any distinctive clinical symptoms, there are distinctive personality characteristics that make psychopathy uniquely different from other personality disorders. The notion of innate antisocial predispositions has been explored in a recent twin study conducted by Blonigen et al. (2003) that provided evidence of a significant genetic influence in psychopathy; the data reveal substantial evidence of a large genetic contribution in monozygotic twins (identical) and relatively little relationship in dizygotic (nonidentical) twins that suggest a genetic rather than an environmental contribution.

Arousal Theory

Low-arousal theory suggests that psychopaths have a pathologically low level of autonomic and cortical arousal, and hyperactivity when compared to nonpsychopathic individuals.

Consequently, the psychopath will be in a chronic state of stimulation and sensation seeking, thus explaining why psychopaths do not become autonomically aroused to stimuli that would otherwise be stressful, exciting, or frightening to non-psychopaths. This results in the psychopath needing a greater variety and intensity of sensory input to increase his or her arousal level to the optimum. The theory assumes that arousal level and sensory intake are dynamically related in such a way that an optimal level of arousal is maintained, in other words, a common level presumed to be functionally desirable for all individuals. However, when that level of arousal falls below the optimum, stimulation and/or sensation-seeking behavior and sensory intake increases dramatically to raise the arousal level to the desired optimum.

Raine, Venables, and Williams (1990) conducted a study looking at the relationship between measures of arousal and criminality at two different age intervals. Arousal was measured at age 15 years, using electrodermal, cardiovascular, and cortical responses and criminality was measured at age 24 years. The findings demonstrated that on all measures of arousal, future criminals showed lower levels of arousal in the experimental situation than did future noncriminals. Therefore, this does seem to provide some evidence that low arousal is associated with future criminality; however, criminality does not equate to psychopathy. Blair, Jones, Clark, and Smith (1997) investigated the psychophysiological responsiveness of psychopathic individuals to distress cues and to threatening stimuli. Electrodermal responses and skin conductance responses were recorded during exposure to the stimuli. The results demonstrated that psychopaths, relative to controls, had reduced electrodermal and skin conductance responses to the threatening and neutral stimuli.

Low arousal makes conditioning less likely to occur, and there's some evidence to indirectly support low conditionability due to low arousal in a study reported by Raine et al. (1990) where it was found that identified low arousal in childhood was related to later levels of adult criminality. Flor et al. (2002) also provided evidence to support the theory that psychopaths have low condition-ability compared to nonpsychopathic controls. Further analysis of the data indicated that psychopaths have a deficit in association formation, and it was concluded that this deficit could be related to deficient interaction between the limbic and cortical systems of the brain.

Neurobiological Theory

One of the basic principles of neurobiology and/or neuropsychology is the notion that certain functions are, to some degree, localized within certain areas of the cerebral hemisphere, whereas others are lateralized to one hemisphere of the brain. There have been strong associations and a belief that psychopathic individuals are biologically different from the norm, in the sense that their brains are structurally different.

Kiehl et al. (2004) used functional magnetic resonance imaging (fMRI) to clarify and characterize the neural architecture involved in the lexico-semantic processing in criminal psychopaths compared with a matched group of controls. Analysis revealed that psychopaths failed to show appropriate neural differentiation between abstract and concrete stimuli in the right anterior temporal gyrus and the surrounding cortex. Kiehl et al. (2004) concluded that semantic processing of abstract material in psychopathy is associated with abnormalities in the right hemisphere. However, Blair (2003) argued that the lifestyle of the psychopath may exacerbate neurobiological impairments, if any, rather than the impairments being defined at birth.

Amygdala and Other Cortical Dysfunction

Dysfunction in emotional processing seems to point toward a neural basis for the development of psychopathy. Herpertz and Sass (2000) provided some evidence of a close association between the difficulties that psychopaths have in emotional processing and poor prefrontal functioning. There is now, to a certain extent, agreement that amygdala dysfunction is the underlying neural structure responsible for the development of psychopathic tendencies. This is because of the association of the amygdala with aversive conditioning and instrumental learning, as well as fearful and sad facial expression responses. Kiehl et al. (2001) used fMRI to examine neural responses in criminal psychopaths during an affective word memory task. The results provided evidence that criminal psychopaths demonstrated significantly less affect-related activity in the amygdala, when compared to criminal nonpsychopaths and noncriminal controls.

Séguin (2004) reviewed the role of the orbitofrontal cortex (OFC) in relation to antisocial behavior. It was concluded that although OFC dysfunction seems to be associated with variations in antisocial problems, they are rarely associated with physical violence. Furthermore, Blair (2004) suggests that OFC dysfunction is involved in the modulation of reactive aggression, which is physical violence elicited in response to frustration and/or threat. Blair went on to distinguish reactive aggression from goal-directed, instrumental aggression, which the individual would engage in just to gratify the self. Psychopaths tend to demonstrate marked levels of instrumental aggression, which is modulated by the amygdala. Therefore, OFC dysfunction may not be evident in psychopaths. However, psychopaths seem to show marked levels of instrumental aggression and reactive aggression; therefore both may play a role (Blair, 2004).

Interhemispheric Deficits Although the amygdala and OFC are the main focus of attention for the neurological cause of psychopathy, there is now evidence to suggest poor interhemispheric integration and lateralization deficits. In a study conducted by Hiatt, Lorenz, and Newman (2003), the assumption was tested that psychopaths have abnormal language lateralization and abnormal processing asymmetries in other domains. They concluded that psychopaths' abnormal processing asymmetries are related to poor interhemispheric integration.

Using structural magnetic resonance imaging (sMRI), Raine et al. (2003) assessed whether psychopaths, and individuals with antisocial personality disorder (APD) show any structural and/or functional impairments in the corpus callosum. A sample of 15 men with high psychopathy scores and APD was compared to a group of matched controls. They observed that psychopaths showed an increase in callosal white matter and an increase in callosal length when compared to controls. Raine et al. concluded that corpus callosum abnormalities in psychopaths might be a reflection of neurodevelopmental processes involving early axonal pruning or increased white myelination, which results in abnormal or poor interhemispheric integration.

Social Learning Theory

Although there is strong evidence linking psychopathy with neural and cerebral dysfunctions, there has been speculation that a certain element of learning is involved in the development of psychopathic personality and tendencies. Lykken (1957) used passive-avoidance learning tasks to demonstrate psychopaths' poor avoidance learning. The tasks required the respondents to learn a "mental maze"; and at specific points the respondent had to choose a response from a possibility of four. The correct response from the four led to progression in the maze; however, one of the four responses led to an electric shock. The main observation of the study was the extent to which control respondents learned to passively avoid the electric shocks, whereas psychopaths made significantly more responses resulting in punishment, thus providing evidence of poor avoidance learning in psychopaths.

According to Eysenck (1996/1998), conscience is a conditioned response (CR), which is developed through socialization to control an individual's behavior. Eysenck gave two possible explanations as to why some individuals do not develop a conscience. First, a permissive society results in conditioning experiences for conscience to either be missing or inadequate and, hence, leading to antisocial behavior. Second, Eysenck suggested that the wrong experiences are reinforced; therefore, the CR to this reinforcement is also wrong.

Moral-Reasoning Deficits

The idea that psychopaths have moral-reasoning deficits stems back to Prichard's (1835) concept of "moral insanity," whereby psychopaths have a diseased moral faculty. In a study conducted by Trevathan and Walker (1989), no significant difference was found between the reasoning levels of psychopaths and nonpsychopathic controls. Rather than a diseased moral faculty, Blair (1995) suggested that psychopaths fail to make moral and/or conventional distinctions because of their lack of a violence inhibition mechanism. Blair tested this assumption and concluded that psychopaths judged all transgressions as moral and not conventional. Although psychopaths judged all transgressions as moral, they did not justify their responses with reference to welfare of others.

Blair, Sellers, et al. (1995) investigated emotional deficiency in terms of a deficit in emotional attribution, rather than failure to mentalize. Twenty-five psychopaths and 25 nonpsychopathic controls were presented with short vignettes or stories depicting happiness, sadness, embarrassment, and guilt. The respondents were then asked to attribute emotion to the story protagonist. Psychopaths, and nonpsychopathic controls, did not differ in the attribution of happiness, sadness, or embarrassment but differed in the attribution of guilt. However, while the controls attributed guilt correctly, psychopaths attributed either happiness or indifference to vignettes depicting guilt. This suggests that psychopaths' deficient moral-reasoning abilities may stem from the inability to experience guilt and anxiety, rather than a deficit in making mental representations.

Selective Attention Deficits

It has been demonstrated that high aversion in psychopaths is facilitated by a weakness in the initial evaluation of the target stimulus (Levenston, et al., 2000). This was demonstrated in a study conducted by Verona et al. (2004) investigating the physiological reactions to emotional sounds, either pleasant or unpleasant, in prisoners. Those who scored high on the psychopathy measure showed diminished skin conductance responses to pleasant and unpleasant sounds. Findings indicate an abnormal reactivity to emotionally positive and negative stimuli in psychopathic individuals, which suggests that there is a deficit in the action-mobilization component of their emotional responses. This would suggest that psychopaths have an inability to differentiate between positive and negative stimuli and this selective attention deficit is manifested in their high aversion, as measured by low skin conductance and normal heart rate.

Developmental Pathways

Although the development of psychopathy in the individual can be based on neurological underpinnings and theoretical explanations, there seems to be a number of different developmental pathways. Saltaris (2002) attributed the development of psychopathy in adulthood to aggressive and antisocial behavior exhibited in childhood. Saltaris acknowledges that not all children with conduct problems will develop psychopathy in adulthood; however, the link between the two must not be ignored, and she concluded that conduct problems coupled with temperamental dispositions and an inability to form attachment bonds with early caregivers are involved in the manifestation of psychopathic traits in adulthood.

This is due to the aversive interpersonal style they engage during childhood and adolescence, which ultimately contributes to negative interactions with family members and others in their immediate environment and continues to deepen throughout the wider environment.

Harris, Rice, and Lalumière (2001) examined the interrelationships and the independent contributions of three major constructs associated with male criminal violence. The constructs examined in this study were neurodevelopmental insults, antisocial parenting, and psychopathy. Results indicated that neurodevelopmental insults and psychopathy were not interrelated but were independently and directly related to criminal violence. Antisocial parenting had no direct relationship with criminal violence however; it was related to neurodevelopmental insults and psychopathy. Hence, Harris et al. concluded that criminal violence has at least two developmental pathways, one originating early in life involving neurodevelopmental insults and one involving psychopathy, with antisocial parenting facilitating the development of both.

Moffit (1993) suggested a theory of life-course-persistent antisocial behavior as a developmental pathway from childhood neuropsychological impairments and adolescence behavioral problems to adult psychopathology. Although this is generally more related to aspects of APD, there are distinct parallels with psychopathy. Moffit suggested that during a life course of an antisocial individual there would be changing manifestations of antisocial behavior. For example, an individual may exhibit behaviors of biting and hitting at age 4 years, shoplifting and truancy by age 10 years, selling drugs and stealing cars at age 16 years, robbery and rape at age 22 years, and fraud and child abuse at age 30 years. The underlying disposition of the individual is the same; however, the expression of behavior changes because of new social opportunities that arise at different points in the individual's life. Although Moffit's (1993) life-course theory does incorporate neurological and social factors into the developmental pathway to psychopathy, this pathway is not dissimilar to persistent adult criminal behavior with the absence of psychopathology. It can be concluded that the developmental pathways of psychopathy in the individual cannot be attributed to any one theory or explanation. One must take into account all the differing, and sometimes even opposing, arguments to understand in more depth the development of psychopathy in the individual, which would then lead us to a better understanding of the etiology of psychopathy.

Measurement of Psychopathy

The PCL-R (Hare, 2003) is currently the most widely used and validated measure of psychopathy. At times when interview information is unattainable, such as research using archival information, clinical assessments of actively psychotic patients or even individuals who refuse to either be interviewed or cooperate with the clinician, it is suggested that ratings can be made solely from file information. PCL-R items are rated based on a person's lifetime functioning as evident in the assessment of the interview and file information. The maximum total score on the PCL-R is 40, and the diagnostic cutoff for psychopathy on the PCL-R is more than 30 (Hare, 2003).

When the PCL-R was first devised, a two-factor model of the measure was proposed. Factor 1 measures and identifies core personality characteristics of psychopathy, and Factor 2 measures antisocial behaviors as assessed by APD. Hall, Benning, and Patrick (2004) suggested the PCL-R would be better accounted for in a three-factor model. They validated this three-factor model on a sample of 310 incarcerated offenders. They concluded that the affective factor was correlated with low social closeness and violent offending and the behavioral factor was associated with negative emotionality, disinhibition, reactive aggression, and poor adaptability. The interpersonal factor was related to social dominance, low stress reactivity, and higher adaptive functioning.

Risk Assessment

Although the PCL-R was not constructed as a risk-assessment tool, it does seem to have an ability to predict risk and recidivism quite accurately and is routinely used as part of risk-assessment, for example in the Violence Risk Assessment Guide (Quinsey, Rice, & Harris, 1995), The link between psychopathy and offending seems to lie within the levels of risk of recidivism for violent sexual and violent nonsexual reoffending and the personality characteristics of psychopathy.

Prediction of Violent Offending

Psychopathy as measured by the PCL-R suggests that individuals with high (over 30) PCL-R scores will commit more violent and/or serious offenses compared to low (below 30) PCL-R scores. Not only are they more likely to commit serious crimes, but they are also at higher levels of risk to recidivism. Hemphill et al. (1998) reported that the rate of violent recidivism for psychopaths is greater than 80% during a 5-year period; therefore, it would be reasonable to conclude that psychopaths are at high levels of risk for future violence and violent recidivism.

In a retrospective study conducted by Hare and McPherson (1984), it was concluded that evidence of psychopathy in offenders increases the likelihood of violent and aggressive behavior. Harpur and Hare (1994) reported that psychopathy scores, in particular Factor 2 scores, decline with age. One possible explanation for this observation is long spells of incarceration whereby they cannot engage in certain antisocial behaviors that is measured by Factor 2 on the PCL-R, for example, Item 9 (Parasitic Lifestyle) and Item 19 (Revocation of Conditional Release). However, evidence for the decline of Factor 2 scores with the increase of age is inconclusive as there is not yet enough data to determine whether this relationship holds true across the life span or whether incarceration was the main reason for a decline in antisocial behaviors. Hare and Jutai (1983) reported from their longitudinal study that psychopaths may, in fact, continue to engage in extensive criminal and antisocial behaviors and activities long after other criminals have shown decline.

Prediction of Sexual Offending

Porter et al. (2000) investigated the contribution of psychopathy to the heterogeneity of sexual violence. The results showed that sex offenders tend to have elevated Factor 1 scores, and nonsexual offenders had higher scores on Factor 2. Therefore, it could be argued that sex offending is related to the interpersonal aspects of psychopathy, whereas nonsexual offences are more related to antisocial behavioral aspects. It was also noted that rapists and mixed rapists and/or molesters scored higher on the PCL-R compared to offenders who victimize only children. Furthermore, offenders who sexually victimize adults and children were between 2 to 10 times more likely to be high PCL-R scorers,

Rice and Harris (1997) reported that sexual recidivism, as opposed to general violent recidivism, was strongly predicted by a combination of a high PCL-R total score and deviant sexual arousal (as defined by a phallometric preference for deviant sexual stimuli, such as pictures of children, rape cues, or even nonsexual violent cues). 70% of those with such a profile recidivated sexually compared to 40% in all other groups. Similarly, Hildebrand, de Ruiter, and de Vogel (2004) examined the roles of the PCL-R and sexual deviancy scores in predicting recidivism in a sample of convicted rapists in a Dutch forensic psychiatric hospital. Survival analysis provided considerable evidence that sex offenders with high PCL-R scores (PCL-R total score about 26 in the case of this sample) who also had high sexual deviancy scores are substantially at a greater risk of committing a new sexual offense.

Psychopathy and White-Collar Crime

Not all psychopaths are violent and incarcerated criminals, some are untrustworthy and unreliable employees or unscrupulous and predatory businessmen, corrupt politicians, or even unethical or immoral professionals, who victimize their clients, patients and the general public. However, little empirical research is available for this group of "white-collar" psychopaths; however, there are indications that the personality structure and propensity for unethical behavior are probably much the same as the criminal psychopath (Hare, 2003). Babiak (1995) used the context of an organization going through chaotic change as the backdrop to an example of a successful industrial psychopath. In such circumstances, these psychopaths use their manipulation skills to effectively manage the discrepant views of supporters and detractors to enhance their movement up the career ladder. Although there is plenty of speculation of the similarities of personality characteristics between the industrial psychopath and the incarcerated offending psychopath, empirical research in this area remains minimal.

Treatment Approaches

Psychopaths present the most serious problems for the Criminal Justice System: not only are they responsible for many types of serious crimes, they are also viewed to be "untreatable" or disruptive in treatment. It is believed that traditional treatment programs have a low success rate in treating psychopaths; however, this does not mean that all treatment would fail. A number of studies do provide empirical evidence as to what can constitute inappropriate treatment methods to use with psychopaths. For example, Rice et al. (1992) reported that psychopaths who had taken part in a patient-run, unstructured therapeutic community program had violently recidivated, on release, at a much higher rate than did psychopaths who had not taken part in the program.

Psychopaths' disruptive behavior during treatment has been attributed to the personality characteristics of glibness and/or superficial charm and grandiose sense of self-worth. However, Skeem, Monahan, and Mulvey (2002) concluded from their study that patients with psychopathic tendencies appeared just as likely to benefit from treatment as nonpsychopathic patients, in terms of becoming less violent. They added that unfavorable treatment effects on psychopaths could be eradicated if the frequency or dosage of treatment is increased for these individuals. Patients who received seven or more sessions of treatment during a 10-week period were approximately 3 times less likely to be violent during a subsequent 10-week period, when compared to those who received six or fewer sessions (Skeem et al., 2002).

New Approaches to Treatment

Wong and Hare (2005) suggested that violence and antisocial behavior are the primary reasons for psychopaths to offend and be incarcerated, and a suitable treatment program should focus on changing behavior rather than trying to change their core personality characteristics. Wong and Hare reported guidelines in their Psychopathy Treatment Program (PTP). The PTP is more a strategy of behavioral self-management rather than a cure for psychopathy; participants of the PTP should be assisted in developing deeper insights into their lifelong psychopathology and to accept the fact that they will require long-term and continuing self-management for most aspects of their lives to keep them from recidivating rather than any quick-fix treatment program. PTP guidelines are similar to many other treatment program guidelines that are available for nonpsychopathic offenders, However, it does document ways in which to approach treatment with psychopaths by using what can be termed as the "what not to do" approach. This has been done by justifying using certain techniques with psychopaths to bypass the obstacle of core personality characteristics rather than head-on challenges.

Pharmacological Approaches

Pharmacological treatment of psychopathy has had little attention in terms of empirical research. The general trend has been identification of medication that is used to treat behavioral symptoms apparent in psychiatric conditions, such as schizophrenia, that are similar to behaviors psychopaths engage in, such as impulsivity and aggression. There are a number of known drugs that seem to demonstrate effectiveness in controlling behavioral symptoms such as aggression and impulsivity in respondents with psychiatric disorders, and the move now would be to see whether the same applies to psychopathy. SSRIs or selective serotonin reuptake inhibitors, such as Prozac, are agents that inhibit the reuptake of serotonin as part of a much more widespread effect on neurotransmitters. Although SSRIs are not directly used as treatment for aggression and impulsivity in psychopaths, they have been found to be useful in the treatment of major depression, anxiety disorders, pain disorders, and premature ejaculation.

Conclusions

Although it can be concluded that the etiology of psychopathy is an interplay between neurobiological, psychological, social and environmental factors, there is still disagreement as to what the underlying cause or developmental pathway for each factor is. The only agreement so far is some kind of amygdala and orbitofrontal cortex dysfunction. Moral reasoning deficits and heightened aversion thresholds do help put into context why psychopaths are responsible for a large majority of serious sexual and violent crimes. It would seem that the pessimistic view of treatment effectiveness on psychopaths has been fuelled by the unsuccessful out-comes of treatment programs inappropriately used. It appears that when an appropriately designed treatment program is implemented, effectiveness of treatment can be demonstrated. A gap in the literature is the pharmacological treatment of psychopaths. Little research has been carried out in this area, and what is available is mostly based on literature written about psychiatric disorders and then transferred to psychopaths.

References

Babiak, P. (1995). When psychopaths go to work: A case study of an industrial psychopath. *Applied Psychology: An International Review, 44,* 171–188.

Blair, R. J. R. (1995). A cognitive developmental approach to morality: Investigating the psychopath. *Cognition, 57,* 1–29.

Blair, R. J. R. (2003). Neurobiological basis of psychopathy. *British Journal of Psychiatry, 182,* 5–7.

Blair, R. J. R. (2004). The roles of orbital frontal cortex in the modulation of antisocial behavior. *Brain and Cognition, 55,* 198–208.

Blair, R. J. R., Jones, L., Clark, F., & Smith, M. (1997). The psychopathic individual: A lack of responsiveness to distress cues? *Psychophysiology, 34,* 192–198.

Blair, R. J. R., Sellers, C., Strickland, I., Clark, F., Williams, A. O., Smith, M., et al. (1995). Emotion attributions in the psychopath. *Personality and Individual Differences, 19,* 431–437.

Blonigen, D. M., Carlson, S. R., Kruegar, R. F., & Patrick, C. J. (2003). A twin study report of self-reported psychopathic personality traits. *Personality and Individual Differences, 35,* 179–197.

Cleckley, H. (1941). *The mask of sanity.* St Louis, MO: C. V. Mosby.

Eysenck, H. J. (1998). Personality and crime. In T. Millon, E. Simonsen, M. Birket-Smith, & R. D. Davis (Eds.), *Psychopathy: Antisocial, criminal and violent behavior* (pp. 40–49). London: Guilford. (Original work published in *Psychology, Crime and Law, 2,* 1996, 143–152).

Flor, H., Birbaumer, N., Hermann, C., Ziegler, S., & Patrick, C. J. (2002). Aversive Pavlovian conditioning in psychopaths: Peripheral and central correlates. *Psychophysiology, 39,* 505–518.

Hall, J. R., Benning, S. D., & Patrick, C. J. (2004). Criterion-related validity of the three-factor model of psychopathy personality: Behavior and adaptive functioning. *Assessment, 11,* 4–16.

Hare, R. D. (1995). *Psychopaths: New trends in research.* Retrieved February 26, 1999, from www.mental_health.com/mag1/p5hpe01.html

Hare, R. D. (2003). *The Hare Psychopathy Checklist–Revised* (2nd ed.). Toronto, Canada: Multi-Health Systems.

Hare, R. D., & Jutai, J. W. (1983). Criminal history of the male psychopath: Some preliminary data. In K. T. Van Dusen & S. A. Mednick (Eds.), *Perspective studies of crime and delinquency* (pp. 225–236). Boston: Kluwer-Nijhoff.

Hare, R. D., & McPherson, L. M. (1984). Violent and aggressive behavior by criminal psychopaths. *International Journal of Law and Psychiatry, 7,* 35–50.

Harpur, T. J., & Hare, R. D. (1994). Assessment of psychopathy as a function of age. *Journal of Abnormal Psychology, 103,* 604–609.

Harris, G. T., Rice, M. E., & Lalumière, M. (2001). Criminal violence: The role of psychopathy, neurodevelopmental insults, and antisocial parenting. *Criminal Justice and Behavior, 28,* 402–426.

Hemphill, J. F., Templemann, R., Wong, S., & Hare, R. D. (1998). Psychopathy and crime: Recidivism and behavior. In D. J. Cooke, A. E. Forth, & R. D. Hare (Eds.), *Psychopathy: Theory, research, and implications for society* (pp. 375–399). Dordrecht, The Netherlands: Kluwer.

Herpertz, S. C., & Sass, H. (2000). Emotional deficiency and psychopathy. *Behavioral Sciences and the Law, 18,* 567–580.

Hiatt, K. D., Lorenz, A. R., & Newman, J. P. (2003). Assessment of emotion and language processing in psychopathic offenders: Results from a dichotic listening task. *Personality and Individual Differences, 32,* 1255–1268.

Hildebrand, M., de Ruiter, C., & de Vogel, V. (2004). Psychopathy and sexual deviance in treated rapists: Association with sexual and nonsexual recidivism. *Sexual Abuse: A Journal of Research and Treatment, 16,* 1–24.

Kiehl, K. A., Smith, A. M., Hare, R. D., Mendrek, A., Forster, B. B., Brink, J., et al. (2001). Limbic abnormalities in affective processing by criminal psychopaths as revealed by functional magnetic resonance imaging. *Biological Psychiatry, 50,* 677–684.

Kiehl, K. A., Smith, A. M., Mendrek, A., Forster, B. B., Hare, R. D., & Liddle, P. F. (2004). Temporal lobe abnormalities in semantic processing by criminal psychopaths as revealed by functional magnetic resonance imaging. *Psychiatry Research-Neuroimaging, 130,* 297–312.

Levenston, G. K., Patrick, C. J., Bradley, M. M., & Lang, P. J. (2000). The psychopath as observer: Emotion and attention in picture processing. *Journal of Abnormal Psychology, 109,* 373–385.

Lykken, D. T. (1957). A study of anxiety in the sociopathic personality. *Journal of Abnormal Psychology, 55,* 6–10.

Moffitt, T. E. (1993). Adolescence-limited and life-course-persistent antisocial behavior: A developmental taxonomy. *Psychological Review, 100,* 674–701.

Porter, S., Fairweather, D., Drugge, J., Hervé, H., Birt, A., & Boer, D. P. (2000). Profiles of psychopathy in incarcerated sexual offenders. *Criminal Justice & Behavior, 27,* 216–233.

Prichard, J. C. (1835). *A treatise on insanity and other disorders affecting the mind.* London: Sherwood, Gilbert & Piper.

Quinsey, V. L., Rice, M. E., & Harris, G. T. (1995). Actuarial prediction of sexual recidivism. *Journal of Interpersonal Violence, 10,* 85–105.

Raine, A., Lencz, T., Taylor, K., Hellige, J. B., Bihrle, S., Lacasse, L., et al. (2003). Corpus callosum abnormalities in psychopathic antisocial individuals. *Archives of General Psychiatry, 60,* 1134–1142.

Raine, A., Venables, P., & Williams, M. (1990). Relationships between central and autonomic measures of arousal at age 15 years, and criminality at age 24 years. *Archives of General Psychiatry, 47,* 1003–1007.

Rice, M. E., & Harris, G. T. (1997). Cross-validation and extension of the Violence Risk Appraisal Guide for child molesters and rapists. *Law and Human Behavior, 21,* 231–241.

Rice, M. E., Harris, G. T., & Cormier, C. A. (1992). An evaluation of a maximum-security therapeutic-community for psychopaths and other mentally disordered offenders. *Law and Human Behavior, 16,* 399–412.

Saltaris, C. (2002). Psychopathy in juvenile offenders: Can temperament and attachment be considered as robust developmental precursors? *Clinical Psychology Review, 22,* 729–752.

Séguin, J. R. (2004). Neurocognitive elements of antisocial behavior: Relevance of an orbitofrontal cortex account. *Brain and Cognition, 55,* 185–197.

Skeem, J. L., Monahan, J., & Mulvey, E. P. (2002). Psychopathy, treatment involvement, and subsequent violence among civil psychiatric patients. *Law and Human behavior, 26,* 577–603.

Trevathan, S. D., & Walker, L. J. (1989). Hypothetical versus real-life moral reasoning among psychopathic and delinquent youth. *Development and Psychopathology, 1,* 91–103.

Verona, E., Patrick, C. J., Curtin, J. J., Bradley, M. M., & Lang, P. J. (2004). Psychopathy and physiological response to emotionally evocative sounds. *Journal of Abnormal Psychology, 113,* 99–108.

Wong, S., & Hare, R. (2005). *Guidelines for a psychopathy treatment program.* Toronto: Multi-Health Systems.

DISCUSSION QUESTIONS

1. Why do you think that treatment often serves to make psychopaths worse?
2. As we saw in the text portion of this chapter, most prominent theorists such as Robert Hare do not believe that environmental factors play a part in the etiology of psychopathy. This article suggests that they might. How would you reconcile these two viewpoints?
3. Go to any search engine and type in "amygdala." What is the function of this organ and how is it related to psychopathy?

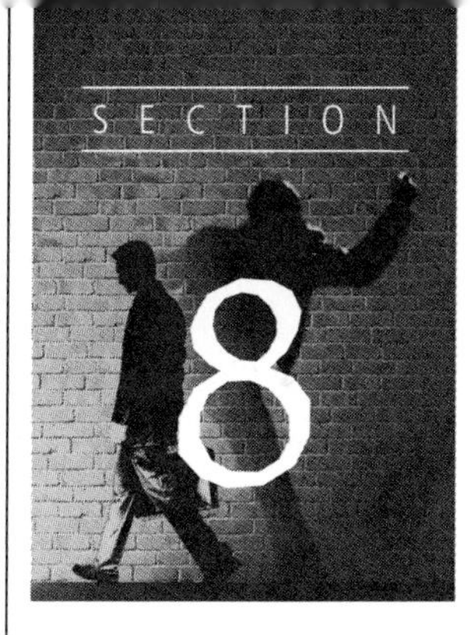

BIOSOCIAL APPROACHES

In February of 1991, Stephen Mobley walked into a Domino's Pizza store in Georgia to rob it. After getting the money, Mobley forced store manager John Collins onto his knees and shot him execution style. After committing several other robberies and bragging to friends about Collins' murder, Mobley was apprehended by Atlanta police, charged with aggravated murder, and sentenced to death. In the automatic appeal to the Georgia Supreme Court to get his sentenced commuted to life in prison, his primary defense boiled down to claiming that his "genes made [him] do it." In support of this defense, Mobley's lawyers pointed to a Dutch study of an extended family in which, for generations, many of the men had histories of unprovoked violence. The Researchers took DNA samples from 24 male members of the family and found that those with violent records had a marker for a mutant or variation of a gene for the manufacture of monoamine oxydase (MAO), an enzyme that regulates a lot of brain chemicals. Mobley's lawyers found a similar pattern of violent behavior and criminal convictions among his male relatives across the generations and asked the court for funds to conduct genetic tests on Mobley to see whether he had the same genetic variant.

The court wisely denied the defense motion. Even if it were found that Mobley had the same genetic variant, it would not show that he lacked the substantial capacity to appreciate the wrongfulness of his acts or to conform to the requirements of the law. Mobley's lawyers were hoping to mitigate his sentence by appealing to a sort of genetic determinism that simply does not exist. As we shall see in this section, genes don't "make" us do anything; they simply bias us in one direction rather than another. Except in cases of extreme mental disease or defect, we are always legally and morally responsible for our behavior. Cases such as Mobley's underline the urgent need for criminologists to understand the role of genes in human behavior as that role is understood by geneticists.

Biosocial criminologists believe that because humans have brains, genes, hormones, and an evolutionary history, they should integrate insights from the disciplines that study these things into their theories and dismiss naïve nature versus nurture arguments in favor of nature via nurture. Any trait, characteristic, or behavior of any living thing is always the result of biological factors interacting with environmental factors (Cartwright, 2000), which is why we call modern biologically informed criminology biosocial rather than biological. Biosocial approaches have three broad complementary areas: behavior genetics, evolutionary psychology, and neuroscience.

Behavior Genetics

Behavior genetics is a branch of genetics that studies the relative contributions of heredity and environment to behavioral and personality characteristics. Genes and environments work in tandem to develop any trait—height, weight, IQ, impulsiveness, blood sugar levels, blood pressure, and so on—the sum of which constitutes the person.

Behavior geneticists stress that genes do not *cause* us to behave or feel; they simply *facilitate* tendencies or dispositions to respond to the environments in one way rather than in another. There are no genes "for" criminal behavior, but there are genes that lead to particular traits such as low empathy, low IQ, and impulsiveness that increase the probability of criminal behavior when combined with the right environments.

Behavior geneticists use twin and adoption studies to disentangle the relative influences of genes and environments, and they tell us that genes and environments are always jointly responsible for any human characteristic. To ask whether genes or environment is most important for a given trait is just as nonsensical as asking whether height or length is most important to the area of a rectangle. Geneticists also tell us that gene expression depends on the environment (think of identical rose seeds planted in an English garden and in the Nevada desert, and then think about where the full potential of the seeds will be realized).

Behavior geneticists quantify the extent to which genes influence a trait with a measure called **heritability** (symbolized as h^2), which ranges between 0 and 1. The closer h^2 is to 1.0, the more the variance in a trait in a population, not in an individual, is due to genetic factors. Since any differences among individuals can only come from two sources—genes or environment—heritability is also a measure of environmental effects ($1 - h^2$ = environmental effects). All cognitive, behavioral, and personality traits are heritable to some degree, with the traits mentioned in the psychosocial section being in the .30 to .80 range (Carey, 2003).

Heritability estimates only tell us that genes are contributing to a trait; they do not tell us which genes; only molecular genetics can tell us this. Fortunately, we can now go beyond heritability estimates and straight to the DNA by genotyping individuals using simple cheek swabs for about $10 each. This eliminates the need for special twin or adoptee samples, but we do need cooperative studies with scientists having access to laboratories. Such studies are being conducted with increasing frequency, with the huge National Longitudinal Study of

Adolescent Health (ADD Health) study being one in which some very important genetic findings (the discovery of specific genes) relating to criminal behavior have been recorded (Beaver, 2006). It is emphasized that any individual gene only accounts for a miniscule proportion of the variance in criminal behavior, and that it contributes to a trait linked to criminality, not to criminality itself. Genes always have *indirect* effects on behavior via their effects on traits.

▲ **Photo 8.1** Former major league baseball player Jose Canseco is sworn in at a U.S. House of Representatives baseball steroids hearing. Canseco presents a fascinating case for biosocial theories. Jose had a fraternal twin brother, Ozzie, who also chose a career in baseball. However, in comparison with Jose's 462 home runs and over 1,400 RBI, Ozzie had only a "cup of coffee" in the major leagues. He came to bat only 65 times over 3 seasons and never hit a home run. Had he been an identical rather than a fraternal twin, might Ozzie have performed more like his brother? After finishing his baseball career, Jose wrote a book (*Juiced*) in which he admitted using steroids for most of his playing career and claimed that 85% of other players in his era did likewise. Because of his steroid use, many baseball experts predict Jose will never be elected to the baseball Hall of Fame, though his career numbers exceed those of many current Hall of Fame players.

Gene/Environment Interaction and Correlation

Gene/environment (G/E) interaction and G/E correlation describe people's active transactions with their environment. **G/E interaction** involves the notion that people are differentially sensitive to identical environmental influences and will thus respond in different ways to them. For instance, a relatively fearless and impulsive child is more likely to seize opportunities to engage in antisocial behavior than is a fearful and constrained child. **G/E correlation** means that genotypes and the environments they find themselves in are related. These concepts enable us to conceptualize the indirect way that genes help to determine what aspects of the environment will and will not be important to us. There are three types of G/E correlation: passive, reactive, and active.

Passive G/E correlation is the positive association between genes and their environments that exists because biological parents provide children with genes for certain traits and an environment favorable for their expression. Children born to intellectually gifted parents, for instance, are likely to receive genes for above-average intelligence and an environment in which intellectual behavior is modeled and reinforced, thus setting them on a trajectory independent (passively) of their actions.

Reactive G/E correlation refers to the way others react to the individual on the basis of his or her evocative behavior. Children bring traits with them to situations that increase or decrease the probability of evoking certain kinds of responses from others. A pleasant and well-mannered child will evoke different reactions than will a bad-tempered and ill-mannered child. Some children may be so resistant to socialization that parents may resort to coercive parenting or simply give up, either of which may worsen any antisocial tendencies and drive them to seek environments where their behavior is accepted. Reactive G/E correlation thus serves to magnify phenotypic differences by funneling individuals into like-minded peer groups ("birds of a feather flock together").

Active G/E correlation refers to the active seeking of environments compatible with our genetic dispositions. Active G/E correlation becomes more pertinent as we mature and acquire the ability to take greater control of our lives because within the range of possibilities available in our cultures, our genes help to determine what features of the environment will and will not be attractive to us. Our minds and personalities are not simply products of external forces, and our choices are not just passive responses to social forces and situations. We are active "niche picking" agents who create our own environments just as they help to create us.

Genes imply human self-determination because, after all, our genes are *our* genes. As Colin Badcock (2000) put it: "Genes don't deny human freedom; they positively guarantee it" (p. 71). Genes are constantly at our beck and call, extracting information from the environment and manufacturing the substances we need to navigate it. They are also what make us uniquely ourselves and thus resistant to environmental influences that grate against our natures. In short, genes do not constrain us, they enable us. This view of humanity is far more respectful of human dignity than the blank slate view that we are putty in the hands of the prevailing environmental winds.

Behavior Genetics and Criminal Behavior

A review of 72 behavior genetic studies conducted up to 1997 found that 93% were supportive of the hypothesis than genes affect antisocial behavior (Ellis & Walsh, 2000). Although there are no behavior genetic theories of criminal behavior per se, behavior genetic studies help us to temper and make more sense of traditional criminological theories. For instance, large behavior genetic studies conducted in the U. S. (Cleveland, Wiebe, van den Oord, & Rowe, 2000) and the U. K. (Moffitt & The E-Risk study team, 2002) showed that genetic factors play a large part in sorting individuals into different family structures (broken vs. intact homes), a variable often linked to antisocial behavior.

A major longitudinal study of child abuse and neglect that integrated genetic (the same MAO gene mentioned in the section vignette) data showed why only about one-half of abused/neglected children become violent adults (Caspi et al., 2002). This study showed that neither genetic risk nor abuse/neglect by themselves have much effect on antisocial behavior, but the odds of having a verified arrest history for a violent crime for subjects with both genetic and environmental risk factors (G/E interaction) were 9.8 times greater than the odds for subjects with neither.

Another study looked at Gottfredson and Hirschi's (1990) assumption that parents are primarily responsible for their children's self-control (Wright & Beaver, 2005). A modest relationship between parental practices and children's self-control was found, but disappeared when genetic information was added. In other words, not using genetically informed methods leads researchers to misidentify important causal influences. Wright and Beaver (2005) concluded, "for self-control to be a valid theory of crime it must incorporate a more sophisticated understanding of the origins of self-control" (p. 1190).

Unlike the relatively strong genetic influences discovered for most human traits, genetic influences on antisocial behavior are modest, especially during the teenage years. A study of 3,226 twin pairs that found genes accounted for only 7% of the variance in antisocial behavior among juvenile offenders, but 43% among adult offenders (Lyons et al., 1995). Heritability coefficients for most traits related to antisocial behavior are typically in the .20 to .80 range, and for antisocial behavior itself, two meta-analyses concluded that they are in the .40 to .58 range (Miles & Carey, 1997; Rhee & Waldman, 2002), with h^2 being higher in adult than in juvenile populations.

What this means is that the majority of delinquents have little, if any, genetic vulnerability to criminal behavior while a small minority may have considerable vulnerability. Pooling these two groups has the effect of elevating estimates of the overall influence of genes while minimizing it for those most seriously involved. For instance, although Mednick, Gabrielli, and Hutchings (1984) found a weak overall pattern of genetic effects for delinquency among a large number of young males, just 1.0% of the cohort (37 males) who had biological fathers with three or more criminal convictions accounted for fully 30% of all the cohort's convictions. Genetic effects on antisocial behavior are most likely to be found among chronic offenders who begin offending prior to puberty and who continue to do so across the life course (Moffitt & Walsh, 2003).

The article by Anthony Walsh in this section further explains behavior genetic concepts and ties them to the anomie/strain theory. Walsh contends that achieving the American Dream, the success goal so central to anomie/strain theory, is a matter of having the requisite cognitive and temperamental characteristics to perform well in school and at work. He shows how behavior genetics can augment sociological criminology theories by illuminating the genetic underpinnings of many of their favored concepts such as SES, occupational mobility, frustration, and so forth.

Evolutionary Psychology

Evolutionary psychology explores human behavior using an evolutionary theoretical framework and seeks to explain human behavior with reference to human evolutionary history. Criminologists operating within the evolutionary framework explore how certain behaviors society now calls criminal may have been adaptive in ancestral environments. Evolutionary psychology complements behavior genetics because it informs us how the genes of interest came to be present in the first place. The primary difference between the two disciplines is that while behavior genetics looks for what makes people different, evolutionary psychology focuses on what makes us all the same. Another basic difference is that evolutionary psychology looks at ultimate level "why" questions (what evolutionary problem did this behavioral mechanism evolve to solve?), and behavioral geneticists look at proximate level "how" questions (to what extent is this behavioral mechanism influenced by genes in this population at this time?).

Evolutionary psychologists agree with most criminologists that although it is morally regrettable, crime is normal behavior for which we all have the potential (Kanazawa, 2003). Evolutionary logic tells us that if criminal behavior is normal, it must have conferred some evolutionary advantage on our distant ancestors. However, because modern environments are so radically different from the hunter/gatherer environments in which we evolved, many traits selected for their adaptive value at the time may not be adaptive today. It is important to realize that it is the *traits* underlying criminal behavior that are the alleged adaptations, not the specific acts; genes do not code themselves for burglarizing a house or stealing a car (Rowe, 1996).

Criminal behavior is a way to acquire resources illegitimately. Evolutionary psychologists refer to such behavior (whether it is defined as criminal or not) as *cheating*, and think of individual traits associated with it, such as impulsiveness, aggression, and low empathy, in terms of normal distributions dispersed around adaptive species averages. Whether exploitation occurs depends on environmental triggers interacting with individual differences and with environmental constraints. Although we all have the potential to exploit and deceive others, we are a highly social and cooperative species with minds forged by evolution to form cooperative relationships built on trust (Barkow, 2006). Cooperation is typically contingent on the reciprocal cooperation of others, and is thus a tit-for-tat strategy favored by natural selection because of the benefits it confers. We cooperate with our fellows because we feel good when we do and because it identifies us as reliable and trustworthy, which confers valued social status on us.

Because cooperation occurs among groups of other cooperators, it creates niches for non-cooperators to exploit others by signaling their cooperation and then failing to follow through (Tibbetts, 2003). In the human species, criminal behavior may be viewed as an extreme form of defaulting on the rules of cooperation. But cheating comes at a cost, so before deciding to do so the individual must weigh the costs and benefits of cooperating versus defaulting. Cheating is rational (not to be confused with moral) when the benefits outweigh the costs. But if cheating is so rational, how did cooperation come to be predominant in social species? The answer is that cheating is only rational in circumstances of limited interaction and communication. Frequent interaction and communication breeds trust and bonding, and cheating becomes a less rational strategy because cooperators remember and retaliate against those who have cheated them. Ultimately, cooperation is the most rational strategy in any social species because each player reaps in the future what he or she has sown in the past.

Yet, we continue to see cheating behavior despite threats of exposure and retaliation. We do so because exposure and retaliation are threats only if cheats are constrained to operate within the same environment in which their reputations are known. Cheats can move from location to location meeting and cheating a series of others who are unaware of their reputation. This is the pattern of many career criminals who move from place to place, job to job, and relationship to relationship, leaving a trail of misery behind them before their reputation catches up. This is why cheats are more likely to prosper in large cities in modern societies than in small traditional communities where the threat of exposure and retaliation is great (Ellis & Walsh, 1997).

The Evolution of Criminal Traits

There are a number of evolutionary theories of crime, all of which focus on reproductive strategies. This is not surprising because from a biological point of view, the evolutionary imperative of all living things is reproductive success. There are two ways that members of any animal species can maximize reproductive success: parenting effort and mating effort. **Parenting effort** is the proportion of reproductive effort invested in rearing offspring, and **mating effort** is that proportion allotted to acquiring sexual partners. Because humans are born more dependent than any other animal, parenting effort is particularly important to our species. The most useful traits underlying parenting effort are altruism, empathy, nurturance, and intelligence (Rowe, 2002).

Humans invest more in parenting effort than any other species, but there is considerable variation within the species. Gender constitutes the largest division due to different levels of obligatory parental investment between the genders. Female parental investment necessarily requires an enormous expenditure of time and energy, but the only *obligatory* investment of males is the time and energy spent copulating. Reproductive success for males increases in proportion to the number of females to whom they have sexual access, and thus males have evolved a propensity to seek multiple partners. Mating effort emphasizes quantity over quality (maximizing the number of offspring rather than nurturing a few), although maximizing offspring numbers is obviously not a conscious motive of any male seeking sex. The proximate motivation is sexual pleasure, with more offspring being a natural consequence (in pre-contraceptive days) when the strategy proved successful.

Reproductive success among our ancestral females rested primarily on their ability to secure mates to assist them in raising offspring in exchange for exclusive sexual access, and thus human females evolved a much more discriminating attitude about sexual behavior (Fisher, 1998; Geary, 2000). According to evolutionary biologists, the inherent conflict between the reckless and indiscriminate male mating strategy and the careful and discriminating female mating strategy drove the evolution of traits such as aggressiveness and the lowering of trait levels (relative to female levels) such as empathy and constraint that help males to overcome both male competitors and female reticence. The important point to remember is that *although these traits were designed by natural selection to facilitate mating effort, they are also useful in gaining non-sexual resources via illegitimate means* (Quinsey, 2002; Walsh, 2006).

The reverse is also true—traits that facilitate parenting effort underlie other forms of prosocial activity: "Crime can be identified with the behaviors that tend to promote mating effort and noncrime with those that tend to promote parenting effort" (Rowe, 1996, p. 270). Because female reproductive success hinges more on parenting effort than mating effort, females have evolved higher levels of the traits that facilitate it (e.g., empathy and altruism) and lower levels of traits unfavorable to it (e.g., aggressiveness) than males. Of course, both males and females engage in both mating and parenting strategies, and both genders follow a mixed mating strategy. It is only claimed that mating behavior is more typical of males and parenting effort is more typical of females.

Empirical research supports the notion that an excessive concentration on mating effort is linked to criminal behavior. A review of 51 studies relating number of sex partners to criminal behavior found 50 of them to be positive, and in another review of 31 studies it was found that age of onset of sexual behavior was negatively related to criminal behavior (the earlier the age of onset, the greater the criminal activity) in all 31 (Ellis & Walsh, 2000). A British cohort study found that the most antisocial 10% of males in the cohort fathered 27% of the children (Jaffee, Moffitt, Caspi, & Taylor, 2003), and anthropologists tell us that there are striking differences in behavior between members of cultures that emphasize either parenting or mating strategies. The world over, cultures emphasizing mating effort exhibit behaviors (low-level parental care, hypermasculinity, transient bonding) considered antisocial in Western societies (Ember & Ember, 1998). Finally, a recent molecular genetic study found the same genes that were significantly related to number of sexual partners were also significantly related to antisocial behavior (Beaver, Wright, & Walsh, in press).

The Neurosciences

Whatever the source of human behavior, it is necessarily funneled through the brain, arguably the most awe-inspiring structure in the universe. Although the brain is only about 2% of body mass, it consumes 20% of the body's energy as it perceives, evaluates, and responds to its environment (Shore, 1997). This three-pound marvel of evolutionary design is the CEO of all that we think, feel, and do. All our thoughts, feelings, emotions, and behavior are the results of communication networks of brain cells called **neurons.** The more "primitive" networks come "hard wired" at birth, but development of the higher brain areas depends a lot on environmental "software" downloaded after birth. The message neuroscience has for us is that the experiences we encounter largely determine the patterns of our neuronal connections, and thus our ability to successfully navigate our lives (Quartz & Sejnowski, 1997).

Neural networks are continually being made and selected for retention or elimination in a use-dependent process governed by the strength and frequency of experience, and it is biased in favor of networks that are most stimulated during early development (Restak, 2001). This is why bonding and attachment are so vital to human beings, and why abuse and neglect is so injurious. Chronic stress can produce neuron death via the production of stress hormones, and children with high levels of these hormones experience cognitive, motor, and social development delays (Gunnar, 1996). As Perry and Pollard (1998) point out: "Experience in adults *alters* the *organized* brain, but in infants and children it *organizes* the *developing* brain" (p. 36). Brains organized by stressful and traumatic events tend to relay events along the same neural pathways laid out by those early events because pathways laid down early in life are more resistant to elimination than pathways laid down later in life. A brain organized by negative events is ripe for antisocial behavior.

▲ **Photo 8.2** Harkening back to the 19th century, when postmortem examinations of the brains of criminals were a frequent phenomenon, the brain of serial killer John Wayne Gacy was dissected after his execution. The attempt to locate an organic explanation of his monstrous behavior was unsuccessful.

Reward Dominance and Prefrontal Dysfunction Theories

Reward dominance theory is a neurological theory based on the proposition that behavior is regulated by two opposing mechanisms, the **behavioral activating system** (BAS) and the **behavioral inhibition system** (BIS). The BAS is associated with the neurotransmitter dopamine and with pleasure areas in the brain (Gove & Wilmoth, 2003). The BIS is associated with serotonin and with brain structures that govern memory (Pinel, 2000). **Neurotransmitters** such as dopamine and serotonin are the chemical messengers that shunt information between neural networks. Dopamine facilitates goal-directed behavior and serotonin generally modulates behavior (Depue & Collins, 1999).

The BAS is sensitive to reward and can be likened to an accelerator motivating a person to seek rewarding stimuli. The BIS is sensitive to threats of punishment and can be likened to a brake that stops a person from going too far too fast. The BAS motivates us to seek whatever affords us pleasure, and the BIS tells us when we have had enough for our own good. A normal BAS combined with a faulty BIS, or vice versa, may lead to a very impulsive person with a "craving brain" that can lead him or her into all sorts of physical, social, moral, and legal difficulties, by becoming addicted to pleasures such as food, gambling, sex, alcohol, and drugs (Ruden, 1997).

While most of us are more or less equally sensitive to both reward and punishment (in state of dopamine/serotonin balance), in some people one system might dominate the other most of the time (Ruden, 1997). The theory asserts that criminals, especially chronic criminals, have a dominant BAS, which tends to make them overly sensitive to reward cues and relatively insensitive to punishment cues (Lykken, 1995). Reward dominance theory provides us with hard *physical* evidence relating to the concepts of sensation seeking, impulsiveness, and low self-control we have previously discussed because each of these traits is underlain by either a sticky accelerator (high dopamine) or faulty brakes (low serotonin).

A third system of behavior control is the **flight/fight system** (FFS) chemically controlled by epinephrine (adrenaline). The FFS is that part of the autonomic nervous system that mobilizes the body for vigorous action in response to threats by pumping out epinephrine. Fear and anxiety at the chemical level is epinephrine shouting its warning: "Attention, danger ahead; take action to avoid!" Having a weak FFS (low epinephrine) that whispers rather than shouts, combined with a BAS (high dopamine) that keeps shouting "Go get it" and a BIS (low serotonin) too feeble to object, is obviously very useful when pursuing all kinds of criminal and antisocial activities.

Another neurologically specific theory of criminal behavior is **prefrontal dysfunction theory**. The **prefrontal cortex** (PFC) is responsible for a number of human attributes such as making moral judgments, planning, analyzing, synthesizing, and modulating emotions. The PFC provides us with knowledge about how other people see and think about us, thus moving us to adjust our behavior to consider their needs, concerns, and expectations of us. These PFC functions are collectively referred to as *executive functions* and are clearly involved in prosocial behavior. If these functions are compromised in some way via damage to the PFC, the result is often antisocial behavior.

Positron emission tomography (PET) and functional magnetic resonance imaging (fMRI) studies consistently find links between PFC activity and impulsive criminal behavior. A PET study comparing impulsive murderers with murderers whose crimes were planned found that the former showed significantly lower PFC and higher limbic system activity (indicative of emotional arousal) than the latter and other control subjects (Raine, Meloy, Bihrle, Stoddard, LaCasse, & Buchsbaum, 1998). Cauffman, Steinberg, and Piquero (2005) combined reward dominance and PFC dysfunctions theories in a large-scale study of incarcerated and nonincarcerated youths in California and found that seriously delinquent offenders have slower resting heart rates and performed poorly relative to nondelinquents on various cognitive functions mediated by the PFC.

Jana Bufkin and Vickie Lettrell's review of 17 neuroimaging studies in this section show conclusively that impulsive violent behaviors are associated with PFC deficits. They interpret the findings in terms of problems with the PFC's regulation of negative emotionality. They also explore activity between the PFC and subcortical structures associated with violence and aggression. Like many biosocial researchers, they make a plea for interdisciplinary studies of criminality so that the biological, psychological, and social factors associated with it can be better understood.

Lee Ellis's evolutionary neuroandrogenic (ENA) theory presented in this section attempts to tie many genetic, evolutionary, and neurological factors together into a unified theory. Ellis claims that his theory can explain why many of the demographic correlates of crime, such as age, sex, and SES, as well as biological correlates, such as mesomorphic body build, heart rate, and brainwave patterns, exist. He begins with two assumptions that are ultimately about mating versus parenting effort: (1) Males have been naturally selected for engaging in status striving and resource procurement because over countless generations, females who have chosen mates based on a male's ability to obtain resources will have left more offspring in subsequent generations than females who use other criteria for selecting mates. (2) Fetal exposure of male brains to testosterone makes them more prone to competitive status striving than females.

Table 8.1 summarizes the strengths and weaknesses of biosocial perspectives and theories.

Evaluation of the Biosocial Perspective

Biosocial theories have never been popular with mainstream social scientists because they were interpreted as implying a Lombrosian biological inferiority of criminals. This kind of thinking is rarer today as social scientists have become more sophisticated about the interaction of biology and environment (Robinson, 2004). There are still people who fear that "biological" theories can be used for racist ends, but as Bryan Vila (1994) remarks, "Findings can be used for racist or eugenic ends only if we allow perpetuation of the ignorance that underlies these arguments" (p. 329). Bigots and hate-mongers will climb aboard any vehicle that gives their prejudices a free ride, and they have done so for centuries before genes were heard of.

The strength of biosocial approaches lies in their ability to incorporate biological factors into their theories and to physically measure many of them via various chemical, electrophysiological, and neuroimaging methods. However, studies are more difficult to do and far more expensive than the typical social science study. If we want genetic information, we cannot simply go to the nearest high school and survey a few hundred students. Genetic studies require comparing samples consisting of pairs of identical and fraternal twins and/or adoptees, and these are difficult to come by. However, new technologies have allowed us to go straight to the DNA, thus eliminating the need for special samples consisting of paired subjects with known degrees of genetic relatedness.

It is difficult to make generalizations from the typical neuroimaging study because many consist of a small number of known offenders matched with a control group. Neuroimaging studies are extremely expensive to conduct (an average of $3,000 per scan), and many of them may only consist of 20 or 30 identified criminals matched with a control group of similar size. Nevertheless, biosocial studies provide criminologists with "harder" evidence than they are typically able to get, and this evidence will help them to more solidly ground their theories, and when numerous small studies tell the same story we are able to place more confidence in their conclusions.

Table 8.1 Summarizing Biosocial Perspectives and Theories

Theory	Key Concepts	Strengths	Weaknesses
Behavior Genetics Perspective	Genes affect behavior in interaction with environmental influences. Heritability estimates the relative contribution of genetic and environmental factor traits affecting criminality. All individual traits are at least modestly influenced by genes.	Looks at both the genetic and environmental risk factors for criminal behavior. Understanding genetic contributions also identifies the complementary contributions of environmental factors.	Requires samples of twins and/or adoptees, which are difficult to come by. However, technology now enables us to eliminate the need for special samples and go straight to the DNA.
Evolutionary Psychology Perspective	Human behavior is rooted in evolutionary history. Natural selection has favored victimizing tendencies in humans, especially males. These tendencies arose to facilitate mating effort, but are useful in pursuing criminal behavior as well. Criminals emphasize mating effort over parenting effort more than males in general.	Ties criminology to evolutionary biology. Mating effort helps to explain why males are more criminal than females and why criminals tend to be more sexually promiscuous than persons in general. Emphasizes that crime is biologically "normal" (although regrettable) rather than pathological.	Gives some the impression that because crime is considered "normal" it is justified or excused. Makes assumptions about human nature which may or may not be true. While recognizing that culture is important, it tends to ignore it.
Neuroscience Perspective	Whatever its origin, all stimuli are channeled through the brain before being given expression in behavior. The development of the brain is strongly influenced by early environmental experiences, especially those involving nurturance and attachment.	Shows how environmental experiences are physically "captured" by the brain. Emphasizes the importance of nurturing for optimal development of the brain. Uses sophisticated technology and provides "harder" evidence.	High cost of neuroimaging studies. Very small samples of known criminals are used, thus limiting generalizations. Linking specific brain areas to specific behaviors is problematic.
Reward Dominance Theory	Behavioral activating (BAS) and behavioral inhibiting system (BIS) are dopamine and serotonin driven, respectively. Among criminals the BAS tends to be dominant over the BIS. This BIS/BAS imbalance can lead to addiction to many things, including crime.	Explains why low serotonin is related to offending (low serotonin = low self-control). Explains why criminality is persistent in some offenders because they develop a taste for the "thrill of it all."	The neurological underpinnings of the BAS and BIS have been difficult to precisely identify. Studies difficult and expensive to conduct.
Prefrontal Dysfunction Theory	Frontal lobes control long-term planning and temper emotions and their expressions. Criminals have frontal lobes that fail to function as they do in most people, especially in terms of inhibiting actions that harm others.	Explains why moral reasoning is inversely related to involvement in persistent criminality Explains why criminality has been linked to frontal lobe damage and to abnormal brain waves.	Dysfunction of the prefrontal lobes remains difficult to measure, even with fMRI scans. Some sampling difficulties noted for the neurosciences in general.

Policy and Prevention: Implications of Biosocial Theories

The policy issues suggested by the biosocial perspective are midway between the macro-level sociological suggestions aimed at whole societies or communities and the micro-level suggestions of psychological theories aimed at already convicted criminals. Mindful of how nurturing affects both gene expression and brain development in humans, many biosocial criminologists have advocated a wide array of "nurturant" strategies such as pre- and postnatal care for all women, monitoring infants and young children through the early developmental years, paid maternal leave, nutritional programs, and a whole host of other interventions (Vila, 1997; Walsh & Ellis, 1997).

Biosocial criminologists are typically in the forefront in advocating treatment over punishment, and toward this end, they have favored indeterminate sentences over fixed sentences (Lanier & Henry, 1998). Pharmacological treatments in conjunction with psychosocial treatments have proven to be superior to psychosocial treatment alone for syndromes (alcoholism, drug addiction, etc.) associated with criminal behavior (Walsh, 2002). Of course, there are always dangers of seeking simple medical solutions to complex social problems. Requiring sex offenders to take anti-androgen treatment to reduce the sex drive raises both medical and legal/ethical issues regardless of how effective the treatment is. Prescribing selective serotonin reuptake inhibitors such as Prozac and Zoloft helps to curb low self-control and irritability, but there is always the temptation to treat everyone the same regardless of their serotonin levels.

Some behavioral scientists tend to feel that identifying biological risk factors will lead to cessation of efforts to reduce crime through environmental improvement because such factors are wrongly thought to be intractable. But biosocial studies provide information about *both* environmental and biological risk factors and, as such, are "more likely to refine social policies by better specification of environmental factors than to divert funds from environmental crime prevention strategies" (Morley & Hall, 2003). In other words, they will enable us to better pinpoint environmental factors that may prove fruitful in our crime prevention efforts.

Summary

- Behavior geneticists study the genetic underpinning of traits and characteristics in populations by calculating heritability coefficients. There are no genes "for" any kind of complex human behavior; genes simply bias trait values in one direction or another. This view is respectful of human dignity because it implies self-determinism because our genes are *our* genes.

- Gene/environment interaction tells us that the impact an environmental situation (e.g., living in a crime ridden neighborhood) has on us depends on who we are, and gene/environment correlation tells us that who we are is a product of our unique genotype and the environments we find ourselves in.

- Genes have practically no influence on juvenile delinquency, probably because of the high base rate of delinquency. There are genetic effects for chronic and serious delinquents, but these few individuals tend to get "lost" in studies that combine them with those who limit their offending to adolescence. Adult criminality is much more influenced by genes. One of the reasons that we find only modest genetic effects in criminality when the traits that underlie it are strongly influenced by genes is that parents have control over their children's behavior, but little or none over the underlying traits.

◆ Evolutionary psychology focuses on why we have the traits we do and is more interested in their universality than in their variability. Crime is viewed as a normal but regrettable response to environmental conditions. By this it is meant that many human adaptations forged by natural selection in response to survival and reproductive pressures are easily co-opted to serve morally wrong purposes.

◆ In common with all sexually producing species, humans are preeminently concerned with their own survival and reproductive success. The traits designed to assist males in their mating efforts include many that can also assist them to secure other resources illegitimately; traits designed to assist females in their parenting efforts are conducive to prosocial behavior. Mating vs. parenting effort is not an either/or thing; males and females engage in both at various times in their lives, it is just that mating effort is more typical of males and parenting effort is more typical of females

◆ Socially cooperating species create niches that cheats can exploit to their advantage by signaling cooperation but then defaulting. Cheating is a rational strategy in the short term, but invites retaliation in the long term. This is why chronic criminals rarely have successful relationships with others and why they typically die broke.

◆ Neuroscience tells us that genes have surrendered control of human behavior to the brain. Following genetic wiring to jump-start the process, the brain literally wires itself in response to environmental input. The softwiring of our brains is an electro-chemical process that depends on the frequency and intensity of early experiences. Adverse experiences can literally physically organize the brain so that we experience the world negatively, which is why nurturing, love, and attachment are so important to the healthy development of humans.

◆ Reward dominance theory informs us that the brain regulates our behavior through the BIS and BAS systems. The BIS and BAS systems (underlain by serotonin and dopamine neurotransmitters, respectively) in most people are balanced, but criminals tend to have either an overactive BAS or an underactive BIS. This means that their behavior is dominated by reward cues and relatively unaffected by punishment cues.

◆ Prefrontal dysfunction theory posits that the brain's prefrontal cortex (PFC) is vital to the so-called executive functions such as planning and modulating emotions. If the PFC is damaged in any way, the individual is deficient in these executive functions and tends to be impulsive.

EXERCISES AND DISCUSSION QUESTIONS

1. If it could be shown with high scientific confidence that some young children inherit genes that put them at 85% risk for developing antisocial proclivities, what do you think should be done? Should their parents be warned to be especially vigilant and to seek early treatment for their children, or would such a warning tend to stigmatize children? What are the costs and benefits of each option?

2. We know that males, especially young males, are more likely to perpetrate and be victimized by violent crimes. Provide a plausible evolutionary explanation for this.
3. How might reward dominance theory add strength and coherence to low self-control theory?

USEFUL WEB SITES

Behavioral Genetics. http://www.ornl.gov/sci/techresources/Human_Genome/elsi/behavior.shtml.

Crime Times. http://crimetimes.org/.

Evolutions Voyage. (Evolutionary Psychology). http://www.evoyage.com/.

The Human Brain. http://www.fi.edu/brain/index.htm.

Naturalistic Fallacy. http://www.iscid.org/encyclopedia/Naturalistic_Fallacy.

Anatomy of the Brain. http://www.neuroguide.com/index.html.

GLOSSARY

Behavior genetics: A branch of genetics that studies the relative contributions of heredity and environment to behavioral and personality characteristics.

Behavioral activating system: A brain reward system associated chemically with the neurotransmitter dopamine.

Behavioral inhibition system: A brain system that inhibits or modulates behavior and is associated with serotonin.

Evolutionary psychology: A way of thinking about human behavior using a Darwinian evolutionary theoretical framework.

Flight/fight system: An autonomic nervous system mechanism that mobilizes the body for action in response to threats by pumping out epinephrine.

G/E correlation: The notion that genotypes and the environments they find themselves in are related because parents provide children with both.

G/E interaction: The interaction of genes and the environment.

Heritability: A concept defined by a number ranging between 0 and 1 indicating the extent to which variance in a phenotypic trait in a population is due to genetic factors.

Mating effort: The proportion of total reproductive effort allotted to acquiring sexual partners; traits facilitating mating effort are associated with antisocial behavior.

Neurons: Brain cells consisting of the cell body, an axon, and a number of dendrites.

Neurotransmitters: Brain chemicals that carry messages from neuron to neuron.

Parenting effort: The proportion of total reproductive effort invested in rearing offspring; traits facilitating parenting effort are associated with prosocial behavior.

Prefrontal cortex: Part of the brain that plays the major integrative and supervisory roles in the brain.

Prefrontal dysfunction theory: A neurological theory of antisocial behavior based on dysfunction of the prefrontal cortex.

Reward dominance theory: A neurological theory based on the proposition that behavior is regulated by two opposing mechanisms, the behavioral activating system (BAS) and the behavioral inhibition system (BIS).

READING

Behavior Genetics and Anomie/Strain Theory

Anthony Walsh

In this article, Anthony Walsh states that criminology is in need of a conceptual boost from the more fundamental sciences such as behavior genetics. He claims that behavior genetics is a biologically friendly environmental discipline because it tells us just as much about the effect of the environment on a trait as it does that of genes. He uses anomie/strain theory to illustrate this. Anomie/strain is about occupational success, or the lack thereof, and Walsh looks at the role of intelligence and temperament, both of which are highly heritable, on occupational success. He also shows how behavior genetics can throw light on many of criminology's favored concepts.

The decade of the 1990s saw numerous reports of crisis within sociology in general as well as in criminology in particular. The substance of most of these reports is that the discipline is moribund and desperately needs something to breathe fresh life into it. The reason most often given for this state of affairs is sociology's dogged refusal to entertain the possibility that biological factors may play an important role in helping us to understand phenomena of concern to us. Perhaps a major reason for this refusal can be found in what is probably the most cited passage in all of sociology: "The determining cause of a social fact should be sought among antecedent social facts and not among the states of the individual consciousness" (Durkheim, 1982:131). This dictum has become a sociological mantra used to assert and defend the ontological autonomy of the discipline. According to Udry (1995:1267), however, although Durkheim clearly meant this statement to be a boundary axiom defining sociology's purview, sociologists came to think of it as "a true statement about the nature of the world instead of a set of deliberate blinders to help them focus their attention."

There are signs that certain criminological theories are becoming more psychologically and biologically informed, and there have been a number of recent encapsulations of biosocial approaches in the criminology literature. Authors

---

Source: Walsh, A. (2000). Behavior genetics and anomie strain theory. *Criminology, 38*(4). Reprinted with permission of the American Society of Criminology.

of such overviews, however, tend not to be mainstream criminologists, but psychologists, psychiatrists, and biologists, and none of them has tried to show the relevance of biosocial variables to mainstream criminological theories. Perhaps biologically informed theories will continue to have minimal impact on mainstream criminology until it can be shown that they are not antithetical to current environmental theories of crime. My primary purpose in this paper is to demonstrate that at least one biologically informed discipline—behavior genetics—is not antithetical to at least one environmental theory—anomie/strain theory.

Behavior Genetics

I suggest that it is time for mainstream criminology to at least pull back its blinders and peek at what behavior genetics has to offer. Very few criminologists have taken biology beyond an introductory class, and even fewer believe that genetic factors are important in explaining criminal behavior, despite the overviews of biosocial theories that have been published in the past decade. The lack of human genetic knowledge is probably not as important as ideology in keeping our blinders on. To acknowledge that genes may play a role in criminality has been virtually taboo for decades, and many criminologists still suffer from "biophobia," to use Ellis's (1996a) expression.

Behavior genetics, a branch of quantitative genetics, is not so much a "biological" discipline as it is a biologically-friendly environmental discipline. Although the discipline's main focus is to understand genetic influences on human behavior, traits, and abilities, behavior geneticists are aware that this cannot be accomplished without understanding the complementary role of the environment. Behavioral genetic studies have typically found that the environment accounts for more variance than does genetics in the most human traits, behaviors, and abilities. Thus, an often unappreciated aspect of behavior genetics is that it tells us as much, or sometimes more, about environmental effects as it does about genetic effects.

Behavior genetics research designs can also address and rectify a major problem with what has been called the standard social science model (SSSM) of socialization (Tooby and Cosmides, 1992). The problem is that the SSSM can never determine if any observed effects are primarily genetically or environmentally driven, i.e., whether parent/child or sibling/sibling similarities are caused primarily by shared genes or shared environments. This has led some socialization researchers to dismiss the entire SSSM of socialization as essentially useless.

We can only disentangle genetic from environmental effects if we can hold one constant. Behavior genetics does this using twin and adoption studies. The twin method compares pairs of individuals reared together (e.g., monozygotic twins) with a known degree of genetic relatedness with other pairs of reared-together individuals with a different degree of genetic relatedness (e.g., dizygotic twins), which holds environments constant and allows genes to vary. The adoption method examines genetically related individuals reared in different homes (or genetically unrelated individuals reared in the same home), which holds genes constant and allows environments to vary. In other words, the effects of shared environment can be determined by studying genetically unrelated individuals raised in the same environment, and the effects of heredity can be gauged by the phenotypical similarities of genetically related individuals reared apart.

Until molecular genetics identifies specific genes underlying specific traits, genetic effects must be inferred rather than directly demonstrated. Genetic effects are inferred by calculating a trait's heritability (h^2). Heritability coefficients range between 0 and 1 and estimate the extent to which variance in a trait in a population is attributable to genes. Heritability is based on the assumption that if genes affect a trait, the more genetically related two individuals are, the more similar they will be on that trait. If genes do not affect a trait, it would be logically and empirically impossible to calculate a heritability coefficient for that trait significantly greater than zero.

Heritability estimates fluctuate among different populations and within the same population as they experience different environments. Knowing what percentage of variance in a trait is attributable to genetic factors does not set limits on creating other environments that may influence the trait. A large heritability coefficient informs us that the present environment at the present time has minimal effect (accounts for little variance) on the trait; it does not tell us what other environments may have appreciably greater effects on the trait. Heritability tells us what *is* affecting trait variance; it does not tell us what *can* affect it. Moreover, the heritability of a trait informs us only about how much of the variance in a trait is attributable to genes, not how much the trait is attributable to genetic influences.

A related misunderstanding about heritability is the assumption that it carves nature neatly at the joint because it apportions traits into "genetic" and "environmental" components (e.g., 60% genetic, therefore, 40% environmental). Although heritability estimates statistically transform genes and environments into components, they are not separable ingredients in the real world. Genes and environments have the same relationship to phenotypic attributes as hydrogen and oxygen have to water and length and width have to area. It takes twice as many hydrogen atoms as oxygen atoms to make a molecule of water and an area's length may be twice its width, but in terms of the wholes they describe, the quantitative differences are meaningless. Without their complements, there would be no water or area, only a gas and a straight line. Likewise, genes and environments in isolation make no sense in terms of the phenotypic wholes they describe. There is no nature versus nurture, there is only nature via nurture.

In addition to apportioning variance into genetic and environmental components, the methods of behavior genetics yield a further benefit to social science in that they allow researchers to break down environmental variance into shared and nonshared components. Shared environments are environments of rearing that serve to make siblings alike (e.g., parental SES, religion, neighborhood, intactness of home), and nonshared environments are microenvironments unique to each sibling (birth order, differential treatment, different peers, etc.) that serve to make them different. A cascade of behavior genetic research has shown unequivocally that although shared rearing effects are real, albeit modest, during childhood and adolescence, they almost completely disappear in adulthood.

Agency and Gene/Environment Correlation

Social scientists are increasingly acknowledging that people's choices are not passive responses to social and cultural situations, and that these choices are to some degree autonomous. Many social scientists have gravitated toward the concept of agency in response to the "oversocialized" conception of human nature that has for so long dominated the social sciences. Agency simply means that as individuals strive for autonomy, they affect their environments just as surely as they are affected by them.

Transformations of self and environment are achieved by the actions and interactions of human subjects propelled by the subjective meanings that different people assign to similar situations. This concept of agency, which has much to do with our wishes, goals, and desires, is highly congenial to behavior genetics, which has always emphasized that people make their environments. That is, people's unique genotypes will largely determine what aspects of the social environment will be salient to them. This is a position considerably more respectful of human dignity (and more scientifically defensible) than is the image of human development that views it as little more than a process of class-, race-, age-, or gender-based adjustment to structural and cultural demands made on us. The concept of agency pulls us away from thinking of socialization as a mechanistic parent-to-child process and provides us with the skeleton of reciprocal effects thinking. Behavior genetics goes a step further to put the flesh on the bones of this thinking.

The concept of gene/environment (G/E) correlation is philosophically related to the concept of agency, but goes beyond it to provide an understanding of the underlying mechanisms that constitute the basis for the subjective meanings agency theorists articulate. G/E correlation essentially means that genotypes and the environments they encounter are not random with respect to one another, and that individuals are active shapers of their lives. The concept enables us to conceptualize the indirect way that genes exert their influence to help to determine the effective environment of the developing individual. Behavior geneticists differentiate between passive, reactive, and active G/E correlation.

Children raised by their biological parents experience passive G/E correlation by virtue of being provided with genes that underlie certain traits and a home environment favorable for their expression. A child born to highly intelligent parents, for example, typically receives genes conducive to high intelligence and a home in which intellectual activity is modeled and reinforced. The synergistic effects of correlated genes and environment will make for the unfolding of the child's intellectual abilities almost independently (passively) of what the child does. This means that the child has simply been exposed to the environment and has not been instrumental in forming it, not that the child does not actively engage it. On the downside, children born to low IQ, impulsive, or bad-tempered parents receive both genes and an environment developmentally biasing them in the same direction. The influence of passive G/E correlation declines dramatically from infancy to adolescence as the scope of environmental interaction widens and the person is confronted with and engages a wider variety of other people and behavioral options.

Reactive G/E correlation picks up the developmental trajectory as children grow older and are exposed to an increasing number of people and situations in their environments and begin to respond more actively to them. Children bring their developing phenotypical characteristics and abilities with them to the interpersonal situations they encounter that increase or decrease the probability of evoking certain kinds of responses from others. A bad-tempered child will evoke less solicitude than will a good-natured child, and a hyperactive, moody, and mischievous child will evoke less benign responses from others than will one who is pleasant and well behaved. Likewise, a child who shows enthusiasm and ability for school work evokes better treatment from teachers and will be afforded greater opportunities for advancement than will a child who shows little enthusiasm and ability for school work.

The important lesson of reactive G/E correlation is that the behavior of others toward the child is as much a function of the child's evocative behavior as it is of the interaction style of those who respond to it. Socialization is not something that others simply do to children; it is a reciprocal process that children and their caretakers engage in together. Some children may be difficult to control for even the most patient and loving parents. Some parents may abuse difficult children in their efforts to make them conform, and other parents may just give up trying to socialize their children. Both abusive and permissive responses to difficult children are likely to exacerbate their antisocial tendencies and drive them to seek others more accepting of their tendencies, presumably because the others toward whom they gravitate harbor such tendencies. Groups of individuals with similar tendencies provide positive reinforcement for each other and provide ample opportunities to exercise these tendencies. It is in this way that the feedback nature of reactive G/E correlation amplifies differences among phenotypes.

The seeking out of environments in which one feels accepted and psychologically comfortable is referred to as active G/E correlation. Large-scale twin studies provide striking evidence that genes play a very important role in "niche-picking." A number of studies have reported that the similarity in intelligence, personalities, attitudes, interests, and constructed environments of monozygotic (MZ) twins is essentially unaffected by whether they were reared together or apart. That is, MZ twins reared apart construct their environments and order their lives about as similarly as they would have had they been raised together, and this similarity is considerably greater than is the similarity between dizygotic (DZ) twins reared together.

Although such findings are robust, they appear counterintuitive to those who believe that the rearing environment mostly determines life outcomes. However, they make perfect sense to behavior geneticists, for whom it would be counterintuitive for people to either accept or to seek out environments incompatible with their genetic dispositions. Because MZ twins share 100% of their genes and DZ twins share, on average, only 50%, it makes sense that MZ twins would order their lives more similarly than would DZ twins. People with genes facilitative of different temperaments, traits, and abilities will seek out environments that mesh well with them (genes and environments will covary positively). It is no surprise to behavior geneticists that within the range of cultural possibilities and constraints, our genes set us on a developmental trajectory that will largely determine what features of the social world will be meaningful and rewarding to us, and what features will not.

Behavior Genetics and Antisocial Behavior

I must strongly emphasize that there is absolutely nothing in the concept of G/E correlation that can in any way be construed as supporting the notion of congenital criminality. Genes are self-replicating slices of DNA that code for proteins, which code for hormonal and enzymatic processes. There is no mysterious cryptography by which genes code for certain kinds of brains, which in turn code for different kinds of behaviors. Genes do not code for feelings or emotions either; what they do is make us differentially sensitive to environmental cues and modulate our responses to them. Genes always exert whatever influence they have on behavior in an environmental context. Although there can be no gene(s) "for" crime, there are genes that, via a number of neurohormonal routes, lead to the development of different traits and characteristics that may increase the probability of criminal behavior in some environments and in some situations.

A recent behavior genetic study illustrates the role of G/E correlation for antisocial behavior in late childhood/early adolescence (O'Connor et al., 1998). A number of adopted children were classified as either being or not being at genetic risk for antisocial behavior on the basis of their biological mothers' self-reported antisocial behavior collected prior to the birth of their children. It was found that from ages 7 to 12, children at genetic risk for antisocial behavior consistently received more negative parenting from their adoptive parents than did children not at genetic risk. This effect was interpreted as reactive G/E correlation in that children's poor behavior was seen as evoking negative parenting. O'Connor et al. (1998) did, however, find an environmentally mediated parental effect not attributable to reactive G/E correlation.

Another adoption study of antisocial behavior focused on G/E interaction (the differential effects of similar environments on different genotypes). Cadoret et al. (1995) examined the antisocial history of adopted children separated at birth from biological mothers with verified antisocial histories, compared with other adoptees with biological mothers with no known history of antisocial behavior. It was found that adverse adoptive home environments (divorce/separation, substance abuse, neglect/abuse, marital discord) led to significant increases in antisocial behavior for adoptees at genetic risk, but not for adoptees without genetic risk. Thus, both genes and environments operating in tandem (interacting) were required to produce significant antisocial behavior, whereas neither seemed powerful enough in this study to produce such effects independent of the other.

Antisocial behaviors, especially adolescent antisocial behaviors, are an interesting exception to the modest shared environmental influences typically discovered for most human characteristics. Shared group influences reflect the socializing influences of peer groups, although such influences cannot be considered apart from the tendency of similar people to befriend one another. The shifting pattern of genetic and environmental effects is readily seen in studies of juvenile and adult offending.

Genetic factors do not have similar explanatory power across all environments and across all developmental periods. Few things point to the vital importance of the environment to gene expression than the fact that heritability coefficients are always higher in advantaged than in disadvantaged environments. Just as a rose will express its fullest genetic potential planted in an English garden and wither when planted in the Nevada desert, human beings will realize their genetic potential to the fullest when reared in positive environments and fall short of doing so when reared in negative environments. High heritabilities for positive traits index how well a society is doing in equalizing environmental opportunities.

Differential genetic effects also apply to delinquent and criminal behaviors. In environments where resistance to crime is low, very little variance in criminal behavior will be attributable to genes; in environments where resistance to crime is high, the genetic contribution will be high. Venables (1987), for example, found that (low) tonic heart rate was a significant predictor of antisocial behavior among high SES children, but not among low SES children. Similarly, Walsh (1992) found that cognitive imbalance (as measured by verbal/performance IQ discrepancy scores) significantly predicted violent delinquency in advantaged environments, but not in disadvantaged environments. This does not mean these studies found that low tonic heart rate or cognitive imbalance was more prevalent in advantaged environments. On the contrary, what it means is that environmental causes tend to overwhelm putative genetic causes in disadvantaged environments.

The environmental complexities illustrated by studies in this section make criminal/antisocial behavior especially appealing for behavior genetic analysis of environmental influence.

Traits Linked to Middle-Class Success

In the following section, I briefly review the literature on the two major individual-level factors (temperament and intelligence) stressed by Agnew [in his general strain theory] (1992, 1997) as they relate to achieving middle-class success. Agnew remarks on a number of occasions that temperament and intelligence bear a strong relationship to problem-solving skills, and that the lower classes feel strain most acutely (1997:111–114). He never tries to make the connection between SES and these correlates of problem solving, although he is more than willing to state that such traits are a function of both biological and social factors. After the discussion of temperament and intelligence, I will attempt to show how other biosocial variables can be usefully integrated with Agnew's (1997) recently formulated developmental version of GST.

Temperament

Temperament refers to an individual's habitual mode of emotionally responding to stimuli. Temperamental style is identifiable very early in life, and it tends to be stable across the life course. Variance in temperamental measures is largely a function of heritable variation in central and autonomic nervous system arousal. Heritability coefficients for the various components of temperament range from about .40 to .80. Temperamental differences are largely responsible for making children differentially responsive to socialization. The unresponsiveness of a bad tempered (sour, unresponsive, quick to anger) child is exacerbated by the fact that temperaments of children and their parents are typically positively correlated. Parents of children with difficult temperaments tend to be inconsistent disciplinarians, irritable, impatient, and unstable, which makes them unable or unwilling to cope constructively with their children. Their children are thus burdened with both a genetic and an environmental liability. A cascade of evidence shows that children with difficult temperaments evoke negative responses from parents, teachers, and peers, and that these children find acceptance only in association with peers with similar dispositions (Moffitt, 1996).

Various physiological measures confirm that disinhibited temperament is associated with central nervous system (CNS) and peripheral nervous system (PNS) underarousal (suboptimal levels of arousal under normal environmental conditions). Individuals differ in the level of environmental stimulation they find optimal because of variation in the CNS's reticular activating system (a finger-size cluster of cells extending from the spinal cord that monitors incoming stimuli for processing by the cerebral cortex), and the PNS's autonomic nervous system. What is optimal for most of us will be stressful for some and boring for others. Suboptimally aroused people are easily bored and continually seek to boost stimuli to more comfortable levels. This search for amplified pleasure and excitement often leads the underaroused person into conflict with the law. A number of studies have shown that relative to the general population, criminals, especially those with the most serious criminal records, are chronically underaroused as determined by EEG brain wave patterns, resting heart rate and skin conductance, and histories of hyperactivity and attention deficit disorders. Individuals who are chronically bored and continually seeking intense stimulation are not likely to apply themselves to school or endear themselves to employers. One study found that various measures of reticular and autonomic nervous system arousal taken at age 15 correctly classified 74.7% of subjects as "criminal" or "noncriminal" at age 24 after controlling for a number of social factors (Raine et al., 1990).

Temperament provides the foundations for personality, which emerges from its interaction with the environment. Conscientiousness, one of the "big five" factors of personality, is a trait that is particularly important to success in the work force and is thus important to anomie/strain theory. Conscientiousness is a continuous trait ranging from well organized, disciplined, reliable, responsible, and scrupulous at one end of the continuum, to disorganized, careless, unreliable, irresponsible, and unscrupulous at the other. As we might expect, variance in conscientiousness is heritable. An analysis of 21 behavior genetic studies of conscientiousness found a median heritability of .66. In short, individuals with temperaments biasing them in the direction of being lacking in conscientiousness typically do not possess the personal qualities needed to apply themselves to the often long and arduous task of achieving legitimate occupational success. They will become "innovators' or "retreatists," not because a reified social structure has denied them access to the race, but because they find the race intolerably boring and busy themselves with more "exciting" pursuits instead.

Intelligence

Intelligence, as operationalized by IQ tests, is another obvious determinant of both occupational success and coping strategy, but also one that is conspicuously absent in sociological discussions of social status. As Lee Ellis has opined: "Someday historians of social science will be astounded to find the word intelligence is usually not even mentioned in late-twentieth-century text books on social stratification" (1996b:28).

Few scientists who study intelligence seriously doubt the importance of genes as well as the environment to explaining IQ variation. Twin studies, adoption studies, position electron tomography and magnetic resonance image scan studies, and even molecular genetic studies all point to substantial genetic effects on intelligence. Contrary to the claims of many social scientists, the National Academy of Sciences and the American Psychological Association's Task Force on Intelligence have concluded that there is no empirical evidence to indicate that IQ tests are biased against any racial/ethnic group or social class.

The litmus test for any assessment tool is its criterion-related validity—its ability to predict outcomes. An examination of 11 meta-analyses of the relationship between IQ and occupational success found that IQ predicted success better than did any other variable in most occupations, particularly in higher status occupations, and that it predicted equally well for all races/ethnic groups. Intelligence is particularly important in open and technologically advanced societies. Although there is no such thing as a totally open society, unlike the rigid caste-like societies of the past. In the rigid and aristocratic caste societies of the past, genes played almost no role in determining social class. They play an increasingly important role in more modern and egalitarian societies, however. Genes, and the individual differences they underlie, become important in determining SES in roughly direct proportion to equalization of environments. Although this may seem paradoxical at first blush, it is a basic principle of genetics: The more homogeneous (or equal) the environment, the greater the heritability of a trait; the more heterogeneous (unequal) the environment, the lower the heritability of a trait. High heritability coefficients for socially important traits tell us that the society is doing a good job of equalizing the environment.

The degree of occupational mobility in the U.S. labor force can be gauged by a major study's finding that 48% of sons of upper white-collar status fathers had lower status occupations than did their fathers, with 17% falling all the way to "lower manual" status, and that 51% of sons of lower manual status fathers achieved higher status, with 22.5% achieving "upper white-collar" status (Hurst, 1995: 270). Given this degree of upward/downward social mobility, and given the degree to which IQ predicts it equally for all races and social classes, it is difficult to maintain that any group is systematically denied access to legitimate opportunities to attain middle-class status.

Social scientists are more prone to attribute IQ level to SES level rather than the other way around. A large number of studies have found the correlation between parental SES and children's IQ to be within the .30 to .40 range (which is predictable from polygenetic transmission models). However, the correlation between individuals' IQ and their attained adult SES is in the .50 to .70 range. As Jensen remarks: "If SES were the cause of IQ [rather than the other way around], the correlations between adults' IQ and their attained SES would not be markedly higher than the correlation between children's IQ and their parents' SES" (1998: 491).

If differential IQ predicts differential adult SES, and if the lack of success leads to a mode of adaptation that includes criminal activity, IQ must be a predictor of criminal behavior. When evaluating the relatively small (eight IQ points) difference said to separate criminals and noncriminals, we must remember that researchers do not typically separate what Moffitt (1993) calls adolescent-limited (AL) and "life-course persistent" (LCP) offenders. As statistically normal individuals responding to the contingencies of their environments, we would not expect AL offenders to be significantly different from nonoffenders on IQ, and they are not. Moffitt (1996) reports a one-point mean IQ deficit between AL offenders and nonoffenders, but a 17-point deficit between LCP offenders and nonoffenders, and Stattin and Klackenberg-Larsson (1993) found a mean deficit of 4 points between nonoffenders and "sporadic" offenders and a 10-point difference between nonoffenders and "frequent" offenders. Aggregating temporary and persistent offenders creates the erroneous perception that IQ has minimal impact on antisocial behavior.

Although these studies separated temporary and persistent offenders, they did not separate IQ subtest scores. This also leads to an underestimation of the effects of IQ by pooling verbal IQ (VIQ), which uniformly shows a significant difference between offenders and nonoffenders, and performance IQ (PIQ), which typically does not. The most serious and persistent criminal offenders tend to have a PIQ score exceeding their VIQ score by about 12 points. Low verbal IQ (about one standard deviation below the mean) indexes poor abstract reasoning, poor judgment, poor school performance, impulsiveness, low empathy, and present orientedness. None of these traits is conducive to occupational success, but it is conducive to antisocial behavior, especially if combined with a disinhibited temperament.

Agnew's Developmental GST and Other Biosocial Processes

Agnew's (1997) developmental strain theory attempts to explain why antisocial behavior peaks during adolescence, focusing on increases in negative relationships with others during this period and the tendency to cope with the resulting strain through delinquency. Thus, a number of situations experienced as aversive are hypothesized to account for the age effect on crime and delinquency. As other theorists have pointed out, however, the age effect tends to remain robust, controlling for a number of demographic and situational variables, indicating that none of the various social correlates of age predict crime as well as age. For instance, Udry's (1990) sociological model found that age remained the strongest predictor of "problem behavior," controlling for a number of other independent variables. However, in his biosocial model that included measures of testosterone (T) and sex hormone binding globulin (SHBG), age dropped out of the equation, prompting him to suggest that age is a proxy for the hormonal changes of puberty.

During puberty, male T-levels increase 10- to 20-fold and SHBG is halved. Although the heritability of base-level T is around .60, the environmental effects on this hormone are well known. It is plain that T/SHBG ratios alone cannot explain the dramatic increase in antisocial behavior during this period, nor can they explain the fairly rapid decline in such behavior in the late teens and early twenties, because the decline does not correspond with a similar decline in T. Whatever the environmental or hormonal correlates of the increase and decline in adolescent antisocial behavior are, they are necessarily mediated by the brain.

Recent MRI studies of brain development confirm that the prefrontal cortex (PFC) is the last part of the human brain to fully mature. The PFC is the part of the brain that serves various executive functions, such as modulating emotions from the more primitive limbic system and making reasoned judgments. The PFC undergoes an intense prepubescent period of synaptogenesis and a period of pruning of excess synapses during adolescence. This process of synaptic elimination may be part of the reason many young persons find it difficult to accurately gauge the meanings and intentions of others and to experience more stimuli as aversive.

In addition to the synaptic pruning process, the adolescent's PFC is less completely myelinated than is the adult PFC. Myelination is important to the speed and conductive efficiency of neurotransmission. The fact that many syndromes associated with delinquency, such as oppositional disorder and conduct disorder/socialized type, first appear during this period may reflect a brain that is sometimes developmentally not up to the task of dealing rationally with the strains of adolescence. For instance, magnetic resonance imaging studies have shown that only the emotional limbic system is activated for many adolescents when shown photographs of frightened people, whereas both the limbic system and the PFC show activity among mature adults. This is another indication that the PFC is probably not performing its reasoning and emotion-modulating duties efficiently for at least some adolescents.

The increase in behavior-activating hormones coupled with an immature brain reflect two biological processes temporarily on conflicting trajectories that may both generate and exacerbate the strains of adolescence. This conjecture is supported by studies indicating that boys entering puberty early throw significantly more temper tantrums than do later maturing boys, and that early-maturing girls engage in significantly more problem behaviors than do later-maturing girls. Also consistent with this is the finding that T-levels predict future problem behavior only for early-pubertal-onset boys, again implying that the immature adolescent brain may facilitate a tendency to assign faulty attributions superimposed on an unfamiliar and diffuse state of physiological arousal.

Many other neurohormonal processes have the potential to assist anomie/strain theorists to understand strain across the life span, particularly during the teenage years. These include the shifting ratios of behavior-facilitating dopamine and the behavior-moderating serotonin during adolescence. The deaminating enzyme monoamine oxydase, which removes excess neurotransmitters at the synapse, is at its lowest during this period. These neurochemical fluctuations suggest that adolescence may be a period in which the brain is particularly sensitive to rewards and relatively insensitive to punishment, a combination facilitative of risk-taking and sensation-seeking behaviors.

It has long been known that humans and other animals in a low serotonin state are prone to violence, impulsiveness, and risk taking, especially if combined with high T. However, none of this means that teenagers are at the mercy of their biology, or that the environment has little impact. Serotonin levels, like T-levels, largely reflect environmental influences, and the lower average levels of serotonin during adolescence are as likely to be the effect of the increased strains of this period as they are to be the cause (Bernhardt, 1997). The stresses of adolescence outlined in Agnew's (1997) theory prompt the release of corticosteroids, which in turn suppress serotonin receptors. Thus, the vicissitudes of biology and social situations during this period of life combine in synergistic fashion to become both the producers and the products of strain. Further exploration of these interesting processes is beyond the scope of this paper.

Conclusion

I have attempted to show how behavior genetics can complement, extend, and add coherence to criminological theory, using one of its most revered and long-lived theories as an illustration. Perhaps all social scientists would acknowledge in the abstract that human behavior is the result of complex interactions between genes and environments, but having done so, most will continue to neglect the genetic 50% of the human behavioral equation in their thinking and research. As I have repeatedly emphasized, behavior genetics is not a "biological" perspective of human behavior. It is a biosocial perspective that takes seriously the proposition that all human traits, abilities, and behaviors are the result of the interplay of genetics and environments, and it is the only perspective with the research tools to untangle their effects. Given the exponential increase in knowledge in genetics over the past decades, criminologists cannot continue to reject insights from behavior genetics if the discipline is to remain scientifically viable. The history of science tells us that cross-disciplinary fertilization by a more mature science has, without exception, proven to be of immense value to the immature science. Criminology will be no exception.

References

Agnew, R. (1992). Foundations for a general strain theory of crime and delinquency. *Criminology, 30,* 47–87.

Agnew, R. (1997). Stability and change in crime over the life-course: A strain theory explanation. In T. Thornberry (Ed.), *Developmental theories of crime and delinquency.* New Brunswick, NJ: Transaction.

Bernhardt, P. (1997). Influences of serotonin and testosterone in aggression and dominance: Convergence with social psychology. *Current Directions in Psychological Science, 6,* 44–48.

Cadoret, R., Yates, W., Troughton, E., Woodworth, G., & Stewart, M. (1995). Genetic-environmental interaction in the genesis of aggressivity and conduct disorders. *Archives of General Psychiatry, 52,* 916–924.

Durkheim, E. (1982). *The rules of sociological method.* New York: Macmillan.

Ellis, L. (1996a). A discipline in peril: Sociology's future hinges on curing its biophobia. *American Sociologist, 27,* 21–41.

Ellis, L. (1996b). Arousal theory and the religiosity-criminality relationship. In P. Cordella & L. Siegel (Eds.), *Readings in contemporary criminological theory.* Boston: Northeastern University Press.

Hurst, C. (1995). *Social inequality: Forms, causes, and consequences.* Boston: Allyn & Bacon.

Jensen, A. (1998). *The g factor.* Westport, CT: Praeger.

Moffitt, T. (1993). Adolescent-limited and life-course persistent antisocial behavior: A developmental taxonomy. *Psychological Review, 100,* 674–701.

Moffitt, T. (1996). The neuropsychology of conduct disorder. In P. Cordella & L. Siegal (Eds.), *Readings in contemporary criminological theory.* Boston: Northeastern University Press.

O'Connor, T., Deater-Deckard, K., Fulker, D., Rutter, M., & Plomin, R. (1998). Genotype-environment correlations in late childhood and early adolescence: Antisocial behavioral problems and coercive parenting. *Developmental Psychology, 34,* 970–981.

Raine, A., Venables, P., & Williams, M. (1990). Relationships between central and autonomic measures of arousal at age 15 years and criminality at age 24 years. *Archives of General Psychiatry, 47,* 1003–1007.

Stattin, H., & Klackenberg-Larsson, I. (1993). Early language and intelligence development and their relationship to future criminal behavior. *Journal of Abnormal Psychology, 102,* 369–378.

Tooby, J., & Cosmides, L. (1992). The psychological foundations of culture. In J. Barkow, L. Cosmides, & J. Tooby (Eds.), *The adapted mind: Evolutionary psychology and the generation of culture* (pp. 19–136). New York: Oxford University Press.

Udry, J. R. (1990). Biosocial models of adolescent problem behaviors. *Social Biology, 37,* 1–10.

Udry, J. R. (1995). Sociology and biology: What biology do sociologists need to know? *Social Forces, 73,* 1267–1278.

Venables, P. (1987). Autonomic nervous system factors in criminal behavior. In S. Mednick, T. Moffitt, & S. Stack (Eds.), *The causes of crime: New biological approaches* (pp. 110–136). Cambridge: University of Cambridge Press.

Walsh, A. (1992). Genetic and environmental explanations of violent juvenile delinquency in advantaged and disadvantaged environments. *Aggressive Behavior, 18,* 187–199.

DISCUSSION QUESTIONS

1. How does the concept of G/E correlation highlight the role of human agency in development?
2. Why is talking about the role of intelligence in occupational success (and crime) controversial?
3. Outline the ways in which behavior genetics has thrown light on the anomie/strain tradition.

READING

Neuroimaging Studies of Aggressive and Violent Behavior

Current Findings and Implications for Criminology and Criminal Justice

Jana L. Bufkin and Vickie R. Luttrell

This article reviews neuroimaging studies with the goal of showing how neuroscience can advance criminological theories and inform criminal justice practice. The availability of new functional and structural neuroimaging techniques has allowed researchers to localize brain areas that may be dysfunctional in offenders who are aggressive and violent. Bufkin and Luttrell's review of 17 neuroimaging studies reveals that the areas associated with aggressive and/or violent behavioral histories, particularly impulsive acts, are located in the prefrontal cortex and the medial temporal regions. These findings are explained in the context of negative emotion regulation, thus supporting many studies outside of neuroscience that link criminality to low self-control and negative emotionality. Bufkin and Luttrell provide suggestions concerning how such findings may affect future theoretical criminology, crime prevention efforts, and treatment in criminal justice.

Aggressive and/or violent behaviors persist as significant social problems. In response, a substantial amount of research has been conducted to determine the roots of such behavior. Case studies of patients with neurological disorders or those who have suffered traumatic brain injury provide provocative insights into which brain regions, when damaged, might predispose to irresponsible, violent behavior. Psychophysiological and neuropsychological assessments have also demonstrated that violent offenders have lower brain functioning than controls, including lower verbal ability and diminished executive functioning. However, until recently it has been impossible to determine which brain areas in particular may be dysfunctional in violent individuals. With the availability of new functional and structural neuroimaging techniques, such as single-photon emission computed tomography (SPECT), positron emission tomography (PET), magnetic resonance imaging (MRI) and functional MRI (fMRI), it is now possible to examine regional brain dysfunction with a higher sensitivity and accuracy than was possible with previous techniques. This newfound ability to view the brain "in action" has broadened our understanding of the neural circuitry that underlies emotional regulation and affiliated behaviors. In particular, evidence suggests that individuals who are vulnerable to faulty regulation of negative emotion may be at increased risk for aggressive and/or violent behavior.

SOURCE: Bufkin, J. L., & Luttrell, V. R. (2005). Neuroimaging studies of aggressive and violent behavior: Current findings and implications for criminology and criminal justice. *Trauma, Violence, & Abuse*, 6(2), 176–191. Reprinted with permission of Sage Publications, Inc.

In this review, we evaluate the proposed link between faulty emotion regulation and aggressive or violent behavior. We define *aggression* as any threatening or physically assaultive behavior directed at persons or the environment. *Violence* refers to behaviors that inflict physical harm in violation of social norms. Specifically, we (a) discuss briefly the neurobiology of emotion regulation and how disruptions in the neural circuitry underlying emotion regulation might predispose to impulsive aggression and violence; (b) summarize the results of modern neuroimaging studies that have directly assessed brain functioning and/or structure in aggressive, violent, and/or antisocial samples and evaluate the consistency of these findings in the context of negative emotion regulation; and (c) discuss theoretical and practical implications for criminology and criminal justice.

Emotion Regulation and Theoretical Links to Impulsive Aggression and Violence

Emotion is regulated by a complex neural circuit that involves several cortical areas, including the prefrontal cortex, the anterior cingulate cortex (ACC), the posterior right hemisphere, the insular cortex, as well as several subcortical structures, such as the amygdala, hippocampus, and thalamus. These cortical and subcortical areas are intricately and extensively interconnected. In this article, we focus on three key elements of this neural circuitry: the prefrontal cortex, the ACC, and the amygdala.

The prefrontal cortex is a histologically heterogeneous region of the brain and has several (somewhat) functionally distinct sectors, including the ventromedial cortex and the orbitofrontal cortex (OFC). Damage to the ventromedial cortex and its behavioral affiliations have been assessed in case studies of individuals who experienced traumatic brain injury, either during childhood or adulthood, and in large, systematic studies on cohorts of war veterans with head injury.

Studies have found that patients with early-onset ventromedial lesions experience an insensitivity to future consequences, an inability to modify so-called risky behaviors even when more advantageous options are presented, and defective autonomic responses to punishment contingencies. Studies have also demonstrated that patients with adult-onset ventromedial damage show defects in real-life decision making, are oblivious to the future consequences of their actions, seem to be guided by immediate prospects only, and fail to respond autonomically to anticipated negative future outcomes.

The OFC, also a part of the prefrontal circuit, receives highly processed sensory information concerning a person's environmental experiences. The OFC is hypothesized to play a role in mediating behavior based on social context and appears to play a role in the perception of social signals, in particular, facial expressions of anger. Blair et al. (1999), using PET scans, assessed 13 male volunteers as they viewed static images of human faces expressing varying degrees of anger. They found that increasing the intensity of angry facial expressions was associated with enhanced activity in participants' OFC and the ACC. Dougherty et al. (1999) used functional neuroimaging and symptom provocation techniques to study the neurobiology of induced anger states and found that imaginal anger was associated with enhanced activation of the left OFC, right ACC (affective division), and bilateral anterior temporal regions. Also using imaginal scenarios, Pietrini et al. (2000) found that functional deactivation of OFC areas was strongest when participants were instructed to express unrestrained aggression toward assailants rather than when they tried to inhibit this imaginal aggression. Taken together, these lines of evidence support the suggestion that heightened activity in the left OFC may prevent a behavioral response to induced anger.

Based on these findings, and consistent with fearlessness theories of human aggression, a logical prediction is that OFC and ACC activity in response to provocation may be attenuated in certain individuals, predisposing them to aggression and violence. Consistent with this prediction, patients with OFC damage tend to exhibit poor impulse control, aggressive outbursts, verbal lewdness, and a lack of interpersonal sensitivity, which may increase the probability of sporadic so-called crimes of passion and encounters with the legal system. In contrast, evidence suggests that the ACC plays a role in processing the affective aspects of painful stimuli, such as the perceived unpleasantness that accompanies actual or potential tissue damage.

In addition to the prefrontal cortex and the ACC, another hypothesized neural component of emotion regulation is the amygdala, a subcortical structure, which is located on the medial margin of the temporal lobes. Similar to the OFC, the amygdala appears to play a role in extracting emotional content from environmental stimuli and may also play a role in individuals' ability to regulate negative emotion. However, neuroimaging studies have found that the amygdala is activated in response to cues that connote threat, such as facial expressions of fear (instead of anger), and that increasing the intensity of fearful facial expressions is associated with an increased activation of the left amygdala.

Davidson, Putnam, et al. (2000) suggested that individuals can typically regulate their negative affect and can also profit from restraint-producing cues in their environment, such as others' facial expressions of fear or anger. Information about behaviors that connote threat (e.g., hostile stares, threatening words, or lunging postures) is conveyed to the amygdala, which then projects to other limbic structures, and it is there that information about social context derived from OFC projections is integrated with one's current perceptions. The OFC, through its connections with other prefrontal sectors and with the amygdala, plays an important role in inhibiting impulsive outbursts because prefrontal activations that occur during anger arousal constrain the impulsive expression of emotional behavior.

Davidson, Putnam, et al. (2000) also proposed that dysfunctions in one or more of these regions and/or in the interconnections among them may be associated with faulty regulation of negative emotion and an increased propensity for impulsive aggression and violence. First, people with prefrontal and/or amygdalar dysfunction might misinterpret environmental cues, such as the facial expressions of others, and react impulsively, as a preemptive strike, to a misperceived threat. The perception of whether a stimulus is threatening is decisive in the cognitive processing leading to the aggressive behavior. Evidence suggests that individuals vary considerably in their ability to suppress negative emotion. Therefore, individuals with decreased prefrontal activity may have greater difficulty suppressing negative emotions than those individuals who have greater prefrontal activation. Finally, although prefrontal activity helps one to suppress negative emotion, this negative emotion is generated by subcortical structures, including the amygdala. Therefore, an individual may be more prone to violence in general, and impulsive violence, in particular, if prefrontal functioning is diminished in relation to subcortical activity.

Research on individuals who have suffered traumatic head injury is of key importance in understanding the neural substrates of aggressive and/or violent behavior; however, Brower and Price (2001) noted many limitations of head injury studies, such as inadequate controls for known risk factors, including prior history of aggressive or violent behavior, socioeconomic status, stability of employment, and substance abuse. Research of behavior following head injury is also one step removed from the question of whether aggressive and/or violent individuals (who may have no history of head trauma) have neurological dysfunction localized to specific areas in the brain.

Studies of aggressive, violent, and/or antisocial offenders using functional (SPECT and PET) and structural (MRI) neuroimaging are beginning to reveal abnormalities in these groups (Raine, Lencz, et al., 2000). Specifically, 17 neuroimaging studies have been conducted on samples derived from forensic settings, prisons, psychiatric hospitals, and on violent offenders who are noninstitutionalized. Our review of these works reveals four consistent patterns: (a) prefrontal dysfunction is associated with aggressive and/or violent behavioral histories; (b) temporal lobe dysfunction, particularly left-sided medial-temporal (subcortical) activity, is associated with aggression and/or violence; (c) the relative balance of activity between the prefrontal cortex and the subcortical structures is associated with impulsive aggression and/or violence; and (d) the neural circuitry underlying the regulation of emotion and its affiliated behaviors is complex. Each of these patterns is described in theoretical context below.

Prefrontal Dysfunction Is Associated With Aggressive and/or Violent Behavioral Histories

Of the 17 studies reviewed, 14 specifically examined possible links between frontal lobe pathology and aggressive and/or violent behavior. In the 10 SPECT and PET studies, 100% reported deficits in either prefrontal (8 of 10 studies) or frontal (2 of 10 studies) functioning in aggressive, violent, and/or antisocial groups compared to nonaggressive patients or healthy controls. Analyses of specific regions in the medial prefrontal cortex revealed that individuals who were aggressive and/or violent had significantly lower prefrontal activity in the OFC (4 of 10 studies), anterior medial cortex (5 of 10 studies), medial frontal cortex (2 of 10 studies) and/or superior frontal cortex (1 of 10 studies). In the four MRI studies, 50% (2 of 4 studies) reported decreased grey matter volume in prefrontal or frontal regions, and 25% (1 of 4 studies) reported nonspecific white matter abnormalities, not localized to the frontal cortex.

The consistency with which prefrontal disruption occurs across studies, each of which investigated participants with different types of violent behaviors, suggests that prefrontal dysfunction may underlie a predisposition to violence. Evidence is strongest for an association between prefrontal dysfunction and an impulsive subtype of aggressive behavior. Empirical findings concerning the regulation of negative emotion suggest that prefrontal sectors, such as the OFC, appear to play a role in the interpretation of environmental stimuli and the potential for danger. Consequently, disruptions in prefrontal functioning may lead individuals who are impulsive and aggressive to misinterpret situations as threatening and potentially dangerous, which in turn increases the probability of violent behavior against a perceived threat.

Nevertheless, four caveats are noteworthy. First, although prefrontal disruption was consistently related to aggressive and/or violent behavior, this association may reflect a predisposition only, requiring other environmental, psychological, and social factors to enhance or diminish this biological risk. Second, prefrontal dysfunction has also been documented in a wide variety of psychiatric and neurological disorders not associated with violence, and it may be argued that frontal hypometabolism is a general, nonspecific finding associated with a broad range of conditions. However, Drevets and Raichle (1995) reported that although frontal deficits have been observed in conditions, such as major depression, schizophrenia, and obsessive-compulsive disorder, the neurological profile for individuals who are aggressive and/or violent is different from these other groups. For example, while murderers exhibit widespread bilateral prefrontal dysfunction, individuals with depression tend to have disruptions localized in the left hemisphere only and to the left dorsolateral prefrontal cortex, in particular. (See Raine, Buchsbaum, et al., 1997, for a discussion of alterations in brain functioning across a variety of psychiatric conditions.)

Of the 10 SPECT and PET studies reviewed, 70% reported temporal lobe dysfunction in aggressive and/or violent groups, with reductions in left temporal lobe activity in 6 of 7 studies. Examination of the medial-temporal lobe, which includes subcortical structures, such as the amygdala, hippocampus, and basal ganglia, revealed that subcortical disruptions also characterized individuals who were aggressive and/or violent (4 of 7 studies). In the six MRI studies that examined the possibility of temporal lobe abnormalities, 100% (6 of 6 studies) reported temporal irregularities, including asymmetrical gyral patterns in the temporal-parietal region, decreases in anterior-inferior temporal lobe volume (including the amygdala-hippocampal region or adjacent areas), increases in left temporal lobe volume, or pathologies specific to the amygdala.

It is important to note that excessive right subcortical activity or abnormal temporal lobe structure was most common in patients with a history of intense violent behavior, such as that seen in those with intermittent explosive disorder rather than in patients who had aggressive personality types or who had high scores on an aggression scale. In humans, right-hemisphere activation has been suggested to play a role in the generation of negative affect. Therefore, increased subcortical activity in the right hemisphere could lead an individual to experience negative affect that promotes aggressive feelings and acts as a general predisposition to aggression and violence. These findings are generally consistent with current conceptions of emotion regulation and its purported relationship to impulsive violence, in particular

The Relative Balance of Activity Between the Prefrontal Cortex and the Subcortical Structures Is Associated With Impulsive Aggression and/or Violence

Previous research has suggested that individuals may be predisposed to impulsive violence if prefrontal functioning is diminished relative to subcortical activity. Raine, Meloy, et al. (1998) found that reduced prefrontal functioning relative to subcortical functioning was characteristic of those who commit impulsive acts of aggression and/or violence. By contrast, aggression and/or violence of a predatory nature was not related to reduced prefrontal and/ or subcortical ratios. They also suggest that although most biological studies of aggression and/or violence have not distinguished between impulsive and premeditated aggression, this distinction is likely relevant to understanding the neuroanatomical and functional underpinnings of these behaviors.

An additional line of evidence that lends support to the impulsive and/or predatory distinction comes from investigations regarding the mechanism underlying the suppression of negative emotion. The neurochemical link mediating prefrontal and/or subcortical interactions is purportedly an inhibitory serotonergic connection from the prefrontal cortex to the amygdala. The prefrontal cortex is a region with a high density of serotonin receptors, which sends efferents to the brainstem where most of the brain's serotonin-producing neurons originate. The prefrontal cortex, amygdala, and hippocampus also receive serotonergic innervation. Therefore, it is logical that dysfunction in the prefrontal and/or subcortical regions disrupts serotonergic activity in the brain. Consistent with this hypothesis, the serotonergic system has been shown to be dysfunctional in victims of violent suicide attempts, impulsive violent offenders, impulsive arsonists, violent offender and arson recidivists, children and adolescents with disruptive behavior disorders, and "acting out" hostility in normal volunteers. In all those studies, low serotonin levels were strongly related to the maladaptive behaviors noted.

Implications for Criminology and Criminal Justice

Historically, paradigms guiding criminological programs of study have tended to bypass complex webs of interconnections that produce and reproduce criminality, favoring instead an emphasis on one dimension or level of analysis. The trend has been to maintain a specialized focus, often within the confines of a sociological or a legalistic model. Attempts to expand the image of crime through theory integration, which have surfaced quite frequently since the mid-1970s, shift attention to different realities of crime. The general emphasis, however, has been on the integration of ideas within and/or across the two dominant paradigms rather than on a broader, interdisciplinary strategy.

Resistance to interdisciplinarity or disciplinary cross-fertilization has not been inconsequential. Failure to incorporate interdisciplinary insights has stifled exploration of the intersections among structure, culture, and the body, leaving a knowledge void where provocative social facts "merely hang in space as interesting curiosities" (Pallone & Hennessey, 2000, chap. 22, p. 11), and critical questions go unanswered. More specifically, lack of an imagination of how nurture and nature interact to affect behavior, or what may be understood as biography in historical context, has resulted in an incapacity to either deal with variability or deal with it well. Some male individuals socialized in a patriarchal society rape and some adolescents from poor, urban, single-families chronically offend; however, most do not. In other words, there is individual variation within social contexts, and those differences may be better understood if criminologists begin to consider all pertinent angles or dimensions.

Although the studies in our review may appear to be firmly planted in the tradition of specialization and unidimensional thinking, they should be interpreted within the framework of Barak's (1998) interdisciplinary criminology, where knowledges relevant to a behavioral outcome are treated as complements in an image expansion project. Understanding that each perspective offers a reality of behavior from a different, though interrelated angle, the objective is to develop a logical network of theories that will capture the most dimensions and provide the most accurate information about phenomena of interest. In this vein, our appraisal of knowledge from the field of neuroscience intends to elucidate the image of aggression and/or violence without supplanting other perspectives and paradigms. Our desire is not to reduce aggression and/or violence to brain functioning but to inform of advances in neurological analyses of emotion regulation and their importance to studies of that behavior.

It should be noted that the more comprehensive, interdisciplinary paradigm has been embraced by some criminologists linked to the biological sciences. In the 17 studies reviewed, researchers attempted to examine (or at least statistically controlled for) a variety of biological, psychological, and social correlates of aggressive and/or violent behavior and, in some cases, analyzed biopsychosocial interactions affecting behavior. Across studies, biological variables included history of head injury, substance use/abuse/dependence, diseases of the nervous system, left-handedness, body weight, height, head circumference, and sex. Psychological variables included the presence of psychological disorders (such as schizophrenia), indices of intellectual functioning (such as IQ scores), and performance-related motivational differences. Social variables included indices of psychosocial deprivation (such as physical and/or sexual abuse, extreme poverty, neglect, foster home placement, being raised in an institution, parental criminality, parental physical fights, severe family conflict, early parental divorce), family size, and ethnicity.

The benefits of integrating ideas or investigating an image from several angles in a research design is demonstrated in the neuroimaging studies provided. Raine, Lencz, et al. (2000), for example, found that prefrontal and autonomic deficits contributed substantially to the prediction of group membership (antisocial personality disorder vs. control group) over and above 10 demographic and psychosocial measures. The 10 demographic and psychosocial variables accounted for 41.3% of the variance. After the addition of three biological variables into the regression equation (prefrontal gray matter, heart rate, and skin conductance), amount of variance explained increased significantly to 76.7%, and the prediction of group membership increased from 73% to 88.5% classified correctly. These findings suggest that a more contextualized theoretical grasp of aggression and/or violence is possible when this behavior is conceptualized as multi-dimensional. When nature-nurture dichotomies are countered by interdisciplinary image expansion, clues about individual variability emerge, and criminologists come closer to understanding the complexity of aggression and/or violence.

The compelling evidence about this behavior revealed in the reviewed neuroimaging studies is valuable, then, not because it allows for completely reliable predictions of behavioral outcomes, but because it makes the image of aggression and/or violence a little less murky. Moreover, when merged with existing knowledges, particularly ideas about social structures and social psychology (sociological model) and rational choice (legalistic model), such findings may spawn new visions of justice centered on prevention and treatment. Within an interdisciplinary framework that values neuroscience, virtually every essential sociological factor elaborated by criminologists, structural and processual, acquires a greater potential to explain aggression and/or violence and influence policy making. According to the works in our review, as well as other research in this area, all forms of child abuse and neglect, direct exposure to violence (including media violence), an unstable family life, poor parenting, lack of prenatal and perinatal services, individual drug use, maternal drug use during pregnancy, poor educational and employment structures, poverty, and even exposure to racism play a vital role in the production of aggression and/or violence. Thus, the inclusion of insights from neuroscience further legitimizes prevention strategies touted by advocates of the sociological paradigm, from social disorganization theory to self-control theory.

When aggression and/or violence is not prevented, the criminal justice system is granted responsibility for social control. . . . Drawing from empirical findings across disciplines and levels of analysis, a vision of therapeutic justice encourages the development of holistic treatment regimens that hold offenders to "scientifically rational and legally appropriate degree[s] of accountability" (Nygaard, 2000, chap. 23, p. 12). The potential for this approach to replace the utilitarian model lies in its continued ability to unveil the often-perplexing ways in which choice is structured. This facilitates an awareness that the legally appropriate and the scientifically rational are in unity. Human creativity is not ignored in this paradigm; however, the clearer image it provides points to an amalgam of limitations. When an individual is brought into the criminal justice system, an inter-disciplinarian seeks to examine those restrictions on behavior and to tailor treatments accordingly.

With varying levels of success, criminologists have sought to qualify choice and diminish the impact of legalistic factors on conceptions of justice since the advent of positivism in the 19th century. Assessments of measures associated with social psychology, psychology, and psychiatry, along with input implicating structural concerns, such as unemployment, have been utilized, and a plethora of interventions have evolved. Thus, cracks in the utilitarian mold of justice have accrued, laying the foundation for interdisciplinarity in thought and in treatment. Applied to aggression and/or violence, this translates into the implementation of treatment plans with multidimensional components, to include neurological techniques that address how brain dysfunction affects choice. Although not the sole neurological strategy, the intervention most consistently promoted is drug therapy. Several types of drugs, such as anti-convulsants, psychostimulants, and serotonergic agents, have been successful in reducing aggressive behavior. Inter-disciplinary thinkers should not be hesitant to consider using these pharmacological remedies when biopsychosocial indicators overwhelmingly suggest that an individual is at risk to violently recidivate, for it is a step in the direction of therapeutic justice.

Other than paradigmatic preferences disallowing an interdisciplinary consideration of aggressive and/or violent crimes and lack of funding, the largest obstacle in attaining therapeutic justice is the inability to predict future behavior. When informed by neuroscience, classification and prediction instruments are fine-tuned. To illustrate, Robinson and Kelley (2000) discovered that, among probationers, indicators of brain dysfunction correlated with repeat violent offending, as opposed to repeat nonviolent offending and first-time offending. Birth complications, family abuse, head injury, parental drug use, abnormal interpersonal characteristics, and offender substance abuse were found to be risk factors for recidivism within this group. Given that it is estimated that less comprehensive prediction models reap false positives in approximately two thirds of all cases, added precision is welcome.

Still, prediction is not foolproof. Shortcomings in this area lead some to conclude that drug therapy and other invasive strategies are unwarranted. Before throwing in the towel, it should be acknowledged that pharmacological remedies already abound in the criminal justice system, along with many other intrusions. Knowledge from neuroscience merely allows for the targeted distribution of services to appropriate populations, a fruitful strategy given the scarcity of resources at the system's disposal. Prevention strategies directed at alleviating environmental conditions that increase the probability for aggression and/or violence are optimal; however, criminologists should not dismiss neuroscientific individual-level interventions in cases where patterns of aggressive and/or violent criminality are detected. Converging lines of evidence suggest that those patterns are produced by a unique combination of external and internal risk factors, each of which is integral to the construction of treatment regimens intended to effect therapeutic justice.

Blind spots in the image of aggression and/or violence should not deter interventions where they hold promise for enhancing quality of life. It is unfair and unjust to those processed in the criminal justice system and to society at large for criminologists to ignore this evidence and the control strategies proposed.

Conclusion

Functional and morphometric neuroimaging have enhanced our understanding of the distributed neural networks that subserve complex emotional behaviors. Research emanating from affective, behavioral, and clinical neuroscience paradigms is converging on the conclusion that there is a significant neurological basis of aggressive and/or violent behavior over and above contributions from the psychosocial environment. In particular, and consistent with modern theories of emotion regulation, reduced prefrontal and/or subcortical ratios may predispose to impulsive aggression and/or violence. Further progress in the study of these behaviors will require a forensically informed, interdisciplinary approach that integrates neuropsychological and psychophysiological methods for the study of the brain, emotional processing, and behavior.

As this line of interdisciplinary research unfolds, it is vital that criminology and criminal justice begin to incorporate what is known about human behavior into its explanatory models, as well as its classrooms. Evidence suggests that brain structure and brain functioning do affect behavior, particularly aggressive and/or violent behavior. It is also the case that neuroscience offers means for curbing aggression and/or violence. Traditional criminology and criminal justice paradigms tend to sidestep these issues because of aversions to less dominant knowledges, especially biological programs of study. Biological insights are often dubbed Lombrosian, suggesting that some behavioral scientists retain notions of a born criminal easily identifiable using some magic test. Continued aversion to anything biological on these grounds is anachronistic and will hamper the development of theory and policy.

The problem is that neurobiological discovery has carried on with little to no input from criminology and criminal justice, and there is every reason to believe that the research will progress. There is also reason to believe that the functioning of the criminal justice system will be affected by the findings produced. The general public is already being widely exposed to such advances through numerous television news clips and articles appearing in newspapers and weekly periodicals. If criminology and criminal justice wants to be relevant in more than a historical sense when it comes to theorizing about aggressive and/or violent behavior and formulating policies accordingly, it is imperative that the field embrace the interdisciplinary model.

Implications for Practice, Policy, and Research

Practice

Bridging the gap between nature and nurture, a biopsychosocial model for understanding aggression enhances the explanatory capacity of sociologically based criminological theories by accounting for individual variability within social contexts.

Insights derived from a biopsychosocial model offer the most promise in the realm of crime prevention, which entails devising holistic treatment strategies for those exposed to numerous risk factors.

Policy

The accuracy of risk classification devices, used extensively throughout the criminal justice process, may be enhanced by incorporating what is known about negative emotion regulation.

Research

Research reveals that other cortical and subcortical structures likely play a role in emotion regulation through their inextricable link to the prefrontal and medial-temporal regions. The complexity of this neural circuitry must be explored with greater precision.

References

Barak, G. (1998). *Integrating criminologies.* Needham Heights, MA: Allyn & Bacon.

Blair, R. J. R., Morris, J. S., Frith, C. D., Perrett, D. I., & Dolan, R. J. (1999). Dissociable neural responses to facial expressions of sadness and anger. *Brain, 122,* 883–893.

Brower, M. C., & Price, B. H. (2001). Neuropsychiatry of frontal lobe dysfunction in violent and criminal behavior: A critical review. *Journal of Neurology, Neurosurgery, and Psychiatry, 71,* 720–726.

Davidson, R. J., Putnam, K. M., & Larson, C. L. (2000). Dysfunction in the neural circuitry of emotion regulation: A possible prelude to violence. *Science, 289*(5479), 591–594.

Dougherty, D. D., Shin, L. M., Alpert, N. M., Pitman, R. K., Orr, S. P., Lasko, M., et al. (1999). Anger in healthy men: A PET study using script-driven imagery. *Biological Psychiatry, 46,* 466–472.

Drevets, W. C., & Raichle, M. E. (1995). Positron emission tomographic imaging studies of human emotional disorders. In M. S. Gazzaniga (Ed.), *The cognitive neurosciences* (pp. 1153–1164). Cambridge, MA: MIT Press.

Nygaard, R. L. (2000). The dawn of therapeutic justice. In D. H. Fishbein (Ed.), *The science, treatment, and prevention of antisocial behaviors: Application to the criminal justice system* (chap. 23, pp. 1–18). Kingston, NJ: Civic Research Institute.

Pallone, N. J., & Hennessy, J. J. (2000). Indifferent communication between social science and neuroscience: The case of "biological brain-proneness' for criminal aggression. In D. H. Fishbein (Ed.), *The science, treatment, and prevention of antisocial behaviors: Application to the criminal justice system* (chap. 22, pp. 1–13). Kingston, NJ: Civic Research Institute.

Pietrini, P., Guazzelli, M., Basso, G., Jaffe, K., & Grafman, J. (2000). Neural correlates of imaginal aggressive behavior assessed by positron emission tomography in healthy subjects. *American Journal of Psychiatry, 157*(11), 1772–1781.

Raine, A., Buchsbaum, M., & LaCasse, L. (1997). Brain abnormalities in murderers indicated by positron emission tomography. *Biological Psychiatry, 42,* 495–508.

Raine, A., Lencz, T., Bihrle, S., LaCasse, L., & Colletti, P. (2000). Reduced prefrontal gray matter volume and reduced autonomic activity in antisocial personality disorder. *Archives of General Psychiatry, 57,* 119–127.

Raine, A., Meloy, J. R., Bihrle, S., Stoddard, J., LaCasse, L., & Buchsbaum, M. S. (1998). Reduced prefrontal and increased subcortical brain functioning assessed using positron emission tomography in predatory and affective murderers. *Behavioral Sciences and the Law, 16,* 319–332.

Robinson, M., & Kelley, T. (2000). The identification of neurological correlates of brain dysfunction in offenders by probation officers. In D. H. Fishbein (Ed.), *The science, treatment, and prevention of antisocial behaviors: Application to the criminal justice system* (chap. 12, pp. 12-1–12-20). Kingston, NJ: Civic Research Institute.

DISCUSSION QUESTIONS

1. How might we make the case that the findings presented in this review support Agnew's "super traits" theory discussed in Section 9?
2. What are the authors' main arguments for "biopsychosocial" integration?
3. What do you think are the main obstacles to biopsychosocial integration?

READING

A Theory Explaining Biological Correlates of Criminality

Lee Ellis

In this article, Lee Ellis proposes a biosocial theory he calls the evolutionary neuroandrogenic theory (ENA), which purports to explain many of the correlates of criminal behavior. Two propositions form the theory's foundation: an evolutionary proposition and a neurohormonal proposition. According to the first proposition, males have been favored for victimizing others because doing so has helped them to compete and acquire resources that they use to attract mates. The second proposition is concerned with identifying the neurochemistry underlying this evolved strategy. It maintains that male sex hormones—particularly testosterone—alter brain functioning in ways that promote competitive and victimizing types of behavior, which includes, but is not limited to, many types of criminality.

Despite growing evidence that biology plays an important role in human behavior, most theories of criminal behavior continue to focus on learning and social environmental variables. This article proposes a biosocial theory of criminality that leads one to expect variables such as age, gender and social status will be associated with offending in very specific ways. According to the theory, androgens (male sex hormones) have the ability to affect the brain in ways that increase the probability of what is termed *competitive/victimizing behavior* (CVB). This behavior is hypothesized to exist along a continuum, with "crude" (criminal) forms at one end and "sophisticated" (commercial) forms at the other. Theoretically, individuals whose brains receive a great deal of androgen exposure will be prone toward CVB. However, if they have normal or high capabilities to learn and plan, they

SOURCE: Ellis, L. (2005). A theory explaining biological correlates of criminality. *European Journal of Criminology, 2*(3), 287–315.

will transition rapidly from criminal to non-criminal forms of the behavior following the onset of puberty. Individuals with high androgen exposure and poor learning and planning capabilities, on the other hand, often continue to exhibit criminality for decades following the onset of puberty.

The Evolutionary Neuroandrogenic Theory of Criminal Behavior

The theory to be presented is called the *evolutionary neuroandrogenic theory* (*ENA*). The main types of offenses it attempts to explain are those that harm others, either by injuring them physically or by depriving them of their property. Two main propositions lie at the heart of ENA theory. The first addresses evolutionary issues by asserting that the commission of victimful crimes evolved as an aspect of human reproduction, especially among males. The second is concerned with identifying the neurochemistry responsible for increasing the probability of criminality among males relative to females. The theory maintains that sex hormones alter male brain functioning in ways that promote CVB, which is hypothesized to include the commission of violent and property crimes.

The concept of CVB is illustrated in Table 8.2. At one end of the continuum are acts that intentionally and directly either injure others or dispossess them of their property. In all societies with written laws, these obviously harmful acts are criminalized. At the other end of the CVB continuum are acts that make no profits on the sale of goods or services, although those who administer and maintain the organizations under which they operate usually receive much higher wages than do those who provide most of the day-to-day labor. In a purely socialist economy, the latter type of minimally competitive activities is all that is allowed; all other forms are criminalized. A capitalist economy, on the other hand, will permit profit-making commerce and often even tolerate commerce that involves significant degrees of deception. With the concept of CVB in mind, the two propositions upon which the theory rests can now be described.

The Evolutionary Proposition

Throughout the world, males engage in victimful crimes (especially those involving violence) to a greater extent than do females. To explain why, ENA theory maintains that female mating preferences play a pivotal role. The nature of this mating preference is that females consider social status criteria much more than males do in making mate choices, a pattern that has been documented throughout the world (Ellis, 2001). From an evolutionary standpoint, this female preference has served to increase the chances of females mating with males who are

Table 8.2 Continuum of Victimizing Behavior (Reflecting Competitive/Victimizing Tendencies)

The Continuum	very crude------------------------------intermediate--------------------------------------very sophisticated				
Probability of Being Criminalized	virtually certain------------------------intermediate-----------------------------------exceedingly unlikely				
Examples	Violent and property offenses ("street crime")	Embezzlement, fraud ("white collar crime")	Deceptive business practices, price gouging	Profit-making commerce	Nonprofit-making commerce

reliable provisioners of resources, allowing females to focus more of their time and energy on bearing offspring. Another consequence has been that female choice has made it possible for males who are status strivers to pass on their genes at higher rates than males who are not. Such female preferences are found in other mammals, as evidenced by their mating more with dominant males than with subordinate males.

According to ENA theory, female preferences for status-striving males have caused most males to devote considerable time and energy to competing for resources, an endeavor that often victimizes others. In other words, natural selection pressure on females to prefer status-striving mates has resulted in males with an inclination toward CVB. ENA theory maintains that the brains of males have been selected for exhibiting competitive/victimizing behavior to a greater extent than the brains of females, and that one of the manifestations of this evolved sex difference is that males are more prone than females toward victimful criminality.

Theoretically, the same natural selection pressure that has resulted in the evolution of CVB has also favored males who flaunt and even exaggerate their resource-procuring capabilities. More unpleasant consequences of the female bias for resource provisioning mates are male tendencies to seek opportunities to circumvent female caution in mating by using deceptive and even forceful copulation tactics. This implies that rape will always be more prevalent among males than among females. ENA theory also leads one to expect complex social systems to develop in order to prevent crime victimization. In evolutionary terms, these systems are known as *counter-strategies.* An example of a counter-strategy to crude forms of CVB is the evolution of the criminal justice system.

As with any theory founded on neo-Darwinian thinking, ENA theory assumes that genes are responsible for substantial proportions of the variation in the traits being investigated. In the present context, the average male is assumed to have a greater genetic propensity toward CVB than is true for the average female. However, this assumption must be compromised with the fact that males and females share nearly all of their genes. Consequently, the only possible way for the theory to be correct is for some of the genes that promote criminality (along with other forms of CVB) to be located on the one chromosome that males and females do not share—the Y-chromosome.

The Neuroandrogenic Proposition

The second proposition of ENA theory asserts that three different aspects of brain functioning affect an individual's chances of criminal offending by promoting CVB. Two additional neurological factors help to inhibit offending by speeding up the acquisition of sophisticated forms of CVB. Testosterone's ability to affect brain functioning in ways that promote CVB is not simple, but most of the complexities will not be considered here. The main point to keep in mind is that testosterone production occurs in two distinct phases: the organizational (or perinatal) phase and the activational (or postpubertal) phase. Most of the permanent effects of testosterone effects occur perinatally. If levels of testosterone are high, the brain will be masculinized; if they are low, the brain will remain in its default feminine mode.

ENA theory asserts that androgens increase the probability of CVB by decreasing an individual's sensitivity to adverse environmental consequences resulting from exhibiting CVB. This lowered sensitivity is accomplished by inclining the brain to be *suboptimally aroused.* Suboptimal arousal manifests itself in terms of individuals seeking elevated levels of sensory stimulation and having diminished sensitivity to pain.

The second way androgens promote CVB according to ENA theory is by inclining the limbic system to seizure more readily, especially under stressful conditions. At the extreme, these seizures include such clinical conditions as epilepsy and Tourette's syndrome. Less extreme manifestations of limbic seizuring are known as *episodic dyscontrol* and *limbic psychotic trigger*. These latter patterns include sudden bursts of rage and other negative emotions, which often trigger forceful actions against a perceived provocateur.

Third, ENA theory asserts that androgen exposure causes neocortical functioning to be less concentrated in the left (language-dominated) hemisphere and to shift more toward a right hemispheric focus. As a result of this so-called *rightward shift in neocortical functioning*, males rely less on language-based reasoning, emphasizing instead reasoning which involves spatial and temporal calculations of risk and reward probabilities. Coinciding with this evidence are intriguing new research findings based on functional magnetic resonance imaging (fMRI) which suggest that empathy-based moral reasoning occurs primarily in the left hemisphere. Predictably, empathy-based moral reasoning seems to be less pronounced in males than in females. Such evidence suggests that empathy-based moral reasoning is more likely to prevent victimful criminality than so-called justice-based moral reasoning.

Theoretically, the three androgen-enhanced brain processes just described have evolved in males more than in females because these processes contribute to CVB. Furthermore, competitive/victimizing behavior has evolved in males more than in females because it facilitates male reproductive success more than it facilitates female reproductive success.

Inhibiting Criminal Forms of Competitive/Victimizing Behavior

Regarding the inhibiting aspects of brain functioning, two factors are theoretically involved. One has to do with learning ability and the other entails foresight and planning ability. According to ENA theory, the ability to learn will correlate with the rapidity of male transitioning from crude to sophisticated forms of CVB. This means that intelligence and other measures of learning ability should be inversely associated with persistent involvement in criminal behavior. Likewise, neurological underpinnings of intelligence such as brain size and neural efficiency should also correlate negatively with persistent offending. These predictions apply only to persistent victimful offending, with a much weaker link to occasional delinquency and possibly none with victimless criminality.

The frontal lobes, especially their prefrontal regions, play a vital role in coordinating complex sequences of actions intended to accomplish long-term goals. These prefrontal regions tend to keenly monitor the brain's limbic region, where most emotions reside. Then the prefrontal regions devise plans for either maximizing pleasant emotions or minimizing unpleasant ones. In other words, for the brain to integrate experiences into well-coordinated and feedback-contingent strategies for reaching long-term goals, the frontal lobes perform what has come to be called *executive cognitive functioning*. Moral reasoning often draws heavily on executive cognitive functioning, since it often requires anticipating the long-term consequences of ones actions.

Factors that can impact executive cognitive functioning include genetics, prenatal complications, and various types of physical and chemical trauma throughout life. According to ENA theory, inefficient executive cognitive functioning contributes to criminal behavior. Similar conclusions have been put forth in recent years by several other researchers.

To summarize, ENA theory asserts that three aspects of brain functioning promote competitive/victimizing behavior, the crudest forms of which are victimful crimes. At least partially counterbalancing these androgen-promoted tendencies are high intelligence and efficient executive cognitive functioning. These latter two factors affect the speed with which individuals quickly learn to express their competitive/victimizing tendencies in sophisticated rather than crude ways. Sophisticated expressions are less likely to elicit retaliation by victims, their relatives, and the criminal justice system than are crude ones. Males with low intelligence and/or with the least efficient executive cognitive functioning will therefore exhibit the highest rates of victimful criminal behavior.

Correlates of Criminal Behavior

Twelve biological correlates of crime with special relevance to ENA theory (testosterone, mesomorphy, maternal smoking during pregnancy, hypoglycemia, epilepsy, heart rate, skin conductivity, cortisol, serotonin, monoamine oxidase, slow brainwave patterns, and P300 amplitude) are discussed below.

Testosterone ENA theory predicts that correlations will be found between testosterone and CVB. However, the nature of these correlations will not involve a simple one-to-one correspondence between an individual's crime probability and the amount of testosterone in his/her brain at any given point in time. Earlier, a distinction was made between the organizational and activational effects of testosterone on brain functioning, and that the most permanent and irreversible effects of testosterone occur perinatally. For this reason alone, testosterone levels circulating in the blood stream or in saliva following puberty may have little direct correlation with neurological levels, especially within each sex. Therefore, one should not expect to find a strong correlation between blood or saliva levels of testosterone among, say, 20-year-old males and the number of offenses they have committed even though testosterone levels in the brain at various stages in development are quite influential on offending probabilities.

Numerous studies have investigated the possible relationship between blood levels or saliva levels of testosterone and involvement in criminal behavior, and most have found modest positive correlations (Maras et al. 2003). Additional evidence of a connection between testosterone and aggressive forms of criminality involves a recent study of domestic violence, where offending males had higher levels of saliva testosterone than did males with no history of such violence (Soler et al. 2000).

Overall, it is safe to generalize that circulating testosterone levels exhibit a modest positive association with male offending probabilities, particularly in the case of adult violent offenses. According to ENA theory, males are more violent than females, not because of cultural expectations or sex role training, but mainly because of their brains being exposed to much higher levels of testosterone than the brains of females.

Mesomorphy Body types exist in three extreme forms. These are sometimes represented with a bulging triangle. Most people are located in the center of the triangle, exhibiting what is termed a basically *balanced body type.* At one corner of the triangle are persons who are extremely muscular, especially in the upper body, called *mesomorphs. Ectomorphs* occupy a second corner. Individuals with this body type are unusually slender and non-muscular. In the third corner, one finds *endomorphs,* individuals who are overweight and have little muscularity.

Studies have consistently revealed that offending probabilities are higher among individuals who exhibit a mesomorphic body type than either of the two other extreme body types (e.g., Blackson & Tarter 1994). ENA theory explains this relationship by noting that testosterone affects more than the brain; it also enhances muscle tissue, especially in the upper part of the body.

Maternal Smoking During Pregnancy There is considerable evidence that maternal smoking may lead to an elevated probability of offspring becoming delinquent (e.g., Rasanes et al. 1998). ENA theory assumes that fetal exposure to carbon monoxide and other neurotoxins found in cigarette smoke disrupt brain development in ways that adversely affects IQ or executive cognitive functioning, thereby making it more difficult for offspring to maintain their behavior within prescribed legal boundaries. However, it is possible that genes contributing to nicotine addiction may also contribute to criminal behavior. In fact, a recent study reported that the link between childhood conduct disorders (a frequent precursor to later criminality) and maternal smoking was mainly the result of mutual genetic influences (Maughan et al. 2004).

Hypoglycemia *Glucose*, a type of natural sugar, is the main fuel used by the brain. The production of glucose is largely regulated by the pancreas in response to chemical messages from a portion of the brain called the *hypothalamus*. When the hypothalamus senses that glucose levels are becoming too high or too low, it sends chemical instructions to the pancreas to either curtail or increase production of glucose by regulating the amount of insulin released into the blood system. In most people, this feedback regulatory process helps to maintain brain glucose at remarkably stable levels. For a variety of reasons, some people have difficulty stabilizing brain glucose levels. These people are said to be *hypoglycemic*. Dramatic fluctuations in brain glucose can cause temporary disturbances in thoughts and moods, with the most common symptoms being confusion, difficulty concentrating, and irritability.

Studies have indicated that hypoglycemia is associated with an elevated probability of crime, especially of a violent nature (e.g., Virkkunen 1986). To explain such a connection, ENA theory draws attention to the importance of maintaining communication between the various parts of the brain in order to control emotionality. In particular, if the frontal lobes receive distorted signals from the limbic system, bizarre types of behavioral responses sometimes result, including responses that are violent and antisocial.

Epilepsy Epilepsy is a neurological disorder typified by *seizures*. These seizures are tantamount to "electrical storms' in the brain. While people vary in genetic susceptibilities, seizures are usually induced by environmental factors such as physical injuries to the brain, viral infections, birth trauma, and exposure to various chemicals.

The main behavioral symptoms of epilepsy are known as *convulsions* (or *fits*), although not all epileptics have full-blown convulsive episodes. Mild epileptic episodes may manifest themselves as little more than a momentary pause in an ongoing activity accompanied by a glazed stare. Seizures that have little to no noticeable debilitating effects on coordinated movement are called *subconvulsive* (or *subclinical*) *seizures*. Studies of human populations have shown that epilepsy affects only about one in every 150 to 200 persons. In prison populations, however, the prevalence of epilepsy is around one in 50, at least three times higher than in the general population (e.g., Mendez et al. 1993).

ENA theory can explain the links between epilepsy and offending by noting that very basic and primitive emotional responses sometimes emanate from the limbic region of the brain. While seizures in [illegible] control centers are most likely to receive a diagnosis of epilepsy, seizures in the limbic region could provoke very basic survival instincts.

Resting Heart and Pulse Rates Heart and pulse rates rise in response to strenuous exercise along with stressful and frightening experiences. Studies have shown that on average, the resting heart rate and pulse rate of convicted offenders are lower than those of persons in general (e.g., Mezzacappa et al. 1997:463). ENA theory would account for these relationships by stipulating that both low heart and low pulse rates are physiological indicators of suboptimal arousal. Such arousal levels should incline individuals to seek more intense stimulation and to tolerate unpleasant environmental feedback to a greater extent than individuals with normal or superoptimal arousal under most circumstances.

Skin Conductivity (Galvanic Skin Response) Sweat contains high concentrations of sodium, which is a good electrical conductor. A device called a *Galvanic Skin Response* (GSR) meter was developed nearly a century ago to monitor palmer sweat. The GSR works by measuring electrical impulses passing through our bodies from one electrode to another. Thus, by putting one's fingers on two unconnected electrodes of a GSR device, one completes an electrical circuit through which imperceptible amounts of electricity flows. Temperature obviously affects how much people sweat, but so too do emotions. The more intense one's emotions become (especially those of fear and anger), the more one will sweat, and thus the stronger will be the readings on the GSR meter.

Numerous studies have examined the possibility that persons with the greatest propensities toward criminal behavior have distinctive skin conductivity patterns. These studies suggest that offenders exhibit lower skin conductivity under standard testing conditions than do people in general (e.g., Buikhuisen et al. 1989; Raine et al. 1996). As in the case of heart and pulse rates, ENA theory can account for such findings by hypothesizing that low GSR readings especially under stressful testing conditions are another indication of suboptimal arousal.

Cortisol So-called *stress hormones* are secreted mainly by the adrenal glands during times of anxiety, stress, and fear. The stress hormone that has been investigated most in connection with criminality is cortisol. Most of these studies have suggested that offenders have below normal levels (e.g., Lindman et al. 1997). As with heart rates and skin conductivity, one could anticipate a low cortisol-high criminality relationship by assuming that low cortisol production even in the face of stress is another indicator of suboptimal arousal. This would suggest that offenders are less intimidated by threatening aspects of their environments than are persons in general.

Serotonin *Serotonin* is an important neurotransmitter. When serotonin is relatively active in the synaptic regions connecting adjacent nerve cells, people typically report feeling a sense of contentment and calm. Several drugs that have been designed to treat depression and anxiety disorders operate by either prolonging the presence of serotonin in the synaptic gaps between neurons or by facilitating the ability of receptor sites on the dendrites to bond to the serotonin that is available. Low serotonin activity has been linked to crime by numerous studies, especially impulsive crimes (e.g., Virkkunen et al. 1996; Matykiewicz et al. 1997). Explaining the link between serotonin and criminality from the perspective of ENA theory draws attention to serotonin pathways connecting the brain's prefrontal areas with the emotion-control centers in the limbic system. Serotonin may facilitate the sort of executive cognitive functioning required to restrain impulsive behavior, especially regarding rage and persistent frustration.

Monoamine Oxidase *Monoamine oxidase* (MAO) is an enzyme found throughout the body. Within the brain, MAO helps to break down and clear away neurotransmitter molecules (including serotonin), portions of which often linger in the synaptic gap after activating adjacent nerve cells. Studies indicate that MAO activity is unusually low among offenders (e.g., Alm et al. 1996; Klinteberg 1996). ENA draws attention to the fact that low MAO activity seems to be related to high levels of testosterone. Furthermore, low MAO brain activity may interfere with the brain's ability to manufacture or utilize serotonin.

Brain Waves and Low P300 Amplitude Brain waves are measured using electrodes placed on the scalp. These electrodes can detect electrical activity occurring close to the surface of the brain fairly clearly. Despite their complexity, brain waves can be roughly classified in terms of ranging from being rapid and regular (alpha brain waves) to being slow and irregular (delta brain waves). Most studies based on electroencephalographic (EEG) readings have found that offenders have slower brain waves than do persons in general (e.g., Petersen et al. 1982).

Unlike traditional brain wave measurement, modern computerized brain wave detection is able to average responses to dozens of identical stimuli presented to subjects at random intervals. This reveals a distinctive brain wave pattern or "signature" for each individual. Nearly everyone exhibits a noticeable spike in electrical voltage, interrupted by a "dip" approximately one-third of a second following presentation of test stimuli. This is called the P300 amplitude of an event-related evoked potential. From a cognitive standpoint, the P300 amplitude is thought to reflect neurological events central to attention and memory.

While research has been equivocal thus far in the case of criminality, several studies have found a greater dip in P300 responses by individuals diagnosed with antisocial personality disorder than is true for general populations (see Costa et al. 2000). ENA theory can account for slower EEG patterns among offenders and a P300 decrement among persons with antisocial behavior by again focusing on suboptimal arousal. From a neurological standpoint, both slow brain waves and a tendency toward a greater than normal P300 decrement can be considered symptomatic of suboptimal arousal. If ENA theory is correct, both of these conditions will be found associated with elevated brain exposure to testosterone.

Summary and Conclusions

Unlike social environmental theories, the evolutionary neuroandrogenic (ENA) theory can account for statistical associations between biological variables and criminal behavior. Furthermore, ENA theory predicts the universal concentration of offending among males between the ages of 13 and 30, patterns that strictly environmental theories have always had difficulty explaining. As its name implies, ENA theory rests on two over-arching assumptions. The first assumption is an extension of Darwin's theory of evolution by natural selection. It maintains that males on average exhibit CVB more than females because females who prefer to mate with such males increase their chances of having mates who are competent provisioners of resources. These female biases have evolved because females who have had the assistance of competent provisioners have left more offspring in subsequent generations than other females. No comparable reproductive advantage comes to males who select mates based on resource procurement capabilities.

Some forms of CVB are crude in the sense of requiring little learning, nearly all of which are either assaultive or confiscatory in nature. Other forms are sophisticated in the sense that they require complex learning and involve much more subtle types of "victimization." A major expression of sophisticated competitive/victimizing behavior involves profitable business ventures and/or the management of large organizations. In most societies, these expressions are tolerated and even encouraged. However, the vast majority of people in all societies condemn the crudest expressions of CVB, and, in all literate societies, the criminal justice system has evolved to punish such behavior.

The theory's second assumption is that genes on the Y-chromosome have evolved which cause male brains to exhibit higher rates of competitive/victimizing behavior than female brains. These genes operate in part by causing would-be ovaries to develop instead into testes early in fetal development. Once differentiated, the testes produce testosterone and other sex hormones, which have three hypothesized effects upon brain functioning, all of which promote CVB. The three effects are termed *suboptimal arousal, seizuring proneness*, and *a rightward shift in neocortical functioning*. Furthermore, two neurological processes are hypothesized to help individuals shift from crude to sophisticated forms of competitive/victimizing behavior. These are learning ability (or intelligence) and executive cognitive functioning (or planning ability). The better one's learning ability or executive functioning, the quicker he/she will transition from crude to sophisticated forms of the behavior.

To illustrate the theory's predictive power, the second portion of this article reviews evidence regarding several biological correlates of criminal behavior. For all of these correlates, the theory was able to explain their apparent relatedness to criminality. Proponents of critical theory, differential association theory, control theory, etc. cannot explain these correlates. They must simply ignore them. Overall, ENA theory should help move criminology beyond strictly social environmental theories toward a new, more variable-rich paradigm. From the theory's perspective, criminality results from a complex interaction of evolutionary, biological, learning, and social environmental factors.

References

Alm, P. O., af Klinteberg, B., Humble, K., Leppert, J., Sorensen, S., Thorell, L. H., et al. (1996). Psychopathy, platelet MAO activity and criminality among former juvenile delinquents. *Acta Psychiatrica Scandinavica, 94*, 105–111.

Blackson, T. C., & Tarter, R. E. (1994). Individual, family, and peer affiliation factors predisposing to early-age onset of alcohol and drug use. *Alcoholism: Clinical and Experimental Research, 18*, 813–821.

Buikhuisen, W., Eurelings-Bontekoe, E. H. M., & Host, K. B. (1989). Crime and recovery time: Mednick revisited. *International Journal of Law and Psychiatry, 12*, 29–40.

Costa, L., Bauer, L., Kuperman, S., Porjesz, B., O'Connor, S., & Hesselbrock, V. M. (2000). Frontal P300 decrements, alcohol dependence, and antisocial personality disorder. *Biological Psychiatry, 47*, 1064–1071.

Ellis, L. (2001). The biosocial female choice theory of social stratification. *Social Biology, 48*, 297–319.

Klinteberg, A. (1996). Biology, norms, and personality: A developmental perspective: Psychobiology of sensation seeking. *Neuropsychobiology, 34*(3), 146–154.

Lindman, R. E., Aromaki, A. S., & Eriksson, C. J. P. (1997). Sober-state cortisol as a predictor of drunken violence. *Alcohol and Alcoholism, 32*, 621–626.

Maras, A., Laucht, M., Gerdes, D., Wilhelm, C., Lewicka, S., Haack, D., et al. (2003). Association of testosterone and dihydrotestosterone with externalizing behavior in adolescent boys and girls. *Psychoneuroendocrinology, 28*, 932–940.

Matykiewicz, L., La Grange, L., Vance, P., Wang, M., & Reyes, E. (1997). Adjudicated adolescent males: Measures of urinary 5-hydroxyindoleacetic acid and reactive hypoglycemia. *Personality and Individual Differences, 22*, 327–332.

Maughan, B., Taylor, A., Caspi, A., & Moffitt, T. E. (2004). Prenatal smoking and early childhood conduct problems: Testing genetic and environmental explanations of the association. *Archives of General Psychiatry, 61*, 836.

Mendez, M. F., Doss, R. C., & Taylor, J. (1993). Interictal violence in epilepsy: Relationship to behavior and seizure variables. *The Journal of Nervous and Mental Disease, 181*, 566–569.

Mezzacappa, E., Tremblay, R. E., Kindlon, D., Saul, J. P., Arseneault, L., Seguin, J., et al. (1997). Anxiety, antisocial behavior, and heart rate regulation in adolescent males. *Journal of Psychiatry, 38*, 457–469.

Petersen, K. G. I., Matousek, M., Mednick, S. A., Volovka, J., & Pollock, V. (1982). EEG antecedents of thievery. *Acta Psychiatrica Scandinavica, 65*, 331–338.

Raine, A., Venables, P. H., & Williams, M. (1996). Better autonomic conditioning and faster electrodermal half-recovery time at age 15 years as possible protective factors against crime at age 29 years. *Developmental Psychology, 32*, 624–630.

Rasanes, P., Hakko, H., Isohanni, M., Hodgins, S., Jarvelin, M.-R., & Tiihonen, J. (1999). Maternal smoking during pregnancy and risk of criminal behavior among adult male offspring in the Northern Finland 1966 birth cohort. *American Journal of Psychiatry, 156*, 857–862.

Soler, H., Vinayak, P., & Quadagno, D. (2000). Biosocial aspects of domestic violence. *Psychoneuroendocrinology, 25*, 721–739.

Virkkunen, M. (1986). Reactive hypoglycemic tendency among habitually violent offenders. *Nutrition Reviews, 44*(Supplement), 94–103.

Virkkunen, M., Eggert, M., Rawlings, R., & Linnoila, M. (1996). A prospective follow-up study of alcoholic violent offenders and fire setters. *Archives of General Psychiatry, 53*, 523–529.

DISCUSSION QUESTIONS

1. All three readings in this section have appealed for the integration of biologically informed and strictly environmental theories of criminology. Given the cascade of evidence supporting the role of biological factors in criminality, why do you think some criminologists still resist integration?
2. How would you go about testing ENA theory?
3. What is the primary role of testosterone vis-à-vis criminal behavior?

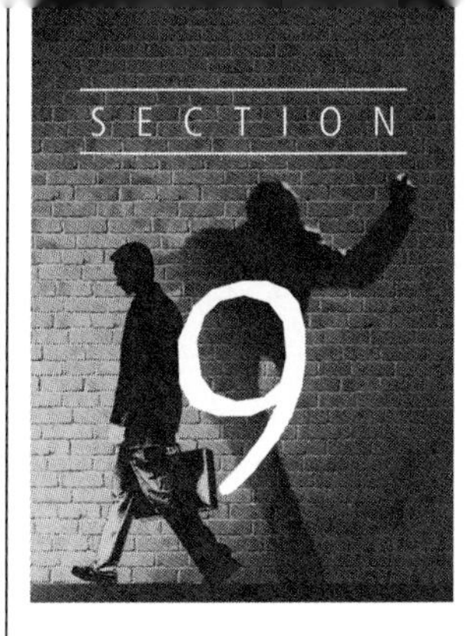

DEVELOPMENTAL THEORIES

From Delinquency to Crime to Desistance

Kathleen Holmes was a sweet child born to an "All-American" family in Boise, Idaho. Her parents sent her to a Catholic girls' school where she did well in her studies. All seemed to be going well for Kathy until she was 16 years old, when she agreed to go to a local air force base with two older friends from the neighborhood to meet one girl's boyfriend. The boyfriend brought along two of his friends, and the six of them partied with alcohol, drugs, and sex. It was Kathy's first time experiencing any of these things, and she discovered that she liked them all. Thus began a nine-year spiral into alcohol, drug, and sex addiction and into all the crimes associated with these conditions, such as drug trafficking, robbery, and prostitution.

When she was 25 years old she was involved in a serious automobile accident in which she broke her pelvis, both legs and an arm, and suffered a concussion. She was charged with drunken driving and possession of methamphetamine for sale. Kathy spent 10 months recuperating from her injuries, during which she was drug, alcohol, and sex free. Because of her medical condition, she was placed on probation. Her P.O. was a real "knuckle-dragger" who, while brooking no nonsense, became something of a father figure to her. While she was recuperating, she was often taken care of by a male nurse she described as "nerdy but nice." Her parents, who had been estranged from her for some time, became reacquainted with her, and her P.O. and nurse taught her to trust men again. She also occupied her time taking on-line college courses on drug addiction and counseling. She eventually married her "nerdy nurse" with her parents' blessing, and one of the honored guests was the "knuckle dragger."

Kathy's story illustrates some core ideas in this section. No matter how low a person sinks into antisocial behavior, he or she is not destined to continue the downward spiral. Certain so-called "turning points" in life can have a dramatic impact on a person's life. The auto accident and meeting the tough P.O. and the tender nurse would certainly qualify as significant turning points, as most certainly would marriage and the decision to continue her education. Before she became involved with "the wrong crowd," she had accumulated what is called "social capital" in the form of a good relationship with her family and a good academic preparation. Although she spent most of her social capital, there was sufficient left to get her back on the right track.

Most theories of crime implicitly assume that their favored causes are applicable across the lifespan and neglect changing social, biological, and psychological contingencies. They also imply that criminal behavior is self-perpetuating once initiated and say little about the process of desisting. In contrast to these static views, **developmental theories** are dynamic in that they emphasize that individuals develop along different pathways, and as they develop, factors that were previously meaningful to them (e.g., acceptance by antisocial peers) no longer are, and factors that previously meant little to them (marriage and a career) become meaningful. These theories are concerned with the onset, acceleration, and deceleration of offending, and finally desisting altogether. This difference notwithstanding, all theories in this chapter tend to be integrative (to varying degrees) in that they look at social, psychological, and biological factors simultaneously.

Offending in the early stages of life is known as **delinquency**, a legal term that distinguishes between youthful (juvenile) offenders and adult offenders. Except in rare instances in which a juvenile commits murder and is waived (transferred) to adult court, juvenile offenders are not referred to as criminals. Acts that are forbidden by law are called delinquent acts when committed by juveniles. The term delinquent comes from a Latin term meaning to "leave undone," with the connotation being that juveniles have not done something that they were supposed to (behave lawfully) rather than done something they were not supposed to. This subtle difference reflects the rehabilitative rather than punitive thrust of the juvenile justice system.

Juvenile delinquency occurs everywhere, and at all times, although the extent and severity of it varies from culture to culture and from time to time. Some criminologists attempt to explain the rise in antisocial behavior in adolescence by the increase in peer involvement at this time, and its decline thereafter by the decreasing influence of peers and the increasing influence of girlfriends, wives, children, and employers (Warr, 2002). However, this does not explain *why* the period between these events is so filled with delinquency or *why* associations with peers so often lead to negative behavior, and as Shavit and Rattner (1988) admit, the age peak in delinquency remains "unexplained by any known set of sociological variables" (p. 1457).

The New York Academy of Sciences confronted this problem in their 2003 conference and concluded that the adolescent brain is undergoing significant changes in directions that lead them to seek high levels of novelty and stimulation to achieve the same feeling of pleasure that children or adults require (White, 2004). In other words, the immature behavior of many adolescents is matched by the immaturity of their brains. The resculpting of the adolescent brain is precipitated by the hormonal surges of puberty, a time when male levels of testosterone (a hormone that underlies and facilitates dominance and aggression) are 10-plus times that of females (Ellis, 2003). Adolescents are also experiencing changes in the ratios of excitatory and inhibitory neurotransmitters; dopamine and other excitatory transmitters are peaking, while the inhibitory transmitter serotonin is reduced (Collins, 2004). All of this neurohormonal preparation leads many theorists (Spear, 2000; White, 2004) to conclude with Martin Daly (1996) that "There are many reasons to think that we've been designed [by natural selection] to be maximally competitive and conflictual in young adulthood" (p. 193).

When adolescence is over and adulthood is attained around the age of 20, the excitatory transmitters start to decrease and the inhibitory transmitters start to increase (Collins, 2004). McCrea and his colleagues (2000, p. 183) report findings from five different countries showing age-related decreases in personality traits positively related to antisocial behavior and increases in personality traits positively related to prosocial behavior. The fine-tuning of neurological and endocrine (hormonal) systems occurs across the lifespan and, thankfully, results in personality traits in adulthood conducive to prosocial behavior, for most individuals.

Risk and Protective Factors for Serious Delinquency

Some individuals possess so many risk factors that they go beyond the normal adolescent hell-raising to commit serious crimes. A **risk factor** is something that increases the probability of offending. These factors are dynamic, meaning that their predictive value changes according to the stage of development they occur in, the presence of other risk and protective factors, and the immediate social circumstances. For instance, low socioeconomic status (SES) is a family risk factor, but a person with a high IQ and warm relationship with both parents is protected from the risks low SES poses. It is typical for risk (and protective) factors to cluster together. A single parent family, for instance, is a risk factor that can lead to low SES and the financial necessity to reside in socially disorganized neighborhoods where children interact with antisocial peers. A report issued by the Office of the Surgeon General of the United States (OSGUS; 2001) indicates that a 10-year-old child with six or more risk factors is approximately 10 times more likely than a 10-year-old child with only one risk factor to be violent by the age of 18. We have already examined many of these risk factors (low IQ, poverty, social disorganization, parental supervision or abuse, etc.), so we will only examine developmental factors here.

Among the risk factors listed by OSGUS are ADHD/impulsivity, restlessness, difficulty concentrating, and aggression, which can be subsumed under the syndromes of **attention deficit with hyperactivity disorder (ADHD)** and conduct disorder (CD). ADHD is a chronic neurological condition manifested as constant restlessness, impulsiveness, difficulty with peers, disruptive behavior, short attention span, academic underachievement, risk-taking behavior, and extreme boredom. Most healthy children will manifest some of these symptoms at one time or another, but they cluster together to form a syndrome in ADHD children (8 out of 14 symptoms are required for diagnosis) and are chronic and more severe than simple high spirits (Durston, 2003). ADHD affects somewhere between 2% and 9% of children and is four or five times more prevalent in males than in females (Levy, Hay, McStephen, Wood, & Waldman, 1997). Brain imaging studies have found differences (albeit small ones) in brain anatomy and physiology between ADHD and non-ADHD children (Raz, 2004; Sanjiv & Thaden, 2004).

Although the precise cause of ADHD is not known, it is known that genes play a large role (Coolidge, Thede, & Young, 2000), with heritability in the .75 to .95 range (Levy et al., 1997). Environmental factors that play a role in the etiology of ADHD are fetal exposure to drugs, alcohol, and tobacco; perinatal complications; and head trauma (Durston, 2003). ADHD symptoms generally decline in their severity with age, although about 90% of ADHD sufferers continue to display some symptoms into adulthood (Willoughby, 2003).

A review of 100 studies found that 99 reported a positive relationship between ADHD and antisocial behavior (Ellis & Walsh, 2000). The probability that ADHD persons will persist in offending as adults rises dramatically if they are also diagnosed with **conduct disorder** (CD). CD is defined as "the persistent display of serious antisocial actions that are extreme given the child's developmental level and have a significant impact on the rights of others" (Lynam, 1996, p. 211). Markus Krueisi and his colleagues (1994) propose that ADHD is a product of a deficient BIS and CD is a product of an oversensitive BAS. ADHD/CD individuals thus suffer a double disability: their oversensitive BAS inclining them to seek high levels of stimulation while their faulty BIS leaving them with little sense of when to put a stop to their search. ADHD and CD are found to co-occur in 30% to 50% of cases (Lynam, 1996). CD has an onset at around five years of age and is a neurological disorder with substantial genetic effects (Coolidge et al., 2000).

The article by David Bennett, Maria Pitale, Vaishali Vora, and Alyssa Rheingold goes a step further than most and examines how ADHD affects children primarily exhibiting two different kinds of antisocial behavior (ASB): proactive and reactive. Proactive ASB is unprovoked person-initiated ASB, such as bullying, stealing, and cheating; reactive ASB is other-initiated ASB, such as becoming angry and hitting in response to the actions of others. ADHD symptoms were significantly related to both forms of ASB; more strongly so for reactive ASB. Children who exhibit proactive ASB tend to befriend others exhibiting the same proactive ASB, but reactive ASB children do not befriend other reactive ASB children, probably because they exhibit high levels of emotional reactivity (they tend to lose their temper easily).

Major Developmental Theories

As we have already noted, developmental theories are dynamic theories concerned with the frequency, duration, and seriousness of offending behavior from onset to desistance. All theories maintain that although a criminal career may be initiated at any time, it is almost always begun in childhood or adolescence, with only about 4% being initiated in adulthood (Elliot, Huizinga, & Menard, 1989). The duration of a criminal career may be limited to one offense or it can last well into old age, with the frequency and seriousness of offending varying widely. Onset, frequency, duration, seriousness, and desistance depend on a variety of interacting individual and situational factors that vary across the life course.

Robert Agnew's General Theory or "Super Traits" Theory

In his **super traits theory**, Robert Agnew identifies five life domains that contain possible crime-generating factors: *personality, family, school, peers,* and *work*. It is a developmental theory because these domains interact and feed back on one another across the lifespan.

Agnew suggests that personality traits set individuals on a particular developmental trajectory that influences how other people in the family, school, peer group, and work domains react to them. In other words, personality variables "condition" the effect of social variables on crime.

Noting that personality traits cluster together, Agnew identifies the latent (underlying) traits of *low self-control* and *irritability* as "super traits" that encompass many of the traits we discussed in previous sections, such as sensation seeking, impulsivity, inattentiveness, and low empathy. People saddled with low self-control and irritable temperaments are likely to evoke negative responses from family members, school teachers, peers, and workmates that feed back and exacerbate those tendencies (reactive gene/environment correlation).

Agnew (2005) states that "biological factors [ANS and brain chemistry] have a direct effect on irritability/low self-control and an indirect effect on the other life domains through [the effects of] irritability/low self-control [on them]" (p. 213). What Agnew calls irritability appears to be analogous to the trait most psychologists call negative emotionality.

The theory can explain gender, racial, and age effects in criminality and can account for the differences between individuals who limit their offending to the adolescent years and those who offend across the life course. In terms of gender differences, Agnew (2005) says that males are more likely to inherit irritability/low self-control than females, perhaps because in evolutionary terms these traits have aided male reproductive success by enhancing male aggressiveness and competitiveness (p. 163). In terms of race differences, Agnew argues that blacks are more likely to be poor and to receive discriminatory treatment, and that this and other factors may significantly increase irritability, and that perceptions of poor job prospects may also lead to the adoption of an impulsive "live-for-the-day" lifestyle among some African Americans.

Agnew (2005) notes that the immaturity of adolescent behavior is tied to the immaturity of the adolescent brain, and that adolescents tend to become more irritable because their brains are undergoing a period of intense "remodeling." At the same time that adolescent brains are changing, adolescents are experiencing massive hormonal surges that tend to facilitate aggression and competitiveness. Thus, the neurological and endocrine changes during adolescence *temporarily* increase irritability/low self-control among adolescents who limit their offending to that period, while for those who continue to offend, irritability/low self-control is a *stable* characteristic.

David Farrington's Integrated Cognitive Antisocial Potential (ICAP) Theory

David Farrington's **ICAP theory** is based on a longitudinal cohort study of boys born in deprived areas (thus putting all boys at environmental risk) of London. The key concepts in the theory are **antisocial potential** (AP), which is a person's risk or propensity to engage in crime, and **cognition**, which is the "thinking or decision-making process that turns potential into actual behavior" (Farrington, 2003, p. 231). AP is ordered on a continuum, with relatively few people with very high levels, but levels vary over time and across life events and peak in adolescence.

Farrington distinguishes between long-term AP and short-term AP. Individuals with long-term AP tend to come from poor families, to be poorly socialized, low on anxiety, impulsive, sensation-seeking, have low IQ, and fail in school (Farrington, 2003, p. 233). Short-term AP individuals suffer few or none of these deficits but may temporarily increase their AP in response to certain situations or inducements. Farrington indicates that we all have "desires for material goods, status among intimates, excitement, and sexual satisfaction," and that people choose illegitimate ways of satisfying them when they lack legitimate means of doing so, or when bored, frustrated, or drunk (p. 231).

Short-term AP may turn into long-term AP over time as a consequence of offending. This can happen if individuals find offending to be reinforcing, either in material terms or in psychological terms by gaining status and approval from peers. Such outcomes lead to changes in cognition such that AP is more likely to turn into actual criminal behavior in the future. Offending can also lead to criminal labeling and incarceration, which limits future opportunities to meet one's needs legitimately. Thus, long-term AP can develop even in the absence of most or all of the risk factors (i.e., for social-situational reasons alone) said to predict chronic offending across the life course.

ICAP theory is also interested in the process of desisting from offending, which occurs for both social and individual reasons and occurs at different rates according to a person's level of AP. As people age they tend to become less impulsive and less easily frustrated. They also experience **turning points** such as marriage, steady employment, and moving to new areas, thus shifting their patterns of interaction from peers to girlfriends, wives, and children. These events (1) decrease offending opportunities by shifting routine activities such as drinking with male peers, (2) increase informal controls in terms of having family and work responsibilities, and (3) change cognition in the form of reduced subjective rewards of offending because the costs of doing so are now much higher than before (the risk of losing their hard-earned stake in conformity); potential peer approval becomes the potential disapproval of wives (Farrington, 2003).

Terrie Moffitt's Dual Pathway Developmental Theory

Terrie Moffitt's **dual pathway** developmental theory is based on findings from an ongoing longitudinal study of a New Zealand birth cohort (in its 30th year as of 2005). It has been called "the most innovative approach to age-crime relationships and life course patterns" (Tittle, 2000, p. 68). The data available to Moffitt and her colleagues come from collaborative efforts by scientists in medicine, neuroscience, and endocrinology, as well as psychologists and other social scientists, enabling them to test biosocial hypotheses.

It has long been known that the vast majority of youth who offend during adolescence desist and that there are a small number of them who continue to offend in adulthood. Moffitt calls the former adolescent limited (AL) offenders and the latter life-course persistent (LCP) offenders. LCP offenders are individuals who begin offending prior to puberty and continue well into adulthood, and who are saddled with neuropsychological and temperamental deficits that are manifested in low IQ, hyperactivity, inattentiveness, negative emotionality, and low impulse control. These problems arise from a combination of genetic and environmental effects on central nervous system development. Environmental risk factors include being the offspring of a single teenage mother, low SES, abuse/neglect, and inconsistent discipline. These related individual and environmental impairments initiate a cumulative process of negative person/environment interactions that result in a life course trajectory propelling individuals toward ever hardening antisocial attitudes and behaviors.

Moffitt (1993) describes the antisocial trajectory of LCP offenders as one of "biting and hitting at age 4, shoplifting and truancy at age 10, selling drugs and stealing cars at age 16, robbery and rape at age 22, fraud and child abuse at age 30; the underlying disposition remains the same, but its expression changes form as new social opportunities arise at different points of development" (p. 679). This age-consistent behavior is matched by cross-situational behavioral consistency. LCP offenders "lie at home, steal from shops, cheat at school, fight in bars, and embezzle at work" (p. 679). Given this antisocial consistency across time and place, opportunities for change and for legitimate success become increasingly unlikely for these individuals. While LCP offenders constituted only 7% of the cohort, they were responsible for more than 50% of all delinquent and criminal acts committed by it (Henry et al., 1996; Jeglum-Bartusch, Lynam, Moffitt, & Silva, 1997). Moreover, whereas AL offenders tend to commit relatively minor offenses such as petty theft, LCP offenders tend to be convicted of more serious crimes against the person such as assault, robbery, and rape.

AL offenders have a different developmental history that places them on a prosocial trajectory that is temporarily derailed at adolescence (Farrington's short-term AP). They are not burdened with the neuropsychological problems that weigh heavily on LCP offenders and they are adequately socialized in childhood by competent parents. AL offenders are "normal" youths adapting to the transitional events surrounding adolescence and whose offending is a social phenomenon played out in peer groups and does not reflect any stable personal deficiencies (Moffitt, 1993). At least 90% of offenders are AL.

According to Moffitt, many more teens than in the past are being diverted from their prosocial life trajectories because better health and nutrition has lowered the average age of puberty while the average time needed to prepare for participation in the economy has increased. These changes have resulted in about a 5- to10-year **maturity gap** between puberty and entry into the job market. Thus, "adolescent-limited offending is a product of an interaction between age and historical period" (Moffitt, 1993, p. 692). Filled with youthful energy, strength, and confidence, and a strong desire to shed the restrictions of childhood, AL offenders are attracted to the excitement of antisocial peer groups typically led by experienced LCP delinquents. Once initiated into the group, juveniles learn the attitudes and techniques of offending through mimicking others and gain reinforcement in the form of much-desired group approval and acceptance for doing so, as social learning theorists argue.

The article by Terrie Moffitt and Anthony Walsh in this section summarizes studies spanning 10 years that have tested Moffitt's AL/LCP theory. They examine the genetic, neurological, personality, and offense pattern differences between the two offender types. They also look at possible reasons why some adolescents abstain from delinquency altogether at a period when delinquency is statistically normative, finding abstainers to be inhibited, extremely self-controlled, and timid. They conclude that these individuals probably possess autonomic nervous systems in the hyperactive range. Gender differences relevant to the AL/LCP dichotomy are also explored. They find that there are about 10 LCP male offenders for every one LCP female, but only 1.5 AL male offenders for every one AL female offender. This developmental theory has perhaps generated more studies since its appearance in 1993 than any other theory.

Sampson and Laub's Age-Graded Theory

All three developmental theories we have thus far discussed posit a set of traits that set individuals on developmentally distinct pro-or anti-social pathways. However, Sampson and Laub (1999, 2005) prefer to call their **age-graded theory** a life-course theory rather than a developmental theory because they deny the notion that people are necessarily locked into developmentally distinct pathways by these traits. Sampson and Laub want to emphasize environmental circumstances and human agency as opposed to individual traits. Their theory is based on data collected in the 1930s through the 1960s by Sheldon and Eleanor Glueck (1950). Age-graded theory is essentially a social control theory extended into adulthood. While the bonds to parents and school are very important during childhood, and to peers during adolescence, they become less important in adulthood when new situations offer opportunities to form new social bonds that constrain offending behavior.

The excerpt from the book *Unraveling Juvenile Delinquency* written by the Gluecks provides an excellent brief introduction to the developmental/life-course approach. The excerpt makes clear the richness of the data collected by them and the sophistication of their research design.

▲ **Photo 9.1** From this series of family images taken over time, can you discern the behavioral and life-course paths of the individuals depicted? One became a teen delinquent but then went on to be a law-abiding adult and, ultimately, a minister. One dropped out of a teen job to move away to marry a federal inmate who had sold a number of types of drugs. Another started a career in sales but became involved in possible "Ponzi" schemes. One became a criminologist.

As with all control theories, the task of age-graded theory is not to explain why people commit crimes but why people do not. The theory assumes that factors such as low IQ, difficult temperament, SES, and broken home have only indirect effects on offending via their influence on the ease or difficulty with which bonding and socialization take place. Thus the theory places emphasis on the process of desisting from offending rather than on risk factors for offending.

People who bond well with conventional others build **social capital**, which is essentially a store of positive relationships built on norms of reciprocity and trust developed over time upon which the individual can draw for support. People who have opened their social capital "accounts" early in life (bonding to parents and school), even though they may spend it freely as adolescents, build quite a nest egg by the time they reach adulthood. They can then gather more interest in the form of a successful career and marriage (bonding to a career and a family of one's own). This accumulation of social capital provides people with a powerful stake in conformity, which they are not likely to risk by engaging in criminal activity.

Life is a series of transitions (or life events) that may change life trajectories for persons lacking much social capital in prosocial directions. Sampson and Laub (2005) call such events turning points and consider this concept the most important concept in their theory (p. 14). Important turning points include getting married, finding a decent job, moving to a new neighborhood, or entering military service. Of course, turning points are processes rather than events, and rather than promoting change may accentuate antisocial tendencies. According to Gottfredson and Hirschi (1990), the problem is that "the offender tends to convert these institutions [marriage, jobs] into sources of satisfaction consistent with his previous criminal behavior" (p. 141). In other words, they expand their antisocial repertoire into domestic abuse and workplace crime. Nevertheless, Laub and Sampson (2005) have shown that obtaining a good job and a good marriage does reduce offending among previously high-rate offenders.

▲ **Photo 9.2** Life-course theories attempt to uncover points of desistance, critical events that may change a person's life path from deviant behavior back to law abiding status. Getting married is considered one of those events. Once one is married, free time to hang out with one's friends tends to disappear.

Although a person may be disadvantaged by the past, he or she does not have to be a prisoner to the past. Age-graded theory strongly emphasizes human agency—"the purposeful execution of choice and individual will" (Sampson & Laub, 2005, p. 37)—and some people freely choose a life of crime because they find it seductive and rewarding despite having full knowledge of the negative consequences. Nevertheless, all members of the

Gluecks' original delinquent sample that Sampson and Laub were able to locate ($N = 52$) had desisted from offending by age 70 regardless of whether they were defined as high-risk or low-risk as children and regardless of the level of social capital they had managed to accumulate.

Robert Sampson and John Laub's article in this section greatly expands on what we have written about their age-graded theory. The authors seek to undermine the idea that developmentally distinct groups of offenders can be explained by unique causal processes. They do this primarily by emphasizing turning points at various points over the life course and the importance of human agency. Sampson and Laub envision development as the result of interaction between individuals and their environment, chance, and purposeful human agency.

Table 9.1 summarizes the strengths and differences of developmental theories.

Evaluation of Developmental Theories

Developmental theories offer many advantages over theories previously discussed because of their dynamic nature. It is not only consideration of the differential impact of risk factors at different junctures across the lifespan that distinguishes developmental theories; it is also their focus on the process of desisting from crime. Both these contributions are extremely important. Developmental theories are mostly based on longitudinal cohort data, which enables theorists to examine the links between risk factors and crime among the same individuals at every developmental stage of their lives. Longitudinal studies also enable theorists to identify causes rather than mere correlates because temporal order (which factor came first) can be established among the correlates, something that cross-sectional studies cannot do.

With the exception of Agnew's theory, all developmental theories we have discussed are based on longitudinal cohort data so that theorists can examine the links between risk factors and crime among the same individuals at every developmental stage of their lives. Such data are hard to come by, and some theorists argue that cross-sectional studies (research that studies a sample at a single moment in time) are adequate and that longitudinal studies are an expensive luxury. Theorists making such claims are those who identify a latent trait (e.g., self-control) that is considered stable throughout life; thus, cross-sectional studies capture characteristics of individuals at any one time, making multi-year studies redundant. Such a position ignores the interaction of this supposed stable latent trait with vastly different life experiences.

In short, we believe that if there is a "gold standard" for criminological theory, developmental theories would have to be it because (1) they generally integrate and consider sociological, psychological, and biological factors as a coherent whole; (2) they follow the same individuals over long periods of time, a strategy that allows for cause/effect analysis; (3) they do not rely on "convenience samples" from high schools and colleges; and (4) they can identify characteristics that lead to onset, persistence, and desistance from crime in the same individuals. One useful addition to all developmental theories considering continuity/desistance of antisocial behavior and the age-crime curve in general is the integration of the neurochemical data because these phenomena "may have a strong basis in normal neurochemistry" (Collins, 2004, p. 12).

Table 9.1 Summary of Key Points, Strengths, and Differences of Developmental Theories

Theory	Key Points	Key Strengths	Key Differences
Agnew's General or Super Traits Theory	Low self-control and irritability set people on a trajectory leading to negative interactions with other in the family, school, peers, work, and in marriage. These different domains interact and feed back on one another.	Parsimoniously integrates concepts from psychology, sociology, and biology and shows how each affects and is affected by all others. Theory states that low self-control and irritability are temporarily increased during adolescence.	Does not address the process of desisting from offending. Does not explicitly address different trajectories as in Moffitt, and to a lesser extent in Farrington and Sampson and Laub.
Sampson and Laub's Age-Graded Theory	Power of informal social controls across the life course. Assumes classical notions of why people commit crimes, therefore no need to dwell too much on risk factors. Turning points in life and human agency are important. These turning points are made easier if one has accumulated significant social capital.	Emphasis on the power of life events to turn trajectories around and to facilitate desistance. Also emphasis on human agency is refreshing. All offenders will eventually desist regardless of risk factors or lack of social capital.	Unlike other theories, there is little emphasis on risk factors setting people on a particular trajectory other than bonding strength. Emphasis on inhibiting (bonding, social capital) rather than facilitating factors.
Farrington's ICAP Theory	People have varying levels of antisocial propensity (AP) due to a variety of environmental and biological factors. Few people have long-term AP, but these people tend to offend across the life course. Short-term AP tends to occur in adolescence and can change to long-term under some circumstances.	Shows how people think (cognition) translates AP into actual offending behavior. As with Moffitt's AL offenders, AP is said to increase temporarily during adolescence, but also it can lead to long-term AP if caught and labeled criminal because such a label limits future opportunities.	Not so much emphasis on desisting as age-graded theory, but more so than Agnew and Moffitt. Less emphasis than Agnew and Moffitt on latent traits, but more than Sampson and Laub. Unlike Moffitt, AP is considered a continuum rather than a distinct two-type typology.
Moffitt's Dual Pathway Theory	There are two main pathways to offending—LCP and AL. LCP offenders have neurological and temperamental difficulties which are exacerbated by inept parenting. LCP offend across time and situations, begin prior to puberty, and continue well into adulthood. AL offenders are "normal" individuals temporarily derailed during adolescence.	Identifies two distinct pathways to offending rather than assuming all people are similarly affected by similar factors. Shows how the social bonds so important to age-graded theory are formed or not formed according to the characteristics of individuals.	Emphasizes a larger number of individual differences that affect offending behavior than all other theories. Also attempts to explain prevalence of offending with reference to the modern maturity gap. Differences between the two trajectory groups are more defined than they are in other theories.

Policy and Prevention: Implications of Developmental Theories

Policies designed to prevent and reduce crime derived from developmental studies do not differ from other theories but rather encompass them all and suggest a broad array of strategies. Developmental theories support the same kind of family-based nurturant strategies supported by biosocial and social- and self-control theories. Regardless of any traits children may bring with them to the socialization process, however, these traits can be muted by patient and loving parenting, and as such many developmental theories suggest family interventions as early as possible to help nurture bonds between children and their parents.

The *Nurse-Family Partnership* program has attempted to do this. The program began in 1978 with a sample of 400 at-risk women and girls and their infants. All mothers were unmarried, most were living in poverty, and 48% were under age 15. The women were randomly assigned to four different groups, with one group (the experimental group) receiving extensive care from nurses in the form of multiple prenatal and postnatal home visitations in which the nurses gave help and advice on a variety of child care matters. The other groups received less comprehensive care for shorter periods. A 15-year follow-up study by Olds, Hill, Mihalic, and O'Brien (1998) found that the program had many beneficial outcomes for the experimental group children and their mothers relative to the subjects in the other groups. For the mothers, there was less substance abuse, fewer subsequent illegitimate births, fewer legal difficulties, and 79% fewer verified instances of child abuse/neglect relative to the control groups. For the children, there was less substance abuse, fewer arrests, better school performance, and better all-around social adjustment.

A very interesting school-based program that has been implemented in several countries and explicitly based on the assumptions of developmental theories is the *Fast Track Project* (2005). Through multistage screening of over 10,000 kindergarten children in 1991–1993, 891 were identified as being at-risk for antisocial behavior. These children were impulsive, had difficult temperaments, and came from unstable families living in low-income high-crime neighborhoods. The children were divided into experimental (n = 445) and control (n = 446) groups. The experimental group was given a curriculum designed to develop social understanding, emotional communication skills, and to improve self-control and problem-solving skills. They were also placed in so-called *friendship groups* and *peer-pairing* designed to increase social skills and enhance friendships. Their parents received parenting effectiveness training and home visits to foster their problem-solving skills and general life management.

The program is evaluated periodically and found to have modest but positive outcomes. By the end of the third grade, 37% of the experimental group was judged to be free of conduct problems, as opposed to 27% of the control group. By the eighth grade, 38% of the experimental group had been arrested as opposed to 42% of the control group. Although these differences are modest at best, they still reflect a good number of people saved from criminal victimization if the improvements hold in the future. If programs such as the Fast Track Project could be combined with programs like the Nurse-Family Partnership, we should see improved results because even kindergarten intervention may be too late, in some cases. Yet, developmental theories tell us that human life is characterized by dynamism and people can change at any time. This is a note of optimism for crime control/prevention strategies.

Summary

- Developmental theories are dynamic and integrative and examine offending across the life course. Juvenile offending has been noted across time and cultures. A sharp rise in offending following puberty and a steady decline thereafter has been noted always and everywhere. Brain and hormonal scientists explain the age effect with respect to the brain and hormonal processes occurring during adolescence. At puberty a huge surge in testosterone levels is experienced, and the brain is undergoing a process of intensive resculpting.

- There are a wide variety of factors that put some teens more at risk for delinquent behavior than others. The factors we focused on were ADHD and CD, which are highly heritable. The co-occurrence of ADHD and CD is a particularly strong risk factor.

- Developmental theories follow individuals across the life course to determine the differential effect of risk factors for offending at different junctures.

- Agnew's super traits theory focuses on how low self-control and irritability (negative emotionality) interact with other life domains (school, work, marriage, etc.) across the life span to impact the probability of offending.

- Farrington's ICAP theory stresses antisocial potential and cognition and how these things are shaped in pro- or antisocial directions at different times and in different situations. Farrington distinguishes between long-term and short-term antisocial potential. Short term antisocial potential occurs primarily during adolescence.

- Moffitt's theory posits a dual pathway model consisting of adolescence-limited (AL) offenders, who limit their offending to the adolescent years, and life-course persistent (LCP) offenders, who offend across the life course. LCP offenders have neurological and temperamental difficulties that set them on a developmental trajectory that leads to antisocial behavior at all ages and social situations. AL offenders do not suffer these disabilities and have accumulated sufficient social capital that they can resettle into a prosocial lifestyle once neurological and social maturity has been reached.

- Sampson and Laub's age-graded theory is concerned with the power of informal social control (bonds) to prevent offending. Turning points in life (marriage, new job, etc.) are important in understanding the process of offenders' desisting from crime.

EXERCISES AND DISCUSSION QUESTIONS

1. Why is it important that we understand what is going on biologically during adolescence?
2. If people age out of crime as well as into it, would it be a good idea to ignore all but the most serious of juvenile crimes so that we don't risk having children gain a "criminal" reputation?
3. What is social capital and how much of it do you believe that you have accumulated?

USEFUL WEB SITES

Conduct Disorder. http://aacap.org/page.ww?section=Facts+for+Families&name=Conduct+Disorder.

Developmental Psychology. http://medicine.jrank.org/pages/455/Developmental Psychology.html.

Moffitt, T. E. (1993). Adolescent-limited and life-course persistent antisocial behavior: A developmental taxonomy. *Psychological Review, 100*(4), 674–701. http://www.soc.umn.edu/~uggen/Moffitt_PR_93.pdf.

National Institute of Mental Heath. http://www.nimh.nih.gov/.

Office of Juvenile Justice and Delinquency Prevention. http://ojjdp.ncjrs.org/.

GLOSSARY

Age-graded theory: Theory stressing the power of informal social controls to explain onset, continuance, and desisting from crime. Emphasizes the concepts of social capital, turning points in life, and human agency.

Antisocial potential: In Farrington's theory, it is a person's risk or propensity to engage in crime. Antisocial potential can be long-term or short-term.

Attention deficit with hyperactivity disorder (ADHD): A chronic neurological condition that is manifested as constant restlessness, impulsiveness, difficulty with peers, disruptive behavior, short attention span, academic underachievement, risk-taking behavior, and extreme boredom.

Cognition: In Farrington's theory, a thinking or decision-making process that turns antisocial potential into actual behavior.

Conduct disorder: The persistent display of serious antisocial actions that are extreme given the child's developmental level and have a significant impact on the rights of others.

Delinquency: A legal term that distinguishes between youthful (juvenile) offenders and adult offenders. Acts forbidden by law are called delinquent acts when committed by juveniles.

Developmental theories: A set of dynamic theories that tend to look at social, psychological, and biological factors simultaneously and are concerned with the onset, acceleration, and deceleration of offending, and finally desisting from crime.

Dual pathway: Theory based on the notion that there are two main pathways to offending: One pathway is followed by individuals with neurological and temperamental difficulties, the other by "normal" individuals temporarily derailed during adolescence.

ICAP theory: Farrington's integrated cognitive antisocial potential theory is based on the notion that people have varying levels of antisocial propensity due to a variety of environmental and biological factors.

Maturity gap: In Moffitt's theory, the widening gap in modern times between onset of puberty and the acquisition of a socially responsible role.

Risk factor: Something in individuals' personal characteristics or their environment that increase the probability of offending.

Social capital: The store of positive relationships in social networks built on norms of reciprocity and trust developed over time upon which the individual can draw for support.

Super traits theory: Developmental theory that identifies five life domains that contain possible crime-generating factors: personality, family, school, peers, and work. Low self-control and irritability are the "super traits."

Turning points: Transitional events in life (getting married, finding a decent job, moving to a new neighborhood) that may change a person's life trajectory in prosocial directions.

READING

Reactive Versus Proactive Antisocial Behavior

Differential Correlates of Child ADHD Symptoms?

David S. Bennett, Maria Pitale, Vaishali Vora, and Alyssa A. Rheingold

This study by Bennett and his colleagues furthers our knowledge of ADHD by examining the relationship between proactive and reactive antisocial behavior and ADHD symptoms. Caregivers of children aged 8 to 15 (n = 84) being evaluated at a child psychiatry outpatient clinic served as participants. Given the conceptual similarity between reactive antisocial behavior (ASB) and particular ADHD symptoms, they hypothesized that ADHD symptoms would be more closely related to reactive than to proactive ASB. Based on peer deviancy training models of ASB, they also hypothesized that the relation between ADHD symptoms and proactive ASB would increase from middle childhood to adolescence. Both hypotheses were supported. These findings suggest that reactive ASB is relatively specific to ADHD symptoms, and that intervening before early adolescence may be critical to prevent the onset of comorbid proactive ASB.

Children with Attention-Deficit/Hyperactivity Disorder (ADHD) symptoms are at risk for elevated rates of antisocial behavior. Furthermore, children with ADHD and comorbid antisocial behavior (ASB) have a poorer prognosis and greater impairment, including increased social problems, than children with only ADHD or only ASB. Among children with ADHD symptoms, the presence of ASB is actually a more robust predictor of adult adjustment than ADHD symptoms themselves (Paternite, Loney, Salisbury, & Whaley, 1999).

Children's ASB [may be divided into] reactive vs. proactive ASB (Dodge & Coie, 1987).

Source: Bennett, D., Pitale, M., Vora, V., & Rheingold, A. (2004). Reactive vs. proactive antisocial behavior: Differential correlates of child ADHD symptoms. *Journal of Attention Disorders, 7*(4), 197–204. Reprinted with permission of Sage Publications, Inc.

Reactive ASB is defined as an angry, defensive response to frustration or perceived provocation. Reactive ASB includes behaviors such as hitting a peer in response to perceived teasing, or becoming angry when corrected by a parent. Proactive ASB is defined as an unprovoked, aversive means of influencing or coercing another person and includes goal-directed behaviors like stealing, bullying, and cheating.

Reactive ASB has been strongly related to possessing a hostile attributional bias, poor social problem-solving skills, low peer acceptance, peer rejection, and victimization by peers; an early onset of problems; physical abuse, and inconsistent parenting. Proactive ASB has been associated with poor parental supervision and belief that it will lead to a positive consequence. Proactive ASB is also associated with an increased risk of future delinquent behavior in childhood and of psychopathy in adulthood and may be particularly resistant to treatment.

Reactive ASB is characterized by deficits in self-regulation and behavioral inhibition. Recently, increased attention has focused on self-regulation and behavioral inhibition deficits as characteristic of children with ADHD and of children with ADHD and comorbid ASB (Melnick & Hinshaw, 2000). Given the conceptual overlap between self-regulation deficits and both reactive ASB and ADHD, we would expect children with ADHD symptoms to be more at risk for reactive than proactive ASB, which is not typically characterized by self-regulation deficits. Yet, although the relation between ADHD symptoms and ASB in general has been well documented, little research has examined whether ADHD symptoms are specifically related to the reactive ASB subtype. Given that reactive ASB is hypothesized to respond to psycho-pharmacological intervention and anger management, whereas proactive ASB is hypothesized to respond to contingency management (Mpofu, 2002), increased understanding of the relation between ADHD symptoms and ASB subtypes may ultimately lead to a better matching of problems to effective interventions for children who have both ADHD symptoms and ASB.

Consistent with our hypothesis that ADHD symptoms are more closely related to reactive ASB, Dodge et al. (1997) found teacher ratings of attention problems and impulsivity to be higher among children with reactive ASB and children with both reactive and proactive ASB than those with only proactive ASB. In a second sample, proactive subtypes, a nonsignificant trend was found for boys who exhibited reactive ASB to be more likely than those who exhibited proactive ASB to receive an attention deficit disorder diagnosis (55% to 37%).

Further support for a relation between ADHD symptoms and the reactive ASB subtype comes from teacher ratings in a community sample (Waschbusch, Willoughby, & Pelham, 1998). Among 70 children who had elevated ratings of both ADHD symptoms and ASB, 22 had high rates of reactive-only ASB, compared to only 3 who had high rates of proactive-only ASB. Forty-one children had high rates of both reactive and proactive ASB, consistent with prior research indicating that children with high levels of one subtype often exhibit high levels of the other. Yet, because reactive ASB was more common than proactive ASB for their overall community sample (Waschbusch et al., 1998), it is not clear from their data that ADHD symptoms are more closely related to the reactive ASB subtype.

A third study, however, supports the hypothesis in that ADHD symptoms (activity level and inattention) assessed at age 6 were greater among children rated as having reactive-only or both reactive and proactive ASB, but not proactive-only ASB, at age 11 (Vitaro, Brendgen, & Tremblay, 2002). Finally, research has found aggressive children with ADHD tend to exhibit a hostile attributional bias in which they attribute malicious intent to others' behavior in ambiguous social situations and are more apt to respond aggressively when provoked (Pelham et al., 1991). Though indirect, given that a hostile attributional bias has also been associated more closely with reactive than proactive ASB, these findings further point to a relatively specific relation between ADHD symptoms and reactive ASB.

Validation of a specific relation between ADHD symptoms and reactive ASB has implications for both clinical work and future research. Clinically, although stimulants have been found to decrease global measures of ASB among children with ADHD (Connor, Barkley, & Davis, 2000), it is unclear how medication might affect the reactive and proactive subtypes. Pelham et al. (1991) found stimulants failed to reduce the number of points taken away from a confederate in response to provocation during a competitive game; however, their lab measure featured elements of both reactive (i.e., response to provocation) and proactive (goal directed removal of points from a competitor) ASB. If reactive ASB is most common in children with ADHD symptoms, then it would be useful to target reactive ASB in medication studies.

Developmentally, research has yet to examine whether the relations between reactive ASB, proactive ASB, and ADHD symptoms change during childhood. ADHD, and particularly hyperactivity-impulsivity symptoms, has an onset in early childhood. Given that self-control and behavioral inhibition deficits appear common to both ADHD and reactive ASB, it is reasonable to hypothesize that reactive ASB may develop at an early age (i.e., preschool or early elementary school years). Given the moderate to high correlation between reactive and proactive ASB, it is also reasonable to hypothesize that initial high levels of reactive ASB precede the development of proactive ASB in older children.

Peer deviancy training provides a model as to how children with ADHD symptoms and reactive ASB may gradually develop increasing rates of proactive ASB. Peer deviancy training is based on the finding that children with ASB tend to develop friendships with peers who also exhibit ASB (Patterson, Reid, & Dishion, 1992). Hence, children with ADHD symptoms and reactive ASB, two problems associated with rejection by the broader peer group, may develop friendships with peers who also exhibit ASB. Given that reactive and proactive ASB tend to co-occur, it is likely that some such peers will exhibit both reactive and proactive subtypes. Increased interaction with antisocial peers during late childhood and early adolescence predicts increased future delinquent behavior, including acts that could be classified as proactive ASB. This peer deviancy training model may be particularly relevant for children with ADHD symptoms, as adolescents with high levels of ADHD symptoms spend more time with friends and less time with family members than do adolescents with low levels of ADHD symptoms.

The purpose of the present study was to extend prior findings suggesting that ADHD symptoms are more strongly related to reactive than proactive ASB. In contrast to the three prior studies that used either the brief 6-item Dodge and Coie (1987) rating scale or a retrospective chart review to assess ASB subtypes, we used a previously validated, more comprehensive measure of reactive and proactive ASB. Whereas prior studies used teacher ratings, we used primary caregiver ratings. Caregiver ratings are only modestly related to teacher ratings, perhaps in part because children's behavior varies by context. Hence, it remains to be established whether ADHD symptoms are specifically related to reactive ASB as perceived by caregivers, who are often the main source of child assessment information for externalizing problems in clinical settings. Finally, we examined whether the relation between ADHD symptoms, reactive ASB, and proactive ASB differs between children in middle childhood and adolescence. Specifically, we tested the hypothesis that the relation between ADHD symptoms and proactive ASB would be highest among the oldest children in our sample.

Participants and Procedure

Participants were caregivers of children ages 8 to 15 years (n = 84) who presented for evaluation at a child psychiatry outpatient clinic. Children were primarily of African American ethnicity (n = 61), male (63 boys, 21 girls), and had a mean age of 11.0 years (SD = 2.0 years). Caregivers were primarily biological mothers. Children were most frequently given diagnoses of ADHD (67%), oppositional defiant disorder (38%) or conduct disorder by the intake clinician.

Measures

Antisocial Behavior Subtyping Scale

This scale contains 25 items rated on a 3-point scale (0 = never, 1 = sometimes, 2 = very often). The scale contains a 6-item reactive ASB factor (e.g., gets mad when corrected; blames others) and a 10-item proactive ASB factor (e.g., tells things that aren't true; takes things from other students; Brown et al., 1996). The scale, initially developed for teacher ratings, was modified such that "student" was changed to "child" in the instructions. Similarly, the item "misbehaves when the teacher's back is turned" was changed to "misbehaves when the parent's back is turned." Subscale means and standard deviations were as follows: reactive ASB, M = 14.6, SD = 2.3; proactive ASB, M = 19.4, SD = 4.4.

Results

ADHD symptoms were significantly correlated with both reactive and proactive ASB (see Table 9.2). However, as hypothesized, ADHD symptoms were more highly correlated with reactive ASB than with proactive ASB (Fisher's r to Z transformation; $Z = 2.95$, $p < .01$, 1-tailed). Given the high correlation between reactive and proactive ASB ($r = .63$, $p < .01$, 1-tailed), partial correlations between ADHD symptoms and each subtype were conducted. As expected, the partial correlation between ADHD symptoms and reactive ASB, controlling for proactive ASB, remained significant ($r = .49$, $p < .01$). However, the correlation between ADHD symptoms and proactive ASB, controlling for reactive ASB, was not significant ($r = -.09$, $p > .10$), indicating that ADHD symptoms were specific to reactive ASB.

Next, we examined whether the correlations between ADHD symptoms, proactive ASB, and reactive ASB varied as a function of age (see Table 9.2). We divided the sample into three groups: children ages 8 to 9 years ($n = 34$), children ages 10 to 11 years ($n = 26$), and children ages 12 to 15 years ($n = 24$). Consistent with hypothesis, correlations were significantly higher between ADHD symptoms and proactive ASB for the oldest ($r = .54$, $p < .01$) than for the youngest ($r = .13$, $p > .10$) group ($Z = 1.65$, $p < .05$). Although not significantly different from the others, the correlation for the middle (10 to 11 years) age group was intermediate to that of the youngest and oldest groups ($r = .25$, $p > .10$). In contrast, the correlation between ADHD symptoms and reactive ASB was already moderately high for the youngest group

Table 9.2 Correlations Between ADHD Symptoms and Antisocial Behavior Subtypes

	Age (Years)			
	8–9 (*n* = 34)	**10–11 (*n* = 26)**	**12–15 (*n* = 24)**	**Total Sample (*n* = 84)**
Pearson correlations				
Proactive ASB	0.13	0.25	0.54**	0.27**
Reactive ASB	0.45**	0.59**	0.63**	0.54**
Partial correlations				
Proactive ASB	–0.07	–0.24	0.11	–0.09
Reactive ASB	0.43**	0.59**	0.41*	0.49**

*$p < .05$; **$p < .01$ (one-tailed).

(r = .45, $p < .01$) and was not significantly higher for the middle (r = .59, $p < .01$) or oldest groups (r = .63, $p < .01$) (e.g., comparing the youngest and oldest groups, Z = 0.95, $p > .10$). Thus, the hypothesis that the relation between ADHD symptoms and proactive ASB, but not reactive ASB, would be higher among the oldest age group was supported.

Finally, we examined the correlation between reactive and proactive ASB as a function of age. If the relation between ADHD symptoms and proactive ASB increases over time, perhaps it is due to a parallel increase in the relation between reactive and proactive ASB. Consistent with this hypothesis, correlations were significantly higher between reactive and proactive ASB for the oldest (r = .76, $p < .01$) than for the youngest (r = .43, $p < .05$) children (Z = 1.93, $p < .05$), with the middle group again intermediate (r = .67, $p < .01$).

Discussion

The present findings indicate that the ASB exhibited by children with ADHD symptoms is somewhat specific to the reactive subtype, consistent with prior research (Vitaro, Brendgen, & Tremblay, 2002). Furthermore, ADHD symptoms do not appear to be associated with proactive ASB until early adolescence. These findings provide further support that reactive and proactive ASB, though correlated, are distinct subtypes.

Developmentally, ADHD symptoms appear to precede the onset of comorbid ASB. Yet, little is known about the onset of ASB subtypes among children with ADHD symptoms. The present findings are consistent with the possibility that reactive ASB may precede proactive ASB among children with ADHD symptoms, as the relation between ADHD symptoms and proactive ASB is modest at best among young children, but is greater in late childhood and early adolescence. Similarly, the relation between reactive ASB and proactive ASB becomes greater during late childhood and early adolescence. Although longitudinal research is needed to examine the relation between ADHD symptoms and ASB subtypes over time, it is possible that the increasing importance of the peer group may explain these findings. Children who exhibit ASB tend to be drawn to each other and, consequently, may become more similar in their acts of ASB. This may be particularly true for proactive ASB, which serves social goals and strategic alliances. Research indicates that children with proactive ASB tend to befriend peers with proactive ASB, but children with reactive ASB do not necessarily befriend peers with reactive ASB (Poulin & Boivin, 2000b). However, another important question to examine is whether children who exhibit reactive ASB befriend peers who exhibit proactive ASB. Given that children who exhibit reactive ASB may be socially rejected by the general peer group, while children who exhibit proactive ASB may exhibit leadership skills, and, as indicated in some studies, greater social acceptance (Poulin & Boivin, 2000a), children who exhibit reactive ASB may be particularly vulnerable to befriending children who exhibit proactive ASB. If supported by longitudinal research, such findings would highlight the importance of identifying and providing interventions for reactive ASB among children with ADHD symptoms at an early age, as such intervention may help to prevent the development of an increasingly antisocial trajectory.

The present findings have several additional implications for clinical work and research. First, intervention studies attempting to decrease comorbid ASB among children with high levels of ADHD symptoms should assess and monitor change in reactive ASB. Second, the increasing association between proactive ASB with both reactive ASB and ADHD symptoms across age groups suggests that intervening before early adolescence may be critical to prevent the onset of comorbid proactive ASB. However, clinicians should recognize that although particular children with ADHD symptoms may be at risk for developing increased proactive ASB, others are not. Additional research is needed to not only replicate the present findings, but also to examine those factors (e.g., poor parental supervision, child belief that antisocial behavior has positive consequences) that may predict which children with ADHD symptoms are at greatest risk for developing proactive ASB. Third, children with reactive ASB and children with ADHD symptoms are both at risk for social problems. Given the existence of specificity between reactive ASB and ADHD symptoms, longitudinal research is needed to test the hypothesis that the relation between ADHD symptoms and social outcomes is mediated by reactive ASB. Consistent with such a model, Hinshaw and Melnick (1995) found aggressive boys with ADHD to exhibit high levels of emotional reactivity, which predicted later peer disapproval.

References

Connor, D. F., Barkley, R. A., & Davis, H. T. (2000). A pilot study of methylphenidate, clonidine, or the combination in ADHD comorbid with aggressive oppositional defiant or conduct disorder. *Clinical Pediatrics, 39,* 15–25.

Dodge, K. A., & Coie, J. D. (1987). Social-information-processing factors in reactive and proactive aggression in children's peer groups. *Journal of Personality and Social Psychology, 53,* 1146–1150.

Dodge, K. A., Lochman, J. E., Harnish, J. D., Bates, J. E., & Pettit, G. S. (1997). Reactive and proactive aggression in school children and psychiatrically impaired chronically assaultive youth. *Journal of Abnormal Psychology, 106,* 37–51.

Hinshaw, S. P., & Melnick, S. M. (1995). Peer relationships in boys with attention-deficit hyperactivity disorder with and without comorbid aggression. *Development and Psychopathology, 7,* 627–647.

Melnick, S. M., & Hinshaw, S. P. (2000). Emotion regulation and parenting in AD/HD and comparison boys: Linkages with social behaviors and peer preference. *Journal of Abnormal Child Psychology, 28,* 73–86.

Mpofu, E. (2002). Psychopharmacology in the treatment of conduct disorder children and adolescents: Rationale, prospects, and ethics. *South African Journal of Psychology, 32,* 9–21.

Paternite, C. E., Loney, J., Salisbury, H., & Whaley, M. A. (1999). Childhood inattention-overactivity, aggression, and stimulant medication history as predictors of young adult outcomes. *Journal of Child and Adolescent Psychopharmacology, 9,* 169–184.

Patterson, G. R., Reid, J. B., & Dishion, T. J. (1992). *Antisocial boys.* Eugene, OR: Castalia.

Pelham, W. E., Milich, R., Cummings, E. M., Murphy, D. A., Schaughency, E. A., & Greiner, A. R. (1991). Effects of background anger, provocation, and methylphenidate on emotional arousal and aggressive responding in attention-deficit hyperactivity disordered boys with and without concurrent aggressiveness. *Journal of Abnormal Child Psychology, 19,* 407–426.

Poulin, F., & Boivin, M. (2000a). Reactive and proactive aggression: Evidence of a two-factor model. *Psychological Assessment, 12,* 115–122.

Poulin, F., & Boivin, M. (2000b). The role of proactive and reactive aggression in the formation and development of boys' friendships. *Developmental Psychology, 36,* 233–240.

Vitaro, F., Brendgen, M., & Tremblay, R. E. (2002). Reactively and proactively aggressive children: Antecedent and subsequent characteristics. *Journal of Child Psychology and Psychiatry, 43,* 495–505.

Waschbusch, D. A., Willoughby, M. T., & Pelham, W. E. (1998). Criterion validity and the utility of reactive and proactive aggression: Comparisons to attention deficit hyperactivity disorder, oppositional defiant disorder, conduct disorder, and other measures of functioning. *Journal of Clinical Child Psychology, 27,* 396–405.

DISCUSSION QUESTIONS

1. Explain the difference between proactive and reactive antisocial behavior.
2. Note the pattern of correlations in the table: Why is it that we see the correlations between ADHD and ASB increasing from age 8 to age 15?
3. What role are peers hypothesized to play in proactive antisocial behavior?

READING

The Adolescence-Limited/Life-Course Persistent Theory of Antisocial Behavior

What Have We Learned?

Terrie E. Moffitt and Anthony Walsh

In this article, Terrie Moffitt and Anthony Walsh review 10 years' worth of research based on the most comprehensive longitudinal cohort study in criminology. This study uses data verified via police, hospital, and social work reports as well as self-report data. It is a collaborative study conducted with medical and various types of biological and social scientists. Moffitt and Walsh first provide a brief overview of the life-course persistent and adolescence-limited offender model and then examine evidence for genetic and neurological origins of persistent antisocial behavior. They also examine personality and offense pattern differences between the two types of offenders, delinquency abstainers, and gender differences.

The life-course persistent (LCP) and adolescence-limited (AL) model of offending is a synthesized theory that draws on biological and social learning variables. The theory posits that the antisocial behavior of LCP offenders has its origins in neurodevelopmental processes, that it begins early in childhood, and continues worsening thereafter. In contrast, the antisocial behavior of AL offenders has its origins in social processes, begins in adolescence, and ceases for the vast majority in young adulthood. LCP offenders are few, persistent, and pathological; AL offenders are common, relatively transient, and are relatively "normal" (Moffitt, 1993).

Life Course Persistent Offenders: Genetic Origins

LCP antisocial behavior originates early in life when the individual characteristics of a high-risk child are exacerbated by a high-risk social environment. The child's risk emerges from inherited or acquired neuropsychological deficits initially manifested as subtle cognitive deficits, difficult temperament, and/or hyperactivity. Environmental risk includes such factors as inadequate parenting, disrupted family bonds, and poverty. The environmental risk domain expands beyond the family as the child ages to include poor relations with peers and teachers. Over the first two decades of development, transactions between an individual and his or her environment gradually construct a disordered personality in which physical aggression and antisocial behavior are prominent features. This antisocial behavior will infiltrate multiple adult life domains: crime, employment problems, and victimization of intimate partners and children. The infiltration into so many domains diminishes the possibility of reform.

Source: Moffitt, T., & Walsh, A. (2003). The adolescence-limited/life-course persistent theory of antisocial behavior: What have we learned? In A. Walsh & L. Ellis (Eds.), *Biosocial criminology: Challenging environmentalism's supremacy* (pp. 125–144). Hauppage, NY: Nova Science. Reprinted by permission.

The AL/LCP theory hypothesizes that genetic processes contribute more to LCP than to AL antisocial development. A number of researchers have observed that adult criminality is more heritable than adolescent delinquency, although a small number of behavioral genetic studies have shown that juvenile antisocial behavior is also heritable, with LCP antisocial behavior more strongly heritable than AL antisocial behavior.

One type of study has taken a behavioral approach, identifying sub-types based on the Aggression and Delinquency scales from the Child Behavior Checklist (CBC). The aggression scale is most strongly associated with LCP offenders because it measures antisocial personality and physical violence and because its scores are stable across developmental stages. The delinquency scale is associated with the AL prototype because it measures rule-breaking and its mean scores rise steeply during adolescence. Of course, both LCP and AL youths engage in the behaviors on the delinquency scale, but because AL offenders are more numerous and are less genetically at risk, we would expect the delinquency scale to yield lower heritability estimates than the aggression scale. In other words, the genetic contribution to the delinquency of the relatively small number of LCP youths will be "swamped" by the much larger group of AL offenders who have little or no genetic risk. Twin and adoption studies using these scales report higher heritability for aggression (around 60%) than delinquency (around 30–40%).

A longitudinal study explored genetic and environmental influences on aggression and delinquency in 1,000 Swedish twin pairs aged 8–9 years and again at 13–14 years. In this study, continuity from childhood to adolescence in the CBCL aggression scale was largely mediated by genetic influences, whereas continuity in the delinquency scale was mediated by both the shared environment and genetic influences. This study suggests that aggression is a stable heritable trait, whereas delinquency (a behavior) is more strongly influenced by the environment and shows less genetic stability over time. A number of other studies have found quite high heritability estimates for traits that contribute to antisocial behavior (.58 to .82) that contrast with the lower estimate of .42 heritability for adolescent and adult antisocial behavior from a recent meta-analysis (Rhee & Waldman, 2002). This contrast is somewhat puzzling.

Lykken (1995:109) is also puzzled as to why heritability is so small for criminality compared to traits such as fearlessness, aggressiveness, impulsiveness, and low IQ that are its constituent parts. He notes that the constituent parts are traits whereas criminality is behavior expressing those traits. Parents have greater control over their offspring's behavior than they do over their personality traits, and thus behavior is more subject to environmental influences than are traits. Lykken asks us to imagine two hypothetical environments in which: (1) all parents are equally feckless and negligent in their parental duties, and (2) all parents are equally diligent and skilled in their parental duties. To the extent that the only environmental factor that mattered was parenting quality, the heritability coefficient would be 1.0 because parenting would be constant across both the feckless/negligent and the diligent/skilled families. Although heritability would be 100% in both cases, there would be much more antisocial behavior in the first case because feckless and negligent parents have created a situation in which only the most fearful, non-aggressive, and conscientious would refrain from antisocial behavior. In the second case, only the most fearless, aggressive, and impulsive children of diligent and skilled parents would become antisocial. In the real world, there is a wide variation in the quality of parenting, and thus wide variation in the control parents have over their children's behavior. This is why the heritability of criminality is less than the basic psychological traits that are its constituent parts.

Sociologically trained criminologists tend to view the socialization process as a one-way street running from parent to child; they rarely stop to evaluate children's effects on parental behavior. Even parental effects are not strictly environmental effects. Parents provide their offspring with a suite of genes as well as an environment. And, given the importance of assortative mating (the tendency for like to seek like), and that there is assortative mating for antisocial behavior, the possibility of intergenerational transmission of criminal behavior via both genetic and environmental routes cannot be ignored. For example, Rowe and Farrington (cited in Rutter, 1996) found a correlation of 0.50 between husbands' and wives' criminal convictions in Britain, and a New Zealand study found substantial correlations (average r = .54) between husbands and wives for a variety of antisocial traits (Krueger et al., 1998). Krueger and his colleagues view this assortative mating phenomenon as creating "criminal families" via the effects of active and passive gene/environment (G/E) correlation. The assortative mating process is active G/E correlation in the parental generation, and passive G/E correlation in the offspring generation. Assortative mating increases the probability of offspring receiving both paternal and maternal genes for traits that make them vulnerable to antisocial behavior, as well as a home environment that facilitates the development of those traits. The AL/LCP model predicts that the intergenerational transmission of antisocial behavior will be observed primarily (almost exclusively) among LCP offenders.

Life-Course Persistent Offenders and Neurological Factors

The AL/LCP theory avers that the biological (not necessarily genetic) variables that set children on a LCP trajectory are neurodevelopmental factors coupled with family adversity. The Dunedin Multidisciplinary Health and Development Study, a 30-year longitudinal study of a birth cohort of 1000 New Zealanders which examined childhood predictors measured between ages 3 and 13, has tested many hypotheses based on this assertion. These studies showed that the LCP path to offending was predicted by individual risk characteristics that included under-controlled temperament, neurological abnormalities, and delayed motor development measured at age 3, low IQ, reading difficulties, poor scores on neuropsychological tests of memory, hyperactivity, and slow heart rate (Jeglum-Bartusch et al., 1997).

The LCP path was also predicted by risk factors such as being the offspring of a teenaged single parent, neglectful and abusive parenting, inconsistent discipline, poor maternal mental health, family conflict, low SES, and rejection by peers. In contrast, study members on the AL path, despite being involved in teen delinquency to the same extent as their LCP counterparts, tended to have normal backgrounds that were sometimes better than the average Dunedin child's. Other studies have reported findings consistent with predictions from the AL/LCP taxonomy. For example, children's hyperactivity interacts with poor parenting skills to predict early onset antisocial behavior that escalates to delinquency (Patterson et al., 2000), an interaction that fits the hypothesized origins of the LCP path. Other studies report that measures of infant nervous-system maldevelopment interact with poor parenting and social adversity to predict chronic aggression from childhood to adolescence, and early-onset violent crime, but not non-violent crime (Arseneault et al., 2000).

Adolescence-Limited Offenders: Origins

AL offenders are much more common than LCP offenders and for a brief period they are almost as antisocial. Nevertheless, AL offenders have a different developmental history that puts them on a prosocial trajectory that is temporarily derailed at adolescence. AL/LCP theory shares with traditional sociological theories the position that individual differences play little or no role in the prediction of short-term adolescent offending and posits that the strongest predictors of AL offending should be peer delinquency, attitudes toward adolescence and adulthood, cultural and historical contexts which influence adolescence, and age. AL offenders are statistically "normal" youths whose offending reflects adaptive responses to conditions and events that temporarily divert them from their prosocial trajectories; it does not reflect any kind of stable personal deficiency. AL offending is motivated by the widening gap between biological and social maturity and is learned by mimicking antisocial peers (LCP offenders).

The "maturity gap" between biological and social maturity is the result of the lower age of puberty experienced by modern adolescents coupled with technological advances that continue to raise the time needed to prepare for participation in today's complex economy. This has resulted in an average of about a five- to10-year gap between puberty and the acquisition of socially responsible roles for youths in modern societies. Having no socially responsible roles while at the same time filled with energy, strength, confidence, and a strong desire to shed the restrictions of childhood, many adolescents gravitate to antisocial peer groups led by LCP youths and the excitement such groups offer. AL offenders learn the techniques of offending and justifications for it via mimicry and the principles of psychological reinforcement.

Antisocial behavior is adaptive in that it offers opportunities to gain valuable resources otherwise temporarily unavailable, the most important of which is mature status. Adult independence seems to be a distant dream to adolescents who are still dependent on parents for almost everything. In the eyes of many young people, LCP offenders have already declared their independence and have obtained a modicum of the resources (i.e., cars, nice clothes, access to sex partners) that signal mature status. The behavior of LCP offenders is viewed by neophyte offenders as bringing positive results, and thus novice offenders are drawn to LCP offenders and mimic their behavior. Developmental research shows that as ordinary young people age into adolescence, they begin to admire good students less and aggressive, antisocial peers more. Popular antisocial youths are rewarded with status among their peers, and their mimics receive vicarious reinforcement by identifying with such youths. These neophyte delinquents internalize the idea that antisocial behavior and popularity go together and thus receive validation for their antisocial behavior.

It has been documented that the AL path is more strongly related to association with delinquent peers than is the LCP path (Jeglum-Bartusch et al., 1997), and that an increase in young teens' awareness of peers' delinquency antedates and predicts the onset of their own delinquency. Other studies have shown that delinquent peer influence directly promotes increases in adolescent-onset delinquency.

The remarkable changes in young people's behavior—moodiness, the rejection of adult values, the sudden popularity of LCP antisocial peers, and most pertinently, the surge of antisocial behavior—indicates that something very interesting is occurring during adolescence. Most criminologists tend to explain the onset of adolescent offending only in terms of structural and situational factors, viewing age as simply a proxy for socially generated factors, not a variable with its own intrinsic properties. They do admit that the age peak in delinquency cannot be explained by any known set of sociological variables, which means that we have to go beyond such models. Perhaps the major factor responsible for the onset of antisocial behavior during adolescence is the fact that the increasing social maturation of children suddenly runs up against the surges of pubertal hormones that herald rapid increases in size and strength and a general increase in novelty-seeking.

The view that adolescence is a period in which physical desires and abilities outrun neurological maturity is supported by several studies showing that the earlier the onset of puberty the greater the level of problem behavior—early maturing boys and girls throw significantly more temper tantrums and engage in a number of problem behaviors more frequently and intensely than late maturing boys and girls. One study found that testosterone levels predicted future problem behavior, but only for early pubertal onset boys (Drigotas & Udry, 1993). Such findings imply that the adolescent brain is physically immature relative to the adult brain. If this is so, it may facilitate a tendency among adolescents to assign faulty attributions to situations superimposed on an unfamiliar and diffuse state of physiological arousal. A brain on "go slow" combined with physiology on "fast forward" may explain why many young persons find it difficult to accurately gauge the meanings and intentions of others and to experience more stimuli as aversive during adolescence than they did in childhood or will in adulthood.

Two magnetic resonance imaging (MRI) studies of brain development confirm that the adolescent brain, particularly the prefrontal cortex (PFC), is immature. The PFC serves various executive functions such as modulating emotions from the limbic system and making reasoned judgments. The PFC undergoes a wave of synaptic overproduction just prior to puberty, which is followed by a period of pruning during adolescence that continues into early adulthood. Selective retention and elimination of synapses depend on input from the environment, making adolescence a "critical stage of development" (Geidd et al., 1999:863).

In short, because their pre-delinquent development was normal, most AL delinquents are able to desist from crime when they age, obtain adult roles, and return gradually to a more conventional lifestyle. They are able to do this because they have accumulated sufficient social capital (any value that can be extracted from social networks that enhances one's prospects of obtaining positive life outcomes such as a positive reputation, education, job, desirable friends and mates/spouses) from previous positive interactions with others. Their recovery may be delayed, however, if their antisocial activities attract factors we called snares, such as a criminal record, incarceration, addiction, or truncated education without credentials. Such snares can compromise the ability to make a successful prosocial transition to adulthood.

The Different Personality Structures of AL and LCP Offenders

Referring to LCP offenders, Moffitt (1993:684) observes, "Over the years, an antisocial personality is slowly and insidiously constructed and accumulating consequences of the youngster's personality problems prune away options for change. . . . A person-environment interaction process is needed to account for emerging antisocial behavior, but after some age, will the 'person' main effect alone predict adult outcome?" In other words, do we arrive at a point in time when environmental circumstances and situations are not just responded to but created by individuals? Do individuals convert such things as jobs, marital relationships, and children into sources of satisfaction consistent with their dispositions, or do these things actually turn their lives around?

Studies of adolescents' personality characteristics measured at age 18 showed that the LCP path was associated with weak bonds to family and psychopathic personality traits of alienation, callousness, and impulsivity. In contrast, the AL path at age 18 was associated with a tendency to endorse unconventional values and with a personality trait called social potency (evidencing self-confidence, social influence, and leadership qualities). Personality traits were assessed again at age 26 using self-reports and reports from informants who knew the study members well (Moffitt et al., 2002). The self-and informant-reports concurred that the LCP men had more negative emotionality (i.e., they were stress-reactive, alienated, and aggressive) and were less agreeable (less social closeness, more callous) compared to AL men. It appears from these longitudinal assessments that the LCP pathway leads to a disordered antisocial personality structure resembling the psychopath: aggressive, alienated, and callous. AL men, in contrast, are unconventional, valuing spontaneity and excitement.

Offense Patterns of AL and LCP Offenders

The AL/LCP theory predicts that LCP offenders would engage in a wider variety of offense types, including "more of the victim-oriented offenses, such as violence and fraud" (Moffitt, 1993:695) than AL offenders. Subsequent research has verified this hypothesis. The Dunedin cohort examined at age 18 showed that the LCP pathway was associated with conviction for violent crimes while the AL pathway was associated with nonviolent offenses (Jeglum-Bartusch et al., 1997; Moffitt et al., 1996). It was also shown that preadolescent antisocial behavior accompanied by neuropsychological deficits predicted a greater persistence of crime and more violence up to age 18. The follow-up of these men at age 26 confirmed that LCP men as a group differed from AL men particularly in the realm of violence including violence against women and children. This finding was corroborated by self-reports, informant reports, and official court conviction records (Moffitt et al., 2002). In a comparison of specific offenses, LCP men tended to specialize in serious offenses such as carrying a concealed weapon, assault, and robbery, whereas AL men specialized in non-serious offenses such as petty theft, public drunkenness, and pirating computer software. LCP men accounted for five times their share of the cohort's violent convictions.

Delinquency Abstainers

If AL delinquency is normative adaptive behavior, then the existence of teens that abstain from delinquency requires an explanation. The original theory speculated that teens abstaining from antisocial behavior would be rare and they must have either structural barriers that prevent them from learning about delinquency, a smaller than average maturity gap, or personal unappealing characteristics that cause them to be excluded from teen social group activities. The Dunedin cohort contained only a small group of males who avoided virtually any antisocial behavior during both childhood and adolescence. These abstainers described themselves at age 18 on personality measures as extremely self-controlled, fearful, interpersonally timid, socially inept, and latecomers to sexual relationships (i.e., virgins at age 18). Dunedin abstainers were the type of good and compliant students who become unpopular with adolescent peers. Follow-up data at age 26 confirmed that abstainers had not become late-onset offenders (Moffitt et al., 2002). Although their teenaged years had been troubling, their lifestyle became more successful in adulthood. As adults they retained their self-constrained personality, had virtually no crime or mental disorder, were likely to have settled into marriage, were delaying children (a desirable strategy for a generation needing prolonged education to succeed), were likely to be college educated, held high-status jobs, and expressed optimism about their own futures.

Dunedin abstainers fit the profile Shedler and Block (1990) reported for youth who abstained from drug experimentation in a historical period when it was normative: i.e., they were over-controlled, not curious, not active, not open to experience, socially isolated, and lacking in social skills. Another study found that drug experimenters and abstainers did not differ in parental attachment, but both differed from frequent users, and that virgins and sexual experimenters did not differ in parental attachment, but both differed from sexually promiscuous subjects (Walsh, 1992). Sexual promiscuity is a powerful marker of antisocial behavior.

The kind of inhibited and introverted personality characteristic of abstainers is consistent with studies of autonomic nervous system (ANS) arousal and criminality. Individuals with a hyperarousable ANS are easily conditioned; those with a hypoarousable ANS are conditioned with difficulty. Abstainers may be individuals located on the "hyper" tail of the ANS arousal distribution, and thus have excessive guilt feelings and excessive fear of the negative consequences of nonconformity.

Gender and Antisocial Behavior

It is universally acknowledged that across time and culture, females commit far less crime and delinquency than males and that the more serious the crime the greater the male/female gap. Much of the gender difference in crime is attributed to sex differences in the risk factors for LCP antisocial behavior. Girls are biologically less likely than boys to encounter the putative neurophysiological links that initiate the causal chain for LCP antisocial development. Girls are at lower risk for symptoms of nervous system dysfunction, difficult temperament, late verbal and motor milestones, hyperactivity, learning disabilities, reading failure, and childhood conduct problems. In other words, more girls than boys lack the congenital elements of the passive, reactive, and active person/environment correlations and interactions that initiate and maintain LCP antisocial behavior.

On the other hand, AL delinquency is open to girls as well as to boys, and girls' AL delinquency should begin soon after puberty as boys' does. The extent to which they do so will depend upon their access to antisocial models, whether or not they perceive the consequences of delinquency to be reinforcing. Exclusion from male antisocial groups, however, may cut off opportunities for girls to engage in delinquent behaviors. Girls are more vulnerable than boys to personal victimization, such as injury from dating violence, or pregnancy, if they affiliate with LCP males. "Thus, lack of access to antisocial models and perceptions of serious personal risk may dampen the vigor of girls' delinquent involvement somewhat. Nonetheless, girls should engage in adolescence-limited delinquency in significant numbers" (Moffitt, 1994:39–40).

In the Dunedin cohort it was found that for every LCP female there were 10 LCP males but the AL difference was negligible at 1.5 male for every one female. Childhood offending-onset females had high-risk neurodevelopmental and family backgrounds, but adolescent-onset females did not. This documents the fact that females and males on the same trajectories share the same risk factors. The Dunedin sample found that each girl's delinquency onset is linked to the timing of her puberty, that delinquent peers are a necessary condition for the onset of delinquency among adolescent girls, and that an intimate relationship with an offender promotes girls' antisocial behaviors. These accumulated findings suggest that the AL/LCP theory of the origins of both LCP and AL offending are explanatory across the genders. Because few females have the risk factors for LCP development, the theory explains the wide sex difference in serious, persistent antisocial behavior. We do not mean to imply that the AL/LCP theory explains every aspect of the gender gap in antisocial behavior. Although the male/female ratio may be as little as 1.5:1 for AL offending, that ratio should be considerably higher for the more serious violent offenses committed by AL offenders. Males seem to be designed by natural selection to be particularly concerned with status and acquiring sexual partners during adolescence, and the proximate mechanism enabling them to engage in this strategy is testosterone, a hormone that is much more abundant in males than in females.

Discussion

Before 1993, virtually no research compared delinquent subtypes defined on a developmental basis, but now this research strategy has become almost commonplace. The idea that antisocial behavior that begins in childhood has a different developmental history from antisocial behavior that begins in adolescence has been codified in the *DSM-IV* of the American Psychiatric Association and has been invoked in the National Institute of Mental Health Factsheet on Child and Adolescent Violence and in the Surgeon General's report on Youth Violence. Many research teams have assessed representative samples with prospective measures of antisocial behavior from childhood to adulthood, and this has enabled comparisons based on age of onset and persistence. Now that the requisite data bases are available, other hypotheses derived from the original AL/LCP theory need to be tested.

All AL offenders do not desist at the same time, and we suggest that snares such as a criminal record, incarceration, addiction, or a truncated education should explain variation in the age at desistence from crime. The theory also asserted that AL and LCP offenders would react differently to turning-point opportunities: LCP offenders would selectively obtain undesirable partners and employment and would expand their antisocial repertoire into domestic abuse and workplace crime: AL offenders would obtain good partners and employment and would desist from almost all antisocial behavior. The theory also speculated that AL offenders must rely on peer support for crime but that LCP offenders should be willing to offend alone. It was also inferred that childhood measures of antisocial behavior in longitudinal studies should be more highly correlated with adult measures than with adolescent measures. These and many other hypotheses derived from the theory remain to be systematically examined.

References

Arseneault, L., Tremblay, R. E., Boulerice, B., Seguin, J. R., & Saucier, J-F. (2000). Minor physical anomalies and family adversity as risk factors for adolescent violent delinquency. *American Journal of Psychiatry, 157,* 917–923.

Drigotas, S., & Udry, J. (1993). Biosocial models of adolescent problem behavior: Extensions to panel design. *Social Biology, 40,* 1–7.

Giedd, J., Blumenthal, J., Jeffries, N., Castellanos, F., Liu, H., Zijenbos, A., et al. (1999). Brain development during childhood and adolescence: A longitudinal MRI study. *Nature Neuroscience, 2,* 861–863.

Jeglum-Bartusch, D., Lynam, D., Moffitt, T. E., & Silva, P. A. (1997). Is age important: Testing general versus developmental theories of antisocial behavior. *Criminology, 35,* 13–47.

Krueger, R., Moffitt, T., Caspi, A., Bleske, A., & Silva, P. (1998). Assortative mating for antisocial behavior: Developmental and methodological implications. *Behavior Genetics, 28,* 173–185.

Lykken, D. (1995). *The antisocial personalities.* Hillsdale, NJ: Lawrence Erlbaum Associates.

Moffitt, T. E. (1993). "Life-course-persistent" and "adolescence-limited" antisocial behavior: A developmental taxonomy. *Psychological Review, 100,* 674–701.

Moffitt, T. E. (1994). Natural histories of delinquency. In E. Weitekamp & H. J. Kerner (Eds.), *Cross-national longitudinal research on human development and criminal behavior* (pp. 3–61). Dordrecht: Kluwer Academic Press.

Moffitt, T. E., Caspi, A., Dickson, N., Silva, P. A., & Stanton, W. (1996). Childhood-onset versus adolescent-onset antisocial conduct in males: Natural history from age 3 to 18. *Development and Psychopathology, 8,* 399–424.

Moffitt, T. E., Caspi, A., Harrington, H., Milne, B. (2002). Males on the life-course persistent and adolescence-limited antisocial pathways: Follow-up at age 26. *Development & Psychopathology, 14,* 179–206.

Patterson, G. R., DeGarmo, D. S., & Knutson, N. (2000). Hyperactive and antisocial behaviors: Comorbid or two points in the same process? *Development & Psychopathology, 12,* 91–106.

Rhee, S., & Waldman, I. (2002). Genetic and environmental influences on antisocial behavior: A meta-analysis of twin and adoption studies. *Psychological Bulletin, 28,* 490–529.

Rutter, M. (1996). Introduction: Concepts of antisocial behaviour, of cause, and of genetic influence. In G. Bock & J. Goode (Eds.), *Genetics of criminal and antisocial behavior* (pp. 1–20). Chichester, England: Wiley.

Shedler, J., & Block, J. (1990). Adolescent drug use and psychological health. *American Psychologist, 45,* 612–630.

Walsh, A. (1992). Drug use and sexual behavior: Users, experimenters and abstainers. *Journal of Social Psychology, 132,* 691–693.

DISCUSSION QUESTIONS

1. How would you describe "peer influence" if you were writing about (1) AL delinquents and (2) LCP delinquents?
2. Why are LCP girls greatly outnumbered by LCP boys when AL girls are only slightly outnumbered by AL boys?
3. What are the major advantages of interdisciplinary criminology as revealed by the findings in this article?

READING

Unraveling Juvenile Delinquency

Sheldon Glueck and Eleanor Glueck

This excerpt from Sheldon and Eleanor Glueck's 1950 book, *Unraveling Juvenile Delinquency*, is the scaffold and template of all subsequent developmental studies. The Gluecks were prolific researchers who were ahead of their time in realizing that any single unit of analysis was far from sufficient to explain crime and juvenile delinquency. They took numerous measures from their experimental and control groups and concluded that delinquency, especially serious delinquency, is the result of a dynamic interplay of bodily, temperamental, intellectual, and sociocultural variables.

Need for Eclectic Approach to Study of Crime Causation

At the present stage of knowledge, an eclectic approach to the study of the causal process in human motivation and behavior is obviously necessary. It is clear that such an inquiry should be designed to reveal meaningful integrations of diverse data from several levels of inquiry. There is need for a systematic approach that will not ignore any promising leads to crime causation, covering as many fields and utilizing as many of the most reliable and relevant techniques of investigation and measurement as are necessary for a fair sampling of the various aspects of a complex biosocial problem. Ideally, the focus in such a study should be upon the selectivity that occurs when environment and organism interact. The searchlight should be played upon the point of contact between specific social and biological processes as they coalesce, accommodate, or conflict in individuals.

But while the most promising areas of research in human conduct and misconduct are to be found in the nexus of physical and mental functions and in the interplay of person and milieu, the complexities of motivation and varieties of behavior compel a division of the field into areas or levels. These must be interpreted serially before arriving at a meaningful pattern.

Control of the Inquiry

In order to arrive at the clearest differentiation of disease and health, comparison must be made between the unquestionably pathologic and the normal. (This metaphor does not carry with it any implication that we view delinquency as a disease.) Therefore non-delinquents as well as delinquents become the subjects of our inquiry. Comparison is a fundamental method of science; and the true value of any phenomenon disclosed by exploration of human behavior cannot be reliably determined without comparing its incidence in an experimental group with that in a control group. This method in the present research should result not only in isolating the factors which most markedly differentiate delinquents from non-delinquents, but in casting light on the causal efficacy of a number of factors generally accepted as criminogenic.

The control group must of course be truly non-delinquent. But this one factor, although basic, is not enough. Some factors must be held constant as a prerequisite to the comparison of delinquents and non-delinquents in respect to still other factors.

In deciding which factors to use in matching delinquents with non-delinquents, we were guided by several aims. First, since the ultimate comparison should cover subtle processes of personality and environment, the more general or cruder factors should be controlled in the matching; second, those traits that typically affect a whole range of factors ought to be held constant; third, those general characteristics that have already been explored sufficiently by other investigators and about which there is much agreement ought to be equalized in the two groups. Overriding all these aims, however, is the practical difficulty of matching two series of hundreds of human beings, while holding several factors constant.

We therefore decided to match 500 delinquent boys with 500 non-delinquent boys in four respects:

1. Age, because it is often asserted that tendencies to maladjustment and misbehavior vary with age (especially puberty and adolescence); also because morphologic and psychologic factors are more or less affected by age.

2. *General intelligence*, because there is considerable claim that intelligence, as measured by standard tests, bears an intimate relationship to varieties of behavior-tendency. From the point of view of the clinical psychologist concerned with determining the extent and quality of intelligence of the individual before him, the intelligence quotient—standing alone—is nowadays regarded as somewhat naive and not always helpful. However, it still forms an important part of the protocol covering the diagnosis of intellectual make-up, and for our purpose of matching delinquents with non delinquents, it is adequate.

3. *National (ethnico-racial) origin*, because there is a school of thought which stresses ethnic derivation and associated culture patterns in accounting for variations in behavior tendencies. Since we are interested, also, in determining whether meaningful differences exist between delinquents and non-delinquents in respect to bodily morphology (somatotypes and bodily disproportions), such differences or similarities as may emerge from the anthropologic analysis should gain in significance by the fact that ethnico-racial derivation has been controlled in the matching process.

4. *Residence in underprivileged neighborhoods*, because, as noted in Chapter 1, the conception is widespread that delinquency is largely bred by the conditions in such areas. In selecting

both delinquents and non-delinquents from unwholesome neighborhoods, we were attempting to control a complex of socio-economic and cultural factors whose similarity would permit us to find out why it is that even in regions of most adverse social conditions, most children do not commit legally prohibited acts of theft, burglary, assault, sexual aggression, and the like.

Levels of the Inquiry

Though we look at the problem of delinquency from the point of view of the integration of the total personality, it is revealing to study discord or conflict and accord or harmony at various levels: (1) the socio-cultural level, where conflict is shown by delinquency, crime, or other forms of maladjustment of the individual to the taboos, demands, conventions, and laws of society; (2) the somatic level, where disharmony is indicated by dysplasias or disproportions between the structure of two or more segments of the physique and by ill health; (3) the intellectual level, where discord may be revealed by contrasts between capacities of abstract intelligence and those of concrete intelligence, or between special abilities and special disabilities, or by excessive variability in types of intellectual capacity; (4) the emotional-temperamental level, where disharmony is shown by mental conflict and by the tensions between repressed and forgotten emotional experiences and more recent experiences or between divergent instinctual energy propulsions typically reflected in the phenomenon of ambivalence.

In gathering and interpreting data on each of these four levels separately, we are not overlooking the fact that we are dealing with the motivations and behavior of the total organism in its milieu. . . .

. . . Always remembering our main objective—to compare boys reared in underprivileged areas of a large city who become serious and persistent delinquents with boys bred in similar neighborhoods who are not antisocial—we finally arrived at what seemed to be as nearly an ideal plan as could be devised; it promised to produce the necessary data by the most direct means and with the least possible inconvenience to the boys, their families, and the many institutional authorities and social workers who would be involved. This plan called for the examination of delinquent boys preponderantly from the Boston area who at the time of selection were inmates of a state correctional school, the Lyman School for Boys in Westboro, Massachusetts, and the examination of a comparable number of proven non-delinquents from public schools in the city of Boston. (Later it proved necessary to include some boys from the Industrial School for Boys in Shirley, Massachusetts.). . . .

Delimiting the Social Inquiry

In deciding which factors to include, we had to keep in mind the limitations of data gathered so many years after the original event, not only regarding purely objective facts but especially regarding qualitative information about the home environment in which the boy had been reared, his early behavioral difficulties, and the like. Long experience in gathering social histories had taught us the dangers and pitfalls in reconstructing the past because of limitations in recorded data and in the memories of those from whom the information must necessarily be derived. It is also difficult to secure any valid qualitative data, such as the emotional atmosphere surrounding past and even current situations, without the most intensive investigations.

The factors which we had sought in connection with our other studies had all been subjected to rigid testing; and as we were interested, among other things, to secure data that would be comparable to those already compiled on the two thousand case histories of our earlier studies, we incorporated all these factors and added others of special pertinence to the current research. The result was a total of 149 social factors.

In general, we set ourselves to gather vital statistics on the boy and the members of his immediate family, as well as his paternal and maternal grandparents and paternal and maternal aunts and uncles. We were concerned also with reconstructing the history of delinquency and securing evidences of excessive drinking, mental deficiency, emotional disturbance, and certain physical ailments, not only among the members of the boy's immediate family but also among grandparents on both sides and paternal and maternal aunts and uncles, in order to determine the differences in the burden of familial defects among the delinquents and non-delinquents. We were interested, further, as part of the exploration of the family history, to determine the education and economic condition of the boy's parents and of his paternal and maternal grandparents.

With respect to the boy himself, one of our major quests was to achieve a chronological picture of his whereabouts from birth until the time we selected him for study and to describe the physical aspects of his present home and environment.

Another major concern in the social investigation was with the kind of cultural, intellectual, and emotional atmosphere in which the boy had been reared. Still another area of exploration dealt with the boy's habits and his use of leisure, our attention being specifically directed to an inquiry concerning his age at the onset of aberrant behavior and the nature of the earliest manifestations of his antisociality. Of special interest to us was a determination of the parents' knowledge of the boy's habits and use of leisure time, as contrasted with the actual facts in the case, and the parents' explanation of the boy's behavior. A history of the boy's schooling was also to be secured, as well as an estimate of his behavior in school. . . .

Testing and Examining Delinquents and Non-Delinquents

As a preliminary to determining the content of the somatic, intellectual, and emotional-temperamental levels of the research project, various tests and examinations were explored. The clinicians on our staff as well as other specialists were consulted, the main focus always being on what a particular test or examination might contribute to answering the basic question of the study: What are the differences between juvenile delinquents and non-delinquents that might throw light on crime causation?

The sum total of our explorations resulted in a plan to photograph each boy for ultimate classification into a somatotype and to make a twenty- to thirty-minute medical examination (somatic level); to administer a Wechsler-Bellevue Test and Stanford Achievement Tests in Reading and Arithmetic (intellectual level); and to administer a Rorschach Test and hold a psychiatric interview (emotional-temperamental level). . . .

Dynamic Pattern of Delinquency

The finding of marked differences between the persistently delinquent group and the non-delinquent group in the incidence of many biologic, sociologic, and biosocial factors reveals the operation of cause and effect by establishing the probability of a functional relationship between the factors taken as a whole and a tendency to antisocial behavior. This does not mean that a boy possessing one or more of the differentiating traits must inevitably become delinquent. A group of differentiative factors derived from any single level of the inquiry is not, standing alone, too likely to bring about delinquency. But when to that cluster of distinguishing traits are added those from one or more of the other levels, the possibilities of delinquent behavior are greatly enhanced. In a case in which many differentiative factors from all levels are present, the impulsion to delinquency is virtually unavoidable.

Three points ought to be borne in mind: (1) The selection and matching of the two groups of boys with respect to ethnic-racial derivation, age, intelligence quotient, and residence in underprivileged areas has, of course, meant the exclusion of these controlled variables from the comparison. (2) In planning the research, we took into account the fact that statistical correlation in itself may not necessarily mean actual functional relationship. Therefore, in choosing factors for comparison, we selected only those which, from our own experience and from the writings of those who have made inquiries into various branches of the behavior disciplines, seemed to us to have a possible functional connection with delinquent or non-delinquent behavior tendencies. (3) Although we included a great many factors, and although, as has been shown in the preceding chapter, we have been able to construct efficient predictive instrumentalities based on the more differentiative among them, further investigations may well disclose still other causal factors.

The Causal Complex

It will be observed that in drawing together the more significant threads of each area explored, we have not resorted to a theoretical explanation from the standpoint, exclusively, of any one discipline. It has seemed to us, at least at the present stage of our reflections upon the materials, that it is premature and misleading to give exclusive or even primary significance to any one of the avenues of interpretation. On the contrary, the evidence seems to point to the participation of forces from several areas and levels in channeling the persistent tendency to socially unacceptable behavior. The foregoing summation of the major resemblances and dissimilarities between the two groups included in the present inquiry indicates that the separate findings, independently gathered, integrate into a dynamic pattern which is neither exclusively biologic nor exclusively socio-cultural, but which derives from an interplay of somatic, temperamental, intellectual, and socio-cultural forces.

We are impelled to such a multidimensional interpretation because, without it, serious gaps appear. If we resort to an explanation exclusively in terms of somatic constitution, we leave unexplained why most persons of mesomorphic tendency do not commit crimes; and we further leave unexplained how bodily structure affects behavior. If we limit ourselves to a socio-cultural explanation, we cannot ignore the fact that socio-cultural forces are selective; even in underprivileged areas most boys do not become delinquent and many boys from such areas do not develop into persistent offenders. And, finally, if we limit our explanation to psychoanalytic theory, we fail to account for the fact that the great majority of non-delinquents, as well as of delinquents, show traits usually deemed unfavorable to sound character development, such as vague feelings of insecurity and feelings of not being wanted; the fact that many boys who live under conditions in which there is a dearth of parental warmth and understanding nevertheless remain non-delinquent; and the fact that some boys, under conditions unfavorable to the development of a wholesome superego, do not become delinquents, but do become neurotics.

If, however, we take into account the dynamic interplay of these various levels and channels of influence, a tentative causal formula or law emerges, which tends to accommodate these puzzling divergencies so far as the great mass of delinquents is concerned: The delinquents as a group are distinguishable from the non-delinquents: (1) physically in being essentially mesomorphic in constitution (solid, closely knit, muscular); (2) temperamentally, in being restlessly energetic, impulsive, extroverted, aggressive, destructive (often sadistic)—traits which may be related more or less to the erratic growth pattern and its physiologic correlates or consequences; (3) in attitude, by being hostile, defiant, resentful, suspicious, stubborn, socially assertive, adventurous, unconventional, non-submissive to authority; (4) psychologically, in tending to direct and concrete, rather than symbolic, intellectual expression, and in being less methodical in their approach to problems; (5) socio-culturally, in having been reared to a far greater extent than the control group in homes of little understanding, affection, stability, or moral fiber by parents usually unfit to be effective guides and protectors or, according to psychoanalytic theory, desirable sources for emulation and the construction of a consistent, well-balanced, and socially normal superego during the early stages of character development. While in individual cases the stresses contributed by any one of the above pressure-areas of dissocial behavior tendency may adequately account for persistence in delinquency, in general the high probability of delinquency is dependent upon the interplay of the conditions and forces from all these areas.

In the exciting, stimulating, but little-controlled and culturally inconsistent environment of the underprivileged area, such boys readily give expression to their untamed impulses and their self-centered desires by means of various forms of delinquent behavior. Their tendencies toward uninhibited energy-expression are deeply anchored in soma and psyche and in the malformations of character during the first few years of life.

DISCUSSION QUESTIONS

1. In what ways was the Gluecks' study of experimental and control groups superior to studies in which variables are controlled only statistically?
2. How do you think that this study emphasizing the role of biology and psychology as well as sociocultural factors was received by the criminologists of the time who only emphasized the latter?
3. In what ways are the Gluecks' explanations of their data consistent with social control theory?

READING

A Life-Course View of the Development of Crime

Robert J. Sampson and John H. Laub

In this article, Sampson and Laub present a life-course perspective on crime with their fundamental argument being that persistent offending and desistance from it can both be understood within their revised age-graded theory of informal social control. The authors examine some of their early work as well as their ongoing work in which they have sought to undermine the idea that developmentally distinct groups of offenders can be explained by unique causal processes. They strongly emphasize the concepts of turning points from a time-varying view of key life events and the all-too-often overlooked importance of human agency in the development of crime. Their life-course theory envisions development as the constant interaction between individuals and their environment, coupled with random developmental noise and a purposeful human agency. They take pains to distinguish human agency from rational choice. Contrary to influential developmental theories in criminology such as those addressed in the text portion of this chapter, Sampson and Laub conceptualize crime as an emergent process reducible neither to individual differences nor the sociocultural environment. Criminal behavior (and other forms of behavior) has an emergent property of both and is not necessarily deducible from either a priori.

In this article, we argue for a life-course perspective on trajectories of crime, focusing on the question of whether (and why) adolescent delinquents persist or desist from crime as they age across the adult life course. The growing tendency in developmental perspectives on crime, often called "developmental criminology," is to subdivide the offender population and assume different causal influences at different stages of the "criminal career." Although at first it may seem counterintuitive, our fundamental argument is that persistent offending and desistance—and hence trajectories of crime—can be meaningfully understood within the same theoretical framework. Our strategy is to start with the assumption of generality and see how far it takes us in understanding patterns of criminal offending across the full age range of the life course. We explore this logic, beginning with a summary of results from our prior research.

Crime in the Making and the Origins of Life-Course Criminology

Unraveling Juvenile Delinquency, along with subsequent follow-ups conducted by Sheldon and Eleanor Glueck of the Harvard Law School, is one of the most influential research projects in the history of criminological research. The Gluecks' data were derived from a three-wave prospective study of juvenile and adult criminal behavior that began in 1940. The research design involved a sample of five hundred male delinquents aged ten to seventeen and five hundred male non-delinquents aged ten to seventeen matched case by case on age, race/ethnicity, IQ, and low-income residence in Boston. Extensive data were collected on the one thousand boys at three points in time—ages fourteen, twenty-five, and thirty-two. Over the period 1987 to 1993, we reconstructed, augmented, and analyzed these longitudinal data that, owing to the Gluecks' hard work over many years, are immensely rich and will likely never be repeated given modern institutional review board restrictions (e.g., wide-ranging interviews with teachers, neighbors, and employers; detailed psychiatric and physical assessments; extensive searches of multiple agency records).

In [our 1993 book] *Crime in the Making,* we developed a theoretical framework to explain childhood antisocial behavior, adolescent delinquency, and crime in early adulthood. The general organizing principle was that crime is more likely to occur when an individual's bond to society is attenuated. Our analysis of the causes of delinquency shared much in common with the focus in classical control theory (Hirschi 1969) on adolescence, but the reality of later life-course milestones required us to develop a modified theoretical perspective. After all, the transition to young adulthood brings with it new social control institutions and potential turning points that go well beyond adolescence. We thus developed an *age-graded* theory emphasizing informal social controls that are manifested in shifting and possibly transformative ways as individuals age. For example, we focused on parenting styles (supervision, warmth, consistent discipline) and emotional attachment to parents in childhood; school attachment and peers in adolescence; and marital stability, military service, and employment in adulthood. Although these are manifestly distinct domains that are age graded, we argued that there are higher-order commonalities with respect to the concept of social connectivity through time.

Stability and Change in Criminal Behavior Over the Life Course

The delinquents and nondelinquents in the Gluecks' study displayed considerable between-individual stability in crime and many problematic behaviors well into adulthood. This stability held independent of age, IQ, ethnicity, and neighborhood SES. Indeed, delinquency and other forms of antisocial conduct in childhood were strongly related to troublesome adult behavior across a variety of experiences (e.g., crime, military offenses, economic dependence, and marital discord). But why? One of the mechanisms of continuity that we emphasized was "cumulative disadvantage," whereby serious delinquency and its nearly inevitable correlates (such as incarceration) undermined later bonds of social control (such as employability), which in turn enhanced the chances of continued offending.

At the same time, we found that job stability and marital attachment in adulthood were significantly related to *changes* in adult crime—the stronger the adult ties to work and family, the less crime and deviance among both delinquents and nondelinquent controls. We even found that strong marital attachment inhibits crime and deviance regardless of that spouse's own deviant behavior and that job instability fosters crime regardless of heavy drinking. Despite differences in early childhood experiences, adult social bonds to work and family thus had similar consequences for the life-course trajectories of the five hundred delinquents and five hundred nondelinquent controls. These results were consistent for a wide variety of crime outcome measures, control variables (e.g., childhood antisocial behavior and individual-difference constructs), and analytical techniques ranging from methods that accounted for persistent unobserved heterogeneity in criminal propensity to analyses of qualitative data.

Taken as a whole, these findings suggested to us that social ties embedded in adult transitions (e.g., marital attachment, job stability) explain variations in crime unaccounted for by childhood propensities. This empirical regularity supports a dual concern with continuity and change in the life course. A fundamental thesis of our age-graded theory of informal social control was that whereas individual traits and childhood experiences are important for understanding behavioral stability, experiences in adolescence and adulthood can redirect criminal trajectories in either a more positive or more negative manner. In this sense, we argue that all stages of the life course matter and that "turning points" are crucial for understanding processes of adult change.

Shared Beginnings, Divergent Lives: An Overview

Crime in the Making raised many unanswered questions, and in its concluding chapter we highlighted directions for future research and theoretical development that appeared fruitful. Two of these directions seemed especially relevant for developmental/life-course theories of crime, namely, the merging of quantitative and qualitative data and further understanding of age and crime. For example, what about crime in middle age? Older age? Is there really such a thing as a lifelong career criminal—or what have been dubbed "life-course persisters" (LCPs)? If so, can this group be prospectively identified? Another set of questions turned on the use of qualitative narratives to delve deeper into a person-based exploration of the life course. Can narratives help us unpack mechanisms that connect salient life events across the life course, especially personal choice and situational context? In our view, life-history narratives combined with quantitative approaches can be used to develop a richer and more comprehensive picture of why some men persist in offending and others stop. We made moves toward a narrative-based inquiry in *Crime in the Making* but were forced to rely on the Gluecks' written records rather than our own original interviews.

These motivations led us to follow up the Glueck men to the present. Our study involved three sources of new data collection—criminal record checks (local and national), death record checks (local and national), and personal interviews with a sample of fifty-two of the original Glueck delinquents. The sample of men to interview was strategically selected to ensure variability in trajectories of adult crime. The combined data represent a roughly fifty-year window from which to update the Glueck men's lives at the close of the twentieth century and connect them to life experiences all the way back to early childhood. We believe these data represent the longest longitudinal study to date in criminology of the same men. The following sections briefly summarize the key findings.

Age and Crime

Our analyses showed that, on one hand, the aggregate age-crime curve is not the same as individual age-crime trajectories, lending apparent support to one of the major claims of the criminal career model. There is enormous variability in peak ages of offending, for example, and age at desistance varied markedly across the Glueck men. On the other hand, we found that crime declines with age even for active offenders and that trajectories of desistance cannot be prospectively identified based on typological accounts rooted in childhood and individual differences. That is, offenses eventually decline for all groups of offenders identified according to extant theory and a multitude of childhood and adolescent risk factors. Whether low IQ, aggressive temperament, or early onset of antisocial behavior, desistance processes are at work even for the highest-risk and predicted life-course persistent offenders. While prognoses from childhood factors such as these are modestly accurate in predicting stable differences in later offending, they did not yield distinct groupings that were valid prospectively for troubled kids. Not only was prediction poor at the individual level, our data raised questions about the sorts of categorically distinct groupings that dominate theoretical and policy discussions (e.g., "life-course persistent offender," "superpredator"). These groupings tended to wither when placed under the microscope of long-term observation (Sampson and Laub 2003).

We thus concluded that a middle-ground position was necessary in the criminal careers debate—yes, there is enormous variability in individual age-crime curves such that it renders the aggregate curve descriptive of few people, and yes, age has a direct effect on offending such that life-course desistance is the more accurate label. We believe this compromise position has general implications for assessing key assumptions of developmental criminology and rethinking its conceptual meaning.

Mechanisms of Desistance

A second goal of our book was to exploit life-history narratives to better understand patterns of stability and change in offending over the life course. In our narrative interviews, we asked the men to describe turning points in their life. We also had the men fill out life-history calendars so that we could more accurately determine the sequencing of major life events. Several turning points were implicated in the process of desistance from crime, including marriage/spouses, military service, reform school, work, and residential change. The mechanisms underlying the desistance process are consistent with the general idea of social control. Namely, what appears to be important about institutional or structural turning points is that they all involve, to varying degrees, (1) new situations that "knife off" the past from the present, (2) new situations that provide both supervision and monitoring as well as new opportunities of social support and growth, (3) new situations that change and structure routine activities, and (4) new situations that provide the opportunity for identity transformation. The lesson we drew is that involvement in institutions such as marriage, work, and the military reorders short-term situational inducements to crime and, over time, redirects long-term commitments to conformity. In making the case for the importance of the adult life course, we have referred to involvement in these institutions as turning points because they can change trajectories over time (Laub and Sampson 1993).

A potential objection, however, is that turning points are a result of selection bias or, put differently, the unobserved characteristics of the person (e.g., Gottfredson and Hirschi 1990). To shed further light on life events, we exploited the longitudinal nature of the long-term data to examine within-individual change, where the unit of variation is across time. As such, stable characteristics of the person are held constant and we can exploit changes in social location, such as marriage, in terms of deviations from a person's expected trajectory. Holding age constant and allowing individual heterogeneity in age effects, we found that *when* in a state of marriage, the propensity to crime was lower for the same person than when not in marriage. Similar results were found for military service and steady employment. Quantitative models of within-individual change thus give statistical evidence of the probabilistic enhancement of desistance associated with life-course events like marriage, military service, and employment. . . .

A Revised View of the Causal Importance of Turning Points

[W]e now turn to the role of "turning points" in development and growth. To date, our work has tended to conceptualize turning points in terms of singular, sometimes rare events (e.g., serving in military during wartime). Recently, we have begun to modify this view in light of the fact that many important life events are repeating in nature. For illustrative purposes, we examine here the institution of marriage.

Why is marriage important in the process of desistance from crime? There appear to be at least five mechanisms of desistance, none of which are to our knowledge limited to the particular historical period or demographic subgroups represented in the Gluecks' data. Consistent with the general turning point processes discussed above, theoretically marriage has the potential to lead to one or more of the following in the lives of criminal men: (1) a "knifing off" of the past from the present; (2) opportunities for investment in new relationships that offer social support, growth, and new social networks; (3) forms of direct and indirect supervision and monitoring of behavior; (4) structured routines that center more on family life and less on unstructured time with peers; and (5) situations that provide an opportunity for identity transformation and that allow for the emergence of a new self or script, what Hill (1971) described as the "movement from a hell-raiser to a family man."

It follows from this theoretical conceptualization that the mechanisms associated with marriage are not a constant once set in motion and thus vary through time. The spousal monitoring of drinking patterns, for example, is predicted to vary over time depending on the state of whether one is in or out of a marital relationship. Consider further the demographic reality that people enter and exit (and often reenter) marriage over time. Sampson, Laub, and Wimer (2005) followed through on this observation by conceptualizing the potential causal effect of being in the state of marriage (which hypothetically could be randomly or exogenously induced) with the state of nonmarriage *for the same person*. In dynamic terms, marriage is thus not seen as a single turning point but as part of a potential causal dynamic over the life course. We further hypothesize that the effect of marriage on desistance from crime is independent of the developmental history of the person—in this sense, the marriage effect is "nondevelopmental."

Causal Effects and the Life Course

The biggest threat to the validity of any analysis claiming causal effects of a social state like marriage is to account for the nonrandom selection of individuals into the state itself. Marriage is not a random event, and homophily in partner characteristics is well established, even though it is simultaneously true that fortuitous events influence mating patterns. To the extent that marriage is influenced by individual self-selection, the marriage-crime relationship is potentially spurious. Indeed, selection is the main critique put forth by those suspicious of social forces (e.g., Gottfredson and Hirschi 1990). Since marriage cannot be randomized in practice, the canonical solution to date has been to "control" for a host of potentially confounding factors, most notably lagged states of crime itself and other factors that may cause both crime and later marriage, such as prior crime and deviance, personality, unemployment, and so on. Instrumental variables are also possible, but in practice they have not proven effective. Moreover, controlling past values of the treatment or outcome results in biased estimates because such a method controls for the very pathways that are hypothesized to lead to crime.

In recent work, we have addressed this conundrum through a multipronged approach that combines a longitudinal fixed-effects analysis of changes in marriage and crime over the life course with recently pioneered methods for identifying causal effects using observational data—what are typically called "counterfactual methods' of causal inference (Sampson, Laub, and Wimer 2005). Drawing from the language of randomized experiments, counterfactual methods conceptualize causal effects as the effect of a definable "treatment" (e.g., marriage) on some outcome (e.g., likelihood of committing a crime). In this case, one would divide the sample population into a treatment group (those who marry) and a control group (those who do not marry). When examining the causal effect of the treatment, counterfactual methods assume that each individual has two "potential outcomes," at least theoretically. The first is the outcome that the individual demonstrates under the treatment condition, which we will call Yi^t. The second is the outcome that the individual demonstrates under the control condition, which we will call Yi^c. For each individual, however, only one of these outcomes can be actually observed at the same time. We can thus recast questions of causality as a "missing data problem" of the unobserved counterfactual, one that is solved in experimentation through randomization. Assuming equivalence of controls and treatments permits the estimation of the causal effect, Yt – Yc.

Observational data are another matter. When dealing with a treatment at one point in time, one statistical solution is propensity score matching. With this technique, one can model the propensity that each individual receives the treatment and then create two groups by matching those who did or did not receive the treatment on this propensity score. This strategy has been shown to yield consistent and unbiased estimates of causal effects, as long as all potential confounding factors are included in the model used to create the propensity score. The surprising outcome is that matching on the propensity score fully balances the treatment and control groups on all of the covariates used in modeling the propensity of receiving the treatment, allowing the identification of the causal effect by Yt – Y.llc

In a recent article, we applied this model, but because of space constraints we note here just the basic results (Sampson, Laub, and Wimer 2005). Our essential strategy was to exploit the rich individual baseline data and time-varying covariates over the full life course to model the propensity to marriage. Rather than control for the proverbial "kitchen sink" in estimating crime, the inverse proportional treatment weighting (IPTW) method forces conceptual clarity in the sense of distinguishing between pretreatment confounders and posttreatment outcomes. From IQ to the cumulative history of both the outcome and treatment itself, we accounted for twenty baseline covariates and approximately a dozen time-varying confounders measured from widely varying sources—many of which predict the course of marriage as theoretically expected. For example, all the childhood and family adversity risk factors noted earlier were considered as baseline (pre-first marriage) covariates, and employment, military service, offspring, and crime itself were modeled as time-varying covariates (cumulative history up to the year before a marriage observation).

To give an example, married men who had a high probability of being married at any given age based on their marital, criminal, employment, military, and offspring history were effectively "downweighted" in the IPTW analysis for that year. Such person-periods reflect a higher degree of "selection" into the observed treatment status given values on confounding covariate histories that make them especially likely to be married (or unmarried). As a result, we do not want them to contribute as much information to the estimation of the causal effect of marriage on crime. On the other hand, married men with low probabilities of being married (but who actually marry) at a given age based on the same histories provide more information, and they are therefore "upweighted" when estimating the final causal effect.

Applying this counterfactual modeling strategy that weights observations by the inverse probability of men being in the state of marriage as predicted by observed covariates and prior treatment history, we found that being married is associated with a 35 percent average reduction in the probability of crime for our sample of fifty-two men assessed from ages seventeen to seventy. This finding was maintained for our full sample of nearly five hundred men examined from ages seventeen to thirty-two. Thus, we view this basic finding as robust and consistent with the notion that marriage causally inhibits crime over the life course. Given the extensive list of baseline and cumulative history covariates, omitted confounders would have to be implausibly large to overturn the basic results we obtained under a number of different model specifications and assumptions.

In sum, in our revised framework we see marriage not as a singular turning point but as a potential causal force in desistance that operates as a dynamic, time-varying process through time. Changes in crime *or* marriage can happen in any year, and the explicit point of the counterfactual model is to estimate these associations with the cumulative history of both outcome and covariates explicitly controlled. Given the nature of the results, we raise the question whether the metaphor of development is the proper one when it comes to understanding time-varying turning points over the adult life course. We return to this issue in the conclusion.

Reflections on the Importance of Agency and Choice

A vital feature that emerged from our life-history narratives was the role of human agency—the purposeful execution of choice and individual will—in the process of desisting from crime. As a result, the men we studied were active participants in the process of going straight. We discovered that personal conceptions about the past and future were often transformed as men maneuvered through the transition from adolescence to adulthood. Cohler (1982) has noted that a subjective reconstruction of self is especially likely at times of transition. Many men engaged in "transformative action" in the desistance process. Although informed by the past, agency points toward the future (and hence a future self). Projective actions in the transition from adolescence to adulthood that we uncovered were the advancement of a new sense of self and identity as a desister from crime or, perhaps more aptly, as a family man, hard worker, and good provider.

It also appears that human agency is vitally important for understanding persistent offending over the life course. Some men simply insist on a criminal lifestyle, not out of impulsivity or lack of knowledge of future consequences, but rather because of the rewards of crime itself or a willful resistance to perceived domination. Persistent offenders knowingly engage in these activities at the expense of a future self. As revealed in many of our life-history narratives, crime was viewed as attractive, exciting, and seductive despite the future pains usually called forth as a result. Calculated and articulated resistance to authority was a recurrent theme in lives of persistent offenders. The men's defiance seemed to have been fueled by a perceived sense of injustice resulting from a pattern of corrosive contacts with officials of the criminal justice system, coupled with a general sense of working-class alienation from elite society. Many persistent offenders see "the system" (criminal justice and work alike) as unfair and corrupt.

In crucial ways, then, persistent crime is more than a weakening of social bonds, and desistance is more than the presence of a social bond, as one might be led to conclude (mistakenly) from *Crime in the Making*. At a meta-theoretical level, our long-term follow-up data direct us to insist that a focus purely on institutional, or structural, turning points and opportunities is incomplete, for such opportunities are mediated by perceptions and human decision making. The process of desistance is complex, and many men made a commitment to go straight without even realizing it. Before they knew it, they had invested so much in a marriage or a job that they did not want to risk losing their investment. Drawing on the work of Becker (1960), this is what we call "desistance by default" (Laub and Sampson 2003). Even if below the surface of active consciousness, actions to desist are in a fundamental sense willed by the offender, bringing a richer meaning to the notion of commitment. Further support for this idea is that the men who desisted from crime, but even those who persisted, accepted responsibility for their actions and freely admitted getting into trouble. They did not, for the most part, offer excuses. Tough times due to the Great Depression, uncaring parents, poor schools, discrimination based on ethnicity and class, and the like were not invoked to explain their criminal pasts. One man captured this opinion the best when he was asked to assess his life and said, "Not because of my mother and father. Because of me. I'm the one that made it shitty."

In ongoing work, we make what we believe is a crucial distinction between human agency and rational choice, one that runs opposite to the recent claim by Paternoster and Bushway (2004, 1) that "if you believe in agency you need to adopt a rational choice perspective." From a rational choice perspective, agency is a matter of preferences (e.g., attitudes toward time and attitudes toward others) and how preferences can be used to change or modify inputs or exogenous events like employment and marriage. In our view, the rational choice approach views agency as a static entity representing the stable part of the person as well as within-individual variation over time that is largely driven by age. What is lacking in rational choice is the recognition that we "construct our preferences. We choose preferences and actions jointly, in part, to discover—or construct—new preferences that are currently unknown" (March 1978, 596). At this time, we know little about how preferences are formed. It is thus not surprising that Hechter and Kanazawa (1997, 195) concluded that "the mechanisms for individual action in rational choice theory are descriptively problematic." Perhaps more important, we argue that human agency cannot be divorced from the situation or context, once again making choice situated or relational rather than a property of the person or even the environment; agency is constitutive of both.

In short, human beings make choices to participate in crime or not, and life-course criminology has been remiss to have left agency—which is essentially human social action—largely out of the theoretical picture. We seek to reposition human agency as a central element in understanding crime and deviance over the life course. To be sure, *Shared Beginnings* is an incomplete response, for we did not develop an explicit theory of human agency replete with testable causal hypotheses. Our theoretical claim here is simply that the data make clear that agency is a crucial ingredient in causation and thus will be a first-order challenge for future work in life-course criminology.

Implications for Developmental (Life-Course?) Criminology

We close by considering the implications of our analyses of group-based theories of crime, turning points, and human agency for a broader understanding of human development over the life course—issues that are at the very heart of developmental criminology. Relying on what Wordsworth argued was a central insight from Shakespeare—that the child is father to the man—criminologists have addressed in intense fashion how developmental processes are linked to the onset, continuation, and cessation of criminal and antisocial behavior. Much has been learned, and it is fair to say that developmental criminology is now ascendant.

In our view, however, the meaning of development in developmental criminology remains fuzzy and has not been subjected to theoretical interrogation. The biologist Richard Lewontin (2000, 5) has argued that "the term *development* is a metaphor that carries with it a prior commitment to the nature of the process." Using the analogy of a photographic image, Lewontin argues that the way the term *development* is used implies a process that makes the latent image apparent. From our perspective, this seems to be what much of developmental criminological theory is all about, that is, offering a perspective wherein the environment offers a "set of enabling conditions" that allow individual traits to express themselves over time. Although reciprocal interactions with the environment are often mentioned, the typical working assumption seems to be that offenders are following a preprogrammed line of development in a crucial respect—an unwinding, an unfolding, or an unrolling of what is fundamentally "already there." The underlying view of development as a predetermined unfolding is ultimately linked to a typological understanding of the world—different internal programs will have different outcomes for individuals of a different type.

Debates about development in the social sciences are not new, and we are not saying that development reduces to biological processes only. Still, while most developmentalists allude to social interactions as real, in the end most embrace a focus that emphasizes the primacy of early childhood attributes that are presumed to be stable over the life course in a between-individual sense. How else can we understand the fixation on the prediction of later crime from childhood characteristics? It is indisputable that throughout the history of criminology, one of the dominant themes is past as prologue. This continues and finds full expression in the area of addiction research, where we seem to have come full circle from the crude biology of Cesare Lombroso to the current fascination with DNA sequencing and brain imaging as the promise of the future.

In our life-course theory of crime, we seek to return development to where it probably should have been all along, conceived as the constant interaction between individuals and their environment, coupled with purposeful human agency and "random developmental noise" (Lewontin 2000, 35–36). According to Elder (1998), human agency is one of the key principles of the life-course perspective. The principle states that "individuals construct their own life course through the choices and actions they take within the opportunities and constraints of history and social circumstances" (p. 4). The recognition of developmental noise implies that "the organism is determined neither by its genes nor by its environment *nor even by interaction between them*, but bears a significant mark of random processes" (Lewontin 2000, 38, italics added). The challenge is that human agency and random processes are ever-present realities, making prediction once again problematic. It further follows that long-term patterns of offending among high-risk populations cannot be divined by individual differences (for example, low verbal IQ, temperament), childhood behavior (for example, early onset of misbehavior), or even adolescent characteristics (for example, chronic juvenile offending).

A key difference between the present life-course perspective and most developmental criminology can be clarified by asking what would happen in an imagined world of perfect measurement. Even if *all* risk factors (including social controls!) were measured without error, our framework posits the continuous influence of human agency and randomizing events, leading again to heterogeneity in outcomes, emergent processes, and a lack of causal prediction. The logic of prediction that drives the search for early risk factors takes nearly the opposite view. Indeed, one gets the sense from "early interveners" that it is just a matter of time before risk factors are measured well enough (from the human genome?) that the false positive problem will finally become ancient history. From the perspective of our theory, this is simply wishful thinking, and we instead predict continued heterogeneity in criminal offending over the life course no matter what the childhood classification scheme of the future. Some "destined" offenders will always start late or refrain from crime altogether, whereas some "innocents" will always start early and continue for long periods of time. And a sizable portion of the offending population will always display a zigzag pattern of offending over long time periods.

Concluding Thoughts

We view this article and our larger project as offering a dual critique of social science theory and current policy about crime over the life course. Developmentalists seem to believe that childhood and adolescent risk characteristics are what *really* matter—witness the undeniable rise and dominance of the "early risk-factor" paradigm. Our work simply pleads for balance in the other direction, but this move in no way denies the reality of the stability of individual differences.

Not to be overlooked and equally important, our work is inherently critical of "structuralist" approaches in sociological criminology wherein it is argued that location in the social structure, namely, poverty and social class, are what really matter. We hardly believe that all bad actors would simply desist from crime if they were given jobs. Pure deprivation or materialist theories are not just antediluvian but wrong by offenders' own accounts. Our recent work even questions the idea that some inferred from *Crime in the Making*—that institutional turning points are purely exogenous events that act on individuals. The men we studied in *Shared Beginnings, Divergent Lives* were not blank slates any more than they were rational actors in an unconstrained market of life chances. They were active participants in constructing their lives—including turning points. We were thus compelled to take seriously purposeful human action under conditions of constraint. At the same time, we did see evidence that certain institutions, such as marriage, predicted crime even when each man served as his own control.

How can these seemingly opposite views be reconciled? Although not readily apparent at first glance, we believe the concept of *emergence* unifies the three themes of this article. By studying the group question, we learned that long-term outcomes cannot be easily predicted. By emphasizing time-varying events, we learned that stability and change do not neatly fit a simple linear "growth" model of development. By listening and taking seriously what the Glueck men told us about their lives, in their own words, we learned that human agency is an important element in constructing trajectories over the life course. Each theme shares in common the idea of criminal behavior as a socially emergent and contextually shaped property.

From our perspective, the implied next step is to reconcile the idea of choice or agency with a structural notion of turning points. We refer to this idea as "situated choice" (Laub and Sampson 2003, 281–82; 2005). As Abbott (1997, 102) has written, "A major turning point has the potential to open a system the way a key has the potential to open a lock . . . action is necessary to complete the turning." In this instance, individual action needs to align with the social structure to produce behavioral change and to maintain change (or stability) over the life course. Choice alone without structures of support, or the offering of support alone absent a decision to desist, however inchoate, seems destined to fail. Thus, neither agency nor structural location can by itself explain the life course of crime. Studying them simultaneously permits discovery of the emergent ways that turning points across the adult life course align with purposive actions and, yes, stable individual differences.

References

Abbott, A. (1997). On the concept of turning point. *Comparative Social Research, 16,* 85–105.

Becker, H. (1960). Notes on the concept of commitment. *American Journal of Sociology, 66,* 32–40.

Cohler, B. (1982). Personal narrative and life course. In P. Baltes & O. Brim Jr. (Ed.), *Life span development and behavior* (Vol. 4, pp. 205–241). New York: Academic Press.

Elder, G. (1998). The life course as developmental theory. *Child Development, 69,* 1–12.

Glueck, S., & Glueck, E. (1950). *Unraveling juvenile delinquency.* New York: The Commonwealth Fund.

Gottfredson, M., & Hirschi, T. (1990). *A general theory of crime.* Stanford, CA: Stanford University Press.

Hechter, M. & S. Kanazawa (1997). Sociological rational choice theory. *Annual Review of Sociology, 23,* 191–214.

Hill, T. (1971). From hell-raiser to family man. In J. P. Spradley & D. W. McCurdy (Eds.), *Conformity and conflict: Readings in cultural anthropology* (pp. 186–200). Boston: Little, Brown.

Hirschi, T. (1969). *The causes of delinquency.* Berkeley: University of California Press.

Laub, J., & Sampson, R. (1993). Turning points in the life course: Why change matters to the study of crime. *Criminology, 31,* 301–325.

Laub, J., & Sampson, R. (2003). *Shared beginnings, divergent lives: Delinquent boys to age 70.* Cambridge, MA: Harvard University Press.

Laub, J., & Sampson, R. (2005). *Human agency in the criminal careers of 500 Boston men, circa 1925–1995.* Paper presented at the workshop on Agency and Human Development under Conditions of Social Change, Jena, Germany, June 4–6.

Lewontin, R. (2000). *The triple helix: Gene, organism, and environment.* Cambridge, MA: Harvard University Press.

March, J. (1978). Bounded rationality, ambiguity, and the engineering of choice. *Bell Journal of Economics, 9,* 587–608.

Paternoster, R., & Bushway, S. (2004). *Rational choice, personal agency, and us.* Paper presented at the annual meeting of the American Society of Criminology, Nashville, TN.

Sampson, R., & Laub, J. (1993). *Crime in the making: Pathways and turning points through life.* Cambridge, MA: Harvard University Press.

Sampson, R., & Laub, J. (2003). Life-course desisters? Trajectories of crime among delinquent boys followed to age 70. *Criminology, 41,* 301–339.

Sampson, R., Laub, J., & Wimer, C. (2005). *Assessing causal effects of marriage on crime: Within-individual change over the life course.* Manuscript, Cambridge, MA: Harvard University.

DISCUSSION QUESTIONS

1. Discuss how Sampson and Laub might be giving short shrift to individual differences such as self-control in accounting for variation in continuity and desistence of criminal activity.
2. Discuss the differences between Sampson and Laub's and rational choices theorists' conceptions of human agency.
3. Sampson and Laub claim that "Pure deprivation or materialist theories are not just antediluvian but wrong by offenders' own accounts"; how might those who claim that poverty is a major cause of crime respond to this claim?

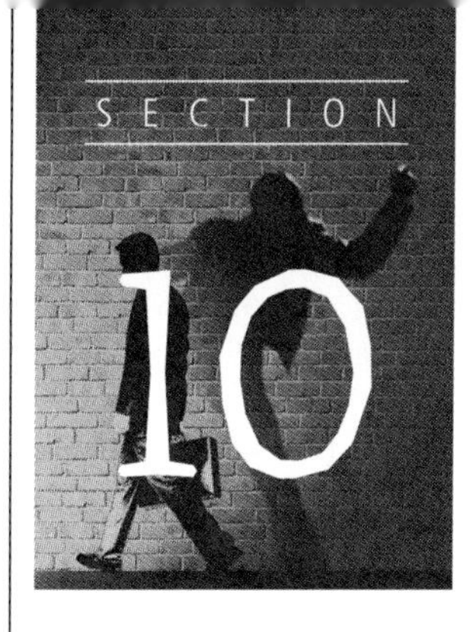

VIOLENT CRIMES

On April 20, 1999, the 110th anniversary of Adolph Hitler's birthday, Columbine High School, located in the small, white middle-class town of Littleton, Colorado, became the site of one of the worst cases of teen terrorism in U.S. history. Eric Harris, 18, and Dylan Klebold, 17, entered the school at 11:35 that morning and began their murderous rampage through the halls and corridors that lasted for 55 minutes. During this period they shot 13 people dead and seriously wounded 25 others. Harris and Klebold hated the world, but they singled out athletes and religious kids as special targets. After running out of targets, they committed suicide as the police closed in.

What made them worshipers of Hitler, guns, and violence? Some accounts of their lives paint them as nerds picked on by jocks, others that they were bullies with a history of threatening violence and of trouble with the police. Diaries and computer Web sites revealed that they had planned the attack to celebrate Hitler's birthday a full year before it occurred. In the meantime, they amassed an arsenal of weapons including handguns, shotguns, a 9mm carbine, and a number of homemade bombs, which they had learned to make on the Internet. They were clearly deeply disturbed young men with a dark fascination with violence.

The potential for violence is in us all, waiting to be ignited by environmental sparks. The vast majority of us will never come into close contact with such sparks, even though they are all around us. We have to ask ourselves what kind of a society we are that glorifies violence in its entertainment, that allows young boys to gain access to deadly weapons and to instructions on how to build bombs, and that does not provide its young with meaningful moral lessons and activities to fill their time. These are among the environmental sparks that should be kept well clear of combustible material like Harris and Klebold.

The history of humankind is a violent one. In 16th-century Rome, "crimes of violence were innumerable. Assassins could be bought almost as cheaply as indulgences. Everyone had a dagger, and brewers of poison found many customers; at last the people of Rome could hardly believe in the natural death of any man of prominence or wealth" (Durant, 1953, p. 590). London in the 14th century had a murder rate of 44 per 100,000 (Hanawalt, 1979, p. 98), which is greater than the murder rate of almost any of American cities today. In Nuremberg, Germany, the murder rate fluctuated between 20 and 65 per 100,000 from 1292 to 1392 (it was only 4.7 for the same city in 1984; Schussler, 1992, pp. 3–4). These high rates in medieval Europe had a lot to do with the combination of the habit of bearing arms, alcohol-induced quarrels, the absence of effective medical treatment for wounds, the absence of effective police power, and the absence of a trusted system of justice.

Murder

Violent crimes are the crimes the public fear most, with **murder**, "the willful (non-negligent) killing of one human being by another" (FBI, 2006, p. 15), being the most feared and most serious. There were 16,692 murders in the United States in 2005, a rate of 5.6 per 100,000, which is almost half of the all-time high rate of 10.2 in 1980. Gary, Indiana (58) had the highest murder rate followed by Detroit (39.3). For known offenders, 90.1% were males, 52.6% were black, 45% white, and 2.4% were of other races. About 90% of murders were intraracial and intrasexual (males killing males).

Figure 10.1 documents the murder rates in the United States during the 20th century and indicates how wildly they fluctuate from year to year. One of the biggest factors in the recent drop in homicide is medical and technological improvements. Cell phones for reporting incidents are everywhere, and emergency medical technicians are alerted and dispatched swiftly, and once hospitalized, victims have the benefit of all that medicine learned about treating violent traumas since the Vietnam War. It is estimated that we would experience five times the murder rate today if medicine and technology were at the same level that they were in 1960 (Harris, Thomas, Fisher, & Hirsch., 2002).

As high as the U.S. murder rate is, it is much lower than in many other countries. According to the World Health Organization's (Krug, Dahlberg, Mercy, Zwi, & Lozano, 2002) homicide data for "youths" 10 to 29 years of age, the United States is about in the middle of the pack, with a rate of 11 homicides per 100,000. Colombia had the highest rate (84.4) and Japan had the lowest (0.4). Countries with low murder rates tend to be politically stable and wealthy democracies; countries with high rates tend to be third-world or developing countries or countries experiencing rapid social and economic changes, such as Russia.

In the United States, young people ages 18 through 24 are most likely to be killed. Young black males are about nine times more likely to be murdered than young white males, and young black females are about six times more likely to be murdered than young white females (Walsh & Ellis, 2007). Males are far more likely to murder than females. When females kill males, it is typically a spouse, ex-spouse, or boyfriend in a self-defense situation (Mann, 1990). Female/female murder is very rare around the world. Daly and Wilson (2000, p. 16) examined data on same-sex non-relative murders from a number of cultures and found that female/female homicide constituted only 2.5% of the total number of murders. Even going back to England in the 13th century, female/female murder accounted for only 4.9% of the total murders (Given, 1977).

Figure 10.1 Murder Rates in the United States in the Twentieth Century

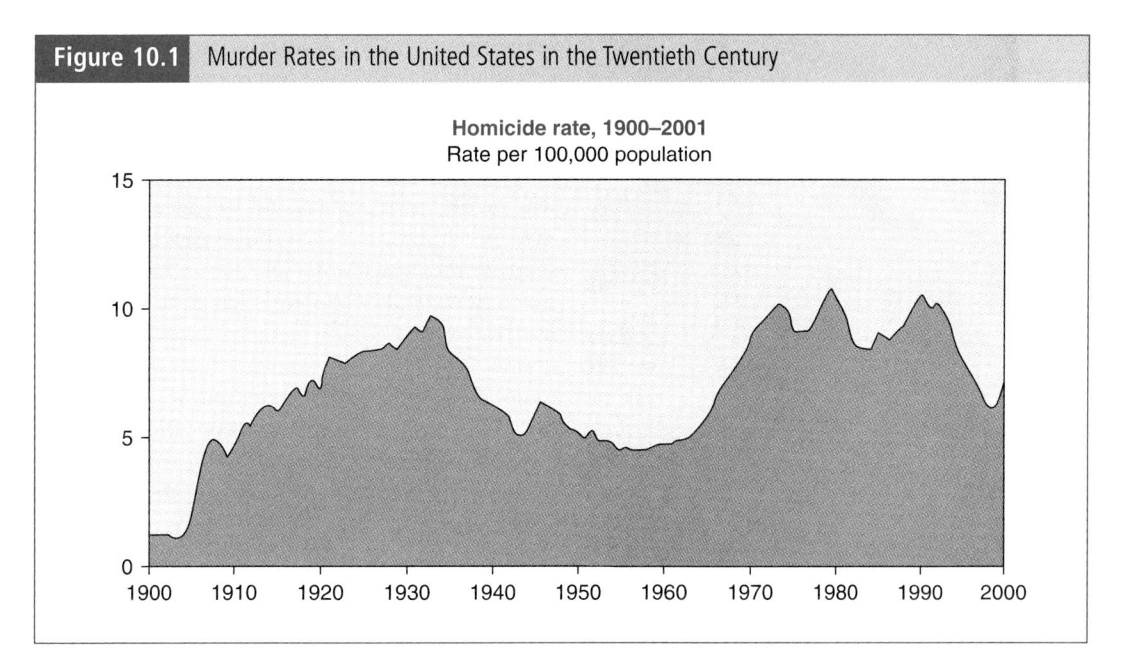

SOURCE: National Center for Health Statistics, Vital Statistics, 2002.

NOTE: The 2001 rate includes deaths attributed to the 9/11 terrorism attacks.

Forcible Rape

Forcible rape is "the carnal knowledge of a female forcibly and against her will" (FBI, 2006, p. 27). This definition includes attempts to commit rape but excludes statutory rape (consensual sex with an underage female) and the rape of males. Rapes of males are classified as either assaults or other sex offenses, depending on the circumstances. According to the 2006 UCR, there were 93,934 reported rapes in 2005, a rate of 62.5 per 100,000 females. By race, 65.3% of those arrested for rape were white, 32.2% black, and the remaining 2.5% were other races. Poor, young, unmarried, non-white females are disproportionately likely to be victimized, and poor, young, unmarried, non-white males are disproportionately likely to be perpetrators (Catalano, 2006).

Some facts about rape from 20 years of victimization surveys on rape:

- About half of all rapes are committed by someone known to the victim.
- The offender was armed in about 20% of the cases. Stranger rapists were more likely to be armed (29%) than were rapists known to the victim.
- Among the victims who fought their attackers or yelled and screamed, more reported that it helped the situation rather than made it worse.
- Slightly more than half of the victims report the assault to the police. Victims are more likely to report the incident if the perpetrator was armed or if they sustained physical injuries (in Ellis & Walsh, 2000).

There is a variety of theories about the causes of rape, with different assumptions. Feminist theories of rape assert that rape is learned and is motivated by power rather than

▲ **Photo 10.1** Some males excuse their inappropriate behavior with women by adopting rape myths. For example, girls who dress provocatively or consume significant amounts of alcohol are sometimes considered to be "asking" to be victimized. Even the police and courts appear to be less sympathetic toward sex crime victims who do not take "all precautions" to avoid situations in which males might assume a female is not really saying "No!"

sexual desire and that all men are capable of it (Brownmiller, 1975). Social learning theorists agree with these assumptions, except that they view rapists as a psychologically unhealthy category of males, not as "normal men." Therapists who work with rapists tend to agree that every man is a potential rapist, but assert that rape is sexually motivated (Mealey, 2003). Evolutionary psychologists also contend that rape is sexually motivated and that it is a maladaptive consequence of a generally adaptive behavior; that is, males seeking as many sexual partners as possible (Thornhill & Palmer, 2000). Other theories have been offered that combine the most empirically supported assumptions of the feminist, social learning, and evolutionary approaches (Ellis, 1991).

Rape victimization produces feelings of depression, guilt, self-blame, lowered self-esteem, shock, anger, and depression, which sometimes translate into *rape trauma syndrome.* This syndrome is similar to post-traumatic stress syndrome (re-experiencing the event via "flashbacks," avoiding anything at all associated with the event, and a general numbness of affect) often suffered by those who have experienced the horrors of war (van Berlo & Ensink, 2000). Victimization may also negatively change victims' perceptions of other people from sources of support to sources of threat.

Robbery

Robbery is "the taking or attempted taking of anything of value from the care, custody, or control of a person or persons by force or threat of force or violence and/or putting the victim in fear" (FBI, 2006, p. 31). In 2005, there were 417,122 reported robberies in the United States, a rate of 150.8 per 100,000, down considerably from the peak rate of 272.7 in 1991. Victims were murdered in 921 of the robberies. Of those robbers arrested, 60.1% were under 25 years of age, and 89.0% were male. By race, 53.6% were black, 43.9 white, and the remaining 1.5% were of other races.

The danger posed by resisting victims, and the severe penalties attached to committing robbery, make it a high-risk crime, which suggests that those who commit it may be among the most daring and dangerous of all criminals. Interviews of active street robbers (Wright & Decker, 1997) reveal them to be the least educated, most fearless, most impulsive, and most hedonistic of criminals. Obtaining legitimate work is simply not an option that robbers

◀ **Photo 10.2** Gang members brandish a handgun. Unlike the days of *West Side Story*, when gangs fought with knives and pipes, from the mid-1980s, urban youth gangs started to carry both handguns and automatic weapons. Partly, this was to protect the drug distribution networks these gangs became part of at that time.

entertain because work would seriously interfere with their "every night is Saturday night" lifestyles.

Robbery, and flaunting the material trappings signaling its successful pursuit, is seen ultimately as a campaign for respect and status in the street culture in which most robbery specialists participate. As James Messerschmidt (1993) puts it,

> The robbery setting provides the ideal opportunity to construct an "essential" toughness and "maleness"; it provides a means with which to construct that certain type of masculinity—hardman. Within the social context that ghetto and barrio males find themselves, then, robbery is a rational practice for "doing gender" and for getting money. (p. 107)

With the exception of rape, robbery is the most "male" of all crimes, but women do engage in it. A favorite ploy for female robbers is to appear sexually available (prostitution or otherwise) to a male victim and then, either alone or with the help of an accomplice, rob him. Female robbers will seldom rob males without an accomplice, but will practice their "art" mostly on other females. Much like their male counterparts, female robbers eschew legitimate work and prefer their hedonistic "money for nothing" lifestyles (Miller, 1998).

The article by Bruce Jacobs and Richard Wright in this section tells the stories of active street robbers in their own words. Their interviewees view robbery as the quickest and safest way to gain quick cash. Burglary takes time and requires the burglar to find buyers for stolen property, and burglars never know who or what they might run into inside a house. Drug selling means dealing with a lot of people, the risk of being robbed oneself, and most importantly, coping with the temptation of being one's own best customer. Robbery, on the other hand, allows the robber to pick the time, place, and victim at leisure, and then complete the job in a matter of minutes and seconds. It is the perfect crime for those with a pressing and constant need for fast cash to feed a hedonistic lifestyle and who enjoy the rush that the crime affords them. Robbery, and flaunting the material trappings signaling its successful pursuit, is seen ultimately as a campaign for respect and status in the street culture in which robbery specialists participate.

Aggravated Assault

Aggravated assault is "an unlawful attack by one person upon another for the purpose of inflicting severe or aggravated bodily injury" (FBI, 2006, p. 37). As opposed to simple assault, aggravated assault is an assault in which a weapon such as a knife or gun is used, although sometimes the use of hands and feet can result in a charge of aggravated assault. Aggravated assault is the most common felony violent crime, accounting for 62.5% of all Part I violent crimes in 2004. Each incident of aggravated assault carries the potential threat of becoming a murder because without the speedy access to modern medicine we enjoy today, many aggravated assaults would have turned into murders.

There were 862,947 aggravated assaults reported in 2006 (FBI, 2006). This rate of 291.1 per 100,000 is a 34% drop from the all-time high of 441.8 in 1992. Personal weapons were used in 26.6% of the cases, firearms in 19.3%, knives or other cutting instruments in 18.6%, and the remaining 35.6% were other miscellaneous weapons. About 40% of those arrested for aggravated assault were under the age of 25, 79.3% were male, 64.5% were white, 33.1% were black, and 2.3% were other races.

Mass, Spree, and Serial Murder

In April of 1973, Edmund Kemper, a six-foot nine-inch, 300-pound 25-year-old hate machine, crept into his mother's bedroom and bludgeoned her to death, decapitated her, had sex with her headless body, and played darts with the head. He then invited his mother's best friend over for a "surprise" dinner to honor his mother and bludgeoned and decapitated her also. He killed at least six other women, whom he also decapitated, and sexually assaulted their headless corpses. He even ate the flesh of some of his victims, cooking it into a macaroni casserole. But Kemper's biggest thrill was not sex, murder, or cannibalism; it was decapitation. In Kemper's own words: "You hear that little pop and pull their heads off and hold their heads up by the hair. Whipping their heads off, their body sitting there. That'd get me off" (in Leyton, 1986, p. 42).

Kemper's behavior belongs to a class of violent behavior we find extremely difficult to understand. We can all imagine circumstances in which we might act violently, and even kill someone, but few of us can imagine ourselves becoming mass, spree, or serial killers. **Mass murder** is the killing of several people at one location that begins and ends within a few minutes or hours. **Spree murder** is the killing of several people at different locations over a period of several days. Research suggests that the time frame involved is the only factor that differentiates mass and spree killers and that both mass and spree killers are different from the serial killer.

Mass murderers are divided into two types: (1) Those who chose specific targets who the killers believe to have caused them stress (e.g., disgruntled workers attacking supervisors), and (2) those who attack targets having no connection with the killer but who belong to groups the killer dislikes. For instance, Marc Lepine killed 14 women and wounded 13 others in an engineering classroom at the University of Montreal in 1989 after ordering all the men out. Most mass murderers are motivated by a hatred that simmers until some specific incident provides the flame that brings it to a boil. Lepine, for instance, had been denied admission to the engineering school and sought revenge on the women who had taken his place in a "man's profession" (Fox & Levin, 2001, p. 119).

◀ **Photo 10.3** Spree killer or serial killer? Lee Malvo leaves court after being convicted of sniper killings in and around the Washington, D.C., region. Ten people were killed and three others critically injured by Malvo and his associate over a several-week period.

Spree killers move from victim to victim in fairly rapid succession, and like mass murderers make little effort to hide their activities or avoid detection, as if driven by some frenzied compulsion. Spree killing is rare, but spree-killing teams are even rarer and are typically composed of a dominant leader and a submissive lover. The most recent spree killer team is the sniper team of John Muhammad and Lee Malvo, who killed 13 random individuals and wounded six others in the Washington, DC area in 2002. Although the team killed without regard to race or gender, Muhammad belonged to the Nation of Islam, which may have provided ideological impetus to his vicious spree (Hurd, 2003).

Serial murder is the killing of three or more victims over an extended period of time (Keeney & Heide, 1995). In addition to this time-frame distinction, there are several generalities that differentiate the serial killer from spree and mass killers, as well as from the "typical" (non-multiple) murderer (Hickey, 2006):

- Whereas spree and mass killers almost invariably use firearms and kill their victims quickly, serial killers prefer to stalk and often torture their victims.
- The typical serial killer is a white male who is older than the typical murderer.
- African Americans are overrepresented among known serial killers and Asian Americans and women of all races are greatly underrepresented.
- Researchers have identified four general types of serial killers: visionary, mission-oriented, hedonistic, and power/control.
- Since 1977, the prevalence of serial murder has been more than twice that of spree or mass murders and has accounted for more than twice the number of victims.

Although the typical serial killer is a white male, African Americans are overrepresented among serial killers relative to their proportion of the population. Hickey (2006, p. 143) claims that about 44% of the serial killers operating between 1995 and 2004 have been black. The only *known* Asian American serial killer operating in the United States during the 20th century is Charles Ng. Ng and his white partner, Leonard Lake, tortured and killed at least 19 people in the early 1980s. Ng was sentenced to death in 1999 (Newton, 2000).

There have been female serial killers, such as nursing home proprietor Amy Archer-Gilligan, who may have murdered up to 100 patients in her charge between 1907 and 1914 (Newton, 2000). Almost all female serial killers, including Archer-Gilligan, kill for instrumental (profit) or affective (emotional) reasons, such as "mercy" killings. The key distinction between male and female serial killers may be that, "There are no female counterparts to a Bundy or a Gacy, to whom sex or sexual violence is a part of the murder pattern" (Segrave, 1992, p. 5).

No one knows exactly how many serial killers there are at any given time or how many murders per year are attributable them. Figure 10.2 is simply an estimate of numbers and rates of serial killers in the United States from 1795 to mid-2004.

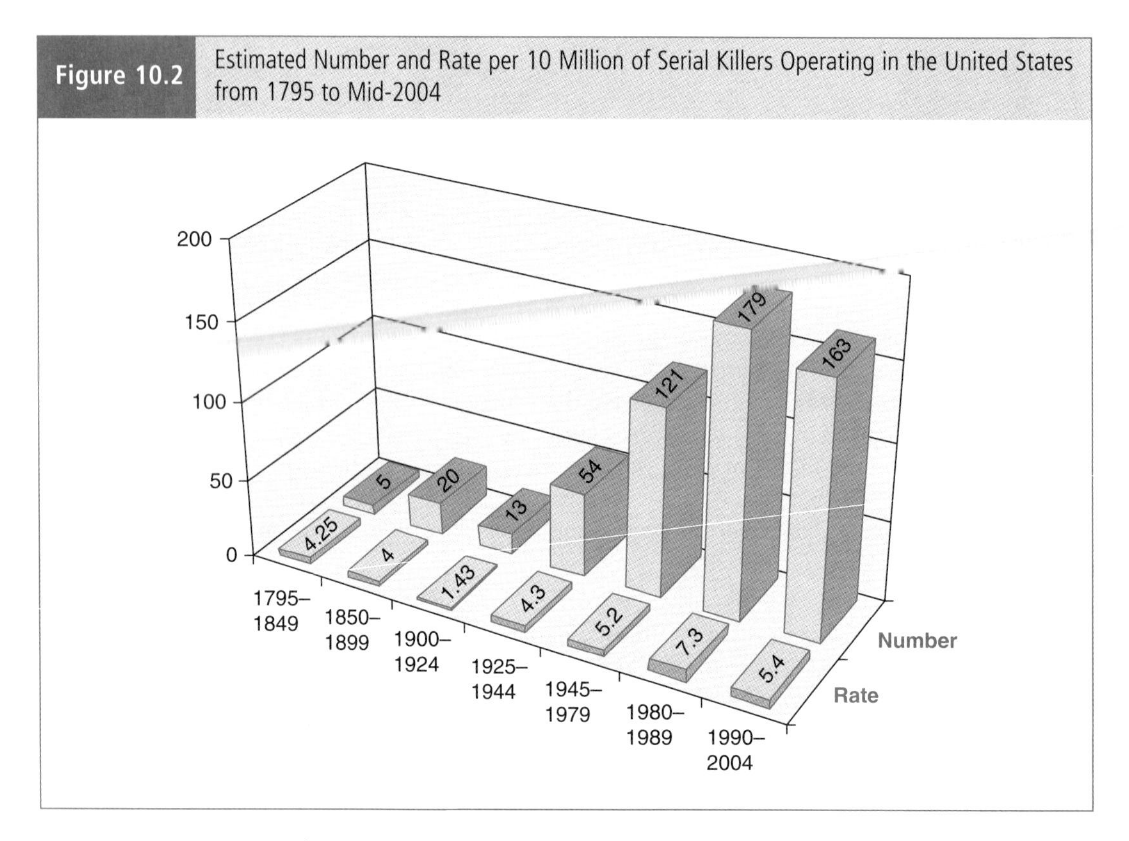

Figure 10.2 Estimated Number and Rate per 10 Million of Serial Killers Operating in the United States from 1795 to Mid-2004

SOURCES: U.S. Justice Department figures as reported by Jenkins (1994); updated figures from Hickey (2006) and Walsh (2005). Rates per 10 million population computed by authors.

A Typology of Serial Killers

Holmes and DeBurger (1998) have provided a typology that divides serial killers into four broad types: *visionary, mission-oriented, hedonistic, and power/control.* These are not definitive categories; many serial killers may evidence aspects of all types at various times.

- **Visionary serial killers** are typically out of touch with reality, may be psychotic or schizophrenic, and feel impelled to commit murder by visions or "voices in my head."
- **Mission-oriented serial killers** feel it is their mission in life to kill certain kinds of people such as prostitutes and homosexuals.
- **Hedonistic serial killers** are the majority of serial killers. They kill for the pure thrill and joy of it, engaging in cruel and perverted sexual activity.
- **Power/control serial killers** gain more satisfaction from exercising complete power over their victims rather than from "bloodlust," although sexual activity is almost always involved.

What Causes Serial Killing?

Becoming a serial killer is a long, drawn-out process, not a discrete event. A theory that has attempted to integrate cultural, developmental, psychological, and biological concepts is Stephen Giannangelo's *diathesis-stress* model (1996). The theory states that all serial killers have a congenital propensity to behave and think in ways that lead to serial killing, if combined with environmental stressors. This combination leads to the development of self-esteem, self-control, and sexual dysfunction problems. These problems feed back on one another and lead to the development of maladaptive social skills, which moves the person to retreat into his private pornographic fantasy world. As he dwells longer and longer in this world, he enters a dissociative process in which he takes his fantasies to their moral limits. At this point, the killer seeks out victims to act out his fantasies, but the actual kill never lives up to his expectations or to the thrill of the hunt, so the whole process is repeated and becomes obsessive-compulsive and ritualistic.

The article by Anthony Walsh in this section examines why African American serial killers do not get the attention that white serial killers do. Walsh indicates that the media gives the impression that only white males commit serial murder, and this becomes a "fact" in the minds of the public. His sample of 413 serial killers operating from 1945 to mid-2004 in the United States, however, found that African Americans constituted 22% of the total. Because blacks have constituted about 10% of the American population averaged over that period, they are represented in the population more than twice as often as expected from their proportion of the population.

Terrorism

On the morning of September 11, 2001, Americans woke to horrifying images that are seared into their memories forever. Nineteen Islamic terrorists led by Mohamed Atta had hijacked four airliners and used them in coordinated attacks against symbols of America's financial and military might. At 8:45 a.m., American Airlines flight 11, with 92 people on board, crashed into the north tower of the World Trade Center. Eighteen minutes later, United flight 175, with 64 people aboard, smashed into the south tower. At 9:40, American Airlines flight 77, carrying 64 people, crashed into the Pentagon. Then at 10 a.m., United flight 93, carrying 45 people, crashed into a Pennsylvania field, having been prevented from accomplishing its mission (apparently to destroy the Capitol Building or the White House) by the courageous actions of its passengers. These actions cost the lives of close to 3,000 people from 78 different countries, making it the deadliest terrorist attack in history anywhere (U.S. Department of State, 2004). What were these people trying to accomplish by such a wanton act, and what drove them to sacrifice their own lives in the process?

The FBI defines **terrorism** as "the unlawful use of force or violence against persons or property to intimidate or coerce a government, the civilian population, or any segment thereof, in furtherance of political or social goals" (Smith, 1994, p. 8). It is estimated that up until 1995, terrorists cost the lives of 500,000 people in the 20th century (Rummel, 1992), with another 1,269 killed in the last five years of the century (U.S. Department of State, 2004). Terrorism has a long history; it is "as old as the human discovery that people can be influenced by intimidation" (Hacker, 1977, p. ix). Terrorism is a tactic used to influence the behavior of others through intimidation, although terrorists typically appeal to a higher moral "good," such as ethnic autonomy or some religious or political dogma, to justify the killing of innocents. Nevertheless, they strike at innocents because the very essence of terrorism is public intimidation, and the randomness of terrorist action accomplishes this better than targeting specific individuals would. Victims are incidental to the aims of terrorists; they are simply instruments in the objectives of (1) publicizing the terrorists' cause, (2) instilling in the general public a sense of personal vulnerability, and (3) provoking a government into unleashing repressive social control measures that may cost it public support (Simonson & Spindlove, 2004).

▲ **Photo 10.4** The September 11, 2001 attack on New York's Twin Towers and the Pentagon, carried out by al-Qaeda operatives, stunned the world. Ultimately, it resulted in passage of a comprehensive Patriot Act, the creation of the Department of Homeland Security, and a war with Iraq.

While terrorist violence is immoral, it is not "senseless" because it has an ultimate purpose: Evil means are justified by the ends they seek. The terrorist attacks on trains in Madrid, Spain on March 11, 2004 (exactly 911 days after the 9/11 attacks in America), which took the lives of at least 200 people, led to the fall of a conservative government that supported the U. S. action in Iraq and the election of a socialist government three days later. The new government immediately pulled Spanish troops out of Iraq, which was evidently the purpose of the bombings. Every time terrorists gain an objective they have sought, the rationality of terrorism is demonstrated along with its immorality.

While terrorism has ancient roots, it is far more prevalent today. Of the 74 groups listed by the U.S. Department of State (USDS) in 2003, only three active at the turn of the 21st century originated before 1960 (IRA, ETA, and the Muslim Brotherhood). Terrorist incidents also rose dramatically after the 1960s, peaked at a high of 665 in 1987, and then dropped to a low of 190 incidents in 2003. The National Counterterrorism Center (NCC) reports 651 terrorist attacks in 2004 that claimed 1,907 victims (NCC, 2005). Of the 74 listed terrorist groups, 39 are Islamic, 18 were Marxist/Maoist, and the remaining 17 were hybrids of Marxist/Islamic groups or nationalist groups.

The Causes of Terrorism

There are as many causes of terrorism as there are terrorist groups because it cannot be understood without understanding the historical, social, political, and economic conditions behind the emergence of each group. Perhaps the one generality we can make is that all groups originated in response to some perceived injustice. Although certain kinds of people may be drawn to terrorism, terrorists are not a bunch of "sicko-weirdos." If they were, we would have defeated terrorism long ago. Terrorist groups take pains not to recruit anyone showing signs of mental instability because such people are not trustworthy and would arouse the suspicion of their intended targets (Hudson, 1999, p. 60).

Many Islamic terrorists are recruited from religious schools known as **madrasas**. Some of these schools teach secular subjects, but they mostly focus on religious texts and stress the immorality and materialism of Western life and the need to convert all infidels to Islam (Armanios, 2003). The madrasas are appealing to poor Muslim families because they offer free room and board as well as free education. Many members of the Afghani *Taliban* regime studied and trained in Pakistani madrasas stressing a strict form of Islam. Children are indoctrinated in these schools with anti-Israeli and anti-American propaganda from the earliest days of their lives. A person nurtured on such material is ideal material for recruitment as a "martyr" to the cause. Martyrdom brings with it the promise of immediate ascension into heaven, where he will find "rivers of milk and wine . . . lakes of honey, and the services of 72 virgins" (Hoffman, 2002, p. 305).

Is There a Terrorist Personality?

Some theorists are of the opinion that we should look at what terrorist groups have to offer if we want to understand why individuals join them: "Terrorism can provide a route for advancement, an opportunity for glamour and excitement, a chance of world renown, a way of demonstrating one's courage, and even a way of accumulating wealth" (Reich, 1990, p. 271). Terrorism is much like organized crime, in that it provides illegitimate ways to get what most of us would like to have—fame and fortune. Terrorists also have a bonus in that they, and their comrades and supporters, see themselves as romanticized warriors fighting for a just and noble cause, and in the case of religious terrorists, the favor of their God and the promise of a rewarding afterlife.

Some scholars view terrorists as people with marginal personalities drawn to terrorist groups because their deficiencies are both accepted and welcomed by the group (Johnson & Feldman, 1992). These scholars also see the terrorist group as made up of three types of individuals: (1) the *charismatic leader,* (2) the *antisocial personality,* and (3) the *follower.* The charismatic leader is socially alienated, narcissistic, arrogant, and intelligent, with a deeply idealistic sense of right and wrong. The terrorist group provides a forum for his narcissistic rage and intellectual ramblings, and the subservience of group members feeds his egoism. Antisocial (or psychopathic) individuals have opportunities in terrorist groups to use force and violence to further their own personal goals, as well as the goals of the group. For the psychopath, the group functions like an organized crime family, providing greater opportunity, action, and prestige than could be found outside the group (Perlman, 2002). The majority of terrorists, however, are simple followers who see the world purely in black ("them") and white ("us") and have deep needs for acceptance, which makes them susceptible to all sorts of religious, ideological, and political propaganda (Ardila, 2002).

The article by Lawrence Miller in this section considerably expands our discussion of terrorism. He examines the major forms of terrorism, the motivations of terrorists, and the various factors that contribute to the causes of terrorism. He is particularly concerned with the motivations of the suicide bomber, and the decision to enter and leave a life of terrorism. He concludes his article with a number of policy recommendations for dealing with this scourge of modern times.

Domestic Violence

Domestic violence refers to any abusive act (physical, sexual, or psychological) that occurs within the family. While not as spectacular or newsworthy as serial killing and terrorism, family violence is included here because it is the most prevalent form of violence in the United States today, and most of that is intimate partner (spouse or lover) violence (Tolan, Gorman-Smith, & Henry, 2006). Except for minor forms of abuse, intimate partner violence is overwhelmingly committed by males against females (Hampton, Oliver, & Magarian, 2003). Just over one-third of all murders of females in the United States are committed by intimate partners, whereas less than 4% of males are killed by intimate partners (Rennison, 2003; see Figure 10.3). Assaults against spouses or lovers are primarily driven by male sexual proprietariness, jealousy, and suspicion of infidelity. Evidence from around the world indicates that the single most important cause of domestic violence (including homicide) is male jealousy and suspicion of infidelity (Lepowsky, 1994). DNA data indicate that between 1% and 30% (depending on the culture or subculture) of children are sired by someone other than the presumed father (Birkenhead & Moller, 1992; Brock & Shrimpton, 1991). The threat of cuckoldry (being fooled into raising someone else's child) is thus real, which suggests that male violence against spouses and lovers should be most common in environments where the threat of infidelity is most real. Such environments would be those in which marriages are most precarious, where moral restrictions on pre- and extra-marital sexual relationships are weakest, and where illegitimacy rates are highest.

Although by no means limited to the lower classes, domestic violence is most often committed by "competitively disadvantaged (CD) males" (Figueredo & McClosky, 1993). CD males have low mate value because they have less to offer in terms of resources or prospects of acquiring them, which should tend to make their mates less desirous of maintaining the relationship with them and to seek other partners. Lacking alternative means of controlling their partner's behavior, CD males may turn to violently coercive tactics to intimidate them. This is one of the reasons that intimate personal violence is 2 to 3 times more prevalent and more deadly among African American males than among males of other races (Hampton

Figure 10.3 Highlights of the 2003 Report on Intimate Partner Violence

Intimate partner violence—by current or former spouses, boyfriends, or girlfriends—made up 20% of all nonfatal violence against females age 12 or older in 2001.

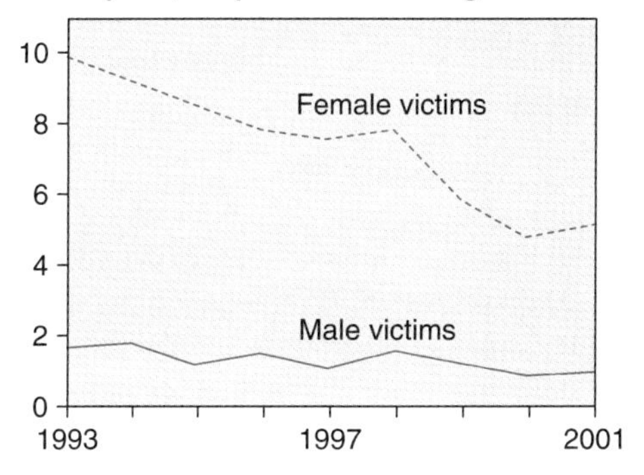

- The number of violent crimes by intimate partners against females declined from 1993 to 2001. Down from 1.1 million nonfatal violent crimes by an intimate in 1993, women experienced about 588,490 such crimes in 2001.
- In 1993 men were victims of about 162,870 violent crimes by an intimate partner. By 2001 that total had fallen to an estimated 103,220 victimizations.
- Intimate partner violence made up 20% of all nonfatal violent crime experienced by women in 2001. Intimate partners committed 3% of the nonfatal violence against men.
- For intimate partner violence, as for violent crime in general, simple assault was the most common type of crime.
- 1,247 women and 440 men were killed by an intimate partner in 2000. In recent years an intimate killed about 33% of female murder victims and 4% of male murder victims.

SOURCE: Rennison (2003).

et al., 2003). Hampton and his colleagues also list the anger and frustration born of poverty and unemployment, the reluctance of black females to report incidents, and the general fractious and antagonistic relationship that seems to exist between black men and women as reasons.

Summary

- Murder rates have been significantly higher in the past than they are today, primarily because of the lack of effective law enforcement and adequate medical attention. Homicide trends in the United States have fluctuated wildly over the years, and the U.S. is situated somewhere in the middle of nations in terms of its homicide rate. In the United States, the typical perpetrator and victim of homicide is a young black male living in an urban center. Female/female homicide is very rare worldwide.

- Poor, young, unmarried, non-white females are disproportionately likely to be victimized by rape, and poor, young, unmarried, non-white males are disproportionately likely to be perpetrators. Feminist theories maintain that all men have a propensity to rape and that the act is about power, not sex. Social learning and feminist theory assert that rape is the result of male socialization, while evolutionary theorist maintains that it is a maladaptive consequence of male reproductive strategy.

- Robbery is a violent crime and robbers tend to be the most impulsive, hedonistic, daring, and dangerous of all street criminals, as well as the least educated and least conscientious. Robbery is also considered an excellent way to prove a certain kind of "manliness" in certain urban areas.

- Aggravated assault is the most frequently committed of the violent Part I crimes. Each such incident carries the threat of ending up as a criminal homicide, and but for speedy access to medical treatment, many of them would have done so.

- Spree, mass, and serial murder have increased dramatically since the 1960s, especially serial murders. Serial murder is the murder of three or more victims over an extended period of time. A popular typology of serial killers contains visionary, mission-oriented, hedonistic, and power/control types. Visionary killers are usually psychotic, and mission-oriented killers feel that it is their duty to rid the world of people they consider undesirable. Hedonistic killers (the most common) kill for the pure joy of it, while power/control killers get more satisfaction from exerting complete control over their victims.

- African Americans are overrepresented in the ranks of serial killers relative to their numbers in the population, and females are even more underrepresented than they are among other kinds of criminals. Asian Americans are extremely underrepresented among serial killers. The diathesis/stress model posits that serial killers have a biological disposition to kill that is exacerbated by severe environmental stress during childhood.

- Terrorism is an ancient method of intimidating the public by the indiscriminate use of violence for social or political reasons. Terrorism increased rather dramatically from the 1960s to the mid-1980s, steadily dropped off in the 1990s, and has again increased in the 21st century.

- There are as many causes of terrorism as there are terrorist groups. Each group has its origins in some perceived injustice, but only a minuscule number of people react to such conditions by joining terrorist organizations.

- Domestic violence (mostly intimate partner violence) is the most prevalent form of violence in the United States today. Much of it is driven by jealousy and real or imagined infidelity and is most likely to be committed by competitively disadvantaged males.

EXERCISES AND DISCUSSION QUESTIONS

1. Why is female/female homicide so rare, and why is it the case around the world?
2. What do you think of the idea that rape is only about violence and not sex?
3. Mass and spree murderers almost by necessity need guns to carry out their activities. Would a ban on private ownership of weapons be an acceptable price to pay to save the lives of victims of these acts (as well as other victims of guns)?
4. Do you agree than one man's terrorist is another man's freedom fighter?

USEFUL WEB SITES

Bureau of Justice Statistics. http://www.ojp.usdoj.gov/bjs/.

Federal Bureau of Investigation. http://www.fbi.gov/.

Rape, Abuse, and Incest National Network. http://www.rainn.org/.

Serial Killers. http://www.karisable.com/crserial.htm.

Subculture of Violence Theory. http://www.criminology.fsu.edu/crimtheory/wolfgang.htm.

Women Organized Against Rape. http://www.woar.org/.

GLOSSARY

Aggravated assault: An unlawful attack by one person upon another for the purpose of inflicting severe or aggravated bodily injury.

Domestic violence: Any abusive act (physical, sexual, or psychological) that occurs within the family setting. Intimate partner violence is the most common form.

Forcible rape: The carnal knowledge of a female forcibly and against her will.

Hedonistic serial killer: A killer who kills for the pure thrill and joy of it.

Madrasas: Islamic religious schools that stress the immorality and materialism of Western life and the need to convert all infidels to Islam.

Mass murder: The killing of several people at one location within a few minutes or hours.

Mission-oriented serial killer: A killer who feels it to be a mission in life to kill certain kinds of people.

Murder: The willful (non-negligent) killing of one human being by another.

Power/control serial killer: A killer who gains most satisfaction from exercising complete power over his victims.

Serial murder: The killing of three or more victims over an extended period of time.

Spree murder: The killing of several people at different locations over several days.

Robbery: The taking or attempted taking of anything of value from the care, custody, or control of a person or persons by force or threat of force or violence and/or putting the victim in fear.

Terrorism: The unlawful use of force or violence against persons or property to intimidate or coerce a government, the civilian population, or any segment thereof, in furtherance of political or social goals.

Visionary serial killer: A killer who feels impelled to commit murder by visions or "voices in my head."

READING

Stick-Up, Street Culture, and Offender Motivation

Bruce A. Jacobs and Richard Wright

In this article, Jacobs and Wright describe interviews conducted with 86 active armed robbers. Rather than looking at the usual background characteristics (age, race, SES, etc.) of the robbers, they looked at the foreground conditions, that is, what robbers were thinking before they committed their crimes. Jacobs and Wright focus on the motivating factors and decision-making processes of the robbers. They find that most robbers decide to commit their crimes impulsively with very little rational thought. Robbers' participation in street culture leads them to be blind to legitimate opportunities, to the point where many seem to believe that they have little choice but to rob. In other words, armed robbers appear to be so overwhelmed by their emotional, financial, and drug problems that they perceive robbery as their only alternative to make money.

In this article we attend to exploring the decision-making processes of active armed robbers in real-life settings and circumstances. Our aim is to understand how and why these offenders move from an unmotivated state to one in which they are determined to commit robbery. We argue that while the decision to commit robbery stems most directly from a perceived need for fast cash, this decision is activated, mediated, and channeled by participation in street culture. Street culture, and its constituent conduct norms, represents an essential intervening variable linking criminal motivation to background risk factors and subjective foreground conditions.

Methods: Money, Motivation, and Street Culture

The study is based on in-depth interviews with a sample of 86 currently active robbers recruited from the streets of St. Louis, Missouri. Respondents ranged in age from 16 to 51. All but 3 were African-American; 14 were female. All respondents had taken part in armed robberies, but many also had committed strong-arm attacks. Respondents did not offend at equal rates, but all (1) had committed a robbery within the recent past (typically within the past month), (2) defined themselves as currently active, and (3) were regarded as active by other offenders.

Sixty-one of the offenders admitted to having committed 10 or more lifetime robberies. Included in this group were 31 offenders who estimated having done at least 50 robberies. Seventy-three of the offenders said that they typically robbed individuals on the street or in other public settings, 10 reported that they usually targeted commercial establishments, and 3 claimed that they committed street and commercial robberies in roughly equal proportions.

Fast Cash

With few exceptions, the decision to commit a robbery arises in the face of what offenders perceive to be a pressing need for fast cash. Eighty of

SOURCE: Jacobs, B., & Wright, R. (1999). Stick-up, street culture, and offender motivation. *Criminology, 37*, 149–174. Reprinted by permission.

81 offenders who spoke directly to the issue of motivation said that they did robberies simply because they needed money. Many lurched from one financial crisis to the next, the frequency with which they committed robbery being governed largely by the amount of money—or lack of it—in their pockets:

> [The idea of committing a robbery] comes into your mind when your pockets are low; it speaks very loudly when you need things and you are not able to get what you need. It's not a want, it's things that you need, . . . things that if you don't have the money, you have the artillery to go and get it. That's the first thing on my mind; concentrate on how I can get some more money.

> I don't think there is any one factor that precipitates the commission of a crime, . . . I think it's just the conditions. I think the primary factor is being without. Rent is coming up. A few months ago, the landlord was gonna put us out, rent due, you know. Can't get no money no way else; ask family and friends, you might try a few other ways of getting the money and, as a last resort, I can go get some money [by committing a robbery].

Many offenders appeared to give little thought to the offense until they found themselves unable to meet current expenses.

> [I commit a robbery] about every few months. There's no set pattern, but I guess it's really based on the need. If there is a period of time where there is no need of money . . . , then it's not necessary to go out and rob. It's not like I do [robberies] for fun.

The above claims conjure up an image of reluctant criminals doing the best they can to survive in circumstances not of their own making. In one sense, this image is not so far off the mark. Of the 59 offenders who specified a particular use for the proceeds of their crimes, 19 claimed that they needed the cash for basic necessities, such as food or shelter. For them, robbery allegedly was a matter of day-to-day survival. At the same time, the notion that these offenders were driven by conditions entirely beyond their control strains credulity. Reports of "opportunistic" robberies confirm this, that is, offenses motivated by serendipity rather than basic human need:

> If I had $5,000, I wouldn't do [a robbery] like tomorrow. But [i]f I got $5,000 today and I seen you walkin' down the street and you look like you got some money in your pocket, I'm gonna take a chance and see. It's just natural . . . . If you see an opportunity, you take that opportunity . . . . It doesn't matter if I have $5,000 in my pocket, if I see you walkin' and no one else around and it look like you done went in the store and bought somethin' and pulled some money out of your pocket and me or one of my partners has peeped this, we gonna approach you. That's just the way it goes.

Need and opportunity, however, cannot be considered outside the open-ended quest for excitement and sensory stimulation that shaped much of the offenders' daily activities. Perhaps the most central of pursuits in street culture, "life as party" revolves around "the enjoyment of 'good times' with minimal concern for obligations and commitments that are external to the . . . immediate social setting" (Shover and Honaker, 1992:283). Gambling, hard drug use, and heavy drinking were the behaviors of choice:

> I [have] a gambling problem and I . . . lose so much so I [have] to do something to [get the cash to] win my money back. So I go out and rob somebody. That be the main reason I rob someone.

> I like to mix and I like to get high. You can't get high broke. You really can't get high just standing there, you got to move. And in

order to move, you got to have some money . . . Got to have some money, want to get high.

While the offenders often referred to such activities as partying, there is a danger in accepting their comments at face value. Many gambled, used drugs, and drank alcohol as if there were no tomorrow; they pursued these activities with an intensity and grim determination that suggested something far more serious was at stake. Illicit street action is no party, at least not in the conventional sense of the term. Offenders typically demonstrate little or no inclination to exercise personal restraint. Why should they? Instant gratification and hedonistic sensation seeking are quite functional for those seeking pleasure in what may objectively be viewed as a largely pleasureless world.

The offenders are easily seduced by life as party, at least in part because they view their future prospects as bleak and see little point in long-range planning. As such, there is no mileage to be gained by deferred gratification:

> I really don't dwell on [the future]. One day I might not wake up. I don't even think about what's important to me. What's important to me is getting mine [now]

The offenders' general lack of social stability and absence of conventional sources of support only fueled such a mindset. The majority called the streets home for extended periods of time; a significant number of offenders claimed to seldom sleep at the same address for more than a few nights in a row. Moving from place to place as the mood struck them, these offenders essentially were urban nomads in a perpetual search for good times. The volatile streets and alleyways that criss-crossed St. Louis's crime-ridden central city neighborhoods provided their conduit:

> I guess I'm just a street person, a roamer. I like to be out in the street . . . Now I'm staying with a cousin . . . That's where I live, but I'm very rarely there. I'm usually in the street. If somebody say they got something up . . . I go and we do whatever. I might spend the night at their house or I got a couple of girls I know [and] I might spend the night at their house. I'm home about two weeks out of a month.

Keeping Up Appearances

The open-ended pursuit of sensory stimulation was but one way these offenders enacted the imperatives of street culture. No less important was the fetishized consumption of personal, nonessential, status-enhancing items. Shover and Honaker (1992:283) have argued that the unchecked pursuit of such items—like anomic participation in illicit street action—emerges directly from conduct norms of street culture. The code of the streets calls for the bold display of the latest status symbol clothing and accessories, a look that loudly proclaims the wearer to be someone who has overcome, if only temporarily, the financial difficulties faced by others on the street corner. To be seen as "with it," one must flaunt the material trappings of success. The quest is both symbolic and real; such purchases serve as self-enclosed and highly efficient referent systems that assert one's essential character in no uncertain terms.

> You ever notice that some people want to be like other people . . . ? They might want to dress like this person, like dope dealers and stuff like that. They go out there [on the street comer] in diamond jewelry and stuff. "Man, I wish I was like him!" You got to make some kind of money [to look like that], so you want to make a quick hustle.

The functionality of offenders' purchases was tangential, perhaps irrelevant. The overriding goal was to project an image of "cool transcendence," (Katz, 1988) that, in the minds of offenders, knighted them members of a mythic street aristocracy. As Anderson (1990:103–104) notes, the search for self-aggrandizement takes on a powerful logic of its own and, in the end, becomes all-consuming. Given the day-to-day desperation that dominates most of these offenders' lives, it is easy to appreciate why they are anxious to show off whenever the opportunity presents itself (particularly after making a lucrative score). Of course, it would be misleading to suggest that our respondents differed markedly from their law-abiding neighbors in wanting to wear flashy clothes or expensive accessory items. Nor were all of the offenders' purchases ostentatious. On occasion, some offenders would use funds for haircuts, manicures, and other mundane purchases. What set these offenders apart from "normal citizens" was their willingness to spend large amounts of cash on luxury items to the detriment of more pressing financial concerns.

Obviously, the relentless pursuit of high living quickly becomes expensive. Offenders seldom had enough cash in their pockets to sustain this lifestyle for long. Even when they did make the occasional "big score," their disdain for long-range planning and desire to live for the moment encouraged spending with reckless abandon. That money earned illegally holds "less intrinsic value" than cash secured through legitimate work only fueled their spendthrift ways (Walters, 1990: 147). The way money is obtained, after all, is a "powerful determinant of how it is defined, husbanded, and spent" (Shover, 1996, p. 104). Some researchers have gone so far as to suggest that through carefree spending, persistent criminals seek to establish the very conditions that drive them back to crime. Whether offenders spend money in a deliberate attempt to create these conditions is open to question; the respondents in our sample gave no indication of doing so. No matter, offenders were under almost constant pressure to generate funds. To the extent that robbery alleviated this stress, it nurtured a tendency for them to view the offense as a reliable method for dealing with similar pressures in the future. A self-enclosed cycle of reinforcing behavior was thereby triggered.

Why Robbery?

The decision to commit robbery, then, is motivated by a perceived need for cash. Why does this need express itself as robbery? Presumably the offenders have other means of obtaining money. Why do they choose robbery over legal work? Why do they decide to commit robbery rather than borrow money from friends or relatives? Most important, why do they select robbery to the exclusion of other income-generating crimes?

That the decision to commit robbery typically emerges in the course of illicit street action suggests that legitimate employment is not a realistic solution. Typically, the offenders' need for cash is so pressing and immediate that legal work, as a viable money-making strategy, is untenable: Payment and effort are separated in space and time and these offenders will not, or cannot, wait. Moreover, the jobs realistically available to them—almost all of whom were unskilled and poorly educated—pay wages that fall far short of the funds required to support a cash-intensive lifestyle:

> Education-wise, I fell late on the education. I just think it's too late for that. They say it's never too late, but I'm too far gone for that. . . . I've thought about [getting a job], but I'm too far gone I guess. . . . I done seen more money come out of [doing stick-ups] than I see working.

Legitimate employment also was perceived to be overly restrictive. Working a normal job requires one to take orders, conform to a schedule, minimize informal peer interaction, show up sober and alert, and limit one's freedom of movement for a given period of time. For many in our sample, this was unfathomable; it cramped the hedonistic, street-focused lifestyle they chose to live:

> I'm a firm believer, man, God didn't put me down on this earth to suffer for no reason. I'm just a firm believer in that. I believe I can have a good time every day, each and every day of my life, and that's what I'm trying to do. I never held a job. The longest job I ever had was about nine months . . . at St. Louis Car; that's probably the longest job I ever had, outside of working in the joint. But I mean on the streets, man, I just don't believe in [work]. There is enough shit on this earth right here for everybody, nobody should have to be suffering. You shouldn't have to suffer and work like no dog for it, I'm just a firm believer in that. I'll go out there and try to take what I believe I got comin' [because] ain't nobody gonna walk up . . . and give it to me. [I commit robberies] because I'm broke and need money; it's just what I'm gonna do. I'm not going to work! That's out! I'm through [with work]. I done had 25 or 30 jobs in my little lifetime [and] that's out. I can't do it! I'm not going to!

The "conspicuous display of independence" is a bedrock value on which street-corner culture rests (Shover and Honaker, 1992:284): To be seen as cool one must do as one pleases. This ethos clearly conflicts with the demands of legitimate employment. Indeed, robbery appealed to a number of offenders precisely because it allowed them to flaunt their independence and escape from the rigors of legal work.

This is not to say that every offender summarily dismissed the prospect of gainful employment. Twenty-five of the 75 unemployed respondents claimed they would stop robbing if someone gave them a "good job"—the emphasis being on good:

> My desire is to be gainfully employed in the right kind of job. . . . If I had a union job making $16 or $17 [an hour], something that I could really take care of my family with, I think that I could become cool with that. Years ago I worked at one of the [local] car factories; I really wanted to be in there. It was the kind of job I'd been looking for. Unfortunately, as soon as I got in there they had a big layoff.

Others alleged that, while a job may not eliminate their offending altogether, it might well slow them down:

> [If a job were to stop me from committing robberies], it would have to be a straight up good paying job. I ain't talkin" about no $6 an hour . . . I'm talkin" like $10 to $11 an hour, something like that. But as far as $5 or $6 an hour, no! I would have to get like $10 or $11 an hour, full-time. Now something like that, I would probably quit doing it [robbery]. I would be working, making money, I don't think I would do it [robbery] no more . . . I don't think I would quit [offending] altogether. It would probably slow down and then eventually I'll stop. I think [my offending] would slow down.

While such claims may or may not be sincere, it is unlikely they will ever be challenged. Attractive employment opportunities are limited for all inner-city residents and particularly for individuals like those in our sample. Drastic changes in the post World War II economy—deindustrialization and the loss of manufacturing jobs, the increased demand for advanced education and high skills, rapid suburbanization and out-migration of middle class residents—have left them behind, twisting in the wind. The lack of legal income options speaks to larger societal patterns in which major changes in the U.S. economy have reduced the number of available good-paying jobs and created an economic underclass with unprecedented levels of unemployment and few options—beyond income-generating crime—to exercise. Governmental directives, such as changes in requirements and reductions in public transfer payments, decidedly reduce the income of already marginalized persons in inner-city communities—those at highest risk for predatory crime. This only intensifies their economic and social isolation, makes their overall plight worse, and their predisposition to criminality stronger. Most offenders realized this and, with varying degrees of bitterness, resigned themselves to being out of work:

> I fill out [job] applications daily. Somebody [always] says, "This is bad that you got tattoos all over looking for a job." In a way, that's discrimination. How do they know I can't do the job? I could probably do your job just as well as you, but I got [these jailhouse] tattoos on me. That's discriminating. Am I right? That's why most people rob and steal because, say another black male came in like me [for a job], same haircut, same everything. I'm dressed like this, tennis shoes, shorts and tank top. He has on (a) Stacy Adams pair of slacks and a button-up shirt with a tie. He will get the job before I will. That's being racist in a way. I can do the job just as well as he can. He just dresses a little bit better than me.

Clearly, these offenders were not poster children for the local chamber of commerce or small business association. By and large, they were crudely mannered and poorly schooled in the arts of impression management and customer relations. Most lacked the cultural capital necessary for the conduct of legitimate business. They were not "nice" in the conventional sense of the term; to be nice is to signal weakness in a world where only the strong survive.

Even if the offenders were able to land a high-paying job, it is doubtful they would keep it for long. The relentless pursuit of street action—especially hard drug use—has a powerful tendency to undermine any commitment to conventional activities. Life as party ensnares street-culture participants, enticing them to neglect the demands of legitimate employment in favor of enjoying the moment. Though functional in lightening the burdensome present, gambling, drinking, and drugging—for those on the street—become the proverbial "padlock on the exit door" (Davis, 1995) and fertilize the foreground in which the decision to rob becomes rooted.

Borrowing

In theory, the offenders could have borrowed cash from a friend or relative rather than resorting to crime. In practice, this was not feasible. Unemployed, unskilled, and uneducated persons caught in the throes of chronically self-defeating behavior cannot, and often do not, expect to solve their fiscal troubles by borrowing. Borrowing is a short-term solution, and loans granted must be repaid. This in itself could trigger robberies. As one offender explained, "I have people that will loan me money, [but] they will loan me money because of the work [robbery] that I do; they know they gonna get their money [back] one way or another." Asking for money also was perceived by a number of offenders to be emasculating. Given their belief that men should be self-sufficient, the mere prospect of borrowing was repugnant:

> I don't like always asking my girl for nothing because I want to let her keep her own money.... I'm gonna go out here and get some money.

The possibility of borrowing may be moot for the vast majority of offenders anyway. Most had long ago exhausted the patience and goodwill of helpful others; not even their closest friends or family members were willing to proffer additional cash:

> I can't borrow the money. Who gonna loan me some money? Ain't nobody gonna loan me no money. Shit, [I use] drugs and they

> know [that] and I rob and everything else. Ain't nobody gonna loan me no money. If they give you some money, they just give it to you; they know you ain't giving it back.

When confronted with an immediate need for money, then, the offenders perceived themselves as having little hope of securing cash quickly and legally. But this does not explain why the respondents decided to do robbery rather than some other crime. Most of them had committed a wide range of income-generating offenses in the past, and some continued to be quite versatile. Why, then, robbery?

For many, this question was irrelevant; robbery was their "main line" and alternative crimes were not considered when the pressing need for cash arose:

> I have never been able to steal, even when I was little and they would tell me just to be the watch-out man. . . . Shit, I watch out, everybody gets busted. I can't steal, but give me a pistol and I'll go get some money. . . . . [Robbery is] just something I just got attached to.

When these offenders did commit another form of income-generating crime, it typically was prompted by the chance discovery of an especially vulnerable target rather than being part of their typical modus operandi:

> I do [commit other sorts of offenses] but that ain't, I might do a burglary, but I'm jumping out of my field. See, I'm scared when I do a burglary [or] something like that. I feel comfortable robbing . . . , but I see something they call "real sweet," like a burglary where the door is open and ain't nobody there or something like that, well. . . .

Many of the offenders who expressed a strong preference for robbery had come to the offense through burglary, drug selling, or both. They claimed that robbery had several advantages over these other crimes. Robbery took much less time than breaking into buildings or dealing drugs. Not only could the offense be committed more quickly, it also typically netted cash rather than goods. Unlike burglary, there was no need for the booty "to be cut, melted down, recast or sold," nor for obligatory dealings with "treacherous middlemen, insurance adjustors, and wiseguy fences" (Pileggi, 1985:203). Why not bypass all such hassles and simply steal cash.

> Robbery is the quickest money. Robbery is the most money you gonna get fast. . . . Burglary, you gonna have to sell the merchandise and get the money. Drugs, you gonna have to deal with too many people, (a) bunch of people. You gonna sell a $50 or $100 bag to him, a $50 or $100 bag to him, it takes too long. But if you find where the cash money is and just go take it, you get it all in one wad. No problem. I've tried burglary, I've tried drug selling . . . the money is too slow.

Some of the offenders who favored robbery over other crimes maintained that it was safer than burglary or dope dealing:

> I feel more safer doing a robbery because doing a burglary, I got a fear of breaking into somebody's house not knowing who might be up in there. I got that fear about house burglary. . . . On robbery I can select my victims, I can select my place of business. I can watch and see who all work in there or I can rob a person and pull them around in the alley or push them up in a doorway and rob them. You don't got [that] fear of who . . . in that bedroom or somewhere in another part of the house.

> [I]f I'm out there selling dope somebody gonna come and, I'm not the only one out

> there robbing you know, so somebody like me, they'll come and rob me. . . . I'm robbin' cause the dope dealers is the ones getting robbed and killed you know.

A couple of offenders reported steering clear of dope selling because their strong craving for drugs made it too difficult for them to resist their own merchandise. Being one's own best customer is a sure formula for disaster, something the following respondent seemed to understand well:

> A dope fiend can't be selling dope because he be his best customer. I couldn't sell dope [nowadays]. I could sell a little weed or something cause I don't smoke too much of it. But selling rock [cocaine] or heroin, I couldn't do that cause I mess around and smoke it myself. [I would] smoke it all up!

Others claimed that robbery was more attractive than other offenses because it presented less of a potential threat to their freedom:

> If you sell drugs, it's easy to get locked up selling drugs; plus, you can get killed selling drugs. You get killed more faster doing that.

> Robbery you got a better chance of surviving and getting away than doing other crimes. . . . You go break in a house, [the police] get the fingerprints, you might lose a shoe, you know how they got all that technology stuff. So I don't break in houses. . . . I leave that to some other guy.

Without doubt, some of the offenders were prepared to commit crimes other than robbery; in dire straits one cannot afford to be choosy. More often than not, robbery emerged as the "most proximate and performable" (Lofland, 1969:61) offense available. The universe of money-making crimes from which these offenders realistically could pick was limited. By and large, they did not hold jobs that would allow them to violate even a low-level position of financial trust. Nor did they possess the technical know-how to commit lucrative commercial break-ins, or the interpersonal skills needed to perpetrate successful frauds. Even street-corner dope dealing was unavailable to many; most lacked the financial wherewithal to purchase baseline inventories—inventories many offenders would undoubtedly have smoked up.

The bottom line is that the offenders, when faced with a pressing need for cash, tend to resort to robbery because they know of no other course of action, legal or illegal, that offers as quick and easy a way out of their financial difficulties. As Lofland (1969:50) notes, most people under pressure have a tendency to become fixated on removing the perceived cause of that pressure "as quickly as possible." Desperate to sustain a cash-intensive lifestyle, these offenders were loathe to consider unfamiliar, complicated, or long-term solutions (Lofland, 1969:50–54). With minimal calculation and "high" hopes, they turned to robbery, a trusted companion they could count on when the pressure was on. For those who can stomach the potential violence, robbery seems so much more attractive than other forms of income-generating crime. Contemplating alternative offenses becomes increasingly difficult to do. This is the insight that separates persistent robbers from their street-corner peers:

> [Robbery] is just easy. I ain't got to sell no dope or nothing, I can just take the money. Just take it, I don't need to sell no dope or work. . . . I don't want to sell dope, I don't want to work. I don't feel like I need to work for nothing. If I want something, I'm gonna get it and take it. I'm gonna take what I want. . . . If I don't have money, I like to go and get it. I ain't got time [for other

offenses]; the way I get mine is by the gun. I don't have time to be waiting on people to come up to me buying dope all day. . . . I don't have time for that so I just go and get my money.

Discussion

The overall picture that emerges from our research is that of offenders caught up in a cycle of expensive, self-indulgent habits (e.g., gambling, drug use, and heavy drinking) that feed on themselves and constantly call for more of the same. It would be a mistake to conclude that these offenders are being driven to crime by genuine financial hardship; few of them are doing robberies to buy the proverbial loaf of bread to feed their children. Yet, most of their crimes are economically motivated. The offenders perceive themselves as needing money and robbery is a response to that perception.

Being a street robber is a way of behaving, a way of thinking, an approach to life. Stopping such criminals exogenously—in the absence of lengthy incapacitation—is not likely to be successful. Getting offenders to "go straight" is analogous to telling a lawful citizen to "relinquish his history, companions, thoughts, feelings, and fears, and replace them with [something] else" (Fleisher, 1995:240). Self-directed going-straight talk on the part of offenders more often than not is insincere—akin to young children talking about what they're going to be when they grow up: "Young storytellers and . . . criminals . . . don't care about the [reality]; the pleasure comes in saying the words, the verbal ritual itself brings pleasure" (Fleisher, 1995:259). Gifting offenders money, in the hopes they will reduce or stop their offending, is similarly misguided. It is but twisted enabling and only likely to set off another round of illicit action that plunges offenders deeper into the abyss of desperation that drives them back to their next crime.

References

Anderson, E. (1990). *Streetwise.* Chicago: University of Chicago Press.

Davis, P. (1995, October 12). *If you came this way* [Interview]. All Things Considered, National Public Radio.

Fleisher, M. S. (1995). *Beggars and thieves: Lives of urban street criminals.* Madison: University of Wisconsin Press.

Katz, J. (1988). *Seductions of crime: Moral and sensual attractions in doing evil.* New York: Basic Books.

Lofland, J. (1969). *Deviance and identity.* Englewood Cliffs, NJ: Prentice-Hall.

Pileggi, N. (1985). *Wiseguy.* New York: Simon & Schuster.

Shover, N. (1996). *Great pretenders: Pursuits and careers of persistent thieves.* Boulder, CO: Westview.

Shover, N., & Honaker, D. (1992). The socially-bounded decision making of persistent property offenders. *Howard Journal of Criminal Justice, 31,* 276–293.

Walters, G. (1990) *The criminal lifestyle.* Newbury Park, CA: Sage.

DISCUSSION QUESTIONS

1. What would you think are the main differences (in degree, not in kind) between individuals who specialize in robbery versus those who specialize in burglary?
2. What do Jacobs and Wright think of the rehabilitative potential of the typical robber? Do you agree?
3. Which criminological theory discussed so far best explains robbers as described here?

READING

African Americans and Serial Killing in the Media

The Myth and the Reality

Anthony Walsh

In this article, Anthony Walsh explores why so few black serial killers are known to the public, or even to many criminologists. There were many expressions of shock and surprise voiced in the media in 2002 when the "D.C. Sniper" turned out to be two black males. Two of the stereotypes surrounding serial killers are that they are almost always white males and that African American males are barely represented in their ranks. In a sample of 413 serial killers operating in the United States from 1945 to mid-2004, it was found that 90 were African American. Relative to the African American proportion of the population across that time period, African Americans were overrepresented in the ranks of serial killers by a factor of about 2. Possible reasons why so few African American serial killers are known to the public are explored.

In the movie *Copycat,* Sigourney Weaver plays a criminal psychologist and expert on serial killing. In the opening scene, she is giving a lecture in which she asks all the males in the audience to stand, emphasizing that serial killing is primarily a male behavior. She then asks all African American and Asian American males to sit down, leaving only White males standing as representative of serial killers. The message that viewers of *Copycat* get is that only White males commit these heinous crimes and that members of other races or ethnicities never do.

This stereotype is pervasive in the United States. A commentator in the *Harlem Times* expressed shock and disbelief when the D.C. Sniper turned out to be two Blacks, because "white guys have pretty much cornered the market on mass murders and serial killing" (Charles, 2002). Psychologist Na'im Akbar stated, "This is not typical conduct for us. I mean Black folks do some crazy stuff, but we don't do anonymous violence. That's not in our history. We just don't do that" (in White, Willis, & Smith, 2002: 2).

It is one of the mysteries of modern criminology that a group responsible for a highly disproportionate number of homicides of all other types has gained a reputation for not producing serial killers. For instance, data from the period encompassing 1976 through 1998 reveal that African Americans committed 51.5% of the recorded homicides in the United States (Fox & Levin, 2001). Between 1946 and 1990, homicide rates among Black males have ranged from 6.56 times the White male rate in 1984 to 15.78 times the White male rate in 1952 (LaFree, 1996).

There is no doubt that White males have constituted the majority of multiple murderers in the United States, but White males have constituted the vast majority of males across that time period. However, recent data for the years

Source: Walsh, A. (2005). African Americans and serial killing in the media: The myth and the reality. *Homicide Studies, 9*(4), 271–291. Reprinted with permission of Sage Publications, Inc.

1976 through 1998 found African American representation among murderers with multiple victims (serial, mass, and spree combined) to have increased to 38.2% of all such offenders (Fox & Levin, 2001). African Americans are thus overrepresented among killers having multiple victims by about 3 times relative to their proportion of the population.

Pre-World War II African American serial killers such as Jarvis Catoe, Jake Bird, and Clarence Hill were among those claiming the largest number of victims. Bird was particularly prolific, with a verified 44 victims, a number just 4 victims short of White killer Gary Ridgeway's (the Green River killer) 48 victims, which is the record number of verified victims in the annals of American serial killing. Coral Watts and Milton Johnson are two of the most notorious examples of Black serial killers of post–World War II years. Watts, known as the "Sunday Morning Slasher," confessed to 13 murders and was linked to at least 8 others between 1978 and 1983, and Milton Johnson was responsible for at least 17 murders in the 1980s. More contemporary African American serial killers include Henry Louis Wallace, who raped and strangled at least nine women from 1993 until his capture in 1996. All of Wallace's victims were acquaintances, which makes him unique among serial killers, who almost always seek strangers. Another recent example is Kendall Francois, who was indicted in 1999 for the murders of eight women, all but one of whom was White. Perhaps the most chilling of recent African American serial killers was Maury Travis, who was arrested in 2002. Travis had a secret torture chamber in his basement, where police found bondage equipment, videotapes of his rape and torture sessions, and clippings relating to police investigations of his murder victims (mostly prostitutes and crack addicts). Travis hanged himself in jail after confessing to 17 murders (Shinkle, 2002). Among the most recent African American serial killers at the time of writing are Derrick Todd Lee, Lorenzo Gilyard, and Daniel Jones. Lee was arrested in Atlanta in 2003 for the murders of five women in Louisiana and is a suspect in many other murders, rapes, and assaults going back to 1992. Gilyard and Jones were both arrested in Kansas City, Missouri, in 2004 for the murder of 12 and 4 women, respectively (Lambe, 2004).

Although the victim counts of these African American serial killers fall short of those attributed to notorious White killers, such as Ted Bundy (20-plus victims) or John Wayne Gacy (33), they exceed the figures attributed to more publicized killers, such as David Berkowitz (6) and Ed Kemper (8). The extensive media coverage of the Bundy, Gacy, and Berkowitz cases have made these killers almost household names, but African Americans such as Watts, Johnson, Francois, and Wallace are practically unknown, despite having operated within the same general time framework (1980s to 1990s). Before the Mohammed-Malvo sniper case, Wayne Williams (the Atlanta child murders) was the only African American serial killer to gain a modicum of the notoriety attached to his White counterparts, and that may be because, as with the sniper case itself, the police believed that they were looking for a White perpetrator or perpetrators.

Race as a Victim Selection Factor

Although most serial killers, like most killers in general, tend to kill within their own racial or ethnic group, there are a number who appear to purposely move across racial lines to secure victims. Given the added risks of apprehension trolling for victims in areas where the killer is likely to stand out, racial bigotry must be considered a partial motive for those who cross racial lines to kill. There are and have been White racist serial killers, but as Newton (1992) states, "Unlike their Caucasian counterparts, Black racists tend to murder in groups—De Mau Mau in Chicago, California's Death Angels, the Yahweh cult of Hebrew Israelites in Florida" (p. 67).

The San Francisco–based Death Angels may have killed more people in the early to mid-1970s than all the other serial killers operating during that period combined (Lubinskas, 2001). In

Clark Howard's (1979) study of these killings (dubbed the zebra killings by police, apparently because the perpetrators were Black and the victims White), identifies 270 alleged victims, although Newton and Newton (1991) indicate that the police believed them to have killed about 80. Whatever the true number is, convictions were obtained for only 23 of the murders. Five Black Muslims apparently carried out the majority of the killings attributed to this group, believing it their Islamic duty to rid the world of "White devils."

The Yahweh Ben Yahweh cult, also sometimes known as the Death Angels, operated in Miami in the 1980s. The cult began by killing ex-members (all Black) for abandoning it but soon began to focus rage on Whites. As with their San Francisco counterparts, members were instructed to kill "White devils" and to bring back various body parts to prove that they had done so. The most prolific killer was ex-NFL football player Robert Rozier, who confessed to seven killings (Freedberg & Gehreke, 1990).

The sibling team of Anthony and Nathaniel Cook also targeted only White victims. Both men pleaded guilty in 2000 in Lucas County, Ohio, to eight murders committed in the late 1970s and early 1980s and were suspected of committing others outside Lucas County's jurisdiction (Emch, 2000). Kendall Francois and Derrick Todd Lee are recent examples of African American solo serial killers who targeted only White victims.

Method

We define a serial killer as a person who kills three or more individuals in separate events and who is motivated by various combinations of hedonism, sexual lust, desire for power and dominance, and/or misplaced missionary zeal to rid the world of "undesirables." In other words, serial killing is viewed as an expressive crime rather than an instrumental crime. Not included in this definition are those whose murders are motivated by financial gain or by various criminal enterprises (such as gangland wars, hit men, revenge killings, "Bluebeard" killings, and so forth), or committed by groups for religious or political reasons, such as those committed by the Death Angels.

The individuals in Table 10.1 are those whose killings encompass the years 1945 up to the first half of 2004 and are males only. The year 1945 was chosen as the beginning year because the end of World War II may be considered a turning point in race relations in the United States and the beginning of the crumbling of Jim Crow barriers (Thernstrom & Thernstrom, 1997). Prior to the war (and certainly to some extent after it), American law enforcement paid little attention to Black crime unless it involved White victims. Given that situation, it is plausible to assume that serial killers operating in the Black community may have gone largely unnoticed and thus unrecorded. Cases were placed in a particular period according to the year the killer killed his first known victim.

The data for this study were derived from a variety of sources. The major sources of information are the encyclopedias written by Newton (1990, 1992, 2000) and by Wilson and Seaman (1983, 1990) as well as the works of other authors cited herein. Newspaper (particularly Newspaper Source) and Internet sources were used for cases that occurred after 1999. Using these resources, 90 African American serial killers were identified. Using the same method, 323 White American serial killers were identified as operating in the same time framework. This provided us with a combined sample of 413 known Black or White serial killers operating in the United States between 1945 and the first 6 months of 2004 (Hispanic and Asian Americans are not included in this study).

Findings

Table 10.1 provides the names of the 90 identified post-World War II African American serial killers and the time period and state(s) in which they operated. Cases ranged from Monroe Hickson in 1945 to Lorenzo Gilyard and Daniel Jones in

Table 10.1 African American Serial Killers 1945 to May 2004

1945 to 1979			1980 to 1989			1990 to May 2004		
Name	Number of Victims	State(s)	Name	Number of Victims	State(s)	Name	Number of Victims	State(s)
CarltonGary	7	GA	Clinton Bankston	5	CA	Benjamin Atkins	11	MI
Vaughn Greenwood	11	CA	Norman Bernand	3	NC,CA	Luscious Boyd	3+	LA
Vincent Groves	14	CO	Vemon Brown	5+	IN,MO	Engene Britt	11	IN
William Hence	3	GA	Nathaniel Code	13	LA	Andre Grawford	10	IL
Anthony Cook, Nathaniel Cook	8+	OH	Alton Coleman	8	MI,OH,IN	Reginald Carr, Jonathon Carr	5+	KS
Lester Harrison	7	IL	Louis Craine	5	CA	Jemme Dennis	5	NJ
Monrce Hickson	4	SC	Lorenzo Fayne	5	MO	Paul Durousseau	6	GA,NL
John Henry	3	FL	Thomson Hawkins	3	PA	James Swan	3	DC
Calvin Jackson	9+	NY	Harrison Graham	7+	PA	Lorenzo Gilyard	12	MO
Deveron LeGrand	6+	NY	Ronald Gray	4	NC	kendall Francois	8	NY
Laskey Posteal	7	OH	Richard Grissom	3	MO	Hubert Geralds	6	GA, FL
Bobby Joe Maxwell	10	CA	Kevin Haley, Reginald Haley	8	CA	Samuel Ivery	4+	CA,AZ,IL,AL
Winston Moseley	3	NY	Ray Jackson	6	MO	Daniel O.Jones	5	MO
David Roberts	4	IN	Wilbur Jennings	4	CA	Henry Lee Jones	4+	Several
Winford Strokes	3	MO	Milton Johnson	17	IL	Arohn kee	3+	NY
Clarance walker	14	TN,OH,MI	Bryan Jones	4	CA	Gregory Klepper	8	IL
Coral Watts	13+	TX,MI	Jeffrey Jones	4	CA	Derrik Todd Lee	10	LA
Robert Williams	3	NE,IA	Anthony Joiner	6	PA	David Middleton	3	CO
Wayne Williams	5+	GA	Horace Kelly	3	CA	Christopher Peterson	7	IN
Ben Cheney, Martin Rutrell, L.L. Thompson	4	FL,SC	Michael Player	10	CA	John Mohammed, Lee Malvo	11+	Several

1945 to 1979			1980 to 1989			1990 to May 2004		
Name	Number of Victims	State(s)	Name	Number of Victims	State(s)	Name	Number of Victims	State(s)
			Anthony McKnight	7	CA	James Pough	11	FL
			Eddie Lee Moseley	16+	FL	Cleophus Prince	6	CA
			Donald Murphy	5	MI	Earl Richmond	5	NJ,NC
			Calvin Perry	5	IN	George Russell	3	WA
			Craig Price	4	RI	Marc Sappington	4	KS
			Michael Player	10	CA	David Selepe	11	OH
			Yusef Rahman	4	KS,NY	Maury Travis	17+	MO,IL
			Robert Rozier	7+	FL,NY,MD	Henry Wallace	9	NC
			Beoria Simmons	3	KY	John Williams	5	NC
			Morris Solomon	7	CA	Nathaniel White	6	NY
			Timothy Spencer	4	VA			
			James Stuard	3	AZ			
			Micheal Terry	6	GA			
			Anthony Wimberly	3	CA			

2004. It is noted that African American serial killers were particularly well represented during the 1980s (as were White serial killers). There were 34 known African American serial killers in the 10 years from 1980 through 1989 (an average of 3.4 per year) compared with only 24 in the 34-year period between 1945 and 1979 (an average of 0.70 per year). Based on these figures, there were approximately 5 times more African American serial killers operating in the 1980s than there were from 1945 to 1979. Likewise, the 13-year period between 1990 and the first half of 2004 revealed 32 African American serial killers—an average of about 2.3 per year.

The 90 identified African American serial killers compose 21.8% of the sample of Black and White killers. This figure approximates the 22% estimate made by Hickey's (1997) sample of 337 serial killers of all races operating in the United States across a 165-year period. Far from being absent or severely underrepresented in the ranks of serial killers, then, African Americans are represented among serial killers at a rate approximately twice one would expect based on the average percentage of African Americans in the population (approximately 10.5%) across the 58-year time period examined. Given these findings, the next task is to try to determine why the reality is so far removed from the myth.

Why So Little Media Coverage of African American Serial Killers?

The mass media are the major sources of public information and perceptions about crime and criminality. The media are the gatekeepers of what the public is entitled to know, and the media are very anxious not to attract accusations of racism by zeroing in on heinous crimes committed by African Americans with the same zealousness it exhibits when such crimes are committed by Whites (Greek, 2001; Perazzo, 1999). Charges of racism and all the negative consequences that accrue when such charges are made may feature prominently in the maintenance of this double standard.

Jenkins (1998) lists three more specific reasons the media ignore or downplay stories about African American serial killers. The weakest reason he gives is that the language often used to describe serial killers (e.g., "primitive," "monsters," "animals") by commentators would be deemed racist if applied to African Americans by a mostly White media. When an official of Nassau County, New York, for instance, called Colin Ferguson (the African American who killed 6 and wounded 17 others on a commuter train in 1993) an "animal," he was soundly rebuked by many civil rights leaders, none of whom have ever been heard to complain when such terms are applied to White killers.

A more convincing argument made by Jenkins (1998) is that until recently, law enforcement agencies were less likely to take Black crimes seriously unless the victims were White. Given this relative lack of interest, Jenkins is of the opinion that African American serial killers may have been more hidden from the mainstream culture and thus more prevalent than the record indicates, especially during earlier periods of the 20th century. Pre-World War II African American Jarvis Catoe, for instance, drew little attention when he concentrated on killing Black women, but his switch to White women in 1941 proved to be his undoing. Likewise, White killer Albert Fish was able to remain at large for about 25 years by concentrating on killing Black children, but when he killed a White girl in 1928, he was arrested. Other serial killers operating exclusively in the Black community may have likewise escaped notice by the police and thus are not known to us today. On the other hand, Jake Bird murdered White women almost exclusively in his years of travel across the United States, until captured in 1947 (Newton, 1992).

The third reason is the entertainment media's (as opposed to the news media) perception that books and movies featuring African American characters are not likely to appeal to mass audiences. There does appear to be a reluctance to cast African Americans in negative roles. A survey of more than 600 prime-time television programs aired across 3 decades found that "nine out of 10 murders on TV were committed by Whites. Only three in 100 murders on TV were committed by Blacks. Blacks are about 18 times less likely to commit homicide on TV than in real life" (Lichter, Lichter, & Rothman, 1991: 198). However, movie producers have no concerns about casting African Americans in the star roles of lead investigators in serial killer movies, such as Morgan Freeman in *Seven* and *Along Came a Spider* or Denzel Washington in *The Bone Collector*. These movies are box office successes, which indicates that Black characters do have appeal to White audiences and that producers are willing to cast them in major roles—at least in "good guy" roles.

It could be argued that fear of being branded racist prevents the depiction of African Americans as villains in the entertainment media rather than their alleged lack of appeal to mass audiences. The TV series *Hill Street Blues*, which realistically featured both Black and White criminals, has been depicted as racist for encouraging the stereotype that African Americans are criminals, as has the *Cosby Show*, because it paints an overly positive picture of Black life in America and thus absolves White society of any responsibility for the welfare of African Americans (D'Sousa, 1995). Given this "damned if you do, damned if you don't" situation, producers might be forgiven for their reluctance to negatively cast African Americans.

We believe that the primary reason for the lack of coverage of African American serial killers, like the lack of coverage of African Americans in other sensitive areas, such as organized crime and hate crime, is that the print and electronic news media (as opposed to the entertainment media) largely ignore them. The media have tended to avoid more than minimal coverage of heinous crimes committed by African Americans at the same time as they extensively publicize the same kinds of crimes committed by Whites. The differential coverage of the trial of the White police officers who beat Rodney King in Los Angeles and the trial of members of the Yahweh Ben Yahweh cult in Miami for killing "White devils" is a case in point. Almost everyone is aware of the first trial because we were bombarded daily with images of the King beating, the riots, and the trial itself, but few outside Miami have heard of the second trial, which took place concurrently with the police officers' trial (Taylor, 1992). African American columnist Armstrong Williams (2002) wrote that these crimes received little national media attention "largely because the victims were White, which meant no Jesse Jackson screaming into his megaphone."

It would be difficult to find a better example of the media's reluctance to portray African American serial killers than the differential coverage of White killer Gary Heidnik and Black killer Harrison Graham. Both men kidnapped and kept a number of women imprisoned in their basements where they raped and tortured them and killed some of them. Although unknown to each other, Heidnik and Graham lived only 3 miles apart in Philadelphia and both were arrested only 5 months apart in 1987 (Jenkins, 1998). Gary Heidnik received widespread national attention, became the subject of books and television shows, and served as a model for the fictitious Buffalo Bill in *Silence of the Lambs*. Harrison Graham received virtually no media attention outside of Philadelphia, despite having been convicted of four more murders than Heidnik (seven vs. three), and despite the obvious public interest such attention would generate, given the almost uncanny demographic and modus operandi similarities involved.

At the intersection of news and entertainment lies the documentary. In a 1998 documentary called *Heidnik and Dahmer: Killers for Company*, Heidnik was compared to Jeffrey Dahmer. The choice of Dahmer to "costar" with Heidnik was a curious one given that Dahmer's crimes occurred in a different city in another decade, that his modus operandi was very different from Heidnik's, and that his victims were male rather than female. Given the similarities of the Graham and Heidnik cases and the differences between the Heidnik and Dahmer cases, it is difficult to see why Dahmer was chosen instead of Graham, except that Dahmer was also White and thus fit the stereotype. A more appropriate counterpart to Jeffrey Dahmer would have been African American Marc Sappington (the "Kansas City Vampire"). Sappington's killing career lasted only about a month (March to April 2001), but before his capture, the 21-year-old Sappington had killed four young males and eaten body parts or drunk the blood of three of them.

African American Serial Killers and Criminology

The overrepresentation of African Americans in serial killing is rarely explicitly (although it may be tacitly) stated in the criminological literature, even by criminologists actively working in the area. Neglecting to point out Black overrepresentation in crimes generally considered a White domain allays any fears of being smeared by allegations of racism. Schatzberg and Kelly (1996) have addressed this concern as it relates to organized crime research and have opined that the primary reason for the academic community's neglect of African American organized crime is that anyone interested in the topic has to "consider the question of race and [accusations of] racism" (p. 21). Likewise, Martens (1990) states that Black involvement in organized crime "is one topic that dare not be discussed, for fear of racism being attributed to the discussants" (p. 43). Other criminologists may practice self-censorship out of a genuine concern that an already disadvantaged group will be further stigmatized if findings pertaining to race are discussed too forthrightly. Whatever the reason may be, Sampson and Wilson (2000) assert that it has resulted in "an unproductive mix of controversy and silence" (p. 149).

Criminologists who specialize in homicide studies are obviously aware that Black serial killers exist and do name them in their work. However, African American serial killers are never represented in the works of these criminologists in proportion to their share of the serial killer population. For instance, of the 19 serial killers discussed in Sears's (1991) book, 2 (10%) were Black; of the 45 discussed in Holmes and Holmes (1998), 2 (4.3%) were Black; Egger (1998) mentions 2 out of 48 serial killers (4.2%); and Fox and Levin (2001) name (but do not discuss) 5 Black serial killers out of 45 (11.1%). What we do see often are statements such as the following: "[Serial killers] were almost never drawn from the ranks of the truly oppressed; there are few women, Blacks or native Americans in our files" (Leyton, 1986, p. 288) and "Serial killing is perpetrated predominantly by white males on white females" (Holmes & Holmes, 1998, p. 31).

Although the words *almost never* and *predominantly* qualify the above statements, making them without further comment gives the reader the impression that White males exhaust the serial killer category. As noted before, it is no surprise that Whites constitute the majority of serial killers in absolute numbers, but it is a surprise that the concepts of proportionality and disproportionality are never mentioned in the works of serial killer researchers. No writer, to our knowledge, has ever pointed out that in relation to their proportion of the population African Americans are overrepresented among serial killers.

Ironically, not focusing on African American serial killers has also been deemed racist because serial killers are seen not only as exclusively White, "but also White and brilliant, thus alluding that Blacks aren't smart enough to carry out organized murder" (Chehade, 2002). An unconscious albeit well-meaning paternalistic racism may thus be another factor in the neglect of African American serial killers among White scholars.

Conclusion

We found that approximately 21.8% of the identified Black or White serial killers in our sample across a 59-year period were African American.

African Americans are thus overrepresented in the ranks of serial killers by a factor of about 2, relative to their proportion of the United States population during the years examined.

Our lack of knowledge of African American serial killers, almost by definition, implies a strong media bias against highlighting it. This may be because of fears about being branded racist, a reluctance to further stigmatize an already stigmatized group, or in the case of the entertainment media, the perception that African American characters have little appeal to White audiences. We can largely dismiss the accusation that a racist police force has little interest in Black crime in modern America when the victims are themselves Black, as evidenced by the 90 African Americans that constitute our sample. If there were still any truth in the accusation, it would mean that even more African Americans than we were able to account for have been engaged in this activity.

Regarding the stereotype that African Americans do not commit serial murder, we concur with Jenkins (1998) that "this apparently favorable stereotype is both inaccurate and as pernicious as any of the more familiar racial slurs" (p. 30). It is pernicious because on one hand, it implies that African Americans lack the requisite "brilliance" to commit such crimes, as Chehade (2002) intimated, and on the other, because it can lead to law enforcement neglecting to protect potential victims in the African American community. One wonders how many additional victims have been lost because law enforcement succumbed to the stereotype and concentrated their efforts on White males.

References

Charles, N. (2002, November 2). Black serial killers: A rare breed. *Harlem Times*, p. 1.

Chehade, C. (2002, November 4). Colorizing crime. *Black Electorate*. Retrieved from http://www.blackelectorate.com/articles.asp?ID=733

D'Sousa, D. (1995). *The end of racism: Principles for a multiracial society*. New York: Free Press.

Egger, S. (1998). *The killers among us: An examination of serial murder and its investigation*. Upper Saddle River, NJ: Prentice Hall.

Emch, D. (2000, April 7). Black brothers admit to murdering eight Whites. *Toledo Blade*, p.1.

Fox, J., & Levin, J. (2001). *The will to kill: Making sense of senseless murder*. Boston: Allyn & Bacon.

Freedberg, S., & Gehreke, D. (1990, December 31). From idealists to "Death Angels"? *Miami Herald*. Retrieved May 30, 2004, from http://www.miami.com/mld/miamiherald/

Greek, C. (2001). *Media crime*. Retrieved May 30, 2004, from Florida State University, School of Criminology and Criminal Justice Web site: http://www.criminology.fsu.edu/crimemedia/lecture4.html

Hickey, E. (1997). *Serial killers and their victims*. Belmont, CA: Brooks Cole

Holmes, R., & Holmes, S. (1998). *Serial murder* (2nd ed.). Thousand Oaks, CA: Sage.

Howard, C. (1979). *Zebra: The true account of the 179 days of terror in San Francisco*. New York: Richard Marek.

Jenkins, P. (1998). African Americans and serial homicide. In R. Holmes & S. Holmes (Eds.), *Contemporary perspectives on serial murder* (pp. 17–32). Thousand Oaks, CA: Sage.

LaFree, G. (1996). Race and crime trends in the United States, 1946–1990. In D. Hawkins (Ed.), *Ethnicity, race, and crime: Perspectives across time and space* (pp. 169–193). Albany: State University of New York Press.

Lambe, J. (2004, May 22). Gilyard connection to suspect possible. *Kansas City Star*, p. A1.

Leyton, E. (1986). *Compulsive killers. The story of modern multiple murder*. New York: Washington Mews.

Lichter, R., Lichter, L., & Rothman, S. (1991). *Watching America*. New York: Prentice Hall.

Lubinskas, J. (2001, Aug. 30). Remembering the zebra killings. *Frontpage Magazin*. Retrieved from http://www.frontpagemag.com/guestcolumnists/lubinskas.

Martens, F. (1990). African American organized crime, an ignored phenomenon. *Federal Probation*, *54*, 43–50.

Newton, M. (1990). *Hunting humans: An encyclopedia of modern serial killers*. Port Townsend, WA: Loompanics.

Newton, M. (1992). *Serial slaughter: What's behind America's murder epidemic?* Port Townsend, WA: Loompanics.

Newton, M. (2000). *The encyclopedia of serial killers*. New York: Checkmark.

Newton, M., & Newton, J. (1991). *Racial and religious violence in America: A chronology*. New York: Garland.

Perazzo, J. (1999). *The myths that divide us: How lies have poisoned American race relations*. Briarcliff Manor, NY: World Studies.

Sampson, R., & Wilson, W. J. (2000). Toward a theory of race, crime, and urban inequality. In S. Cooper (Ed.), *Criminology* (pp. 149–160). Madison, WI: Coursewise.

Schatzberg, R., & Kelly, R. (1996). *African American organized crime: A social history.* New Brunswick, NJ: Rutgers University Press.

Sears, D. (1991). *To kill again: The motivation and development of serial murder.* Wilmington, DE: Scholarly Resources.

Shinkle, P. (2002, June 17). Serial killer caught by his own Internet footprint. *St. Louis Post-Dispatch.* Available from http://www.rense.com/general26/serial.htm

Taylor, J. (1992). *Paved with good intentions. The failure of race relations in contemporary America.* New York: Carroll & Graff.

Thernstrom, S., & Thernstrom, A. (1997). *America in black and white: One nation indivisible.* New York: Simon & Schuster.

White, T., Willis, L., & Smith, L. (2002, October 25). African Americans grapple with race of sniper suspects: Relief at capture, worry about repercussions [Electronic version]. *Baltimore Sun.*

Williams, A. (2002, October 23). *Hate crime reversed.* Retrieved from http://www.townhall.com/columnists/Armstrongwilliams/aw20021023.shtml

Wilson, C., & Seaman, D. (1983). *The encyclopedia of modern murder.* London: Barker.

Wilson, C., & Seaman, D. (1990). *The serial killers.* New York: Carol.

DISCUSSION QUESTIONS

1. What other reason(s) might account for our lack of knowledge about African American serial killers?
2. Is it racist to write about black serial killers or racist not to write about them?
3. Walsh doesn't provide an explanation for why blacks are overrepresented in serial killing. What is your explanation for this phenomenon of black overrepresentation?

READING

The Terrorist Mind I

A Psychological and Political Analysis

Laurence Miller

Miller describes the major forms of terrorism, the motivations of the perpetrators, and the psychological, social, and political forces that contribute to this most particular expression of violence. The article addresses the question of whether all terrorists are sick or evil and considers the possibility that some forms of terrorism, however odious their result, can be a rational response to a situation of perceived intolerable injustice. The article examines what motivates

SOURCE: Miller, L. (2006). The Terrorist Mind: I. A psychological and political analysis. *International Journal of Offender Therapy and Comparative Criminology, 50*(2), 121–138. Reprinted with permission of Sage Publications, Inc.

people to join terrorist groups and what may later move them to leave the terrorist lifestyle. Special consideration is given to the psychological and religious dynamics of suicide terrorism and what might motivate some people to give their lives for their cause. Finally, the article offers recommendations for a multipronged approach to dealing with this modern yet ageless scourge.

The Nature and Purposes of Terrorism

The word *terrorism* derives from the Latin *terrere*, which means to frighten, and the first recorded use of the term as it is currently understood derives from the 18th-century "Reign of Terror" associated with the French Revolution. Although we may think of it as a recent crisis in this country, terrorism is as old as civilization, as timeless as human conflict, and it has existed ever since people discovered that they could intimidate the many by targeting the few. However, terrorism has achieved special prominence in the modern technological era, beginning in the 1970s as international terrorism, continuing in the 1980s and 1990s as American domestic terrorism, and apparently coming full circle in the 21st century with mass terror attacks on United States soil by foreign nationals. Arguably, the two culmination points of domestic and international terrorism in the past decade have been Oklahoma City and the World Trade Center.

The Federal Bureau of Investigation defines terrorism as "the unlawful use of force or violence against persons or property to intimidate or coerce a government, the civilian population, or any segment thereof, in furtherance of political or social objectives" (Seger, 2003). According to the U.S. Department of Defense, "terrorism is the calculated use of violence or threat of violence to instill fear, intended to coerce or try to intimidate governments or societies in the pursuit of goals that are generally political, religious, or ideological" (Seger, 2003). The operative terms in both of these definitions are coercion and intimidation and the ability to convert weakness of numbers into strength of impact.

Almost all conventional warfare contains a terroristic element. Why threaten war at all unless the goal is to intimidate your enemy into complying with your demands? And if they resist, your strategy is to instill as much fear as possible to increase the likelihood of their surrender with minimal casualties on your part, the whole rationale behind "shock and awe"-type campaigns.

On the other hand, where one side's formal battlefield armies are deficient, terrorism puts disproportionate psychological power into the hands of small groups of ideologues or opportunists. Historically, a terrorist act is rarely an end in itself but is rather designed to instill fear in whole populations by targeting a small, representative group; Mao Zedong spoke of "killing one to move a thousand" (Bolz et al., 1996). However, this may be changing. A major characteristic of mass terrorism such as the World Trade Center attack and the much feared potential nuclear-biological-chemical terrorism of the future (Romano & King, 2002) is the terrorists' apparent desire to wreak maximum destruction as an end in itself, going far beyond the symbolic value of the act and turning terrorism into a veritable war of annihilation.

Along these lines, Butler (2002) divides terrorism into two broad categories. *Instrumental terrorism* describes terrorist acts carried out to coerce a group into taking some action or complying with a demand. The perpetrators are usually political terrorists who want to effect a tangible result, such as the IRA's desire to end British control of Northern Ireland or the political faction of the PLO that wants to drive out the Israeli presence in disputed territories. Theoretically, at least, the terror will end if and when the demands are met or a compromise is forged.

By contrast, there is little that may be done to appease the perpetrators of *retributive terrorism;* perpetrators who are primarily interested in destroying, not influencing, their enemies. Here, the target is hated not because of what they do but for the very fact that they exist, so nothing less than their complete eradication will suffice. Often, instrumental-type and retributive-type terrorist groups are admixed and ill defined even among themselves, which further complicates negotiations and compromises.

Several elements appear to be almost universal in modern terrorist activities. The first is the use of violence itself as a primary methodology of influence, persuasion, or intimidation. In this sense, the true target of the terrorist act extends far beyond those directly affected. An Israeli pizza parlor is blown up to effect withdrawal of settlements in the West Bank. The USS Cole is torpedoed in Yemen to drive the infidel from the holy lands. The goal of these activities is to use threats, harassment, and violence to create an atmosphere of fear that will eventually lead to some desired behavior on the part of the larger target population or government. This is instrumental terrorism.

Second, victims are usually selected for their maximum propaganda value, usually ensuring a high degree of media coverage. A great deal of thought may go into the symbolic value of the attacks, or the victims may simply be targets of opportunity. This approach may backfire if the goal is to garner public sympathy and the result is that noninvolved innocents, especially children, are killed along with the symbolic targets. Alternatively, if the aim is to inflict as much pain and panic as possible, then indiscriminate slaughter may serve only too well: The target population had better comply because the terrorists are desperate enough to "do anything." Traditionally, the aim of most terrorist acts has been to achieve maximum publicity at minimum risk, yet the recent spate of suicide bombings in the Middle East and elsewhere shows that fanaticism will often trump caution, and this lack of restraint even in the service of self-preservation is what makes suicide terrorism so frightening.

Third, unconventional military tactics are used, especially secrecy and surprise ("sneak attacks"), as well as targeting civilians, including women and children. This is a commonly cited distinction between a terrorist and a soldier or guerrilla. Again, if the goal is to inflict maximum horror, then it makes sense to choose locations that contain the largest number of victims from all walks of life. Everyone is a target. No one is safe.

In fact, Nacos (2003) points out that the September 11 terrorists precisely calculated their acts to achieve the maximum amount of publicity. She cites an Afghan Jihad terror manual advising holy warriors to target sentimental landmarks such as the Statue of Liberty in New York, the Big Ben in London, and the Eiffel Tower in Paris because of the intense publicity their destruction would generate. Moreover, terrorists hope to stick a thumb in the eye of hypocritical Western civil libertarian values by forcing democratic governments to defensively crack down on terrorism by adopting all kinds of repressive antiterrorism measures.

Fourth, intense and absolutist loyalty to the cause of the organization characterizes most terrorist groups. Although there are exceptions, the bulk of hard-core terrorist members are not typically part-timers or mercenaries. In general, the ability to commit otherwise unspeakable acts—not to mention giving one's own life—necessitates an unshakable belief that these acts are somehow in the cause of some absolute and worthy purpose.

The Changing Demographics of International Terrorism

Most of the international terrorist groups of the 1960s and 1970s consisted of well-educated, well-trained, well-traveled, multilingual, and reasonably sophisticated middle-class men and women. This individual tended to be intelligent, disciplined, and sufficiently resourceful to deal with unforeseen circumstances or last-minute changes in plans to successfully complete his or her mission. New members were typically recruited from among the ranks of university students or within urban cultural centers.

In the 1980s, 1990s, and today, the prototypical foreign terrorist is likely to be a poorly educated, unemployed, and ill-trained male refugee of Middle Eastern origin. These are teenagers or young men who have grown up as members of street gangs, and what formal education they have received has been steeped in extreme religious and political doctrine. They have been taught to hate Western society, and to especially resent those who have been able to escape their drab life and make successes of themselves. Psychologically, this is a defensive reaction formation against the despair of never being able to partake of the bounties they may secretly covet—so these are now viewed as evil temptations, unholy excrescences of Western decadence, to be expunged and destroyed. A smaller group of terrorists may retain the educational status and cosmopolitanism of their relatively more privileged background but find the clash of values threatening to their religious and cultural self-identity.

Another difference between 30 years ago and today is that the regional and international support structure of today's terrorist is not nearly as extensive as it was in the past. Despite the current politicized fears about massive conspiratorial funding and logistical support of worldwide networks of terrorist cells sponsored by powerful rogue nations, the more common trend today is for local terrorist groups to act in relative isolation or with only loose coordination, and their successes depend not so much on paramilitary precision as in their focus on relatively unprotected targets, taking full advantage of weaknesses in the system and the element of surprise.

Today's terrorists actually spend less time and money on training than in the past. Popular media accounts often portray battalions of terrorist recruits receiving the equivalent of a graduate university education in terror technology. Although a few facilities of this type may exist, today's terrorists typically do not receive the type of broad paramilitary training geared for a range of tactics, strategies, and contingencies. More common, they narrowly prepare for a specific mission, the approach being to train fast and hit hard. One reason for this is the availability and expendability of young recruits. Especially for those missions that involve suicide—whether a lone backpacker blowing up a bus or a hijack team turning a plane into a bomb—there is obviously no need to train the perpetrators beyond the operation itself because nobody is coming back.

Becoming a Terrorist

The first issue concerns viewing terrorism as something one does, as opposed to something one is, which relates to personal and group identity. What we believe and what we do comprise the individual threads of our self-definition that together weave the broad cloak of our identity. Different elements of that fabric will have different proportions of meaning for each of us. For some, our vocation may be the most important element; for others, our politics or religion. For most of us in the Western democratic societies, the fabric of our identity is stitched together from a range of textures and colors. We of the Western world are able to weave so richly a textured raiment of identity because our economy and culture give us so wide a range of choices.

But for most of the poorer and less-advantaged peoples of the world, menial work gives little satisfaction, political freedom is sparse or nonexistent, avenues of recreational escapism are few, and social mobility and hope for a better life is little more than a fantasy. From such existential remnants, these people are forced to stitch together a patchwork quilt of meaning that is stiff and irrefutable, to shield them from the harsh climate of their daily lives. For such people, ideologies become the guardians of identity (Gibbs, 2005). If people have already got next to nothing, the one thing you cannot take away from them is their religious or political or philosophical belief, especially if that belief tells them that all their travails and deferred dreams are for a reason, a loftier purpose, God's will, or the purity of the race. If doubts arise, or circumstances occur to challenge any of the meager securities in their lives, imagine how they will fight to preserve the few knotted cords of identity that keep their worldview from unraveling. Add to this the fact that adolescence and young adulthood are the critical periods for identity formation, and the reason that many terrorists come from the ranks of the young and disaffected becomes clear.

In areas where this deprivation is combined with overt political persecution, a collective sense of injustice swells and simmers. Combine this with the typical period of adolescent angst and adrenaline and embed it in an eye-for-an-eye religious culture, and terrorism or some other extreme action may be seen as a perfectly legitimate means of striking back. What better way to defuse one's desires for the comforts of the material life than to sublimate one's yearnings into 180-degree antipathy, to brand that subconsciously coveted life as categorically evil and try to destroy it? And far better than eruptions of individual rebellion that may be condemned as mere street crime, politically or religiously guided terrorism has the added benefit of receiving the sacred imprimatur of a respected community, if not the society as a whole. The nascent terrorist thereby focuses the frustration, channels his or her aggression, and gains respect.

Although the causes and motivations for terrorism are complex and there is no set formula for creating a terrorist group, certain regularities in the psychological and sociopolitical dynamics of such groups have been identified. In one representative model, the evolution of the terrorist mind-set is divided into four stages (Borum, 2003). Stage 1 ("It's not right") begins with an individual or group identifying some set of conditions in their life that is unpleasant, undesirable, or unacceptable. This can be poverty, political repression, runaway immorality, or anything else that produces confusion, discomfort, or distress.

Stage 2 ("It's not fair") involves a basis of comparison. Not only do we—through no fault of our own—have it bad, others—through no credit of theirs—have it better. This breeds resentment and a desire to find a cause of this gross injustice.

This leads to Stage 3 ("It's your fault"), in which the cause of the injustice is projected onto a vilifiable out-group, alien culture, or corrupt regime. All the sociopolitical complexities of the situation are homogenized and distilled into a single, all-purpose explanation for the in-group's travail: The White supremacists' problem is the Blacks; the Palestinians' problem is the Zionists; the Chechens' problem is the Russians; the Northern Irish's problem is the British; the Muslim fundamentalists' problem is the whole Western world. And if your group or society or way of life is persecuting us, tormenting us, keeping us down, and laughing in our faces, then: "You're evil" is the logical 4th stage of the process, in which the purported exploiters and tormenters are dehumanized and demonized. By this logic, any aggressive corrective action on the part of the in-group is justified and elevated to noble resistance and freedom fighting. If you're evil and I'm good, then I'm entitled—indeed, obligated—to destroy you in the name of righteousness, to make the ends justify the means and to dismiss as collateral damage any innocent bystanders who happen to get in the way.

Assuming one makes the decision to become a terrorist, how does one start? Much depends on the nature of the political or religious organization one wishes to join. Some terrorist organizations are entities unto themselves, whereas others exist as splinter groups of larger, more mainstream political organizations. Some groups will be eager to attract new members, whereas others will screen prospective applicants carefully. In some communities, terrorist group members may be well known to the general populace; in others, the neighbors may be surprised when the double life is eventually revealed.

Typically, the would-be terrorist approaches some larger legitimate group associated with the terrorist cause, starts working with it, and may eventually express an interest in, or be selected for, more dangerous work. The recruit may be subjected to successive tests of loyalty and commitment by proving oneself through increasingly dangerous acts. Some individuals join mainstream organizations just to do conventional work but are eventually socialized into the more radical aspects of the group's activities.

In other cases, indigenous or expatriate members of the population, often itinerant workers, come in contact with political group members and are befriended by them. The participant is typically not aware at first that their new friends have any association with an extremist organization. A soft-pedal approach to recruitment then ensues, by way of discussions and commiseration with their mutual plight. When the participant expresses a strong desire and willingness to do something, the new friends then decide that the time is right to make the appropriate introductions and suggest to the participant that there is indeed a way to put his beliefs into action.

Suicide Terrorists

What is worth giving up your life? That may depend on what you think your life is worth. For some individuals, death in the name of a noble cause may be the one act that gives life its ultimate meaning—a paradox that most of us have immense difficulty comprehending.

The first problem concerns the term "suicide." In the Western mind, suicide is invariably associated with despair, capitulation, depression, and a disordered mind. Western religions generally discourage purposeless suicide and are not entirely comfortable with the idea of giving one's life even as an affirmative act of faith. We tend to take religion as one component of a full life, not the end-all and be-all of life itself, and we recoil at the idea of willingly giving our lives for a religious principle, even if some of us might have less compunction about killing others, as the actions of both Medieval crusaders and modern self-appointed slayers of abortion doctors have demonstrated. But we look on suicide bombers in about the same way as we view self-immolating Buddhist monks—with a sense of bemused but itchy revulsion: We just don't get it.

This kind of understanding requires an appreciation of the role death plays in many people's conception of life. For most religions, death may be the entree into one or another form of eternal life. Becker (1973, 1975) pointed out how annealing oneself to something that transcends one's own life can give a person what he called "immortality power." Religion is the most obvious choice because most faiths promise an afterlife of some sort. Whether you accept this literally as playing a harp on a cloud or more metaphorically as a melding with some great universal consciousness, it boils down to the essential reassurance that when you die, you do not really die. Somehow, in some form, you defeat death by continuing on.

But immortality power is not denied to the secularist or atheist, either. Indeed, the messianic zeal that has characterized proponents of socialism, fascism, humanism, or any of the sweeping sociopolitical movements of the past century illustrates their power to grant their adherents at least historical and philosophical immortality. I may be the smallest cog in the grand engine of historical destiny, but I am thus part of the whole machine and thereby derive both limitless power and eternal existence through my connection with it. The reason that religion and politics are such loaded topics is because to challenge someone's belief system is to threaten not just their life's meaning but their very (eternal) life itself.

Add to this uncounted millennia of human evolution within close-knit, intra-dependent, insular tribal clans, each guided by its own totemic deities, and it is easy to see why we quite naturally gravitate to the beliefs of our in-group, and why, especially under conditions of stress, scarce resources, and conflict, my in-group will be elevated to absolute righteous goodness and your out-group will be loathed and demonized. This conceptualization explains the paradoxical acts of altruism that seem to fly in the face of self-preservation, acts that are familiar to every war movie buff who has gotten choked up at the brave soldier throwing himself on a grenade to save his buddies. He willingly dies so that his companions might live. Why?

In Becker's (1973, 1975) interpretation, what the soldier is really saving is his transcendent alliance with his warrior clan and, by extension, the survival and immortality of his nation and cultural heritage. He physically dies, but his people live on, and by extension, so does he. It certainly helps him to know that God is on his side, too, as this gives him the added bonus of real immortality and honor in heaven. But even this spiritual perk may not be necessary, as history has shown millions willing to sacrifice themselves, or at least put themselves in grave danger, for the sake of political beliefs or even the chance of personal wealth, honor, or power—a species of material and temporal immortality to supplement the spiritual and eternal kind.

Accordingly, it makes sense that suicide attacks seem to spring most readily from cultures that condone and encourage self-sacrifice, especially in the context of long-running conflicts that have endured extensive and repeated casualties on both sides. Suicide attacks, in this view, may emerge from a sense of desperation but not despair or depression. Suicide attackers are not killing themselves as a way of "going out and taking as many of you as I can" in the kind of final exit that marks the suicidal workplace violence perpetrator. This is not reckless self-destruction but a noble act, the ultimate sacrifice—gift, even—that the suicide attacker can give for his cause and his people.

The few psychological analyses that have been carried out on this subject have found that Middle Eastern suicide terrorists are rarely the wild-eyed crazies caricatured in the Western media. Rather, these are typically young men in their late teens and early 20s who have been generally well-behaved youth in their communities, good students, and regarded as helpful and generous. They come from relatively stable, religious homes, often with large extended families. But like many terrorists, at least part of their decision to sacrifice themselves comes from the rage and resentment at what they perceive to be an endless onslaught of unjust persecutions and humiliations at the hands of the out-group. Thus, it is not depression and despair that fuels their self-sacrificial impulse but the assertive, energetic desire to fuse themselves with something greater and stronger, to become one with an eternal and omnipotent vindicating force.

That this is not the purposeless throwing away of lives is further illustrated by the simple secular fact that, culture notwithstanding, the usefulness of suicide missions is often dictated by the numbers. Considering the sacrifice involved, terrorist organizations seek to maximize the deadly payoff, and suicide bombings that kill only one or two other persons represent a poor investment return. In the grim calculus of Mideast terror, for example, Palestinians typically glean a respectable margin on their bombings, favoring targets that concentrate their victims tightly, such as loaded buses or markets: The average ratio for Palestinian suicide terrorists is 7 dead and 30 injured per bombing.

In addition to return on investment, suicide bombing is relatively easy. A certain level of training, preparation, and precision timing is required to pull off an attack if the aim is to complete the mission and then escape to fight another day. In contrast, all the suicide bomber has to do is literally show up and pull a cord or press a cell phone button. As Ismail Abu Shanab, the assassinated Hamas leader, is quoted as saying, all one needs to qualify as a suicide bomber is "a moment of courage" (Van Natta, 2003). Indeed, this is as close as anyone ever gets to actually, literally, going out in a "blaze of glory."

Glory, however, may sometimes need a little outside help, and so many terrorist organizations build into the process a number of fail-safe measures to ensure that the suicide mission is completed. For one, emphasizing indoctrination of the subject and leaving the tactical details to others limits the number of psychological junctures at which doubts or wavering motivation might intrude. As part of the rigorous indoctrination process, there occur a number of "point-of-no-return rituals" to ensure compliance. These include having members write last letters to friends and relatives, videotaping a goodbye narrative, saying final prayers, and so on. Once a person has pledged himself or herself to a suicide mission, groups such as Hamas and Islamic Jihad cement that commitment by thenceforth referring to the member as "the living martyr." In essence, the person has already left the physical world and exists in a temporary corporeal state solely to carry out his last mission on earth.

Even so, although rare, there are a few isolated cases in which suicide attackers have changed their minds at the last minute. As a final safeguard against such last-moment derelictions, some groups take the decision out of the bomber's hands and arrange for remote control detonation (Silke, 2003a). In some cases, this is part of the indoctrination ritual: Like any successful operation, the fewer surprises, the better.

Leaving Terrorism

Like the Mafia, it is generally assumed that the only way out of a terrorist organization is feet first. The terrorist lifestyle is not an easy one, and although most members stay committed, some actually do leave. The motivations for quitting may vary, depending on the personality of the terrorist and the nature of the organization. Some terrorist organizations seem to rely on a kind of freelance subcontractor system involving operatives who perform a specific task and then drop out of sight until they are needed next time. The advantage for the organization is that this looseness makes the ties hard to trace. The drawback is lack of control, putting the organization at the mercy of even one blabbermouth or loose cannon.

At the other extreme are terrorist organizations that use only a select cadre of dedicated operatives, carefully selected, screened, and indoctrinated, who have earned the right to carry out missions through a hierarchical progression of skill and loyalty tests. Certainly, all gradations of commitment are seen in various extremist groups. In some cases, once you are in, there is only one terminal opt-out policy. In other cases, members who do not wish to carry out violent missions may be assigned intelligence, technical, and other support roles. In the latter case, the member's distaste for personally carrying out violence does not diminish his commitment to the broader goals of the group. In still other cases, the roles are fluid and members are cross-trained, alternating between direct attack and behind-the-scenes support operations from mission to mission.

It is difficult to just quit something to which you have devoted your life, heart, and soul for any length of time, for which you have made sacrifices and burned bridges. The theory of *cognitive dissonance* states that, when confronted with information that disconfirms your beliefs or devalues your actions, the first impulse is to dig in your heels and cling ever more desperately to the crumbling belief structure. People want to be existentially consistent and will struggle and fight to retain this consistency in their beliefs. Only when overwhelmed by the sheer volume of contrary evidence does the belief system topple, and in the tailspin of what Horgan (2003a) calls the "spiraling of commitment," the individual may then attempt to squeeze out one desperate last act of violent activity to prove the rightness of the cause before then being snapped 180 degrees and coming to revile the very group for which the person was once willing to give his or her life. Alternatively, he or she may react to this shattering of his worldview by simply slinking away in a state of burned-out anomie.

That is, assuming that any form of elective retirement is even an option. Although giving one's notice may not automatically be fatal, the decision to leave a terrorist organization is rarely as simple as a goodbye and a handshake. The organization may have a lot invested in the terrorist member in terms of time and training. There is always the security issue, and sometimes vows to keep silent are not enough. In religious groups, there may be a great deal of social stigma and ostracism by the supporting community for betraying the ideals of one's faith. Once having left the group, any protection from rival groups, law enforcement, or government agencies is gone: "Once you're out, you're on your own." This, plus the reactive antipathy the former member may now feel for his erstwhile comrades, makes the individual a prime security risk.

The good news for the rest of us is that even a few exceptions to the "once-a-terrorist-always-a-terrorist" rule may provide insights and tools for deterring terrorism. If we can learn more about what makes terrorists give up the life, perhaps we can encourage more of them to seek alternatives to destruction for expressing their concerns, whatever we may think of the merits of their cause. But inasmuch as the best intervention is prevention, far better still would be to work the path backward to the causes of terrorism, to find a way to pull the fuse from the bomb before it is lit.

Stopping Terrorism

Currently, the conventional antiterrorism methodology consists of a surgical version of shock and awe. After a terrorist act occurs, find out who and where the perpetrators are, hit them soon, hit them hard, and thereby teach a lesson to other miscreants who may be thinking of hatching similar plots. This approach, which is the standard response of antiterrorism units throughout the world, may be fundamentally misguided. In fact, systematic analysis shows that the standard retaliatory approach to terrorism not only fails to deter and discourage it but in fact actually increases the violence by encouraging retribution. This leads to more counterattacks and counter-counterattacks, resulting in the now clichéd "cycle of violence."

The problem with most military reactions to terrorist attacks is that they rarely are the total, all-cleansing, scorched-earth responses that the military forces imagine them to be. In fact, part of the standard doctrine of retaliation consists of a proportional or step-wise approach to retaliation, as if terrorist acts were equivalent to workplace rule infractions or schoolyard hijinks. A graded response defeats the whole purpose of deterrence completely: If you want to create habituation, what better way to do it than administer carefully titrated doses of punishment, progressively inoculating your enemy to further retaliation and thus emboldening him by your perceived impotence.

Indeed, factions at war with each other seem to engage in a strange kind of folie-à-deux mis-logic (Silke, 2003b) that goes something like this: If they kill our people, this shows just how evil they are, and their aggression will only stiffen our resolve in our righteous cause and motivate us to fight to the bitter end. On the other hand, if we kill their people, that will show them we mean business, teach them a lesson, and they will melt into submission at our mighty force. The result? Escalating violence on both sides. Perhaps unwilling to be seen as utter barbarians, the stronger side fails to wage a war of total destruction, diluting the effect of whatever half-hearted retaliation it applies and thereby achieving not subjugation but further rebellion.

So why do these ineffectual actions continue? One reason is that they make good press for politicians and generals who generally prefer doing something to doing nothing. Even if total victory is elusive, who in power wants to be the one that backs down a step, who calls a truce, who searches for dialogue? What leader wants to be seen as "soft on terrorism"? Indeed, public opinion consistently shows that, in the face of attack, populations of even the most civilized Western democracies generally view negotiations and diplomacy as hapless dithering and prefer their leaders to act forthrightly and aggressively, especially when dealing with foreign enemies.

To the extent that terrorism is at least partly motivated by legitimate social and political grievances of the host population, any comprehensive antiterrorism approach must inevitably deal with these factors. Unfortunately, this is often misinterpreted as giving in or coddling terrorists. But engagement is not the same thing as surrender, and many an adversary has been defanged by being given the courtesy of just being heard. A key principle of all forms of *active listening*—whether in criminal interrogation, business negotiation, or psychological crisis intervention—is to engage your opposite number by taking the time to listen to his or her point of view, demonstrating that you have done your best to understand it, and then presenting your own point of view. You do not have to agree on the issue. You may even go to war over it later. But once you have established a dialogue, it is hard to completely forget or discount it, and the possibility of further communication will hang over the smoke of battle, even as the conflict rages on.

Talking will not stop all terrorists, of course. Where terrorism represents an act of desperation to achieve otherwise legitimate rights and freedoms by a marginalized, disenfranchised, or persecuted group, the ruling power's willingness to put at least some issues on the table might well be effective in stemming further terrorist acts because the subgroup now feels it has something to gain, at least for now, from the dialogue. But with many domestic or international extremists, fueled by religious or racial fanaticism and programmed to destroy their enemy at all costs, no amount of either accommodation or forceful counterreaction is likely to deter them. They want to kill us because they hate us, period. Indeed, habituation can also occur to well-meaning measures, as when small concessions are mocked and reviled as being too little too late, mere window dressing or cosmetic changes, or worse, taken as signs of weakness to be exploited. Just as the doctrine of proportional, step-by-step aggressive retaliatory response to terrorism must be carefully rethought, so must the form and amount of agreement and concessions so that making peace is not mistaken for giving in or selling out.

For peace to truly work, justice must be provided for victims of terror. As Truth and Reconciliation Commissions and the so-called Restorative Justice Movement illustrate, evildoers—whatever their original motivation and rationalization—must undergo some societally sanctioned consequences for their actions. Also, formal social arrangements must be made that acknowledge and validate the unacceptable suffering that was imposed on victims, and that also makes future harm-doing less likely. Simply arresting and releasing violent terrorists as political bargaining chips sends a corrosive message to victims and to society that we do not really take this matter seriously, that it is all a political game. For in our fascination with the terrorist mind-set—no matter how clinical or lurid that interest may be—we cannot and must not overlook or abandon the victims of terror, because, if for no other reason, the next victims may be us.

References

Becker, E. (1973). *The denial of death.* New York: Free Press.

Becker, E. (1975). *Escape from evil.* New York: Free Press.

Bolz, F., Dudonis, K. J., & Schultz, D. P. (1996). *The counterterrorism handbook: Tactics, procedures, and techniques.* Boca Raton, FL: CRC Press.

Borum, R. (2003, July). Understanding the terrorist mindset. *FBI Law Enforcement Bulletin,* 7–10.

Butler, P. (2002). Terrorism and utilitarianism: Lessons from, and for, criminal law. *Journal of Criminal Law and Criminology, 93,* 1–22.

Gibbs, S. (2005). Islam and Islamic extremism: An existential analysis. *Journal of Humanistic Psychology, 45,* 156–203.

Horgan, J. (2003a). Leaving terrorism behind: An individual perspective. In A. Silke (Ed.), *Terrorists, victims and society: Psychological perspectives on terrorism and its consequences* (pp. 109–130). Chichester, UK: Wiley.

Jenkins, P. (1994). *Using murder: The social construction of serial homicide.* New York: Aldine De Gruyter.

Nacos, B. L. (2003). The terrorist calculus behind 9-11: A model for future terrorism? *Studies in Conflict and Terrorism, 26,* 1–16.

Romano, J. A., & King, J. M. (2002). Chemical warfare and chemical terrorism: Psychological and performance outcomes. *Military Psychology, 14,* 85–92.

Seger, K. A. (2003). Deterring terrorists. In A. Silke (Ed.), *Terrorists, victims and society: Psychological perspectives on terrorism and its consequences* (pp. 257–269). Chichester, UK: Wiley.

Silke, A. (2003a). The psychology of suicidal terrorism. In A. Silke (Ed.), *Terrorists, victims and society: Psychological perspectives on terrorism and its consequences* (pp. 93–108). Chichester, UK: Wiley.

Silke, A. (2003b). Retaliating against terrorism. In A. Silke (Ed.), *Terrorists, victims and society: Psychological perspectives on terrorism and its consequences* (pp. 215–231). Chichester, UK: Wiley.

Van Natta, D. (2003, August 24). Terror's ultimate weapon. *The New York Times,* pp. E1, E7.

DISCUSSION QUESTIONS

1. According to Miller, why does something as bad as terrorism exist?
2. Explain the difference between instrumental and retributive terrorism.
3. Explain Miller's conception of altruistic suicide. Can you see yourself, under any circumstances at all, doing what a suicide bomber does?

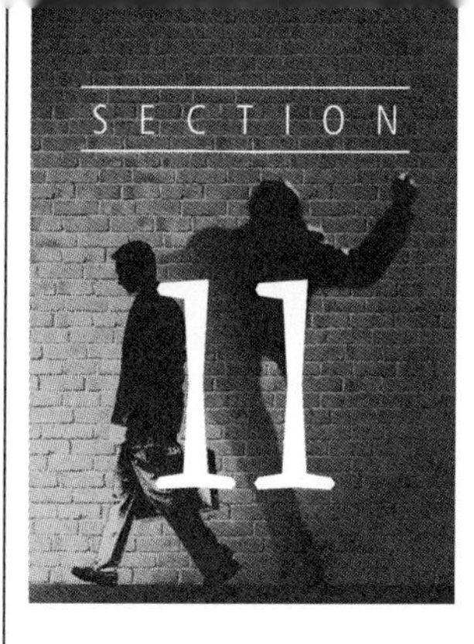

PROPERTY CRIME

Jay Scott Ballinger was a property offender on a dark mission. He admitted in court to setting fire to between 30 and 50 churches in 11 states between 1994 and 1999. A volunteer firefighter was killed in one of the blazes, which makes Ballinger a murderer, too. Ballinger did not set these fires for profit or because he got some weird sexual kick from watching them burn; he did so on an anti-Christian mission. Ballinger, his girlfriend Angela Wood, and accomplice David Puckett traveled around the country seeking converts to Satanism, and churches were the sanctuaries of the enemy. The trio would burglarize and shoplift, and Woods would work as a stripper, to finance their travels around the Midwest and South.

It all came to an end when Ballinger was arrested after paramedics treating him for severe burns wondered why he waited two days to seek treatment (he had burns to 40% of his body and had to receive four skin grafts). A police officer who remembered Ballinger's name from a previous investigation questioned him and then summoned ATF (Alcohol, Tobacco, and Firearms) agents, who found fire-setting paraphernalia and satanic literature at Ballinger's home. Among the writings agents found were 50 "contracts" signed by teenagers in their own blood, pledging their souls to the devil and to do "all types of evil," for which they would be rewarded with wealth, power, and sex, the perennial male motivators. Ballinger, aged 36 at the time of his arrest, was described as a misfit loner and high school dropout who was more comfortable with teens than people his own age. He was sentenced to 42 years in prison for his multi-state arson spree (Ross, 1999). As the Ballinger case shows, crimes against property can sometimes morph into something much more deadly.

While violent crime gets the lion's share of media attention, 88.3% of the 11,695,264 offenses reported to the police in 2005 were property crimes. Property crimes either involve the illegal acquisition of someone else's property (money or goods) or the malicious destruction of property (sometimes including one's own). Just about everyone will be victimized by a property offense at some time in life, but few of us will be victimized by a serious violent crime. It is also true that while the vast majority of us will never commit a violent crime, most of us have committed, or will commit, a property offense of some kind, such as pilfering items from work or shoplifting.

Figure 11.1 shows property crime trends in the United States over the history of the National Crime Victimization Surveys (Catalano, 2006), showing how it has been dropping dramatically over the past two decades. Recall that the NCVS only includes offenses against individuals and households; it does not include the huge number of offenses committed against commercial establishments (stores, garages, warehouses, factories, and so forth), thus leaving out a very large number of property offenses.

Figure 11.1 Thirty-Year Trends in U.S. Property Crime

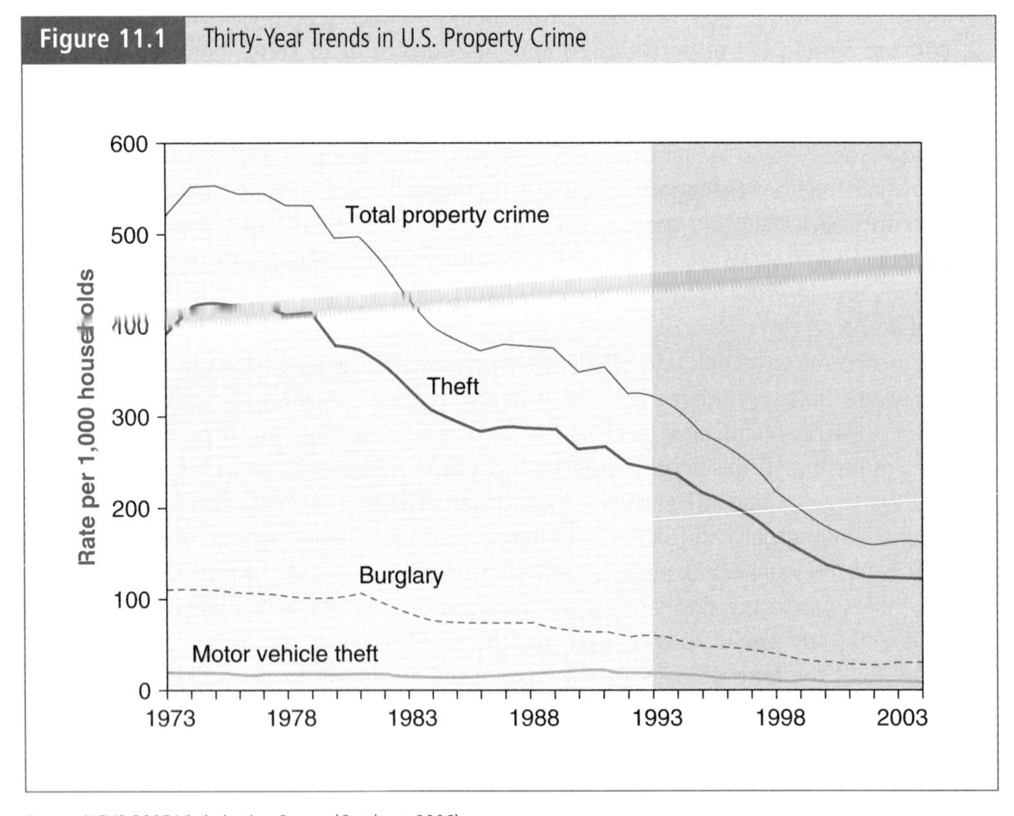

Source: NCVS 2005 Victimization Survey (Catalano, 2006).

Larceny/Theft

Larceny/theft is the most common property crime committed in the United States and is defined as "the unlawful taking, leading, or riding away from the possession or constructive possession of another" (FBI, 2006, p. 49). The number of larceny/thefts reported in the United States in 2005 was 6,776,807, for a rate of 2,286.3 per 100,000 U.S. residents. This crime constituted 66.7% of property crimes in 2005. The average loss due to larceny/theft was $764, for a total national loss of approximately $5.2 billion. Of those arrested for larceny/theft in 2005, 61.7% were males. Whites (a category which includes non-black Hispanics) were 69.3% of larceny/theft arrests, blacks 28%, and other races 2.7%.

In early English law, larceny only applied to persons who achieved possession of goods belonging to others by stealth or force; it did not cover persons who abused their victim's trust to steal from them. For instance, if person A gave person B a sheep to graze in B's field, but B killed and ate the sheep, no larceny was committed because A voluntarily handed over the sheep to B. You can see the problems this would cause in today's society where every day numerous people hand over money to bank tellers and vehicles and appliances to mechanics. Lawmakers responded to this by enlarging the definition of larceny to include taking by fraud or false pretenses as well as by stealth and force. Taking by stealth has evolved into other crimes such as burglary or embezzlement, and taking by force has evolved into the crime of robbery.

▲ **Photo 11.1** Actress Winona Ryder reacts to her sentencing for shoplifting at a Saks Fifth Avenue store on Rodeo Drive. On December 6, 2002, she was sentenced to three years probation, 480 hours of community service, fined $11,300, and ordered to undertake drug and psychological counseling.

Today larceny/theft covers most types of theft that do not include the use of threats or force; excluded is theft of a motor vehicle, forgery, passing bad checks, and embezzlement. Larceny/theft includes grand theft (a felony) and petty theft (a misdemeanor), with the distinction depending on the value of the asset stolen. The cutoff value varies from state to state, but presently it is under $1,000 in every state. Whether a person is charged with grand or petty theft depends on the value of the item at the time it was stolen, not its replacement value. A stolen computer bought for $700 but now worth $100 is a petty theft, but a stolen guitar bought for $100 in 1955 which is now a classic valued at $10,000 is classified as a grand theft. Because the grand theft/petty theft distinction varies across states, the UCR includes both grand theft and petty theft as the same thing in its yearly larceny/theft tally.

Larceny/theft is subclassified into shoplifting, pocket picking, purse snatching, thefts from motor vehicles (except for parts and accessories), theft of motor parts and accessories, theft of bicycles, and theft from buildings. Theft from motor vehicles is the most common type and includes thefts from just about any type of motorized vehicle, such as automobiles, trucks, buses, or motor homes.

Shoplifting—theft by a person other than an employee of goods exposed for sale in a store—is the most studied of the subcategories of larceny/theft. Compulsive shoplifting is sometimes defined as a psychiatric problem (kleptomania, Greek for "stealing madness"). For instance, actress Winona Ryder was arrested in 2001 for shoplifting about $5,000 worth of goods from a Saks Fifth Avenue store in Beverly Hills (Mowbray, 2002). Ryder's "five-finger discount" is one of millions occurring each year, costing the retail industry approximately $31 billion in 2004 (Orlando Business Journal, 2005). This loss is passed on to customers in the form of increased prices.

About 8% of shoplifters say they engage in the practice as a primary source of income (Moore, 1984). These are the individuals with the greatest level of expertise who know how to minimize the risk of being caught, target the most expensive items, and steal almost exclusively for resale. Most shoplifters, however, are impulsive amateurs who shoplift relatively inexpensive items on the spur of the moment and for their own gratification (Lamontagne, Boyer, Hetu, & Lacerte-Lamontagne, 2000). One self-report study is consistent with rational choice theory in that shoplifters said that they engaged in shoplifting simply because it is an easy, low-risk crime for which there are abundant opportunities (Tonglet, 2001).

Jack Katz's article in this section shows that the reward received from such crimes as shoplifting often go far beyond any material gain or peer recognition. He argues that a big part of the reward for engaging in crime comes from the sheer thrill of getting away with something in the face of the possibility that one could get caught. He points out that for many crimes, there is little economic gain; it is, rather, the "seduction" of crime itself, the conquering of fear, and finally the euphoric thrill of completion that is the real payoff. Like Albert Cohen in Section 4, Katz believes that such rewards are underappreciated and not understood by most criminologists. Katz's "sneaky thrills" proposition is consistent with findings that criminals tend to score high on impulsiveness, risk taking, and sensation seeking.

Burglary

Burglary is "the unlawful entry of a structure to commit a felony or theft" (FBI, 2006, p. 45). Burglary has always been considered a very serious offense under common law, dating back many centuries, because of the importance attached to the sanctity of the home. Victims of residential burglary experience feelings of anger, fear, and a profound sense of invasion of privacy and vulnerability, in addition to financial loss. The original common law definition involved breaking and entering at nighttime with the intention of committing a felony, but now includes any unlawful entry whether forceful or not and regardless of the time of day.

The number of burglaries reported to the police in 2005 was 2,154,126, for a rate of 730.5 per 100,000. Of these, 62.4% were residential with an average loss of $1,725. Of those arrested for burglary in 2005, 85.7% were male, 69.6% were white, 28.5% black, and the remaining 1.9% were of other races.

Burglars and Their Motives

The "typical" burglar is a young male firmly embedded in street culture. They are perhaps a little less daring than robbers because there is less chance of victim contact, injury, and identification for burglary than for robbery, and the penalties and probability of arrest are lower (the national clearance rate for burglary in 2005 was 12.7 versus 25.4 for robbery; FBI, 2005). Almost all of the 105 active burglars interviewed by Wright and Decker (1994) admitted numerous other offenses they had committed. They evidenced pride in their ability to exploit the range of criminal opportunities that come their way, but most of them considered burglary their preferred crime because it offers the greatest chance of success and reward with the least amount of risk.

Demographically, Wright and Decker's (1994) burglars came from poor run-down and socially disorganized neighborhoods where unemployment was rife. They were poorly educated, unreliable, resistant to taking orders, and most came from single-parent homes. Consistent with Walter Miller's focal concerns concept addressed in Section 4, there was a strong sense of toughness and masculine independence, fate ("I had little choice but to burgle"), excitement (sexual activity, drugs, alcohol), autonomy ("As a burglar, I'm my own man"), and smartness (outwitting the law; getting something for nothing).

Wright and Decker (1994, p. 58) state that burglars constantly need money quickly to finance their high living, but many of them also reported that they found the psychic rewards of committing burglaries a secondary benefit, describing the act as "an adventure," "a challenge," "fun," "exciting," and "thrilling." Given the lack of legitimate skills and general trustworthiness they see among burglars, Wright and Decker are dubious about the possibility that job creation programs can change burglars into law-abiding citizens because most burglars see burglary as being far more profitable than working. Rengert and Wasilchick (2000, p. 47) also reject the notion that burglary is a default option of the jobless, claiming that many burglars give up jobs to concentrate on burglary: "Unemployment is not what caused crime. Crime caused the unemployment." Legitimate employment simply would not fit into these people's lives because for them, party time is all the time. Almost all of the proceeds of Wright and Decker's burglars were spent on drugs, alcohol, and sex, and legitimate jobs provide neither adequate time nor money to engage in these pleasures to the extent desired.

In 2004, 14.3% of persons arrested for burglary were females. Females overwhelmingly commit burglaries in mixed gender teams, and thus obviously share most of the demographic characteristics of their male partners (Mullins & Wright, 2003). Mullins and Wright found that most of the women were initiated into burglary by their boyfriends, and some were coerced ("if you love, me you'll do it") against their will. Mullins and Wright also found that female burglars capitalized on their sexuality to locate potential targets and gain access to homes that they and their partners would burglarize later. Once inside a target's home, they could assess potential valuables and entry points and perhaps even discover where their victim kept spare keys. They could also elicit other important information, such as the target's schedule, so that they and their partner could be sure to enter the house when the target was not at home.

Choosing Burglary Targets

Selecting a suitable home for burglary is an obvious concern for burglars. The four most important considerations in target selection, according to Mawby (2001, p. 29) are *target exposure, guardianship, target attractiveness,* and *proximity.*

- Target exposure refers to the visibility and accessibility of the home, i.e., isolation from other homes and easy access via side and back doors shielded by abundant trees and shrubs. Can the premises be seen by neighbors and passers-by?

- Guardianship refers to how well the home is protected. Does the home show signs of occupancy, such as cars in the driveway, lights on, music playing; is there a burglar alarm or dog present, or is there mail in the mailbox, newspapers in the foyer, and a general silence about the place?
- Target attractiveness refers to signs that there should be rich pickings in the house. Previous surveillance may have revealed high-priced cars in the driveway or delivery trucks delivering expensive items.
- Proximity refers to the distance between the target home and the burglar's home.

All these concerns are relative to the "professionalism" of the burglar. High-level burglars may travel miles to a particularly attractive target after very careful surveillance and planning, but the majority of burglars are low- to mid-level opportunists who engage only in rudimentary, even spur-of-the-moment, planning. For these individuals, the "planning" of a burglary is little more than opportunism. Proximity is important both because burglars are most familiar with their own areas and because many of them lack transportation. As one of Wright and Decker's (1994, p. 86) subjects put it, "I ain't gonna go no further than 10 blocks; that's a ways to be carryin' stuff. . . . Since I'm on foot, I got to keep walkin' back and forth until I get it all." As is the case with murderers, robbers, and rapists, the great majority of low- and mid-level burglars prey on residents in the same neighborhoods in which they reside. Target exposure and attractiveness is simply making the best of a bad deal for such burglars because the pickings are pretty slim in their neighborhoods.

Guardianship is the most important consideration for low- and mid-level burglars, with many choosing homes occupied by individuals known to them, such as neighbors, acquaintances, and even friends (Mawby, 2001, Wright & Decker, 1994). Wright and Decker (1994, p. 70) report the statement of one of their respondents: "I be knowin' what house I'm going to hit. It could be a friend of mine, I could be over at his house all last week, I know he got a new VCR, we be lookin' at movies. I know what time they work, I know where his wife at or he stay by himself." Typically, the only planning such individuals do is to call and see whether their intended victims are home. Some of those who victimize friends and acquaintances do report occasional pangs of guilt but justify their actions as the result of a desperate need to get money for another drug fix. Given their willingness to criminally exploit almost anyone, we can easily see why Wright and Decker (1994, p. 72) characterized their sample of burglars (who obviously lack the social emotions) as "self-centered individuals without notably strong bonds to other human beings; their allegiance seemed forever to be shifting to suit their own needs."

Disposing of the Loot

The most immediate pressure facing burglars after a successful burglary is to convert the stolen goods into cash. Burglars turn to a variety of sources to dispose of the loot, with fences being the preferred method. A fence is a person who regularly buys stolen property and who often has a legitimate business to cover his activities. Only a minority of burglars (the high-level burglar) use a fence because fences prefer to deal only with people they trust. Fences are valued by burglars who use them because it is the fastest way to get rid of "hot" property, and they can be trusted to be discreet. Fencing is a UCR Part II crime, which is formally known as receiving stolen property. Anyone knowingly buying or possessing stolen property can be charged with this crime.

▲ **Photo 11.2** One way the police respond to street crime is to create maps detailing locations where reported crimes have occurred. Note in this photo that, as of 2002, the NYPD was still using pushpins rather than sophisticated GIS software to research crime location patterns.

Burglars without connections to a professional fence must turn to other outlets. One method is a pawnshop, but this is not a very popular outlet for most burglars because pawnbrokers must ask for identification, take pictures of people selling to them, and possess "hot sheets" of stolen goods. Some burglars who have developed a trusting relationship with certain pawnbrokers are able to sell "off camera," but because pawnbrokers always have the upper hand in negotiations and offer very little for the goods, only 13 of Wright and Decker's (1994) sample said that they regularly used them. A more popular outlet was the drug dealer because it can entail a strict "drugs-for-merchandise" deal without involving any middleman. Others regularly sold to relatives, friends, and acquaintances because few people can resist buying merchandise at below even "fire-sale" prices. Because of the high value of the property they go after, high-level burglars would never use any of these alternatives to the professional fence.

Richard Wright's article in this section explains burglary from the point of view of burglars. As in the Katz article, Wright describes the emotions felt when planning the burglary, being in the burgled home, and the aftermath of the burglary. Mostly, however, it explores the A to Z of committing a burglary once entrance to the house has been made. It talks primarily of where, when, and how to search a residence and how burglars feel and act during the search phase of the burglary.

Motor Vehicle Theft

Motor vehicle (MV) **theft** is simply "the theft or attempted theft of a motor vehicle" (FBI, 2006, p. 55), that is, any motorized land vehicles such as motorcycles, buses, automobiles, trucks, and snowmobiles, although the vast majority of thefts involve automobiles. There were an estimated 1.2 million MV thefts in 2005, for a rate of 416.7 per 100,000. The nationwide loss attributable to MV theft was approximately $7.6 billion. Of those arrested for MV theft, 82.9% were males, 62.8% white, 34.8% black, and other races the remainder. Table 11.1 presents the top 10 most stolen vehicles in 2005 and the cities where they are most likely to be stolen. The older vehicles are stolen primarily for their parts.

Most MV thefts are committed by juveniles strictly for fun. Juveniles will spot a "cool" car with the keys in the ignition, steal it, drive it around until it runs out of gas, and then abandon it. Some of the more malicious joyriders will get an additional kick by smashing it up a little first (Rice & Smith, 2002). The high recovery rate of stolen vehicles (about 62%) indicates that most MV thefts are for expressive (to show off, to get some kicks) rather than for instrumental reasons (financial gain; Linden & Chaturvedi, 2005). Motor vehicles are also obviously stolen for profit. Most vehicles stolen for profit are taken to so-called "chop shops" where they are stripped of their parts and accessories. These items are easily sold to auto supply stores, repair shops, and individuals who get faster delivery at a cheaper price than they would from legitimate suppliers. Other stolen vehicles may be shipped abroad where they are

Table 11.1 Ten Most Stolen Vehicles and 10 Cities with Highest MV Theft Rates, 2005

Vehicle	Rating	City
1991 Honda Accord	1	Modesto, CA
1995 Honda Civic	2	Las Vegas/Paradise, NV
1989 Toyota Camry	3	Stockton, CA
1994 Dodge Caravan	4	Phoenix/Mesa/Scottsdale, AZ
1994 Nissan Sentra	5	Visalia/Porterville, CA
1997 Ford F150 Series	6	Seattle/Tacoma/Bellevue, WA
1990 Acura Integra	7	Sacramento/Arden-Arcade/Roseville, CA
1986 Toyota Pickup	8	San Diego/Carlsbad/San Marcos, CA
1993 Saturn SL	9	Fresno, CA
2004 Dodge Ram Pickup	10	Yakima, WA

SOURCE: National Insurance Crime Bureau. (2006).

worth more than they are in the United States. Some professional auto thieves (called *jockeys*) even steal particularly high-value vehicles "to order" for specific customers.

The most serious form of MV theft is **carjacking** (the theft or attempted theft of a motor vehicle from its occupant by force or threat of force). Carjacking thus involves multiple crimes for which the offender may be charged, including robbery, assault, and MV theft. Carjacking has increased significantly recently (an average of 34,000 incidents per year over the past decade; Klaus, 2004). Media accounts of carjacking have sparked many copycat offenders: "Hey, I can steal any vehicle I want without damaging it, I get the car keys, and I can rob the owner too; what a concept!" (McGoey, 2005, p. 1).

Victims report that 93% of the carjackers were male, 3% involved a male/female team, and 3% were lone females. By race/ethnicity, 56% were African American, 21% white, with the remainder not identified by race. Interviews with active carjackers find that they are more like street robbers in terms of motivations and lifestyles than they are like professional car thieves (Jacobs, Topoli, & Wright, 2003). Many times carjacking is a spur-of-the-moment thing: "I was broke. I didn't have enough bus fare. I'm walking down the street, there's a guy sitting in his car. I go ask him for change. He was going for his pocket. I just grabbed him outta his car. Why just take his change when I can take his car and get a little bit more?" (Jacobs et al., 2003).

Arson

Arson is "any willful or malicious burning or attempting to burn, with or without intent to defraud, a dwelling house, public building, motor vehicle or aircraft, personal property of another, etc." (FBI, 2005, p. 61). Arson was added to the UCR in 1979, and there still exists a great deal of difficulty in gathering statistics from reporting agencies because of the problem of deciding whether a "suspicious" fire was arson. Only fires that have been determined by investigators to have been willfully and maliciously set are classified as arsons. For these reasons, the UCR does not provide an estimated national rate for arson, but did report a total of 67,504 arsons in 2005. The average loss for property destroyed or damaged in 2005 was $14,910. Fully 50.5% of all persons arrested for arson in 2005 were juveniles. Whites accounted for 76.6% of arson arrests and blacks for 21.2%. Males accounted for 83.7% of arson arrests.

Arson can have a variety of instrumental motivations, such as financial gain, revenge, and intimidation, or expressive motivations that may signal psychopathology of some sort. For instance, an owner of a failing business may hire a professional arsonist (a *torch*) to burn down his or her place of business, a person may set fire to the property of another because of some perceived wrong suffered, or labor unionists may set fires in a labor dispute to intimidate management, as was the case with the massive Dupont Hotel fire in Puerto Rico in 1986, which led to the death of 97 people and injured 140 others.

Because juveniles comprise only about 6% of the American population but are arrested in 50% of the arsons, expressive motivations are of great interest. Juvenile fire-setting may be the result of curiosity and may never be repeated if the juvenile is caught and dealt with, but persistent fire-setters are another matter. A variety of studies have shown that persistent fire-setters have higher levels of other antisocial behaviors, and higher levels of hostility and impulsiveness, and lower levels of sociability and assertiveness than youths in general (Brett, 2004; Hakkanen, Puolakka, & Santilla, 2004).

Crimes of Guile and Deceit: Embezzlement, Fraud, and Forgery/Counterfeiting

The crimes we have discussed thus far are "physical" rather than "cerebral," but many crimes are committed each day purely by guile and deceit. The UCR lists three Part II property crimes of this type: embezzlement, fraud, and forgery/counterfeiting.

Embezzlement is the misappropriation or misapplication of money or property entrusted to the embezzler's care, custody, or control. Embezzlement is the rarest of property crimes, with only 12,087 cases being reported in the UCR in 2005. Females (50.5%) were actually arrested more often than males in 2005 (FBI, 2005). Whites constituted 67% and blacks 31% of arrests. Most embezzlers do what they do because they have some pressing financial problem, or simply because they have access to money and the ability to hide any discrepancies for some time. After being exposed, many insist that they were only "borrowing the money" and that they fully intended to pay it back.

Banks have long been embezzlement targets, but the advent of computers has made it both easier to commit embezzlement and more lucrative. In the first decade of the "computer revolution" in banking, arrests for embezzlement rose 56%, with the average loss to banks per computer embezzlement crime being as high as $500,000, compared with the average loss per armed bank robbery of just over $3,000 (Rosoff, Pontell, & Tillman, 1998). A favorite method of stealing via the computer is known as the salami ("slicing off") technique, whereby the embezzler will open up "phantom accounts" in his or her name and slice off a few cents from a large number of accounts whose owners are hardly likely to notice. This technique can garner the embezzler large sums of money over a period of time (Rosoff et al., 1998).

Fraud is theft by trick, namely, obtaining the money or property of another through deceptive practices such as false advertising and impersonation. The FBI (2006) reported 217,650 arrests for fraud in 2005, of which 55.1% were males and 44.9% were females. In terms of race, 68.7% were white and 30.1% were black.

Obtaining resources by fraudulent means probably began when the first human being realized that he or she could obtain them with less risk and effort by using brains rather than brawn. Examples of fraud include dishonest telemarketing, quack medical cures, phony faith healers, "cowboy" home repair companies, price gouging, and diploma mills promising "accredited" college degrees for a lot of money and little study.

Forgery is the creation or alteration of documents to give them the appearance of legality and validity with the intention of gaining some fraudulent benefit from doing so. Strictly speaking, forgery is the "false writing" of a document and *uttering* is the passing of that document to another with knowledge of its falsity and with intent to defraud. One can thus commit a forgery without uttering (passing the document on) and can utter (passing a forged document on he or she did not forge) without committing a forgery.

Counterfeiting, the creation or altering of currency, is a special case of forgery. In most states, forgery and counterfeiting are allied offenses, which is the reason that they appear that way in the UCR. Would-be counterfeiters no longer need the engraver's fine craftsmanship to produce quality plates for professional counterfeiters. Printing currency with copiers available today has become an amateur's do-it-yourself enterprise (just feed in a $20 bill and press the button). There were 83,747 arrests for forgery/counterfeiting in 2005, of which 60.2% were males and 39.8% were females. By race, 70.7% were white and 27.6% black (FBI, 2006).

Cybercrime: Oh What a Tangled World Wide Web We Weave

Cybercrime is the use of computer technology to criminally victimize unwary individuals or groups. Any invention that can be used by criminals to exploit others has been, but few of these have been as useful as the computer. Now even the weak and timid who would never dream of using a gun to rob or otherwise victimize someone can steal, assault, and harass in the comfort of his or her home with little or no risk involved. Everyone who enters cyberspace, uses a credit card, and/or has a social security number—which means just about everybody—is a potential victim of cybercrime. We have seen that conventional criminals, such as robbers and burglars, typically operate in their own or nearby neighborhoods, but the global reach of the Internet now allows someone in Birmingham, England to victimize someone in Birmingham, Alabama, or vice versa. The number of offenses it is possible to classify under cybercrime are legion, ranging from terrorism (the targeting of a country's computer-run infrastructures such as air traffic control and power grid systems) to simple e-mail harassment. We can concentrate only on the most common ones here, beginning with identity theft.

Identity Theft

Identity theft is the use of someone else's personal information without their permission to fraudulently obtain goods and services. According to a Federal Trade Commission Report, over 9.9 million Americans were victims of identity theft in 2003, with an estimated total loss to consumers of $5 billion (Stafford, 2004). Identity theft can range from a criminal's short-term use of a stolen or lost credit card to the long-term use of a person's complete biographical information (name, social security number, and other identifiers) to "clone" the victim's identity and to commit multiple crimes that may be attributed to the victim.

Criminals gain access to the personal information of others in a variety of ways. They can steal it, buy it, or simply be given it by their unwary victims. People are continually providing confidential information to all sorts of businesses and agencies that goes into huge data banks that may be legitimately accessed by employees who may steal it, or they can be "hacked" into and information stolen. Credit card numbers can be copied during a financial transaction, such as when a restaurant server takes your card for processing, or they can be surreptitiously recorded on a skimming device, which is typically a cigarette-pack-size device that is run across a credit card to record the electronic information in the magnetic strip. This information is then used to make duplicate cards. Thieves can also steal original checks left in mailboxes for pickup, copy the information on them, and buy duplicate checks from mail-order firms.

Another method is *phishing*, which, as the name implies, involves thieves casting thousands of fraudulent e-mails into the cyberpond asking for personal information and waiting for someone to bite. Phishers may send out official looking e-mails with a bank logo asking recipients to "update" their information or telling them that their account may have been fraudulently used and that the bank needs to "verify" their personal information. One study indicated that 40% of recipients of one fraudulent bank e-mail believed it to be real (Kshetri, 2006). A victim may also be literally scared into providing his information. Imagine receiving an e-mail from "Lolita Productions" telling you that your credit card has been billed $99.95 for the first two child pornography CDs and that it will be automatically billed $49.95 each month for further CDs. The message also says that if you want to cancel membership you should e-mail back with full credit card details "for verification." Knowing the penalties for possessing child porn, you may be anxious to do anything to free yourself from the electronic embrace of Lolita Publications.

Perhaps the most notorious phishers are the so-called "Nigerian frauds" run by Nigerian organized crime groups. E-mails have been received by millions of people the world over. A small number of people fall for them. These people are first asked to send a small amount of money (perhaps $200 or less) to "cover expenses," but are suckered into sending ever larger amounts as "complications" arise. Some of the more gullible have even been lured to Nigeria with their cash and have been killed (Baines, 1996).

Most stolen identity information is not for the personal use of the thief but for sale to others. An organization of about 4,000 individuals called the Shadowcrew stole large volumes of personal information for many years and arrogantly advertised and sold it on Web sites worldwide. If you wanted to buy card numbers with security codes, you could get 50 of them for $200; if you wanted the same thing complete with the original owner's social security number and date of birth, you would only have to pay $40 each (Levy & Stone, 2005). Leading members of the Shadowcrew were arrested by the U.S. Secret Service in 2004, effectively closing the business that authorities estimated had trafficked at least 1.5 million credit and bank cards, account numbers, and other counterfeit documents such as passports and driver's licenses (Department of Justice, 2004).

Denial of Service Attack: Virtual Kidnapping and Extortion

Denial of service (DoS) attacks occur when criminals "kidnap" a business Web site or threaten to kidnap it so that business cannot be conducted. DoS attacks are accomplished by overloading the computational resources of the victim's system by flooding it with millions of bogus messages and useless data. Sometimes an attack is simply malicious mischief carried out by computer-savvy disgruntled employees, customers, or just someone who has a bone to pick with the services the company provides. Other times DoS attacks are committed by criminals who demand ransom. On-line gambling sites are prime targets for cyber-extortionists because a "kidnapped" Web site cannot accept bets and stand to lose millions. Paying the ransom is cheaper than losing the business, especially if the threat comes during peak operation times. Millions of dollars have been paid to cyber-extortionists, with only a miniscule few ever reported to the police (Kshetri, 2006). Not that the police could do much anyway, as many of these attacks originate overseas.

Who Are the Hackers?

A **hacker** may be simply defined as someone who illicitly accesses someone else's computer system. Hackers may be seen as the upscale version of Albert Cohen's lower-class delinquents we met in Section 4, who engaged in malicious, destructive, and non-utilitarian vandalism "just for the heck of it." We do not include in this categorization people who hack into computers for instrumental or political reasons, such as professional criminals or cyberterrorists.

There are many kinds of hackers, some of whom are purely interested in the intellectual challenge of breaking into difficult systems and do so without damaging them, and others (sometimes known as cyberpunks or virtual vandals) who break into systems and implant viruses to destroy data. Most, however, appear to be intellectual thrillseekers who enjoy the challenge of doing something illegal and getting away with it (Voiskounsky & Smyslova, 2003). There is something of a counterculture among these people analogous to graffiti artists and gang members. They take on cybernames such as Nightcrawler and Kompking and romanticize and tell stories about their accomplishments, as well as the accomplishments of "legendary" hackers. Gaining the respect of fellow hackers serves as a source of psychological reinforcement for them in ways similar to ordinary street delinquents (Kshetri, 2006). Hackers tend to be young white males, loners, "nerdy," high IQ, and idealistic, but also unpopular with others, prone to lying and cheating, and perhaps prone to drug and/or alcohol abuse (Power, 2000; Voiskounsky & Smyslova, 2003).

Software Piracy

Software piracy is illegally copying and distributing software for free or for sale. The Business Software Alliance (BSA) has estimated that the worldwide cost of software piracy in 2004 was $31 billion, and that although the United States has the lowest piracy rate (ratio of legitimate to pirate market) in the world, it leads the world in losses to piracy—about $7 billion annually (BSA, 2005). In the U.S., the illegal market is 21% of the total market, whereas in countries such as China, Vietnam, and Russia, the illegal market is around 90% of the total. The BSA estimates that worldwide, for every two dollar's worth of software purchased, one dollar's worth was illegally obtained.

Software piracy is a crime, but few people see it as such unless multiple copies are made and sold for profit. Many view making copies for friends the same way they view loaning books to them—"I bought it; shouldn't I be able to give it to whomever I please?" Having purchased something legally, they see no reason why the law should mandate that he or she only should be allowed rights to it. A large survey of university employees found that this was indeed the attitude of many. Forty-four percent of the respondents said that they had obtained unauthorized copies of software and 31% said that they had made such copies (Seale, Polakowski, & Schneider, 1998). The cost of piracy to the economy, however, has moved the Congress to authorize heavy penalties for offenders. Infringement of copyright can land a first-time offender in prison for 5 years if 10 or more copies are made in a 6-month period, or 10 years for subsequent offenses (Cornell Law School, 2005).

Majid Yar's article on cybercrime examines the phenomenon through the lens of rational choice theory. He looks at the question of whether cybercrime is a completely new form of criminal behavior requiring its own theoretical explanation, or whether those who commit "virtual" crimes are no different (perhaps a little brighter) than those who commit "terrestrial" crimes and thus cybercrime does not require special theoretical treatment. Yar concludes that there are significant similarities between the two forms of crime, but also there are significant differences that limit the utility of rational choice theory to explain cybercrime.

Summary

◆ Property crimes constitute the vast majority of crimes committed in the U.S. Larceny/theft is the most common of these crimes and is divided into misdemeanor and felony categories depending on the value of the stolen property. Shoplifting has received the most attention of all the subcategories of larceny/theft, largely because some people supposedly suffer from a psychiatric condition known as kleptomania.

◆ Burglary is a more serious property offense because it violates victims' homes and could lead to personal confrontation. Studies of burglars find them to be motivated by the need to get quick and easy cash to finance a hedonistic lifestyle. They are typically members of the lower class with the focal concerns of that class, such as seeking excitement and autonomy. Female burglars typically work as auxiliaries to male partners and they, too, spend much of their money on alcohol and drugs.

◆ Motor vehicle theft is a serious larceny often committed by joyriding juveniles, although many vehicles are stolen for profit. Carjacking is a relatively new method of stealing cars made "necessary" by the improvement of anti-theft devices. There is a high degree of injury to the victim inherent in this crime, which is essentially a robbery.

◆ Arson is a particularly dangerous property crime because it can lead to deaths. Juveniles commit the majority of arsons, suggesting that it serves some expressive function for them. Compulsive fire-setting signals some very serious underlying psychological problems.

◆ Crimes of guile and deceit included in the UCR property crime classification are embezzlement, fraud, and counterfeiting/forgery. There are only small gender differences in the commission of these crimes.

◆ Cybercrime is the use of computer technology to criminally victimize unwary individuals or groups. There are many different forms of cybercrime ranging from the use of the Internet to steal identity information to the destruction of whole computer networks. Other offenses include denial of services "kidnapping" and the very common crime of software piracy.

EXERCISES AND DISCUSSION QUESTIONS

1. Survey several friends or classmates and ask them whether they have ever shoplifted and how they felt afterwards. What were their motivations; were some guilty, some proud about getting way with it; did some get a thrill out of it?

2. According to the burglary researchers discussed in this chapter, why or why not would it be wise policy to implement job training and job creation programs for convicted burglars?

3. Go to http://www.sosfires.com/new.html and view the various youth fire-setting intervention programs listed there. Click on two or three of the research reports listed there and report about what you have learned about juvenile fire-setting and its treatment.

USEFUL WEB SITES

FTCs Identity Theft Site. http://www.ftc.gov/bcp/edu/microsites/idtheft/.

InterFire. http://www.interfire.org/.

Motor Vehicle Theft Statistics. http://www.fbi.gov/ucr/cius_04/offenses_reported/property_crime/motor_vehicle_theft.html.

Property Crime Statistics. http://www.fbi.gov/ucr/cius_04/offenses_reported/property_crime/index.html.

Property Crime Trends. http://www.ojp.gov/bjs/glance/house2.htm.

GLOSSARY

Arson: Any willful or malicious burning or attempting to burn, with or without intent to defraud, a dwelling house, public building, motor vehicle or aircraft, or personal property of another.

Burglary: The unlawful entry of a structure to commit a felony or theft.

Carjacking: The theft or attempted theft of a motor vehicle from its occupant by force or threat of force.

Counterfeiting: The creation or altering of currency.

Cybercrime: The use of computer technology to criminally victimize unwary individuals or groups.

Embezzlement: The misappropriation or misapplication of money or property entrusted to the embezzler's care, custody, or control.

Forgery: The creation or alteration of documents to give them the appearance of legality and validity with the intention of gaining some fraudulent benefit from doing so.

Fraud: Obtaining the money or property of another through deceptive practices such as false advertising, impersonation, and other misrepresentations.

Hacker: Someone who illicitly accesses someone else's computer system.

Identity theft: The use of someone else's personal information (credit card, social security number, etc.) without his or her permission to fraudulently obtain goods and services.

Larceny/theft: The unlawful taking, leading, or riding away from the possession or constructive possession of another.

Motor vehicle theft The theft or attempted theft of a motor vehicle.

Shoplifting: Theft by a person other than an employee of goods exposed for sale in a store.

Software piracy: Illegally copying and distributing software for free or for sale.

READING

Sneaky Thrills

Jack Katz

Jack Katz's book, *Seductions of Crime* (1988), from which this excerpt is taken, emphasizes the emotional aspect of criminal activity rather than the rational. He contends that crime has its own intrinsic rewards in the form of an emotional thrill independent of any material reward it may bring. He shows how the crime of shoplifting is fraught with anxiety and danger, and the thrill comes from overcoming that and getting away with the merchandise, which is often of little monetary value. The real reward is neurological; the buildup of tension followed by the relief once out of the store—stealing for the sake of stealing, and nothing more. For Katz, the real seduction of crime is the challenge of overcoming obstacles to the commission of a crime and the emotional high that accompanies the successful accomplishment of a crime.

Various property crimes share an appeal to young people, independent of material gain or esteem from peers. Vandalism defaces property without satisfying a desire for acquisition. During burglaries, young people sometimes break in and exit successfully but do not try to take anything. Youthful shoplifting, especially by older youths, often is a solitary activity retained as a private memory. "Joyriding" captures a form of auto theft in which getting away with something in celebratory style is more important than keeping anything or getting anywhere in particular.

In upper-middle-class settings, material needs are often clearly insufficient to account for the fleeting fascination with theft, as the account by one of my students illustrates:

> I grew up in a neighborhood where at 13 everyone went to Israel, at 16 everyone got a car and after high school graduation we were all sent off to Europe for the summer. . . . I was 14 and my neighbor was 16. He had just gotten a red Firebird for his birthday and we went driving around. We just happened to drive past the local pizza place and we saw the delivery boy getting into his car. . . . We could see the pizza boxes in his back seat. When the pizza boy pulled into a high rise apartment complex, we were right behind him. All of a sudden, my neighbor said, "You know, it would be so easy to take a pizza!" . . . I looked at him, he looked at me, and without saying a word I was out of the door. . . . got a pizza and ran back. . . . (As I remember, neither of us was hungry, but the pizza was the best we'd ever eaten.)

It is not the taste for pizza that leads to the crime; the crime makes the pizza tasty.

Qualitative accounts of *initial* experiences in property crime by the poorest ghetto youths also show an exciting attraction that cannot be explained by material necessity. John Allen, whose career as a stickup man living in a Washington, D.C., ghetto . . . recalled his first crime as stealing comic books from a junkyard

Source: Katz, J. (1988). *Seductions of crime.* New York: Basic Books. Reprinted by permission.

truck: "we destroyed things and took a lot of junk—flashlights, telephones." These things only occasionally would be put to use; if they were retained at all, they would be kept more as souvenirs, items that had acquired value from the theft, than as items needed before and used after the theft.

What are these wealthy and poor young property criminals trying to do? A common thread running through vandalism, joyriding, and shoplifting is that all are sneaky crimes that frequently thrill their practitioners. Thus I take as a phenomenon to be explained the commission of a nonviolent property crime as a sneaky thrill.

In addition to materials collected by others, my analysis is based on 122 self-reports of university students in my criminology courses. Over one-half were instances of shoplifting, mostly female; about one-quarter described vandalism, almost all male; and the rest reported drug sales, nonmercenary housebreaking, and employee theft. In selecting quotations, I have emphasized reports of female shoplifters, largely because they were the most numerous and sensitively written.

The sneaky thrill is created when a person (1) tacitly generates the experience of being seduced to deviance, (2) reconquers her emotions in a concentration dedicated to the production of normal appearances, (3) and then appreciates the reverberating significance of her accomplishment in a euphoric thrill. After examining the process of constructing the phenomenon, I suggest that we rethink the relationships of age and social class to devious property crime.

Flirting With the Project

In the students' accounts there is a recurrent theme of items stolen and then quickly abandoned or soon forgotten. More generally, even when retained and used later, the booty somehow seems especially valuable while it is in the store, in the neighbor's house; or in the parent's pocketbook. To describe the changing nature of the object in the person's experience, we should say that once it is removed from the protected environment, the object quickly loses much of its charm.

During the initial stage of constructing a sneaky thrill, it is more accurate to say that the objective is to be taken or struck by an object than to take or strike out at it. In most of the accounts of shoplifting, the shoplifters enter with the idea of stealing but usually do not have a particular object in mind. Indeed, shoplifters often make legitimate purchases during the same shopping excursions in which they steal. The entering mood is similar to that which often guides juveniles into the short journeys or sprees that result in pranks and vandalism. Vandals and pranksters often play with conventional appearances; for example, when driving down local streets, they may issue friendly greetings one moment and collectively drop their pants ("moon") to shock the citizenry the next. The event begins with a markedly deviant air, the excitement of which is due partly to the understanding that the occurrence of theft or vandalism will be left to inspirational circumstance, creative perception, and innovative technique. Approaching a protected property with disingenuous designs, the person must be drawn to a particular object to steal or vandalize, in effect, inviting particular objects to seduce him or her. The would-be offender is not hysterical; he or she will not be governed by an overriding impulse that arises without any anticipation. But the experience is not simply utilitarian and practical; it is eminently magical.

Magical Environments

In several of the students' recountings of their thefts, the imputation of sensual power to the object is accomplished anthropomorphically. By endowing a thing with human sensibilities, one's reason can be overpowered by it. To the *Alice in*

Wonderland quoted below, a necklace first enticed—"I found the one that outshone the rest and begged me to take it"—and then appeared to speak.

> [15] There we were, in the most lucrative department Mervyn's had to offer two curious (but very mature) adolescent girls: the cosmetic and jewelry department. . . . We didn't enter the store planning to steal anything. In fact, I believe we had "given it up" a few weeks earlier; but once my eyes caught sight of the beautiful white and blue necklaces alongside the counter, a spark inside me was once again ignited. . . . Those exquisite pukka necklaces were calling out to me, "Take me! Wear me! I can be yours!" All I needed to do was take them to make it a reality.

Another young shoplifter endowed her booty, also a necklace, with the sense of hearing. Against all reason, it took her; then, with a touch of fear, she tossed it aside in an attempt to exorcise the black magic and reduce it to a life-less thing.

> [56] I remember walking into the store and going directly to the Jewelry stand. . . . This is very odd in itself, being that I am what I would consider a clothes person with little or no concern for accessories. . . . Once at home about 40–45 minutes after leaving the store, I looked at the necklace. I said "You could have gotten me in a lot of trouble" and I threw it in my jewelry box. I can't remember the first time I wore the necklace but I know it was a very long time before I put it on.

The pilferer's experience of seduction often takes off from an individualizing imputation. Customers typically enter stores, not to buy a thing they envisioned in its particularities but with generic needs in mind. A purchased item may not be grasped phenomenally as an individualized thing until it is grasped physically. Often, the particular ontology that a possession comes to exhibit—the charm of a favorite hat or an umbrella regarded as a treasure—will not exist while the item sits in a store with other like items; the item will come to have charm only after it has been incorporated into the purchaser's life—only when the brim is shaped to a characteristic angle or the umbrella becomes weathered. But the would-be thief manages to bring the particular charm of an object into existence before she possesses it. Seduction is experienced as an influence emanating from a particular necklace, compact, or chopstick, even though the particular object one is drawn to may not be distinguishable from numerous others near it.

In some accounts, the experience of seduction suggests a romantic encounter. Objects sometimes have the capacity to trigger "love at first sight." Seduction is an elaborate process that begins with enticement and turns into compulsion. As a woman in her mid-thirties recalled:

> A gold-plated compact that I had seen on a countertop kept playing on my mind. Heaven knows I didn't need it, and at $40 it was obviously overpriced. Still, there was something about the design that intrigued me. I went back to the counter and picked up the compact again. At that moment, I felt an overwhelming urge.

Participant accounts often suggest the image of lovers catching each other's eyes across a crowded room and entering an illicit conspiracy. The student next quoted initially imagines herself in control and the object as passive—she is moving to put it in her possession; but at the end of her imagining, the object has the power to bring her pocket to life.

> [67] I can see what I want to steal in plain sight, with no one in the aisle of my target. It would be so easy for me to get to the Chapstick without attracting attention and

> simply place it in my pocket. . . . I'm not quite sure why I must have it, but I must. . . .

Some of the details that would make the deviant project hard or easy really are not up to the would-be shoplifter. In part the facility of the project is a matter of environmental arrangements for which she has no responsibility. While she is appreciating the object as a possible object of theft, she considers it at a particular angle. She will approach it from this side, with her back to that part of the scene, taking hold of it at just that part of its surface. In her experience that "It would be so easy," she is mobilizing herself to concentrate on the tangible details of the object. Thus the would-be shoplifter's sense of the facility of the project is constituted not as a feature of her "intent" or mental plan, but as a result of the position of the object in the store and the posture the object takes toward her.

To specify further how the would-be shoplifter endows the inanimate world with a real power to move her, we might consider why the initial stage of magical provocation is part of the project of sneaky thrills and not of other, equally fascinating, forms of deviance. Not all projects in deviance begin with the seductive sense, "It would be so easy." Indeed, some projects in deviance that are especially attractive to young people begin with an appreciation of the difficulties in becoming seduced to them.

As the budding shoplifting project brings the object of deviance to life, the person and the object enter a conspiratorial relationship. "It would be so easy" contains a touch of surprise in the sudden awareness that no one else would notice. The tension of attraction/hesitation in moving toward the object is experienced within a broader awareness of how others are interpreting one's desires. For all *they* know, one's purposes are moral and the scene will remain mundane. The person's situational involvement in sneaky property crimes begins with a *sensual concentration on the boundary between the self as known from within and as seen from without.*

The Reemergence of Practiced Reason

Independent of the would-be shoplifter's construction of a sense that she might get away with it are any number of contingencies that can terminate the process. For example, the sudden attentions of a clerk may trigger an intimidating awareness of the necessity to produce "normal appearances."

At some point on the way toward all sneaky thrills, the person realizes that she must work to maintain a conventional, calm appearance up to and through the point of exit. The timing of this stage, relative to others in the process, is not constant. The tasks of constructing normal appearances may be confronted only after the act is complete; thus, during the last steps of an escape, vandals may self-consciously slacken their pace from a run to normal walking, and joyriders may slow down only when they finally abandon the stolen car.

In shoplifting, the person occasionally becomes fascinated with particular objects to steal only after appreciating an especially valuable resource for putting on normal appearances. In the following recollection of one of my students, the resource was a parent:

> [19] I can clearly remember when we coaxed my mom into taking us shopping with the excuse that our summer trip was coming up & we just wanted to see what the stores had so we could plan on getting it later. We walked over to the section that we were interested in, making sure that we made ourselves seem "legitimate" by keeping my mom close & by showing her items that appealed to us. We thought "they won't suspect us, two girls in school uniforms with their mom, no way." As we carried on

> like this, playing this little game, "Oh, look how pretty, gee, I'll have to tell Dad about all these pretty things." Eventually a necklace became irresistible.

Whichever comes first, the pull of the person toward the object to be stolen or the person's concentration on devices for deception, to enact the theft the person must bracket her appearance to set it off from her experience of her appearance, as this student's account shows:

> [19] My shoplifting experiences go back to high school days when it was kind of an adventurous thing to do. My best friend & I couldn't walk into a store without getting that familiar grin on our faces. . . . Without uttering a word, we'd check out the place. . . . The whole process pretty much went about as if we were really "shopping" except in our minds the whole scene was different because of our paranoia & our knowledge of our real intentions.

Sensing a difference between what appears to be going on and what is "really" going on, the person focuses intently on normal interactional tasks. Everyday matters that have always been easily handled now rise to the level of explicit consciousness and seem subtle and complex. The thief asks herself, "How long does a normal customer spend at a particular counter?" "Do innocent customers look around to see if others are watching them?" "When customers leave a store, do they usually have their heads up or down?" The recognition that all these questions cannot possibly be answered correctly further stimulates self-consciousness. As one student expressed it,

> [19] Now, somehow no matter what the reality is, whether the salesperson is looking at you or not, the minute you walk in the store you feel as if it's written all over your face "Hi, I'm your daily shoplifter."

Unless the person achieves this second stage of appreciating the work involved—if she proceeds to shoplift with a relaxed sense of ease—she may get away in the end but not with the peculiar celebration of the sneaky thrill. Novice shoplifters, however, find it easy to accomplish the sense that they are faced with a prodigious amount of work. "Avoiding suspicion" is a challenge that seems to haunt the minute details of behavior with an endless series of questions—How fast should one walk? Do customers usually take items from one department to another without paying? and so on.

To construct normal appearances, the person must attempt a sociological analysis of the local interactional order. She employs folk theories to explain the contingencies of clerk-customer interactions and to guide the various practical tasks of the theft. On how to obscure the moment of illicit taking:

> [44] The jewelry counter at Nordstrom's was the scene of the crime. . . . I proceeded to make myself look busy as I tried on several pairs of earrings. My philosophy was that the more busy you look the less conspicuous.

On where to hide the item:

> [15] Karen and I were inside the elevator now. As she was telling me to quickly put the necklace into my purse or bag, I did a strange thing. I knelt down, pulled up my pants leg, and slipped the necklace into my sock! I remember insisting that my sock was the safest and smartest place to hide my treasure. I knew if I put it in an obvious place and was stopped, I'd be in serious trouble. Besides, packages belonging to young girls are usually subject to suspicion.

Some who shoplift clothes think it will fool the clerks if they take so many items into a dressing room that an observer could not easily keep count, as this student recalled:

> [5] We went into a clothing shop, selected about six garments apiece (to confuse the sales people), entered separate dressing rooms and stuffed one blouse each into our bags.

Others, like the following student, think it sufficiently strategic to take two identical garments in, cover one with the other, and emerge with only one visible:

> [46] We'd always take two of the same item & stuff one inside the other to make it seem like we only had one.

Many hit on the magician's sleight of hand, focusing the clerk's attentions on an item that subsequently will be returned to hide their possession of another.

> [56] [While being watched by a clerk] I was now holding the green necklace out in the open to give the impression that I was trying to decide whether to buy it or not. Finally, after about 2 minutes I put the green necklace back but I balled the brown necklace up in my right hand and placed my jacket over that hand.

In its dramatic structure, the experience of sneak theft has multiple emotional peaks as the thief is exposed to a series of challenges to maintaining a normal appearance. The length of the series varies with the individual and the type of theft, but, typically, there are several tests of the transparency of the thief's publicly visible self, as one student indicated:

> [122] I can recall a sneak theft at Penney's Dept. store very well. I was about 12 years old. . . . I found an eye shadow kit. I could feel my heart pounding as I glanced around to make sure that others weren't watching. I quickly slipped the eye shadow in my purse and sighed heavily with relief when I realized that no one had seen. I nervously stepped out of the aisle and once again was relieved when I saw that there was no one around the corner waiting to catch me. I caught my friend's eye; she gave me a knowing glance and we walked to the next section in self-satisfaction for having succeeded so far.

When the person devises the deviant project in advance, even entering the store normally may be an accomplishment. Having entered without arousing suspicion, the would-be shoplifter may relax slightly. Then tension mounts as she seizes the item. Dressing rooms provide an escape from the risk of detection, but only momentarily, as in this student's account:

> [19] So, here we were, looking at things, walking around & each time getting closer to the dressing room. Finally we entered it & for once I remember feeling relieved for the first time since I'd walked into the store because I was away at last from those "piercing eyes" & I had the merchandise with me. At this point we broke into laughter. . . . We stuffed the items in our purses making sure that they had no security gadgets on them & then we thought to ourselves "well we're halfway there." Then it hit me, how I was safe in the dressing room, no one could prove anything. I was still a "legitimate" shopper.

Then a salesperson may come up and, with an unsuspecting remark, raise the question of transparency to new heights:

> [19] I remember coming out of the dressing room & the sales lady looking at me & asking me if I had found anything (probably concerned with only making her commission). I thought I would die.

Finally there is the drama of leaving the store:

> [19] Walking out the door was always a big, big step. We knew that that's when people get busted as they step out & we just hoped & prayed that no one would run up to us & grab us or scream "hey you"! The whole time as we approached the exit I remember looking at it as a dark tunnel & just wanting to run down it & disappear as I hung on to my "beloved purse."

Once they have hidden the booty and so long as they are in the store, the would-be-shoplifters must constantly decide to sustain their deviance. Thus, the multiple boundaries of exposure offer multiple proofs not only of their ability to get away with it but of their will toward immorality:

> [5] We went into the restroom before we left and I remember telling Lori, "We can drop all this stuff in here and leave, or we can take it with us." Lori wanted to take everything, and as we neared the exit, I began to get very nervous.

Many of these shoplifters understand that clerks or store detectives may be watching them undercover, in preparation for arresting them at the exit door. They also believe that criminal culpability is only established when they leave with the stolen goods. As they understand it, they are not irretrievably committed to be thieves until they are on the other side of the exit; up to that point, they may replace the goods and instantly revert from a deviant to a morally unexceptional status. Were they to believe that they were criminally culpable as soon as they secreted the item, they would continue to face the interactional and emotional challenges of accomplishing deception. But because they think they are not committed legally until they are physically out of the store, they experience each practical challenge in covering up their deviance as an occasion to reaffirm their spiritual fortitude for being deviant. One student described the phenomenon this way:

> [56] I guess I had been there so long that I started to look suspicious. I was holding a bright lime green necklace in my left hand and a brown Indian type necklace in my right. A lady, she must have been the store manager, was watching me. She was about 20 ft. away from me and on my left. I could feel her looking at me but I didn't look directly at her. . . . I remember actually visualizing myself putting back both the necklaces and walking out the store with pride and proving to this bitch she was wrong and that I was smarter than her, but I didn't. . . . I started out the store very slowly. I even smiled at the lady as I passed by the cash register. It was then that she started toward me and my mind said okay T. what are you going to do now. There was a table full of sweaters on sale near me and I could have easily dropped the necklace on the table and continued out the door. I knew I could and I considered it but I wouldn't do it. I remember just holding the necklace tighter in my right hand. As she was coming toward me I even thought of dropping the necklace and running out the door but I continued in a slow pace even though the thought of them calling my mother if I was caught and what she would do to me was terribly frightening.

In addition to focusing on the practical components of producing a normal appearance, the would-be shoplifter struggles not to betray the difficulty of the project. This is the second layer of work—the work of appearing not to work at practicing normal appearances. The first layer of work is experienced as the emergence of a novel, analytical attention to behavioral detail; the second, as a struggle to remain in rational control, as the following statement by a student illustrates:

> [19] You desperately try to cover it up by trying to remember how you've acted before but still you feel as if all eyes are on you! I think, that's the purpose of settling in one area & feeling everything & everyone out. It's an attempt to feel comfortable so that you don't appear obvious. Like maybe if I'm real cool & subtle about it & try on a few things but don't seem impressed w/ anything, I can just stroll out of here & no one will notice. . . .

Being Thrilled

Usually after the scene of risk is successfully exited, the third stage of the sneaky thrill is realized. This is the euphoria of being thrilled. In one form or another, there is a "Wow, I got away with it!" or an "It was so easy!" A necklace shoplifter stated:

> [56] Once outside the door I thought Wow! I pulled it off, I faced danger and I pulled it off. I was smiling so much and I felt at that moment like there was nothing I couldn't do.

After stealing candy with friends, another student recalled:

> [87] Once we were out the door we knew we had been successful! We would run up the street . . . all be laughing and shouting, each one trying to tell just how he pulled it off and the details that would make each of us look like the bravest one.

The pizza thief noted:

> [82] The feeling I got from taking the pizza, the thrill of getting something for nothing, knowing I got away with something I never thought I could, was wonderful. . . . I'm 21 now and my neighbor is 23. Every time we see each other, I remember and relive a little bit of that thrill. . . .

In a literal sense, the successful thieves were being thrilled: they shuddered or shook in elation, often to the rhythms of laughter. For many, whether successful or not, the experience of youthful shoplifting was profoundly moving, so moving that they could vividly recall minute details of the event years later.

DISCUSSION QUESTIONS

1. Does the idea of "sneaky thrills" help you to make sense of why some people are willing to risk trouble for even the smallest material reward?
2. Why is the emotional side of criminal behavior often ignored in favor of such explanations as poverty or discrimination?
3. Does Katz's study help to make sense of kleptomania? How would one become a kleptomaniac (think nucleus accumbens)?

READING

Searching a Dwelling

Deterrence and the Undeterred Residential Burglar

Richard Wright

In the book *Burglars on the Job* (1994), from which this excerpt is taken, Wright and Decker report on interviews with 105 residential burglars in an attempt to understand the strategies and tactics they use to search a dwelling for items of value. In doing so, of course, they also learn something of the characters of these individuals. Burglars know from experience that they have to get into and out of a residence as quickly as possible. To do this, they must have a fair knowledge about where the most valuable property is to be found, what rooms to avoid, and what to do in an emergency. They learn all this through experience and are often able to easily, quickly, and methodically ransack a home on "autopilot." The burglars discuss with Wright their motivations and fears, as well as their search techniques.

In most jurisdictions, a residential burglary has been completed in the eyes of the law the moment an offender enters a dwelling without permission, intending to commit a crime therein. But seen through the eyes of the burglars themselves, a break-in is far from complete at this point. Indeed, the offense has just begun. They must still transform their illicit intentions into action—which, in practice, almost invariably involves searching for goods and stealing them—and escape from the scene without getting caught, injured, or killed. But in doing this, offenders are on the horns of a dilemma. On the one hand, the more time they spend searching a residence, the better chance they stand of realizing a large financial reward. On the other hand, the longer they remain inside a target, the greater risk they run of being discovered. Having entered a dwelling, then, offenders must strike a deceptively complex, subjective balance that maximizes reward within the limits of acceptable risk. How is such a balance actually struck? Criminologists interested in the decision-making of residential burglars have devoted almost no attention to this process, despite the fact that such offenders obviously continue to make decisions throughout the commission of their break-ins. An examination of this matter is crucial to the development of a fuller understanding of the decision-making calculus of property offenders.

In an attempt to learn more about the tactics used by offenders to search dwellings, a colleague, Scott Decker, and I located and interviewed 105 currently active residential burglars in St. Louis, Missouri. The residential burglars were recruited through the efforts of a field-based informant—an ex-offender with a solid reputation for integrity and trustworthiness in the criminal underworld. Working through chains of street referrals, the field recruiter contacted active residential burglars, convinced

Source: Wright, R. T., & Decker, S. H. (1994). *Burglars on the job: Streetlife and residential break-ins.* Boston: Northeastern University Press. Reprinted by permission.

them to take part in the project, and sat in on interviews that lasted two hours or more. In the pages that follow, I report just a small portion of what the offenders said during those interviews, focusing on how they search dwellings once they have broken into them.

Once inside a target, the first concern of most offenders is to reassure themselves that no one is at home. They do this in a variety of ways. Some run through the dwelling and take a quick glance into every room. Others remain still and silent, listening for any sound of movement. Still others call out something along the lines of "Is anybody home?" More than anything, such actions probably represent an attempt by offenders to put worries about being attacked by an occupant behind them so that they can devote their attention to searching for cash and goods: "When you first get inside, you go through all the rooms to make sure no one's home. Once [you see] there's no one home, that's when you start gettin' busy, doin' your job."

Thus having reassured themselves, offenders often experience a sudden realization that everything inside the residence is theirs for the taking. One female offender said this realization made her feel as if she was in Disneyland—calling to mind a magical world in which fantasy had become reality. Another offender likened the feeling he got inside an unoccupied dwelling to being in a fashionable shopping mall, "except you don't have to pay for stuff; just take it." Such comments suggest that, during this phase of the burglary process, offenders perceive themselves to be operating in a world that is qualitatively different from the one they inhabit day-to-day. Jack Katz refers to this world as "an enchanted land," the phenomenological creation of a mind bent on crime (Katz 1988).

Inside dwellings, shielded from public view, many offenders also experience a marked reduction in anxiety. They already have broken in, there is no turning back, and it makes little sense to agonize over the potentially negative consequences of their actions. Recognizing this, offenders have a tendency to settle down and turn their attention to searching the residence. This is not to suggest that the burglars stop worrying about the risks altogether. Most of them continue to be somewhat fearful, with the length of time they are willing to spend inside targets providing a rough indication of the gravity of their concern.

The Brief Search

The outside world does not stand still while burglars are searching dwellings and, as noted above, their vulnerability to discovery increases the longer they remain inside them. Occupants may return unexpectedly. Neighbors may become suspicious and call the police. Patrolling officers may spot a broken window and stop to investigate. Burglars are well aware of these risks, and the vast majority of them try to limit the time they spend in places: "When you first get in [to a dwelling], do whatever you gon do; do it quick and get on out of there! When you doin' [a burglary], you work fast. You go straight to what you want to get and then you come out of there. I know three minutes don't seem like a long time, but in a house that's long! You just go straight and do what you got to do; [three minutes is] a long time."

The ability of the burglars to locate goods without undue delay is facilitated by strict adherence to a cognitive script that guides their actions almost automatically as they move through dwellings; this script allows them to flow through the search process without periodically having to stop to calculate their next step. Virtually all of the burglars reported having a tried-and-true method of searching residences which, they believe, produces the maximum yield in cash and goods per unit of time invested. The search pattern varies somewhat from offender to offender, largely as a function of the time an individual is prepared to remain inside a given target. With few exceptions, however, the burglars agree that, upon entering a dwelling, one should make a beeline for the master bedroom; this is where cash, jewelry, and guns are most likely to be found. These items are highly prized because they are light, easy to conceal, and represent excellent pound-for-pound value.

> I'm hittin' that bedroom first. I'd rather hit that bedroom than the living room cause it's more valuables in the bedroom than it is in the living room—jewelry, guns and money. So everything is in the back [of the dwelling] somewhere; most likely in the bedroom. The first thing you always do when you get to a house is you always go to the bedroom. That's your first move. . . . [b]ecause that's where the majority of people keep they stuff like jewelry or cash. You know it's gon be a jewelry box in the bedroom; you know you ain't gon find it in the living room. Guns, you ain't gon find that too much in the living room.

The burglars believed that searching the master bedroom first also enhances their personal safety, especially when this room is located on the second floor. They feel particularly vulnerable upstairs because their only escape route—the stairway—can easily be cut off should an occupant return unexpectedly. As a result, most of them want to begin their search on the second floor and get downstairs as quickly as possible.

The burglars reported searching four main places within the master bedroom. The first stop for most of them is the dresser, where they quickly go through each drawer—often dumping the contents onto the floor or bed—looking primarily for cash and jewelry: "[Y]ou got to look around, you got to ransack a little. You got to realize too that you don't have very much time to ransack neither. . . . You always start in the bedrooms, in the drawers; that's where they keep the money and the jewelry."

They typically turn next to the bedside table, hoping to find a handgun and, perhaps, some cash and jewelry. From there, many of them search the bed itself because some people, especially those who are elderly or poor, continue to keep their savings under the mattress. Lastly, the burglars usually will rummage through the bedroom closet, looking mainly for cash and pistols hidden in a shoebox or similar container.

If the search of the main bedroom has been moderately successful, some of the burglars will not bother to look through the remaining rooms, preferring simply to make good their escape. They know full well that a more exhaustive search of the premises could net them a larger financial return, but are unwilling to assume the increased risk entailed in spending longer inside the target. As one burglar who typically searches only the master bedroom explained: "You miss a lot, but it's all gravy if you get away." A majority of the burglars, however, conduct at least a cursory search of the rest of the dwelling before departing. A number of them said they usually have a quick look around the kitchen. Surprisingly, most of these burglars are not searching for silverware or kitchen appliances so much as for cash and jewelry; they claimed that such items sometimes can be found hidden in a cookie jar or in the refrigerator/freezer: "[The valuables are located in] either the freezer, the icebox, in they bedroom . . . in the dresser drawer or under the mattress. . . . People put money in plastic bags behind the meat in they freezer."

Some of the burglars also search the bathroom, concentrating on the medicine cabinet where they hope to find not only psychoactive prescription drugs (e.g., valium), but perhaps hidden money and valuables: "Like a lot of times I've [gone through] the medicine cabinets, quite a few people leave money in the medicine cabinet. I don't know if you've ever heard of that, but I . . . found about forty dollars in there once and the second time I found about twenty-five dollars."

Few burglars, however, bother to rummage about in bedrooms occupied by young children because, in their view, such rooms are unlikely to contain anything worth stealing.

> I ain't gon even worry about the kids' bedrooms. Mom and Dad have all the jewelry and stuff in they room. I know the little kids ain't got nothing in they room. Little kids' rooms, I don't usually go in there because they don't usually have much. I don't have any kids so ain't no sense me goin' in there unless one of my little nieces or nephews might want something and I might keep an eye out for them.

Most of the burglars usually search the living room just prior to leaving the target because the items kept there tend to be heavy or bulky (e.g., television sets, videocassette recorders, stereo units); hence they best are left to the last minute. Indeed, some offenders do not bother to search the living room because carrying out the cumbersome goods located therein is likely to draw the attention of neighbors. As one offender explained: "I don't carry no TVs . . . because that's an easy bust." And another added: "You never take nothing big. You look for [something small]; something won't nobody see you bringin' out."

By employing such strategies to search dwellings, the burglars usually can locate enough cash and other valuables to meet their immediate needs in a matter of minutes. As one said: "You'd be surprised how fast a man can go through your house." Occasionally, however, the predetermined search strategies used by offenders fail to yield the expected results. This may cause some of them to depart from their normal modus operandi, remaining in the residence for longer than they feel is safe in order to find something of value.

> [I] go straight to the main spots where I think the main stuff is in. Never mess around with, well, at least try not to, stay out the petty spots. Bathrooms, you know, what you goin' in there for? I may go in there if the house ain't no good, you know, [isn't] what I thought it was; then I start gettin' desperate and stuff and lookin' everywhere. Gotta get somethin'. Let me see if they got a gold toothbrush, you know, just anything.

In breaking into the dwelling, these offenders already have assumed considerable risk and they are determined to locate something worth stealing, if only to justify having taken that risk in the first place. Add to this the fact that most of them are under pressure to obtain money quickly, and their decision to carry on searching despite the increasing risk of being discovered seems more sensible still; to abandon the offense would require quickly finding and breaking into an alternative target, with all of the attendant hazards.

On other occasions, offenders are tempted to linger in residences when they discover something that convinces them that an especially desirable item must be hidden elsewhere in the building. Put differently, they are enticed into accepting a higher level of risk, believing that the extra time devoted to searching is justified by the potential reward: "Then when I'm goin' through the dresser drawer or somethin', I might find shell boxes; they got a gun! Definitely! And I'm gon find it."

It is at this point that we can begin to glimpse the danger of allowing oneself to be seduced by the possibility of realizing a large financial gain. As noted above, the burglars are operating in a world that is qualitatively different from the one they inhabit day-to-day; a world in which they can take whatever catches their fancy. It is easy to see how they could get carried away by the project at hand, trying to take everything and disregarding the risks altogether. Generally speaking, the burglars are aware of this threat and try their best to avoid becoming so wrapped up in the offense as to throw caution to the wind. They do this by focusing on the items they originally intended to steal, resolving not to get greedy.

> [M]ost of the time you want to get in and get out as quick as possible. See, that's how a lot of people get caught; they get greedy. You go in and get what you first made up your mind to get. When you take the time to ramble for other things, and look through this and look through that, you taking a chance.

> I put my mind on one thing. That's what I'm a get. I ain't gon be ransackin' all through there . . . See, I don't get greedy once I go in and do a burglary.

From the perspective of these burglars, it makes little sense to steal more than necessary to meet their immediate needs because doing so involves additional risk. Besides, most of them could not transport more than they already had. As one put it: "I don't want to be in [the dwelling] that long. I ain't gon be able. To carry all that stuff anyway."

Not surprisingly, some offenders did report getting carried away during particular offenses, being seduced by the allure of the available goods such that they forgot all about the risks of lingering in the target.

> The stuff that was in there, it just had this attraction to your eyes. It made you feel like, "God, I need that!," you know. So that's what, so we just kept on . . . everything just attracted our eyes.

> I was downstairs just lookin' around cause I was real choicy; I was real choicy this last job I had. [My partner] came downstairs and said, "Man, you better get what you gon get and come on!" "Man, what time is it?" I said, "I got about fifteen minutes." He said, "Man, you better hurry up!" You know, he was rushin' me then cause usually I be rushin' them. "Alright then, I'm a take this VCR, this TV, let's go."

Cases such as these, though, are the exception rather than the rule; almost all of the burglars usually adhere to a well-rehearsed cognitive script in searching targets. Admittedly, not all of them enter dwellings intending to steal specific items, preferring instead to allow themselves to be seduced by whatever catches their eye. But even these offenders typically stick to a set pattern in moving through residences.

> When you go in [to a target], the first thing you do is go straight to the back. As you go to the back, you already lookin'. While you lookin', you pick certain things out that you gon take with you that you know you can get. You can pick them out just by lookin' as you walkin'. Then, when you turn around to come out of there, you already know what you gon get. . . . It depends on what I spot and if I think it's of value; not no particular things.

Many residential burglars, of course, commit their offenses acting in concert with others. As Neal Shover has observed, co-offending appeals to many burglars because "it facilitates management of the diverse practical demands of stealing (Shover 1991). Foremost among these demands is avoiding detection. Working with others allows offenders to locate and transport goods more quickly, thereby reducing the risk of being caught in the act.

> [I have searched dwellings more extensively], but that was only when I do it with friends. Cause I have more time; while I'm downstairs, he's upstairs. It's all timing. Fast as you do it and then get out of there.

> I guess [I work with others] because it would take so much time for me to have to look for everything all by myself. If it's two or three of us it will be that much quicker. A lot of times the places that I normally pick, it's quite a few items there. . . . But

then I use someone else so we can get the job done and move on out; get away as soon as possible instead of making a lot of trips to get everything [out of the dwelling]. [I commit burglaries with a partner] just in case they have lots of stuff in there I want. We can hurry up an' get it out.

A number of the burglars also reported that working with one or more accomplices is safer than operating alone because it permits them to post a look-out.

It's always good to go in a house with at least three dudes. You know, two of you'll get the stuff together. The other one look out the window; he be the watch.

[I]t's almost always a little safer to have someone else with you. . . . Because if you got someone outside, they can always give a little signal and let you know when someone's coming or whatever. If you're alone, you can't hear these things.

These offenders believe that using a look-out not only is objectively safer, it also carries the subjective benefit of eliminating concerns about being caught by surprise and thus enhances their ability to concentrate on searching for goods to steal. As one of them noted: "By yourself, you never know who behind you."

Several offenders pointed out that accomplices represent ready assistance should they encounter unanticipated resistance.

Sometime you want somebody with you. . . . Sometimes I like to do [residential burglaries] with someone because I like to have protection in case something do happen. [I work with someone else] because we spent five and a half years together and we can handle ourselves real well. I can trust him; if I get in a tough situation, he would kick ass for me. If a guy with a big baseball bat is going to come kill me, he'll come to the rescue type of thing.

A few offenders chose to work with others in the belief that, should the police arrive unexpectedly, this increases the odds of at least one of them getting away.

[Working with others decreases the risk] because if it's more than one [offender] and the police do come, then somebody is bound to get away. I guess if you get caught, you know, [the police] gonna catch one of yas, to put it that way, if you're gonna get caught, they're gonna catch one of ya, one of yas always gotta chance of getting away.

The logic underlying this belief is that two or more offenders can split up, running in different directions so that officers have to decide who to chase and who to allow to escape. The delusional aspect of this position is self-evident; why should the police elect not to pursue these offenders in favor of catching their accomplices? Here we are confronting a force that transcends rationality. The burglars acknowledge the risk of apprehension but believe that, because luck is on their side, they personally will avoid such a fate. That said, it remains true that by working with others, offenders may well reduce their individual risk of being caught during a police chase. One subject referred specifically to this fact: "I work with some partners because it's a better chance of gettin' away. They might get caught and I might get away." Along the same lines, another burglar said that he liked working with others because witnesses have a tendency to confuse the features of multiple offenders, making it difficult for them to provide the police with a good description of individual suspects.

Quite a few of the burglars who commit their break-ins with others reported doing so just in case they do get caught. Most of them simply want the reassurance of knowing that, should they be arrested, their co-offenders will be there to share the guilt and shame (Shover 1991)

> [I work with others] cause I feel if I get caught, I want them to get caught with me. I mean, I don't want to get caught by myself. . . . They gon get caught with me. . . . I feel I won't be so guilty if I have somebody with me. Then I won't feel so guilty, I'll feel kind of safe.

Two experienced female offenders, however, do not want to share the blame with accomplices, but rather hope to shift it entirely onto their associates. Both of these offenders work exclusively with men, believing that the police will show leniency toward a woman who claims that she was coerced into an offense by her male partner: "I think down in my mind, when I first started doing burglary I saw a show and in it the woman claimed mental incompetent; that she'd been brainwashed. And I guess I feel like if we ever got caught that I could blame it on him. That's a pretty shit attitude, but . . . I don't know him and kind of feel like I'm smarter than [the male burglars] are."

There may be a grain of truth in this belief. Be that as it may, the important point for our purposes is that these women are convinced this is the case, and thus are able to mentally discount the threat of arrest and punishment. This allows them to get on with the business of searching dwellings, unimpeded by concerns about getting caught.

In short, the burglars who choose to work with others in committing their residential break-ins do so not only for practical reasons, but for psychological ones as well; the company of co-offenders dampens their fear of apprehension and bolsters their confidence.

> Like I told you, I know it sounds strange, but I be scared when I do [a residential burglary]. Then if I have somebody with me and they say, "Ah, you can do it," they boost me up and I go on and do it.

> [I work with someone else] when I don't really know about that place, you know, I'm kind of nervous about it. So I feel like I wouldn't be as nervous by me havin' somebody with me on a place that I don't know too much about.

While co-offending has a number of potential advantages, working with other criminals inevitably entails certain risks. In the best of circumstances, such individuals are of dubious reliability, and the pressures inherent in offending can undermine their trustworthiness still further. Several of the burglars said they are becoming increasingly reluctant to work with others, having been let down in the past by co-offenders who failed to carry their weight during offenses.

> See, you can't depend on no one else. That's why I'm goin' to court now. . . . I had this so-called buddy of mine supposed to be watchin' this house for me [while I was inside]. And I told him to stand across the street so when I come out I can look across the street and see him. When I came out, he was gone and I had merchandise up under my arm. So I said, "Let me get on out of here." I don't know what happened, he might have just left me. So I was gettin' nervous and I just went on and left. And there the police was! Walked right into they arms!

Even those who continue to work with others typically believe that, should something go wrong, their crime partners might well let them down. Many expressed doubts about the ability of accomplices to withstand police interrogation without naming them as co-participants.

> You never know, it just ain't no sure thing. I just say, "Do unto others what's done unto you." So I'm not banking on [my partner's ability to remain silent]; if he get caught, then maybe that's damn near my ass is probably caught too. Cause I know if the

> police say, "Was somebody with you?" he'll probably say, "Yeah, yeah, oh yes." Police talk to you, you know; [partners] start spillin' they guts. They scared and then all they thinkin' about is themselves. So to be truthful with you, yeah, I never bank on [silence].

Most of these offenders appeared to accept the potential for duplicity among their colleagues with equanimity. As they see it, the police put pressure on arrestees to inform on their partners, and it is naive to expect them to remain silent. One put it this way: "[My co-offenders] would probably tell on me, but, to be honest, I'd probably tell on them too."

Perhaps the aspect of co-offending that the burglars find most irksome involves what they perceive as a tendency for accomplices to "cuff" (that is, neglect to tell them about) some of the loot found during the search of a target. In their eyes, it is bad enough having to split the proceeds of their crimes, without having to deal with the possibility that their co-workers will try to cheat them. Some offenders attempt to reduce the risk of being cheated by working only with those they know well. For a few, this strategy seems to payoff.

> Well, like they say, "There's no honor among thieves." That's what they say, [but] I believe that they really are wrong about that because this guy really loyal. We done did one house, man, where this guy had [a pistol he found during the search] in his pocket already and we didn't know. But he came out and told us when we was splittin' everything up. He took it out and he put it with the rest of the shit to be split up. I wouldn't a did it; I would've kept that.

But most of them continue to believe that even close acquaintances might try to deceive them by cuffing booty: "That's why I work with the same people, you know what I mean? He don't know what I got from downstairs. I might have found a ten thousand dollar diamond ring. Of course, you got to trust these damn fools, you know? It's easy to do man; it's easy for somebody to rip you off."

One offender reported that he and his usual burglary partner have an agreement whereby they always search the master bedroom together. The logic behind their agreement is that the most easily concealed valuables tend to be found in this room; going through it together represents a means of "keeping each other honest."

Despite the risks of co-offending, the fact remains that a majority of the burglars continue to work with others. As much as anything, their decision to do so undoubtedly reflects the powerful influence of routine. The burglars are used to co-offending. Many of them work with regular partners, and have developed cognitive scripts that incorporate roles for their accomplices. These roles are well understood by their co-offenders and therefore break-ins can be carried out efficiently, with a minimum of confusion or conflict.

> I work with him because when you get a [regular] partner, it's like two pieces of a machine; two gears clicking together and that's the way me and him work. Everybody has they routine. I check upstairs and then they stay downstairs and get the VCRs and everything. That's the routine we been doing for years.

When the pressures that give rise to burglaries intensify, then, it makes little sense for offenders to deviate from their typical modus operandi and set off alone. These pressures often arise in the context of partying—a group activity—where a shared desire to obtain fast cash for more alcohol or drugs precipitates a decision among those present to commit an offense. As one burglar explained: "We be gettin' high together anyway, I might as well go [with] them. I come back by myself, they gon get high with me anyway."

The Leisurely Search

On any given residential burglary, the safest course of action for offenders is to search the target quickly and then leave without delay. Adopting this approach, however, means that they seldom will come away from offenses having stolen more than is necessary to meet their immediate financial needs; there is unlikely to be anything left over to help them deal with their next monetary crisis. The price of reducing the risks in the short term, therefore, may well be a foreshortening of the time between break-ins and a consequent increase in the frequency with which those risks must be taken. Some burglars, albeit a small minority, are unwilling to pay this price, preferring instead to remain in targets long enough to make certain they have found everything of value: "[I take] the whole day going through the whole house, sitting down and eating and things of that nature . . . On a burglary, you get all that you can. Some people will just go get certain items, [but] I can just take everything cause everything has a value."

These burglars claim to understand the schedules kept by occupants of their targets and to have a clear idea about how long residents will be out Thus they can proceed unimpeded by concerns about being discovered in the act of searching places: "When I do a burglary, I don't go in there and come back out. I go in there and stay! I go in there and stay for a couple of hours. I know these people won't be back home until about five in the evening if they leave at seven in the morning. I be done ransacked the house by then."

Even offenders who do not routinely linger in targets occasionally succumb to the temptation to stay longer when they know that the occupants will be away for some time. Their reasons for doing so, however, often seem to transcend the desire for greater financial rewards. Indeed, many devote this extra time almost wholly to relaxation and entertainment: "I usually go straight to the bedroom and then I walk around to the living room. I have set at people's house and cooked me some food, watched TV and played the stereo. . . . I knew they wouldn't be there. But I usually go straight to the bedroom."

The offenders recognize that these offenses are special, being largely free of the temporal constraints that circumscribe most of their break-ins. They respond by taking full advantage of the situation and making themselves at home. In effect, they are acting out the widely held adolescent fantasy of having the run of a place without the obligation to answer to anyone. Some of them, of course, are adolescents. But even among those who are older, few have any experience of being in full control of their living space; the majority still stay with their mothers or have no fixed address. It is easy to appreciate why they enjoy having someplace to themselves. The irony is that they seldom do anything very outrageous: Like teenagers left alone for the weekend by their parents, most simply help themselves to whatever alcohol and food is available and take pleasure in not having to clean up afterward. One, for instance, reported: "Sometimes I cook me some breakfast, but I never wash the dishes."

A number of the burglars said that they sometimes urinate or defecate inside their targets. They attributed their need to do so to the emotional pressures involved in offending. Contrary to popular media accounts, however, they generally do not use the carpet for this purpose; most of them reported using the toilet, sometimes not flushing it afterwards because the resulting noise can drown out the warning sounds of approaching danger. A couple of the burglars did admit to sometimes relieving themselves in rooms other than the bathroom, but they explained this action in terms of safety—they do not want to get cornered, literally with their pants down, in a small space with just one exit—rather than attributing it to any special contempt for the residents. At the same time, these burglars seemed untroubled about the distress this might cause their victims; their sole concern is for their own well-being.

Summary

The vast majority of residential burglars want to search dwellings as quickly as possible in the belief that the longer they remain inside, the more chance they stand of being discovered. They do this by adhering to a cognitive script developed through trial and error to assist them in locating the maximum amount of cash and goods per unit of time invested. Using this script, burglars can proceed almost automatically, without having to make complicated decisions at each stage of the search process. Although the script varies from one offender to the next, it usually calls for them to search the master bedroom first; this is where money, jewelry, and guns are most likely to be found. The living room typically is searched last because the items kept there tend to be difficult to carry and hence are best left until the last minute. Many burglars work with others to expedite the search process; co-offenders can explore one part of the residence while they look through another. By employing a consistent, well-rehearsed modus operandi in searching dwellings, burglars often can locate enough valuables to meet their immediate needs in a matter of minutes. Having successfully done so, most leave without delay.

By the time offenders have entered a target with the intention to steal, a burglary has been committed; the offense can no longer be deterred or prevented. It is at this point that criminologists and crime prevention experts show a marked tendency to lose interest, ceding the field to victimologists and police investigators. This is unfortunate because the activities of offenders during break-ins also may have implications both for decision-making theory, especially in regard to deterrence, and for crime, prevention policy. To be sure, the burglars have not been deterred by the threat of sanctions, but that threat nevertheless seems to have a pronounced effect on their actions as they search targets. Most are unwilling to remain inside for long, forgoing the possibility of greater rewards in favor of reducing the risk of being discovered. In fact, for actual offenders this may be where deterrence operates most effectively. And while it is too late to prevent the burglaries, there is still an opportunity to limit the loss of cash and goods if we can understand the cognitive scripts used by burglars to search dwellings well enough to be able to disrupt those scripts.

References

Katz, J. (1988). *Seductions of crime: Moral and sensual attractions in doing evil.* New York: Basic Books.

Shover, N. (1991). Burglary. In M. Tonry (Ed.), *Crime and justice: An annual review of research.* Chicago: University of Chicago Press.

DISCUSSION QUESTIONS

1. In what ways might the emotional rewards of burglary be analogous to those of shoplifting described by Katz in the previous reading?
2. In what ways are these burglars similar to Jacobs and Wright's robbers discussed in Section 10?
3. Wright talks about preventing burglary by disrupting burglars' scripts. What does he mean by that, and how can it be done?

READING

The Novelty of "Cybercrime"

An Assessment in Light of Routine Activity Theory

Majid Yar

Majid Yar's article explores the possibility that the concepts of routine activities theory can be fruitfully applied to cybercrime given the apparent novelty of the phenomenon. Some criminologists claim that cybercrime is no different from "terrestrial crime" and can be analysed and explained using established theories of crime causation, particularly by routine activity theory. Yar's examination concludes that although some of the theory's core concepts can indeed be applied to cybercrime, there remain important differences between "virtual" and "terrestrial" worlds that limit the theory's usefulness. These differences, he claims, give qualified support to the suggestion that cybercrime does indeed represent the emergence of a new and distinctive form of crime. The one thing that remains the same, however, is that motivated offenders may be considered cut from the same cloth, whether they operate in the terrestrial or virtual spheres.

Introduction

Does cybercrime denote the emergence of a "new" form of crime and/or criminality? Would such novelty require us to dispense with (or at least modify, supplement or extend) the existing array of theories and explanatory concepts that criminologists have at their disposal? Unsurprisingly, answers to such questions appear in positive, negative and indeterminate registers. Some commentators have suggested that the advent of "virtual crimes" marks the establishment of a new and distinctive social environment (often dubbed "cyberspace," in contrast to "real space") with its own ontological and epistemological structures, interactional forms, roles and rules, limits and possibilities. In this alternative social space, new and distinctive forms of criminal endeavour emerge, necessitating the development of a correspondingly innovative criminological vocabulary. Sceptics, in contrast, see "cybercrime" at best as a case of familiar criminal activities pursued with some new tools and techniques. If this were the case, then cyber-crime could still be fruitfully explained, analysed and understood in terms of established criminological classifications and aetiological schema. Grabosky (2001) nominates Cohen and Felson's "routine activity theory" (RAT) as one such criminological approach, thereby seeking to demonstrate "that 'virtual criminality' is basically the same as the terrestrial crime with which we are familiar" (2001: 243). Others, such as Pease (2001: 23), have also remarked in passing upon the helpfulness of the RAT approach in discerning what might be different about "cybercrime," and how any such differences (perhaps ones of degree, rather than kind) present new challenges for governance, crime control and crime prevention.

Source: Yar, M. (2005). The novelty of "cybercrime": An assessment in light of routine activity. *European Journal of Criminology, 2*(4), 407–427. Reprinted with permission of Sage Publications, Ltd.

Cybercrime: Definitions and Classifications

Cybercrime might best be seen to signify a *range* of illicit activities whose common denominator is the central role played by networks of information and communication technology (ICT) in their commission. The specificity of cybercrime is therefore held to reside in the newly instituted interactional environment in which it takes place, namely the "virtual space" ("cyberspace") generated by the interconnection of computers into a worldwide network of information exchange, primarily the Internet. Thus Wall (2001: 3–7) subdivides cybercrime into four established legal categories:

1. Cyber-*trespass*—crossing boundaries into other people's property and/or causing damage, e.g., hacking, defacement, viruses.
2. Cyber-*deceptions* and *thefts*—stealing (money, property), e.g., credit card fraud, intellectual property violations (a.k.a. "piracy").
3. Cyber-*pornography*—activities that breach laws on obscenity and decency.
4. Cyber-*violence*—doing psychological harm to, or inciting physical harm against others, thereby breaching laws pertaining to the protection of the person, e.g., hate speech, stalking.

This classification is certainly helpful in relating cybercrime to existing conceptions of proscribed and harmful acts, but it does little in the way of isolating what might be qualitatively *different* or *new* about such offences and their commission when considered from a perspective that looks beyond a limited legalistic framework. Borrowing from sociological accounts of globalization as "time-space compression," theorists of the new informational networks suggest that cyberspace makes possible near-instantaneous encounters and interactions between spatially distant actors, creating possibilities for ever-new forms of association and exchange. Criminologically, this seemingly renders us vulnerable to an array of potentially predatory others who have us within instantaneous reach, unconstrained by the normal barriers of physical distance.

Moreover, the ability of the potential offender to target individuals and property is seemingly amplified by the inherent features of the new communication medium itself—computer-mediated communication (CMC) enables a single individual to reach, interact with and affect thousands of individuals simultaneously. Thus the technology acts as a "force multiplier," enabling individuals with minimal resources (so-called "empowered small agents") to generate potentially huge negative effects (mass distribution of email "scams" and distribution of viral codes being two examples). Further, great emphasis is placed upon the ways in which the Internet enables the manipulation and reinvention of social identity—cyberspace interactions afford individuals the capacity to reinvent themselves, adopting new virtual personae potentially far-removed from their "real world" identities. From a criminological perspective, this is viewed as a powerful tool for the unscrupulous to perpetrate offences while maintaining anonymity through disguise and a formidable challenge to those seeking to track down offenders.

Delimiting the Routine Activity Approach: Situational Explanation, Rationality and the Motivated Actor

Birkbeck and LaFree (1993: 113–14) suggest that the criminological specificity of routine activity theory (RAT) can be located via Sutherland's (1947) distinction between "dispositional" and "situational" explanations of crime and deviance. Dispositional theories aim to answer the question of "criminality," seeking some causal mechanism (variously social, economic, cultural, psychological or biological) that might account for why *some* individuals or groups come to possess an inclination toward law- and rule-breaking behaviour. Dispositional theories comprise the standard reference points of criminological discourse.

In contrast, routine activity theorists take criminal inclination as given, supposing that there is no shortage of motivations available to all social actors for committing law-breaking acts. They do not deny that motivations can be incited by social, economic and other structural factors, but they insist that any such incitements do not furnish a *sufficient* condition for actually following through inclinations into law-breaking activity. Rather, the *social situations* in which actors find themselves crucially mediate decisions about whether or not they will act on their inclinations (whatever their origins). Consequently, routine activity theorists choose to "examine the manner in which the spatio-temporal organization of social activities helps people translate their criminal inclinations into action" (Cohen and Felson 1979: 592). Social situations in which offending becomes a viable option are created by the routine activities of other social actors; in other words, the routine organizational features of everyday life create the conditions in which persons and property become available as targets for successful predation at the hands of those so motivated. For routine activity theorists, the changing organization of social activities is best placed to account for patterns, distributions, levels and trends in criminal activity. If this is the case, then the emergence of cybercrime invites us to enquire into the routine organization of *online* activities, with the aim of discerning whether and how this "helps people translate their criminal inclinations into action." More broadly, it invites us to enquire whether or not the analytical schema developed by RAT—in which are postulated key variables that make up the criminogenic social situation; what Felson (1998) calls "the chemistry for crime"—can be successfully transposed to cyber-spatial contexts, given the apparent discontinuities of such spaces vis-à-vis "real world" settings.

For example, Felson has elaborated and refined his original "chemistry for crime" over a 25-year period by introducing additional mediating variables into what is an ever-more complex framework. Here I discuss RAT in something like its "original" formulation. This statement of the theory hypothesizes that "criminal acts require the convergence in space and time of *likely offenders, suitable targets* and the *absence of capable guardians*" (Cohen and Felson 1979: 588). This definition has the virtue of including the central core of three concepts which appear as constant features of all routine activity models.

Routine activity approaches are generally held to be consistent with the view that actors are free to choose their courses of action, and do so on the basis of anticipatory calculation of the utility or rewards they can expect to flow from the chosen course. Felson, for example, has made explicit this presupposition and his work has been marked by a clear convergence with "rational choice theory" (Clarke and Felson 1993). One common objection raised in light of this commitment is the theory's potential inability to encompass crimes emanating from non-instrumental motives. Similar objections can be raised by proponents of "cultural criminology" who highlight the neglect of emotional and affective "seductions" that individuals experience when engaged in criminal and deviant activity. I would suggest, however, that the basic difficulty here arises not so much from the attribution to actors of "rationality" per se, but from taking such rationality to be necessarily of a limited, economic kind. It may be a mistake to view affective dispositions as inherently devoid of rationality; rather, as Archer (2000) argues, emotions can better be seen as responses to, and commentaries upon, situations that we encounter as part of our practical engagements with real-world situations. Particular emotional dispositions (such as fear, anger, boredom, excitement) are not simply random but "reasonable" responses to the situations in which we as actors find ourselves. My point here is that, by adopting a more capacious conception of rationality (which includes aesthetic and affective dimensions), the apparent dualism between "instrumental" and "expressive" motivations can be significantly overcome.

Convergence in Space and Time: The Ecology and Topology of Cyberspace

At heart, routine activity theory is an *ecological* approach to crime causation, and as such the spatial (and temporal) localization of persons, objects and activities is a core presupposition of its explanatory schema. The ability of its aetiological formula (offender + target - guardian = crime) to explain and/or anticipate patterns of offending depends upon these elements converging in space and time. Routine activities, which create variable opportunity structures for successful predation, always occur in particular locations at particular times, and the spatio-temporal accessibility of targets for potential offenders is crucial in determining the possibility and likelihood of an offence being committed. Cohen and Felson (1979) suggest that the postwar increases in property crime rates in the United States are explicable in terms of changing routine activities such as growing female labour force participation, which takes people increasingly out of the home for regularized periods of the day, thereby increasing "the probability that motivated offenders will converge in space and time with suitable targets in the absence of capable guardians" (1979: 593). Similarly, they argue that "proximity to high concentrations of potential offenders" is critical in determining the likelihood of becoming a target for predation (1979: 596). Thus, at a general level, the theory requires that targets, offenders and guardians be located in particular places, that measurable relations of spatial proximity and distance pertain between those targets and potential offenders, and that social activities be temporally ordered according to rhythms such that each of these agents is either typically present or absent at particular times. Consequently, the transposability of RAT to virtual environments requires that cyberspace exhibit a *spatiotemporal ontology* congruent with that of the "physical world," i.e., that place, proximity, distance and temporal order be identifiable features of cyberspace.

Spatiality

Discourses of cyberspace and online activity are replete with references to space and place. There are purported to exist "portals," "sites" complete with "back doors," "chat rooms," "lobbies," "classrooms," "cafes," all linked together via "superhighways," with "mail" carrying communications between one location and another. Such talk suggests that cyberspace possesses a recognizable geography more or less continuous with the familiar spatial organization of the physical world to which we are accustomed. The virtual environment is seen as one in which there is "zero distance" between its points, such that entities and events cannot be meaningfully located in terms of spatial contiguity, proximity and separation. Everyone, everywhere and everything are always and eternally "just a click way." Consequently, geographical rules that act as a "friction" or barrier to social action and interaction are broken. If this is true, then the viability of RAT as an aetiological model for virtual crimes begins to look decidedly shaky, given the model's aforementioned dependence on spatial convergence and separation, proximity and distance, to explain the probability of offending. To take one case in point, if all places, people and objects are at "zero distance" from all others, then how is it possible meaningfully to operationalize a criterion such as "proximity to a pool of motivated offenders"? Despite these apparent difficulties, I would suggest that all is not lost—that we can in fact identify spatial properties in virtual environments *that at least in part* converge with those of the familiar physical environment.

Positions that claim there is no recognizable spatial topology in cyberspace may be seen to draw upon an absolute and untenable separation of virtual and non-virtual environments—they see these as two ontologically distinct orders or experiential universes. However, there are good reasons to believe that such a separation is overdrawn, and that the relationships between these domains are characterized by both similarity and dissimilarity, convergence and divergence. I shall elaborate two distinctive ways in which cyberspace may be seen to retain a spatial geometry that remains connected to that of the "real world."

First, cyberspace may be best conceived not so much as a "virtual reality," but rather as a "real virtuality," a socio-technically generated interactional environment rooted in the "real world" of political, economic, social and cultural relations. Cyberspace stands with one foot firmly planted in the "real world," and as a consequence carries non-virtual spatialities over into its organization. This connection between virtual and non-virtual spatialities is apparent along a number of dimensions. The virtual environments (websites, chat rooms, portals, mail systems, etc.) that comprise the virtual environment are themselves physically rooted and produced in "real space." The distribution of capacity to generate such environments follows the geography of existing community relations and hierarchies. Access to the virtual environment follows existing lines of social inclusion and exclusion, with Internet use being closely correlated to existing cleavages of income, education, gender, ethnicity, age and disability. Consequently, presence and absence in the virtual world translate "real world" marginalities, which themselves are profoundly spatialized ("first world" and "third world," "urban" and "rural," "middle-class suburb" and "urban ghetto," "gated community" and "high-rise estate"). In short, the online density of both potential offenders and potential targets is not neutral with respect to existing social ecologies, but translates them via the differential distribution of the resources and skills needed to be present and active in cyberspace.

A second way in which cyberspace may exhibit a spatial topology refers to the purely internal organization of the information networks that it comprises. Many commentators see the Internet and related technologically generated environments as heralding "the death of distance" and the collapse of spatial orderings, such that all points are equally accessible from any starting point (Dodge and Kitchin 2001: 63). However, reflection on network organization reveals that *not* all "places" are equidistant—proximity and distance have meaning when negotiating cyberspace. This will be familiar to all students and scholars who attempt to locate information, organizations and individuals via the Internet. Just because one knows, suspects or is told that a particular entity has a virtual presence on the Net, finding that entity may require widely varying expenditures of time and effort. Arriving at a particular location may require one to traverse a large number of intermediate sites, thereby rendering that location relatively distant from one's point of departure; conversely, the destination may be "only a click away."

Temporality

The ability to locate actors and entities in particular spaces/places *at particular times* is a basic presupposition of RAT. The explanatory power of the theory depends upon routine activities exhibiting a clear temporal sequence and order. It is this temporal ordering of activities that enables potential offenders to anticipate when and where a target may be converged upon; without such anticipation, the preconditions for the commission of an offence cannot be fulfilled, nor can criminogenic situations be identified by the analyst.

The temporal structures of cyberspace, I would argue, are largely devoid of the clear temporal ordering of real-world routine activities. Cyberspace, as a *global* interactional environment, is populated by actors living in different real-world time zones, and so is populated

"24/7." Moreover, online activities span workplace and home, labour and leisure, and cannot be confined to particular, clearly delimited temporal windows. Consequently, there are no *particular* points in time at which actors can be anticipated to be *generally* present or absent from the environment. From an RAT perspective, this means that rhythm and timing as structuring properties of routine activities become problematic—for offenders, for potential targets and for guardians. Given the "disordered" nature of virtual spatiotemporalities, identifying patterns of convergence between the criminogenic elements becomes especially difficult.

Thus far, I have largely focused on the question of cyber-spatial convergence between the entities identified as necessary for the commission of an offence. Now I turn to consider the properties of those entities themselves, in order to reflect upon the relative continuity or discontinuity between their virtual and non-virtual forms. The "motivated offender," is assumed rather than analysed by RAT . . . [and] I shall follow RAT in focusing upon the other two elements of the criminogenic formula, namely "suitable targets" and "capable guardians."

Targets in Cyberspace: VIVA la Difference?

For routine activity theory, the suitability of a target (human or otherwise) for predation can be estimated according to its four-fold constituent properties—value, inertia, visibility and accessibility, usually rendered in the acronym VIVA.

Value The valuation of targets is a complicated matter, even when comparing "like with like," e.g., property theft. This complexity is a function of the various purposes the offender may have in mind for the target once appropriated—whether it is for personal pleasure, for sale, for use in the commission of a further offence or other non-criminal activity, and so on. Equally, the target will vary according to the shifting valuations attached socially and economically to particular goods at particular times—factors such as scarcity and fashion will play a role in setting the value placed upon the target by offenders and others. Most cybercrime targets are *informational* in nature, given that all entities that exist and move in cyberspace are forms of digital code. Prime targets of this kind include the various forms of "intellectual property," such as music, motion pictures, images, computer software, trade and state secrets, and so on. In general terms it may well be that, in the context of an information economy, increasing value is attached to such informational goods, thereby making them increasingly valued as potential targets. The picture becomes more complex when the range of targets is extended—property may be targeted not for theft but for trespass or criminal damage (a cybercriminal case in point being "hacking," where computer systems are invaded and websites are "defaced," or "malware" distribution, where computer systems are damaged by "viruses," "Trojan horses" and "worms"); the target may be an individual who is "stalked" and "abused"; or members of a group may be subjected to similar victimization because of their social, ethnic, religious, sexual or other characteristics; the target may be an illicit product that is traded for pleasure or profit (such as child pornography). Broadly speaking, we can conclude that the targets of cybercrime, like those of terrestrial crime, vary widely and attract different valuations, and that such valuations are likely to impact on the suitability of the target when viewed from the standpoint of a potential offender.

Inertia This term refers to the physical properties of objects or persons that might offer varying degrees of resistance to effective predation: a large and heavy object is relatively difficult to remove, and a large and heavy person is relatively difficult to assault. Therefore, there is (at least for terrestrial crimes against property and persons) an inverse relationship between inertia and suitability, such that the greater the inertial resistance the lower the suitability of the target, and vice versa. The operability of the inertial criteria in cyberspace, however, appears more problematic, since the targets of cybercrime do not possess physical properties of volume and mass— digitized information is "weightless" and people do not carry their physical properties into the virtual environment. This apparent "weightlessness" seemingly deprives property in cyberspace of any inherent resistance to its removal. Information can be downloaded nearly instantaneously; indeed, it can be infinitely replicated thereby multiplying the offence many-fold (the obvious example here being media "piracy"). However, further reflection shows that even informational goods retain inertial properties to some degree. First, the volume of data (e.g., file size) impacts upon the portability of the target— something that will be familiar to anyone who has experienced the frustration of downloading large documents using a telephone dial-up connection. Secondly, the technological specification of the tools (the computer system) used by the "information thief" will place limits upon the appropriation of large informational targets; successful theft will require, for example, that the computer used has sufficient storage capacity (e.g., hard drive space or other medium) to which the target can be copied. Thus, although informational targets offer *relatively* little inertial resistance, their "weightlessness" is not absolute.

Visibility RAT postulates a positive correlation between target visibility and suitability: Property and persons that are more visible are more likely to become targets. Conceptualizing visibility in cyberspace presents a difficult issue. Given that the social raison d'etre of technologies such as the Internet is to invite and facilitate communication and interaction, visibility is a ubiquitous feature of virtually present entities. The Internet is an inherently *public*. Moreover, since the internal topology of cyberspace is largely unlimited by barriers of *physical* distance, this renders virtually present entities *globally* visible, hence advertising their existence to the largest possible "pool of motivated offenders."

Accessibility This term denotes the "ability of an offender to get to the target and then get away from the scene of a crime" (Felson 1998: 58). Again, the greater the target's accessibility, the greater its suitability, and vice versa. Given that traversal of cyberspace is "non-linear," and it is possible to jump from any one point to any other point within the space, it is difficult to conceive targets as differentiated according to the likelihood of accessibility to a potential offender in this manner. Similarly, the availability of egress from the "scene of the crime" is difficult to operationalize as a discriminating variable when applied to cyberspace. The ability to "get away" in cyberspace can entail simply severing one's network connection, thereby disappearing from the virtual environment altogether. It is, of course, possible that an offender may be noticed during the commission of the offence (e.g., by an "Intrusion Detection System") and subsequently "trailed" back to his/her "home" location via electronic tracing techniques. However, such tracing measures can be circumvented with a number of readily available tools, such as "anonymous re-mailers," encryption devices, and the use of third-party servers and systems from which to launch the commission of an offence; this brings us back to the problem of *anonymity*, noted earlier. The one dimension in which accessibility between non-virtual and virtual targets might most closely converge is that of security devices that prevent unauthorized access. Cohen and Felson (1979: 591) note the significance of "attached or locked features of property inhibiting its illegal removal." The cyber-spatial equivalents of such features include passwords and other authentication measures that restrict access to sites where vulnerable targets are stored (e.g., directories containing proprietary information). Such safeguards can, of course, be circumvented with tools such as "password sniffers," "crackers" and decryption tools, but these can be conceived as the virtual counterparts of lock-picks, glass-cutters and crowbars.

Are There "Capable Guardians" in Cyberspace?

"Capable guardianship" furnishes the third key aetiological variable for crime causation postulated by routine activity theory. Guardianship refers to the capability of persons and objects to prevent crime from occurring. Guardians effect such prevention "either by their physical presence alone or by some form of direct action" (Cohen et al. 1980: 97). Although direct intervention may well occur, routine activity theorists see the simple presence of a guardian in proximity to the potential target as a crucial deterrent. Such guardians may be "formal" (e.g., the police), but RAT generally places greater emphasis on the significance of "informal" agents such as homeowners, neighbours, pedestrians and other "ordinary citizens" going about their routine activities.

How does the concept of guardianship transpose itself into the virtual environment? The efficacy of the concept as a discriminating variable between criminogenic and non-criminogenic situations rests upon the guardian's co-presence with the potential target at the time when the motivated offender converges upon it. In terms of formal social guardianship, maintaining such co-presence is well nigh impossible, given the ease of offender mobility and the temporal irregularity of cyber-spatial activities (it would require a ubiquitous, round-the-clock police presence on the Internet). However, in this respect at least, the challenge to formal guardianship presented by cyberspace is only a more intensified version of the policing problem in the terrestrial world; as Felson (1998: 53) notes, the police "are very unlikely to be on the spot when a crime occurs." In cyberspace, as in the terrestrial world, it is often only when private and informal attempts at effective guardianship fail that the assistance of formal agencies is sought. The cyber-spatial world, like the terrestrial, is characterized by a range of such private and informal social guardians: these range from in-house network administrators and systems security staff who watch over their electronic charges, through trade organizations oriented to self-regulation, to "ordinary online citizens" who exercise a range of informal social controls over each other's behaviour (such as the practice of "flaming" those who breach social norms on offensive behaviour in chat rooms). In addition to such social guardians, cyberspace is replete with "physical" or technological guardians, automated agents that exercise perpetual vigilance. These range from "firewalls," intrusion detection systems and virus scanning software, to state e-communication monitoring projects such as the US government's "Carnivore" and "ECHELON" systems. In sum, it would appear that RAT's concept of capable guardianship is transposable to cyberspace, even if the structural properties of the environment (such as its variable spatial and temporal topology) amplify the limitations upon establishing guardianship already apparent in the terrestrial world.

Conclusion

The impetus for this article was provided by the dispute over whether or not cybercrime ought to be considered as a new and distinctive form of criminal activity, one demanding the development of a new criminological vocabulary and conceptual apparatus. I chose to pursue this question by examining if and to what extent existing aetiologies of crime could be transposed to virtual settings. I have focused on the routine activity approach because this perspective has been repeatedly nominated as a theory capable of adaptation to cyberspace; if such could be established, this would support the claim of *continuity* between terrestrial and virtual crimes, thereby refuting the "novelty" thesis. If not, this would suggest *discontinuity* between crimes in virtual and non-virtual settings, thereby giving weight to claims that cybercrime is something criminologically new. I conclude that there are both significant continuities and discontinuities in the configuration of terrestrial and virtual crimes.

With respect to the "central core of three concepts," I have suggested that "motivated offenders" can be treated as largely homologous between terrestrial and virtual settings. The construction of "suitable targets" is more complex, with similarities in respect of value but significant differences in respect of inertia, visibility and accessibility. The concept of "capable guardianship" appears to find its fit in cyberspace, albeit in a manner that exacerbates the possibilities of instituting such guardianship effectively. However, these differences can be viewed as ones of *degree* rather than *kind*, requiring that the concepts be adapted rather than rejected wholesale.

A more fundamental difference appears when we try to bring these concepts together in an aetiological schema: whereas people, objects and activities can be clearly located within relatively fixed and ordered spatio-temporal configurations in the "real world," such orderings appear to destabilize in the virtual world. In other words, the routine activity theory holds that the "organization of time and space is central" for criminological explanation (Felson 1998: 148), yet the cyber-spatial environment is chronically spatio-temporally *disorganized*. The inability to transpose RAT's postulation of "convergence in space and time" into cyberspace thereby renders problematic its straightforward explanatory application to the genesis of cybercrimes. Routine activity theory (and, indeed, other ecologically oriented theories of crime causation) thus appears of limited utility in an environment that defies many of our taken-for-granted assumptions about how the socio-interactional setting of routine activities is configured.

References

Archer, M. (2000). *Being human: The problem of agency*. Cambridge: Cambridge University Press.

Birkbeck, C., and LaFree, G. (1993). The situational analysis of crime and deviance. *Annual Review of Sociology, 19*, 113–137.

Clarke, R., and Felson, M. (Eds.). (1993). *Routine activity and rational choice*. London: Transaction Press.

Cohen, L., & Felson, M. (1979). Social change and crime rate trends: A routine activity approach. *American Sociological Review, 44*, 588–608.

Cohen, L., Felson, M., & Land, K. (1980). Property crime rates in the United States: A macrodynamic analysis, 1947–1977; with ex ante forecasts for the mid-1980s. *American Journal of Sociology, 86*, 90–118.

Dodge, M., & Kitchin, R. (2001). *Mapping cyberspace*. London: Routledge.

Felson, M. (1998). *Crime and everyday life* (2nd ed.). Thousand Oaks, CA: Pine Forge Press.

Grabosky, P. (2001). Virtual criminality: Old wine in new bottles? *Social & Legal Studies, 10*, 243–249.

Pease, K. (2001). Crime futures and foresight: Challenging criminal behaviour in the information age. In D. Wall (Ed.), *Crime and the internet*. London: Routledge.

Sutherland, E. (1947). *Principles of criminology*. Philadelphia: Lippincott.

Wall, D. (2001). Cybercrimes and the internet. In D. Wall (Ed.), *Crime and the internet*. London: Routledge.

DISCUSSION QUESTIONS

1. If routine activities theory cannot explaining cybercrime, which theory can?
2. Why can we consider "motivated offenders" to be homologous regardless of whether they operate in the terrestrial or virtual spheres?
3. Isn't predatory crime always predatory crime regardless of the tools used to commit it? If so, why Yar's fuss about needing special explanations?

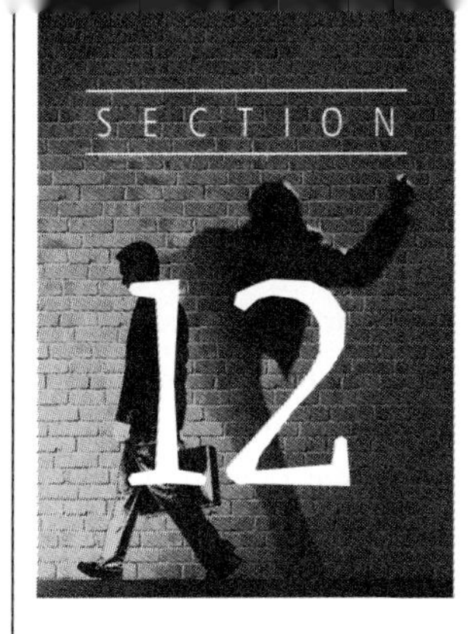

PUBLIC ORDER CRIME

Joe Alladyce and Jared Livingston were both literally born drunk. Their mothers were heavy drinkers who continued to drink during their pregnancies, and if mothers drink, so do their fetuses. If the fetus survives this assault, it is highly likely to be born with a condition called fetal alcohol syndrome (FAS), symptoms of which include neurological abnormalities, intellectual impairment, behavioral problems, and various bodily and facial imperfections. Joe and Jared were made wards of the court and sent to a special institution where staff did their best to educate and care for them. The boys formed a bond with each other and soothed each other's feelings of anger and depression. When they were both 17 they walked away from the home and made their way to the nearest town, where they robbed a liquor store and went on a drinking binge. Walking down the street in a stupor, they came across Mr. and Mrs. Whelan and little 7-year-old Angela walking toward them. Angela made a remark about their behavior and appearance and started to giggle. Enraged, Jared smashed Angela over the head with the beer bottle he was carrying and Sam did the same thing to her father when he tackled Jared. Both boys mercilessly beat and kicked all three family members to death.

This tragic story illustrates the insidious nature of alcohol abuse. Joe and Jared didn't ask to be born with incurable disabilities, and according to many FAS experts they could no more be held responsible for their actions than a blind person could for not recognizing faces. They have brains incapable of appreciating right from wrong and of linking cause and effect. Their mothers not only ruined their own lives, but also the lives of their sons and the lives of surviving members of the Whelan family. There is a huge cost to society caused by what has been aptly named "the beast in the bottle" and by other substances that tear the rationality from our brains and replace it with all manner of monsters.

Public order crimes are a smorgasbord of offenses, some of which have been variously called vice offenses, consensual offenses, victimless crimes, or even nuisance offenses. Some public order crimes are considered very serious (the sale of drugs), and some are dismissed with a shrug of the shoulders or a look of disgust (drunken and disorderly behavior). Public order crimes are of the "moving target" type—legal in some places and at some times (prostitution in Nevada, drugs in Amsterdam, gambling in London) and illegal at other times and in other places. All public order offenses cause some social harm, but whether or not the harm is great enough to warrant siphoning off criminal justice resources that could be applied to more serious crimes is a matter of debate. For instance, the debate about whether the use of mind-altering drugs should be legalized is not about the effects of these drugs—everyone realizes that they are harmful—but rather it is about whether legalization or decriminalization would be the lesser of two evils.

The notion that offenses categorized as public order offenses are "victimless" has been rejected by most criminologists today because there are always secondary victims (family members, friends, etc.) who may be profoundly harmed by the actions of the offender. The "victimless" act by the mothers of Joe and Jared of drinking alcohol to excess started a horrible chain of events that took three lives and ruined many others. These two mothers caused more social, financial, and emotional harm than any two burglars or thieves probably ever did.

The Scope of the Alcohol/Crime Problem

Humans have a love of ingesting substances that alter their moods. We swallow, sniff, inhale, and inject with a relish that suggests that sobriety is a difficult state for us to tolerate. Alcohol has always been humans' favorite way of temporarily escaping reality. We drink this powerful drug to loosen our tongues, to be sociable, to liven up our parties, to feel good, to sedate ourselves, and to anesthetize the pains of life. Many centuries before the birth of Christ, the ancient Sumerians and Egyptians were singing the praises of beer, wine, and the various spirits, but also warning about the consequences of excessive use (Burns, 2004, p. 2).

Of all the substances used to alter mood and consciousness, alcohol is the one most directly linked to crime, especially violent crime (Martin, 2001). It has been estimated that at least 70% of American prison inmates (Wanberg & Milkman, 1998) and 60% of British inmates (McMurren, 2003) are alcohol and/or drug addicted. Alcohol is linked to about 110,000 deaths a year versus the "mere" 19,000 fatalities attributable to other drugs (Robinson, 2004), although this should be interpreted in light of the fact that many more people drink alcohol than take illicit drugs.

Police officers spend more than half of their time on alcohol-related offenses, and it is estimated that one-third of all arrests (excluding drunk driving) in the United States are for alcohol-related offenses (Mustaine & Tewkesbury, 2004). About 75% of robberies and 80% of homicides involve a drunken offender and/or victim, and about 40% of other violent offenders in the United States had been drinking at the time of the offense (Martin, 2001). The U.S. Department of Health and Human Services (2002) estimates the cost of alcohol abuse to society to be a staggering $185 billion. Of the total costs to society of both drug and alcohol abuse, the National Institute on Alcohol Abuse and Addiction (1998) estimates that alcohol abuse accounts for 60% and drug abuse the remaining 40%.

The Effects of Alcohol and Context on Behavior

The effects of alcohol (or any other drug) on behavior is a function of the interactions of the pharmacological properties of the substance, the individual's physiology and personality, and the social and cultural context in which the substance is ingested. Pharmacologically, alcohol is a depressant drug that inhibits the functioning of the higher brain centers. As more and more alcohol is drunk, behavior becomes less and less inhibited as the rational cortex surrenders its control of the drinker's demeanor to the more primitive limbic system. What's going on in the drinker's brain to cause this? Although alcohol is a brain-numbing depressant, at low dosages it is actually a stimulant because it raises dopamine levels (Ruden, 1997). Alcohol also reduces inhibition by affecting a neurotransmitter called GABA, which is a major inhibitor of internal stimuli such as fear, anxiety, and stress (Buck & Finn, 2000). Additionally, alcohol decreases serotonin, reduces impulse control, and increases the likelihood of aggression (Martin, 2001). Alcohol's direct effects on the brain can thus help us to reinvent ourselves as "superior" beings: the fearful become more courageous, the self-effacing become more confident, and the timid become more assertive.

As powerful a behavioral disinhibitor as alcohol is, it is not sufficient by itself to change anyone's behavior in the direction of serious law violations. Most people don't become violent or commit criminal offenses when drinking, or even when they are "over the limit." Alcohol is a releaser of behaviors that we normally keep under control but which we may be prone to exhibit when control is weakened. Hence, we may become silly, amorous, melancholic, maudlin, and even aggressive and violent when our underlying propensity to be these things is facilitated by alcohol and the social context in which it is drunk. In some social contexts drinking may lead to violence, but not in others. Many violent incidents between strangers take place in or around drinking establishments in which both victims and perpetrators had been drinking (Richardson & Budd, 2003).

Groups of young men assembled in bars is a recipe for trouble. Experimental research has shown that drinking increases fantasies of power and domination, and that men who are the heaviest drinkers were the most likely to have them (Martin, 2001). With loosened inhibitions, such fantasies might lead to males flirting with the girlfriends of males from another group and then not backing off when challenged or interpreting some comment or gesture as threatening. If a male values his reputation as a macho tough guy, aggressive responses are more likely when his friends are present and he is looking to validate his reputation. There's an old saying among heavy drinkers: "It's not how many beers you drink, it's who you drink them with."

There are also cultural factors to be considered when evaluating the alcohol/crime relationship. Two of the major cultural factors influencing the relationship between alcohol consumption and criminal behavior are "defining a drinking occasion as a 'time-out' period in which controls are loosened from usual behavior and a willingness to hold a person less responsible for their actions when drinking than when sober by attributing the blame to alcohol" (Martin, 2001, p. 146). If one's culture defines alcohol as a good time elixir, the unfortunate (but often subjectively experienced as enjoyable) byproduct of which is a loss of control over behavioral inhibitions, then one is granted cultural "permission" to do just that.

Binge drinkers frequently consume anywhere between 5 and 10 drinks in a few hours' time and are particularly likely to define drinking as a time-out period. Binge drinkers are typically college-age single young adults who drink solely to get drunk. An American study found

◀ **Photo 12.1** Many alcohol researchers compare societies in which drinking alcohol is considered normal behavior from an early age with American society, in which youth alcohol drinking is illegal. In the United States, many teens drink anyway, and binge drinking rather than moderate use is a potential problem.

that 40% of college students reported at least one episode of binge drinking in the previous two weeks (Johnson, O'Malley, & Bachman, 2000), and a Russian nationwide study found that almost one-third of the men admitted binge drinking at least once a month (Pridemore, 2004). The cultures of both American college students and Russians in general have a high level of **tolerance** for engaging in heavy drinking. Richardson and Budd's (2003) British study found that 39% of binge drinkers admitted a criminal offense in the previous 12 months, whereas 14% of other regular drinkers and 8% of occasional or nondrinkers did. The corresponding percentages for a violent crime were 17, 4, and 2. A survey of 180,155 male and 3,664 female arrestees in major U.S. cities found that 47.9% of the males and 34.9% of the females reported that that they had engaged in binge drinking on at least one occasion in the 30 days preceding their arrest (National Institute of Justice, 2004).

But does heavy drinking plus social context per se cause increased antisocial behavior? It could well be that antisocial individuals are more prone to drink heavily and to be attracted to social contexts in which violence is most likely to occur. In this view, antisocial propensities are simply exacerbated under the influence of alcohol and social setting (Bartol, 2002). Heavy alcohol intake certainly has a greater disinhibiting effect on behavior than heavy tea intake, so alcohol-induced disinhibition may be considered a cause of antisocial acts. Likewise, violence and other antisocial behaviors are assuredly more likely to occur in a biker bar than in a tearoom, and thus social context may be considered a cause as well. But a stricter standard of causation may want to consider that perhaps the substance and the setting are secondary in causal importance to the traits of individuals drinking the beverage of their choice in the settings of their choice.

The article by Rosemary and Bud Ballinger in this section focuses on the role of alcohol and context in aggressive behavior aimed at partner, stranger, and general violence. Not surprisingly, males with the highest degree of alcohol problems also had the highest level of violence against both partners and strangers, although much of this "violence" was verbal rather than physical aggression. Surprisingly, among females, alcohol was not a significant predictor of committing or being victimized by violence.

Alcoholism: Type I and Type II

Alcoholism is a chronic disease condition marked by progressive incapacity to control alcohol consumption despite psychological, social, or physiological disruptions. It is a state of altered cellular physiology caused by chronic consumption of alcohol that manifests itself in physical disturbances (**withdrawal** symptoms) when alcohol use is suspended. While most alcoholics do not get into serious trouble with the law, numerous theorists have hypothesized that alcoholism and criminality are linked because they share a common cause (Fishbein, 1998; Gove & Wilmoth, 2003). Recall that the behavioral activating system (BAS) is dopamine driven, and the behavioral inhibiting system (BIS) is serotonin driven. Alcoholics have a saying that one drink is one too many and a hundred drinks are not enough. This seemingly contradictory statement tells us that a single drink activates the brain's pleasure centers and leads to a craving for more that a hundred drinks will not satiate. Thus, both alcoholics and serious criminals are "reward dominant" in terms of their neurophysiology.

There are two different types of alcoholics: Type I and Type II. Crabbe (2002) describes the two types in this way: "**Type I alcoholism** is characterized by mild abuse, minimal criminality, and passive-dependent personality variables, whereas **Type II alcoholism** is characterized by early onset, violence, and criminality, and is largely limited to males" (p. 449). Type II alcoholics start drinking (and using other drugs) at a very early age and rapidly become addicted and have many character disorders and behavioral problems that precede their alcoholism. Type I alcoholics start drinking later in life than Type II's and progress to alcoholism slowly. Type I's typically have families and careers, and if they have character defects, these are induced by their alcohol problem and are not permanent (DuPont, 1997).

Heritability estimates for Type II alcoholism are about 0.90, and about 0.40 for Type I's (McGue, 1999), indicating that environmental factors are more important to understanding Type I alcoholism than Type II alcoholism (Crabbe, 2002). The genetic influence on alcoholism reflects genetic regulation of neurotransmitters such as GABA, dopamine, and serotonin (Buck & Finn, 2000) and/or their enzyme regulation (Demir, Ucar, G., Ulug, B., Ulosoy, S., Sevinc, I., & Batur, 2002).

Illegal Drugs and Crime

The Extent of the Illicit Drug Problem

Alcohol use is a legal and socially acceptable way of drugging oneself, but substances discussed in this section are not. This was not always the case, for many of these drugs have been legitimately used in religious rituals, for medical treatment, and for recreational use around the world and across the ages. Up until 1914, drugs now considered illicit were legally and widely used in the United States for medicinal purposes. Many substances were openly advertised and sold as cures for all sorts of ailments and for refreshing "pick-me-ups" because people were not fully aware of the dangers of addiction. The most famous of these was Coca-Cola, which was made with the coca leaf (used to process cocaine) and kola nuts (hence the name) until 1903. Many patented medicines, such as Cocaine Toothache Drops and Mother Barley's Quieting Syrup, used to "soothe" infants and young children, contained cocaine, morphine, or heroin.

Attitudes toward drug usage in America gradually began to change as awareness of the addictive powers of many of these substances grew. The **Harrison Narcotic Act** of 1914 was the benchmark act for changing America's concept of drugs and their use. According to Richard Davenport-Hines (2002), "by the early 1920s, the conception of the addict changed from that of a middle-class victim accidentally addicted through medicinal use, to that of a criminal deviant using narcotics (or stimulants) for pleasure" (p. 14). The Harrison Act did reduce the number of addicts (estimated at around 200,000 in the early 1900s), but it also spawned criminal black market operations (as did the Volstead Act prohibiting the production and sale of alcohol in 1919) and ultimately many more addicts (Casey, 1978, p. 11).

Figure 12.1 shows percentages of individuals participating in the 2004 National Household Survey on Drug Abuse (NHSDA) who admitted the use of any illicit drug during the month prior to being interviewed. As with delinquency and crime, drug use rises to a peak in the age 18–20 category and then drops precipitously. The use of illicit drugs by most adolescents probably reflects experimentation (adolescence-limited use) while their continued use in adulthood (life-course persistent use) reflects a far more serious antisocial situation.

Figure 12.1 Illicit Drug Use in Past Month by Age: 2004

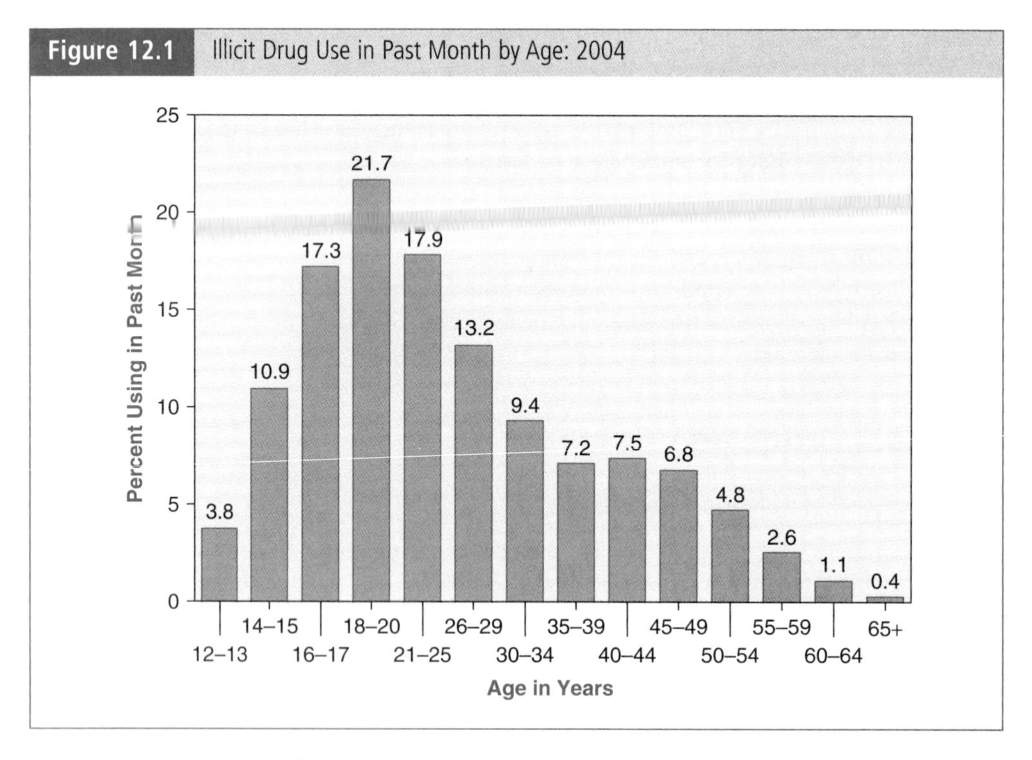

SOURCE: United States Department of Health and Human Services (2005).

Drug Addiction

All addictive drugs mimic the actions of normal brain chemistry by inhibiting or slowing down the release of neurotransmitters, stimulating or speeding up their release, preventing their reuptake after they have stimulated neighboring neurons, or by breaking transmitters down more quickly. Drugs hijack the brain and produce more powerful, rapid, and predictable effects on our pleasure centers than are naturally obtained by the action of neurotransmitters in response to non-drug-induced pleasant experiences.

People turn to illegal drugs for many of the same reasons that people turn to alcohol—to be "with it," to be sociable, to conform, to induce pleasure, to escape stress, or to escape chronic boredom. Among those who experiment with drugs, there are some who are genetically predisposed to develop addiction to their substance(s) of choice, just as others are "sitting ducks" for alcoholism (Robinson and Berridge, 2003).

The Drug Enforcement Administration (DEA) defines **drug addiction** as "compulsive drug-seeking behavior where acquiring and using a drug becomes the most important activity in the user's life," and estimates that five million Americans suffer from drug addiction (2003, p. 13). **Physical dependence** on a drug refers to changes to the body that have occurred after repeated use of it and necessitate its continued administration to avoid withdrawal symptoms. Physical dependence is not synonymous with addiction, as is commonly thought, but **psychological dependence** (the deep craving for the drug and the feeling that one cannot function without it) is synonymous with addiction.

Regardless of the type of drug, addiction is not an invariable outcome of drug usage any more than alcoholism is an invariable outcome of drinking. The DEA (2003) estimates that about 55% of today's youth have used some form of illegal substance, but few descend into the hell of addiction (Kleber, 2003). Genetic differences are undoubtedly related to people's chances of becoming addicted given identical levels of usage and an identical period of time using.

Drug Classification

There are several drug classification schemes determined by the purpose for which the classification is being made. The DEA schedule classification scheme divides chemical substances into five categories, or schedules. Schedule I substances are those that have high abuse liability and no medical use in the United States, such as heroin, peyote, and LSD. Schedule II substances have equally high (or higher) abuse liability, but have some approved medical usage, such as opium or cocaine. Schedule III and IV substances have moderate to moderately high abuse liability and are legally available with prescription, and schedule V substances can be purchased without prescription. The three major types of drugs defined in terms of their effects on the brain are the narcotics, stimulants, and hallucinogenics.

- *Narcotics:* Narcotics drugs are those that reduce the sense of pain, tension, and anxiety and produce a drowsy sense of euphoria. Heroin is an example.
- *Stimulants:* The stimulants have effects opposite to of the narcotics. Stimulants such as cocaine and methamphetamine keep the body in an extended state of arousal.
- *Hallucinogenics:* Hallucinogenic drugs are mind-altering drugs such as lysergic acid diethylamide (LSD) and peyote.

The Drugs/Violence Link

Illegal drugs are associated with violence in three ways: (1) pharmacological, (2) economic-compulsive, and (3) systemic (Goldstein, 1985). **Systemic violence** is violence associated with "doing business" (the growing, processing, transporting, and selling of drugs) in the criminal drug culture. There is so much systemic violence because the drug business is tremendously lucrative for those involved in it, and there is much competition for a slice of that business. The United Nations estimates the annual worth of the international illicit drug trade at $400 billion (Davenport-Hines, 2002, p. 11).

Systemic violence and other criminal activity begins with the bribery and corruption of law enforcement officials and political figures, or their intimidation and assassination, in the countries where raw materials are grown and through which the processed product is transported. On the streets of the United States, systemic violence is most closely linked with gang battles over control of territory (control of drug markets). Goldstein and his colleagues (1989) found that just over one-half of a sample of 414 murders committed in New York in 1988 was drug related, with 90% of them involving cocaine.

Economic-compulsive violence is violence associated with efforts to obtain money to finance the high cost of illicit drugs. The drugs most associated with this type of activity are heroin and cocaine because they are the drugs most likely to lead to addiction among their users and the most expensive (Parker & Auerhahn, 1998). Crimes committed to obtain drug money run the gamut from shoplifting, robbery, prostitution, to trafficking in the very substance the addict craves. A study of newly incarcerated drug users found that 72% claimed that they committed their latest crime to obtain drug money (Lo & Stephens, 2002).

Pharmacological violence is violence induced by the pharmacological properties of the drug itself. Violence induced by illicit drug uses is rare compared with violence induced by alcohol, the legal drug. A criminal victimization survey found that less than 5% of victims of violent crimes perceived their assailants to be under the influence of illicit drugs versus 20% who perceived them to be under the influence of alcohol (Parker & Auerhahn, 1998).

What Causes Drug Abuse?

Sociological explanations of drug abuse mirror almost exactly their explanations for crime. Erich Goode illustrates the almost indistinguishable explanations offered for the causes of crime and drug abuse in his book *Drugs in American Society* (2005, pp. 62–74). In anomie terms, drug abuse is a retreatist adaptation of those who have failed in both the legitimate and illegitimate worlds, and drug dealing is an innovative adaptation. In social control terms, drug abusers lack social bonds; in self-control terms, drug abuse is the hedonistic search for immediate pleasures; and in social learning terms, drug abuse reflects differential exposure to individuals and groups in which it is modeled and reinforced.

Goode (2005) favors conflict theory most as an explanation. As the rich get richer, the poor poorer, and economic opportunities are shrinking for the uneducated and the unskilled, drug dealers have taken firm root among the increasingly demoralized, disorganized, and politically powerless "underclass." He notes that most members of this class do not succumb to addiction, but enough do "to make the lives of the majority unpredictable, insecure, and dangerous" (p. 77). Goode maintains that conflict theory applies "more or less exclusively to heavy chronic, compulsive use of heroin or crack" (p. 74).

Does Drug Abuse Cause Crime?

Table 12.1 shows the percentage of male and female adult arrestees in some of our largest cities who tested positive for illicit drugs (National Institute of Justice, 2004). Clearly, these data show that illicit drug abuse is strongly *associated* with criminal behavior, but is the association a *causal* one? A large body of research indicates that drug abuse does not appear to initiate a criminal career, although it does increase the extent and seriousness of one (McBride & McCoy, 1993; Menard, Mihalic, & Huizinga, 2001). Drug abusers are not "innocents" driven into a criminal career by drugs, although this might occasionally be true. Rather, chronic drug abuse and criminality are part of a broader propensity of some individuals to engage in a variety of deviant and antisocial behaviors (Fishbein, 2003; McDermott, Alterman, Cacciola, Rutherford, Newman, & Mulholland, 2000). The reciprocal (feedback) nature of the drugs/crime connection is explained by Menard et al. (2001) as follows:

> Initiation of substance abuse is preceded by initiation of crime for most individuals (and therefore cannot be a cause of crime). At a later stage of involvement, however, serious illicit drug use appears to contribute to continuity in serious crime, and serious crime contributes to continuity in serious illicit drug use. (p. 295)

Trevor Bennett and Katy Holloway's examination of the drugs/crime connection among arrestees with drug misuse in England and Wales finds that the connection is hard to pin down. There is certainly a connection, and the greater the number of drugs used by arrestees,

Table 12.1 Male and Female Adult Arrestees Testing Positive for Various Drugs

Males			Females		
City	Any of 5 Drugs[a]	Multiple Drugs[b]	City	Any of 5 Drugs[a]	Multiple Drugs[b]
Atlanta	72.4	73.5	Albany, NY	60.9	65.2
Chicago	86.0	86.0	Chicago	61.1	66.7
Dallas	62.3	63.8	Denver	69.1	24.9
Houston	61.7	61.9	Honolulu	74.5	27.7
Los Angeles	68.6	68.9	Los Angeles	59.3	63.0
New York	67.7	72.7	New York	67.7	72.7
Philadelphia	67.0	68.8	New Orleans	58.8	17.8
Phoenix	74.1	76.8	Phoenix	74.6	78.5
San Diego	66.8	71.2	San Diego	69.1	72.6
Washington	65.6	65.8	Washington	61.1	66.7

SOURCE: Adapted from the Arrestee Drug Abuse Monitoring Program (National Institute of Justice, 2004).

a. The five drugs are cocaine, marijuana, methamphetamine, opiates, and phencyclidine (PCP).

b. Multiple drugs are any of nine drugs which include the basic five plus barbiturates, methadone, benzodiazepines, and propoxyphene.

the greater the mean number of offenses over a 12-month period. They conclude that there are three possible explanations for the connection: (1) multiple drug use causes high rates of offending, (2) high rates of offending cause multiple drug use, and (3) there is no causal connection, that is, certain individuals are predisposed to high levels of involvement in both drugs and crime. While they concede that their data do not allow them to choose between these possibilities, they prefer the first explanation, arguing that the more drugs a person uses, the more the "necessity" to commit crimes to finance the habit.

Prostitution and Commercialized Vice

No other crime has been subjected to shifts of attitudes and opinions across the centuries and across different cultures more than prostitution. The FBI defines **prostitution** and commercialized vice in such a way as to cover people who sell their sexual services (prostitutes), those who recruit (procure) them, those who solicit clients (pander) for them, and those who house them. The common term for a procurer and panderer is a pimp and for the keeper of a bawdy house (a brothel) is a madam. There were 62,501 arrests for prostitution and commercialized vice in 2005. This is obviously only the tiniest fraction of all such offenses (FBI, 2005).

▲ **Photo 12.2** A female prostitute hitchhikes along a road at night.

Exchanging sexual favors for some other valued resource is as old as the species, and prostitution has long been referred to as the world's oldest profession. It has not always had the same sordid reputation that is attached to it today, however. Many ancient societies employed prostitutes in temples of worship with whom worshipers "communed" and then deposited a sum of money into the temple coffers according to their estimation of the worth of the communion. In ancient Greece, many women of high birth who had fallen on hard times became courtesans called hetaerae, who supplied their wealthy clients with stimulating conversation and other cultured activities as well as sexual services. The lower classes had to content themselves with the brothel-based pornae or the prettier and more entertaining auletrides, who would make house calls (Bullough & Bullough, 1994).

This ancient Greek hierarchy of sex workers (as most prostitutes like to be called) is mirrored in modern American society. The modern American hetaerae belong to the elite escort services and call houses and tend to be much better educated, more sophisticated, and better looking than other sex workers because they cater to a wealthy clientele who want to be made to feel special as well as sexually satisfied.

These women (and sometimes men who cater to a gay clientele) can earn six-figure incomes annually and are able to sell their "date books" for thousands of dollars upon retiring (Kornblum & Julian, 1995, p. 109).

Brothel prostitutes are the modern auletrides. The only legal brothels in the United States are in certain counties of Nevada, but illegal brothels probably exist in every town of significant size in the United States although they are not as prominent a part of community life as they used to be. Brothel prostitutes must accept whatever client comes along, but may make from $50 to $100 from each client. The streetwalker is the lowest member of the sex worker hierarchy. These prostitutes solicit customers on the streets and may charge only about $20 a trick (typically a quick act of fellatio).

Becoming a Prostitute

It has been estimated that prostitution is the primary source of income for over one million women in the United States, many of whom view sex work as the most financially lucrative option open to them (Bartol, 2002, p. 369). Many brothel and streetwalker prostitutes typically progressed from casual promiscuity at an early age to reasoning that they could sell what they were giving away under the influence of peer pressure from more experienced girls and from pimps (Kornblum & Julian, 1995, p. 112). Pimps exploit the strong need for love and acceptance among vulnerable girls. The pimp frequently takes on the roles of father, protector, employer, lover, husband, and often drug supplier, thus making the girl totally dependent on him (Tutty & Nixon, 2003). The girls most vulnerable to pimps and other pressures to enter prostitution are those who have experienced high rates of physical, sexual, and emotional abuse at home and who are drug abusers (Bartol, 2002).

The article by Shu-ling Hwang and Olwen Bedford in this section picks up this thread in a Chinese context. Hwang and Bedford interviewed a number of prostitutes in a Taiwanese remand center and asked them about their motives for remaining in prostitution. Unlike many Western prostitutes, very few cited economic motives for entering the profession. As is the case in the United States, many of these Taiwanese prostitutes ran away from home to escape abuse and were befriended by pimps who supplied them with drugs and a modicum of affection. A certain number of the prostitutes were indentured to a brothel by their parents who were in desperate need of money. Interestingly, some prostitutes cited professional pride in their work as a reason for continuing in it.

Should Prostitution Be Legalized?

Prostitution is one of those things that we can never really stop, although the AIDS epidemic has greatly reduced it (a 1989 study found that about 40% of streetwalkers and 20% of call girls were HIV positive; in Kornblum & Julian, 1995, p. 109). If we can't stop it, should we legalize it and therefore make it safer? When the ancient Greek lawmaker Solon (638–559 BC) legalized and taxed prostitution, he was widely praised: "Hail to you Solon! You bought public women [prostitutes] for the benefit of the city, for the benefit of the morality of a city that is full of vigorous young men who, in the absence of your wise institution, would give themselves over to the disturbing annoyance of better women" (Durant, 1939, p. 116). Taxes on prostitution enabled Athens to build the temple to Aphrodite and provided its "vigorous young men" safe outlets for their urges. To borrow a term from sociology, the citizens of Athens found prostitution to be "functional," meaning that it had a socially useful role to play. Such an attitude, however, ignores the important (functional) role of morality to society, and the issue of legalization becomes how much morality are we willing to sacrifice for the sake of expediency.

Driving Under the Influence

Driving under the influence is defined by the FBI as "driving or operating a motor vehicle or common carrier while mentally or physically impaired as the result of consuming an alcoholic beverage or using a drug or narcotic" (FBI, 2005, p. 144). In state statutes this crime is typically referred to as driving under the influence (DUI) or driving while intoxicated (DWI). DUI is far from being a victimless offense. According to the Insurance Information Institute (2005), 16,694 people in the United States died in alcohol-related crashes in 2004, which is 557 more than the number of murder victims reported by the FBI in that year. In 2004 there were 805,854 DUI arrests, 81.4% of whom were males and 85% white (FBI, 2005, p. 14).

Many people used to consider deaths due to drunk drivers as "accidents" rather than "crimes" and penalties were relatively light. Attitudes began to change with the founding of MADD (Mothers Against Drunk Driving) in 1980, an organization that has effectively lobbied for legislation nationwide to increase the legal drinking age and for stricter penalties for drunk drivers. MADD also lobbied to lower the blood alcohol count (BAC) level that defines intoxication from 0.10 to 0.08 grams per deciliter of blood. Every state in the union has now enacted all these measures. As we see in Figure 12.2 from the National Highway Traffic Safety Commission (2003), these combined measures reduced alcohol-related traffic fatalities by 62.5% from 1982 to 2002.

Figure 12.2 Alcohol-Related Fatality Rate per 100 Million Vehicle Miles Traveled

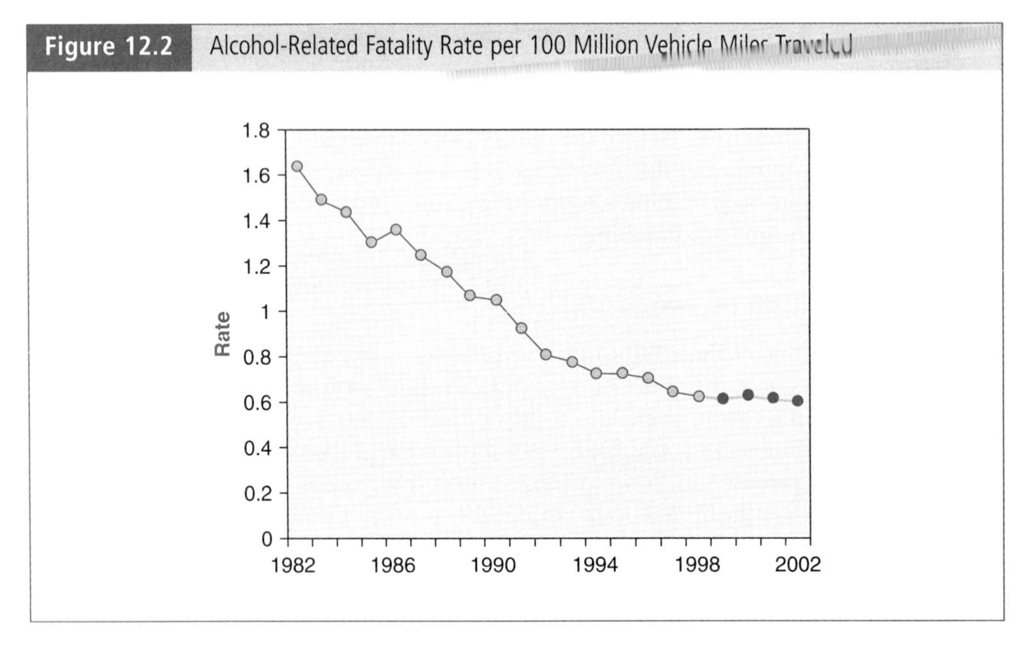

SOURCE: National Highway Traffic Safety Commission (2003).

A 1999 nationwide study of DUI offenders found that the average BAC at arrest was 0.24, or three times the legal limit. The average time for DUI offenders in jail was 11 months, and for offenders sent to prison it was 49 months (Maruschak, 1999). The significant drop in alcohol-related traffic fatalities following a lowered tolerance and increased penalties shows that we can indeed legislate morality, if only in the case of drunken driving.

Gambling

The FBI defines gambling as "to unlawfully bet or wager money or something else of value; assist, promote, or operate a game of chance for money or some other stake; possess or transmit wagering information; manufacture, sell, purchase, possess, or transport gambling equipment, devices or goods; or tamper with the outcome of a sporting event or contest to gain a gambling advantage (2005, p. 143). The FBI (2005, p. 14) reports only 6,365 arrests for gambling in 2004 (90.2% males), which, given the widespread popularity of gambling, points to the lack of seriousness with which the police view it and the lack of public support for enforcement.

We have ambivalent attitudes toward gambling in the United States. For instance, the Federal Council of Churches affirms its "vigorous opposition to gambling which we regard as an insidious menace to personal character and morality. By encouraging the idea of getting something for nothing, of getting a financial return without rendering any service, gambling tends to undermine the basic idea of public welfare" (in McCaghy & Cernkovich, 1987, p. 434). On the other hand, wanting something for nothing is hardly a strange desire. The biggest problem with gambling is not the person who wants the occasional flutter on the lottery (the fact that these lotteries are state supported reveals our moral ambivalence) or on the horses, but the person who becomes addicted and gambles away everything he or she owns. Pathological gambling is thus far from victimless. It is similar to drug, sex, food, and alcohol addiction in terms of what is going on in the reward centers of the brains of addicted gamblers (Ruden, 1997).

Summary

- Public order offenses are sometimes dismissed as minor nuisance offenses, but they can be quite serious. Criminologists now use this term, rather than victimless, with the realization that there are always secondary victims.

- Alcohol is humankind's favorite way of drugging itself and has always been associated with criminal and antisocial behavior. It reduces the inhibiting neurotransmitters and thus reduces impulse control. Contextual factors also play their part in producing the kinds of obnoxious behavior associated with drinking too much alcohol. So-called binge drinking is a major contextual problem.

- There are two types of alcoholism: Type I and Type II. Type I is associated with mild abuse, minimal violence, moderate heritability, and character disorders that result from alcoholism. Type II is characterized by early onset, violence, criminality, high heritability, and character disorders that precede alcoholism. Type II alcoholics may have inherited problems that drive both their alcoholism and their criminality.

- Illicit drug use is also a major problem. Like delinquency, drug usage increases at puberty and drops off in early adulthood to almost zero by the age of 65. Drug addiction is fairly similar to alcoholism in terms of brain mechanisms. Drugs hijack the pleasure centers in the brain and make addicts crave drugs to gain any sort of pleasures at all. Most people who try drugs do not become addicted.

- Drugs are associated with violence in the following ways: pharmacological, economic-compulsive, and systemic, with the latter having the strongest association. It does so because violence is part of "doing business" in the lucrative illicit drug business. The economic-compulsive link with violence is the result of addicts' efforts to gain money to purchase drugs, and the rarest link, pharmacological, is violence induced by ingested drugs.

- Most people arrested for a crime test positive for drugs, but this does not mean that drugs cause crime. Drug abuse is part of a broader propensity of some individuals to engage in all kinds of antisocial behavior, and such behavior is usually initiated before drug abuse behavior. Drug abuse does exacerbate criminal behavior, however.

- Prostitution is as old as the species. While many individuals are coerced into prostitution, others become prostitutes because it is a lucrative business. There are many arguments for and against the legalization of prostitution.

- Driving under the influence is the most serious Part II offense because of its sometimes deadly consequences. It was pointed out that more people are killed by drunken drivers in a typical year than are murdered by other means. Activism and legislation since the 1980s have succeeded in significantly reducing drunk driving.

- Gambling is often viewed as a harmless pastime engaged in by millions of people, but some people take it to pathological extremes and gamble away everything they own, thus victimizing loved ones as well as themselves.

EXERCISES AND DISCUSSION QUESTIONS

1. Discuss with classmates how each of you act—silly, aggressive, lusty, maudlin, etc.—when you have "gone over the limit" drinking alcohol. Why do you think that the same substance "makes" different people react differently?

2. Given what you know about the history of drug laws in the U.S. and the link between drug abuse and violence (and crime in general), would legalizing drugs be the lesser of two evils? Give reasons why or why not.

3. Give your reasons why we should or why we should not legalize prostitution.

USEFUL WEB SITES

Crimes Against Public Order and Morals. http://faculty.ncwc.edu/TOCONNOR/293/293lect13.htm.

Maintaining Public Order. http://police.homeoffice.gov.uk/operational-policing/crime-disorder/public-order.html.

Mothers Against Drunk Driving. http://www.madd.org/.

Paraphilias. http://www.athealth.com/Consumer/disorders/Paraphilias.html.

Prostitution Research and Education. http://www.prostitutionresearch.com/.

GLOSSARY

Alcoholism: A chronic disease condition marked by progressive incapacity to control alcohol consumption despite psychological, spiritual, social, or physiological disruptions.

Binge drinkers: People who frequently consume anywhere between 5 and 10 drinks in a few hours time (go on a binge).

Drug addiction: Compulsive drug-seeking behavior where acquiring and using a drug becomes the most important activity in the user's life.

Economic-compulsive violence: Violence associated with efforts to obtain money to finance the cost of illicit drugs.

Harrison Narcotic Act: A 1914 Congressional Act that criminalized the sale and use of narcotics.

Pharmacological violence: Violence induced by the pharmacological properties of a drug.

Physical dependence: The state in which a person is physically dependent on a drug because of changes to the body that have occurred after repeated use of it and necessitate its continued administration to avoid withdrawal symptoms.

Prostitution: The provision of sexual services in exchange for money or other tangible reward as the primary source of income.

Psychological dependence: The deep craving for a drug and the feeling that one cannot function without it; psychological dependence is synonymous with addiction.

Systemic violence: Violence associated with aggressive patterns of interaction within the system of drug distribution and use.

Tolerance: The tendency to require larger and larger doses of a drug to produce the same effects after the body adjusts to lower dosages.

Type I alcoholism: A form of alcoholism characterized by mild abuse, minimal criminality, and passive-dependent personality.

Type II alcoholism: A form of alcoholism that is characterized by early onset, violence, and criminality and is largely limited to males.

Withdrawal: A process involving a number of adverse physical reactions that occur when the body of a drug abuser is deprived of his or her drugs.

READING

Alcohol Problems and the Differentiation of Partner, Stranger, and General Violence

Rosemary Cogan and Bud C. Ballinger III

In this article, Rosemary Cogan and Bud Ballinger explore the effect of alcohol on violence with partners and strangers. They studied 457 college men and 958 college women with low, intermediate, or high scores on the Short Michigan Alcohol Screening Test. Respondents reported conflict tactics on the Conflict Tactics Scale in the past year to and by partners and strangers. Not surprisingly, they found that more men than women had high alcohol problem scores and that men with alcohol problems were more likely than other men to commit violence against strangers or against partners and strangers. However, men with alcohol problems were not more likely than other men to commit violence against their partners only. Among women, alcohol problems had little relationship to either committing violence or being the victim of violence.

Several relationships between alcohol use and the commission of physical aggression have been reported with some consistency. Reviewing the literature concerning the relationship between alcohol and several types of physical aggression, Lipsey, Wilson, Cohen, and Derzon (1997) commented that "no firm conclusion can be drawn about whether alcohol plays a causal role in [violent] behavior" (p. 245). The relationship between alcohol and the violent victimization of men, particularly by partners, has received less attention. Graham and West (2001) noted that alcohol is related to the violent victimization of men by strangers. Porcerelli et al. (2003) reported more violent victimization by partners and strangers among men and women with alcohol problems than men and women with no alcohol problems among family practice patients.

Women's violence and alcohol use has been studied less often than men's violence. The relationship between alcohol problems and women's violence toward partners is described as equivocal by Riggs et al. (2000). Women with alcohol problems are more likely to be violently victimized by their partners than other women, perhaps because their partners are also more likely to have alcohol problems and/or perhaps because women who have been violently victimized by partners develop alcohol use problems (e.g., Tollestrup et al., 1999). However, some studies have found no relationship between alcohol problems and violent victimization among women (e.g., Plichta, 1996). With respect to violence toward strangers, Bushman (1997) concluded in a meta-analysis of the literature that alcohol has a larger effect on aggression by men than by women.

There is growing recognition in the partner violence literature that men who are violent toward their partners are not a homogenous

Source: Cogan, R., & Ballinger, B. (2006). Alcohol problems and the differentiation of partner, stranger, and general violence. *Journal of Interpersonal Violence, 21*(7), 924–935. Reprinted with permission of Sage Publications, Inc.

group but include some men who are violent only toward their partners and some who are violent toward their partners and toward nonpartners. Men who are violent toward partners and nonpartners commit more severe physical aggression toward their partners than men who are violent only toward their partners. Some research shows that general and nonpartner physical aggression by men are related to antisocial features and alcohol problems, whereas partner physical aggression is not (Cogan et al., 2001).

We planned the current work to directly compare differences between men and women low, intermediate, or high in alcohol problems and the commission of violence toward partners only, strangers only, or partners and strangers. We also compare differences between men and women low, intermediate, or high in alcohol problems and violent victimization by partners only, strangers only, or partners and strangers. Our intent is to explore alcohol problems as a feature that may differentiate between violence toward partners only and violence toward partners and strangers.

Method and Materials

The participants included 457 men and 958 women enrolled in beginning psychology classes at a large southwestern university during three academic semesters who were between age 18 and 24 years, had been in a relationship with a partner during the previous year, and completed the measures described below. The mean age of the participants was 18.7 years. Most participants were White non-Hispanic (85.5%), 12% were Hispanic, 2.2% were Black, and less than 1% were Other. Most were single (94.4%), 2.7% were married, 1.2% were cohabiting, 1.0% were separated or divorced, and 0.6% described themselves as Other. Most were freshmen (78.0%), 14.6% were sophomores, 5.4% were juniors, 2.8% were seniors, and 0.6% were classified as Other.

Each participant completed a brief demographic measure with questions about sex, age, race and ethnicity, marital status, and academic classification and completed the SMAST, the Conflict Tactics Scale—Partners, and the Conflict Tactics Scale—Strangers, described below.

Short Michigan Alcohol Screening Test (SMAST). The SMAST is a 13-item self-report questionnaire for alcoholism screening. The SMAST is an often-used screening measure with good psychometric properties. Respondents indicate yes or no to each item (e.g., "Are you able to stop drinking when you want to?"). Participants responding to 0 or 1 item in the direction of alcohol problems were classified as being "low" in alcohol problems. Participants responding to 2 items in the direction of alcohol problems were classified as "intermediate" with respect to alcohol problems. Participants responding to 3 or more items in the direction of alcohol problems were classified as "high" in alcohol problems.

Conflict Tactics Scales (CTS). The CTS is an 18-item self-report measure of things that might happen between people in conflict situations. Respondents report what they have done to others and what others have done to them. Items range from Item 1 ("Discussed the issue calmly") to 18 ("Used a knife or gun"). Respondents indicate whether each item occurred never, once, twice, 3 to 5 times, 6 to 10 times, 11 to 20 times, or more than 20 times in the past year. Ballinger (2000) concluded, based on the factor analysis, that the CTS to Strangers and CTS by Strangers to and by men and women each includes four factors: Reasoning (Items 1 to 3; e.g., Item 1, "Discussed the issue calmly"), Verbal Aggression (Items 4 to 10; e.g., Item 4, "Insulted or swore at the other one,"), Physical Aggression–mild (Items 11 to 13; e.g., Item 11, "Threw something at the other one"), and Physical Aggression–severe (Items 14 to 18; e.g., item 14, "Kicked, bit, or hit with a fist"). We have combined the two physical aggression subscales into one scale of physical violence.

Four versions of the CTS were included in the current study: CTS-to Partners, CTS-by Partners, CTS-to Strangers, and CTS-by Strangers. The most extreme item that the respondent reported on each version was identified and grouped as

Reasoning (Items 1 to 3), Verbal Aggression (Items 4 to 10), or Physical Aggression (Items 11 to 18). Respondents were grouped into those who reported that they committed no violence toward either partners or strangers, committed violence toward partners only, committed violence toward strangers only, or committed violence toward partners and strangers. Respondents were similarly grouped into four groups based on reports of violent victimization by others.

Results

There were differences between men and women in the frequency of SMAST groups, $\chi^2(3, N = 1{,}415) = 30.76$, $p < .0001$. Follow-up tests showed no differences in the number of men and women with low or intermediate SMAST scores, $\chi^2(1, N = 1{,}193) = 2.12$, $p = .15$. More men than women had high than low SMAST scores, $\chi^2(1, N = 1{,}132) = 30.78$, $p < .0001$. The SMAST information is shown in Table 12.2.

For CTS-to Partners, $\chi^2(2, N = 1{,}415) = 25.89$, $p < .0001$. Follow-up tests showed no significant differences between the number of men and women who were violent toward their partners versus those who were not, $\chi^2(1, N = 1{,}415) = 2.65$, $p = .10$. For CTS-by Partners, $\chi^2(2, N = 1{,}415) = 17.1$[illegible], $p < .0001$. A follow-up test showed that more men than women reported being the victims of violence by their partners, $\chi^2(1, N = 1{,}415) = 13.43$, $p = .0002$. For CTS-to Strangers, $\chi^2(2, N = 1{,}415) = 124.93$, $p < .0001$. A follow-up test showed that more men than women reported directing violence toward strangers, $\chi^2(1, N = 1{,}415) = 124.72$, $p < .0001$. For CTS-by Strangers, $\chi^2(2, N = 1{,}415) = 95.77$, $p < .0001$. A follow-up test showed that more men than women reported being the victims of violence by strangers, $\chi^2(1, N = 1{,}415) = 105.38$, $p < .0001$). The CTS information is shown in Table 12.3.

Table 12.2 Short Michigan Alcohol Screening Test (SMAST) Scores Reported by Men (n = 457) and Women (n = 958)

SMAST Alcohol Problems Items	Men n (%)	Women n (%)
Low (0 or 1 item)	258 (56.5)	652 (68.1)
Intermediate (2 items)	93 (20.4)	190 (19.8)
High (3 or more items)	106 (23.2)	116 (12.1)

Alcohol and Conflict Tactics To extend consideration of partner versus general violence, the percentage of men and women low, intermediate, and high in SMAST alcohol problems reporting committing no violence toward either partners or strangers (C-None), violence toward partners only (C-PO), violence toward strangers only (C-SO), or violence toward partners and strangers (C-PS) were compared, shown in Table 12.4. The percentages of men and women low, intermediate, or high in alcohol problems who were the victims of violence by neither partners nor strangers (V-None), violence by partners only (V-PO), violence by strangers only (V-SO), or violence by partners and strangers (V-PS) are shown in Table 12.5. Differences in the overall frequencies were significant for men for violence directed toward others, $\chi^2(3, N = 457) = 27.09$, $p < .0001$, and for men's victimization by others, $\chi^2(3, N = 457) = 1$[illegible]$.34$, $p = .004$. Follow-up chi-square analyses were carried out comparing the number of men low, high, or intermediate in alcohol problems who committed no violence (C-None), or who committed violence toward partners only (C-PO), strangers only (C-SO), or partners and strangers (C-PS). The number of men in the C-None and C-PO groups did not differ as a function of being low, high, or intermediate in alcohol problems, $\chi^2(2, N = 248) = 1.32$, $p = .52$. The number of men in the C-None and C-SO groups did differ as a function of being low, high, or intermediate in alcohol problems, $\chi^2(2, N = 336) = 14.58$, $p = .0007$, and as can be seen in Table 12.5, men were more likely to commit violence toward strangers as alcohol problems increased. The number of men in the C-None and C-PS groups did differ as a function of being low, high, or intermediate in alcohol

Table 12.3 Most Extreme Conflict Tactics Scale (CTS) Factor Reported by Men (n = 457) and Women (n = 958)

	To Partners*		By Partners*		To Strangers*		By Strangers*	
	Men	Women	Men	Women	Men	Women	Men	Women
Reasoning	48 (10.5)	36 (3.8)	46 (10.1)	67 (7.0)	63 (13.8)	188 (19.6)	61 (13.4)	184 (19.2)
Verbal Aggression	288 (63.0)	628 (65.6)	249 (54.5)	642 (67.0)	185 (40.5)	601 (62.7)	179 (39.2)	574 (59.9)
Physical Aggression	121 (26.5)	294 (30.7)	162 (35.4)	249 (26.0)+	209 (45.7)	169 (17.6)+	217 (47.5)	200 (20.9)+

*$p < .0001$ for each of the four chi-square tests comparing differences between men and women; + $p < .0002$ for chi-square tests comparing differences between men and women for physical versus not physical as the most extreme report.

Table 12.4 Number and Percentage of Men (n = 457) and Women (n = 958) With High, Intermediate, and Low Short Michigan Alcohol Screening Test (SMAST) Alcohol Problem Scores Who Did Not Commit Violence (C-None), Committed Violence Toward Partners Only (C-PO), Committed Violence Toward Strangers Only (C-SO), Toward Partners and Strangers (C-PS)

	Short Michigan Alcohol Screening Test Score					
	Men			Women		
Commission	Low n (%)	Intermediate n (%)	High n (%)	Low n (%)	Intermediate n (%)	High n (%)
C None	150 (58.1)	38 (40.9)	35 (33.0)	422 (64.7)	113 (59.5)	66 (56.9)
C-PO	16 (6.2)	3 (3.2)	6 (5.7)	113 (17.3)	47 (24.7)	28 (24.1)
C-SO	53 (20.5)	25 (26.9)	35 (33.0)	45 (6.9)	12 (6.3)	6 (5.2)
C-PS	39 (15.1)	27 (29.0)	30 (28.3)	72 (11.0)	18 (9.5)	16 (13.8)

problems, $\chi^2(2, N = 319) = 20.05$, $p < .0001$, and as can be seen in Table 12.5, men intermediate or high in alcohol problems were more likely to commit violence toward both partners and strangers (C-PS).

Discussion

Several findings in the current work replicate findings often reported by others. More men than women had high SMAST alcohol problems scores. Most of the men and women reported that reasoning or verbal aggression was the most extreme conflict reported to and by partners and strangers. The differences between men and women in reporting the commission of violence toward their partners were not significant; however, more men than women reported being the victims of violence by their partners. Women are more often than men physically injured by partner violence, and comparisons of more female-to-male and male-to-female partner violence are not without controversy. Men's report of more violent victimization by partners than women's

Table 12.5 Number and Percentage of Men (n = 457) and Women (n = 958) With High, Intermediate, and Low Short Michigan Alcohol Screening Test (SMAST) Alcohol Problem Scores Who Were Not Victims of Violence by Others (V-None), Victims of Violence Only by Partners (V-PO), Victims of Violence Only by Strangers (V-SO), or Victims of Violence by Partners and Strangers (V-PS)

	Short Michigan Alcohol Screening Test Score					
	Men			Women		
Victimization	Low	Intermediate	High	Low	Intermediate	High
V-None	127 (49.2)	27 (30.1)	32 (30.2)	413 (63.3)	117 (61.6)	69 (59.5)
V-PO	29 (11.2)	12 (12.9)	12 (11.3)	104 (16.0)	35 (18.4)	20 (17.2)
V-SO	51 (19.8)	29 (31.2)	28 (26.4)	76 (11.7)	21 (11.0)	13 (11.2)
V-PS	51 (19.8)	24 (25.8)	34 (32.1)	59 (9.0)	17 (9.0)	14 (12.1)

report of violent victimization by partners in the present data is in harmony with the research literature (see Archer, 2000). Far more men than women reported committing violence toward strangers and being the victims of violence by strangers, which is commonplace in the literature (Graham & West, 2001).

We carried out the current work to explore the possibility that alcohol problems might be related to violence with strangers and not related to violence with partners. The hypothesis of differential involvement of alcohol problems in stranger and partner violence led us to separate two types of partner violence that are often confounded. Some men are violent toward their partners only, and some men are violent toward their partners and strangers. The current findings show, as predicted, that more men who commit violence toward strangers or toward partners and strangers have alcohol problems than men who are not violent, whereas few men who committed violence toward their partners only had alcohol problems. Among men, alcohol problems successfully differentiate between the two types of partner violence. Among women, on the other hand, alcohol-use problems and the commission of violence toward strangers occurred less often than they did among men, and alcohol problems did not differentiate between women who are violent toward their partners and women who are violent toward partners and strangers. The current findings extend earlier work on violence by men to the less-often-explored area of violence by women.

References

Ballinger, B. C. (2000). *Factor analysis of the partner and stranger versions of the Conflict Tactics Scale.* Unpublished doctoral dissertation, Texas Tech University, Lubbock.

Bushman, B. J. (1997). Effects of alcohol on human aggression: Validity of proposed explanations. In M. Galanter (Ed.), *Recent developments in alcoholism, Vol. 13: Alcohol and violence: Epidemiology, neurobiology, psychology, family issues* (pp. 227–243). New York: Plenum.

Cogan, R., Porcerelli, J., & Dromgoole, K. (2001). Dynamics of partner, stranger, and generally violent college student men. *Psychoanalytic Psychology, 18,* 515–533.

Graham, K., & West, P. (2001). Alcohol and crime: Examining the link. In N. Heather, T. J. Peters, & T. Stockwell (Eds.), *International handbook of alcohol dependence and problems* (pp. 439–470). New York: John Wiley.

Lipsey, M. W., Wilson, D. B., Cohen, M. A., & Derzon, J. H. (1997). Is there a causal relationship between alcohol use and violence? A synthesis of evidence. In M. Galanter (Ed.), *Recent developments in alcoholism, Volume 13: Alcoholism and violence* (pp. 245–282). New York: Plenum.

Plichta, S. B. (1996). Violence and abuse: Implications for women's health. In M. M. Falik & K. S. Collins (Eds.),

Women's health: The Commonwealth Fund Survey (pp. 236–270). Baltimore: Johns Hopkins University Press.

Porcerelli, J. H., Cogan, R., West, P. P., Rose, E. A., Lambrecht, D., Wilson, K. E., et al. (2003). Violent victimization of women and men: Physical and psychiatric symptoms. *Journal of the American Board of Family Practice, 16*, 32–39.

Riggs, D. S., Caulfield, M. B., & Street, A. E. (2000). Risk for domestic violence: Factors associated with perpetration and victimization. *Journal of Clinical Psychology, 56*, 1289–1318.

Tollestrup, K., Sklar, D., Frost, F. J., Olson, L., Weybright, J., Sandvig, J., et al. (1999). Health indicators and intimate partner violence among women who are members of a managed care organization. *Preventive Medicine: An International Journal Devoted to Practice and Theory, 29*, 431–440.

DISCUSSION QUESTIONS

1. What was the most surprising finding of this study for you?
2. Why would alcohol have different effects on men and women vis-à-vis violence?
3. What is the difference between violence against women by their partners and violence against men by their partners?

❖

READING

The Association Between Multiple Drug Misuse and Crime

Trevor Bennett and Katy Holloway

In this article, Bennett and Holloway examine aspects of multiple drug use and how such usage is associated with crime. They assert that previous research that has investigated the association between specific drug types and crime has tended to focus on the specific drug type and ignored multiple usage. This is problematic because it cannot be assumed that the relationship between use of a specific drug and crime will be the same regardless of the additional drugs consumed. Bennett and Holloway investigated whether there was a correlation between number and type of drugs used and involvement in crime among arrestees taking part in the New English and Welsh Arrestee Drug Abuse Monitoring program in the United Kingdom. Their results showed that both the number of drug types consumed and the particular drug type combinations used explained offending rate. For instance, multiple drug users reported an average of twice as many crimes as single-drug offenders. The authors conclude that their research of the links between multiple drug use and crime might help inform anti-drugs strategies and treatment services.

Source: Bennett, T., & Holloway, K. (2005). The association between multiple drug misuse and crime. *International Journal of Offender Therapy and Comparative Criminology, 49*(1), 63–81. Reprinted with permission of Sage Publications, Inc.

Research on the connection between drug misuse and crime has tended to focus on either aggregated measures of drug misuse and criminal behavior or specific types of drugs and specific types of offences. Little attention has been paid to the extent to which combinations of drug misuse might be connected to crime. This is surprising for at least two reasons. First, there has been considerable attention paid in the research literature to the existence and role of what is sometimes referred to as polydrug or multiple drug misuse. However, much of this discussion has focused on explaining the phenomenon of drug misuse or the implications of multiple drug misuse for treatment. Second, there are a number of plausible reasons to suspect that drug use combinations might be important in explaining crime. These include direct effects, such as the potential interactive or additive effects of drug mixing on judgment or behavior, and indirect effects, such as the potential amplifying effect of involvement in drug misuse on offending (or vice versa).

Multiple Drug Use

There are a number of examples in the literature of the concept of multiple drug use being linked to specific contexts. Some writers have defined polydrug use in the context of subsidiary drug use among users in treatment. The term polydrug user is used to refer to users who supplement their prescribed drugs with additional nonprescribed drugs [and for] users in treatment whose consumption of heroin is supplemented with a combination of prescribed and nonprescribed drugs. There are also examples of variations in the time scale over which the drugs are consumed. Wilkinson et al. (1987) use the term polydrug use to mean use of more than one drug type, either over time or at a point in time. In this article, we have used the words multiple drug use to mean use of two or more drugs over a 12-month period of time. It is used in preference to polydrug use, which tends to be associated with treatment diagnoses and is sometimes used to refer to the use of subsidiary drugs in addition to prescription drugs. It is also used in preference to simultaneous drug use, as the article aims to investigate drug use and criminal behavior over the same time period and over longer periods of time.

Multiple Drug Use and Crime

The main research interest of this article is whether multiple drug use in general and specific kinds of multiple drug use in particular are associated with criminal behavior. The research on this topic can be divided into three main groups: (a) the prevalence of multiple drug use and crime, (b) the number of drug types used and crime, and (c) combinations of drug types and crime.

Prevalence of Multiple Drug Use and Crime

Some studies have looked at the prevalence of criminal behavior among multiple drug users from data derived from general population surveys. Data from the U.S. National Youth Survey show that crime commission rates per year were between 10 and 20 times higher among multiple drug users (who used alcohol, marijuana, and other drugs four or more times each) than among nonusers. Other studies have investigated the various measures of criminal behavior among multiple drug users within criminal populations. Some of the most detailed findings on multiple drug use and crime have come from studies based on arrestee surveys. Smith and Polsenberg (1992) found, in a study based on adult arrestee data for the District of Columbia, that 81% of arrestees testing positive for two or more drugs had a prior criminal record, compared with 71% of those who tested positive for one drug and 52% of those who tested positive for no drugs.

Number of drug types used and crime. Smith and Polsenberg (1992) explored the relationship between number of positive tests for different drug types among a sample of arrestees and the average number of prior arrests. They found that the average number of prior arrests increased with the number of positive tests. Those who tested positive for no drug type recorded an average of 1.95 prior arrests, those who tested positive for just one drug type had an average of 2.75 prior arrests, and those who tested positive for two or more drug types had an average of 4.64 prior arrests. Bennett (2000) used data from the second developmental stage of the New English and Welsh Arrestee Drug Abuse Monitoring (NEW-ADAM) program in the United Kingdom to explore the relationship between number of drug types used and self-reported offending. Arrestees who used one drug type in the past 12 months reported an average of 26 acquisitive offences during the past 12 months. Arrestees who used two drug types reported an average of 95 offences, and those who used three or more drug types reported an average of 176 offences.

Combinations of drug types used and crime. The available research to date has tended to focus on the effect of different combinations of heroin, crack, and cocaine, plus subsidiary drugs, on crime. Shaw et al (1999) found, from among a sample of arrestees in Los Angeles, that those who had used cocaine only or crack only in their lifetimes had lower prevalence rates of criminal activities (10% and 14%, respectively) than those who used both cocaine and crack (16% among those who used cocaine first and 24% among those who used crack first). Other research has confirmed the effect of heroin, crack, and cocaine combining on crime. Sanchez, Johnson, and Israel (1985) found among a sample of incarcerated females that those who used heroin and cocaine in the past year had higher mean rates of drug and prostitution offences than users of heroin only.

Theories of Multiple Drug Use and Crime

Economic Explanations Economic theories of the association between drug use and crime are based on the idea that greater involvement in drug use leads to greater expenditure on drugs and greater involvement in acquisitive crime to pay for these drugs. Some writers have made generalized statements about the relationship that include multiple drug use. Leri et al. (2003) argue that opioid users who also use cocaine will have drug habits that are even more expensive, which, in turn, might lead some individuals to engage in income-generating crime. They also note that opioid addicts sometimes use amphetamines to sustain the activity level needed to hustle the necessary funds to pay for their opioid habit. The main principle of economic theory is that regular drug use is expensive and that some users will seek funds for their drug use from illegal sources. This argument is usually made in relation to heroin addiction and the costs of habitual drug use. Users of multiple drugs (especially when two or more of them are expensive drugs) may face additional financial pressures to commit acquisitive crime.

Psychopharmacological Explanations Psychopharmacological explanations are based on the idea that drugs can have a direct or indirect effect on behavior as a result of their chemical properties. These explanations are typically directed at drug use and violent crime and in most cases refer to the effects of individual drugs. However, some writers have discussed the interaction, protective, or additive effects of multiple drugs on the nature or rate of criminal behavior. Hammersley and Morrison (1987) believe that multiple drugs used simultaneously may increase intoxication. One reason for this is that drug combinations might create unique metabolites that are absent when the drugs are used individually. These metabolites may have greater toxicity than those formed when the drugs are used individually. Pennings et al. (2002) argue that there has been much theorizing about the possible mechanism by which the alcohol and cocaine combination might lead to greater violence than from either drug alone. These include the idea that alcohol and cocaine each elevate extraneuronal dopamine and serotonin levels, which may lead to deficits in impulse control and to violent behavior.

Lifestyle Explanations Lifestyle explanations are sometimes referred to as systemic explanations in that crime is seen as an intrinsic (or systemic) part of the drug-using lifestyle. Lifestyle explanations are also sometimes referred to as spurious explanations in that there may be no direct causal connection between drug use (including multiple drug use) and crime. Instead, they both coexist within the same lifestyle context. The lifestyle perspective rejects the view that drug use can be seen as a cause of crime or that crime can be seen as the cause of drug use. However, some writers conceive of lifestyles as a common cause that explains both drug use and crime. Walters (1998) argues that lifestyles evolve out of predisposing factors, initiating factors, and maintenance factors. The maintaining factors help reinforce and escalate forms of behavior. In the case of drugs and crime, common maintaining factors encourage the convergence and reinforcement of both drug using and criminal lifestyles.

Method

The analysis is based on data collected as part of the New English and Welsh Arrestee Drug Abuse Monitoring (NEW-ADAM) program over the period from May 2000 to March 2002. A total of 9,499 arrestees were processed through the 16 custody suites during the 30-day periods covered by the research. The majority of arrestees were male (86%) and age 25 or older (51%).

The questionnaire included questions on the use of 19 illicit drug types and offending behavior in relation to 10 types of acquisitive crime. The drug types included heroin, crack and cocaine, and a range of other drugs, including cannabis, amphetamines, ecstasy, diazepam, and temazepam. The offence types included vehicle crime, shoplifting, burglary, robbery, theft person, handling, fraud, and drug supply offences. Arrestees were asked about drug use during their lifetime, in the past 12 months, in the past 30 days, and in the past 3 days. They were asked about their offending behavior during their lifetime and in the past 12 months. The following analysis is based on drug use and offending behavior in the past 12 months.

Findings

Prevalence of Multiple Drug Use and Crime

There is a significant difference in the proportion of multiple drug users and nonmultiple drug users who reported acquisitive offending in the past 12 months. About one third of single drug users compared with two thirds of multiple drug users said that they had committed one or more acquisitive crimes in the past 12 months. There is also a significant difference in the mean number of offences reported among multiple and nonmultiple drug users. Multiple drug users reported on average twice as many offences as single drug users over the past 12 months.

Number of Drug Types Used and Crime

The association between multiple drug use and crime can be revealed more fully by examining variations in the number of offences committed and the number of drug types used over the same period of time. About one fifth of arrestees reported that they consumed no drugs, one fifth used just one drug type, one fifth used two or three drug types, and two fifths used four or more drug types (Table 12.6). The mean number of reported offences was 14 among nonusers, compared with 36 among users of one drug type, 128 among users of five drug types, and 306 among users of 11 drug types.

Discussion: Explaining the Connection

The study has shown a correlation between number of drug types used and rate of offending. Almost twice as many multiple drug users as single drug users reported offending in the

Table 12.6 Mean Number of Offenses by Number of Drug Types Used in the Past 12 Months

Number of Drug Types	Mean Number of Offences	Number of Cases	Percentage of Cases	Standard Deviation
0	13.6	691	22.3	96
1	36.4	588	19.0	148
2	47.7	329	10.6	154
3	95.4	329	10.6	218
4	133.0	255	8.2	257
5	128.2	247	8.0	227
6	206.2	202	6.5	301
7	225.3	147	4.8	290
8	222.1	130	4.2	293
9	293.1	81	2.6	314
10	268.9	57	1.8	345
11	306.1	38	1.2	333
Total		3,094	100.0	

previous 12 months. Multiple drug users who offended reported on average twice as many offences as single drug users who offended. Multiple drug users who used a large number of drug types reported committing a greater number of offences than multiple drug users who used a small number of drug types. Multiple drug users who included heroin, crack, and cocaine in their drug combinations committed a greater number of offences on average than multiple drug users who used only recreational drugs. Multiple drug users who used heroin and crack and who also used heroin substitutes, recreational drugs, and tranquilizers had higher offending rates than multiple drug users who used heroin and crack without these additional drug types.

There are at least three broad explanations for a relationship between multiple drug use and crime: (a) Multiple drug use causes high rates of offending, (b) high rates of offending cause multiple drug use, and (c) the relationship is spurious.

Multiple drug use causes high rates of offending. There are at least two possible explanations based on the view that multiple drug use causes high rates of crime. The first is the economic argument that greater involvement in drug use leads to greater involvement in acquisitive crime as a means of financing drug use. One explanation is that multiple drug use leads to high-offending rates simply because a greater number of drug types cost more to finance than a smaller number of drug types. Users might have limited legitimate sources of income and may resort to illegal means to finance the shortfall. Another explanation is that multiple drug use leads to high-offending rates because multiple drug users are more likely to include use of the more expensive drugs, such as heroin, crack, and cocaine, which again places pressure on them to offend to raise funds for drugs. A third explanation, which combines the previous two explanations, is that the most heavily involved multiple drug users are likely to experience a funding shortfall from both the use of more expensive drug types and the range of other supporting drugs used (such as heroin or crack substitutes).

The second is the psychopharmacological argument that multiple drug use can lead to high rates of offending because of the effects of certain types of drugs or certain types of drug interactions. Multiple drug users are more likely than single drug users to be at risk of using these criminogenic drug types or drug type combinations. The research on the psychopharmacology of the drugs-crime connection tends to be based on violence and the links between multiple drug use and violent crime. However, it can be argued that the addictive properties of certain drug types can lead to greater drug use and greater offending to finance the drug use. It can also be argued that certain drug combinations might moderate or enhance the compulsive effects of addictive drug use, which might in turn moderate or enhance offending behavior.

High rates of offending cause multiple drug use. Another explanation given for the drugs-crime connection is the idea that crime might cause drug use. It has been argued that drugs might be one of the items that are purchased from the proceeds of crime. This explanation has sometimes been described as the "life as a party" or "hedonistic pursuits' explanation (Wright & Decker, 1994). Offenders might find that they have surplus funds following a successful crime spree and might choose to use these funds to finance pleasurable pursuits. High-rate offenders might achieve high levels of illegal funds, which might in turn finance high levels of drug use, including multiple drug use.

The relationship is spurious (there is no causal connection). There are two explanations that argue that the links between drug use and crime are not directly causal. The first is the common cause explanation that argues that the drug use and crime are linked by a third or common variable. It is possible that various dispositional factors linked to early development or recent history (e.g., family absence or family breakdown) might predispose certain individuals to high levels of involvement in both drug use and crime. The second is sometimes described as the lifestyles explanation and argues that many forms of behavior, including excessive and problem behavior, are a systemic part of the broader lifestyle. It cannot be argued that any one of these problem behaviors causes the other. In the case of multiple drug use and crime, it could be argued that excessive offending and excessive drug use simply coexist in a broader framework of problem behavior.

Strictly speaking, it is not possible to use the research findings to support or reject these explanations as they are based largely on mechanisms that are beyond the scope of the current research to investigate. The economic argument, for example, requires knowledge about the motivation to offend, and the hedonistic pursuits argument requires knowledge about consumption choices. However, it is possible to make some attempt at comparing the results and the explanations.

Research Support

The research findings are broadly consistent with the economic argument. However, they offer slightly stronger support for the third version of the argument (in the sense that the results are more consistent with the theory) than the first or second version.

The first version (a large number of drug types costs more to finance than a small number) is based on the principle that multiple drug use would lead to higher offending rates regardless of the types of drugs consumed. However, the research shows that multiple drug use involving certain combinations of drugs is associated with higher levels of offending than multiple drug use involving other combinations of drugs. Hence, the type of drugs consumed appears to be important in the explanation. The second version (multiple drug users are more likely to use expensive drugs) is based on the view that multiple drug use leads to higher offending rates because users consume expensive drug types, such as heroin, crack, and cocaine. The research offers some support for this explanation in that it shows that the use of drug combinations involving heroin, crack, and cocaine are associated with higher offending rates than use of combinations based on recreational drugs only. However, the results are most consistent with the third economic explanation that argues that the total number of drug types used and the inclusion of heroin, crack, or cocaine in the drug type combinations are both important in explaining offending rates. The research shows that multiple drug users who use heroin, crack, and cocaine and a large number of other drug types have higher offending rates than those who use heroin, crack, and cocaine and a small number of other drug types.

The research findings are less consistent with the hedonistic pursuits explanation that argues that high offending rates cause multiple drug use. It is difficult to see why high-rate offenders would adopt the drug use patterns shown solely as a result of a desire for hedonistic pursuits following a successful crime spree. The pattern of multiple drugs for the higher offending rate groups is fairly specific and includes heroin, crack, and cocaine as well as a range of heroin substitutes and support drugs. This pattern of drug use is more typical of someone with a substantial drug habit rather than someone who wants to have a party. However, the research offers only indirect insight into the mechanisms that link multiple drug misuse and crime, and hedonistic explanations cannot be ruled out.

The research offers slightly more support for the lifestyles explanation and the idea that multiple drug use and crime are not causally linked but are both systemic to a certain kind of problematic lifestyle. The research shows that multiple drug users who are excessive in terms of the number of drug types they consume are also excessive in terms of the number of offences they commit. Multiple drug users who commit the greatest number of offences are also those who include in their repertoire of drugs the greatest number and most serious drugs. However, the research cannot comment on whether the coexistence is causal or spurious.

Overall, the strongest conclusions that can be drawn from the current research is that the research findings are consistent with the causal explanation that the connection between multiple drug use and high rates of criminal behavior is the product of users consuming a large number of expensive drug types and the noncausal explanation that the connection is the product of overlapping problematic and excessive lifestyles. However, the research findings cannot rule out the possibility that other explanations might also play a part. It is possible, for example, that different types of explanations might apply to different types of individuals or to different times in relation to the same individual.

Policy Implications

The connection between multiple drug use and crime is relevant to policy on drugs and crime. First, drug strategies aimed at reducing drug-related crime typically focus on users of heroin and crack without taking into account use of other drugs. However, the results of the current research have shown that heroin and crack use is not universally associated with high rates of offending. Heroin or crack users who use few other drugs (according to our results) are likely to have lower offending rates than those who use many other drugs. Further, even among heroin and crack users who are multiple drug users, there are likely to be variations in their offending rates depending on the particular multiple drug type combination. Heroin and crack users who use only recreational drugs committed offences at about half the rate of heroin and crack users who used heroin substitutes, recreational drugs, and tranquilizers. Hence, the nature of multiple drug use (in particular the nature of other drugs used by heroin and crack users) might be useful in guiding intervention strategy.

Second, knowledge about multiple drug use is relevant to treatment services provided as part of an antidrug strategy. Heroin users who also use crack might require a different treatment approach than heroin users who use only heroin. Similarly, heroin users who use many other drug types might require a different approach from those who use just a few. In many countries (notably the United Kingdom and the United States), treatment services have grown up around the treatment of heroin addiction. Knowledge about the treatment of cocaine and crack addiction and multiple drug dependency is less well developed. Future research might develop further the concept of multiple drug use and investigate further its value in explaining criminal behavior. The results of this research might help inform national drug prevention strategies and help direct treatment provision and services.

References

Bennett, T. H. (2000). *Drugs and crime: The results of the second developmental stage of the NEWADAM programme* (Home Office Research Study No. 205). London: Home Office.

Hammersley, R., & Morrison, V. (1987). Effects of polydrug use on the criminal activities of heroin-users. *British Journal of Addiction, 82,* 899–906.

Leri, F., Bruneau, J., & Stewart, J. (2003). Understanding polydrug use: Review of heroin and cocaine co-use. *Addiction, 98,* 7–22.

Pennings, J. M., Leccese, A. P., & De Wolff, F.A. (2002). Effects of concurrent use of alcohol and cocaine. *Addiction, 97,* 773–783.

Sanchez, J. E., Johnson, B. D., & Israel, M. (1985). *Drugs and crime among Riker's Island women.* Paper presented at the Annual Conference of the American Society of Criminology, San Diego, CA.

Shaw, V. N., Hser, Y., Anglin, D. M., & Boyle, K. (1999). Sequences of powder cocaine and crack use among arrestees in Los Angeles county. *American Journal of Drug and Alcohol Abuse, 25,* 47–66.

Smith, D. A., & Polsenberg, C. (1992). Specifying the relationship between arrestee drug test results and recidivism. *The Journal of Criminal Law and Criminology, 83,* 364–377.

Walters, G. D. (1998). *Changing lives of crime and drugs: Intervening with substance abusing offenders.* Chichester, UK: Wiley.

Wilkinson, D. A., Leigh, G. M., Cordingley, J., Martin, G. W., & Lei, H. (1987). Dimensions of multiple drug use and a typology of drug users. *British Journal of Addiction, 82,* 259–273.

Wright, R. T., & Decker, S. H. (1994) *Burglars on the job: Street life and residential break-ins.* Boston: Northeastern University Press.

DISCUSSION QUESTIONS

1. Which of the three explanations of Bennett and Holloway's findings do you find most plausible?
2. Why would treatment for multiple drug use differ from treatment for single drug use?
3. If drug addicts engage in large amounts of crime to pay for their drugs, would the decriminalization (not legalization) of drugs so that addicts could get them free (or very cheaply) from their physicians help to solve a big part of the crime problem?

READING

Juveniles' Motivations for Remaining in Prostitution

Shu-Ling Hwang and Olwen Bedford

In this article, Shu-Ling Hwang and Olwen Bedford use qualitative data from in-depth interviews collected in 1990–1991, 1992, and 2000 from 49 juvenile prostitutes remanded to two rehabilitation centers in Taiwan. They analyze these data to explore Taiwanese prostituted juveniles' feelings about themselves and their work, their motivations for remaining in prostitution, and their difficulties leaving it. Hwang and Bedford suggest that juveniles have four major motivations for remaining in prostitution: financial, emotional, drug, and identity-related factors. Analysis of interviews with juveniles working in confined (forced) prostitution suggested four stages of attitude change: resistance, development of interpersonal connections, self-injury and loss of hope, and acceptance of prostitution. Their concluding discussion includes comparison of motivations to remain in prostitution with reasons for entry, relation of findings to prostituted juveniles in other countries, and directions for future research.

Regardless of race or class, prostitutes (both juveniles and adults) tend to have a history of parental abuse and neglect, incest, rape, disruptive school activity, running away, and early sexual experiences. Although an understanding of the reasons for entering prostitution is important for identifying those at risk and designing effective intervention aimed at preventing entry into prostitution, it may not provide sufficient information to develop programs aimed at assisting those already in prostitution to leave it if they want to. Thus, although there is a body of research available to direct efforts in protecting children from ever entering prostitution, little research has been conducted relevant to assisting already prostituted juveniles with reintegration into society. The main goal of this article is to contribute to the international body of research relevant to this topic by identifying juveniles' major motivations for remaining in prostitution as well as the differences between the motivations to remain and the reasons for entry into prostitution.

The majority of research has been conducted in Western countries. Although many aspects of the circumstances and conditions accompanying juvenile prostitution may be shared across cultures, there are also likely to be aspects that differ. For example, family relations play a different role in the identity of people from Confucian and Western cultures. Western studies focusing on juveniles with a relatively individualized sense of identity may not be able to reflect the realities of Chinese juveniles whose identities to a greater extent are integrated with their family members. For example, the Confucian value of filial piety may still play a role in pathways to prostitution for some prostituted juveniles in Taiwan. Another important cultural difference is that the basis of moral behavior

Source: Hwang, S., & Bedford, O. (2004). Juveniles' motivations for remaining in prostitution. *Psychology of Women Quarterly, 28,* 136-146. Reprinted with permission.

differs greatly between Western and Chinese cultures. Social perception of the acceptability of juvenile prostitution and the stigma associated with such behavior may therefore also differ. Solutions aimed at identifying Taiwanese juveniles at risk or at rehabilitating Taiwanese juveniles back into society would likely need to include an understanding of cultural factors. The second goal of this article is to expand understanding of the broad themes that may be common to juvenile prostitution in a Confucian context.

Prostitution in Taiwan

Filial piety is a central concept in Confucianism containing important ethical requirements concerning children's responsibility to their parents. Children are expected to provide for parents in their old age, and are required to submit to parental authority without exception out of respect for their parents and appreciation for their lives and upbringing. In past decades, unmarried women and girls in Taiwan have voluntarily entered prostitution to support parents and siblings, and were socially ratified as "filial daughters." Poor families also indentured their daughters for survival, emergency needs, and debts from gambling. To indenture a girl, a contract was signed with a brothel and the girl handed over. Payment was made in full to the family, and the girl became collateral and lost her rights as a free person. The blend of filial and gender ideology that dictates parents have absolute rights over their children and that devalues daughters while emphasizing the importance of sons is the foundation of indentured prostitution in Taiwan.

In this study we examine contemporary Taiwanese prostituted juveniles' feelings about themselves and their work, and their major motivations for remaining in prostitution and difficulties leaving it. Emphasis is placed on understanding these motivations from the perspective of the girls themselves. The analysis focuses on identifying the reasons the girls tell themselves and each other for remaining in the sex industry. Particular attention is paid to the following research questions: Do the motivations of aboriginal and Han girls differ? Do the motivations of those who have worked in indentured prostitution differ from those who have not? And, have the motivations to remain in prostitution changed in the past decade? Discussion includes relation of findings to reasons for entry into prostitution in Taiwan, relation of the findings in Taiwan to findings on prostituted juveniles in other countries, and directions for future research.

For the purposes of this paper, *prostitution* is defined as exchange of personal interaction of a sexual nature for payment. This personal interaction may range from flirting, dancing, and drinking to sexual intercourse. The exchange may be voluntary or forced, and the individuals engaging in the behavior are not necessarily the ones making or receiving payment. *Special service clubs* are the most dominant form of prostitution in Taiwan. These clubs are staffed by women (called *hostesses*) who provide companionship and exchange their attentions for tips, and encourage men to drink so that the club makes money. Attentions can range from playing drinking games, singing and flirting, to any kind of physical contact, such as kissing, fondling, stripping, or even more graphic behavior. Special service clubs do not usually provide intercourse as a part of the club's offering, but many do have rooms for customers and hostesses to conduct their own transactions. They can also agree on sexual transactions outside the clubs.

Method

The data set is composed of detailed case studies from interviews with 49 girls arrested for prostitution and remanded to two government-run rehabilitation centers in Taiwan. Juveniles detained at these centers are required to go through rehabilitation programs including counseling and/or education through the 9th grade and vocational programs, which may last from

1 month to 2 years. Length of stay in the rehabilitation center depended on the juvenile's family situation and whether or not she had been indentured. In order to check for possible changes in the motivations for juveniles to remain in prostitution over the past decade (as might be expected with implementation of the Child and Teenage Sex Trade Prevention Act in 1995, which has implications for indenture of children by their parents), 8 additional girls were interviewed in 2000 to check whether major changes in motivations might be evident. Participants' ages at the time of the interview ranged from 13 to 28 years old (M = 16.41, mode = 16). Age at first entry into prostitution ranged from 9 to 17 years old (M = 13.4; mode = 13), and number of months in prostitution ranged from 1 month to 10 years. Years of education ranged from 1 to 11 (M = 7.5).

Results

Two main categories with respect to motivation to remain in prostitution immediately emerged from analysis of the interviews. The juvenile herself either believes she has a choice about remaining a sex worker and chooses to do so, or she believes she does not have a choice, and is forced to remain a prostitute. These two categories constitute a very basic distinction on a commonsense level as to why a juvenile might remain in prostitution. They are also the main categories the juveniles themselves use to categorize themselves and other girls.

The first category is made up of juveniles who were not physically coerced into prostitution and have freedom of movement and the possibility of standing up for their rights. Generally, they can decide when they would like to work, and can leave the establishment at any time they wish. This category is termed *free prostituted juveniles.* The term free is selected not because in some objective sense it is evident that the juveniles are free to choose whether they enter or remain in prostitution, but rather because the juveniles in this category perceive themselves as having the option of whether or not to remain in prostitution. That is, they feel they have the power to change their status as a prostitute if they want to. It also reflects the terminology the juveniles themselves use *(zuo zi you de,* literally "doing it with freedom," or alternatively, *zi yuan de* "voluntary"). In contrast, *confined prostituted juveniles* are imprisoned, constantly watched while with customers, and allowed to go outdoors only when they are ill and need medical treatment. They may be forced to take 20–30 clients 15–17 hours a day and given only one day off every month during their menstrual periods. They have little right to protect their health and are beaten for disobedience. They cannot refuse customers, but instead work on demand.

In the following, the two main categories of prostituted juveniles in Taiwan, the free and the confined, are examined to discover differences between and within the groups in motivation to remain in prostitution, and attitudes toward self and prostitution. Although a previous study noted differences between Han and aboriginal juveniles in Taiwan with respect to pathways to prostitution (Hwang & Bedford, 2003), the categorical analysis of data in this study revealed no differences between Han and aboriginal girls in terms of attitudes toward prostitution and self, lifestyle, or motivation for remaining in prostitution. Furthermore, no differences were found between the 1990–1992 and 2000 groups in terms of motivations for remaining in prostitution. In the following analyses, all juveniles, both Han and aborigine, and those interviewed in the different years are discussed as a group.

Free Prostituted Juveniles

Thirty-three (out of 49, 67%) prostituted juveniles in this study described themselves as working in free prostitution for a period of time ranging from 3 months to over 10 years. Types of free prostitution engaged in included full-time call girls or streetwalkers (who generally go to hotels to meet johns for transactions), full-time hostessing (working at special service clubs) with or without clothes and with or without intercourse as part of the hostess job, and hostessing without intercourse while moonlighting as a call girl. In contrast to the West where streetwalking is common among prostituted juveniles, hostessing was most common among the girls in this study. Twenty of the 33 (61%) juveniles in free prostitution began as hostesses, and 31 of the 49 participants (63%) in this study had experience with hostessing. However, no clear pattern of movement from one type of prostitution to another was evident. That is, although the majority of girls had experience with hostessing of some type, there was no clear progression of movement from hostessing to other forms of prostitution, or from other forms into hostessing. Examination of motivations by type of free prostitution engaged in did not appear to be a meaningful distinction.

Analysis of the interviews with the 33 juveniles working in free prostitution revealed four types of motivations contributing to the decision to remain in prostitution: financial/lifestyle, emotional, drug-related, and identity-related. Nearly all girls in free prostitution directly or indirectly expressed the first three motivations, although they varied in which motivation they most emphasized. The fourth motivation, identity, was not as common, and was only indirectly stated.

Finance- and lifestyle-related motivations were mentioned by all 33 participants in this category, and 24 (73%) said they were the most important motivations for remaining in prostitution. Accustomed to high income and flexible working hours, a low-paying, rigidly scheduled job seemed unbearable to many girls. For example, Mei stated she had become addicted to easy money, and that "you work so hard but gain so little back while working in other jobs." June described herself as "totally controlled by money." Wen-wen felt "very happy" in prostitution because she "had no need to worry about money or to borrow it from here and there." Although they reported that money was "easy and ample," they also reported being unable to save it. Most runaways said they spent all their earnings on hotels, clothing, transportation, entertainment, and video games, or gave it away to friends. In Yun-Yun's words: "I think the difficulty [leaving prostitution] is that you adopt some habits in the occupation, for example, the way of spending money, your personality, your language, your behavior, and your appearance . . . it would be difficult to change in a short time. Once you are in this occupation, you have a kind of mentality—you tell yourself 'I'll leave when I make enough money.' But when you have earned the money, you have another kind of mentality—you feel the money is so easy, and it does not do harm to you. You'll continue, continue, and continue . . . you cannot leave it."

Second, 24 girls (73%) described emotional and social support or lack thereof as a major factor keeping them in prostitution, and 9 of them (27%) considered it to be the most important reason. Clients, pimps, and friends in prostitution supplied many juveniles with companionship, attention, affection, excitement, and a sense of accomplishment. For example, 13-year-old Hui-hui said that she did not want to leave her friends—the customers, hostesses, clerks, and waiters in the teahouse. Li-li planned to return to work for a female brothel owner whom she considered to have been very nice to her. Feng-feng said that after having a family full of "blood violence," she was "thirsty for affectionate attachment" and remained in prostitution to help her boyfriend and a sister-like friend. In fact, a lack of support at home was a common theme among 90% of the participants, and most described a history of physical (73%) and/or sexual (55%) abuse at home. For example, an aboriginal girl stated, "Only if my mother took me back home, and their attitudes changed for the better and would not beat me without any reason would I give it up." Eight girls said they "were on the run and did not know where else to go."

Drug abuse (mostly amphetamines and hallucinogens) was a third factor that 21 girls (64%) emphasized as a reason for not leaving prostitution. All 33 free prostituted juveniles had at least tried using amphetamines. The problem was not that they needed money to buy drugs; the amphetamines supplied by customers and friends were not costly. The problem was that the habit created a sense of powerlessness to leave prostitution, and a sense of being unworthy to do anything else. Ying-ying explained: "They were like a tool to help me be depraved and gave me an excuse. For example, when I was thinking of getting out of this circle, I'd tell myself that I had used amphetamines for so long, I was a wastrel. What did it matter if I continued to be bad? . . . I had no goals, I felt like a walking corpse."

Finally, the fourth set of motivations expressed related to identity. Only 11 participants of the 33 (33%), all of whom had been in prostitution for 2 or more years, made statements relating to this motivation. *Identity* refers to the juveniles' acceptance of prostitution as their profession no matter whether they liked it or not. They did not move in and out of prostitution or think of quitting. More important, these 11 girls felt some measure of pride in their work, and considered themselves to be good at prostitution or felt that they had a personality that was suited for it. For example, 4 girls expressed pride in their skill at hostessing, or being the most popular on the street. Christa defended the legitimacy of her occupation: "We should be proud of ourselves, because what we can do TV actresses can't necessarily do. We suppress our resentment with smiles, and nobody can tell." Finny, who had worked in a massage parlor for 3 years, said she did not feel ashamed of her occupation because she "was paid for her labor" and because she considered prostitution to be helpful to society: "Without us, many girls on the streets would be raped."

In sum, when asked why they remained in prostitution, most participants in free prostitution emphasized positive reasons reflecting perceived benefits, such as financial rewards and friendships. They also pointed out the detriments of leaving, such as violence at home or inability to find another job, or cited lack of motivation due to drug use. Many conveyed the overall impression that in their first few months of work, prostitution provided fast money, excitement, companionship, attention, affection, and a sense of accomplishment—things that were unattainable at home, school, and other occupations.

Attitudes Toward Prostitution and Customers

The free prostituted juveniles' attitudes toward prostitution appeared to undergo a process of change throughout their employment. Nineteen girls (58%) recalled how much they enjoyed "the chance to meet new and exciting people," and how excited they were to go to work regularly when they first started. The financial and lifestyle motivation was an especially powerful aspect of this positive attitude in the beginning. As they remained longer in prostitution, they adapted to both the exciting and negative aspects of their work, experienced dramatic mood swings, felt reluctant to go to work, and went irregularly. However, for 28 of the 33 girls (85%), the perceived benefits combined to create the feeling that staying in prostitution was better than getting out, even after many months or even years in prostitution. The other 5 (15%) said that after the first few months they started to feel their lifestyle was "abnormal," and felt they were "drinking too much" and "living a life of debauchery," and that thoughts of leaving "would flash once in a while."

Eight girls (24%) said their personalities were positively influenced by their interactions with customers, and that they had become more "mature" and "experienced." Ling-Ling said the people she met in bars and dance halls influenced her a great deal. "I had been very impulsive, simple, childish, and innocent. . . . I became more sensible, diplomatic, and not as stubborn." About 75% of the girls felt that although some customers were "really malicious," most treated them with politeness and would stop fondling them when they were told. Nine girls (27%) felt that customers treated them with respect. That is, younger men treated them equally as friends, while older customers were kind, telling them to quit prostitution. These girls generally had a positive attitude toward their customers and believed that they had true friendships with them. Two girls (who were both abused by their mothers) even felt their customers treated them more nicely than their mothers. Eight girls (24%) saw no real friendships with customers, only superficial ones. They considered both their own actions and those of their customers to be calculated: All they wanted from customers was money; all the customers wanted from them was sex. Ying-ying described clients as "gentlemen in appearance but beasts in conduct."

Although all juveniles experienced changes in their attitudes toward prostitution and their customers, it was difficult to identify a common pattern in terms of the timing of these changes, even among girls for whom the period of time they remained in prostitution was identical. A number of girls seemed to cycle through these feelings. The lack of pattern is likely due to the interaction of several factors, such as type of prostitution, the degree of abuse by and alienation from their families, severity of early sexual and physical abuse, interaction with customers, relationship with colleagues, and age.

Filial Daughters Filial daughters (6 of 33, 18%) entered prostitution in order to meet their family's financial needs. Runaways and filial daughters alike expressed all four motivations outlined previously for remaining in prostitution. However, there were slight differences relating to the first (financial) and fourth (identity) motivations. Filial daughters did not place greater emphasis on financial considerations. Rather, it was the way they disposed of their money, and consequently their lifestyle, that differed from runaways. Unlike the runaways, who, according to filial daughter Ya-lu, "hardly saved anything and squandered money on clothing and entertainment," filial daughters gave a large part of their earnings to their families. The lifestyle of filial daughters appeared to be more reserved than that of the runaways. The filial daughters' descriptions of their expenditures and participation in nightlife and drug use seemed to be more conservative than that reported by runaways. Furthermore, 2 of the 6 filial daughters in this study would not mix with runaways socially because of "the runaways' wild ways." Thus, although they were similarly motivated to remain in prostitution for financial reasons, the financial rewards were not so much for themselves as for their family members.

The fourth motivation, identification with prostitution, also had a different significance for filial daughters than it did for the runaways. In the filial daughters' view, runaways were "fun-loving girls who rambled in prostitution without a goal," while they themselves were "different because we know what we are doing in prostitution." The filial daughters identified themselves primarily as filial daughters, and not as prostitutes. However, it was this very identification that bound them tightly to prostitution and would not allow them to consider leaving. All 6 filial daughters claimed that prostitution was their choice to support their families. Two even did so without their mother's knowledge.

Because there were only 6 of them, it was difficult to assess whether differences in attitudes toward prostitution and customers existed between the runaways and filial daughters. Only 2 of the filial daughters (33%) had any sort of positive orientation toward the work or the customers, which contrasts with the 58% of the nonfilial juveniles that had at least experienced a positive stage when starting the work. Of these 2 girls, one had a boyfriend who once was a customer. The other, Winnie, said, "I do my best at my job. Even my boss begs me to take more breaks." She also said, "Some customers really dote on me."

A final difference between filial daughters and runaways is that filial daughters worried about their family due to the loss of their incomes during their rehabilitation. All 6 filial daughters clearly expressed they would go back to prostitution if they had to help their families, although all

6 also felt the rehabilitation center provided a chance to inspect their life in prostitution and "to reclaim my other self." Talking about future plans, 2 filial girls said they would go back to school, one planned to open a restaurant, and another 2 "wanted to find a normal job."

In sum, juveniles working in free prostitution expressed four main motivations for remaining in the work. Implications of two of these motivations differed somewhat for filial daughters and runaways. No clear pattern of changing attitudes toward prostitution or clients over the course of their experience with prostitution could be identified among free prostituted juveniles.

Confined Prostituted Juveniles

In this study, 23 (of the 49, 47%) prostituted juveniles said they had experience working as prostitutes with no possibility of leaving or changing their condition. Most worked in confinement under slave-like conditions. Fifteen were indentured by their families, who received the money for their indenture. Five had been kidnapped and confined to a brothel where they were forced to work, and 3 were taken around to hotels and forced to work.

Motivation to Remain in Prostitution Some difficulty exists in discussing motivation for remaining in prostitution for juveniles forced to work against their will. Obviously, they have no choice but to remain, so it would seem motivation does not enter into it. However, 11 of the 18 (61%) girls who had been freed from confined prostitution prior to their arrest and detention at the center returned to prostitution of their own volition. Seven of the 11 (64%) moved into free prostitution. The other 4 actually returned to the brothels where they had been confined; 2 worked under no contract, and 2 gave consent to their parents to be indentured again. Examination of their reasons for returning to prostitution revealed a similarity to the four motivations described for free prostituted juveniles, with an emphasis on the fourth motivation—identification with prostitution. Examination of confined girls' attitudes toward prostitution sheds light on their motivations to return to the work.

Attitudes Toward Prostitution Unlike the free prostituted juveniles, the 23 confined girls in this study did appear to share a clear pattern of change in attitudes toward prostitution and toward clients. The pattern may have emerged because the majority of the girls were confined under quite similar conditions and for a long period of time (12 girls over 3 years, 7 between 1 and 2 years, and 4 for a few months). This change occurred over the duration of their term as confined prostitutes regardless of the degree of their willingness to be confined, and regardless of their pathway into prostitution the first time. (Some had worked in non-confined prostitution prior to being confined.) Four main stages of adaptation in attitudes were identified: resistance, development of interpersonal connections, self-injury and loss of hope, and acceptance of prostitution. These stages were not discrete—they often overlapped, and the length of time for each stage varied among the girls.

In the first stage, with the loss of freedom, unwanted violation of their bodies, and no control of customers' behavior, all but 2 girls (91%) recalled feeling "overflowing with pent-up bitterness and rage." All of them fought against customers as a way to show "resistance against their fate." Those who were kidnapped (8 girls) or who had not given their consent to be indentured (4 girls) especially emphasized their violent struggles to refuse. In return they were "tortured viciously." This period of resistance lasted for days, weeks, or months, depending on the individual. When asked whether her character had changed because of her life in a brothel, Mori explained, "My temper became very bad because each person [customer] was so different. In normal days, we repressed our tempers inside; when we had bad customers, we exploded quickly." When the girls blew up, they struck out verbally, and in many cases, also physically. This behavior, however, was regarded by customers as merely a negative trait shared by all confined prostitutes.

Gradually all of them entered a second stage in which they began to form attachments with the people around them. This second stage is characterized by significant changes in the girl's view of prostitution and her customers. Some of the early resentment against customers, who were their only connections to the outside world, gradually abated. For instance, Didi, an indentured girl, did not cooperate for 2 months. She recalled the reasons she gave up violent resistance, "I finally realized that if I treated customers nicely, the boss wouldn't beat me. . . . If I treated them mean, my life would not change because of it. I'd rather be nice to them and make my life easier."

"Customers were not always bad." All 15 indentured girls and 3 of the kidnapped girls (60%) felt that most of their customers were nice to them, although some were mean. They also found in customers friendship, romantic fantasies, and the dream that a customer would eventually "rescue me." Occasionally a customer who particularly favored them did a great deal to bolster their self-esteem. They especially preferred customers who "came just to chat with me and tell me interesting things." Sayi gave a rather typical description of the changes in her relationship with clients. "In the beginning, I hated customers. When I saw them, I did not like talking to them. I would tell them to come up [to her bed] and just did it [had sex]. Later on, I learned to chat with them." In addition to relationships with customers, friendships with other girls were another important source of emotional support.

In the third stage (reached by all but one girl) they abandoned hope that they could soon leave the brothel, experienced increasing depression, and fell into drug use and self-abuse. After learning that her father had indentured her for 5 years, Min-li collapsed with hopelessness. "I began hurting myself, abusing alcohol, cigarettes, and amphetamines." Substance abuse helped her forget where she was, what she was doing, and her pain and anger. "I had no other way to vent my anger" was a common theme in the girls' descriptions. With drugs, they could retreat to a world where they were safe and where the pain stopped. Moreover, amphetamines had an additional function: They helped indentured girls stay up and meet the demands of long work. "I had to work 17 hours; how could I prop myself up?" said Mori, who was quite offended when asked why she used amphetamines. Many other girls used the term "hurting myself" to mean not only substance abuse, but also physical self-torture and suicide attempts. "I like hurting myself," said Sayi, "I don't like hurting others." Li-li had large, appalling scars on her wrist and had thought about killing herself many times. Lisa attempted suicide twice by swallowing amphetamines.

In stage four the girls began to "accept prostitution as their destiny" and started to see the way they lived as "easy." All but 2 girls (91%) reached this stage. Such attitudes were exactly what their owners tried to instill in them. Rewarded with abundant material gifts, all they had to do was to obey, forget about the outside world, and accept that the essence of their lives reduced to a few acts: eating, sleeping, lying down, and getting up. Sayi recalled: "I was very disgusted at first . . . then gradually it seemed I liked it better. It was not very tiring, yet not that easy, either. Customers came, I lay down, got up, lay down, got up. It's not a big deal. . . . The advantage was that I had plenty of food and clothes. As long as I obeyed and did what I was told, I could have whatever I wanted."

Brothel owners deliberately reinforced filial ideology in the 15 indentured girls to make them accept their fate and prevent them from escaping. Indentured girls were told (by brothel owners and their parents, and even by other indentured juveniles) that their families had the right to sell them because they had reared them and given them their lives. All but one of the filial girls had no strong intention to escape after a period of time. Chen-chen explained, "Since my family has already received the payment, it would be unfair to my owner if I did not fulfill my contract." The indentured girls did not turn their pain outward to their families, and most claimed not to resent or hate their families. "I would tell myself they are my parents; hating them won't do any good," said Li-li. The girls expressed their feelings more as being tragic and wounded, rather than full of hatred. Min-li was the only indentured girl who had bitterly argued with her owner, other prostitutes, and customers who told her that her hatred of her father was immoral. Min-li was able to avoid the dictates of filial ideology because her father only "gave me my life but did not rear me." Min-li was raised by her paternal grandparents. Her father never lived with her or paid her his fatherly duty.

Kidnapped girls came to accept prostitution in a somewhat different way. Their living conditions seemed to be worse than those of the indentured girls. They did not have the support of their families, and they were constantly threatened. "If I escaped they would kill my family," said Olive. But after being freed from confinement, 5 of the 8 (63%) reentered prostitution. "I am not suited to other jobs," said Pai-Pai explaining why she had returned to prostitution. She considered prostitution to be "more satisfying" and was "quite content" with her life as a free prostitute. However, she emphasized, "I would not recommend it to other girls; it would not do any good for them. . . . But, I am already used to it." After Teh-Lung escaped from confinement, she took a job in a teahouse as a hostess. She thought the work "fun, like doing a stage show" and considered it to be a "legitimate job." It appears that after the slave-like conditions of confined prostitution, these girls saw only advantages to free prostitution.

In sum, juveniles in confined prostitution brought up the same four main motivations for moving into free prostitution as the free prostituted juveniles did for remaining in prostitution. They differed from the free prostituted juveniles in that their attitudes toward prostitution and their clients followed a clear progression of development while in confinement.

Discussion

In the West, economic motivations have been cited as a reason for entry into prostitution. It may be that runaways in the States have trouble finding other types of jobs because running away is considered a juvenile offense, or because of laws limiting the age at which a person can be hired. In contrast, 80% of the runaways in this study did not enter prostitution deliberately to fulfill an economic need. Most had other employment at the time they entered prostitution. In Taiwan, financial needs play a direct role in entry into prostitution mainly for filial daughters. However, financial needs can still be considered to play a role in juvenile entry into prostitution in Taiwan. Most free prostituted juveniles in this study reported being attracted to the possibility of a high income for little work (as they perceived it).

Lack of support at home, particularly in the form of parental abuse or neglect, has been clearly indicated as a precursor to prostitution in the West. The case is no different in Taiwan. As mentioned, 73% of the participants in this study had experienced physical abuse, and 55% had been sexually abused. In this aspect, motivations to enter and to remain in prostitution completely overlap. However, whereas strong emotional connections to people involved in prostitution might motivate a juvenile to stay in the work, they are not a major factor in the decision to enter prostitution in Taiwan. The case may be somewhat different in the West, where many studies have indicated that emotional support from pimps plays an important role in introducing juveniles to prostitution. That is, a pimp may provide emotional (and financial) support in the form of comfort, protection, and understanding to a runaway juvenile. Once the juvenile is financially and emotionally dependent on the pimp, he introduces the juvenile to prostitution. Only 2 girls in this study were introduced to prostitution by a boyfriend and only one was forced into prostitution by her boyfriend.

The fourth and final motivation relates to identity and requires special consideration. In this study, some of the free juveniles had come to see prostitution as their occupation. As they explained, part of what kept them in prostitution was their perception that it is what they are good at and what they are suited for; they felt pride in their work. Pride arises from favorable comparison of oneself with others, especially when one makes an internal attribution for success. The girls who expressed this motivation in some way identified with prostitution—does this mean that they had formed an identity as a prostitute? The process of identity formation is a complex issue that may involve not only feelings of pride and self-efficacy, but also such factors as learned-helplessness or rationalization. It has been examined from numerous perspectives. For example, previous studies have examined how construction of identity as a prostitute relates to self-esteem, psychological paralysis, identity disorders, normalization of stigmatized activities, or by using a career model and demonstrating that juveniles may employ a set of occupational ideologies and model older adult colleagues.

[R]esults from this study strongly suggest a number of important relevant questions for future research on prostitution and identity. First, what is the relation between attitudes and identity as a prostitute? In the fourth stage of attitude change, the confined girls no longer thought of leaving in the near future, but accepted prostitution as their destiny or thought of it as easy. Those who returned to prostitution after being released from confinement *all* made statements reflecting the fourth motivation when asked why they chose to return to prostitution. In contrast, only 33% of the free prostituted juveniles made statements relating to the fourth motivation. This difference points to a need for further research on the relation between trauma and identity. Although all of the girls in this study may be considered to have experienced some level of trauma in their lives, there is a sense in which the confined girls experienced trauma different from that of the others. In situations of captivity, the perpetrator's ultimate goal is not merely compliance, but rather "the creation of a willing victim" (Herman, 1992, p. 75). The method for achieving this goal is repetitive infliction of psychological trauma designed to "destroy the victim's sense of self" (p. 77).

Two more aspects of identity formation specifically in relation to Taiwanese culture deserve consideration. Research on American prostitution "has consistently emphasized the importance of family disaffiliation as a career contingency" (McCaghy & Hou, 1994). Likewise, the majority of prostituted juveniles in Taiwan have run away from their families. However, it is clear from this study that family ties can play a greater span of roles in prostitution in Confucian than Western cultures. First, a number of prostituted juveniles in Taiwan not only retain close ties with their families, they enter prostitution out of a sense of filial obligation to help them. Psychological aspects of filial piety such as its role in the formation of a girl's identity and self-esteem deserve attention in future research, particularly as filial piety may possibly play a contradictory role, both in helping the girl to preserve a sense of identity other than as a prostitute and in committing her to prostitution. Second, family ties can influence girls in yet another way. Both the group of filial daughters and all the indentured girls in this study expressed very strong fatalism (a component of Chinese traditionalism). In contrast, no runaways expressed fatalism. It is possible that this fatalism is related to the fact that many of the filial daughters and indentured girls had a mother or sister involved in prostitution (12 of 21, 57%). Far fewer runaways had a mother or sister in prostitution (5 of 28, 18%). The link between fatalism and identity deserves future investigation.

References

Herman, J. (1992). *Trauma and recovery.* New York: Basic Books.

Hwang, S. L., & Bedford, O. (2003). Precursors and pathways to juvenile prostitution in Taiwan. *Journal of Sex Research, 40,* 201–210.

McCaghy, C., & Hou, C. (1994). Family affiliations and prostitution in a cultural context: Career onsets of Taiwanese prostitutes. *Archives of Sexual Behavior, 23,* 251–265.

DISCUSSION QUESTIONS

1. What does this article tell you about cultural differences between the U.S. and Taiwan that leads to different conclusions about prostitution?
2. Does this article change your mind about legalizing (yes or no) prostitution?
3. Explain why a filial daughter would remain in prostitution.

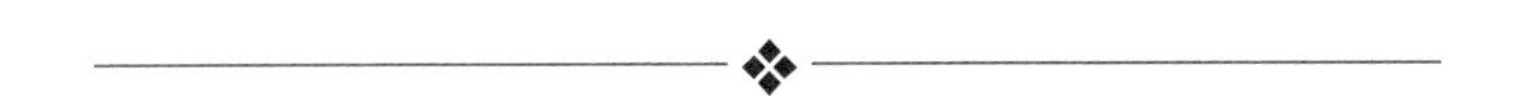

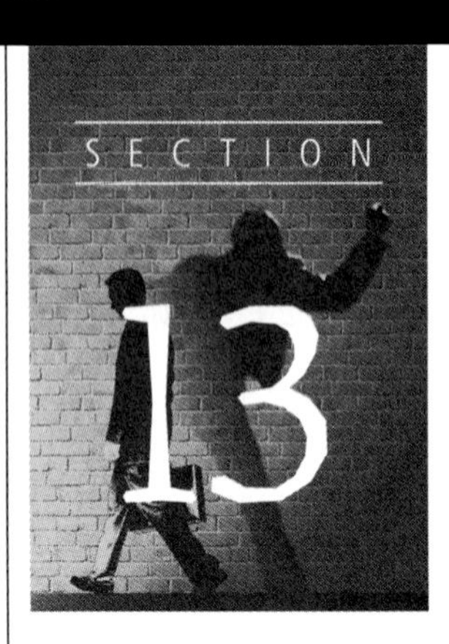

WHITE-COLLAR AND ORGANIZED CRIME

On August 10th, 1978, teenage sisters Judy and Lyn Ulrich and their cousin Donna were on a 20-mile journey to Goshen, Indiana in their Ford Pinto when they were rear-ended by another car. As a result, gas spilled onto the highway, caught fire, and all three trapped girls died horrible fiery deaths. Pintos were fitted with gas tanks that easily ruptured and burst into flames in rear-end collisions of over 25 miles per hour. The problem could have been fixed at a cost of $11 per vehicle, but with 11 million Pintos and 1.5 million light trucks with the problem, Ford accountants calculated that it would cost $137 million to fix. It was calculated that not fixing it would result in 180 burn deaths, 180 serious burn injuries, and 2,100 burned vehicles, which they estimated would cost about $49.5 million dollars in lawsuits and other claims. Comparing those two figures, it was determined that in light of the $87.5 million Ford would save by not fixing the gas tank problem, to fix it would be unprofitable and irrational.

The consciences of Ford executives did not bother them because they openly used these figures to lobby against federal fuel leakage standards to show how unprofitable such standards would be! According to a Ford engineer, 95% of the 700 to 2,500 people who died in Pinto crashes would have survived if the problem had been fixed. If this is an accurate estimate of deaths caused by the defect, the executives who conspired to ignore it may be the worst multiple murderers in U.S. history. Yet no Ford executive was ever imprisoned, and many went on to bigger things. Lee Iacocca, whose personal maxim, "safety doesn't sell," was still in evidence in 1986 when he opposed mandatory airbags for automobiles, went on to become president of Chrysler Corporation and to chair the committee for the Centennial celebrations for the Statue of Liberty.

The Concept of White-Collar Crime

What images pop into your head when you hear the word *crime?* Whatever images you conjured up, we wager that it was not of a well-dressed, middle-aged person sitting in a leather recliner dictating a memo authorizing the marketing of defective automobiles or the dumping of toxic waste. We seldom think that the chain of events set into motion by a business memo may do more harm than the activities of any "street punk." Yet more money is stolen and more people die every year as the result of scams and willful illegal corporate activity than as the result of the activities of street criminals. Kappeler, Blumberg, and Potter (2000) estimate that crimes committed by corporations result in economic losses that total between 17 and 31 times greater than losses resulting from street crimes, and the losses due to non-corporate white-collar crime are roughly the same.

The term *white-collar crime* was coined in the 1930s by Edwin Sutherland (1940), who defined it as crime "committed by a person of respectability and high social status in the course of his occupation" (p. 9). In its Administration Improvement Act (AIA) of 1979, the U.S. Congress defined **white-collar crime** as "an illegal act or series of illegal acts committed by non-physical means and by concealment or guile, to obtain money or property, or to obtain business or personal advantage" (Weisburd, Wheeler, Waring, & Bode, 1991). This definition focuses on characteristics of the offense as opposed to Sutherland's focus on the offender as a high-status person because most white-collar crime is not committed by "high-status persons." The AIA definition, however, fails to differentiate between persons who commit crimes for personal gain and those who do so primarily on behalf of an employer. Our analysis of white-collar crime will differentiate between individuals who steal, defraud, and cheat both in and out of an occupational context and those who commit the variety of offenses attributed to business corporations. There are definitional difficulties in the area of white-collar crime, but rather than presenting endless subcategorizations, we follow Rosoff, Pontel, and Tillman (1998) in using the term *occupational* crime for the former and *corporate* crime for the latter.

Occupational Crime

Occupational crime is crime committed by individuals in the course of their employment. Such crimes might range from the draining of company funds by sophisticated computer techniques to stealing pens and paper clips or vandalizing company property by scrambling computer data or scratching graffiti on newly painted walls. Although such activities may seem relatively mundane to most of us, according to the business insurance industry, employee activities such as theft, fraud, and vandalism cost American businesses a total of $660 billion in 2003 (Parekh, 2004). Just as we all pay for street crime through taxes that support the criminal justice system, we all pay for employee crime because companies merely pass on their losses to their customers. Employee crime may also lead to businesses going bankrupt and employees losing the jobs these businesses provided.

Professional Occupational Crime Occupational crimes are crimes committed by professionals such as physicians and lawyers in the course of their practices. Frauds committed by physicians include practices such as filing insurance claims for tests or procedures not performed, performing unnecessary operations, steering patients to laboratories or pharmacies in which the doctor has financial stakes, and referrals to other doctors in return for kickbacks. Anyone who has been charged $8 for an aspirin, $500 for a nursing bra, or $200 for a pair of crutches on their hospital bill knows how hospitals rip patients off. In many respects, Medicare and Medicaid programs are welfare programs for physicians. Fraud within these programs has been estimated to cost between $50 and $80 billion per year, and overall, medical fraud is estimated to be $100 billion, or about 10% of the total U.S. health care bill (Rosoff et al., 1998).

Most lawyers also work on a fee-for-service basis, thus generating the same temptations to increase their incomes by fraudulent means. Frauds perpetrated by lawyers can include major embezzlement of clients' funds, bribery of witnesses and judges, persuading clients to pursue fraudulent or frivolous lawsuits, billing clients for hours not worked, filing unnecessary motions, and complicating a simple legal matter to keep clients on the hook—"I will defend you all the way to your last dollar." It seems that every occupational category generates a considerable number of criminals, and the higher the prestige of the occupation, the more their criminal activities cost the general public.

Causes of Occupational White-Collar Crime: Are They Different?

According to Hirschi and Gottfredson (1987), occupational crime differs from common street crime only in that it is committed by people in a position to do so—Medicaid fraud can typically only be committed by physicians and bank embezzlement by bank employees in positions of trust. The motives of occupational criminals are the same as those of street criminals—to obtain benefits quickly with minimal effort—and the age, sex, and race profiles of occupational criminals are not that much different from those of street criminals. Hirschi and Gottfredson concluded that "When opportunity is taken into account, demographic differences in white collar crime are the same as demographic differences in ordinary crime" (p. 967).

The article by Glen Walter and Matthew Geyer in this section examines Hirschi and Gottfredson's assertion. Using a number of indicators of criminal thinking patterns and attitudes, they found that white-collar criminals with prior arrests for non-white-collar crimes were largely indistinguishable from street criminals in their demographics, lifestyle, and endorsement of criminal thinking patterns, but were quite different from white-collar criminals with no history of arrest for non-white-collar crimes. Walters and Geyer cite a study of federal white-collar criminals that showed that most of them were much less involved in crime and deviance than street offenders, but that chronic white-collar offenders (about 16% of the sample) were similar to street criminals in their patterns of prior deviance (Benson & Moore, 1992).

Corporate Crime

Corporate crime is criminal activity on *behalf* of a business organization committed during the course of fulfilling the legitimate role of the corporation, and in the name of corporate profit and growth. During much of American history, the primary legal stance relating to the activities of business was *laissez-faire* (leave it alone to do as it will). American courts traditionally adopted the view that government should not interfere with business, so for a very long period in our history, victims of defective and dangerous products could not sue corporations for damages because the guiding principle was *caveat emptor* (let the buyer beware). Unhealthy and dangerous working conditions in mines, mills, and factories were excused under the freedom of contract clause of the Constitution (Walsh & Hemmens, 2000).

▲ **Photo 13.1** Corporate crime cases are typically not handled by local or even county prosecutors. Typically, these cases are moved up to the state attorney general's office, which is better equipped with investigators to prepare these cases for prosecution.

Although attitudes in the 20th century changed considerably, the public continued to be victimized by crimes of fraud and misrepresentation committed by businesses. For instance, the savings and loan (S & L) scandal of the 1980s amounted to one of the most costly crime sprees in American history and is likely to cost the U.S. taxpayer (because the Federal Deposit Insurance Corporation insures all bank savings up to $100,000) up to $473 billion. This staggering amount is many times greater than losses from all the "regular" bank robberies in American history put together (Schmalleger, 2004, p. 375).

Most of the looting took the form of extravagant salaries, bonuses, and perquisites that executives awarded themselves as their banks sank ever further into debt. Other methods involved selling land back and forth ("land flipping") within a few days until its paper value far exceeded its real value, and then finding "sucker" institutions to buy it at the inflated price, and loans made back and forth between employees of different banks with the knowledge that the loans would never be called in (Calavita & Pontell, 1994).

The first major corporate scandal of the first decade of the 21st century is that of Enron Corporation. This crime has been called "one of the most intricate pieces of financial chicanery in history," and for investors in its stock and its employees, "the financial disaster of a lifetime, a harrowing, nerve-racking disaster from which they may never recover" (English, 2004, p. 1). The Enron scandal (and other similar scandals in the first two years of the 21st century) did tremendous damage to the economy and "created a crisis of investor confidence the likes of which hasn't been seen since the Great Depression" (Gutman, 2002, p. 1).

Theories About the Causes of Corporate Crime

Classical theorists, with their assumption that human behavior can be understood in terms of striving to maximize pleasure and minimize pain, have no difficulty appreciating the motivation for individuals to engage in corporate crime. Opportunities abound in corporate America to gain wealth beyond what individuals could earn legitimately. The prevalence of corporate crime can also be appreciated given the weakness of formal and informal controls over business activities and the lenient penalties imposed. We may expect business executives to have assimilated the egoism, acquisitiveness, and competitiveness fostered by capitalism more completely than most.

But all companies and their executives are exposed to the capitalist ethos, the strain of seeking their versions of the American Dream, and the widespread opportunities for illegal gains. We might say that corporate criminals are "high-class innovators," to borrow a phrase from strain theory. "Strain" can be evoked as a *motive* for corporate crime but not for the *choice* to engage in it. So what differentiates "innovators" from the other modes of adaptation available in the corporate world?

Just as there are criminogenic neighborhoods, there are criminogenic corporations with a "tradition" of wrongdoing. One of the first systematic examinations of corporate recidivism found that 98% of the nation's 70 largest corporations were recidivists with an average of 14 regulatory or criminal decisions against them (Sutherland, 1956). A study of 477 major U.S. corporations found that 60% of them were known to have violated the law, and that 13% of the violator companies accounted for 52% of all violations, with an average of 23.5 violations per company (Clinard &Yeager, 1980). And a study of brokerage firms found that many of the biggest names in the business, such as Prudential, Paine Webber, and Merrill Lynch, have had an average of two or more serious violations *per year* since 1981 (Wells, 1995). "Three-strikes-and-you're-out" laws evidently do not apply in the world of pinstripe suits.

Newcomers entering such corporate environments are socialized into the prevailing "way of doing things." If the newcomer does not "fit in" and conform to the prevailing company ethos, he or she is not likely to remain employed there very long. This process can produce a sort of moral apathy in well-socialized executives striving to do their jobs in their own and their company's best interests. If each tiny bending of the rules that brings profit to the company brings the rule bender appreciation and bonuses, the individual's behavior will be almost imperceptibly molded in the direction of ever greater wrongdoing. If he or she is rewarded through the usual stock-based executive compensation, there is further incentive to engage in illegal behavior. Stock-based compensation is designed to increase executives' focus on stockholders' profits, but it is a situation ripe for finagling of stocks through insider trading, for falsifying accounts, for fraudulent trading, and for an emphasis on short-term earnings rather than the long-term success of the corporation (Thorburn, 2004, p. 81).

In terms of the characteristics of individuals linked to corporate crime, it is obvious that things such as low IQ and low self-control do not apply to people who have spent many years of disciplined effort to achieve status. But there are many other factors linked to crime that could apply to corporate crooks as easily as to street criminals. For instance, it appears that people who choose business careers tend have lower ethical and moral standards than people who choose other legitimate careers. A number of studies have concluded that business students are, on average, less ethical than students in other majors (Useem, 1989), and that MBA students make fewer decisions judged as ethical and moral than do law students (McCabe, O'Reilly, & Pfeffer, 1991).

Other studies have focused on *locus of control, moral reasoning,* and *Machiavellianism.* People with an internal locus of control believe that they can influence life outcomes and are relatively resistant to coercion from others. People with an external locus feel that circumstances have more influence over situations than they themselves do. Those who engage in corporate crime tend to have an external locus of control, and whistle blowers tend to have an internal locus of control (Travino & Youngblood, 1990). Furthermore, people with an internal locus of control operate at higher stages of cognitive moral development and tend to behave according to their own beliefs about right and wrong. People at lower stages of moral development tend emphasize conformity to group norms (external locus of control; Weber, 1990).

Machiavellianism is the unprincipled manipulation of others for personal gain. People high on this trait are shallow individuals who exploit superiors and equals by deceit and ingratiation and subordinates by bullying. Simon (2002) describes those who make it to the top of bureaucratic organizations in almost psychopathic terms: such an individual "exudes charisma via a superficial sense of warmth and charm," and he or she exhibits "free floating hostility, competitiveness, a high need for socially approved success, unbridled ambition, aggressiveness [and] impatience" (p. 277).

Law Enforcement Response to Corporate Crime

Corporate crime is monitored and responded to by a variety of criminal, administrative, and regulatory bodies, but very few corporate crooks in the past received truly meaningful sanctions. Of the 1,098 defendants charged in the S & L cases, only 451 were sentenced to prison,

▲ **Photo 13.2** While most cases, even those that involve corporate crime, never come to trial and are settled or plea bargained instead, some do end up in court. Here a prosecutor makes his opening statement to the jury.

with the majority (79%) sentenced to less than five years, with the average sentence being 36.4 months (Calavita, Pontell, & Tillman, 1999, p. 164). The Golden Rule ("those with the gold make the rules") may explain the differences in punishment between street and corporate criminals because great wealth does confer a certain degree of immunity from prosecution and/or conviction. This power may be gauged by the 1990 U.S. Justice Department's withdrawal of its support for proposed tougher sentences for corporate offenders in response to heavy lobbying by many prominent industries, the targets of the proposal (Hagan, 1994). We doubt that street criminals would get very far lobbying against proposals for stricter penalties for them.

The cascade of corporate scandals and failures occurring between 2001 and 2003 may have finally awakened American law enforcement to the realities of the harm done by elite criminals. As a result of Congressional hearings and public outcry, Congress passed the **Sarbanes-Oxley Act** (SOA) of 2002. The SOA increased penalties for corporate criminals, increased the budget of agencies charged with investigating corporate crime, and made prosecution easier. Because of the SOA, it has been said that "prosecutors are driven to go after corporate fraud with an almost evangelical zeal" (Burr, 2004, p. 10). Perhaps the days of leniency for white-collar criminals are over. In 2005, Bernard Ebbers, ex-WorldCom CEO, was sentenced to 25 years in prison for his role in the $11 billion WorldCom fraud; John Rigas, founder of Adelphia Communications Corp., was sentenced to 15 years and his son Timothy to 20 years for their roles in yet another massive fraud; and Tyco executives Dennis Koslowski and Dennis Swartz were sentenced to 8 and 25 years, respectively. Finally, Enron executive Jeffery Skilling was sentenced to 24 years in October, 2006.

Organized Crime

What Is Organized Crime?

Criminologists have had a difficult time deciding what organized crime is and how it differs (if it does) from corporate crime. Some argue that "any distinction between organized and white collar crime may be artificial inasmuch as both involve the important elements of organization and the use of corruption and/or violence to maintain immunity" (Albanese, 2000, p. 412). Corporate crime is "organized crime" in some senses (corporations are organized, and when their members commit illegal acts they are engaging in crime), but there are major differences. Corporate criminals are created from the opportunities available to them in companies organized around doing legitimate business, whereas members of organized crime must be accomplished criminals before they enter groups organized around creating criminal opportunities. Thus, the former make a crime out of business and the latter make a business out of crime.

According to the President's Commission on Organized Crime (PCOC; 1986), **organized crime** is crime committed by structured criminal enterprises that maintain their activities over time by fear and corruption. PCOC concentrated on **La Cosa Nostra** (literally, "our thing"), also commonly referred to as the Mafia, but it should not be inferred from this that La Cosa Nostra (LCN) and organized crime are synonymous. There are many other organized crime groups in the United States and around the world.

LCN groups are structured in hierarchical fashion reflecting various levels of power and specialization. There are 24 LCN families with a national ruling body known as the **Commission**, which is a kind of "board of directors" and consists of the bosses of the five New York families and four bosses from other important families located in other cities.

Figure 13.1 Hierarchical Structure of a Typical Organized Crime Family

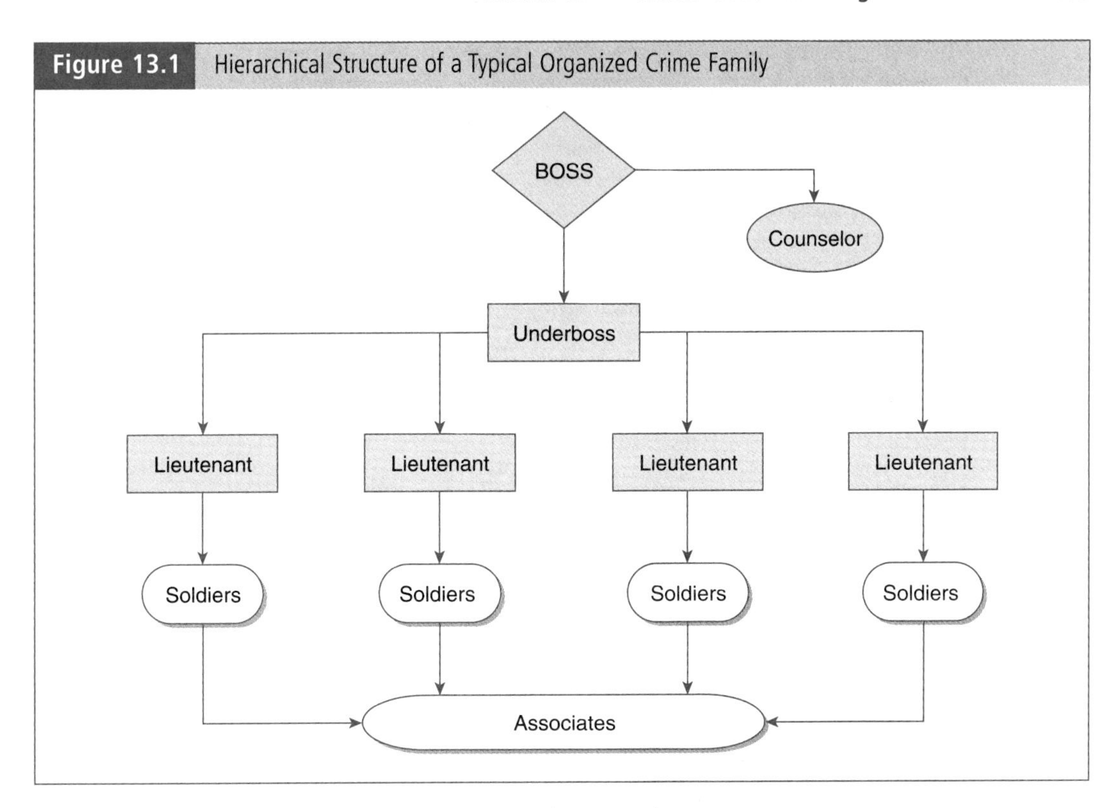

SOURCE: Adapted from President's Commission on Organized Crime (1986, p. 469).

The commission functions to arbitrate disputes among the various families and facilitate joint ventures, approve of new members, and to authorize the executions of errant members (Lyman & Potter, 2004). Members respect the hierarchy of authority in their organization just as corporate executives respect ordered ranks of authority in their corporations. Although there are occasional family squabbles and coups that remove individuals from the hierarchy, the structure remains intact. The formal structure of a LCN Family can be diagramed as in Figure 13.1.

At the top is the boss (the *don* or *capo*), beneath whom is a counselor (*consigliere*) or advisor, and an underboss (*sotto capo*). The counselor is usually an old family member (often a lawyer) wise in the ways of crime; the underboss is being groomed for succession to the top position. Beneath the underboss are the lieutenants (*caporegimas*) who supervise the day-to-day operation of the family through their soldiers (*soldati*), known as *made men, wiseguys,* or *button men.* Although soldiers are the lowest ranking members of the family, they may each run their own crew of nonmember associates.

LCN members are not employees who earn regular incomes from the family. Membership simply entitles the member to run his own rackets using the family's connections and status. A percentage of a soldier's earnings are paid to his lieutenant, who also has his own enterprises on which he pays a percentage to the underboss, and so on up the line.

This model is known as the **corporate model** because of its similarity to corporate structure, but some academics favor the **feudal model**. The feudal model of LCN views it as a loose collection of criminal groups held together by kinship and patronage. The commission may be seen as the king and his ministers (who rarely, if ever, interfered with their nobles), the family bosses may be seen as the lords, the lieutenants as the lesser nobility, and the made men and associates as the peasants. The autonomous operations of each family, the semi-autonomous operations of the soldiers and lieutenants, and the provision of status and protection by the family in exchange for a cut of their earnings, provides evidence that LCN bosses are more like feudal lords than corporate CEOs. Other scholars see evidence for both the corporate and the feudal models, namely, seeing LCN as a highly structured feudal system (Firestone, 1997, p. 78).

Organized crime is like a corporation (or a feudal system, for that matter) in that it continues to operate beyond the lifetime of its individual members. It does not fall apart when key leaders are arrested, die, or are otherwise absent. The criminal group takes on a life of its own, and members subordinate their personal interests to the group's interests, making organized crime different from gangs that spring up and die with their leaders.

LCN is not an equal opportunity employer; it is restricted to males of Italian descent of proven criminal expertise. Prospective members must be sponsored by *made guys* (established members of the family), and applicants are screened carefully for their criminal activity and loyalty before being allowed to apply. A lifetime commitment to the family is required from the newcomer, and in return, he receives a guaranteed and rather lucrative criminal career as part of an organization of great prestige and respect in the underworld. A promising criminal who is not of Italian descent, but who has qualities useful to the organization, may become an associate member of LCN.

The Origins of Organized Crime in the United States

Organized crime groups existed long before there was any major Italian presence in the United States, and many organized crime scholars believe that it is a "normal" product of the competitive and free wheeling nature of American society (Bynum, 1987). Scholars place a variety of dates on the beginnings of organized crime in America as we have defined it, but the two major candidates are the founding of the Society of Saint Tammany in the late 18th century and Prohibition in the early 20th century.

The **Tammany Society** was a corrupt political machine associated with the Democratic Party that ran New York City well into the 20th century from the "Hall" (Tammany Hall). The society made use of street gangs to threaten and intimidate political rivals. Prominent among these gangs were the vicious *Whyos* and the *Five Points* gangs. In order for a new member to be accepted by the Whyos, which at its peak had over 500 members, he had to have killed at least once. The Whyos plied their trade among New York's citizenry by passing out price lists on the streets for the services they provided (ranging from $2 for punching to $100 for murder) as casually as pizza vendors (Browning & Gerassi, 1980). The Five Points gang was a confederation of neighborhood gangs and was said to have had more than 1,500 members at one time.

In 1920, the United States Congress handed every petty gang in America an invitation to unlimited expansion and wealth with the ratification of the Eighteenth Amendment (the Volstead Act, or **Prohibition**), which prohibited the sale, manufacture, or importation of intoxicating liquors within the United States. Prohibition ushered in a vicious 10-year period of crime, violence, and political corruption as gangsters fought over the right to provide the drinking public with illicit alcohol.

The most successful gangster of the era was Al "Scarface" Capone of Chicago. Capone (a former member of the Whyos) was a ruthless criminal and a flamboyant man who established a criminal empire that at its height consisted of over 700 gunmen (Abadinsky, 2003). The wealth Capone accumulated from his bootlegging and prostitution enterprises got him into the *Guinness Book of Records* as having the highest gross income ($105 million) of any private citizen in America in 1927. Capone met his demise when the Supreme Court ruled that unlawful income was subject to taxation. Capone had not paid taxes on his criminal income, so he was sentenced to 11 years in prison in 1931. He died in 1947 of pneumonia following a stroke at age 48.

With the repeal of Prohibition in 1933, organized crime entered a quieter phase, but the modern face of LCN was already beginning to take form in New York. There were two main factions in Italian organized crime in New York at this time, one headed by Giuseppe Masseria and the other by Salvatore Maranzano, who were struggling for supremacy. This struggle, known as the **Castellammarese War**, ended with the deaths of both leaders in 1931. The war saw the end of the old Sicilian leaders and the emergence of an Americanized LCN. The Americanization of LCN saw the emergence of the five New York LCN families. It also saw Lucky Luciano (another former Whyo) set up the organization's national commission and to claim the title of founding father of Italian-American organized crime (Lupsha, 1987).

Interest in organized crime activities waned considerably after World War II, and many law enforcement officials did not acknowledge its existence. Three events affirmed the reality of its existence. In 1950, the (Senator Estes) **Kefauver Committee** was formed to investigate organized crime's involvement in interstate commerce. The Kefauver Committee hearings called in to testify such important organized crime figures as Meyer Lansky, Frank Costello, and Bugsy Siegel, exposing them to the public for the first time, although nothing significant came out of it.

The 1957 **Apalachin meeting**, held in Apalachin, New York, once again riveted national attention on organized crime. The police (who stumbled on the meeting purely by accident) raided the meeting and arrested 63 people, including the bosses and underbosses of New York's LCN families. All arrestees refused to answer questions about the purpose of the meeting and were indicted for obstruction of justice. No convictions came out of these indictments, but it destroyed LCN's hope that it could stay out of the public spotlight. According to the PCOC (1986), it finally confirmed the existence of organized crime, and law enforcement began to focus more seriously on combating it.

The (Senator John L.) **McClellan Commission** was formed in 1956 to look into financial irregularities in the Teamster's Union, but the star witness was a made man in the Genovese Family named Joe Valachi. While in prison, Valachi was accused of being an informer, which meant that he was marked for death. Rather than face this prospect, Valachi decided to testify about the Mafia in front of the McClellan Commission. He revealed much about the operation of the Mafia, including the fact the organization called itself Cosa Nostra, a term unfamiliar to law enforcement officials up to then. Despite the fact that Valachi was only a low-level soldier (albeit with 30 years experience), commission member Senator Robert Kennedy called Valachi's testimony the "biggest intelligence breakthrough yet in combating organized crime" (in Wilson, 1984, p. 566).

Foreign Organized Crime Groups: The Russian "Mafiya" and the Japanese Yakuza

The **Russian "Mafiya"** is considered to be the most serious organized crime threat in the world today (Rush & Scarpitti, 2001). There has been an explosion of crime in Russia since the breakup of the Soviet Union; the crime, bribery, political, and police corruption in Russia make the Prohibition period in America look positively benign. As James Finckenauer (2004, p. 62) put it, "Organized crime has been able to penetrate Russian businesses and state enterprises to a degree inconceivable in most other countries."

The major group in Russian organized crime is known as the *vory v zakone* (thieves-in-law), which began as a large group of political prisoners imprisoned following the Bolshevik Revolution in 1917. The Soviet prison system used this group to maintain order over the general prison population, in exchange for many favors. These "elite" prisoners developed their own structural hierarchy and strict code of conduct or "laws" (hence thieves-"in-*law*"). One of their strictest rules was that there is to be absolutely no cooperation with legitimate authority for any reason.

Russian organized crime is the biggest factor threatening Russia's democratization, economic development, and security. It threatens democratization because if a democratic government cannot control it, an authoritarian one will. It threatens the economy because foreign companies are reluctant to make the much-needed investments in an economy rife with the murder and extortion of business leaders. It threatens public security because many police officers and KGB personnel have left public service for the more lucrative opportunities available with organized crime (Carter, 1994).

Russian organized crime metastasized to the United States with the influx of Russian immigrants in the 1970s and 1980s. Unlike the largely uneducated Mafioso who came to the U.S., many Russian émigré criminals are highly educated individuals driven out by economic

▶ **Photo 13.3** With the demise of the Soviet Union, the West was exposed to Russian and Ukrainian mafiya for the first time on a large scale. This grave of an assassinated mafiya leader is in Odessa, Ukraine. Note the gold leaf and prominent photo on the 15-foot-tall grave memorial

hardship. This level of intelligence and expertise should make them a real threat. Indeed, Rush and Scarpitti (2001) write that

> It has been speculated by intelligence agencies such as the IRS, FBI, and CIA that because of their higher level of criminal sophistication Russian organized crime groups will present a greater overall threat to American society than the traditional Italian-American crime families ever have. (p. 538)

The Japanese Yakuza Japanese organized crime (JOC) groups are probably the oldest and largest in the world, with total membership larger than LCN and estimated at 90,000 (Lyman & Potter, 2004). JOC is commonly believed to have evolved from *ronin,* or masterless samurai warriors, who contracted their services out for assassinations and other illegal purposes. They also protected the peasants from other marauding bandits as a sort of vigilante/law enforcement group. The defeat of Japan in WW II and the ensuing chaos provided the catalyst for the growth of JOC. This period saw many gang wars erupt over control of lucrative illicit markets. As in the United States, these gang wars led to the elimination of some gangs and to the consolidation and strengthening of others. The Kobe-based *Yamaguchi-gumi,* with an estimated membership of over 10,000, is the largest of these groups (Iwai, 1986).

Members of JOC groups are recruited heavily from the two outcast groups in Japanese society—the *burakumin* (outcasts because their ancestors worked at trades that dealt with dead flesh, such as butchery, tanning, grave digging, etc., which was seen as unclean in the Buddhist religious tradition) and Japanese-born Koreans. Once admitted, a *kobun* must pledge absolute loyalty to his superiors, and like his LCN counterpart, must generate his own income and contribute part of it to the *ikka* (the family).

JOC enjoys a unique position in Japanese society. Their historical connection with the samurai; their espousal of traditional norms of duty, loyalty, and manliness; their support for nationalistic programs; and their "law enforcement" functions (yakuza neighborhoods are safe from common criminals) endow them with a certain level of respect and admiration among the Japanese. Furthermore, the yakuza are not shadowy underworld figures; their affiliations are proudly displayed on insignia worn on their clothes and on their offices and buildings, and they publish their own newsletter. The headquarters of one crime group, complete with the gang emblem hanging proudly outside, is only three doors away from the local police station (Johnson, 1990).

The police have tended to tolerate yakuza activity in certain areas as long as it involves only the provision of certain illicit goods and services demanded by the public, but they have cracked down hard when firearms and drugs are trafficked, or when innocent civilians are harmed. With the introduction of the Boryokudan Countermeasures Law of 1992, however, the relationship between the police and the yakuza has become more antagonistic, and there have been many police crackdowns (Hill, 2003).

Theories About the Causes of Organized Crime

Early theories of organized crime relied on the anomie/strain tradition to explain organized crime, describing the gangster as "a man with a gun, acquiring by personal merit what was denied him by complex orderings of stratified society," and saw each successive wave of immigrants ascending a "queer ladder of social mobility" in American society (Bell, 1962). According to this **ethnic succession theory**, upon arrival in the United States, each ethnic group was faced with discriminatory attitudes that denied them legitimate means to success. The Irish, Jews, and Italians were each prominent in organized crime before they became assimilated into American culture and gained access to legitimate means of social mobility. African Americans, Russians, and Asians have been prominent in organized crime, and according to this view, may have to climb their own "queer ladder" until they gain full acceptance in American society.

The idea of opportunity denial implies that those allegedly denied them would have gladly taken advantage of them if they existed and would have spurned crime. In this view, a criminal career is simply the default option undertaken by the downtrodden, with the unspoken corollary being that no one would actively seek criminal opportunities if legitimate options were available to him. However, memoirs of a number of LCN figures show that they had received good educations, came from involved and intact families, and had many opportunities to enter legitimate careers but saw crime as a more lucrative career than any legitimate alternative (Firestone, 1997).

Many gang members grew up hero-worshiping the neighborhood made men. It was the mobster who had the beautiful women, the sleek cars, the fancy clothes, and the respect, not the legitimate "working stiff" (Firestone, 1997, p. 73). Such neighborhoods proved to be fertile ground for the constant cultivation of new batches of criminals because they provided their young inhabitants with exposure to an excess of definitions favorable to law violation. The joy and enthusiasm with which made men describe their acceptance into the gang makes nonsense of the idea that gangsters are deprived individuals making the best of a bad deal: "Getting made is the greatest thing that could ever happen to me . . . I've been looking forward to this day ever since I was a kid." And a made man in the Columbo family gushes, "Since I got made I got a million fuckin' worshipers hanging around" (Abadinsky, 2003, pp. 23–24). As with common street robbers, the attitude toward legitimate employment and those who pursued it among made men is "Anyone who stood waiting his turn on the American pay line" writes Pileggi (1985, p. 37), "was beneath contempt. . . . To wiseguys, 'working guys' were already dead."

The final two articles in this section delve much deeper into the fascinating issue of the causes of organized crime. The article by Jay Albanese asks whether those who are already criminals organize around opportunities available for crime, or do these opportunities create criminals to take advantage of them? He finds evidence for both hypotheses, using case studies of organized crime groups. Albanese also provides extended discussions about the definitions of organized crime and asserts that there are more similarities than differences between organized crime and white-collar crime.

Hung-En Sung's article also pits alternative hypotheses against one another to understand the existence of organized crime across cultures ranging from Argentina to Zimbabwe. He wants to know if state failure (the failure to provide freedoms, rights, and personal security to citizens) or economic failure (inefficient use of labor and capital) is most important in explaining organized crime. Sung's findings show that state failure is the most important factor. Interestingly, however, while judicial independence is the most powerful predictor among eight others (the greater the judicial independence, the lower the level of organized crime), the provision of individual rights was the next most powerful, but in the opposite direction (the greater the rights, the higher the level of organized crime).

Summary

- White-collar crime is our costliest and most deadly form of crime. White-collar crime is divided into occupational and corporate crime. Occupational crime is crime committed against an employer or the general public in the course of an individual's employment.

- Except for crimes requiring high-status occupation for their performance, most white-collar criminals are not all that different from street criminals; they occupy a middle position between street criminals and "respectable" people in terms of criminal convictions. It may, therefore, be possible to explain occupational white-collar crime with the same theories used to explain street crime.

- Corporate crime is criminal activity on behalf of the organization. Corporate crimes involve multiple individuals both as perpetrators and victims, the loss of billions of dollars, and in some cases, of hundreds of lives.

- Corporate crime is explained by a variety of factors, including the juxtaposition of lucrative opportunities and lenient penalties. Personal characteristics associated with corporate criminality include an external locus of control, a low level of cognitive moral development, and a high level of Machiavellianism.

- Corporate wrongdoing is typically investigated and punished by administrative agencies and has typically been treated leniently. New weapons in the fight against white-collar crime, such as the Sarbanes-Oxley Act, have resulted in meaningful penalties imposed upon individual executives of corporations involved in corporate crime.

- Organized crime (OC) is defined by its formal structure, its continuity, and restricted membership. La Cosa Nostra (LCN) is a confederation of families, the leaders of which are of Italian heritage, and which restricts membership to ethnic Italians.

- American OC grew out of the corrupt political machine and its supporting street gangs known as Tammany Hall and received its biggest boost from Prohibition. Many of the gangsters who rose to national prominence during this period got their start in the variety of gangs that supported Tammany Hall.

- The repeal of Prohibition ushered in a quiet period in OC's history, particularly after the founding of the "commission" by Lucky Luciano as a judicial body to settle interfamily disputes without resorting to war.

- Three events—the Kefauver Committee, the Apalachin meeting, and the McClellan Commission—put an end to denials that OC existed and helped to launch major assaults on LCN.

- The Russian "Mafiya" is considered to be the biggest organized crime threat in the world today. The widespread chaos and corruption following the breakup of the U.S.S.R. allowed Russian OC to come out of the closet and proliferate. Not only is Russian OC preventing the democratization and economic stability of Russia, it is spreading its influence to many Western nations. The special danger of Russian OC is that many of its members are highly intelligent and educated men who held professional jobs in the old Soviet Union. Because of this, many government agencies are of the opinion that they will be a much greater threat to the United States than LCN ever was.

- The Japanese yakuza is the oldest OC group in the world. Having evolved from masterless samurai warriors, it received a major boost from the chaos existing in Japan after its defeat in WW II. The yakuza has many characteristics in common with LCN, but it operates openly and proudly, even publishing its own newsletter. Until recently, the police have not interfered with the yakuza unless innocent people are harmed or if drugs and guns are being trafficked.

- The American, Japanese, and Russian examples indicate that OC proliferates in times of social chaos, in a process in which some gangs are weeded out and others consolidate their power. The "queer ladder of success" idea posits that each ethnic group engages in crime until it is fully assimilated into American society. However, many Mafia memoirs show that they had not been denied legitimate opportunities, but rather that they simply preferred crime over other opportunities as easier and more lucrative.

EXERCISES AND DISCUSSION QUESTIONS

1. Do you think there is really any moral difference between (a) setting off a bomb outside a building for some political reason knowing that a certain number of people will be killed and (b) marketing 11 million defective automobiles knowing that a certain proportion of them will explode into flames when rear-ended and burn the occupants alive?
2. Do you think that white-collar crime (occupational and corporate) can be explained by the same principles as street crime? Read one or two of the relevant cited articles for guidance.
3. Looking back at all the theories presented in this book, make a case for one of them as the best in terms of explaining organized crime.
4. Go to http://glasgowcrew.tripod.com/index.html and click on "wiseguy tales." Choose a wiseguy and write a one-page report on him.

USEFUL WEB SITES

Criminal Justice Resources-Organized Crime. http://www.lib.msu.edu/harris23/crimjust/orgcrime.htm.

FBI-Organized Crime. http://www.fbi.gov/hq/cid/orgcrime/ocshome.htm.

FBI-White Collar Crime. http://www.fbi.gov/whitecollarcrime.htm.

National White Collar Crime Center. http://www.nw3c.org/.

RICO Act. http://www.ricoact.com/.

United Nations Office on Drugs and Crime-Organized Crime. http://www.unodc.org/unodc/organized_crime.html.

GLOSSARY

Apalachin meeting: Meeting of the bosses and underbosses of New York's LCN families in 1957 in Apalachin, New York. The police (who stumbled on the meeting purely by accident) raided the meeting and arrested 63 people.

Castellammarese war: A struggle between the two main factions in Italian organized crime in New York during Prohibition, one headed by Giuseppe Masseria and the other by Salvatore Maranzano. The struggle ended with the deaths of both leaders in 1931.

Commission: A national ruling body of La Cosa Nostra consisting of the bosses of the five New York families and four bosses from other important families.

Corporate crime: Criminal activity on behalf of a business organization.

Corporate model: A model of La Cosa Nostra that sees it as similar to corporate structure, i.e., a formal hierarchy in which the day-to-day activities of the organization are planned and coordinated at the top and carried out by subordinates.

Ethnic succession theory: Theory about the causes of organized crime that posits that upon arrival in the United States, each ethnic group was faced with prejudicial and discriminatory attitudes that denied them legitimate means to success in America.

Feudal model: A model of La Cosa Nostra that sees it as similar to the old European feudal system based on patronage, oaths of loyalty, and semi-autonomy.

Kefauver committee: Formed in 1950 to investigate organized crime's involvement in interstate commerce.

La Cosa Nostra: (literally, "our thing") Also commonly referred to as the Mafia; an organized crime group of Italian/Sicilian origins.

McClellan commission: Formed in 1956 to look into financial irregularities in the Teamster's Union; the star witness was a made man in the Genovese Family named Joe Valachi.

Occupational crime: Crime committed by individuals in the course of their employment.

Organized crime: A continuing criminal enterprise that works rationally to profit from illicit activities that are often in great public demand. Its continuing existence is maintained through the use of force, threats, and/or corruption of public officials.

Prohibition: Common term for the Volstead Act, which prohibited the sale, manufacture, or importation of intoxicating liquors within the United States.

Russian "Mafiya": Considered to be the most serious organized crime threat in the world today.

Sarbanes-Oxley Act (SOA): An act passed in 2002 in response to numerous corporate scandals. The provisions of this act include increased funding for the Security Exchange Commission, penalty enhancement for white-collar crimes, and the relaxing of some legal impediments to gaining convictions.

Tammany society: A corrupt political machine associated with the Democratic Party that ran New York City well into the 20th century from the "Hall" (Tammany Hall).

White-collar crime: An illegal act or series of illegal acts committed by non-physical means and by concealment or guile, to obtain money or property, or to obtain business or personal advantage.

READING

Criminal Thinking and Identity in Male White-Collar Offenders

Glenn D. Walters and Matthew D. Geyer

Walters and Geyer attempt in this study to see whether white-collar criminals evidence different thinking patterns compared to street criminals. To this end, 34 male white-collar offenders without a prior history of non-white-collar crime, 23 male white-collar offenders with at least one prior arrest for a non-white-collar crime, and 66 male non-white-collar offenders housed in a minimum security federal prison camp completed the Psychological Inventory of Criminal Thinking Styles and Social Identity as a Criminal scale and were rated on the Lifestyle Criminality Screening Form–Revised. Significant group differences were noted on the Psychological Inventory of Criminal Thinking Styles Self-Assertion/Deception scale, Social Identity as a Criminal Centrality subscale, Social Identity as a Criminal In-Group Ties subscale, and Lifestyle Criminality Screening Form–Revised, which showed that white-collar offenders with no prior history of non-white-collar crime registered lower levels of criminal thinking, criminal identification, and deviance than white-collar offenders previously arrested for non-white-collar crimes.

When Edwin Sutherland coined the term *white-collar crime* in 1939, one of his chief goals was to expose the inadequacies of traditional theories of crime causation (e.g., biological and sociological determinism) in modeling the antisocial behavior of the well to do. Sutherland (1949/1983) would later define white-collar crime as "crime committed by a person of respectability and high social status in the course of his occupation" (p. 7). Although some scholars took issue with Sutherland's definition, choosing to define white-collar crime according to the offense rather than the offender, there is no disputing the fact that white-collar crime, however defined, threatens the social fabric of modern-day society, as evidenced by the recent Enron and WorldCom scandals. Surveys indicate that businesses in the United States incur losses of U.S.$1 billion per annum from employee theft of pens, pencils, paper clips, postage, and stationery (Wells, 1994), and health care fraud, abuse, and waste are estimated to run as high as $100 billion a year, approximately 10% of the total U.S. health care budget (Andrews, 1994). Computer crime, embezzlement, corporate crime, and fraud may have an even more devastating effect on society. Whereas the cost of white-collar crime is undeniable, debate continues to rage over whether white-collar offending should be considered distinct from other categories of criminal conduct.

Most scholars conceptualize white-collar and non-white-collar crime as discrete clinical and theoretical entities. Adopting a contrary view, Gottfredson and Hirschi (1990) posited that the differences between white-collar and non-white-collar crime are more apparent than real based on the assertion that all crime is a product of low self-control. In their general or low self-control theory of white-collar crime, Gottfredson and Hirschi argued that white-collar offenders are just

SOURCE: Walters, G. D., & Geyer, M. D. (2004). Criminal thinking and identity in male white-collar offenders. *Criminal Justice and Behavior, 31*(3), 263–281. Reprinted with permission of Sage Publications, Inc.

as criminally versatile and deviant as their non-white-collar counterparts. What this means is that white-collar offenders do not specialize in white-collar crime anymore than robbers confine themselves to robbery or thieves restrict themselves to theft. In addition, white-collar and non-white-collar offenders are equally likely to own a prior record of criminality and poor social adjustment. There is research that corroborates aspects of Hirschi and Gottfredson's general theory of white-collar crime. Nagin and Paternoster (1994), for instance, uncovered a significant relationship between white-collar crime and low self-control. Weisburd, Waring, and Chayet (1995), in another study that supports Gottfredson and Hirschi's position, determined that imprisonment may be no more effective in deterring white-collar crime than it is in deterring other forms of criminality.

Weisburd, Chayet, and Waring (1990) tested Gottfredson and Hirschi's theory of white-collar crime in a large group of federal offenders restricting their sample to the most chronic white-collar offenders (three or more prior arrests). They discovered that the career pattern of crime was hard to distinguish from that of the average street criminal. Benson and Moore (1992) compared federal white-collar offenders with persons convicted of narcotics violations, bank robbery, and postal forgery. The results revealed that white-collar offenders were 4 times more likely to have been previously arrested for a white-collar crime than non-white-collar criminals, thereby contradicting Gottfredson and Hirschi's versatility hypothesis in the sense that white-collar offenders maintained a higher level of specialization than non-white-collar offenders. Furthermore, the non-white-collar offenders were significantly more deviant than the white-collar offenders on indices of past problem drinking, drug use, poor grades, and social maladjustment. By the same token, a subsample of high-rate white-collar criminals, each with four or more prior arrests, displayed a level of versatility and prior deviance that approached the level attained by non-white-collar offenders. Benson and Moore also uncovered two separate pathways to white-collar crime, one marked by low self-control and prior non-white-collar offending and the other characterized by high self-control and no prior non-white-collar offending.

As the studies reviewed in this section suggest, there are at least two groups of white-collar offender. One group may be indiscernible from the common street criminal, a finding congruent with Gottfredson and Hirschi's (1990) low self-control theory of white-collar crime. The other group, by comparison, is significantly more specialized and less deviant than the first. To the extent that white-collar and non-white-collar crimes are divergent, it would make sense that these offenses are perpetrated by individuals who differ in their thoughts, identifications, and actions toward crime, a possibility that may reflect divergent programming needs for white-collar offenders with and without a history of non-white-collar crime.

The purpose of the current study was to ascertain whether prisoners who have only ever been arrested for a white-collar crime deviated from white-collar offenders previously arrested for a non-white-collar crime and inmates confined for non-white collar offenses on measures of criminal thinking, criminal identity, and criminal lifestyle involvement. Research assessing Gottfredson and Hirschi's (1990) theory of white-collar crime denotes that white-collar crime occurs along two distinct and separate lines, one of which (white-collar offenders with a prior history of non-white-collar crime) may be largely indistinguishable from non-white-collar crime and the other of which (white-collar offenders without a prior history of non-white-collar crime) is less versatile and deviant than the pattern traditionally found in non-white-collar offenders. Employing an offense-based definition of white-collar crime, it was predicted that (a) white-collar offenders with no history of prior non-white-collar crime would receive significantly lower scores than white-collar offenders with a history of non-white-collar crime and non-white-collar offenders on measures of criminal thinking (PICTS factor scores), criminal identity (Self-Identity as a Criminal), and criminal lifestyle involvement (Lifestyle Criminality Screening Form [LCSF]) and

(b) white-collar offenders with a history of prior non-white-collar crime and non-white-collar offenders would perform similarly on these criminal thinking, identity, and lifestyle measures.

Method, Measures, and Procedures

Comparing inmates serving time for white-collar and non-white-collar crimes, white-collar offenders (M = 47.20, SD = 10.78) were significantly older, t(325) = 4.28, $p < .001$, than non-white collar offenders (M = 41.41, SD = 10.79) and significantly more likely to be White, $\chi^2(3, N = 327) = 23.56, p < .001$, than the non-white-collar offenders.

The final sample was composed of 57 white-collar and 66 non-white-collar offenders. All inmates participating in the study were administered the PICTS, an 80-item self-report measure which taps a variety of criminal thinking patterns such as hostility, cutoff, entitlement, power orientation and cognitive Indolence. The 12 items SIC scale was used to assess participants' social identity as a criminal, and the LCSF-R was used to assess the four interactive styles associated with lifestyle patterns of criminal conduct (irresponsibility, self-indulgence, interpersonal intrusiveness, and social rule breaking) that is scored from information found in an inmate's presentence investigation report.

Results

White-collar offenders without a history of non-white-collar crime (WC-only) were significantly older, more highly educated, and serving shorter sentences than the non-white-collar control group (NWC). The WC-only group also possessed significantly more years of education than white-collar offenders with one or more prior arrests for a non-white-collar crime (WC-prior).

As seen in the table below, WC-only inmates attained significantly lower scores on the PICTS Self-Assertion/Deception scale and SIC In-Group Ties subscale relative to participants in the WC-prior and NWC conditions, whereas WC-prior and NWC inmates failed to differ on these two measures. Second, WC-prior inmates scored significantly higher than WC-only and NWC inmates on the SIC Centrality subscale. Finally, all three groups varied on the LCSF-R, with NWC participants scoring significantly higher than WC-only and WC-prior inmates and WC-prior participants scoring significantly higher than WC-only inmates on this measure.

Discussion

Congruent with past research, the current study identified two general categories of white-collar offender: a larger group of white-collar specialists who had never before been arrested for a non-white-collar crime and a smaller group of versatile white-collar offenders who had been arrested at least once for a nonwhite collar crime. When these two groups were compared to a control group of inmates convicted of nonviolent, non-white-collar crimes on measures of criminal thinking, identity, and lifestyle, it was discovered that the white-collar offenders with no history of nonwhite-collar crime (WC-only) were less inclined to endorse criminal thoughts, identify with other criminals, and exhibit signs of a criminal lifestyle than white-collar offenders with prior arrests for non-white-collar crime (WC-prior) and non-white-collar offenders (NWC). The WC-prior and NWC groups, as predicted, failed to differ from each other on measures of criminal thinking and identity, except for the Centrality subscale of the SIC, where WC-prior inmates scored significantly higher than WC-only and NWC participants. On the other hand, NWC inmates registered significantly higher scores on a measure of criminal lifestyle (LCSF-R) than participants in the WC-prior group. White-collar offenders with a history of prior non-white-collar crime were largely indistinguishable from NWC offenders but featured stronger criminal thinking, identity, and general deviance than white-collar offenders with no history of non-white-collar crime.

Demographic variables such as age, education, sentence length, and ethnic status discriminated

Table 13.1 Group Performance on the Psychological Inventory of Criminal Thinking Styles (PICTS), Social Identity as a Criminal (SIC) scale, and Lifestyle Criminality Screening Form–Revised (LCSF-R)

	Group			
Variable	**WC-Only**	**WC-Prior**	**NWC**	**F (2, 120)**
PICTS factor scales				
Problem avoidance				
M	14.59	18.26	16.30	2.65
SD	5.72	7.37	5.49	
Interpersonal hostility				
M	10.97	11.57	11.58	0.88
SD	1.77	2.43	2.41	
Self-assertion/deception				
M	12.65_a	16.09_b	15.30b	4.32*
SD	4.06	7.22	4.38	
Denial of harm				
M	21.79	23.96	23.44	1.65
SD	4.60	5.83	4.89	
Social identity as criminal subscales				
In-group ties				
M	7.09_a	9.48_b	9.61_b	5.58**
SD	2.49	3.58	4.21	
Centrality				
M	11.24_a	13.43_b	10.59_a	3.92
SD	4.78	3.67	4.05	
In-group affect				
M	6.06	6.52	7.11	0.87
SD	3.34	4.81	3.70	
LCSF-R total score				
M1.12_a	4.22_b	5.76_c	37.09***	
SD	1.32	2.61	2.97	
LCSF-R score (w/o arrest items)				
M	1.12_a	2.74_b	3.88_c	27.79***
SD	1.32	1.63	1.98	

*$p < .05$; **$p < .01$; ***$p < .001$.

NOTE: WC-only = white-collar offender with no prior history of non-white-collar crime; WC-prior = white-collar offender with at least one prior arrest for a non-white-collar crime; and NWC = non-white-collar, non-violent, non-white-collar offender. Subscripts (following the mean of each group) represent the results of the Duncan Multiple Range Test. Means with different subscripts differ significantly at $p < .05$.

between the three groups of inmates, and except for ethnic status, correlated with at least one of the eight dependent measures. Group differences on the SIC In-Group Ties, LCSF-R total score, and the LCSF-R persisted even after controlling for the five demographic variables. It may well be that the significant group contrast on the Self-Assertion/Deception scale is a consequence of the WC-only inmates being older than participants in the WC-prior and NWC conditions, coupled with the fact that age was negatively correlated with Self-Assertion/Deception. Group variations on the SIC In-Groups Ties subscale and LCSF-R (with and without the arrest items) appeared to be more robust because they displayed minimal change after age, education, sentence, ethnic status, marital status, and Df-r [defensiveness-revised; a validation scale] scale scores were accounted for. Differences on the SIC Centrality subscale remained stable after controlling for initial group differences on the Df-r scale, however, not when the five demographic variables were entered as covariates in an ANCOVA design.

One could argue that the disparities in criminal thinking and identity found to exist between WC-only and WC-prior participants were a function of the greater number of arrests attained by the WC-prior group. After all, the WC-prior group (M = 1.41, SD = 1.14) had accrued a substantially greater number of prior arrests than the WC only group (M = .24, SD = .43, t(53) = 5.50, $p < .001$). There are two problems with this argument. First, there were no significant group differences found for any of the dependent measures when first-time offending WC-only participants were compared with WC-only inmates who had a previous arrest for a white-collar crime. Second, when the number of prior arrests was employed as a covariate in an ANCOVA of the dependent measures achieving significant ANOVA results (i.e., PICTS Self Assertion/Deception, SIC In-Group Ties, SIC Centrality, LCSF-R) only the PICTS Self-Assertion/Deception scale fell to nonsignificance, as it did when the five demographic measures and PICTS Df-r scale were incorporated as covariates in an ANCOVA of the dependent measures. These findings suggest that there is more to the outcome of the current study than a simple confounding of demographic/criminal arrest variables and white-collar status (WC-only and WC-prior), even though the power of the statistical tests was hindered by a relatively small number of participants in the WC-prior condition and moderately low internal consistency in the SIC subscales.

An interesting and unanticipated finding was the moderately strong inverse relationship that surfaced between In-Group Ties and marital status, (r = –24, $p < .01$). Seeing as marital status was coded 1 for single individuals and 2 for nonsingle inmates, a significant negative correlation indicates that single participants scored higher on the In-Group Ties subscale than persons currently or previously married. Furthermore, this significant effect persisted even after age was controlled for with the aid of a partial correlation (r = –.21, $p < .05$). Hence, relationships or the capacity to form relationships, whether with a person or a profession, in the past or in the present, may protect against the construction of a criminal social identity, or conversely, a criminal social identity may inhibit the development of relationships with conventional people and professions.

Another unanticipated finding from the current study was that inmates from the WC-prior group achieved significantly higher scores on the SIC Centrality subscale than WC-only or NWC inmates. This outcome suggests that a criminal social identity is more central to the self-views of white-collar offenders with a prior history of non-white-collar crime than it is to white-collar offenders with no record of non-white-collar crime and persons convicted of non-white-collar offenses. Perhaps white-collar offenders previously arrested for nonwhite-collar crimes experience a greater sense of cognitive dissonance between their self-view and actual behavior than participants in the other two groups, thereby encouraging them to become preoccupied with a criminal social identity. A second possibility is that WC-prior inmates were more criminally versatile than men in the other two conditions because they committed an average of 1.7 white-collar crimes and 3.7 non-white-collar crimes, compared to a mean of 1.2 white-collar crimes and 0.0 non-white-collar crimes for WC-only inmates and 0.1 white-collar crimes and 6.0 non-white-collar crimes for NWC participants. Thus, relative to inmates in the WC-only and NWC groups, WC-prior participants were dually deviant (white-collar and non-white-collar crime), which, in turn, may have served to inflate the salience of their deviant self-views.

An underpinning assumption of Gottfredson and Hirschi's (1990) general theory of white-collar crime is that white-collar criminals are as versatile and deviant as non-white-collar offenders. Using prior arrests for non-white-collar crime as a proxy for versatility, we uncovered support for Gottfredson and Hirschi's position in the smaller group of white-collar offenders who possessed at least one prior arrest for a non-white collar crime. This group of white-collar offenders displayed criminal thinking and identity on par with inmates serving time for non-white-collar crimes and posted levels of prior non-white-collar arrest (M = 3.74, SD = 3.92) comparable to the non-white-collar group (M = 4.97, SD = 5.12), $t(87) = 1.05$, $p > .10$. Be this as it may, 60% of the white-collar criminals participating in the current study had only ever been arrested for a white-collar crime. Therefore, Gottfredson and Hirschi's versatility hypothesis is only partially supported by the results of this investigation. Seeing as the LCSF-R assesses prior substance misuse, general occupational/school adjustment, and marital/relationship difficulties, removing the three arrest items would seem to make the LCSF-R the ideal proxy for general deviance. When the modified LCSF-R was used for this purpose, it was determined that even less of a relationship existed between deviance and white-collar crime than between criminal versatility and white-collar crime, as the more versatile white-collar offenders (WC-prior) showed greater deviance than the non-versatile white-collar offenders (WC-only) but were themselves less deviant than the non-white-collar control group (NWC).

Outcomes from the current investigation indicate that white-collar offenders can be grouped into two general categories: a larger group of individuals, constituting 60% of the current sample of white-collar offenders, with no history of non-white-collar crime and a smaller group of individuals serving time for a white-collar offense, but with at least one prior arrest for a non-white-collar crime. These findings further denote that the three groups differed modestly on measures of criminal thinking, moderately on measures of criminal identity, and widely on measures of lifestyle involvement. Although not all the comparisons were statistically significant, they were reasonably consistent with the hypotheses established at the onset of the current study. It will still be necessary to determine how well these findings generalize to other groups, particularly female inmates, because research suggests that a fair number of white-collar offenders are women (Daly, 1989). This notwithstanding, one implication of the current findings is that white-collar offenders do not form a homogeneous group with respect to their pattern of offending, level of deviance, attitudes toward crime, or social identity.

References

Andrews, J. H. (1994, August). Health-industry fraud eats up billions yearly. *Christian Science Monitor, 4,* 3.

Benson, M. L., & Moore, E. (1992). Are white-collar and common offenders the same? An empirical and theoretical critique of a recently proposed general theory of crime. *Journal of Research in Crime and Delinquency, 29,* 251–272.

Daly, K. (1989). Gender and varieties of white-collar crime. *Criminology, 27,* 769–794.

Gottfredson, M. R., & Hirschi, T. (1990). *A general theory of crime.* Stanford, CA: Stanford University Press.

Nagin, D. S., & Paternoster, R. (1994). Personal capital and social control: The deterrence implications of a theory of individual differences in criminal offending. *Criminology, 32,* 581–606.

Sutherland, E. H. (1983). *White collar crime: The uncut version.* New Haven, CT: Yale University Press. (Original work published 1949)

Weisburd, D., Chayet, E. F., & Waring, E. J. (1990). White-collar crime and criminal careers: Some preliminary findings. *Crime & Delinquency, 36,* 342–355.

Weisburd, D., Waring, E., & Chayet, E. (1995). Specific deterrence in a sample of offenders convicted of white-collar crimes. *Criminology, 33,* 587–607.

Wells, J. T. (1994, October). The billion dollar paper clip. *Internal Auditor,* pp. 32–37.

DISCUSSION QUESTIONS

1. In what ways does this study support or fail to support Gottfredson and Hirschi's contention that there are really no differences between street and white-collar criminals?
2. Is the similarity/dissimilarity of criminal thinking patterns adequate to test the hypothesis of "no difference"?
3. How would labeling theorists explain the patterns found by Walters and Geyer?

READING

The Causes of Organized Crime

Do Criminals Organize Around Opportunities for Crime or Do Criminal Opportunities Create New Offenders?

Jay S. Albanese

In this article, Jay Albanese examines two competing assumptions about entry into organized crime. The first is that crime-prone individuals or organizations organize around exploiting ever-changing criminal opportunities; the second assumption is that criminal opportunities provide motivation for individuals and organizations formerly not connected with criminal activity. He examines these assumptions with the goal of developing a model of organized crime that employs both offender propensity and opportunity factors to predict the incidence of organized crime activity. Albanese proposes that his model should help law enforcement and public policy decisions on where resources should be focused to reduce criminal opportunities that constantly arise out of social and technological changes.

John "Junior" Gotti, 35-year-old son of the infamous John Gotti, was sentenced in 1999 to more than 6 years in prison for bribery, extortion, gambling, and fraud charges. Many of the charges stemmed from an extortion racket at a Manhattan topless club. Prosecutors claimed that Junior Gotti inherited a leadership position in the Gambino crime family in New York City,

SOURCE: Albanese, J. S. (2000). The causes of organized crime: Do criminals organize around opportunities for crime or do criminal opportunities create new offenders? *Journal of Contemporary Criminal Justice, 16*(4), 409–423. Reprinted with permission of Sage Publications, Inc.

after his father received a life-in-prison sentence for racketeering and murder in 1992. The sentencing judge was amazed that Junior Gotti would repeat his father's mistakes. "You know the toll that kind of behavior [by your father] took on the family and children. Yet the pattern, for reasons I'm unable to fathom, is duplicated" (Associated Press, 1999, p. 1). John "Junior" Gotti and his wife have four children, and his wife was pregnant with a fifth.

Do criminals organize around available opportunities for crime, as it appears in the case of the Gotti family, or do criminal opportunities create new offenders? It appears that examples of each of these types can be found, but it is important to know which is most common and under what circumstances they occur. Such knowledge would go a long way in plotting law enforcement strategy and public policy designed to control organized crime.

Six of the world's largest manufacturers of vitamins agreed to pay more than $1 billion in 1999 to settle a class-action lawsuit claiming that the companies artificially raised the prices of vitamins by forming an international cartel that met in hotel rooms, sometimes under fictitious names. These meetings were designed to divide up the global vitamin market and keep the prices of vitamins artificially high. Two foreign executives were expected to serve jail sentences in the United States as part of the settlement in this case (Barboza, 1999; O'Donnell, 1999). Is this a case of criminals searching for an opportunity or a criminal opportunity that was exploited by unscrupulous people? To make sense of these apparently disparate cases of crimes, criminals, and opportunities, it is necessary to be clear about three precursors: what constitutes organized crime, what constitutes an organized crime group, and what is a criminal opportunity.

Definition of Organized Crime

Definitions of organized crime are numerous but often vague. The U.S. General Accounting Office, the investigative arm of Congress, concluded years ago that the absence of a consensus in the U.S. Justice Department about the fundamental definition of organized crime has hampered the potential success of crime control programs designed to combat it (U.S. Comptroller General, 1977). The President's Commission on Organized Crime (1987), appointed by Ronald Reagan, also did not offer any clear definition of organized crime. Rather, it described a series of characteristics of "criminal groups," "protectors," and "specialist support" necessary for organized crime (President's Commission on Organized Crime, 1987). This apparent confusion over what constitutes organized crime is puzzling, given the long history of interest in the subject. Key words such as *mafia, mob, syndicate, gang,* and *outfit* are often used to characterize it, but the precise meaning of these terms is often lost in discussions of the appearances and earmarks of organized crime.

There is an emerging consensus about what actually constitutes organized crime. The bad news is that 11 different aspects of organized crime have been included in the definitions by the various authors, but few were mentioned often. It appears that a definition of organized crime, based on a consensus of writers over the course of the past three decades, would read as follows: Organized crime is a continuing criminal enterprise that rationally works to profit from illicit activities; its continuing existence is maintained through the use of force, threats, monopoly control, and/or the corruption of public officials. There are, of course, some confounding factors to be addressed. That is, how does an otherwise legitimate corporation, such as the vitamin makers in the example noted above, fit into this definition? What about a licensed massage parlor that also offers sex for money to some customers? As many investigators have recognized, perhaps organized crime does not exist as an ideal type, but rather as a degree of criminal activity or as a point on the spectrum of legitimacy. That is to say, is not the fundamental difference between loan-sharking and a legitimate loan, the interest rate charged (within or in excess of the legal limit)? Is not the important difference between criminal and noncriminal distribution of narcotics whether or not the distributor is licensed (i.e., doctor or pharmacist) or unlicensed by the state? Organized crime, therefore, is actually a type of a larger category of behavior that may be called organizational crime.

It is apparent that crimes by corporations during the course of business, or crimes by politicians or government agencies, can also be considered part of organized crime.

For example, official misconduct by government officials, obstruction of justice, and commercial bribery are all types of organized criminal behavior. If they fulfill the requirements of the definition above, they constitute a part of what is known as organized crime. As the U.S. National Advisory Committee on Criminal Justice Standards and Goals (1976) has recognized, there are more similarities than differences between organized and the so-called white-collar crimes: "Accordingly, the perpetrators of organized crime may include corrupt business executives, members of the professions, public officials, or members of any other occupational group, in addition to the conventional racketeer element" (p. 213). Therefore, any distinction between organized and white-collar crime may be artificial inasmuch as both involve the important elements of organization (i.e., multiple participants), rational criminal motive, and the use of corruption and/or violence to maintain immunity.

Definition of Organized Crime Group

An organized crime group need not be large to achieve this status. Two or more participants suffice, and it is not necessary for these participants to be part of a preexisting organized crime group. Therefore, a conspiracy within or by a corporation is part of organized crime when it fulfills the other elements of the definition explained above. Likewise, crime can be organized (such as instances of illegal immigrant smuggling into other countries) without being part of a larger criminal group enterprise. Organized crime groups are not necessarily culturally driven, as was the case with the traditional Mafia, in which associations existed that performed both legal and illegal functions. Organized crime groups may also be product driven, as has been seen in the case of some illegal narcotics distributors, such as Jamaican posses, who appear to organize simply to commit the crimes and have little or no group loyalty. The President's Commission on Organized Crime (1987) described organized crime as 11 different groups:

1. La Cosa Nostra (Italian)
2. outlaw motorcycle gangs
3. prison gangs
4. Triads and Tongs (Chinese)
5. Vietnamese gangs
6. Yakuza (Japanese)
7. Marielitos (Cuban)
8. Colombian cocaine rings
9. Irish organized crime
10. Russian organized crime
11. Canadian organized crime

This curious mixture includes groups defined in terms of ethnic or national origin, those defined by the nature of their activity (i.e., cocaine rings), those defined by their geographic origin (i.e., prison gangs), and those defined by their means of transportation (i.e., motorcycle gangs). Such a haphazard approach to defining and describing organized crime does little to help make sense of its causes, current events, or how policies against organized crime should be directed.

There is even evidence, as both the President's Commission and independent researchers have pointed out, that these groups and others (such as Jewish gangs) have worked with each other in the past and continue to do so in the present. As a consequence, ethnicity is not a very powerful explanation for the existence of organized crime due to the large number of ethnic groups involved, their interaction with each other in criminal undertakings, and the fact that ethnicity is probably no more a causal factor than are motorcycles. This has been called "the ethnicity trap" (Albanese, 1996). Biographical attributes, like methods of transportation, may help to describe a particular person or group, but they do little to explain that person's or group's behavior (especially when compared to other members of that ethnic group who do not engage in organized crime activity).

A look at several investigations of ethnically based organized crime reveals why it is a weak descriptor. In addition to the fact that no single or multiple ethnic combination accounts for organized crime, ethnicity also has been found to be secondary to local criminal opportunities in explaining organized crime. A study of the early 20th-century illicit cocaine trade in New York by historian Alan Block (1979) found major players with Jewish backgrounds, but also "notable is the evidence of interethnic cooperation" between New York's criminals. He found evidence in Italian, Greek, Irish, and Black involvement of people who did not always work within their own ethnic group. Instead, he found these criminals to be "in reality criminal justice entrepreneurs," whose criminal careers were not within a particular organization but were involved in a "web of small but efficient organizations" (Block, 1979, p. 95).

An ethnography of the underground drug market by Patricia Adler (1985) in "Southwest County" found the market to be "largely competitive" rather than "visibly structured." She found that participants "entered the market, transacted their deals, [and] shifted from one type of activity to another," responding to the demands of the market rather than through ethnic structures or concerns (Adler, 1985, p. 80). Similarly, a study of illegal gambling and loan-sharking by Peter Reuter (1983) in New York found that economic considerations dictated entry and exit from the illicit marketplace. He found that the criminal enterprises he studied in three areas to be "not monopolies in the classic sense or subject to control by some external organization" (Reuter, 1983, pp. 175–176). Like the other investigations of organized crime groups, Reuter found that local market forces shaped the criminal behavior more so than ethnic ties or other characteristics of the criminal groups.

Definition of Criminal Opportunity

Criminal opportunities are of two types: those that provide easy access to illicit funds without incurring high risk and those that are created by motivated offenders. The easy-access type includes the traditional provision of illicit goods and services that are in high public demand: gambling, pornography, and narcotics. Added to these are new criminal opportunities that are made possible by social or technological change. These would include misuse of the Internet, cell phones, and companies or banks for money laundering, among others. The precise type of crime or product is not as important as the use of illegal means for its use, acquisition, or exploitation.

Those criminal opportunities that are created by offenders often involve bribery or extortion. Examples would include protection rackets and schemes to defraud that involve the manufacture of a criminal opportunity in an otherwise legitimate business enterprise.

Model and Method

To examine the relationship between organized crimes, groups, and opportunities, a model is proposed here that will be applied against several case studies of organized crime to assess its power. It is hypothesized that three major elements are important to predicting the incidence

Figure 13.2 Model of Criminal Opportunity: Organized Crime Group Interaction

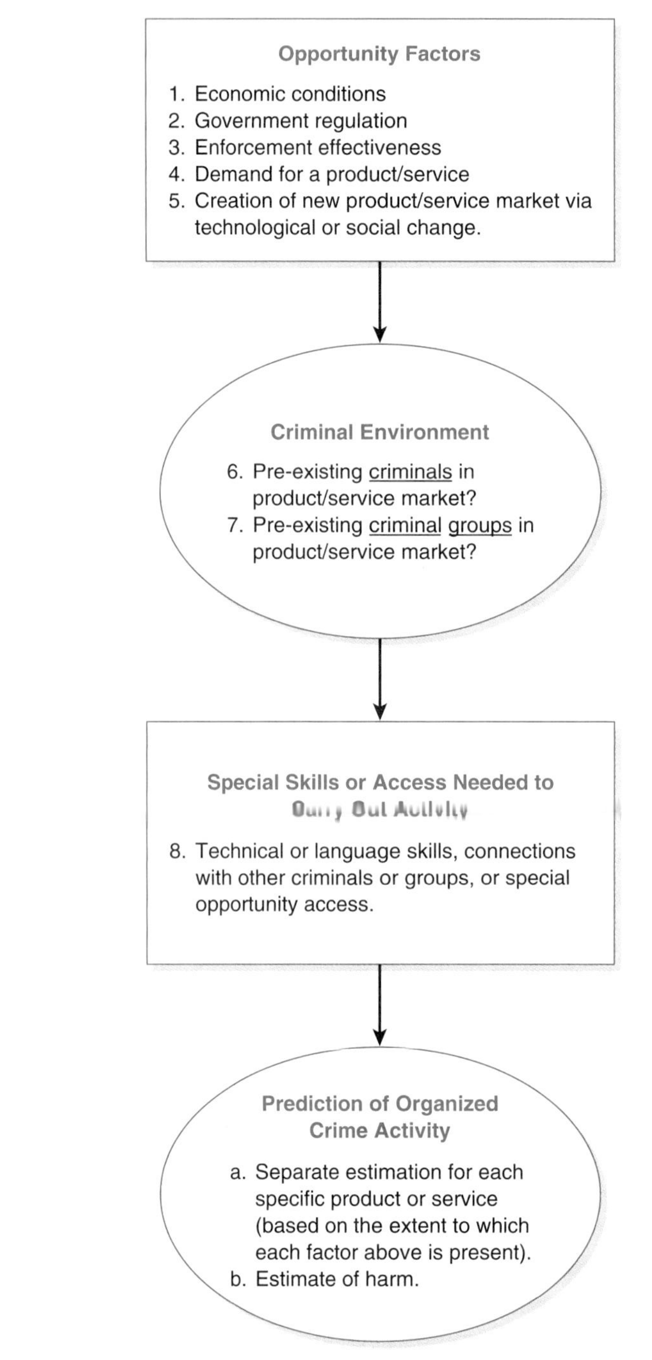

Opportunity Factors

1. Economic conditions
2. Government regulation
3. Enforcement effectiveness
4. Demand for a product/service
5. Creation of new product/service market via technological or social change.

↓

Criminal Environment

6. Pre-existing criminals in product/service market?
7. Pre-existing criminal groups in product/service market?

↓

Special Skills or Access Needed to Carry Out Activity

8. Technical or language skills, connections with other criminals or groups, or special opportunity access.

↓

Prediction of Organized Crime Activity

a. Separate estimation for each specific product or service (based on the extent to which each factor above is present).
b. Estimate of harm.

of organized crime: opportunity factors, the criminal environment, and the skills or access required to carry out the criminal activity. The model builds on earlier work in this area by further specifying the circumstances under which organized crime thrives (Albanese, 1995). Figure 13.2 summarizes this relationship.

As Figure 13.2 illustrates, opportunity factors consist of five major types. These include economic conditions, government regulation, enforcement effectiveness, demand for a product or service, and new product or service opportunities that are created by social or technological changes or by the criminal group itself. The criminal environment is assessed by the extent to which individual offenders and preexisting crime groups are available to exploit these opportunities. Third, technical, language, special access skills, or connections (with other criminals or crime groups) are needed to accomplish certain types of organized crime activity (e.g., narcotics importation requires a manufacturing and distribution capability). This three-factor model (composed of eight different variables) is designed to quantify the relationships among opportunity, criminals, and skills to predict the incidence of organized crime in various markets. It is assumed that a separate version of this model will be needed for each type of criminal activity. This is because the opportunity, criminal environment, and skills required are likely to be different for various types of organized crime, such as narcotics distribution, human smuggling, money laundering, stolen cars, and so on.

Due to the lack of systematic data on the nature and incidence of organized crime, researchers must be resourceful in obtaining means to test models and hypotheses. The use of detailed case studies is one way to learn about how past cases of organized crime activity reflected factors in a proposed model. There are, of course, limitations of such a method. These include representativeness (past cases may not reflect what occurs in undetected activity), generalizability (organization of human smuggling in one location may be different in another location), and measurement (the factors above are not easily measured or quantified). Although these limitations are significant, the use of case studies to predict important factors in the genesis of organized crime may form the basis for future validation efforts. In addition, even a model with relatively low predictive efficiency can be useful in case screening for police and in estimating the true extent of organized crime in various markets.

Testing the Model

Five recent case studies of organized crime were selected for application to the proposed model. They were selected because they are recent, focus on different kinds of offenses, and were conducted in different locations. Each case study was reviewed and the presence or absence of each variable in the model was evaluated, based on the observations of the authors of the studies. A summary of this analysis is presented in Table 13.2.

As Table 13.2 indicates, Gary Potter (1994) conducted a study in an unidentified U.S. city he called "Morrisburg." Illegal gambling was identified as the central organized crime activity, followed by drug sales and prostitution. Morrisburg is an economically depressed area with high unemployment. The laws are in place to prohibit the illicit activities, but authorities were found to be indifferent toward enforcement. The high demand for this vice activity remains high, and there have been no major changes to this market. Potter found that there is a tradition of organized vice and corruption in Morrisburg and that this activity was operated primarily by three separate organized crime groups. The groups endured over the years due to the need for an organization to operate the illegal gambling activity and provide lay-off and banking services for bookies taking bets. The Gianellis (the largest of the three organized crime groups) were "able

Table 13.2 Applying Case Studies to the Model

Focus of Study	Local, Indigenous	Chinese Gangs	Indigenous	Indigenous and Immigrant	Soviet Émigrés
Location	Morrisburg, United States	New York City Chinatown	Former Soviet Union	The Netherlands	New York City (Brooklyn)
Author	Potter (1994)	Chin (19[illegible])	Lee (1997)	Fijnaut, Bovenker, Bruinsma, and van de Bunt (1998)	Finckenauer and Waring (1998)
Nature of primary crime	Illegal gambling (drugs, prostitution)	Extortion (protection)	Nuclear smuggling	Drugs (arms, women)	Fraud (especially fuel tax evasion)
Economic	High unemployment[a]	Businesses benefit in some way[a]	Economic hardship in former USSR[a]	Strong economy	Good immigrant conditions
Government regulation or law	Laws in place	Laws in place	Little international co-operation between republics[a]	Organized crime property ownership in red light areas[a]	Laws in place
Enforcement	Authorities indifferent[a]	Reporting to police unlikely[a]	Security of materials weak[a]	Illicit money in legitimate sector[a]	Long paper trail enforcement difficult[a]
Demand	Demand for vices high[a]	Long tradition of business extortion[a]	Demand for nuclear materials not high	Demand for vices high[a]	Crimes against government tax laws
New opportunities	Unchanged market	Unchanged market	Lax oversight due to political upheaval[a]	Traditional market	Market expands by state law changes[a]
Individual offenders	Local tradition of organized vice activity[a]	Gangs dominate over individual extortion	Insiders are likely offenders[a]	Individual entrepreneurs	Expectation of government corruption, smuggling[a]
Organized crime groups	Three primary organized crime groups in control[a]	Preexisting gangs[a]	Fluid networks not organized crime groups	No sign of organized crime group control	As competitors only
Special skills or access required	Need organized crime groups to handle gambling lay offs[a]	Restricted to Chinatown by language and culture[a]	Special access needed to steal nuclear material[a]	Special skills not required	Few needed, many émigrés have English skills
Total (out of 8 possible factors)	6	5	6	3	3

a. The superscripts indicate positive factors of the prediction model.

to provide their associates with a full range of banking and money-laundering services plus political protection through a highly structured payoff system that, over the years, has simply become part of the normal way of doing business in 'Morrisburg'" (Potter, 1994, pp. 74–75). Six of the eight factors of the prediction model (see Table 13.2) are present in Morrisburg, suggesting a high probability of widespread illegal gambling in this market.

Ko-lin Chin (1996) conducted a study of Chinese gangs in New York City's Chinatown. A significant focus of his study was on the extortion of Chinese-owned businesses. He found this to be a common and long-standing practice—nearly 70% of businesses in Chinatown had been approached by gangs for money, goods, or services. He found that most victims were extorted three or four times per year. The businesses received a benefit from paying off gangs because this protected them from being shaken down by other individuals and gangs. Chin (1996) found that "most Chinese business owners comply with gang extortion demands because such practices are considered consistent with Chinese customs and not worth resisting. Business people are generally willing to pay the gangs some money to avoid further, more significant problems' (p. 97). The gangs were exclusively Chinese (they spoke Cantonese 89% of the time), and their common heritage was central to the extortion efforts. For example, gangs often exploited the Chinese custom of lucky money, in which money is given away on holidays such as the Chinese New Year. These circumstances result in five of the eight factors in the organized crime prediction model (see Table 13.2) being present. This suggests a high probability of continued organized crime activity in New York City's Chinatown.

Rensselaer Lee (1997) studied the incidence of smuggling of nuclear materials out of Russia after the collapse of the Soviet Union. Severe economic hardship, insiders in both government and the nuclear industry, and few international smuggling agreements between the new republics combine to make nuclear smuggling a potential threat. This is counteracted by an uncertain international market for this material (most of it possesses little military significance) and the need for special access to obtain it. Although there have been a number of thefts of nuclear materials, the author found "only one case of an organized crime connection to radioactive smuggling" (p. 111). The continued lax oversight of this market, however, could produce significant organized crime activity in the future (six of eight prediction factors in the model are indicated, suggesting that it is at high risk for organized crime activity).

Fijnaut, Bovenkerk, Bruinsma, and van de Bunt (1998) conducted a comprehensive study of organized crime in the Netherlands. The purpose of the study was to provide an assessment of the nature, seriousness, and scale of organized crime in the country and the effectiveness of investigative methods. It was found that organized crime in the Netherlands was "mainly confined to the traditional illegal supply of certain goods and services—the list being headed by drugs followed by arms and women—and financial and economic fraud" (Fijnaut et al., 1998, p. 203). Although no criminal groups were found to have gained control by infiltrating legitimate business, they found that a significant amount of property is owned by criminal groups in the Red Light District in Amsterdam, which is "cause for great concern" (Fijnaut et al., 1998, p. 204). The continued high demand for the vices is reason for caution, but the strong economy in the Netherlands and the lack of an existing significant presence of organized crime groups make the risk low (three of eight possible factors) according to the prediction model (Table 13.2).

A study by Finckenauer and Waring (1998) of Soviet émigré crime in New York City found fraud to be the most common offense (especially fuel tax evasion). The opportunity factors for this offense are few. Economic conditions for these recent immigrants are generally good, laws are in place to prohibit the behavior in question, the demand for fuel tax evasion is low (because this is a crime against government tax laws), and it does not represent a particularly new opportunity. On the other hand, the complicated paper trail from wholesaler to retail outlet makes enforcement of existing tax laws difficult. With regard to the criminal environment, there was found to be an expectation of government corruption with long-standing practices of smuggling activity in the former Soviet Union by many Soviet émigrés to the United States. Finckenauer and Waring did not find preexisting Soviet organized crime groups in their study, however. Few technical skills are needed to carry out the activity, many of the Soviet émigrés could speak English, and connections with other criminal groups were incidental rather than part of the fuel tax conspiracy. In sum, Soviet émigré organized crime in New York City fulfilled only three of the eight variables in the model, suggesting that its incidence is low. In a survey that was part of their study, Finckenauer and Waring found little first-hand experience with Soviet organized crime of any sort. More importantly, they found that the fuel frauds "do not demonstrate that the involved individuals had a sophisticated criminal organization with dominating harm capacity." They found there to be a significant difference between "the repeated involvement of specific individuals and the repeated involvement of a single criminal organization or network in a range of criminal enterprises" (p. 241). They concluded that "the facts do not support the proposition that Soviet émigrés currently constitute an organized crime threat to the United States" (p. 254). This conclusion concurs with the prediction of the proposed model in Table 13.2, which indicates that there are relatively few opportunity, criminal, and skills factors present in the case of fuel fraud.

Conclusions

This study is a preliminary effort to develop a prediction model that specifies the relationship between criminal opportunities, the criminal (offender) environment, and the skills required to carry out organized crime activity. An effort was made to assess this model through the application of five recent case studies of organized crime in different geographic areas and focusing on different crimes. A few conclusions can be drawn:

- The model shows some ability to distinguish between high-incidence and low-incidence forms of organized crime (by jurisdiction). It will be necessary to develop better indicators to more accurately measure the presence or absence of the variables in the model in order to make more sophisticated predictions.
- The unit of analysis in organized crime studies varies considerably. Some focus on the type of crime activity, others on particular organized crime groups, and still others on a geographic area. Such differences in unit of analysis may obscure facts due to the perspective taken by the researcher.
- Additional case studies of organized crime will be needed that aim to measure its nature and environment with greater precision in order to develop further the model proposed here.
- It is likely that the model would have to be applied to each specific type of crime in each specific geographic area to be useful. This is because the opportunities, offenders, and skills required will likely vary for each kind of offense and in each jurisdiction. Economic conditions, government regulation, enforcement capabilities, demand for products and services, new opportunities for crime, and the available skills and motivated offenders will be different for every offense and between jurisdictions.
- There is a need to be sensitive to changes over time in both the type of illicit activity

and the areas where it occurs because of changes in social conditions, technology, and government policies.

- The model may have utility as a screening device to assess which jurisdictions and potential criminal markets (e.g., drugs, tax fraud, gambling, etc.) are most prone to exploitation. This may be useful in the future in allocating police and other government resources toward areas that are identified to be at greatest risk of organized crime infiltration.

References

Adler, P. A. (1985). *Wheeling and dealing: An ethnography of an upper-level drug dealing and smuggling community.* New York: Columbia University Press.

Albanese, J. S. (1995). Predicting the incidence of organized crime. In J. S. Albanese (Ed.), *Contemporary issues in organized crime* (pp. 35–60). Monsey, NY: Willow Tree.

Albanese, J. S. (1996). *Organized crime in America* (3rd ed.). Cincinnati, OH: Anderson.

Associated Press. (1999, September 3). "Junior" Gotti gets nearly 6 1/2 years. *The New York Times,* p. A1.

Barboza, D. (1999, September 8). Vitamin makers said to agree to pay $1.1 billion to settle suit. *The New York Times,* p. B1.

Block, A. A. (1979). The snowman cometh: Coke in progressive New York. *Criminology, 17,* 75–99.

Chin, K. L. (1996). *Chinatown gangs: Extortion, enterprise and ethnicity.* New York: Oxford University Press.

Fijnaut, C., Bovenkerk, F., Bruinsma, G., & van de Bunt, H. (1998). *Organized crime in the Netherlands.* The Hague: Kluwer Law International.

Finckenauer, J. O., & Waring, E. J. (1998). *Russian Mafia in America: Immigration, culture, and crime.* Boston: Northeastern University Press.

Lee, R. (1997). Recent trends in nuclear smuggling. In P. Williams (Ed.), *Russian organized crime: A new threat?* (pp. 137–145). London: Frank Cass.

O'Donnell, J. (1999, May 21). U.S. fines drug companies $725M for price fixing. *USA Today,* p. B1.

Potter, G. W. (1994). *Criminal organizations: Vice, racketeering, and politics in an American city.* Prospect Heights, IL: Waveland.

President's Commission on Organized Crime. (1987). *The impact: Organized crime today.* Washington, DC: U.S. Government Printing Office.

Reuter, P. (1983). *Disorganized crime: The economics of the visible hand.* Cambridge, MA: MIT Press.

U.S. Comptroller General. (1977). *War on organized crime faltering–Federal strike forces not getting the job done.* Washington, DC: General Accounting Office.

U.S. National Advisory Committee on Criminal Justice Standards and Goals. (1976). *Report of the task force on organized crime.* Washington, DC: U.S. Government Printing Office.

DISCUSSION QUESTIONS

1. Do you think that the distinction between organized crime and corporate white-collar crime is an artificial one, or are there real distinctions?

2. Albanese writes about individuals taking advantage of criminal opportunities as a big factor in organized crime. What do you think Gottfredson and Hirschi would say about that?

3. Making a clear distinction between corporate white-collar crime and organized crime, which of the two is the bigger threat to society at large?

READING

State Failure, Economic Failure, and Predatory Organized Crime

A Comparative Analysis

Hung-En Sung

According to this article by Hung-En Sung, organized crime research gathered momentum in the post-Cold War era, but the field remains dominated by single-society studies of low generalizability. By way of contrast, Sung's study reports findings from comparative analysis of perceptions of predatory organized crime conducted in 59 countries. Sung evaluated two hypotheses of predatory organized crime. First, the state failure hypothesis argues that the failure of the state to deliver key political goods such as security, justice, and stability encourages criminal groups to perform state functions. Second, the economic failure hypothesis holds that poor economic outcomes such as high unemployment, low standards of living, and a reliance on an underground economy stimulates the growth of criminal syndicates as suppliers of demanded goods, services, and jobs. Sung's results provided general support to both hypotheses. Judicial independence and black market activities were the strongest political and economic correlates of predatory organized crime. He concludes by discussing policy implications for organized crime control in developing countries.

Predatory organized crime feeds on developing countries that are hungry for sustainable economic development. Unlike illicit vice industries (e.g., prostitution, gambling, etc.), which promote banned goods and services among consensual individuals, predatory criminal organizations that engage in extortion, fraud, and shakedowns impose burdensome financial costs and physical risks on legal businesses, and figure among the most detrimental obstacles to economic growth. These predatory criminal syndicates extract profits through the use of violence, or its threat, against legal businesses and make it difficult for host countries to attract the domestic and foreign investment enjoyed by less afflicted countries.

This study tests the generality of two observations frequently reported in past single-society research in a sample of 59 countries. The key explanatory constructs examined in the study are state failure and economic failure. Single-country observations have shown that a failure by the state to provide physical safety and to enforce commercial contracts and property rights creates a social climate permeated by a lack of trust. Protection and enforcement rackets develop as a substitute for trust. Criminal organizations act as an authoritarian shadow state, and in the absence of anything else, can be perceived as legitimate by the citizenry. Criminal elements can infiltrate governmental agencies as well through an explicit abuse of state powers. When

Source: Sung, H. (2004). State failure, economic failure, and predatory organized crime: A comparative analysis. *Journal of Research in Crime and Delinquency, 41*(2), 111–129. Reprinted with permission of Sage Publications, Inc.

no action is taken to effectively control profitable criminal behaviors against businesses, extortion is accepted as part of the costs of doing business, thus wearing down the efficiency of the economy as well as the competitiveness of the country in the global economy. It has also been accepted that pervasive organized crime evolves from a climate of changing economic regulation, deteriorating economic conditions, and growing dependence on an underground economy, into a self-perpetuating economic system that performs some of its own productive and distributive functions. The goal of this study is to test to which degree the relationship between state and/or economic failure and predatory organized crime can be generalized beyond the extant scatter of single-nation observations.

State Failure

States fail when the central government ceases to provide political freedoms, civil rights, criminal and civil justice, personal safety, and collective security in an efficient and just manner. The transparent and effective delivery of these political goods legitimizes the political system. Electoral processes, separation of powers, and maintenance of a strong civil society have proved to be the best mechanisms to insure the achievement of these political goals. Failed states are characterized by very high levels of crime and violence, rampant corruption, inability of the rulers to exercise sovereignty without brutal force, absence of consent among the governed, an atrophied public opinion, and a pervasive atmosphere of uncertainty and instability. Foreign interventions in internal crises are also typical among the most serious cases of state failure.

Although countries such as Afghanistan and Somalia are current and unambiguous examples of failed states, the concept of state failure can also be viewed as a quantitative continuum that ranges from high state functioning to complete state failure, according to the degree of successful delivery of political goods. Actually, no state has ever achieved complete control of its jurisdiction, as crime, popular grievances, and abuse of state authority exist in every society to varying degrees. Extreme forms of state failures often lead to tragedies such as virulent civil wars or genocides (e.g., Rwanda), whereas some failed states (e.g., Colombia) manage to survive amidst perpetual political instability as no viable alternatives are available. Failures of the state to implement desirable public policy or government reform are also ubiquitous among advanced constitutional democracies. The failure of reforming campaign finance regulations in the United States that has reinforced the leverage of large contributors of both hard and soft money in the American special-interest politics is just an example. Perfect state functioning is an ideal stage of political development that no country has ever achieved.

Economic Failure

Economic failure occurs when there exists an acutely inefficient use of capital and labor, and/or when there is an exceedingly unjust distribution of basic goods and services for subsistence. Unlike the concept of market failure, which focuses on the breakdown of the fundamental mechanisms of a market economy (i.e., competition, perfect information, property rights, etc.), economic failure refers to the unsuccessful outcomes of a national economy. Rising national income is neither the only nor the most important parameter of a functioning economy, which is best described as an economic system that allocates material and financial resources in such a way that people can lead lives according to their needs and interests. In fact, social factors such as high levels of schooling and good health have been found to be powerful propellers of economic growth, which in turn further improve social well being (Barro 1997).

Financial crisis, skyrocketing unemployment, and scarcity of basic goods and services signal a severe breakdown of the economic system. These economic problems reduce people's material comfort and steal their sense of worth and dignity. The institutional infrastructure of failed or failing economies is likely to be composed of fragile banking and financial sectors, an inefficient tax system, and heavy government deficit. The most extreme expressions of economic failure include famine, hyperinflation, and high long-term unemployment, which if left unresolved can relegate a population to permanent penury. When a distressed economy is not aided with the timely and adequate welfare spending required to erect an effective social safety net, a substantial portion of the population may be diverted into the underworld and the prison system as individuals are relocated to criminal life-course paths. A society weakened by chronic joblessness and poverty puts its youth at a very high risk of being recruited by drug dealing, human trafficking, or extortion syndicates, affecting its health for generations.

Commonly emerging from economic failure is an underground economy, where the shortage of consumer goods and services is ameliorated outside government supervision; meanwhile an underground labor market, associated with both the informal and criminal sectors, emerges to supply employment. The extreme form of an underground economy is the black market, where activities not only are hidden from fiscal authorities, but also involve the illicit trade in goods and services that are contrary to government regulations. Buyers and sellers in the underground economy make conscious efforts to avoid official detection. Therefore, an underground economy predisposes citizens to illegality by training large sections of the population in illegal transactions, "creating human capital specific to illegality and a social morality supportive of activities outside formal legality" (Lotspeich 1995:571). Vice industries and trafficking of illegal merchandises (critical trade components of an underground economy) encroach on economically depressed regions and countries as alternative strategies of recapitalization and growth. Like state failure, economic failure exists in every society in varying degree. For instance, even in countries as prosperous as the United States and the United Kingdom, high unemployment and a thriving illicit drug economy are staples of impoverished inner cities (Foster 2000).

Predatory Organized Crime

State failure fosters predatory organized crime by raising the general levels of violence and fear of violence, which in turn raise the popular demand for security. When the state succumbs to crime and civil strife and ceases to guarantee internal security as a public good to its citizens, personal safety becomes a scarce commodity to be traded. In places where the weakened state has not been further crippled by a parallel economic collapse, security firms and private policing blossom to supply specialized technology and manpower to the market still regulated by the state. But when both the state and the economy fall short of maintaining a minimum coordination, safety and security develop into an elusive good manufactured and manipulated by criminal organizations. Extortion is a core predatory activity of mafia power: it creates jobs for the lower cadres, offers a lucrative source of cash to the organization and, crucially, allows for the exercise of economic and political influence over a given territory. Politically organized crime represents an illegitimate form of authoritarian governance that straightjackets business firms and individual citizens in an extortion-protection trap (Shelley 1999).

An extortion-protection trap is a paradoxical condition in which a criminal organization is capable of providing effective protection simply because it poses a credible threat of illegal violence. Extortion and protection rackets are especially detrimental to legal business activities as both the demand for and the supply of protection services are tightly controlled by the same criminal elements. In a society where a functioning state exists, organized violence is only applicable in the regulation of illegal markets of outlawed goods and services (drugs, prostitution, gambling, etc.) as civil and criminal courts are not sought as mechanisms of mediation and social control. But when institutional anarchy reigns, organized violence permeates legal and illegal economic activities. Although predatory organized crime may deter random violence and reduce government expenditures on law enforcement under certain circumstances, it invariably erodes economic production and destroys properties in the long run. This situation underlies the breakdown of political institutions, their failure to secure public safety, and the failure of the economy to adequately respond to the demand for private protection. Predatory criminal organizations then act as de facto government and as monopolistic firms; they compete with or replace the legitimate government as guarantors of public safety and compete with or drive out legal private firms in the security market.

Even in societies where the extortion-protection trap is not complete, predatory criminal organizations can still be highly demanded as enforcers of legal business contracts or deterrents of competitors in bidding tenders. Market conditions that characterize an industrializing, yet still inefficient, economy—such as low technology, unskilled labor, and unionization—also facilitate the entry of organized crime into the legitimate economy. Advanced criminal organizations that have garnered enough financial capital, professional expertise, and political connections often metamorphose into legal business firms or infiltrate legal markets of commercial goods and services through skillful displays of violence. Violent entrepreneurship provides organized crime-related businesses with a comparative advantage in converting organized violence into wealth through a successful integration of action strategies and commercial decisions.

The success of criminal organizations is also measured by their degree of collusion with the government, which allows them to control entry into and conditions of operation of determined industries. While the application or threat of violence exacts exorbitant costs from the economy, resulting in its inefficiency, business firms can still find it in their economic interests to cooperate with criminal gangs. Given these conditions, predatory criminal organizations are in a position to monopolize niche industries, restrain entry and maximize "revenue by extracting monopoly profits as protection payments; new investment may be discouraged and old investment driven out" (Anderson, 1995:35). As a direct result, organized predatory criminal activities reduce competition in the market, corrupt government officials and the banking system, distort the outcomes of macroeconomic policies, and undermine the efficient performance of the economy. A recent review of Italian and American case studies concluded that past targeted law enforcement initiatives against predatory activities by criminal organizations mostly failed because efficiency and competition in the economy did not improve, and the institutional frameworks of social interactions were left untouched (Gambetta & Reuter, 1997). Coordinated state and economic reforms are always needed.

Hypotheses

Based on the conceptual argument stated in the previous sections, two complementary hypotheses that deal with the institutional sources of predatory organized crime are formulated.

State Failure Hypothesis The level of predatory organized crime is directly and positively correlated with the failure in the delivery of basic political goods by the state.

Three specific aspects of state functioning—institutional stability, judicial independence, and political rights—and their relationship to organized crime were evaluated. The stability of state institutions grants predictability to political processes, an impartial and effective judiciary guarantees individual rights, and the protection of political rights allows citizens to vote and to compete for public offices. The degree to which these political goals are accomplished is hypothesized to be inversely correlated with organized crime activities.

Economic Failure Hypothesis The level of predatory organized crime is directly and positively correlated with the failure of the attainment of general material well-being by the population under the economic system.

The effects of unemployment, underground economy, and economic productivity were assessed. Unemployment is a particular type of inefficiency that exacts important social costs; an underground economy signals allocative inefficiency and weakens public finance; and economic productivity and output determine a country's standard of living. The first two economic maladies are expected to directly vary with changes in the level of organized crime, while productivity and output is anticipated to be inversely connected to organized crime.

Sample, Data, and Method

The present study represents the first statistical cross-national analysis of organized crime ever conducted. A total of 59 countries covered in the 1999 and 2000 annual economic competitiveness evaluations conducted by researchers at Harvard University and the World Economic Forum formed the sample of analysis. Each country was rated on a number of economic and political factors, first in 1999 and then in 2000; thus the resulting sample is a collection of 118 year-country units.

Predatory organized crime is tapped by the 1-to-7 scale survey question that asked whether the respondent perceived that racketeering and extortion had imposed significant costs on businesses. The concept of state failure is operationalized by institutional stability, judicial independence, and political rights; whereas economic failure is measured by unemployment, black-market activities, and gross domestic product per capita. The sizes of the population and the economy were controlled for in the multivariate test.

Results

State Failure Hypothesis

Both political factors maintained firm negative associations with perceptions of organized crime activities. The multivariate state failure equation accounted for more than half (57%) of the variance in the outcome variable (see Table 13.3). Both institutional stability and the independence of the judiciary retained their negative correlations with organized crime, although the effect of institutional stability became weaker when judicial independence was held constant. Indeed the latter contributed most (beta coefficient = .64) to the explanatory power of the model. It strongly suggests that a credible and impartial justice system upholding the rule of law was more important than the predictability of political processes in effectively deterring organized crime for this sample of countries. The inversion of the impact of political rights on organized crime (from negative at the bivariate level to positive at the multivariate level) represented a counterintuitive finding (b coefficient = .196). Societies that permitted greater citizen input in government affairs and policy making suffered higher levels of racketeering and extortion. This finding might have reflected a phenomenon particular to the post-cold war era. During the 1990s, more authoritarian countries often successfully stifled the development of criminal organizations, whereas the sudden political liberalization in other transitional societies created opportunities for the growth of organized crime. The building of electoral democracies in the developing world has not been accompanied by a concomitant rise in the rule of law. In fact, in many emerging democracies the freely elected executive and legislative leaders have little effective influence in improving the rule of law, and some have even colluded with criminal and corrupt interests. The abundance of electoral kleptocracies and illiberal democracies that practice free elections without a true commitment to the rule of law, protection of individual rights, and restriction of state powers demonstrates the vulnerability of populist democracies to infringement by criminal syndicates and corrupt politicians.

Results from the full model with measures of economic failure corroborated findings from the partial model. Judicial independence remained the strongest (negative) correlate of organized

Table 13.3 Results From Least Squares Dummy Variable Models (n = 118)

	State Failure Model			Economic Failure Model			Full Model		
	b	**SE**	**Beta**	**b**	**SE**	**Beta**	**b**	**SE**	**Beta**
State failure measures									
Institutional stability	–.350**	.119	–.226**	—	—	–.167		.121	–.127
Judicial independence	–.661***	.098	–.644***	—	—	–.516***		.105	–.503***
Political rights	.196**	.062	.239**	—	—	.255***		.060	.311***
Economic failure measures									
Unemployment	—	—		–.005	.018	–.021	.011	.015	.051
Black market	—	—		.378**	.123	.376**	.240*	.109	.238*
GDP per capita	—	—		–.000***	.000	–.456***	–.000	.000	–.181
Control variables									
Population size	.000	.000	.056	–.000	.001	–.139	–.000	.000	–.005
Economy size	.000	.000	.025	.000**	.000	.213**	.000	.000	.119
Unit effect									
Year (1999 vs. 2000)	.125	.172	.045	.000	.193	.015	.000	.164	.028
Adjusted R^2		.568****			.452****			.619****	

**$p < .01$; ***$p < .001$; ****$p < .0001$ (two-tailed tests).

crime. Again, when judicial independence and institutional stability were held constant, the expansion of political rights facilitated the growth of predatory organized crime. This result indicates that the bivariate positive association between political rights and organized crime was largely confounded and caused by the high correlation between judicial independence and political rights. Therefore, this unexpected finding reveals that state failure is a multidimensional construct, and, as such, maintains a complex relationship with predatory organized crime. While binding government by the law and making justice available to every citizen are political achievements that directly restrict opportunities for criminal groups, opening up public offices to electoral competition without effective judicial checks and balances risks infiltration by organized crime.

Economic Failure Hypothesis

Higher levels of unemployment and underground economic transactions accompanied more widespread perceptions of racketeering and extortion activities. GDP per capita was in turn inversely linked to organized crime: commercial firms in richer countries were less troubled by criminal syndicates. Black-market activities and GDP per capita significantly correlated with the organized crime measure as expected in both partial and full models, although with varying relative significance. When political factors were held constant in the full model, the effect of underground economy on the outcome variable surpassed that of GDP per capita, to become the driving economic force in determining the perceptions of organized crime. A tested view that democracy nurtures economic prosperity in underdeveloped and developed countries may account for the reduction in the effect of GDP per capita, after controlling for the influence of political freedoms. Black market activities, which generate work opportunities to members of social margins, socialize participants into the subculture of illegality and nurture secretive skills valued in criminal transactions. Because proceeds from both informal and criminal transactions have to be concealed and reinvested into the legitimate financial system, money laundering and government corruption become satellite industries of the underground economy. Most important, as participants are precluded from resorting to the arbitration of formal legal mechanisms to settle disputes and to enforce contracts, credible criminal organizations become the only viable structures to regulate interactions in the underground economy. Violence is particularly likely in unstable markets dominated by freelance entrepreneurs who compete with market players for customers and resources. In sum, underground economies erode the integrity of the institutional fabric of a society and legitimize a syndicate of criminal enterprises as a viable shadow state.

Unemployment was largely irrelevant in the determination of organized crime in this sample of nations; thus the argument that joblessness turns criminal organizations into major employers in social peripheries is unsupported. However, the data did not provide longitudinal information on the influence of sudden or radical changes in unemployment trends on organized crime activities. [U]nlike unemployment and street crime rates that frequently show rapid fluctuations, systemic organized crime is more often a stable feature in the organization of local society. While involvement in street crime can generate income of last resort during extreme economic adversity, participation in criminal organizations generally requires long-term committed involvement in an illegal industry. Individuals recruited by criminal syndicates are not the most disadvantaged members of the population; they are usually quite efficiently employed in racketeering activities in which their relative efficiencies are superior to others.

The larger adjusted R^2 of the state failure model when compared with the economic failure model indicates that the quality of political institutions exercised a stronger influence on the level of organized crime than the performance of the economy. The larger standardized regression coefficients for state failure measures in the full model further confirmed the comparative pre-eminence of political factors over economic forces in the causation of perceived victimization of commercial firms by criminal syndicates. It seems that the strengthening of political institutions in general, and the maintenance of an independent and accountable judicial system in particular, plays a central role in the prevention and control of organized crime activities. As such, organized crime should be dealt with first and foremost as an illicit government to be overthrown, and only secondarily as a competitive firm to be driven out of business.

Conclusion and Discussion

Important lessons have been learned from the World Economic Forum data. State failure and economic failure proved essential elements of an organized crime–infested society. A corrupt judiciary deprives a nation of effective institutional defense against organized crime, and an active underground economy provides criminal syndicates with ample opportunities to expand their influence and legitimacy among ordinary citizens. However, denial of political freedoms and high unemployment were not contributors of organized crime activities as first expected. Democratic participation in political processes per se appears to permit more widespread infiltration by the criminal organizations into a nation's legitimate social life; and higher joblessness in itself does not increase the levels of racketeering and extortion in the two cross-sections of countries examined in this study. It is still plausible that variations in long-term trends of unemployment can have genuine impact on organized crime activities, but only studies with panel or longitudinal design can address this issue.

Unambiguous is the implication that combating organized crime invariably requires the reassertion of judicial sovereignty and integrity, as well as continuous efforts at improving the productive, allocative, and distributional efficiencies of the economy. On the one hand, an independent judiciary made up of well-trained and well-paid judges limits government power, controls state corruption, and protects individual rights from governmental abuses and systemic criminal victimization. Judicial excellence and professionalism multiply the benefits of an open democracy and empower national economy. These qualities also foster the efficient allocation of social resources by reducing the costs and increasing the benefits of using the legal system. On the other hand, the underground economy, which constitutes the power base of criminal organizations, must be carefully dismantled through a revitalization of economic productivity and competitiveness. If a failed or failing state with a bankrupt judiciary and a troubled economy relying on black markets to circulate wealth are the common denominators of societies afflicted with predatory organized crime, antimafia strategies should always include elements of judicial reform and economic liberalization.

Findings from this study warn against the popular notion that there are technical and managerial solutions, under the control of the law enforcement establishment, to organized crime and the improvement of the business environment. Law enforcement agencies are not autonomous social actors, but are constrained by the institutional and economic structures of the community in which businesses and criminal groups exist. In short, making a society more investment-friendly and less vulnerable to predatory organized crime requires reform of more than law enforcement.

References

Anderson, A. (1995). Organized crime, mafia and governments. In G. Fiorentini & S. Peltzman (Eds.), *The economics of organized crime* (pp. 33–60). New York: Cambridge University Press.

Barro, R. J. (1997). *Determinants of economic growth: A cross-country empirical study*. Cambridge, MA: MIT Press.

Foster, J. (2000). Social exclusion, drugs and crime. *Drugs, Education, Prevention and Policy, 7*, 317–330.

Gambetta, D., & Reuter, P. (1997). Conspiracy among the many: The Mafia in legitimate industries. In G. Fiorentini & S. Peltzman (Eds.), *The economics of organized crime* (pp. 116–139). New York: Cambridge University Press.

Lotspeich, R. (1995). Crime in the transition economies. *Europe-Asia Studies, 47,* 555–589.

Shelley, L. I. (1999). Transnational organized crime: The new authoritarianism. In H. R. Friman & P. Andreas (Eds.), *The illicit global economy and state power* (pp. 25–51). New York: Rowman & Littlefield.

DISCUSSION QUESTIONS

1. Why is organized crime a threat to democracy?
2. If the existence of organized crime is positively related to the extent of civil liberties afforded citizens, does this argue for fewer civil liberties?
3. Why is judicial independence more important than a robust economy in controlling organized crime?

VICTIMOLOGY

Exploring the Experience of Victimization

John Sutcliff's entire adult life has been devoted to the sexual seduction of teenage boys. At the age of 33 he was arrested and sentenced to prison for sexually assaulting a 13-year-old boy who was a member of his "Big Brothers Club." By his own admission, he had been sexually active with over 200 "members" of his club. John's favorite activity with these boys was giving and receiving enemas, a paraphilia known as klismaphilia. John became involved with the fetish while enrolled in a residential boys' school where many of the boys were subjected to enemas administered in front of the entire dormitory.

After his release from prison, John became much more "scientific" in his efforts to procure victims. A "theoretical" paper he wrote indicated that father-absent boys were "ripe" for seduction, and he would entice them with his friendly ways and with a houseful of electronic equipment he would teach the boys to repair and operate. He weeded out boys with a father in the home and would spend at least six weeks grooming each victim. He used systematic desensitization techniques, starting with simply getting the boys to agree to type answers to innocuous questions, and escalating to having them view pornographic homosexual pictures, giving them "pretend" enemas, actual enemas, and enemas accompanied by homosexual activity. With each successive approximation toward John's goal, the boys were reinforced by material and nonmaterial rewards (friendship, attention, praise) that made the final events seem almost natural.

John's activities came to light when U.S. postal inspectors found a package containing pictures, letters, and tapes John exchanged with like-minded individuals. On the basis of this evidence, the police raided John's home and found neatly catalogued files detailing 475 boys that he had seduced. His methods were so successful that his actions were never reported to the authorities (indeed, some of the boys were recruited for him by earlier victims). Some of his earlier victims still kept in touch with him and were victimizing boys themselves. Only one victim agreed to testify, but John was allowed to plead to one count of lewd and lascivious conduct and received a sentence of one year and was paroled after serving 10 months, thus serving 15.7 hours for each of his 475 known victims. This case illustrates how victims (totally innocent as children) can be turned into victimizers (totally responsible as adults) and how the distinction between victim and perpetrator can sometimes be blurred.

The Emergence of Victimology

Except for minor public order crimes, for every criminal act there is necessarily at least one victim. Criminologists have spent decades trying to determine the factors that contribute to making a person a criminal, but it wasn't until the German criminologist Hans von Hentig's (1941) work that they began seriously thinking about the role of the victim. It turned out that although victimization can be an unfortunate random event in which the victim was simply in the wrong place at the wrong time, in many, perhaps even in most, cases of victimization, there is a systematic pattern if one looks closely enough.

Victimology is a subfield of criminology that specializes in studying the victims of crime. Criminologists interested in perpetrators of crime ask what are the risk factors for becoming involved in crime; criminologists interested in victims of crime ask pretty much the same questions: Why are some individuals, households, groups, and other entities targeted and others are not (Doerner & Lab, 2002)? The labels "offender" and "victim" are sometimes blurred distinctions that hide the details of the interactions of the offender/victim dyad. Burglars often prey on their own kind, robbers prey on drug dealers, and homicides are frequently the outcome of minor arguments in which the victim was the instigator. As victimologist Andrew Karmen (2004) put it: "Predators prey on each other as well as upon innocent members of the public. . . . When youth gangs feud with each other by carrying out 'drive-by' shootings, the young members who get gunned down are casualties of their own brand of retaliatory street justice" (p. 14). Of course, there are millions of innocent victims who in no way contribute to their victimization, and even lawbreakers can be genuine victims deserving of protection and redress in the criminal courts.

Who Gets Victimized?

Victimization is not a random process. Becoming a victim is a process encompassing a host of systematic environmental, demographic, and personal characteristics. According to the 2005 NCVS study (Catalano, 2006), the individual most likely to be victimized is a young black unmarried male living in poverty in an urban environment. Victimization, like criminal behavior, drops precipitously from 25 years of age onwards; it also drops with increasing household income, and being married is as much a protective factor against becoming a victim as it is against becoming an offender.

Victim characteristics also differ according to the type of crime. Females were 33 times more likely than males to be victimized by rape/sexual assault, but males were 2.1 times more likely to be victimized by aggravated assault. Females are more likely to be victimized by someone they know and males by strangers. Blacks were 3.7 times more likely than "other races" (Asian, American Indian/Alaskan Native) to be victims of aggravated assault, but slightly less likely than whites to be victims of simple assault. Individuals 65 or older were 20 times less likely than individuals 20–24 to be victimized by any type of violent crime, but slightly more likely to be victimized by a personal theft.

Jennifer Shaffer and R. Barry Ruback's article supports the above contention by Karman that the risk factors for offending and being victimized are much the same thing. In other words, we see that violent offenders, all other things being statistically controlled, are about twice as likely as nonviolent offenders to be victimized themselves. Furthermore, past victimization is the best predictor of future victimization (odds ratio = 5.7, which means that if you were victimized in year 1, the odds of you being victimized in year 2 are 5.7 time greater than they are if you were not victimized in year 1). The article also shows that the predictors of violent offending are the same as the predictors for violent victimization.

Victimization in the Workplace and School

Two important demographic variables not included in the 2006 NCVS study are victimization at work and at school. It is important to consider these variables because most of us spend the majority of our waking hours either on the job or at school. The last systematic effort by the Department of Justice to assess the level of workplace violence in the United States was in 1998. This report dealt with workplace violence taking place between 1992 and 1996 and found that on average, over two million incidents take place annually (Warchol, 1998). Table 14.1 provides the average annual level of workplace violence over the five-year period. Two-thirds of the victims in the survey were male, almost 90% were white, and the age category most likely to be victimized was the 35–49 category. The three occupations most at risk were police officers (a rate of 306 per 1,000 workers), corrections officers (217.8), and taxi drivers (183.8); taxi drivers were the most likely to be killed, however. Taxi drivers had an astounding homicide victimization rate of 26.9 per 100,000 workers; all protective service workers had a rate

Table 14.1 Average Annual Number of Violent Victimizations in the Workplace 1992–1996

Offense	Annual Average	Percentage
Homicide	1,023	0.05
Rape/Sexual Assault	50,500	2.50
Robbery	83,700	4.20
Aggravated Assault	395,500	19.70
Simple Assault	1,480,000	73.60
Total	2,010,723	100.00

Source: Warchol (1998).

Figure 14.1 Percentage of Students Ages 12–18 Who Reported Being Bullied During Previous Six Months, by Grade: 1999 and 2001

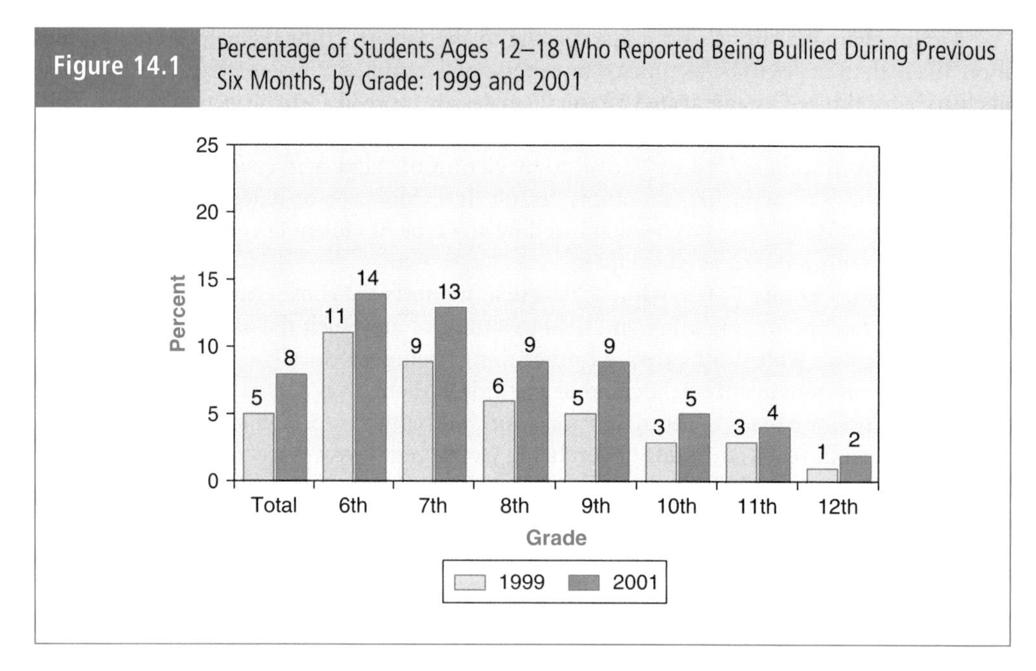

SOURCE: DeVoe et al. (2003).

more than five times lower than taxi drivers (5.0 per 100,000). The most dangerous jobs are those in which the workers must deal with the public in a protective/supervisory capacity, or work alone and relatively isolated from others, work at night, and work with money. The safest job category was university professor (a rate of 2.5).

Public perceptions of victimization at the nation's schools are unfortunately fueled by isolated but spectacular events such as the Columbine school massacre and other similar incidents in the 1990s. The truth is that our schools are some of the safest places we can be. DeVoe et al.'s (2003) study of school crime and safety found that less than 1% of all juvenile homicides and suicides occurred at school during the period studied. Bullying, which also gets a lot of press, seems surprisingly rare as well. Figure 14.1 shows the percentage of students from 6th to 12th grade who reported being bullied in 1999 and 2001. The graph looks a lot like the graphs for delinquency, in that bullying is most frequent at age 12 and levels off thereafter. The somewhat dramatic increase in each grade from 1999 to 2001 probably reflects a growing awareness and willingness to report bullying rather than a real increase in bullying behavior.

Child Molestation: Who Gets Victimized?

Child molestation is perhaps the most prevalent crime against the person in the United States, with approximately two-thirds of incarcerated sex offenders having offended against children (Talbot, Gilligan, Carter, & Matson, 2002). It is more problematic to accurately gauge the prevalence of child molesting, with rates depending on how broad or how narrowly molesting is defined. A "best guess" arrived at from a variety of sources is that the percentage of children in the United States experiencing sexual abuse sometime during their childhood is 25% for girls and 10% for boys (Knudson, 1991). Girls are more likely to be abused within the family, and boys are more likely to be victimized by acquaintances outside of the family and by strangers (Walsh, 1994). The strongest single predictor of victimization for girls is having a stepfather. Stepfathers are about five times more likely to sexually abuse their daughters than are biological fathers (Glaser & Frosh, 1993). The strongest predictor for boys is growing up in a father-absent home (Walsh, 1988). There are many other factors predictive of child sexual abuse, and the more that are present, the more likely abuse is to occur.

Finkelhor (1984) developed a risk factor checklist for the likelihood of girls' victimization containing the following predictors:

1. Living with a stepfather.
2. Living without biological mother.
3. Not close to mother.
4. Mother never finished high school.
5. Sex-punitive mother.
6. No physical affection from (biological) father.
7. Family income under $10,000 (in 1980 dollars; $26,000 in 2006 dollars).
8. Two friends or fewer in childhood.

Finkelhor (1984) found that the probability of victimization was virtually zero among girls with none of the predictors in their background and rose steadily to 66% among girls with five. Given the large number of divorces, out-of-wedlock births, and reconstituted families we are seeing in the United States, these risk factors for sexual abuse will be experienced by an increasing number of children.

Victimization Theories

Victimization can occur at any time or place without warning. Who could have predicted someone gassing her car at the filling station would be gunned down by the Washington, D.C. snipers in 2002, or the typist at his desk in the World Trade Center would be obliterated seconds later by a passenger jet on 9/11/2001? There is no systemic way to evaluate events such as these from a victimology perspective. But as previously noted, most victimizing events are not random or unpredictable. Criminologists no longer view victims as simply passive players in crime who were unfortunate enough to be in the wrong place at the wrong time. In the majority of cases, victims are now seen as individuals who in some way, knowingly or unknowingly, passively or actively, influenced their victimization. Obviously, the role of the victim, however provocative it may be, is never a necessary and sufficient cause of his or her victimization and therefore cannot fully explain the actions of the person committing the criminal act.

Victim Precipitation Theory

Victim precipitation theory was first promulgated by von Hentig (1941) and applies only to violent victimization. Its basic premise is that by acting in certain provocative ways, some individuals initiate a chain of events that lead to their victimization. Most murders of spouses and boyfriends by women, for example, are victim precipitated in that the "perpetrator" is defending herself from the victim (Mann, 1990). Likewise, serious delinquent and criminal behavior and serious victimization are inextricably linked. A study using data from the longitudinal Pittsburgh and Denver studies of delinquency risk factors (e.g., low SES, single-parent household, hyperactivity, impulsiveness, drug usage) showed that the same factors predicted victimization as well (Loeber, Kalb, & Huizinga, 2001). Overall, 50% of seriously violent delinquents were themselves violently victimized compared with 10% of nondelinquents from the same neighborhoods.

Victim precipitation theory has been most contentious when it is applied to rape, ever since Menachem Amir's (1971) study of police records found that 19% of forcible rapes were victim precipitated (defined by Amir as agreeing to sexual relations and then reneging). A number of surveys of high school and college students have shown that a majority of males and a significant minority of females believe that it is justifiable for a man to use some degree of force to obtain sex if the victim had somehow "led him on" (Herman, 1990). This attitude appears to indicate that some people believe that there could be an act labeled "justifiable rape," in the same sense that there is justifiable homicide. These same surveys also indicate that many people continue to believe that rape victims are often at least partially responsible for their rape because of such factors as dress and lifestyle and because of the belief that "nice girls don't get raped" (Bartol, 2002, p. 295). It is for this reason that many criminologists have disparaged victim precipitation theory as victim-blaming, although it was never meant to be that. Hopefully, the attitudes revealed in these 1980s surveys have diminished with the greater awareness of the horrible nature of this crime in evidence today.

Figure 14.2 provides four scenarios illustrating various levels of victim/offender responsibility from this perspective. In the first scenario, the woman who stabbed her husband after suffering years of abuse is judged blameless, although some lacking a little in empathy and understanding of the psychology of domestic abuse may argue that she must take some responsibility for remaining in the relationship. In the second scenario, both the offender and the victim were engaging in a minor vice crime and both are judged equally responsible for the crime (morally he should not have been there and was careless with his wallet). In the third scenario, the victim facilitated the crime by carelessly leaving his keys in the car. In the last scenario, the child is totally innocent of any responsibility for what happened to her. We want to strongly emphasize that whatever the degree of responsibility, "responsibility" does not mean "guilt."

Routine Activities/Lifestyle Theory

Routine activities and lifestyle theories are separate entities, but in victimology they are similar enough to warrant being merged into one (Doerner & Lab, 2002, p. 273). Routine activities theory stresses that criminal behavior takes place via the interaction of three variables that reflect individuals' everyday routine activities: (1) the presence of motivated offenders, (2) the availability of suitable targets, and (3) the absence of capable guardians. The basic idea of lifestyle theory is that there are certain lifestyles (routine activities) that disproportionately

Figure 14.2 Four Scenarios Illustrating the Degree of Victim/Offender Responsibility According to Victim Precipitation Theory

Degree of Criminal Intent of the Perpetrator			
None →	*Some →*	*More →*	*Much*
Victim Provocation A woman who has suffered years of abuse stabs and kills her husband in self-defense as he is beating her again.	Equal Responsibility Victim using the services of a prostitute leaves his wallet on the night stand and leaves. She decides to keep the money in his wallet.	Victim Facilitation Victim leaves keys in his car while he runs into a store. A teenager impulsively steals the car and wrecks it.	Victim Innocent A sex offender kidnaps a screaming young girl from a playground and molests her.
Much	*← More*	*← Some*	*← None*
Degree of Victim Facilitation or Provocation/Precipitation			

expose some people to high risk for victimization. Lifestyles are the routine patterned activities that people engage in on a daily basis, both obligatory (e.g., work-related) and optional (e.g., recreational). A high-risk lifestyle may be getting involved with deviant peer groups or drugs, just "hanging out," or frequenting bars until late into the night and drinking heavily. **Routine activities/lifestyle theory** explains some of the data relating to demographic profiles and risk presented by Loeber, Kalb, and Huizinga (2001) discussed earlier. Males, the young, the unmarried, and the poor are more at risk for victimization than females, older people, married people, and more affluent people because they have riskier lifestyles. On average, the lifestyles of the former are more active and action-oriented than the latter.

These lifestyles sometimes lead to repeat victimization. Prior victimization has been called "arguably the best readily available predictor of future victimization" and it "appears a robust finding across crime types and data sources" (Tseloni & Pease, 2003, p. 196). Lisa Bostaph (2004) reviews the literature on what she calls "career victims" and among the various interesting research findings on this phenomenon, she lists the following attributable to lifestyle patterns:

- A British crime survey that found that 20.2% of the respondents were victims of 81.2% of all offenses.
- A study that found 24% of rape victims had been raped before.
- A study of assault victims in the Netherlands that found 11.3% of victims accounted for 25.3% of hospital admissions for assault over 25 years.
- A study reporting that 67% of sexual assault victims had experienced prior sexual assaults.

Most of the research in routine activities/lifestyle theory has been done on rape victimization. Bonnie Fisher and her colleagues' (2001, p. 15) national sample of college women

◀ **Photo 14.1** Gary Ridgway became known as the Green River Killer for his habit of depositing victims' bodies along this waterway. Serial killers frequently victimize marginalized groups, such as prostitutes. Some of his victims' bodies were only discovered years after their untimely deaths, by searchers such as these, revisiting kill sites.

found that 2.8% had been raped, although 46.5% of these women said that they did not experience the event as rape. Fisher and colleagues (p. 23) report that four lifestyle factors are consistently found to increase the risk of sexual assault: (1) frequently drinking enough to get drunk, (2) being unmarried, (3) having previously been a victim of sexual assault, and (4) living on campus (for on-campus victimization only).

Is Victimology "Blaming the Victim?"

Some victim advocates reject victimology theories as "victim blaming." Victimologists do not "blame" victims; they simply explore the process of victimization with the goal of understanding it and *preventing* it. Although victimology research is used to develop crime prevention strategies, not to berate victims, some victim advocates even reject "as ideologically tainted" crime-prevention tips endorsed by victimologists (Karmen, 2004, p. 129). Crime-prevention tips and strategies are ignored at our peril. We all agree that we *should* be able to leave our cars unlocked, sleep with the windows open in summer, leave our doors unlocked, frequent any bar we choose, walk down any alley in any neighborhood at any time we please, but we cannot. Common sense demands that we take what steps we can to safeguard ourselves and our property in this imperfect world. Crime prevention tips are really no different from tips we get all the time about staying healthy: eat right, exercise, and quit smoking if you want to avoid health problems. Similarly, avoid certain places, dress sensibly, don't provoke, take reasonable precautionary measures, and don't drink too much if you want to avoid victimization.

Victims deserve our sympathy even if they somehow provoked or facilitated their own victimization. Victimologists do not "blame"; they simply remind us that complete innocence and full responsibility lie on a continuum.

The Consequences of Victimization

Some crime victims suffer lifelong pain from wounds and some suffer permanent disability, but for the majority of victims, the worst consequences are psychological. We all like to think that we live in a safe, predictable, and lawful world in which people treat one another decently. When we are victimized, this comfortable "just world" view is shattered. With victimization comes stressful feelings of shock, personal vulnerability, anger, and fear of further victimization, and suspicion of others.

Victimization also produces feelings of depression, guilt, self-blame, and lowered self-esteem and self-efficacy. Rape in particular has these consequences for its victims ("Did I contribute to it?" "Could I have done more to prevent it?"). The shock, anger, and depression that typically afflicts a rape victim is known as **rape trauma syndrome**, which is similar to post-traumatic stress syndrome (re-experiencing the event via "flashbacks," avoiding anything at all associated with the event, and a general numbness of affect) often suffered by those who have experienced the horrors of war (van Berlo & Ensink, 2000). Victimization "also changes one's perceptions of and beliefs about others in society. It does so by indicating others as sources of threat and harm rather than sources of support" (Macmillan, 2001, p. 12).

Victims of property crimes, particularly burglary, also have the foundations of their world shaken. The home is supposed to be a personal sanctuary of safety and security, and when it is "touched" by an intruder, some victims describe is as the "rape" of their home (Bartol, 2002,

▲ **Photo 14.2** Efforts to better recognize victims and their rights have become more common over the past 20 years. This photo memorializes the victims of Columbine High School, including the killers themselves, who committed suicide.

p. 336). A British study of burglary victims found that 65% reacted with anger, 30% with fear of re-victimization, and 29% suffered insomnia as a consequence. The type and severity of these reactions were structured by victims' place in the social structure, with those most likely to be affected being women, older and poorer individuals, and residents of single-parent households (Mawby, 2001).

The first article by Hans Joachim Schneider in this section provides an excellent overview of the international data on victimization. He shows that people are at risk for different kinds of crimes in different areas of the world. People in rich modern societies are more likely to be victims of property crimes than violent crimes, whereas in poorer countries it is the reverse. Schneider also provides additional information on some of the leading theories of victimology.

In Schneider's second article, he concentrates on the physical, psychological, and social effects of victimization. He also looks at the indirect effect of victimization on the victim's loved ones (co-victims) and on "recidivist" victimization. Schneider then examines treatment and support programs for victims, as well as victims' rights in the domains of law and criminal proceedings (the right to be informed and consulted regarding sentencing and parole and the right to restitution).

Victimization and the Criminal Justice System

Until fairly recently, the victim was the forgotten party in the criminal justice system. In the United States, crime is considered an act against the state rather than against the individual who was actually victimized. In 2004, the Senate passed a crime victims' bill of rights (see below) that has gone some considerable way to recognizing the previously discounted victim.

Crime Victims' Bill of Rights

1. The right to be reasonably protected from the accused
2. The right to reasonable, accurate, and timely notice of any public proceeding involving the crime or of any release or escape of the accused
3. The right not to be excluded from any such public proceeding
4. The right to be reasonably heard at any public proceeding involving release, plea, or sentencing
5. The right to confer with the attorney for the Government in the case
6. The right to full and timely restitution as provided in law
7. The right to proceedings free from unreasonable delay
8. The right to be treated with fairness and with respect for the victim's dignity and privacy.

SOURCE: Senate Bill S2329, April 21, 2004.

Although the above rights apply only to victims of federal crimes, all 50 states have implemented constitutional amendments or promulgated bills guaranteeing similar rights.

Crime victims are eligible for partial compensation from the state to cover medical and living expenses incurred as a result of their victimization. All 50 states and all United States protectorates have established programs that typically cover what private insurance does not, assuming the state has sufficient funds. According to the National Association of Crime Victim Compensation Board (NACVCB), in 2004, victims of violent crime nationwide received a total of $426 million in compensation, with the majority (51%) going for medical expenses (NACVB, 2005).

Summary

◆ Victimology is the study of the risk factors for and consequences of victimization and criminal justice approaches dealing with victims and victimization. The risk factors for victimization are basically the same as the risk factors for victimizing in terms of gender, race, age, SES, personal characteristics, and neighborhood.

◆ Theories of victimization such as victim precipitation theory and routine activities/lifestyle theory examine the victim's role in facilitating or precipitating his or her victimization. This is not "victim blaming," but rather an effort to understand and prevent victimization. Victimologists apportion responsibility within the victim/offender dyad on a continuum from complete victim innocence to victim precipitation.

◆ The consequences of victimization can be devastating both physically and psychologically. Although the severity of the psychological consequences of the same sort of victimization can vary widely according to the characteristics of the victim, consequences can range from short-lived anger to post-traumatic stress syndrome, especially for victims of rape.

◆ Until fairly recently, victims were the forgotten party in a criminal justice system that tended to think of them only as "evidence" or witnesses. Things have changed over the last 25 years with the passage of victims' rights bills by the federal government and all 50 states. There are also various victim-centered programs designed to ease the pains of victimization, such as victim compensation.

EXERCISES AND DISCUSSION QUESTIONS

1. Interview a willing classmate or friend who has been victimized by a serious crime and ask about his or her feeling shortly after victimization and now. Did it change his or her attitudes about crime and punishment?

2. Is it a surprise to you that perpetrators of crimes are more likely to also be victims of crime than people in general? Why or why not?

3. Go to your state's official Web site and find out funding levels and what services are available to crime victims.

USEFUL WEB SITES

American Society of Victimology. http://www.american-society-victimology.us/.

International Victimology Website. http://www.victimology.nl/.

National Crime Victimization Survey Resource Guide. http://www.icpsr.umich.edu/NACJD/NCVS/.

National Incident-Based Reporting System Resource Guide. http://www.icpsr.umich.edu/NACJD/NIBRS/.

The World Society of Victimology. http://www.worldsocietyofvictimology.org/.

GLOSSARY

Rape trauma syndrome: A syndrome sometimes suffered by rape victims that is similar to post-traumatic stress syndrome (re-experiencing the event via "flashbacks," avoiding anything at all associated with the event, and a general numbness of affect).

Routine activities/lifestyle theory: A victimization theory that states that there are certain lifestyles (routine activities) that disproportionately expose some people to high risk for victimization.

Victim precipitation theory: A theory in victimology that examines how violent victimization may have been precipitated by the victim by his or her acting in certain provocative ways.

Victimology: A subfield of criminology that specializes in studying the victims of crime.

READING

The Criminal and His Victim

Hans von Hentig

In this excerpt from his 1948 book, Hans von Hentig introduces the study of victimology as a subfield of criminology and lays out the basis of his victim precipitation theory. von Hentig insists that in many cases of criminal victimization, we cannot understand the behavior of the criminal without understanding the concomitant behavior of the victim. Sometimes, he asserts, the victim sets in motion a series of events in which he or she had intended to be the perpetrator of a criminal act rather than the victim. Note that he is not trying to blame the victim, only to understand the victim's role.

The Duet Frame of Crime

Crime, for the most part, is injury inflicted on another person. Setting aside felonies directed against fictitious victims, the state, order, health, and so forth, there are always two partners: the perpetrator and the victim.

This doer-sufferer relation is put by our codes in mechanical terms. A purse is snatched, bodily harm is done. The sexual self-determination of a woman is violated. Mental factors are, of course, taken into account. So is felonious intent or malice aforethought. The "consent" of an adult woman changes the otherwise criminal act of rape into a lawful occurrence, or at least a happening in which the law is not very much interested. No one can complain of injury to which he has submitted willingly. In many other instances, consent changes the legal aspect while the factual situation remains unaltered. By his or her decision the victim can, in spite of loss and pain endured, turn factual crime into a situation devoid of legal significance.[1] Non-complaint after the event practically stands on a par with consent.

Yet experience tells us that this is not all, that the relationships between perpetrator and victim are much more intricate than the rough distinctions of criminal law.[2] Here are two human beings. As soon as they draw near to one another male or female, young or old, rich or poor, ugly or attractive—a wide range of interactions, repulsions as well as attractions, is set in motion. What the law does is to watch the one who acts and the one who is acted upon. By this external criterion, a subject and object, a perpetrator and a victim are distinguished. In sociological and psychological quality the situation may be completely different. It may happen that the two distinct categories merge. There are cases in which they are reversed, and in the long chain of causative forces, the victim assumes the role of a determinant.[3]

We are wont to say and to think that the criminal act is symptomatic for the lawbreaker, as a suicide would he, or a red rash on the skin. We have gone to great lengths, in studying our society, to classify and reclassify groups. Among common situations usually enumerated, however, we do not find the evildoer-evil-sufferer group. It is not always true that common interests give rise to a group; the problem presents many more depths. I maintain that many criminal deeds are more indicative of a subject- object relation than of the perpetrator alone. There is a definite mutuality of some sort. The mechanical outcome may be profit to one party, harm to another, yet the psychological interaction, carefully observed, will not submit to this kindergarten label. In the long process leading gradually to the unlawful result, credit and debit are not infrequently indistinguishable.

Victim Precipitated Criminal Homicide

In a sense the victim shapes and moulds the criminal.[4] The poor and ignorant immigrant has bred a peculiar kind of fraud. Depressions and wars are responsible for new forms of crimes because new types of potential victims are brought into being.[5] It would not be correct or complete to speak of a carnivorous animal, its habits and characteristics, without looking at the prey on which it lives. In a certain sense the animals which devour and those that are devoured complement each other. Although it looks one-sided as far as the final outcome goes, it is not a totally unilateral form of relationship. They work upon each other profoundly and continually, even before the moment of disaster. To know one, we must be acquainted with the complementary partner.

Notes

1. Although morally dubious and under censure of other social controls.

2. It is therefore rather naive to maintain, "All sight has been lost of the fact that for every criminal there must be an innocent sufferer . . . " C. R. Cooper, *Here's to Crime* (Boston, Little, Brown & Co., 1937), p. 434. Read in the "Notable British Trials" series the case of the lawyer

H. R. Armstrong by Filson Young, *Trial of Herbert Rowse Armstrong* (Edinburgh, William Hodge & Co., 1927).

3. The title of a well-known novel by Werfel, *Der Ermordete ist schuld,* expresses the moral transposition. It is already met in the legal notion of self-defense and the concept of grave provocation. The idea of the responsible victim is belatedly conveyed by decisions of the pardoning power; it cannot easily be rendered vocal by laws and court sentences.

4. In others it determines the question of guilt or the degree of penalty. In explaining a sentence of first degree murder and the death penalty in a case of manslaughter, Warden Lawes writes, "the deceased had been well liked and there was a certain amount of prejudice against Chapeleau's foreign origin"—the perpetrator being a Canadian. *Meet the Murderer,* p. 4.

5. A wave of war bond rackets swept the United States in 1944. Utilizing and shrewdly increasing the fear of many bondholders that the bonds might not be redeemed at full value after the war, and playing upon their patriotic reluctance to sell, these frauds proposed to lonely housewives to take orders for merchandise or other transactions, making payment in the bonds. See chapter on war and crime in Hentig, *Crime: Causes and Conditions.*

DISCUSSION QUESTIONS

1. What does von Hentig mean when he writes, "In a certain sense the animals which devour and those that are devoured complement each other"?
2. Is there any sense that von Hentig's tone can be taken as victim blaming?
3. Give an example (either known to you personally or from the news) of victim-precipitated homicide.

READING

Violent Victimization as a Risk Factor for Violent Offending Among Juveniles

Jennifer N. Shaffer and R. Barry Ruback

In this study, Shaffer and Ruback examine data from a large longitudinal study of adolescent health to determine risk factors for violent victimization. They found that violent offenders and victims of violent offenders are basically birds of the same feather. That is, both victims and offenders share the same risk factors, such as neighborhood and drug usage. The second best predictor of violent offending for juveniles (the first being previous violent offending) was previous violent victimization, and the best predictor for violent victimization was previous violent victimization.

Source: Shaffer, J., & Ruback, R. (2002, December). Violent victimization as a risk factor for violent offending among juveniles. *Juvenile Justice Bulletin* (NCJ 195737). Retrieved January 11, 2008, from the U.S. Department of Justice Web site at http://www.ncjrs.gov/html/ojjdp/jjbul2002_12_1/contents.html

Introduction

As a group, juveniles have high rates of violent victimization and violent offending, a pattern suggesting that some juveniles are both victims and perpetrators of violence. To explore that hypothesis, we analyze the relationships between violent victimization and violent offending across a 2-year period, using data for 5,003 juveniles who participated in the National Longitudinal Study of Adolescent Health. We looks at victimization and offending experiences in subgroups of juveniles classified by age, gender, race, and level of physical development [and] identify risk and protective factors for victimization and offending. Key conclusions and policy implications include the following:

- Violent victimization is indeed a warning signal for future violent offending among juveniles. Protecting juveniles against violent victimization may, therefore, reduce overall levels of juvenile violence.
- Because some groups are at higher risk than others for violent victimization, policies and programs aimed at preventing victimization may be most effective if they are focused on these groups.
- Violent victimization and violent offending share many of the same risk factors, and many of these risk factors suggest opportunities for intervention.

Background

Statistical evidence suggests disproportionately high rates of violence by and against juveniles. This evidence comes both from surveys that ask about behaviors and victimization experiences and from official records. Surveys of self-reported behaviors of adolescents and young adults indicate high rates of offending among these age groups (Lauritsen, Sampson, & Laub, 1991). Similarly, surveys of victims' perceptions of offender characteristics indicate that the most common age group for offenders committing rape, robbery, and assault is youth ages 18–20, followed by juveniles ages 15–17 (Hindelang, 1981). Furthermore, Uniform Crime Report data show that arrest rates for murder, forcible rape, robbery, and aggravated assault are higher for older teens than for any other age group.

Theoretical Perspective

According to both lifestyle theory and routine activities theory, individuals' risk of criminal victimization depends on their exposure or proximity to offender populations, and exposure, in turn, depends on individuals' lifestyles and routine activities. Because individuals are most likely to interact with those who are similar to themselves, individuals' victimization risk is directly proportional to the number of characteristics they share with offenders. That is, offenders are more likely than nonoffenders to become victims, because their lifestyles frequently bring them in contact with other offenders. Offenders are also more likely than nonoffenders to use alcohol or illegal drugs, which lowers their ability to protect themselves and their property, and to live in neighborhoods characterized by high levels of population mobility, heterogeneity, and social disadvantage (e.g., poverty and unemployment), which increases their exposure to other offenders (Sampson & Lauritsen, 1994).

Offenders are also likely to be attractive targets for crime because they can be victimized with little chance of legal consequences. Offenders are probably less likely than nonoffenders to report victimization to the police because they do not want to draw attention to their own illegal behavior (e.g., starting the altercation in question or carrying illegal drugs) and because, if they do file a report, the police probably perceive them as less credible than nonoffenders. Offenders' reluctance to report their own victimization might be especially true for violent juvenile offenders, because juveniles in general are less likely than adults to report violent victimization to the police.

Research findings are consistent with these theoretical reasons for expecting that the same individuals are often both victims and offenders. Studies using British Crime Survey data have found a strong positive association between offending and personal victimization among adults (Sampson & Lauritsen, 1990). Studies of juveniles in the United States also show that the individuals most likely to be victims of personal crime are those who report the greatest involvement in delinquent activities. In addition, the greater the variety of delinquent activities, the greater the risk of victimization (e.g., Lauritsen, Sampson, & Laub, 1991).

Data and Methods

Although earlier studies suggest that criminal victimization and criminal offending are related, the nature of the relationship is ambiguous. The present study investigates the nature of the relationship in a sample of juveniles ages 11–17, addressing three issues:

- How are violent victimization and violent offending related over time? Does prior victimization predict subsequent offending, does prior offending predict subsequent victimization, or do they both predict each other? In particular, is victimization a significant risk factor for subsequent offending?
- What individual-level factors might explain the relationship between victimization and offending? Do the same factors predict both violent victimization and violent offending?
- Does drug use affect the relationship between victimization and offending?

The study focuses on violence among juveniles for three reasons. First, from a policy standpoint, it makes sense to concentrate on the most serious offenses, particularly since less is known about the violent victimization of juveniles than about the violent victimization of adults. Second, because many fewer juveniles engage in violence than in property offending and in minor deviant acts, it would be easier to target interventions at this smaller group. Third, the data source for the analyses include measures of nonviolent offending but not of nonviolent victimization.

Data Source

The dataset [consists of] the first two waves of the National Longitudinal Study of Adolescent Health (known as the Add Health Study), which is a longitudinal study of a representative national sample of juveniles in grades 7 through 12. The analyses reflect interview data for 5,003 juveniles: 2,402 males and 2,601 females; 2,768 non-Hispanic white juveniles and 2,235 minority juveniles; 1,147 juveniles ages 11–14 and 3,856 ages 15–17 at the time of the second interview.

Measures

The measures of offending and victimization used in the analyses were dichotomous measures based on juveniles' yes/no responses to multiple items. The measures of violent offending included five items reflecting serious physical offenses against other persons:

- Got into a serious physical fight.
- Hurt someone badly enough to need bandages or care from a doctor or nurse.
- Used or threatened to use a weapon to get something from someone.
- Shot or stabbed someone.
- Pulled a knife or gun on someone.

The measures of violent victimization included four items reflecting serious physical violence:

- Someone pulled a knife or gun on you.
- You were shot.
- You were cut or stabbed.
- You were jumped.

Juveniles were categorized as offenders if they reported committing any of the listed offenses and as nonoffenders if they reported not committing any of these offenses. Juveniles were similarly categorized as victims or nonvictims, based on whether they reported having any of the listed acts committed against them. Juveniles were also categorized on the basis of their reports of using any one of the following drugs: Marijuana, Cocaine, Inhalants, Other drugs, including LSD,

PCP, ecstasy, ice (crystal methamphetamine), heroin, mushrooms, speed (amphetamines), or pills without a doctor's prescription.

Juveniles were categorized into one of four groups: nonusers (no reported use of any drug at either of the two interviews), desisters (reported use at first interview but not at the second interview), new users (reported use at the second interview but not the first), and consistent users (reported use at both interviews). Juveniles were similarly categorized on the basis of reported alcohol use.

Findings

Forty percent of juveniles reported violent offending in year 1, 23% in year 2, and 17% in both years. Nineteen percent reported violent victimization in year 1, 15% in year 2, and 9% in both years. Fifteen percent of juveniles reported both committing and being the victim of a violent crime in year 1, 10% in year 2, and 6% in both years. Generally, the percentages of juveniles reporting offending and victimization were greater in year 1 than in year 2. This decline is likely related to "telescoping," or the tendency of survey respondents to report events that occurred outside the time period about which they were asked (in this study, prior to year 1).

There was a strong link between violent offending and violent victimization within each year. Within year 1, juveniles who offended were 5.3 times more likely than nonoffenders to be victimized (37% versus 7%), and those who were victimized were 2.4 times more likely than nonvictims to offend (78% versus 32%). Within year 2, juveniles who offended were 6 times more likely than nonoffenders to be victimized (42% versus 7%), and those who were victimized were 4 times more likely than nonvictims to offend (66% versus 16%). It should be kept in mind that, although the Add Health Study data indicate a close temporal proximity of offending and victimization, temporal ordering of events within a year—i.e., whether a particular youth first was victimized and then offended, or vice versa—cannot be determined from the data. Because of this limitation, the rest of the analyses will focus on the relationships between violent victimization and offending across years.

In the total sample of 5,003 juveniles, those who committed a violent offense or were victims of violence in year 1 had a significantly increased likelihood of offending or being victimized in year 2. Juveniles who offended in year 1 were 4.4 times more likely than nonoffenders to offend in year 2 (44% 10%) and 4.7 times more likely to be victimized in year 2 (28% versus 6%). Juveniles who were victimized in year 1 were 3 times more likely than nonvictims to offend in year 2 (52% versus 17%) and 6 times more likely to be victimized in year 2 (47% versus 8%).

By Age Juveniles were divided into two groups according to their age in year 2: 11–14 and 15–17. The total percentage of juveniles who were victims of violence in year 2 was significantly greater for the older group; however, there was no significant difference between the two age groups in the percentages who committed a violent offense in year 2. For both age groups, juveniles who offended in year 1 were significantly more likely than nonoffenders to be victimized in year 2, and those who were victimized in year 1 were significantly more likely than nonvictims to offend in year 2. These findings are consistent with the pattern for all juveniles.

By Gender Males were significantly more likely than females to commit a violent offense in year 2 and to be victims of violence in year 2. For both males and females, juveniles who offended in year 1 were significantly more likely than nonoffenders to be victimized in year 2, and those who were victimized in year 1 were significantly more likely than nonvictims to offend in year 2.

By Race The data were analyzed separately for white and minority juveniles. Both the total percentage of juveniles who committed a violent offense in year 2 and the total percentage who were victims of violence in year 2 were greater for the minority category. For both minorities and whites, juveniles who offended in year 1 were significantly more likely than nonoffenders to be victimized in year 2 and to offend in year 2, and juveniles who were victimized in year 1 were significantly more likely than nonvictims to offend in year 2 and to be victimized in year 2.

By Level of Physical Development The data were also analyzed separately by juveniles' level of physical development. Juveniles were categorized as "more physically developed" or "less physically developed," based on their responses to the following questions at the second interview:

> **For males:** How much hair is under your arms? How thick is the hair on your face? How much lower is your voice than when you were in grade school? How advanced is your physical development compared to other boys your age?
>
> **For females:** How much more developed are your breasts than when you were in grade school? How much more curvy is your body compared to when you were in grade school? How advanced is your physical development compared to other girls your age?

The total percentage of juveniles who committed a violent offense in year 2 and the total percentage who were victims of violence in year 2 were significantly greater for more physically developed juveniles. For both groups, juveniles who offended in year 1 were significantly more likely than nonoffenders to be victimized in year 2, and those who were victimized in year 2 were significantly more likely than nonvictims to offend in year 2. Again, these findings are consistent with the pattern for all juveniles.

Effects of Drug Use

Drug use and its influence on the relationship between violent victimization and offending were also examined. In both years, the percentage of juveniles reporting drug use was significantly greater among juveniles ages 15–17 than juveniles ages 11–14 but was not significantly different for males and females or for minorities and whites. To determine the influence of drug use on the relationship between victimization and offending, multivariate analysis was performed within each of the four defined drug use categories: nonusers (n = 3,270), desisters (n = 481), new users (n = 455), and consistent users (n = 797). In general, drug use did not influence the victimization-offending relationship: violent victimization increased the risk of violent offending and violent offending increased the risk of violent victimization, regardless of drug use.

Risk and Protective Factors

Multivariate analyses were also used to identify risk and protective factors for violent offending and victimization (i.e., factors that independently predict offending or victimization in year 2 after statistical controls for other factors are introduced into the model). Tables 14.2 and 14.3 present the results of the analyses.

Violent Offending The results presented in Table 14.2 indicate that, even after other factors related to violent offending were controlled statistically, being a victim of a violent crime in year 1 was still a significant risk factor for committing a violent offense in year 2. Only violent offending in year 1 had a greater influence. The analysis also revealed an important protective factor against violent offending in year 2: juveniles who reported greater support from important people in their lives, such as friends, parents, and teachers, were less likely to commit a violent offense in year 2.

Violent Victimization The results presented in Table 14.3 indicate that, when all other risk factors were controlled statistically, committing a violent offense in year 1 was still a significant risk factor for being the victim of a violent crime in year 2. Only the effects of violent victimization in year 1, being male, being a consistent drug user, and being a new drug user had a greater influence. Finally, violent victimization was significantly less likely among white juveniles than among minority juveniles, among juveniles who resided in two-parent households than among

Table 14.2 Factors Predicting Violent Offending in Year 2

Predictor[a]	Logistic Coefficient[b]	Odds Ratio[c]
Violent offending in year 1	1.39 (.09)	4.01
Violent victimization in year1	0.86 (.11)	2.36
Male	0.71 (.11)	2.03
Consistent drug user	0.62 (.16)	1.86
New alcohol user	0.59 (.15)	1.80
Consistent alcohol user	0.56 (.13)	1.75
New drug user	0.36 (.17)	1.43
More physically developed	0.30 (.10)	1.35
Depression	0.23 (.11)	1.26
Support from significant others	−0.26 (.09)	0.77
Household socioeconomic status	−0.19 (.06)	0.83

a. Only significant ($p < .05$) predictors are reported here.

b. The logistic coefficient represents the effect of a given predictor variable (e.g., violent victimization in year 1) on the log odds of the outcome (i.e., violent victimization in year 2). Positive numbers indicate risk factors; negative numbers indicate protective factors. Standard errors are in parentheses.

c. The odds ratio indicates the proportional change in the odds of violent victimization in year 2, per one-unit increase in the predictor variable. The greater the difference from one, the greater the effect of the variable on violent victimization.

Table 14.3 Factors Predicting Violent Victimization in Year 2

Predictor[a]	Logistic Coefficient[b]	Odds Ratio[c]
Violent victimization in year 1	1.74 (.13)	5.70
Male	0.91 (.13)	2.48
Consistent drug user	0.85 (.18)	2.34
New drug user	0.77 (.19)	2.16
Violent offending in year 1	0.69 (.12)	1.99
Depression	0.52 (.12)	1.68
Consistent alcohol user	0.47 (.16)	1.60
Easy access to gun in home	0.40 (.16)	1.49
More physically developed	0.29 (.13)	1.34
Time spent hanging out with friends	0.17 (.06)	1.19
White	−0.53 (.14)	0.59
Two-parent household	−0.43 (.13)	0.65
Household socioeconomic status	−0.23 (.07)	0.79

a. Only significant ($p < .05$) predictors are reported here.

b. The logistic coefficient represents the effect of a given predictor variable (e.g., violent victimization in year 1) on the log odds of the outcome (i.e., violent victimization in year 2). Positive numbers indicate risk factors; negative numbers indicate protective factors. Standard errors are in parentheses.

c. The odds ratio indicates the proportional change in the odds of violent victimization in year 2, per one-unit increase in the predictor variable. The greater the difference from one, the greater the effect of the variable on violent victimization.

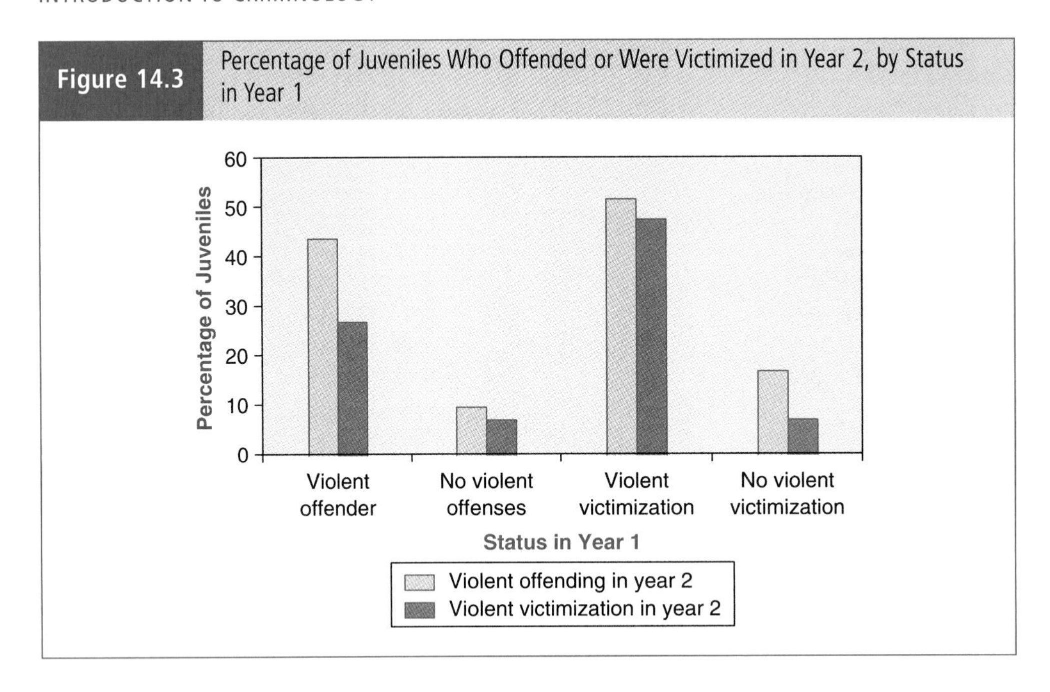

Figure 14.3 Percentage of Juveniles Who Offended or Were Victimized in Year 2, by Status in Year 1

those with other family structures, and among juveniles who resided in households with higher socioeconomic status than among those with lower socioeconomic status.

Conclusions

The analyses suggest three major conclusions:

1. **Violent victimization is an important risk factor for subsequent violent offending.** The percentage of year 1 victims who committed a violent offense in year 2 (52 percent) was significantly higher than the percentage of year 1 nonvictims who committed a violent offense in year 2 (17 percent). Figure 14.3 illustrates this finding.

2. **Repeat offending is more common than repeat victimization.** The relationship of violent offending in year 1 to violent offending in year 2 was stronger than the relationship of violent victimization in year 1 to violent victimization in year 2. Approximately twice as many juveniles committed an offense in both years as were victimized in both years.

3. **Violent victimization and violent offending share many of the same risk factors.** Shared risk factors include previous violent victimization and offending, use of drugs or alcohol, being male, depression, and having a high level of physical development.

References

Hindelang, M. J. (1981). Variations in sex-race-age-specific incidence rates of offending. *American Sociological Review, 46*(4), 461–474.

Lauritsen, J. L., Sampson, R. J., & Laub, J. H. (1991). The link between offending and victimization among adolescents. *Criminology, 29*(2), 265–292.

Sampson, R. J., & Lauritsen, J. L. (1990). Deviant lifestyles, proximity to crime, and the offender-victim link in personal violence. *Journal of Research in Crime and Delinquency, 27*(2), 110–139.

Sampson, R. J., & Lauritsen, J. L. (1994). Violent victimization and offending: Individual-, situational-, and community-level risk factors. In A. J. Reiss & J. A. Roth (Eds.), *Understanding and preventing violence* (Vol. 3, pp. 1–114). Washington, DC: National Academy Press.

DISCUSSION QUESTIONS

1. Why are violent victimization and violent offending so closely related?
2. Are there any policy implications suggested by this study that could be realistically implemented?
3. Which criminological theory you have learned best explains the close relationship between offending and victimization?

READING

Victimological Developments in the World During the Past Three Decades (1)

A Study of Comparative Victimology

Hans Joachim Schneider

In this first article, Hans Schneider presents a comprehensive overview of the results of 10 symposia on crime victims in 10 different countries. He also surveys and discusses the most important literature contributions to criminological victimization research over the past three decades. In the first part of this article, he discusses the distribution of victimization in the population and compares victimization risk across various areas of the world (Asia, Africa, Europe, and North America). Schneider then discusses the causes of victimization, which he divides into social structural, cultural, and institutional victimization. Finally, he discusses the concept of victim precipitation, and describes the lifestyle-routine-opportunity model and the routine-activity model of victimization.

In the past, criminology was centered primarily on the offender. Within the frame of the social psychological interaction theory, over the past three decades interest has turned to the victim of crime. This interest was of course not restricted to criminological issues alone but included the fields of psychology, sociology, and social psychology in particular. The interest in victims of crime shown by the scientific community and the general population stems from the civilization process that the world is undergoing in the long term. People are becoming more aware of what it means to become a victim. Their awareness of the problems associated with violence, particularly of the everyday forms of violence in the close-range social environment, has developed at an increasing rate.

Source: Schneider, H. (2001). Victimological developments in the world during the past three decades (1): A study of comparative victimology. *International Journal of Offender Therapy and Comparative Criminology, 45*(4), 449–468. Reprinted with permission of Sage Publications, Inc.

Differential Victimization: The Distribution of Victimization Within the Population

Victimization surveys belong to the "most exciting developments in criminology" (Fattah, 1991, p. 30) and have turned out to be the "most substantive empirical research concepts of the last two decades" (Kaiser, 1993, p. 6). These victimization surveys, which were first initiated in the United States at the end of the 1960s, have in the meantime spread throughout the world. In these studies, representative samples of the population are asked if they have become victims of selected crimes within a certain period of time and if they reported these crimes to the police. There are four different types of victimization studies.

- International victimization studies are based on representative random samples taken from the populations of numerous countries. They allow a comparison of the extent and structure of victimization in the different societies.
- National victimization surveys are based on a representative random sample taken from the population of a single country. Partly, these surveys are repeated annually and thus allow an analysis of the development of victimization caused by offenders.
- Local victimization studies are confined to the representative population sample of a region or a city.
- Specialized victimization surveys focus on victimization in the close-range social environment, for example family violence, school violence, and victimization related to a specific offense, for example, rape or sexual child abuse.

Empirical victimological research, victimization studies, and victimization surveys have derived a wealth of new information on the frequency of victimization, the extent of bodily and psychical injury, the extent of material loss, and the feeling of (in)security and fear of crime elicited in victims and nonvictims. These victimization studies are landmarks in the study of crime, but still, their limitations are clearly visible.

- They do not assess criminal reality. They merely report on the objective and subjective state of (in)security from the viewpoint of the crime victim and the potential crime victim; they fathom the experiences of the population with victimization.
- Their significance is limited by errors in remembering, memory deficits, a lack of willingness to report, and dubious credibility (exaggeration or withholding information) on the part of the respondents.
- The results of victimization studies are codetermined by the methodical framework of the survey instrument (the questionnaire) and by effects related to the person conducting the survey (e.g., lack of motivation of the interviewer).
- Many offenses, especially those involving violence and sexual abuse, remain in a double dark field: They are communicated neither to the police nor to the interviewer of the victimization study.

The willingness of respondents to disclose information to the interviewer in the interview situation is low when it comes to delicate issues such as female abuse, rape involving marriage partners or acquaintances, bodily child abuse, or sexual abuse of children within the family. These offenses are often not viewed as crimes by the victims and their next of kin. This reflects the social stereotype notion, which is constantly being amplified by the mass media, that crime occurs only between strangers. Offenses in the close-range social environment are also not reported by victims in dark-figure investigations, because the victims are dependent on and in need of the offender, and because they feel obliged or forced to show consideration for his interests.

International Victimization Risk

In 1989, 1992, and 1996, three international victimization studies were carried out in a total of 52 industrialized countries, countries in transition, and developing countries. More than 133,821 inhabitants of different countries and large cities were interviewed as to 11 offenses (property offenses, sexually motivated offenses, and crimes of violence).

On the basis of these studies on victimization by offenders, the international distribution of crime can be outlined as follows.

- Although the legal definitions and crime survey methods vary in the different countries, the basic understanding of the manifestations and the assessment of basic concepts like robbery, burglary, and rape is, on the whole, the same worldwide.
- The rates of victimization determined for the past five years are the highest in Latin America (74.5%) and Africa (74%). These values have reached an intermediate level in the New World (65.3%), eastern Europe (62.2%), and western Europe (61.2%). In Asia (51.4%), the victimization rate is the lowest.
- From the three studies on victimization, it can be concluded that offenses, including crimes of violence, are by no means seldom occurrences but rather, at least in urban centers, statistically normal events.
- In western countries (Europe and the New World), violence against women and men is fairly evenly distributed. The problem of violence in the developing countries of Africa and Latin America is nevertheless characterized mainly by sexual and nonsexual violence against women. Violence against women reaches the highest level in countries in which women hold a low social status. The extent of violence in general is closely related to the frequency of possession of a weapon.
- In most industrialized countries, the number of property offenses is declining, a fact attributed to the improved protection of personal property in these countries. In contrast, property offenses are increasing in the developing countries.
- Victimization related to corruption of government officials is the highest in Latin America (21.3%), Africa (18.8%), and Asia (14.6%). In eastern Europe, its frequency is moderate (10.7%). The figures in countries of the New World (1%) and western Europe (0.7%) are the lowest.
- Reporting rates are symptoms of the perceived level of personal security and sensitivity toward crime within the population. They also reflect the efficiency of the criminal justice system. The figures are the highest for the countries of the New World (54%) and western Europe (52%). They decline significantly in the case of Africa (40%) and eastern Europe (35%) and are the lowest in Asia (31%) and Latin America (27%). Property offenses are reported at a higher rate than crimes of sexual abuse and violence. Crimes against women show the largest dark figure.
- Satisfaction with the performance of the police is higher in the Western countries than in the developing countries; satisfaction is higher in the case of property offenses than in the case of crimes of sexual abuse and violence. Victims complain that the police are not sufficiently active and show no interest. The victims want to be treated with due respect. Victims of sexual abuse and crimes of violence encounter a greater lack of respect than victims of property offenses. Female victims of crimes of violence most frequently experience an impolite and disrespectful response. Satisfaction with police performance

correlates directly with the reporting behavior of the population.

- There is a major gap between the needs of victims and the support they actually receive. Worldwide, only 4% of male victims and 10% of female victims of violent crimes receive assistance. This percentage is higher in the New World (29%) and in western Europe (22%). Two thirds of the victims of severe crimes who had reported the crime to the police need help, which they have failed to receive, however.

Victimological Theories

Social Structural, Cultural, and Institutional Victimization

According to the social structural victimization theory, victimization reflects the economic and the power structures of a society. Marginalized, powerless minorities that have been pushed toward the edge of society are often forced into becoming victims. The social pressure imposed on marginalized minorities leads to social disorganization and the decay of relations and communities, causing a propensity to become a victim. The Australian aborigines who have been deprived of their culture and identity and whose numbers have been decimated by one half can be given as an example here. Another example of social structural victimization is the killing of female infants and dowry homicide in India (Chockalingam, 2000). In India's rural as well as urban areas, the family of the bride owes the family of the groom a substantial dowry in the form of money, jewelry, and household possessions. From childhood, the young girl is neglected bodily and emotionally because she causes considerable dowry expenses to the family. Young boys are welcome because they furnish their family with a dowry. Many families even kill their female infants because they are unable to provide a suitable dowry or do not want to. If the firstborn is a girl, she is left to live. If, however, further children that are born turn out to be girls, they are killed.

Cultural victimization, which is based on customs, tradition, religion, and the ideology of a society, is the subjective form of social structural victimization, as the structure of the economy and the system of power eminently influence views, value concepts, and the stereotypes of a society. Hate crimes that are characterized by the symbolic status of the crime victim constitute an example. The victim belongs to an outsider group symbolizing that which the insider group, to which the offender belongs, does not want to be. The offenses serve to affirm the solidarity and identity of the insider group and at the same time to strengthen the feeling of self-assurance of the group members. This is illustrated by the physical attacks on homosexual men and lesbian women. Homosexuals are beaten and even killed because of their sexual inclinations ("gay bashing"). One is dealing here with an ideology of suppression, which manifests itself in social customs, religious and legal institutions, individual views, and styles of conduct. Homosexuality is stigmatized as being illegal, sinful, and morbid. The heterosexual ideology defines what is to be viewed as masculinity and femininity. As the concepts associated with masculinity and femininity are learned at a very early age, they appear "natural" to an adult; homosexuality is felt to be abnormal.

Institutional victimization not only encompasses victimization within an institution but also victimization by the institution itself. Here, the term institution designates a facility that fulfills certain tasks according to certain rules that govern work procedures and the distribution of tasks among staff members who are working together. A subcategory of institutional victimization is victimization by an enterprise (corporate victimization). Examples of institutional victimization include violence in nursing homes, in schools, and in prisons. The cause of violence can be found in the staff members of the institution and the inmates of the institution, but it can also lie in the structure of the institution. The institution can be socially isolated. Bureaucratic, formalistic impersonality can prevail within the institution; the persons within the institution can be without mutual relations. The inmates can become mere work objects, lifeless abstractions for which the staff feels no more personal responsibility. There is too great a gap between the low number of staff members and the large (in terms of numbers) group of inmates. Power is unilaterally distributed: The decision-making and control authority is completely on the side of the staff members who always feel superior and in the right. The group of inmates is powerless: The inmates are largely deprived of their sphere of privacy and their personal belongings (loss of identity-forming possessions). They are even excluded from decisions affecting their own personal destiny. This group has no more control over itself. Everything is regimented. Its role is ultimately a purely reactive one; it can no longer develop initiatives on its own.

Situation-Oriented Theories

The concept of victim precipitation, which sees the origin of victimization in a misguided offender-victim interaction, was developed in the context of research on homicidal criminality (Wolfgang, 1958); 26% of cases of homicide have been coprecipitated by victims. This model is particularly controversial where the crime of rape is concerned. It is occasionally misinterpreted as victim coresponsibility, responsibility assignment, and blaming the victim. The dynamic, interactionist perspective of victim precipitation does not, however, appraise victim behavior. In this kind of interpretation model, there is no room for normative or value judgments such as guilt or responsibility. The model ultimately describes only the misinterpretation of victim behavior by the offender. The illusionary misinterpretation of the situation by the offender, which is evoked by the victim's behavior, is merely a substantiation of rape-supporting stereotypes in the rape situation. The concept of victim precipitation, which is based on the theory of symbolic interaction and which does not in any way dispense the offender of his exclusive responsibility, thus only marks the application of social structural theory and cognitive social-learning theory in the rape situation. The denial of an offender-victim relationship in the rape situation and of a potential victim precipitation reinforces the questionable proposition that rape is an uncontrollable event and that the victim cannot take any preventive action. This promotes the learned helplessness of the potential rape victim who refrains from developing self-protection measures, because of their assumed futility, and succumbs to his or her fearful, self-defined, victim attitude.

Victimization is associated with a certain lifestyle, with a constantly recurring behavior in which one is exposed to situations bearing a high risk of victimization. The lifestyle-routine-opportunity model focuses on the probability with which individuals are found at certain locations at certain times and under certain circumstances to come into contact with certain people. The risk of becoming a victim strongly depends on the number of hours spent outside one's home, on the frequency of going out in the evening and coming back late at night, on the frequency of going to pubs and discos, and on the closeness of neighborhood contacts.

The lifestyle-routine-opportunity model has been developed into the routine-activity theory. According to this theory, three elements are essential for victimization: the existence of motivated offenders, the presence of a suitable target object of criminal action (i.e., a person or a thing), and the absence of persons effectively able to protect the target object against a violation of the law. Such an effective protector is rarely a policeman but is much more likely to be a housewife, a brother, a friend, or a passerby. Although routine-activity theory proceeds from the assumption of a universal presence of motivated offenders, it places primary emphasis on situational elements, that is to say, the opportunity of committing a crime and the lack of informal control by potential victims and their personal environment. The rise in property offenses in the developed, Western, welfare societies is assessed as follows by routine-activity theory. Many social changes, which have recently improved the quality of life and equality within the population—for example, increased employment, more advanced conditions of academic education, expanded recreational opportunities—form the same factors that have also brought about an increase in crime. Through their mass production, valuable goods have become socially more visible and attainable (e.g., cameras, radio and television sets, dictation equipment, video recorders, and computers). Informal control over them by the potential victims is becoming increasingly more difficult, as these valuable goods are becoming increasingly lighter in weight and smaller in size. Thus, they are easier to remove and transport. At the same time, more people are working outside their homes at an increasing rate. Professional activities outside the home are steadily increasing in the case of women. Vacations are being taken by people outside the towns they live in at a growing rate as well (mass tourism). On working days and during vacation periods, their houses and apartments remain unsupervised and unguarded. Burglars thus find worthwhile targets. Numerous attempts have been undertaken to empirically prove the routine-activity theory. In this context, the proximity and accessibility of the desired target object of the criminal action and the attraction and shelteredness (protection) have been repeatedly emphasized. The nature of professional activity is a determinant factor governing the degree of victimization. More and more households with children and only one child-rearing parent are becoming victims of crime at an increasing rate. Recreational activities outside the home such as participation in sports events and going to movies, theaters, bars and night clubs, restaurants, and discos increase the victimization risk. Consumption of alcoholic beverages in bars and nightclubs can enhance the propensity to victimization. Homeless persons and street children are particularly prone to becoming victims. A deviant lifestyle (e.g., alcoholism, drug consumption, prostitution, homosexuality) evokes a particularly high risk of victimization. Members of youth gangs lead a risk-burdened lifestyle that frequently offers the opportunity of becoming an offender and/or victim. Persons with delinquent and criminal conduct are frequently and continually prone to becoming crime victims.

References

Chockalingam, K. (2000). Female infanticide—A victimological perspective. In P. C. Friday & G. K. Kirchhoff (Eds.), *Victimology at the transition from the 20th to the 21st century* (pp. 273–287). Moenchengladbach, Germany: World Society Of Victimology.

Fattah, E. A. (1991). *Understanding criminal victimization.* Scarborough, ON: Prentice Hall.

Kaiser, G. (1993). Viktimologie [Victimology]. In P.-A. Albrecht, A.P.F. Ehlers, F. Lamott, C. Pfeiffer, H. -D. Schwind, & M. Walter (Eds.), *Festschrift fuer Horst Schueler-Springorum* (pp. 3–17). Munich, Germany: Heymanns.

Wolfgang, M. E. (1958). *Patterns in criminal homicide.* Philadelphia, PA: University of Pennsylvania.

DISCUSSION QUESTIONS

1. What do you think accounts for the fact that victimization is highest in Latin America and lowest in Asia?
2. What does Schneider mean by "cultural victimization"?
3. What is Schneider's opinion of victim precipitation applied to rape? Do you agree?

READING

Victimological Developments in the World During the Past Three Decades (II)

A Study of Comparative Victimology

Hans Joachim Schneider

In this second of a two-part article, Hans Schneider surveys victimology studies and theories emanating from different parts of the world over the past 30 years. In the first part of the article, Schneider discusses the direct, indirect, and secondary damage suffered by victims, with the negative psychic and social effects being emphasized. Damage suffered by what he calls the "indirect victim" (co-victim), such as spouses and other family members, is also highlighted. Noting that the best predictor of future victimization is past victimization, Schneider develops a model of recidivist victimization. The constitutional and procedural legal rights of the victims and potential victims are then elucidated, and he argues for comprehensive treatment programs along restorative justice lines, in an effort to make victims whole again.

Damage to Victims

In national victimization surveys, most respondents report on relatively minor and transient damage. The more recent victim surveys restricted to particular types of offenses have, however, shown that damage to victims is substantially more extensive and lasting than initially assumed. Although victims of crimes of violence clearly suffer the most, psychical injury to victims of property offenses is nonetheless severe. Criminal victimization entails depression, fear, hostility, somatic symptoms, fear of crime, avoidance behavior, reduced self-confidence, increased estrangement, and the need for formal and informal social support. Assurance and self-confidence are shattered by the victimization experience. The victims alter their views about the world and themselves. The victims of bodily and sexual attacks have to bear physical injury. Psychical and social disadvantages generally cause short-, medium-, and long-term pain to a far higher degree. The scope and nature of suffering depends on a variety of influences, such as the degree of inflicted violence, the closeness of the offender-victim relationship, and the extent of support and understanding and the psychological assistance that the victim of a violent or sexually motivated crime receives from his or her family, relatives, neighbors, and friends. The damage suffered by recidivist victims is the most difficult and cumbersome to treat.

The description and substantiation of the "rape-trauma syndrome" (Burgess & Holmstrom, 1985) marked a decisive advance in the development of a critical awareness of the destructive psychical and social effect of victimization: Memories of the victimization experience recur continually and traumatically. Sleep disturbances and nightmares are attempts of the subconscious to cope

Source: Schneider, H. (2001). Victimological developments in the world during the past three decades (2): A study of comparative victimology-Part 2. *International Journal of Offender Therapy and Comparative Criminology, 45*(5), 539–585. (2001) Reprinted with permission of Sage Publications, Inc.

with the traumatizing experience. The victim is in a state of terrible and excruciating fear. Long-lasting nervous tension and emotional disturbances, depression and suicidal thoughts, and sexual dysfunction torment the victim. The female victim's relation to the male sex is persistently disrupted. Her feeling of safety and self-confidence is decisively disturbed. Her social relations toward others are strongly obstructed for months and years. The psychical damage suffered by victims of child abuse can be divided into the following four categories:

1. In psychodiagnostic test methods, increased values of depression and fear have been found in sexually abused children. Moreover, emotional disturbances manifest themselves as sleep disorders and concentration deficits.
2. Child victims of sexual abuse have behavioral problems to an increased degree. Traumatic sexualization is a process through which the sexuality of the child victim is inadequately and dysfunctionally formed with regard to social interrelationships.
3. Sexual victimization results in cognitive distortions (e.g., self-accusation, loss of confidence, internal acceptance of social stigmatization, and negative self-incrimination) that develop in the psyche of the child victim.
4. Sexual abuse of children has an effect on the development process of their self-conception, which coordinates and integrates the various elements of their personality. Psychical tensions arise from damage to their self-conception. These can lead to self-inflicted injuries and to self-mutilation.

Indirect and Secondary Damage to Victims

Whereas the immediate psychical and social damage inflicted on victims had already been established in the 1970s and 1980s, the insight has been gained only during the 1990s that indirect and secondary victims (*covictims*), such as intimate and marriage partners, family members of the victims, and persons with a closer relationship, suffer the same damage to their personality as the victims themselves. For example,

- Family members of homicide victims suffer one of the most profound types of psychical trauma that can be inflicted on victims of crime. After the sudden traumatic death of a marriage partner or child, the surviving victim feels completely helpless, confused, lacking self-control, and unable to comprehend the effect of such a traumatizing experience. Due to the traumatic death, a personality transformation of the surviving covictim often occurs: The covictim completely alters his or her lifestyle.
- Husbands and the intimate partners of women who have been raped by another man react with anger or suffer from feelings of powerlessness, vulnerability, and guilt. Their feeling of self-confidence is heavily affected. They accuse themselves of having failed in protecting their partner.
- The disclosure of sexual child abuse precipitates the victim's mother into an identity crisis characterized by severe self-appreciation problems and great self-doubt. The mother develops feelings of uncertainty and doubt as to her maternal competence. The term *secondary victimization* refers to the phenomenon of renewed victimization due to an inadequate response to the primary act of victimization, namely, becoming a victim by the criminal act itself. Both the victim and the covictim can be affected by secondary victimization. The concept had indeed been developed at quite an early stage, and its validity is still recognized today. However, the double damage suffered by the victim has been somewhat overemphasized by the early victimological research. But still, today as before, it is established knowledge that an inappropriate response to the offense can lead to both victim and covictim becoming victimized a second time.

◆ Exaggerated emotional reactions by the family members of the victim or by persons in his or her close-range social environment can aggravate the victim's psychical coping process. Dramatizing reactions by the social control institutions, such as the police or courts, can also aggravate the psychical and social damage to victims.

◆ It is quite frequent that the formalistic routine and indifference shown by large bureaucratic bodies, such as the police and hospitals, lead to a renewed depersonalization of the victim. The victim feels lost and neglected. Doctors and police officers ask injuring questions of doubtful significance. Skepticism and doubts as to the victim's credibility are difficult to bear by the victim.

The Recidivist Victim

The probability of revictimization increases with each victimization incident. The major share of victimization by offenders is concentrated within a comparatively small segment of the population. Multiple victimization (revictimization) within a fairly short period of time is not rare. A mere 14% of adults experience about 70% of all self-reported cases of victimization (Farrell, 1992). And only 17% of the population is the target of 45% of crimes of violence (Mayhew, Maung, & Mirrlees-Black, 1993). Similar victimization distributions have been reported by a number of researchers over the entire spectrum of different offense categories. Thus, one event of household burglary increases the probability of revictimization of this household by a factor of 4.

The model of the victim career can serve to explain chronic victimization. The model proceeds from the assumption of an initial vulnerability and propensity to becoming a victim that increases the probability of victimization. The model further postulates that each victimization incident in itself forms an important causative factor for further victimization. Traumatic childhood experiences such as separation from one's parents, especially one's mother, fear of separation, victimization during childhood (e.g., physical abuse and sexual child abuse), and disturbed personal relationships (e.g., neglect, rejection, and pampering) can also create a propensity to victimization. Children who long for affection, attention, concern, appreciation, and recognition and children who are fearful, depressive, shy, and socially isolated tend to become victims of recurrent offenses with a higher probability. Shyness and fear can make children vulnerable to victimization. Victimization experiences can further amplify these attitudes and thus increase the future victimization risk in the victim-definition process.

Victimization seriously questions the belief in the benign nature and safety of the world. In this way it damages the self-conception of the victim. He or she develops a victimal self-image and self-understanding. Patterns of subordinate behavior are closely linked with chronic victimization; they are a consequence of earlier victimization experiences. Such experiences during the days of childhood and youth are perpetuated at an adult age. Women who have been sexually abused as children exhibit a high victimization risk during their adult life. The reason for this is that their sexual victimization elicits feelings of a reduced self-appreciation and a lower degree of authority and self-assertion; their psychical self-defense mechanisms are weakened. In consequence of his or her sexual victimization, the child victim has inherently accepted the futility of targeted resistance. The child's thus developed propensity to victimization can then evoke multiple victimization. As a result of the victim's repeated experience of not having been able to evade harmful events (learned helplessness), he or she begins to feel weak, helpless, and needy. Traumatization appears as uncontrollable and inevitable.

An example of such a victim career is the bullying occurring in schools. Victimization-prone children are beaten, knocked around, kicked, and spat at; money is forced out of them; their valuables are stolen, or expensive pieces of clothing are taken away from them; their school materials, homework, or bicycles are damaged or destroyed.

They are forced into committing socially deviant and delinquent acts and into self-degradation. Offenders are mostly boys who have developed an aggressive self-conception in the family home. They pursue violent behavioral patterns and show a pronounced need for and pleasure in subjugating and dominating fellow students. The victims, the ones that are beaten up, exhibit a low and powerless feeling of self-appreciation and develop a subservient self-image. Through overcontrol and overprotective pampering in the family, the victim-prone children have learned to subjugate themselves and to be obedient and submissive. They develop shyness, timidity, fearfulness, insecurity, reservedness, silent behavior, and social unobtrusiveness. By their behavior, they indicate that they will not oppose violence but will rather succumb to attacks and will tolerate insults and harassment. They will readily and without resistance hand over their private possessions to aggressive children. The continual harassment by children of the same age significantly intensifies their fears and insecurity and lowers their self-appreciation. If victim-prone children reward the aggressive attacks of their peers by subjugation, they run the risk of being viewed as vulnerable targets of chronic victimization. Victimization-prone children emanate an aura of fearful vulnerability; they appear as worthwhile victims to their offenders. Due to their submissive manner of behavior, they have a strong attraction for aggressive children who in the process of interaction are solely responsible for their tyrannical and harassing conduct.

Victims' Rights

International criminal policy has made the following two major advances, which had been prepared for many years by the experts of the World Society of Victimology: (a) the declaration by the General Assembly of the United Nations concerning the basic principles of justice for victims of crime and abuse of power and (b) the recommendations by the Committee of Ministers of the Council of Europe (1985) concerning the improvement of the legal position of the victim in the framework of criminal law and procedure. Both documents contain essentially the following four reform proposals:

1. The criminal justice system must conform to the concept of restitution. Here, restitution must be seen as "a creative process, a personality-related and social service" (Schneider, 1998, p. 418) by which the offender accepts his or her responsibility for the criminal act with regard to both the victim and society.

2. The victim of a crime must be granted the right of participation in the criminal proceedings and the right to exert an influence on the course of the proceedings; the victim must be given the status of a legal subject, a *third party*. The criminal law must be repersonalized.

3. A mediation, restitution, and compensation procedure should be introduced as a preliminary procedure before the actual criminal proceedings take place. In such a procedure, the victim and the offender sit at the same table and try to resolve their conflict themselves in an informal framework under the supervision of and mediation by the court and with the assistance of a district attorney, the defense counsel, and possibly a legal expert.

4. A network of professional, state-provided victim assistance and treatment programs must be established. In this kind of institution, it is not only damage inflicted on victims by the victimization event as such that should be treated by appropriate psychological methods to avoid revictimization. It should also be ensured that the victim does not suffer psychical and social damage a second time due to inappropriate responses by the criminal justice system and persons belonging to the victim's close-range social environment.

Victims' Rights in the Domain of Criminal Law, Criminal Proceedings, and Corrections

The victim suffers psychosocial damage as a consequence of [victimization]. A conservative, hard-line criminal policy again makes use of the victim as a mere means to an end. The victim's agony is not appeased by the suffering of the offender; the victim's debasement is not compensated by degradation of the offender. Victims are not hungry for revenge; their main desire is restitution. Victims' needs are not suited for justifying a retaliation or satisfaction ideology. Victims do not want decision-making powers, but they wish to be heard before a court ruling is pronounced. Victims are discontent with their role in criminal proceedings. They want the court to acknowledge the damage they have suffered; they do not want the offender to talk his or her way out of things but to bear the full responsibility for the offense he or she has committed. The victim wants to be recognized as a fully valid participant in the criminal proceedings. He or she wants to play a role that is fully accepted by the criminal justice system.

The issue at hand is that of a repersonalization of the criminal law, criminal procedure, and corrections so as to prevent revictimization. The criminal act is not merely an abstract violation of an object of legal protection; it also constitutes the infliction of a concrete psychosocial injury to the person of the victim. The violation of an object of legal protection must not be negated, as advocated by the abolitionists. But, it has to come second to the damage suffered by the victim. Before implementing its monopoly of power, the state must strive to peacefully settle the conflict in cooperation with the parties directly involved. Formal and informal control must be integrated. It is the responsibility of the criminal justice system to reestablish and support the system of informal control that has been damaged by civilization factors. Restitution reduces the recurrence of crime. The criminal act shall, however, not be attributed to a personality defect for which the offender does not feel responsible. The latter has to be enabled to take on himself or herself the responsibility for his or her misconduct in the face of the victim and legal community without losing his or her self-respect.

On the basis of this concept, the following three main criminal policy proposals have been discussed over the past decade. Restitution must be established as an independent and autonomous central instrument of judicial sanctioning practice. Restitution must be understood as a process of interaction between the victim, the offender, and society that allays the criminal conflict and establishes peace between the parties involved. By restitution, by healing of the criminal conflict, and by restoring peace between offender, victim, and society, adherence to the law is practiced and awareness of the role of justice in society is strengthened. The moral-emotional process of overcoming the psychosocial injuries inflicted on the person of the victim is a socially constructive effort that is able to create internal involvement on the side of the offender. Restitution is a symbolic gesture of reconciliation and a prerequisite for reintegration and acceptance of the offender by society as a whole.

Victim Support and Treatment Programs

Victims of crime need help and psychosocial support. A modern criminal policy cannot deny them this assistance. [W]hat is needed is the establishment of a network of professional, state-provided victim support and treatment centers. Victim support and treatment programs are needed to prevent revictimization and to heal the psychosocial and psychosomatic traumatization caused by the victimization experience. Not only reasons of a humane and just criminal policy advocate that the victim is no more left alone to cope with his or her injuries that he or she has rendered as a special service for crime control. Many victims are so heavily shattered in their self-conception by their victimization experience that they are predisposed—as the result of a process of learned helplessness—to be again chosen as a victim by the criminal on account of their weakened self-assertion (victim vulnerability). In a safe and understanding environment, the victim must systematically, through various treatment methods, expose himself or herself to the traumatic memory and must overcome his or her feelings of self-accusation and stigmatization by the method of cognitive restructuring.

Self-assertion training and supportive group therapy belong to the traditional forms of treatment. In self-assertion training, the determination and the power of the victim to assert himself or herself are reinforced. Group therapy contributes to victims of crime giving each other mutual support by communicating their experiences.

The systematic desensitization serves to promote the process of emotional coping with victimization. In a secure and tension-free atmosphere, the victims have to relive the criminal act in their imagination with all the emotions and fears experienced during their victimization. The crime victim is instructed to imagine the victimization scene as vividly as possible, loudly describe the event in the present tense, and express his or her emotions and fears. Cognitive restructuring helps crime victims to become aware of distortions of their perception and to psychically cope with these. Certain persistent conceptions (victim neutralizations) can give rise to a propensity to anxiety and depression. The victim can thus build on, for example, distorted perceptions as to his or her inadequacy, inability, and helplessness. The victim can hold the view that he or she is responsible for the personal victimization suffered, and the victim can feel worthless as a result of the attack. Cognitive techniques help the victim to recognize such distorted convictions, contrast them with reality, and psychically come to terms with them.

Victims of crime are not mentally disturbed and do not require psychiatric treatment. Their victimization experience has, however, inflicted severe psychical injury on them. The victims can be helped to a great extent by psychological treatment, by trauma-specific therapy, focusing specifically on the victim's traumatization, as schematically outlined earlier. A successful healing process requires that the persons in the victim's close-range environment provide moral assurance and support. If necessary, these persons have to receive psychological instruction as well. It is essential that the therapy be aimed at the victimization experience itself. Victim support programs have proven helpful if conducted with the necessary victimological background knowledge and expertise. Psychological treatment programs for victims of crime were launched only in the 1990s. In view of clinical experience, these programs appear to have a successful outcome. It appears certain that the symptoms evoked by victimization ultimately disappear in numerous crime victims once they have received psychological training centered on their victimization. The psychological training concentrated on victimization has a highly beneficial effect on many victims (Cohen & Mannarino, 2000).

References

Burgess, A. W., & Holmstrom, L. L. (1985). Rape trauma syndrome and post traumatic stress response. In A. W. Burgess (Ed.), *Rape and sexual assault* (pp. 46–60). New York: Garland.

Cohen, J. A., & Mannarino, A. P. (2000). Predictors of treatment outcome in sexually abused children. *Child Abuse and Neglect, 24,* 983–994.

Council of Europe. (1985). *The position of the victim in the framework of criminal law and procedure.* Strasbourg, France: Author.

Farrell, G. (1992). Multiple victimization: Its extent and significance. *International Review of Victimology, 2,* 85–102.

Mayhew, P., Maung, N. A., & Mirrlees-Black, C. (1993). *The 1992 British crime survey.* London: HMSO.

Schneider, H. J. (1998). Viktimologie [Victimology]. In R. Sieverts & H. J. Schneider (Eds.), *Handwoerterbuch der kriminologie* (2nd ed., Vol. 5, pp. 405–425). New York: Aldine de Gruyter.

DISCUSSION QUESTIONS

1. Schneider claims that victims are not "hungry for revenge." Do you believe that this is true and that victims would feel justice has been done using the kinds of methods he advocates?
2. Criminals in the United States have many protected rights; why aren't the same rights extended to victims as victims?
3. What would you say is the most important right that we should afford victims?

Glossary

Actus reus: Literally, *guilty act;* it refers to the principle that a person must commit some forbidden act or neglect some mandatory act before he or she can be subjected to criminal sanctions.

Adaptations: The products of the process of natural selection. Adaptations may be anatomical, physiological, or behavioral.

Aggravated assault: Defined by the FBI as "an unlawful attack by one person upon another for the purpose of inflicting severe or aggravated bodily injury."

Agnew's super traits theory: A developmental theory that asserts that five life domains interact over the life course once individuals are set on a particular developmental trajectory by their degree of low self-control and irritability.

Alcoholism: A chronic disease condition marked by progressive incapacity to control alcohol consumption despite psychological, spiritual, social, or physiological disruptions.

al-Qaeda: Umbrella organization formed by Osama bin Laden that is the "base" for a number of Sunni Muslim terrorist groups.

Altruism: The action component of empathy (i.e., an *active* concern for the well-being of others).

Anomie: A term meaning "lacking in rules" or "normlessness" used by Durkheim to describe a condition of normative deregulation in society.

Antisocial personality disorder: A psychiatric label described as "a pervasive pattern of disregard for, and violation of, the rights of others that begins in childhood or early adolescence and continues into adulthood."

Arousal theory: A theory of crime based on the idea that in identical environmental situations, some people are underaroused and other people are overaroused, and both levels are psychologically uncomfortable. Some people who are underaroused try to raise their level of arousal through antisocial behavior.

Arraignment: A court proceeding in which the defendant answers to the charges against him or her by pleading guilty, not guilty, or no contest (*nolo contendere*).

Arrest: The act of being legally detained to answer criminal charges on the basis of an arrest warrant or the belief of a law enforcement officer that he or she has probable cause to believe that the person arrested has committed a felony crime.

Arson: Defined by the FBI as "any willful or malicious burning or attempting to burn, with or without intent to defraud, a dwelling house, public building, motor vehicle or aircraft, personal property of another, etc."

Atavism: Cesare Lombroso's term for his "born criminals," meaning that they are evolutionary "throwbacks" to an earlier form of life.

Attachment: One of the four social bonds in social bonding theory; the emotional component of conformity refers to one's attachment to others and to social institutions.

Attention deficit with hyperactivity disorder: A chronic neurological condition that is manifested as constant restlessness, impulsiveness, difficulty with peers, disruptive behavior, short attention span, academic underachievement, risk-taking behavior, and extreme boredom.

Autonomic nervous system: Part of the body's peripheral nervous system that carries out the basic housekeeping functions of the body by funneling messages from the environment to the various internal organs; the physiological basis of the conscience.

Behavior genetics: A branch of genetics that studies the relative contributions of heredity and environment to behavioral and personality characteristics.

Behavioral activating system: A reward system associated chemically with the neurotransmitter dopamine and anatomically with pleasure areas in the limbic system.

Behavioral inhibition system: A system that inhibits or modulates behavior and is associated with serotonin.

Belief: In social control theory, belief is one of the four social bonds. It refers to the ready acceptance of the correctness of prosocial values and attitudes.

Binge drinkers: People who frequently consume anywhere between 5 and 10 drinks in a few hours' time (go on a binge).

Biogenetic law: Ernst Haeckel's law, which stated that ontogeny (individual development) recapitulates phylogeny (evolutionary development of the species).

Biosocial theory of rape: A theory that incorporates the most empirically supportable claims of feminist, social learning, and evolutionary theories and adds neurohormonal variables to the mix.

Bipolar disorder: A disorder in which individuals alternate between the poles of extreme elation or euphoria (mania) and deep depression.

Bourgeoisie: The wealthy owners of the means of production.

Burglary: Defined by the FBI as "the unlawful entry of a structure to commit a felony or theft."

Carjacking: The theft or attempted theft of a motor vehicle from its occupant by force or threat of force.

Cartographic criminologists: Criminologists who employ maps and other geographic information in their research to study where and when crime is most prevalent.

Causation: A legal principle stating that there must be an established proximate causal link between the criminal act and the harm suffered.

Cheats: Individuals in a population of cooperators who gain resources from others by signaling their cooperation and then defaulting.

Chicago Area Project: A project designed by Clifford Shaw to "treat" communities from which most delinquents came.

Choice structuring: A concept in rational choice theory referring to how people decide to offend and defined as "the constellation of opportunities, costs, and benefits attaching to particular kinds of crime."

Class struggle: A Marxist principle that there is continuous conflict between political and economic groups (e.g., the bourgeoisie and the proletariat) for power. All history is the history of class struggles.

Classical conditioning: A mostly passive visceral form of learning depending on autonomic nervous system (ANS) arousal that forms an association between two paired stimuli.

Classical school: The classical school of criminology was a nonempirical mode of inquiry similar to the philosophy practiced by the classical Greek philosophers.

Cleared offenses: A crime is cleared by the arrest of a suspect or by exceptional means (cases in which a suspect has been identified but he or she is not immediately available for arrest).

Cognitive dissonance: A form of psychological discomfort resulting from a contradiction between a person's attitudes and his or her behavior.

Collective efficacy: The shared power of a group of connected and engaged individuals to influence an outcome that the collective deems desirable.

Commission: A national ruling body of La Cosa Nostra consisting of the bosses of the five New York families and four bosses from other important families.

Commitment: One of the four social bonds in social bonding theory; the rational component of conformity referring to a lifestyle in which one has invested considerable time and energy in the pursuit of a lawful career.

Concurrence: The legal principle stating that the act (*actus reus*) and the mental state (*mens rea*) concur in the sense that the criminal intention actuates the criminal act.

Conduct disorder: The persistent display of serious antisocial actions that are extreme given the child's developmental level and have a significant impact on the rights of others.

Conformity: The most common of Merton's modes of adaptation (i.e., the acceptance of cultural goals and of the legitimate means of obtaining them).

Conscience: A complex mix of emotional and cognitive mechanisms that we acquire by internalizing the moral rules of our social group in the ongoing socialization process.

Conscientiousness: A personality trait composed of several secondary traits such as well organized, disciplined, scrupulous, responsible, and reliable at one pole and disorganized, careless, unreliable, irresponsible, and unscrupulous at the other.

Consensus or **functionalist perspective:** A view of society as a system of mutually sustaining parts and characterized by broad normative consensus.

Constrained vision: One of the two so-called ideological *visions* of the world. The constrained vision views human activities as constrained by an innate human nature that is self-centered and largely unalterable.

Containment theory: A social control theory asserting that people are prevented from committing crimes by both inner and outer controls.

Contrast effect: The effect of punishment on future behavior depends on how much the punishment and the usual life experience of the person being punished differ or contrast.

Corporate crime: Criminal activity on behalf of a business organization.

Corporate model: A model of La Cosa Nostra that sees it as similar to corporate structure (i.e., a formal hierarchy in which the day-to-day activities of the organization are planned and coordinated at the top and carried out by subordinates).

Corpus delicti: A Latin term meaning "body of the crime" and referring to the elements of a given act that must be present to legally define it as a crime.

Correlates: Factors that that are linked or related to the phenomenon a scientist is interested in.

Counterfeiting: The creation or altering of currency.

Crime: An intentional act in violation of the criminal law committed without defense or excuse and penalized by the state.

Crime rate: The rate of a given crime is the actual number of reported crimes standardized by some unit of the population.

Criminality: A continuously distributed trait composed of a combination of other continuously distributed traits that signals the willingness to use force, fraud, or guile to deprive others of their lives, limbs, or property for personal gain.

Criminaloid: One of Lombroso's criminal types. They had none of the physical peculiarities of the born or insane criminal and were considered less dangerous.

Criminology: An interdisciplinary science that gathers and analyzes data on crime and criminal behavior.

Critical criminology: An umbrella term for a variety of theories united only by the assumption that conflict and power relations between various classes of people best characterize the nature of society.

Cybercrime: A wide variety of crimes committed with computer technology.

Dark figure of crime: The dark (or hidden) figure of crime refers to all of the crimes committed that never come to official attention.

Decommodification: The process of freeing social relationships from economic considerations.

Definitions: Term used by Edwin Sutherland to refer to meanings our experiences have for us, our attitudes, values, and habitual ways of viewing the world.

Delinquency: A legal term that distinguishes between youthful (juvenile) offenders and adult offenders. Acts forbidden by law are called delinquent acts when committed by juveniles.

Department of Homeland Security: Established after the 9/11 attack, its mission is to detect, prevent, prepare for, and recover from terrorist attacks within the United States.

Deterrence: The prevention of criminal acts by the use or threat of punishment; deterrence may be either *specific* or *general.*

Diathesis/stress model: A biosocial model applied to exploring the causes of schizophrenia and bipolar disorder that maintains that mental illnesses reflect an underlying genetic vulnerability (diathesis) interacting with multiple stressful environmental factors.

Differential association theory: Criminological theory devised by Edwin Sutherland asserting that criminal behavior is behavior learned through association with others who communicate their values and attitudes.

Differential detection hypothesis: The hypothesis that low IQ is related to offending because low-IQ offenders are easier to detect.

Differential social organization: Phrase used by Edwin Sutherland to describe lower-class neighborhoods that others saw as disorganized or pathological.

Discrimination: A term applied to stimuli that provide clues that signal whether a particular behavior is likely to be followed by reward or punishment.

Disintegrative shaming: Shaming that results in criminals being shunned and alienated from society.

Drug addiction: Compulsive drug-seeking behavior where acquiring and using a drug becomes the most important activity in the user's life.

Ecological fallacy: The process of making inferences about individuals and groups on the basis of information derived from a larger population of which they are a part.

Economic-compulsive violence: Violence associated with efforts to obtain money to finance the cost of illicit drugs.

Emancipation hypothesis: The assumption that as women become freer to move into male occupations, they will find and take advantage of more criminal opportunities.

Embezzlement: The misappropriation or misapplication of money or property entrusted to the embezzler's care, custody, or control.

Empathy: The emotional and cognitive ability to understand the feelings and distress of others as if they were your own—to be able to "walk in another's shoes."

Enlightenment: A major intellectual shift in the way people viewed the world and their place in it, questioning traditional religious and political values and substituting humanism, rationalism, and naturalism over supernaturalism.

Ethnic succession theory: Theory about the causes of organized crime that posits that, upon arrival in the United States, each ethnic group was faced with prejudicial and discriminatory attitudes that denied them legitimate means to success in America.

Evolutionary psychology: A way of thinking about human behavior using a Darwinian evolutionary theoretical framework.

Evolutionary theory of rape: This theory of rape traces the male propensity to engage in it to evolutionary selection for aggressive tactics in the pursuit of sexual outlets.

Exhibitionism: The exposure of one's genitals to a stranger of either gender for the offender's sexual pleasure.

Experience-dependent brain development: The development of the brain that reflects each person's unique developmental history.

Experience-expected brain development: Hardwired brain development that reflects the evolutionary history of the human species.

Farrington's integrated cognitive antisocial potential (ICAP) theory: Theory based on the notion that people have varying levels of antisocial propensity due to a variety of environmental and biological factors.

Felony: The most serious form of crime; it carries a maximum penalty of greater than 1 year of imprisonment.

Feminist theory of rape: A theory of rape that asserts that rape is motivated by power, not sexual desire, and that it is a crime of violence and degradation designed to intimidate and keep women in their place.

Fence: A person who regularly buys stolen property for resale and who often has a legitimate business to cover his activities.

Feudal model: A model of La Cosa Nostra that sees it as similar to the old European feudal system based on patronage, oaths of loyalty, and semi-autonomy.

Flight/fight system: An autonomic nervous system mechanism that mobilizes the body for action in response to threats by pumping out epinephrine.

Flynn effect: The upward creep in average IQ scores that has been taking place across the past three or four generations in all countries examined.

Focal concerns: Miller's description of the value system and a lifestyle of the lowest classes; they are trouble, toughness, excitement, smartness, fate, and autonomy.

Forcible rape: Defined by the FBI as "the carnal knowledge of a female forcibly and against her will."

Forgery: The creation or alteration of documents to give them the appearance of legality and validity with the intention of gaining some fraudulent benefit from doing so.

Fraud: Obtaining the money or property of another through deceptive practices such as false advertising, impersonation, and other misrepresentations.

Free will: That which enables human beings to purposely and deliberately choose to follow a calculated course of action.

Frotteurism: The desire to press the penis against unsuspecting persons.

Gender ratio problem: An issue in feminist criminology that asks what explains the universal fact that women are far less likely than men to involve themselves in criminal activity.

Gene-environment correlation: The notion that genotypes and the environments they find themselves in are related because parents provide children with both.

Gene-environment interaction: The interaction of a genotype with its environment: people are differentially sensitive to identical environmental influences because of their genes and will thus respond in different ways to them.

General deterrence: The assumed preventive effect of the threat of punishment on the general population (i.e., *potential* offenders).

General strain theory: Agnew's extension of anomie theory into the realm of social psychology stressing multiple sources of strain and how people cope with it.

Generalizability problem: An issue in feminist criminology that asks whether traditional male-centered theories of crime apply to women.

Genes: Strands of DNA that code for the amino acid sequences of proteins.

Genotype: A person's genetic makeup.

Grand jury: An investigatory jury composed of 7 to 23 citizens before which the prosecutor presents evidence that sufficient grounds exist to try the suspect for a crime. If the prosecutor is successful, he or she obtains an indictment from the grand jury listing the charges a person is accused of.

Hacker: A person who illicitly accesses someone else's computer system.

Harm: The legal principle that states that a crime must have a negative impact on either the victim or the general values of the community to be a crime.

Harrison Narcotic Act: A 1914 congressional act that criminalized the sale and use of narcotics.

Hedonism: A doctrine assuming that the achievement of pleasure or happiness is the main goal of life.

Hedonistic calculus: Combining hedonism and rationality to logically weigh the anticipated benefits of a given course of action against its possible costs.

Hedonistic serial killer: A killer who kills for the pure thrill and joy of it.

Heritability: A concept defined by a number ranging between 0 and 1 indicating the extent to which variance in a phenotypic trait in a population is due to genetic factors.

Hierarchy rule: A rule requiring the police to report only the most serious offense committed in a multiple-offense single incident to the FBI and to ignore the others.

Hizballah: A state-funded Shi'ite terrorist organization organized by the Iranian religious leader Ayatollah Khomeini.

Honor subcultures: Communities in which young men are hypersensitive to insult, rushing to defend their reputation in dominance contests.

Hypotheses: Statements about relationships between and among factors we expect to find based on the logic of our theories.

Identity theft: The use of someone else's personal information without their permission to commit an illegal act.

Ideology: A way of looking at the world, a general emotional picture of "how things should be" that forms, shapes, and colors our concepts of the phenomena we study.

Impulsiveness: A personality trait reflecting people's varying tendencies to act on matters without giving much thought to the possible consequences (not looking before one leaps).

Insane criminal: One of Lombroso's criminal types. Insane criminals bore some stigmata but were not born criminals. Among their ranks were alcoholics, kleptomaniacs, nymphomaniacs, and child molesters.

Institutional anomie theory: Messner and Rosenfeld's extension of anomie theory, which avers that high crime rates are intrinsic to the structural and cultural arrangements of American society.

Institutional balance of power: The notion that there is an imbalance of power among American institutions because all noneconomic institutions are subservient to the economy.

Intellectual imbalance: A significant difference between a person's verbal and performance IQ scores.

Involuntary manslaughter: A criminal homicide where an unintentional killing results from a reckless act.

Involvement: In social control theory, involvement is one of the four social bonds. It refers to a pattern of involvement in conventional activities that prevents one's involvement in criminal activities.

Italian school of criminology: Positivist school of criminology associated with Cesare Lombroso, Raffael Garofalo, and Enrico Ferri.

Kleptomania: "Stealing madness": repetitive impulsive stealing for the thrill of stealing and getting away with it.

La Cosa Nostra: Literally, *our thing;* also commonly referred to as the Mafia, an organized crime group of Italian/Sicilian origins.

Larceny-theft: Defined by the FBI as "the unlawful taking, leading, or riding away from the possession or constructive possession of another."

Latent trait: An assumed "master trait" said to influence behavioral choices across time and situations.

Laws of imitation: A set of "laws" devised by Gabriel Tarde to understand the processes whereby people learn criminal behavior.

Left realist criminology: An approach to crime that maintains that although inequality is a cause of crime, the best solution is to work within the system to prevent and control crime.

Level of analysis: That segment of the phenomenon of interest that is measured and analyzed (i.e., individuals, families, neighborhoods, states, etc.).

Lifestyle theory: A theory stressing that crime is not just a behavior but a general pattern of life.

Linkage blindness: The problem of making connections between murders committed in various police jurisdictions.

Lumpenproletariat: The lower classes; the criminal class.

Madrasas: Islamic religious schools that stress the immorality and materialism of Western life and the need to convert all infidels to Islam.

Mala in se: Universally condemned crimes that are "inherently bad."

Mala prohibita: Crimes that are "bad" simply because they are prohibited.

Masculinization hypothesis: The assumption that as females adopt "male" roles and masculinize their attitudes and behavior, they will commit as much crime as men.

Mass murder: The killing of several people at one location within a few minutes or hours.

Mating effort: The proportion of total reproductive effort allotted to acquiring sexual partners; traits facilitating mating effort are associated with antisocial behavior.

Maturity gap: In Moffitt's theory, the gap between the average age of puberty and the acquisition of socially responsible adult roles.

Mechanical solidarity: A form of social solidarity existing in small, isolated, prestate societies in which individuals sharing common experiences and circumstances share common values and develop strong emotional ties to the collectivity.

Mens rea: Literally, *guilty mind;* refers to whether the suspect had a wrongful purpose in mind when carrying out the *actus reus.*

Mental disorder: A clinically significant condition characterized by alterations in thinking, mood (emotions), or behavior associated with personal distress and/or impaired functioning.

Middle-class measuring rods: According to Cohen, because low-class youths cannot measure up to middle-class standards, they experience status frustration, and this frustration spawns an oppositional culture.

Misdemeanor: A less serious crime than a felony; it carries a maximum penalty of less than 1 year in jail.

Mission-oriented serial killer: A killer who feels it to be a mission in life to kill certain kinds of people.

Mobilization for Youth: A delinquency prevention project design by Cloward and Ohlin that concentrated on expanding legitimate opportunities for disadvantaged youths via a number of educational, training, and job placement programs.

Modes of adaptation: Robert Merton's concept of how people adapt to the alleged disjunction between cultural goals and structural barriers to the means of obtaining them. These modes are conformity, ritualism, retreatism, innovation, and rebellion.

Moffitt's dual-pathway developmental theory: Theory based on the notion that there are two main pathways to offending: One pathway is followed by individuals with neurological and temperamental difficulties that are exacerbated by inept parenting, the other by "normal" individuals temporarily derailed during adolescence.

Moral reasoning: The ability to use conscious thought processes to arrive at solutions to a problem that is in accordance with praiseworthy virtuous standards.

Motor vehicle theft: Defined by the FBI as "the theft or attempted theft of a motor vehicle."

Murder: Defined by the FBI as "the willful (nonnegligent) killing of one human being by another."

National Crime Victimization Survey: A biannual survey of a large number of people and households requesting information on crimes committed against individuals and households (whether reported to the police or not) and for circumstances of the offense (time and place it occurred, perpetrator's use of a weapon, any injuries incurred, and financial loss).

National Incident-Based Reporting System: A comprehensive crime statistic collection system that is currently a component of the *UCR* program and is eventually expected to replace it entirely.

Natural selection: The evolutionary process that selects genetic variants that best fit organisms in their present environments and preserves them in later generations.

Naturalistic fallacy: The fallacy of confusing what *is* (a fact) with what *ought* to be (a moral judgment).

Necessary cause: A factor that *must* be present for something to occur and in the absence of which it has never occurred.

Negative emotionality: A personality trait that refers to the tendency to experience many situations as aversive and to react to them with irritation and anger more readily than with positive affective states.

Negligent manslaughter: An unintentional homicide that is charged when a death or deaths arise from some negligent act that carries a substantial risk of death to others.

Net advantage theory: A theory based on the idea that any choice we make rests on cognitive and emotional calculations. Criminals lack the ability to appreciate the long-term consequences of their behavior.

Neural Darwinism: The process by which synapses are selected or eliminated based on those used most often.

Neurons: Brain cells consisting of the cell body, an axon, and a number of dendrites.

Neurotransmitters: Brain chemical that carry messages from neuron to neuron across the synaptic gap.

Occupational crime: Crime committed by individuals in the course of their employment.

Operant psychology: A perspective on learning that asserts that behavior is governed and shaped by its consequences (reward or punishment).

Opportunity: In self-control theory, opportunity refers to a situation conducive to antisocial behavior presenting itself to a person with low self-control. Low self-control and a criminal opportunity are necessary for crime to occur.

Opportunity structure theory: An extension of anomie theory claiming that lower-class youths join gangs as a path to monetary success.

Organic solidarity: A form of social solidarity characteristic of modern societies in which there is a high degree of occupational specialization and a weak normative consensus.

Organized crime: A continuing criminal enterprise that works rationally to profit from illicit activities that are often in great public demand. Its continuing existence is maintained through the use of force, threats, and/or corruption of public officials.

Palestine Liberation Organization: An umbrella organization for several terrorist groups serving a variety of ideologies and agendas united by Palestinian nationalism.

Parenting effort: The proportion of total reproductive effort invested in rearing offspring; traits facilitating parenting effort are associated with prosocial behavior.

Parole: A conditional release from prison granted to inmates some time prior to the completion of their sentences.

Part I offenses (or **Index Crimes**): The four violent (homicide, assault, forcible rape, and robbery) and four property offenses (larceny/theft, burglary, motor vehicle theft, and arson) reported in the *Uniform Crime Reports*.

Part II offenses: The less serious offenses reported in the *Uniform Crime Reports* and recorded based on arrests made rather than cases reported to the police.

Peacemaking criminology: A humanistic approach to crime that claims punitive approaches are counterproductive.

Personality: The relatively enduring, distinctive, integrated, and functional set of psychological characteristics that results from people's temperaments interacting with their cultural and developmental experiences.

Pharmacological violence: Violence induced by the pharmacological properties of a drug.

Phenotype: The observable and measurable behavioral and personality characteristics of any living thing that are the result of genes interacting with the environment.

Physical dependence: The state in which a person is physically dependant on a drug because of changes to the body that have occurred after repeated use of it and necessitate its continued administration to avoid withdrawal symptoms.

Policy: A course of action designed to solve some problem that has been selected by appropriate authorities from among alternative courses of action.

Positivism: An extension of the scientific method—from which more *positive* knowledge can be obtained—to social life.

Postmodernist criminology: A critical theory/radical tradition in that it views the law as an oppressive instrument of the rich and powerful, but unlike other critical approaches, it rejects the "modernist" view of the world.

Power/control serial killer: A killer who gains the most satisfaction from exercising complete power over his victims.

Prefrontal cortex: Part of the brain that occupies about one-third of the front part of the cerebrum. It has many connections with other brain structures and plays the major integrative and supervisory roles in the brain.

Preliminary arraignment: The presenting of suspects in court before a magistrate or municipal judge to advise them of their constitutional rights and of the tentative charges against them, as well as to set bail.

Preliminary hearing: A proceeding before a magistrate or municipal judge in which three major matters must be decided: (1) whether or not a crime has actually been committed, (2) whether or not there are reasonable grounds to believe that the person before the bench committed it, and (3) whether or not the crime was committed in the jurisdiction of the court.

Primary deviance: In labeling theory, the initial nonconforming act that comes to the attention of the authorities resulting in the application of a criminal label.

Primitive rebellion hypothesis: The Marxist hypothesis that crime is the product of unjust, alienating, and demoralizing social conditions that denied productive labor to masses of unemployed.

Principle of utility: A principle that posits that human action should be judged moral or immoral by its effect on the happiness of the community and that the proper function of the legislature is to promulgate laws aimed at maximizing the pleasure and minimizing the pain of the largest number in society—"the greatest good for the greatest number."

Probable cause: Legal standard for making a warrantless arrest (i.e., the officer must possess a set of facts that would lead a reasonable person to conclude that the arrested person had committed a specific felony crime).

Probation: A probation sentence is a suspended commitment to prison that is conditional on the offender's good behavior.

Profiling: A method used to develop a typology of serial killers and other violent offenders based on personality and other offender characteristics to narrow the range of suspects.

Prohibition: Common term for the Volstead Act, which prohibited the sale, manufacture, or importation of intoxicating liquors within the United States.

Proletariat: The working class.

Prostitution: The provision of sexual services in exchange for money or other tangible reward as the primary source of income.

Psychological dependence: The deep craving for a drug and the feeling that one cannot function without it; psychological dependence is synonymous with addiction.

Psychopathy: A syndrome characterized by the inability to tie the social emotions with cognition. Psychopaths come from all social classes and may or may not be criminals.

Puberty: A developmental stage that marks the onset of the transition from childhood to adulthood and prepares us for procreation.

Punishment: A process that leads to the weakening or eliminating of the behavior preceding it.

Rape trauma syndrome: A syndrome sometimes suffered by rape victims that is similar to post-traumatic stress syndrome (reexperiencing the event via "flashbacks," avoiding anything at all associated with the event, and a general numbness of affect).

Rational: Rational behavior is behavior consistent with logic; a logical "fit" between the goals people strive for and the means they use to achieve them.

Rational choice theory: A neoclassical theory asserting that offenders are free actors responsible for their own actions. Rational choice theorists view criminal acts as specific examples of the general principle that all human behavior reflects the rational pursuit of benefits and advantages. People are conscious social actors free to choose crime, and they will do so if they perceive that its utility exceeds the pains they might conceivably expect if discovered.

Recidivism: Refers to "falling back" into criminal behavior after having being punished.

Reinforcement: A process that leads to the strengthening of behavior.

Reintegrative shaming: A form of shaming that condemns the offender's *acts* without condemning his or her personhood, designed to reintegrate the offender into society.

Restorative justice: A system of mediation and conflict resolution oriented toward repairing the harm that has been caused by crime through face-to-face meetings between offender and victim.

Reward dominance theory: A neurological theory based on the proposition that behavior is regulated by two opposing mechanisms, the behavioral activating system (BAS) and the behavioral inhibition system (BIS).

RICO statutes: Statutes that specifically target the continuing racketeering activities of organized criminals and provide for more severe penalties for the same crimes that fall under traditional criminal statutes and for the seizure of property and assets obtained from or involved in illegal activities.

Risk factor: Something in individuals' personal characteristics or their environment that increases the probability of offending.

Robbery: Defined by the FBI as "the taking or attempted taking of anything of value from the care, custody, or control of a person or persons by force or threat of force or violence and/or putting the victim in fear."

Routine activities theory: A neoclassical theory pointing to the routine activities in that society or neighborhood that invite or prevent crime. Routine activities are defined as "recurrent and prevalent activities which provide for basic population and individual needs." Crime is the result of (a) *motivated offenders* meeting (b) *suitable targets* that lack (c) *capable guardians.*

Routine activities/lifestyle theory: A victimization theory that states that there are certain lifestyles (routine activities) that disproportionately expose some people to high risk for victimization.

Sampson and Laub's age-graded developmental theory: Theory stressing the power of informal social controls to explain onset, continuance, and desisting from crime. Emphasizes the concepts of social capital, turning points in life, and human agency.

Sarbanes-Oxley Act: An act passed in 2002 in response to numerous corporate scandals. The provisions of this act include increased funding for the Securities and Exchange Commission, penalty enhancement for white-collar crimes, and the relaxing of some legal impediments to gaining convictions.

Schizophrenia: A group of mental disorders involving auditory and visual hallucinations and general psychosocial deterioration.

Secondary deviance: Deviance that results from society's reaction to offenders' primary deviance.

Self-concept: How people view themselves. In containment theory, it is an important source of social control.

Self-report surveys: The collecting of data by criminologists themselves asking people to disclose their delinquent and criminal involvement on anonymous questionnaires.

Sensation seeking: The active desire for novel, varied, and extreme sensations and experiences often to the point of taking physical and social risks to obtain them.

Serial murder: The killing of three or more victims over an extended period of time.

Short-run hedonism: The seeking of immediate gratification of desires without regard for any long-term consequences.

Social bonding theory: A social control theory focusing on a person's bonds to others.

Social capital: The store of positive relationships in social networks built on norms of reciprocity and trust developed over time upon which the individual can draw for support.

Social control: Any action on the part of others, deliberate or not, that facilitates conformity to social rules.

Social defense: A theory of punishment promulgated by the Italian school of criminology asserting that its purpose is not to deter or to rehabilitate but to defend society against criminals.

Social disorganization: The central concept of the Chicago school of social ecology. It refers to the breakdown or serious dilution of the power of informal community rules to regulate conduct in poor neighborhoods.

Social ecology: Term used by the Chicago school to describe the interrelations of human beings and the communities in which they live.

Social learning theory: A theory designed to explain how people learn criminal behavior using the psychological principles of operant conditioning.

Social learning theory of rape: A theory of rape that asserts that rape is caused by differences in the way women and men are sexually socialized.

Social structure: How society is organized by social institutions—the family and educational, religious, economic, and political institutions—and stratified on the basis of various roles and statuses.

Software piracy: Illegally copying and distributing computer software.

Specific deterrence: The effect of punishment on the future behavior of the person who experiences the punishment.

Spree murder: The killing of several people at different locations over several days.

Status frustration: A form of frustration experienced by lower-class youths who desire approval and status but who cannot meet middle-class criteria and thus seek status via alternative means.

Staying alive hypothesis: The idea that women are less criminal than men because they have evolved a propensity to experience more situations as fearful than men do. This fear keeps women and their children away from danger and thus aids their reproductive success.

Subculture of violence: A part of a larger culture in which the norms, attitudes, and values of its people legitimize the use of violence.

Sufficient cause: A factor that is able to produce an effect without being augmented by some other factor.

Symbolic interactionism: A perspective in sociology that focuses on how people interpret and define their social reality and the meanings they attach to it in the process of interacting with one another via language (symbols).

Synapse: The gap separating the axon of the sending neuron and the axon of the receiving neuron across which neurotransmitters travel.

Synaptogenesis: The process of "soft wiring" the brain via experience.

Systemic violence: Violence associated with aggressive patterns of interaction within the system of drug distribution and use.

Techniques of neutralization: Techniques by which offenders justify their behavior as "acceptable" on a number of grounds.

Temperament: An individual characteristic identifiable as early as infancy that constitutes a habitual mode of emotionally responding to stimuli.

Terrorism: The FBI defines terrorism as "the unlawful use of force or violence against persons or property to intimidate or coerce a government, the civilian population, or any segment thereof, in furtherance of political or social goals."

Theory: A set of logically interconnected propositions explaining how phenomena are related and from which a number of hypotheses can be derived and tested.

Thinking errors: Criminals' typical patterns of faulty thoughts and beliefs.

Thomas theorem: A statement that summarized the symbolic interactionist position in sociology: If people define situations as real, they become real in their consequences.

Tolerance: The tendency to require larger and larger doses of a drug to produce the same effects after the body adjusts to lower dosages.

Toucheurism: The desire to intimately touch women who are strangers to the perpetrator.

Transition zone: An area or neighborhood in the process of being "invaded" by members of "alien" racial or ethnic groups bringing with them values and practices that conflict with those established by the "natural" inhabitants of the area.

Trial: An adversarial process in which the prosecutor must prove beyond a reasonable doubt that the defendant committed the crime the state accused him or her of committing.

Turning points: Transition events in life (getting married, finding a decent job, moving to a new neighborhood) that may change a person's life trajectory in prosocial directions.

Type I alcoholism: A form of alcoholism characterized by mild abuse, minimal criminality, and passive-dependent personality.

Type II alcoholism: A form of alcoholism characterized by early onset, violence, and criminality and largely limited to males.

Unconstrained vision: One of the two so-called ideological *visions* of the world. The unconstrained vision denies an innate human nature, viewing it as formed anew in each different culture.

Uniform Crime Reports: Annual report compiled by the Federal Bureau of Investigation (FBI) containing crimes known to the nation's police and sheriff's departments, the number of arrests made by these agencies, and other crime-related information.

USA Patriot Act: Passed after the 9/11 attack, it grants federal agencies greater authority and power to combat domestic and foreign terrorists.

Victim precipitation theory: A theory in victimology that examines how violent victimization may have been precipitated by the victim when he or she acts in certain provocative ways.

Victim-offender reconciliation programs: Programs designed to bring victims and offenders together in face-to-face meetings in attempts to iron out ways in which the offender can make amends for the hurt and damage caused to the victim.

Victimology: A subfield of criminology that specializes in studying the victims of crime.

Visionary serial killer: A killer who feels compelled to commit murder by visions or "voices in my head."

Voluntary manslaughter: The intentional killing of another human being without malice and forethought, often in response to the mistaken belief that self-defense required the use of deadly force or to adequate provocation while in the heat of passion.

Voyeurism: The act of secretly observing unsuspecting persons who are naked, in the process of disrobing, or engaging in sexual activity.

White-collar crime: Defined by the U.S. Congress as "an illegal act or series of illegal acts committed by non-physical means and by concealment or guile, to obtain money or property, or to obtain business or personal advantage."

Withdrawal: A process involving a number of adverse physical reactions that occur when the body of a drug abuser is deprived of his or her drugs.

Witness Protection Program: Program administered by the U.S. Marshals Service that provides around-the-clock protection while witnesses are awaiting court appearances and new identification documents, employment, housing, and other assistance after testifying.

Photo Credits

Section 1: Introduction and Overview of Crime and Criminology

Photo 1.1: © Getty Images

Photo 1.2: © RichardMilner/Handout/epa/Corbis

Section 2: Measuring Crime and Criminal Behavior

Photo 2.1: Provided by Cecil Greek

Photo 2.2: © Getty Images

Section 3: The Early Schools of Criminology and Modern Counterparts

Photo 3.1: Provided by Cecil Greek

Photo 3.2: Library of Congress

Section 4: Social Structural Theories

Photo 4.1: © Bettmann/Corbis

Photo 4.2: © gettyimages/Robert Yager.

Section 5: Social Process Theories

Photo 5.1: © Getty Images

Photo 5.2: Provided by Cecil Greek.

Section 6: Critical Theories: Marxist, Conflict, and Feminist

Photo 6.1: © PhotriMicroStock™/L.Balterman

Photo 6.2: @ Associated Press

Section 7: Psychosocial Theories: Individual Traits and Criminal Behavior

Photo 7.1: Provided by Cecil Greek.

Photo 7.2: © Corbis

Section 8: Biosocial Approaches

Photo 8.1: © JASON REED/Reuters/Corbis

Photo 8.2: @ Associated Press

Section 9: Developmental Theories: From Delinquency to Crime to Desistance

Photo 9.1: Provided by Cecil Greek.

Photo 9.2: Provided by Cecil Greek.

Section 10: Violent Crimes

Photo 10.1: Provided by Cecil Greek.

Photo 10.2: © gettyimages/Jack Star.

Photo 10.3: @ Corbis

Photo 10.4: None needed

Section 11: Property Crime

Photo 11.1: © Getty Images

Photo 11.2: Provided by Cecil Greek

Section 12: Public Order Crime

Photo 12.1: © Renee Lee/Istockphoto

Photo 12.2: © Michael Newman/PhotoEdit Inc.

Section 13: White-Collar and Organized Crime

Photo 13.1: Provided by Cecil Greek.

Photo 13.2: Provided by Cecil Greek.

Photo 13.3: Provided by Cecil Greek.

Section 14: Victimology: Exploring the Victimization Experience

Photo 14.1: © Associated Press

Photo 14.2: © Associated Press

References

Section 1

Anderson, D. (1999). The aggregate burden of crime. *Journal of Law and Economics, 42,* 611–642.

Cullen, F. (2005). Challenging individualistic theories of crime. In S. Guarino-Ghezzi & J. Trevino (Eds.), *Understanding crime: A multidisciplinary approach* (pp. 55–60). Cincinnati, OH: Anderson.

Daly, M., & Wilson, M. (1988). *Homicide.* New York: Aldine De Gruyter.

Ellis, L. (1994). *Research methods in the social sciences.* New York: McGraw Hill-Brown & Benchmark.

Gottfredson, M., & Hirschi, T. (1990). *A general theory of crime.* Stanford: Stanford University Press.

Hagan, J. (1985). *Modern criminology: Crime, criminal behavior and its control.* New York: McGraw-Hill.

Hawkins, D. (1995). Ethnicity, race, and crime: A review of selected studies. In D. Hawkins (Ed.), *Ethnicity, race, and crime: Perspectives across time and space* (pp. 11–45). Albany: State University of New York Press.

INTERPOL. (1992). *International crime statistics.* Lyon, France: Author.

Kuhn, T. (1970). *The structure of scientific revolutions.* Chicago: University of Chicago Press.

Menard, S., & Mihalic, S. (2001). The tripartite conceptual framework in adolescence and adulthood: Evidence from a national sample. *Journal of Drug Issues, 31,* 905–940.

Moffitt, T. (1993). Adolescent-limited and life-course-persistent antisocial behavior: A developmental taxonomy. *Psychological Review, 100,* 674–701.

O'Manique, J. (2003). *The origins of justice: The evolution of morality, human rights, and law.* Philadelphia: University of Philadelphia Press.

Patterson, O. (1998). *Rituals of blood: Consequences of slavery in two American centuries.* Washington, DC: Civitas Counterpoint.

Sowell, T. (1987). *A conflict of visions: Ideological origins of political struggles.* New York: William Morrow.

Tappan, P. (1947). Who is the criminal? *American Sociological Review, 12,* 96–112.

Udry, J. R. (2003). *The National Longitudinal Study of Adolescent Health (Add Health).* Chapel Hill, NC: University of North Carolina at Chapel Hill, Carolina Population Center.

Walsh, A. (2000). Evolutionary psychology and the origins of justice. *Justice Quarterly, 17,* 841–864.

Walsh, A., & Ellis, L. (2004). Ideology: Criminology's Achilles' heel? *Quarterly Journal of Ideology, 27,* 1–25.

Section 2

Catalano, S. (2006). *Criminal victimization, 2005.* Washington, DC: Bureau of Justice Statistics.

Cernkovich, S., Giordano, P., & Rudolph, J. (2000). Race, crime, and the American dream. *Journal of Research in Crime and Delinquency, 37,* 131–170.

Dunworth, T. (2001, September). Criminal justice and the IT revolution. *Federal Probation,* 55–65.

Farrington, D. (19[illegible]). Longitudinal analyses of criminal violence. In M. Wolfgang & N. Weiner (Eds.), *Criminal violence* (pp. 171–200). Beverly Hills, CA: Sage.

Federal Bureau of Investigation. (2005). *Crime in the United States, 2004.* Washington, DC: U.S. Government Printing Office.

Federal Bureau of Investigation. (2006). *Crime in the United States, 2005: Uniform Crime Reports.* Washington, DC: U.S. Government Printing Office.

Finkelhor, D., & Ormrod, R. (2004, June). Prostitution of juveniles: Patterns from NIBRS. *Juvenile Justice Bulletin.*

Hindelang, M., Hirschi, T., & Weis, J. (1981). *Measuring delinquency.* Beverly Hills, CA: Sage.

Kim, J., Fendrich, M., & Wislar, J. (2000). The validity of juvenile arrestees' drug use reporting: A gender comparison. *Journal of Research in Crime and Delinquency, 37,* 429–432.

National Center for Policy Analysis. (1998b). *Falsified crime data.* Retrieved from http://www.ncpa.org/pi/crime/aug98a.html.

Nettler, G. (1984). *Explaining crime* (3rd ed.). New York: McGraw-Hill.

O'Brien, R. (2001). Crime facts: Victim and offender data. In Sheley, J. (Ed.). *Criminology: A contemporary handbook* (pp. 59–83). Belmont, CA: Wadsworth.

Section 3

Beccaria, C. (1963). *On crimes and punishments* (H. Paulucci, Trans.). Indianapolis: Bobbs-Merrill. (Original work published 1764)

Bentham, J. (1948). *A fragment on government and an introduction to the principles of morals and legislation* (W. Harrison, Ed.). Oxford: Basil Blackwell. (Original work published 1789)

Boudon, R. (2003). Limitations of rational choice theory. *American Journal of Sociology, 104*(3), 817–828.

Catalano, S. (2005). *Criminal victimization, 2004.* Washington, DC: Bureau of Justice Statistics.

Clarke, R., & Cornish, D. (1985). Modeling offenders' decisions: A framework for research and policy. In M. Tonry & N. Morris (Eds.), *Crime and justice annual review of research.* Chicago: University of Chicago Press.

Cohen, L., & Felson, M. (1979). Social change and crime rate trends: A routine activities approach. *American Sociological Review, 44,* 588–608.

Cornish, D., & Clarke, R. (Eds.). (1986). *The reasoning criminal.* New York: Springer-Verlag.

Cornish, D., & Clarke, R. (1987). Understanding crime displacement: An application of rational choice theory. *Criminology, 25,* 933–947.

Curren, D., & Renzetti, C. (2001). *Theories of crime* (2nd ed.). Boston: Allyn & Bacon.

Ferri, E. (1917). *Criminal sociology.* Boston: Little, Brown. (Original work published 1897)

Garofalo, R. (1968). *Criminology.* Montclair, NJ: Patterson Smith. (Original work published 1885)

Levin, Y., & Lindesmith, A. (1971). English ecology and criminology of the past century. In H. Voss & D. Petersen (Eds.), *Ecology, crime, and delinquency* (pp. 47–76l). New York: Appleton-Century-Crofts.

Lombroso, C. (1876). *Criminal man.* Milan: Hoepli.

Lombroso-Ferrero, G. (1972). *Criminal man according to the classification of Cesare Lombroso.* Montclaire, NJ: Patterson Smith. (Original work published 1911)

Mealey, L. (1995). The sociobiology of sociopathy: An integrated evolutionary model. *Behavioral and Brain Sciences, 18,* 523–541.

McCarthy, B. (2002). New economics of sociological criminology. *Annual Review of Sociology, 28,* 417–442.

Nagin, D. (1998). Criminal deterrence research at the onset of the twenty-first century. In M. Tony (Ed.), *Crime and justice: A review of research* (Vol. 23, pp. 1–42). Chicago: University of Chicago Press.

Newman, O. (1972). *Defensible space.* New York: Macmillan.

Robinson, M. (2005). *Justice blind: Ideals and realities of American criminal justice.* Upper Saddle River, NJ: Prentice Hall.

Vold, G., & Bernard, T. (1986). *Theoretical criminology.* New York: Oxford University Press.

Williams, F., & McShane, M. (2004). *Criminological theory* (4th ed.). Upper Saddle River, NJ: Prentice Hall.

Section 4

Agnew, R. (2002). Foundation for a general strain theory of crime. In S. Cote (Ed.), *Criminological theories: Bridging the past to the future* (pp. 113–124). Thousand Oaks, CA: Sage.

Anderson, E. (1999). *Code of the street: Decency, violence, and the moral life of the inner city.* New York: W.W. Norton.

Bartol, C., & Bartol, A. (1989). *Juvenile delinquency: A systems approach.* Englewood Cliffs, NJ: Prentice-Hall.

Bing, L. (1991). *Do or die.* New York: Harper Collins.

Cloward, R., & Ohlin, L. (1960). *Delinquency and opportunity.* New York: Free Press.

Cohen, A. (1955). *Delinquent boys.* New York: Free Press.

Durkheim, E. (1951). *The division of labor in society.* Glencoe, IL: Free Press.

Durkheim, E. (1982). *Rules of sociological method.* New York: Free Press.

Egley, A., & Major, A. (2004). *Highlights of the 2002 National Youth Gang Survey.* OJJDP Fact Sheet, U.S. Department of Justice, Office of Juvenile Justice and Delinquency Prevention.

Kornhauser, R. (1978). *Social sources of delinquency: An appraisal of analytical methods.* Chicago: University of Chicago Press.

LaFree, G., Drass, K., & O'Day, P. (1992). Race and crime in postwar America: Determinants of African-American and white rates. *Criminology, 30,* 157–185.

Mallon, R. (2007). A field guide to social construction. *Philosophy Compass, 2,* 93–108.

Merton, R. (1938). Social structure and anomie. *American Sociological Review, 3,* 672–682.

Messner, S., & Rosenfeld, R. (2001). *Crime and the American dream* (3rd ed.). Belmont, CA: Wadsworth.

Miller, W. (1958). Lower-class culture as a generating milieu of gang delinquency. *Journal of Social Issues, 14,* 5–19.

Moore, J., & Hagedorn, J. (2001). *Female gangs: Focus on research.* OJJDP Juvenile Justice Bulletin, U.S. Department of Justice.

Oberwittler, D. (2004). A multilevel analysis of neighborhood contextual factors on serious juvenile offending. *European Journal of Criminology, 1,* 201–235.

Rosenbaum, D., Lurigio, A., & Davis, R. (1998). *The prevention of crime: Social and situational strategies.* Belmont, CA: West/Wadsworth.

Sampson, R. (2004). Neighborhood and community: Collective efficacy and community safety. *New Economy, 11*, 106–113.

Sampson, R., Raudenbush, S., & Earls, F. (1997). Neighborhoods and crime: A multilevel study of collective efficacy. *Science, 277*, 918–924.

Shaw, C., & McKay, H. (1972). *Juvenile delinquency and urban areas* (Rev. ed.). Chicago: University of Chicago Press.

Shelden, R., Tracy, S., & Brown, W. (2001). *Youth gangs in American society* (2nd ed.). Belmont, CA: Wadsworth.

Spergel, I. (1995). *The youth gang problem: A community approach.* New York: Oxford University Press.

Stiles, B., Liu, X., & Kaplan, H. (2000). Relative deprivation and deviant adaptations: The mediating effects of negative self-feelings. *Journal of Research in Crime and Delinquency, 37*, 64–90.

van Kesteren, J., Mayhew, P., & Nieuwbeerta, P. (2000). *Criminal victimization in seventeen industrialised countries: Key findings from the 2000 international crime victims survey.* The Hague, Netherlands, Ministry of Justice.

Webster, C., MacDonald, R., & Simpson, M. (2006). Predicting criminality? Risk factors, neighborhood influence and desistance. *Youth Justice, 6*, 7–22.

Wilson, J. (1987). *The truly disadvantaged.* Chicago: University of Chicago Press.

Section 5

Akers, R. (2002). A social learning theory of crime. In S. Cote (Ed.), *Criminological theories: Bridging the past to the future* (pp. 135–143). Thousand Oaks, CA: Sage.

Bernhardt, P. (1997). Influences of serotonin and testosterone in aggression and dominance: Convergence with social psychology. *Current Directions in Psychological Science, 6*, 44–48.

Burgess, R., & Akers, R. (1966). A differential association-reinforcement theory of criminal behavior. *Social Problems, 14*, 128–147.

Glueck, S. (1956). Theory and fact in criminology: A criticism of differential association theory. *British Journal of Criminology, 7*, 92–109.

Goldman, D., Lappalainen, J., & Ozaki, N. (1996). Direct analysis of candidate genes in impulsive behavior. In G. Bock & J. Goode (Eds.), *Genetics of criminal and antisocial behaviour* (pp. 183–195). Chichester, England: Wiley.

Gottfredson, M., & Hirschi, T. (1990). *A general theory of crime.* Stanford, CA: Stanford University Press.

Gottfredson, M., & Hirschi, T. (1997). National crime control policies. In M. Fisch (Ed.), *Criminology 97/98* (pp. 27–33). Guilford, CT: Dushkin.

Grasmick, H., Tittle, C., Bursik, R., & Arneklev, B. (1993). Testing the core empirical implication of Gottfredson and Hirschi's general theory of crime. *Journal of Research in Crime and Delinquency, 30*, 5–29.

Harris, J. (1998). *The nurture assumption: Why children turn out the way they do.* New York: Free Press.

Hirschi, T. (1969). *The causes of delinquency.* Berkeley: University of California Press.

Hur, Y., & Bouchard, T. (1997). The genetic correlation between impulsivity and sensation-seeking traits. *Behavior Genetics, 27*, 455–463.

Lemert, E. (1974). Beyond Mead: The societal reaction to deviance. *Social Problems, 21*, 457–468.

Moffitt, T., & Walsh, A. (2003). The adolescent-limited/life-course persistent theory of antisocial behavior: What have we learned? In A. Walsh & L. Ellis (Eds.), *Biosocial criminology: Challenging environmentalism's supremacy* (pp. 123–144). Hauppauge, NY: Nova Science.

Nettler, G. (1984). *Explaining crime* (3rd ed.). New York: McGraw-Hill.

Rodkin, P., Farmer, T., Pearl, R., & Van Acker, R. (2000). Heterogeneity of popular boys: Antisocial and prosocial configurations. *Developmental Psychology, 36*, 14–24.

Sampson, R., & Laub, J. (1999). Crime and deviance over the lifecourse: The salience of adult social bonds. In F. Scarpitti & A. Nielsen (Eds.), *Crime and criminals: Contemporary and classical readings in criminology* (pp. 238–246). Los Angeles: Roxbury.

Sutherland, E., & Cressey, D. (1974). *Criminology* (9th ed.). Philadelphia: J.B. Lippincott.

Sykes, G., & Matza, D. (2002). Techniques of neutralization: A theory of delinquency. In S. Cote (Ed.), *Criminological theories: Bridging the past to the future* (pp. 144–150). Thousand Oaks, CA: Sage.

Tannenbaum, F. (1938). *Crime and community.* New York: Columbia University Press.

Topalli, V. (2005). When being good is bad: An extension of neutralization theory. *Criminology, 43*, 797–835.

Warr, M. (2002). *Companions in crime: The social aspects of criminal conduct.* New York: Cambridge University Press.

Section 6

Adams, J. (1971). *In defense of the constitution of the United States* (Vol. 1). New York: De Capo Press. (Original work published 1778)

Adler, F. (1975). *Sisters in crime: The rise of the new female criminal.* New York: McGraw-Hill.

Adler, F., Mueller, G., & Laufer, W. (2001). *Criminology and the criminal justice system.* Boston: McGraw-Hill.

Akers, R. (1994). *Criminological theories: Introduction and evaluation.* Los Angeles: Roxbury.

Amateau, S., & McCarthy, M. (2004). Induction of PGE2 by estradiol mediates developmental masculinization of sex behavior. *Nature Neuroscience, 7*, 643–650.

Archer, J. (1996). Sex differences in social behavior: Are the social role and evolutionary explanations compatible? *American Psychologist, 51*, 909–917.

Barash, D., & Lipton, J. (2001). Making sense of sex. In D. Barash (Ed.), *Understanding violence* (pp. 20–30). Boston: Allyn & Bacon.

Bartollas, C. (2005). *Juvenile delinquency* (7th ed.). Boston: Allyn & Bacon.

Betzig, L. (1999). When women win. *Behavioral and Brain Sciences, 22,* 217.

Bohm, R. (2001). *A primer on crime and delinquency* (2nd ed). Belmont, CA: Wadsworth.

Bonger, W. (1969). *Criminality and economic conditions.* Bloomington: Indiana University Press. (Original work published 1905)

Butler, W. (1992). Crime in the Soviet Union: Early glimpses of the true story. *British Journal of Criminology, 32,* 144–159.

Campbell, A. (1999). Staying alive: Evolution, culture, and women's intrasexual aggression. *Behavioral and Brian Sciences, 22,* 203–214.

Chambliss, W. (1976). *Criminal law in action.* Santa Barbara, CA: Hamilton.

Champion, D. (2005). *Probation, parole, and community corrections* (5th ed.). Upper Saddle River, NJ: Prentice Hall.

Chesney-Lind, M., & Shelden, R. (1992). *Girl's delinquency and juvenile justice.* Pacific Grove, CA: Brooks/Cole.

Curran, D., & Renzetti, C. (2001). *Theories of crime* (2nd ed.). Boston: Allyn & Bacon.

Currie, E. (1989). Confronting crime: Looking toward the twenty-first century. *Justice Quarterly, 6,* 5–25.

Daly, K., & Chesney-Lind, M. (1996). Feminism and criminology. In P. Cordella & L. Siegel (Eds.), *Readings in contemporary criminological theory* (pp. 340–364). Boston: Northeastern University Press.

Daly, K., & Chesney-Lind, M. (2002). Feminism and criminology. In S. Cote (Ed.), *Criminological theories: Bridging the past to the future* (pp. 207–284). Thousand Oaks, CA: Sage.

Durrant, W., & Durrant, A. (1968). *The lessons of history.* New York: Simon and Schuster.

Ellis, L. (2003). Genes, criminality, and the evolutionary neuroandrogenic theory. In A. Walsh & L. Ellis (Eds.), *Biosocial criminology: Challenging environmentalism's supremacy* (pp. 12–34). Hauppauge, NY: Nova Science.

Fishbein, D. (1992). The psychobiology of female aggression. *Criminal Justice and Behavior, 19,* 99–126.

Geary, D. (1998). Functional organization of the human mind: Implications for behavioral genetic research. *Human Biology, 70,* 185–198.

Gottfredson, M., & Hirschi, T. (1990). *A general theory of crime.* Stanford, CA: Stanford University Press.

Greenberg, D. (1981). *Crime and capitalism: Readings in Marxist criminology.* Palo Alto, CA: Mayfield.

Harris, K. (1991). Moving into the new millennium: Toward a feminist view of justice. In H. Pepinsky & R. Quinney (Eds.), *Criminology as peacemaking* (pp. 83–97.). Bloomington: Indiana University Press.

Kanazawa, S. (2003). A general evolutionary psychological theory of criminality and related male-typical behavior. In A. Walsh & L. Ellis (Eds.), *Biosocial criminology: Challenging environmentalism's supremacy* (pp. 37–60). Hauppauge, NY: Nova Science.

Kimura, D. (1992). Sex differences in the brain. *Scientific American, 267,* 119–125.

Lanier, M., & Henry, S. (1998). *Essential criminology.* Boulder, CO: Westview.

Laub, J., & McDermott, M. (1985). An analysis of serious crime by young black women. *Criminology, 23,* 89–98.

Leonard, E. (1995). Theoretical criminology and gender. In B. Price & N. Sokoloff (Eds.), *The criminal justice system and women: Offenders, victims, and workers* (pp. 54–70). New York: McGraw-Hill.

Lytton, H., & Romney, D. (1991). Parents differential socialization of boys and girls: A meta-analysis. *Psychological Bulletin, 109,* 267–296.

Marx, K., & Engels, F. (1948). *The communist manifesto.* New York: International.

Marx, K., & Engels, F. (1965). *The German ideology.* London: Lawrence and Wishart.

Mears, D., Ploeger, M., & Warr, M. (1998). Explaining the gender gap in delinquency: Peer influence and moral evaluations of behavior. *Journal of Research in Crime and Delinquency, 35,* 251–266.

Siegel, L. (1986). *Criminology.* Belmont, CA: Wadsworth.

Simon, R. (1975). *Women and crime.* Lexington, MA: Lexington Books.

Taylor, I. (1999). Crime and social criticism. *Social Justice, 26,* 150–168.

Taylor, I., Walton, P., & Young, J. (1973). *The new criminology.* New York: Harper & Row.

Triplett, R. (1993). The conflict perspective, symbolic interactionism, and the status characteristics hypothesis. *Justice Quarterly, 10,* 541–556.

Vold, G., & Bernard, T. (1986). *Theoretical criminology.* New York: Oxford University Press.

Vold, G., Bernard, T., & Snipes, J. (1998). *Theoretical criminology.* New York: Oxford University Press.

Walsh, A. (2002). *Biosocial criminology: Introduction and integration.* Cincinnati, OH: Anderson.

Walsh, A., & Ellis, L. (2007). *Criminology: An interdisciplinary approach.* Thousand Oaks, CA: Sage.

Walsh, A., & Hemmens, C. (2000). *From law to order: The theory and practice of law and justice.* Lanham, MD: American Correctional Association.

Section 7

Adler, F., Mueller, G., & Laufer, W. (2001). *Criminology and the criminal justice system.* Boston: McGraw Hill.

Akers, R. (1994). *Criminological theories: Introduction and evaluation.* Los Angeles: Roxbury.

American Psychiatric Association. (1994). *Diagnostic and statistical manual of mental disorders* (4th ed.). Washington, D.C.: American Psychiatric Association.

Andrews, D., & Bonta, J. (1998). *The psychology of criminal conduct.* Cincinnati, OH: Anderson.

Barber, N. (2004). Single parenthood as a predictor of cross-national variation in violent crime. *Cross Cultural Research, 38,* 343–358.

Bartol, C. (2002). *Criminal behavior: A psychosocial approach* (6th ed.). Upper Saddle River, NJ: Prentice Hall.

Brennan, P., Raine, A., Schulsinger, F., Kirkegaard-Sorenen, L., Knop, J., Hutchings, B., et al. (1997). Psychophysiological protective factors for male subjects at high risk for criminal behavior. *American Journal of Psychiatry, 154,* 853–855.

Carey, G. (2003). *Human genetics for the social scientists.* Thousand Oaks, CA: Sage.

Caspi, A. (2000). The child is the father of the man: Personality continuities from childhood to adulthood. *Journal of Personality and Social Psychology, 78,* 158–172.

Caspi, A., Moffitt, T., Silva, P., Stouthamer-Loeber, M., Krueger, R., & Schmutte, P. (1994). Are some people crime-prone? Replications of the personality-crime relationship across countries, genders, races, and methods. *Criminology, 32,* 163–194.

Cleveland, H., Wiebe, R., van den Oord, E., & Rowe, D. (2000). Behavior problems among children from different family structures: The influence of genetic self-selection. *Child Development, 71,* 733–751.

Covell, C., & Scalora, M. (2002). Empathetic deficits in sexual offenders: An integration of affective, social, and cognitive constructs. *Aggression and Violent Behavior, 37,* 251–270.

Ellis, L. (2003). Genes, criminality, and the evolutionary neuroandrogenic theory. In A. Walsh & L. Ellis (Eds.), Biosocial criminology: Challenging environmentalism's supremacy (pp. 12–34). Hauppauge, NY: Nova Science.

Ellis, L., & Walsh, A. (2000). *Criminology: A global perspective.* Boston: Allyn & Bacon.

Ellis, L., & Walsh, A. (2003). Crime, delinquency and intelligence: A review of the worldwide literature. In H. Nyborg (Ed.), *The scientific study of general intelligence: A tribute to Arthur Jensen* (pp. 343–365). Amsterdam: Pergamon.

Fishbein, D. (2001). *Biobehavioral perspectives in criminology.* Belmont, CA: Wadsworth.

Gibson, M. (2002). *Born to crime: Cesare Lombroso and the origins of biological criminology.* Westport, CT: Praeger.

Gottfredson, M., & Hirschi, T. (1997). National crime control policies. In M. Fisch (Ed.), *Criminology 97/98* (pp. 27–33). Guilford, CT: Dushkin.

Hare, R. (1993). *Without conscience: The disturbing world of the psychopaths among us.* New York: Pocket Books.

Harris, G., Skilling, T., & Rice, M. (2001). The construct of psychopathy. In M. Tonry (Ed.), *Crime and justice: A review of research* (pp. 197–264). Chicago: University of Chicago Press.

Herrnstein, R., & Murray, C. (1994). *The bell curve: Intelligence and class structure in American life.* New York: Free Press.

Kochanska, M. (1991). Socialization and temperament in the development of guilt and conscience. *Child Development, 62,* 1379–1392.

Lykken, D. (1995). *The antisocial personalities.* Hillsdale, NJ: Lawrence Erlbaum.

Lynam, D., Moffitt, T., & Stouthamer-Loeber, M. (1993). Explaining the relation between IQ and delinquency: Class, race, test motivation, school failure, or self control? *Journal of Abnormal Psychology, 102,* 187–196.

Matarazzo, J. (1976). *Weschler's measurement and appraisal of adult intelligence.* Baltimore, MD: Williams & Wilkins.

Maynard, R., & Garry, E. (1997). Adolescent motherhood: Implications for the juvenile justice system. *OJJDP Fact sheet #50.* U.S. Department of Justice.

McGue, M., Bacon, S., & Lykken, D. (1993). Personality stability and change in early adulthood: A behavioral genetic analysis. *Developmental Psychology, 29,* 96–109.

Mealey, L. (1995). The sociobiology of sociopathy: An integrated evolutionary model. *Behavioral and Brain Sciences, 18,* 523–541.

Miller, L. (1987). Neuropsychology of the aggressive psychopath: An integrative review. *Aggressive Behavior, 13,* 119–140.

Moffitt, T. (1993). Adolescent-limited and life-course-persistent antisocial behavior: A developmental taxonomy. *Psychological Review, 100,* 674–701.

Moffit, T., & The E-Risk Study Team. (2002). Teen-aged mothers in contemporary Britain. *Journal of Child Psychology and Psychiatry, 43,* 1–16.

Neisser, U., Boodoo, G., Bouchard, T., Boykin, A., Brody, N., Ceci, S., et al. (1995). *Intelligence: Knowns and unknowns: Report of a task force established by the board of scientific affairs of the American Psychological Association.* Washington, DC: American Psychological Association.

Pinel, J. (2000). *Biopsychology* (4th ed.). Boston: Allyn and Bacon.

Quinsey, V. (2002). Evolutionary theory and criminal behavior. *Legal and Criminological Psychology, 7,* 1–14.

Raine, A. (1997). Antisocial behavior and psychophysiology: A biosocial perspective and a prefrontal dysfunction hypothesis. In D. Stoff, J. Breiling, & J. Maser (Eds.), *Handbook of antisocial behavior* (pp. 289–304). New York: John Wiley.

Rosenbaum, D., Lurigio, A., & Davis, R. (1998). *The prevention of crime: Social and situational strategies.* Belmont, CA: West/Wadsworth.

Rothbart, M., Ahadi, A., & Evans, D. (2000). Temperament and personality: Origins and outcomes. *Journal of Personality and Social Psychology, 78,* 122–135.

Rowe, D. (2001). *Biology and crime.* Los Angeles: Roxbury.

Rowe, D. (2002). On genetic variation in menarche and age at first sexual intercourse: A critique of the Belsky-Draper hypothesis. *Evolution and Human Nature, 23,* 365–372.

Scarpa, A., & Raine, A. (2003). The psychophysiology of antisocial behavior: Interactions with environmental experiences. In A. Walsh, & L. Ellis (Eds.), *Biosocial criminology: Challenging environmentalism's supremacy* (pp. 209–226). Hauppaugue, NY: Nova Science.

Seligman, D. (1992). *A question of intelligence: The IQ debate in America.* New York: Birch Lane

Sherman, L., Gottfredson, D., McKenzie, D., Eck, J., Reuter, P., & Bushway, S. (1997). *Preventing crime: What works, what doesn't, what's promising.* Washington, DC: U.S. Department of Justice.

Vold, G., Bernard, T., & Snipes, J. (1998). *Theoretical criminology* (4th ed.). New York: Oxford University Press.

Walsh, A. (2003). Intelligence and antisocial behavior. In A. Walsh & L. Ellis (Eds.), *Biosocial criminology: Challenging environmentalism's supremacy* (pp. 105–124). Hauppauge, NY: Nova Science.

Walters, G. (1990). *The criminal lifestyle.* Newbury Park, CA: Sage.

Walters, G., & White, T. (1989). The thinking criminal: A cognitive model of lifestyle criminality. *Criminal Justice Research Bulletin.* Sam Houston State University.

Ward, D., & Tittle, C. (1994). IQ and delinquency: A test of two competing explanations. *Journal of Quantitative Criminology, 10,* 189–212.

Weibe, R. (2004). Psychopathy and sexual coercion: A Darwinian analysis. *Counseling and Clinical Psychology Journal, 1,* 23–41.

Zuckerman, M. (1990). The psychophysiology of sensation-seeking. *Journal of Personality, 58,* 314–345.

Section 8

Badcock, C. (2000). *Evolutionary psychology: A critical introduction.* Cambridge, England: Polity Press.

Barkow, J. (Ed.). (2006). *Missing the revolution: Darwinism for social scientists.* Oxford: Oxford University Press.

Beaver, K. (2006). *The intersection of genes, the environment, and crime and delinquency: A longitudinal study of offending.* Unpublished doctoral dissertation, Division of Research and Advanced Studies of the University of Cincinnati, Ohio.

Beaver, K., Wright, J., & Walsh, A. (in press). A gene-based evolutionary explanation for the association between criminal involvement and number of sex partners. *Social Biology.*

Carey, G. (2003). *Human genetics for the social sciences.* Thousand Oaks, CA: Sage.

Cartwright, J. (2000). *Evolution and human behavior.* Cambridge, MA: MIT Press.

Caspi, A., McClay, J., Moffitt, T., Mill, J., Martin, J., Craig, I., et al. (2002). Evidence that the cycle of violence in maltreated children depends on genotype. *Science, 297,* 851–854.

Cauffman, E., Steinberg, L., & Piquero, A. (2005). Psychological, neuropsychological and physiological correlates of serious antisocial behavior in adolescence: The role of self-control. *Criminology, 43,* 133–175.

Cleveland, H., Wiebe, R., van den Oord, E., & Rowe, D. (2000). Behavior problems among children from different family structures: The influence of genetic self-selection. *Child Development, 71,* 733–751.

Depue, R., & Collins, P. (1999). Neurobiology of the structure of personality: Dopamine, facilitation of incentive motivation, and extraversion. *Behavioral and Brain Sciences, 22,* 491–569.

Ellis, L., & Walsh, A. (1997). Gene-based evolutionary theories in criminology. *Criminology, 35,* 229–276.

Ellis, L., & Walsh, A. (2000). *Criminology: A global perspective.* Boston: Allyn & Bacon.

Ember, M., & Ember, C. (1998, October). Facts of violence. *Anthropology Newsletter,* 14–15.

Fisher, H. (1998). Lust, attraction, and attachment in mammalian reproduction. *Human Nature, 9,* 23–52.

Geary, D. (2000). Evolution and proximate expression of human paternal investment. *Psychological Bulletin, 126,* 55–77.

Gottfredson, M., & Hirschi, T. (1990). *A general theory of crime.* Stanford: Stanford University Press.

Gove, W., & Wilmoth, C. (2003). The neurophysiology of motivation and habitual criminal behavior. In A. Walsh & L. Ellis (Eds.), *Biosocial criminology: Challenging environmentalism's supremacy* (pp. 227–245). Hauppauge, NY: Nova Science.

Gunnar, M. (1996). *Quality of care and the buffering of stress physiology: Its potential in protecting the developing human brain.* University of Minnesota Institute of Child Development.

Jaffee, S., Moffitt, T., Caspi, A., & Taylor, A. (2003). Life with (or without) father: The benefits of living with two biological parents depend on the father's antisocial behavior. *Child Development, 74,* 109–126.

Kanazawa, S. (2003). A general evolutionary psychological theory of criminality and related male-typical behavior. In A. Walsh & L. Ellis (Eds.), *Biosocial criminology: Challenging environmentalism's supremacy* (pp. 37–60). Hauppauge, NY: Nova Science.

Lanier, M., & Henry, S. (1998). *Essential criminology.* Boulder, CO: Westview.

Lykken, D. (1995). *The antisocial personalities.* Hillsdale, NJ: Lawrence Erlbaum.

Lyons, M., True, W., Eusen, S., Goldberg, J., Meyer, J., Faraone, S., et al. (1995). Differential heritability of adult and juvenile antisocial traits. *Archives of General Psychiatry, 53,* 906–915.

Mednick, S., Gabrielli, W., & Hutchings, B. (1984). Genetic influences in criminal convictions: Evidence from an adoption cohort. *Science, 224,* 891–894.

Miles, D., & Carey, G. (1997). Genetic and environmental architecture of human aggression. *Journal of Personality and Social Psychology, 72,* 207–217.

Moffit, T., & The E-Risk Study Team. (2002). Teen-aged mothers in contemporary Britain. *Journal of Child Psychology and Psychiatry, 43,* 1–16.

Moffitt, T., & Walsh, A. (2003). The adolescence-limited/life-course persistent theory and antisocial behavior: What have we learned? In A. Walsh & L. Ellis (Eds.), *Biosocial criminology: Challenging environmentalism's supremacy* (pp. 125–144). Hauppauge, NY: Nova Science.

Morley, K., & Hall, W. (2003). Is there a genetic susceptibility to engage in criminal acts? *Trends and Issues in Crime and Criminal Justice, 263,* 1–10.

Perry, B., & Pollard, R. (1998). Homeostasis, stress, trauma, and adaptation: A neurodevelopmental view of childhood trauma. *Child and Adolescent Psychiatric Clinics of America, 7,* 33–51.

Pinel, J. (2000). *Biopsychology* (4th Ed.). Boston: Allyn & Bacon.

Quartz, S., & Sejnowski, T. (1997). The neural basis of cognitive development: A constructivist manifesto. *Behavioral and Brain Sciences, 20,* 537–596.

Quinsey, V. (2002). Evolutionary theory and criminal behavior. *Legal and Criminological Psychology, 7,* 1–14.

Raine, A., Meloy, J., Bihrle, S., Stoddard, J., LaCasse, L., & Buchsbaum, M. (1998). Reduced prefrontal and increased subcortical brain functioning assessed using positron emission tomography in predatory and affective murderers. *Behavioral Sciences and the Law, 16,* 319–332.

Restak, R. (2001). *The secret life of the brain.* New York: Dana Press and Joseph Henry Press.

Rhee, S., & Waldman, I. (2002). Genetic and environmental influences on antisocial behavior: A meta-analysis of twin and adoption studies. *Psychological Bulletin, 128,* 490–529.

Robinson, M. (2004). *Why crime? An integrated system theory of antisocial behavior.* Upper Saddle River, NJ: Prentice Hall.

Rowe, D. (1996). An adaptive strategy theory of crime and delinquency. In J. Hawkins (Ed.), *Delinquency and crime: Current theories* (pp. 268–314). Cambridge: Cambridge University Press.

Rowe, D. (2002). *Biology and crime.* Los Angeles: Roxbury.

Ruden, R. (1997). *The craving brain: The biobalance approach to controlling addictions.* New York: Harper Collins.

Shore, R. (1997). *Rethinking the brain: New insights into early development.* New York: Families and Work Institute.

Tibbetts, S. (2003). Selfishness, social control, and emotions: An integrated perspective on criminality. In A. Walsh & L. Ellis (Eds.), *Biosocial criminology: Challenging environmentalism's supremacy* (pp. 83–101). Hauppauge, NY: Nova Science.

Vila, B. (1994). A general paradigm for understanding criminal behavior: Extending evolutionary ecological theory. *Criminology, 32,* 311–358.

Vila, B. (1997). Human nature and crime control: Improving the feasibility of nurturant strategies. *Politics and the Life Sciences, 16,* 3–21.

Walsh, A. (2002). *Biosocial criminology: Introduction and integration.* Cincinnati, OH: Anderson.

Walsh, A. (2006) Evolutionary psychology and criminal behavior. In J. Barkow (Ed.), *Missing the revolution: Darwinism for social scientists* (pp. 225–268). Oxford: Oxford University Press.

Walsh, A., & Ellis, L. (1997). The neurobiology of nurturance, evolutionary expectations, and crime control. *Politics and the Life Sciences, 16,* 42–44.

Wright, J., & Beaver, K. (2005). Do parents matter in creating self-control in their children? A genetically informed test of Gottfredson and Hirschi's theory of low self-control. *Criminology, 43,* 1169–1202.

Section 9

Agnew, R. (2005). *Why do criminals offend? A general theory of crime and delinquency.* Los Angeles: Roxbury.

Collins, R. (2004). Onset and desistence in criminal careers: Neurobiology and the age-crime relationship. *Journal of Offender Rehabilitation, 39,* 1–19.

Coolidge, F., Thede, L., & Young, S. (2000). Heritability and the comorbidity of attention deficit hyperactivity disorder with behavioral disorders and executive function deficits: A preliminary investigation. *Developmental Neuropsychology, 17,* 273–287.

Daly, M. (1996). Evolutionary adaptationism: Another biological approach to criminal and antisocial behavior. In G. Bock & J. Goode (Eds.), *Genetics of criminal and antisocial behaviour* (pp. 183–195). Chichester, England: Wiley.

Durston, S. (2003). A review of the biological bases of ADHD: What have we learned from imaging studies? *Mental Retardation and Developmental Disabilities, 9,* 184–195.

Elliot, D., Huizinga, D., & Menard, S. (1989). *Multiple problem youth: Delinquency, substance abuse, and mental health problems.* New York: Springer-Verlag.

Ellis, L. (2003). Genes, criminality, and the evolutionary neuroandrogenic theory. In A. Walsh & L. Ellis (Eds.), *Biosocial criminology: Challenging environmentalism's supremacy* (pp. 13–34), Hauppauge, NY: Nova Science.

Ellis, L., & Walsh, A. (2000). *Criminology: A global perspective.* Boston: Allyn & Bacon.

Farrington, D. (2003). Developmental and life-course criminology: Key theoretical and empirical issues—The 2002 Sutherland Award address. *Criminology, 41,* 221–255.

Fast Track Project. (2005). *Fast track project overview.* Retrieved from http://www.fasttrackproject.org/fasttrackoverview.htm

Glueck, S., & Glueck, E. (1950). *Unraveling juvenile delinquency.* New York: Commonwealth Fund.

Gottfredson, M., & Hirschi, T. (1990). *A general theory of crime.* Stanford: Stanford University Press.

Henry, B., Caspi, A., Moffitt, T., & Silva, P. (1996). Temperament and familial predictors of violent and non-violent criminal convictions: From age 3 to age 18. *Developmental Psychology, 32,* 614–623.

Jeglum-Bartusch, D., Lynam, D., Moffitt, T., & Silva, P. (1997). Is age important? Testing general versus developmental theories of antisocial behavior. *Criminology, 35,* 13–48.

Krueisi, M., Leonard, H., Swedo, S., Nadi, S., Hamburger, S., Lui, J., et al. (1994). Endogenous opioids, childhood psychopathology, and Quay's interpretation of Jeffrey Gray. In D. Routh (Ed.), *Disruptive behavior disorders in childhood* (pp. 207–219). New York: Plenum.

Levy, F., Hay, D., McStephen, M., Wood, C., & Waldman, I. (1997). Attention-deficit hyperactivity disorder: A category or a continuum? Genetic analysis of a large- scale twin study. *Journal of the American Academy of Child and Adolescent Psychiatry, 36,* 737–744.

Lynam, D. (1996). Early identification of chronic offenders: Who is the fledgling psychopath? *Psychological Bulletin, 120,* 209–234.

McCrae, R., Costa, P., Ostendorf, F., Angleitner, A., Hrebickova, M., Avia, M., et al. (2000). Nature over nurture: Temperament, personality, and life span development. *Journal of Personality and Social Psychology, 78,* 173–186.

Moffitt, T. (1993). Adolescent-limited and life-course-persistent antisocial behavior: A developmental taxonomy. *Psychological Review, 100,* 674–701.

Office of the Surgeon General. (2001). *Youth violence: A report of the Surgeon General.* Washington, DC: U.S. Department of Health and Human Services. Retrieved from www.surgeongeneral.gov/library/youthviolence.

Olds, D., Hill, P., Mihalic, S., & O'Brien, R. (1998). *Blueprints for violence prevention, book seven: Prenatal and infancy home visitation by nurses.* Boulder, CO: Center for the Study and Prevention of Violence.

Raz, A. (2004, August). Brain imaging data of ADHD. *Neuropsychiatry,* 46–50.

Sampson, R., & Laub, J. (1999). Crime and deviance over the lifecourse: The salience of adult social bonds. In F. Scarpitti & A. Nielsen (Eds.), *Crime and criminals: Contemporary and classical readings in criminology* (pp. 238–246). Los Angeles: Roxbury Press.

Sampson, R., & Laub, J. (2005). A life-course view of the development of crime. *Annals of the American Academy of Political and Social Sciences, 602,* 12–45.

Sanjiv, K., & Thaden, E. (2004, January). Examining brain connectivity in ADHD. *Psychiatric Times,* 40–41.

Shavit, Y., & Rattner, A. (1988). Age, crime, and the early life-course. *American Journal of Sociology, 93,* 1457–1470.

Spear, L. (2000). Neurobehavioral changes in adolescence. *Current Directions in Psychological Science, 9,* 111–114.

Tittle, C. (2000). Theoretical developments in criminology. *National Institute of Justice 2000, Vol.1. The nature of crime: Continuity and change.* Washington, DC: National Institute of Justice.

Warr, M. (2002). *Companions in crime: The social aspects of criminal conduct.* New York: Cambridge University Press.

White, A. (2004). *Substance use and the adolescent brain: An overview with the focus on alcohol.* Duke University Medical Center.

Willoughby, M. (2003). Developmental course of ADHD symptomology during the transition from childhood to adolescence: A review with recommendations. *Journal of Child Psychology and Psychiatry, 43,* 609–621.

Section 10

Ardila, R. (2002). The psychology of the terrorist: Behavioral perspectives. In C. Stout (Ed.), *The psychology of terrorism* (Vol. I, pp. 9–15). Westport, CT: Praeger.

Armanios, F. (2003). *Islamic religious schools, madrasas: Background.* (Congressional Research Service, Report RS21654). Washington, DC: Library of Congress.

Birkenhead, T., & Moller, A. (1992, July). Faithless females seek better genes. *New Scientist,* 34–38.

Brock, D., & Shrimpton, A. (1991). Nonpaternity and prenatal genetic screening. *Lancet, 388,* 1151–1153.

Brownmiller, S. (1975). *Against our will: Men, women, and rape.* New York: Simon & Schuster.

Catalano, S. (2006). *Criminal victimization, 2005.* Washington, DC: Bureau of Justice Statistics.

Daly, M., & Wilson, M. (2000). Risk-taking, intersexual competition, and homicide. *Nebraska Symposium on Motivation, 47,* 1–36.

Durant, W. (1953). *Caesar and Christ.* New York: Simon & Schuster.

Ellis, L. (1991). A synthesized (biosocial) theory of rape. *Journal of Consulting and Clinical Psychology, 59,* 631–642.

Ellis, L., & Walsh, A. (2000). *Criminology: A global perspective.* Boston: Allyn & Bacon.

Federal Bureau of Investigation. (2006). *Crime in the United States, 2005: Uniform Crime Reports.* Washington DC: U.S. Government Printing Office.

Figueredo, A., & McCloskey, L. (1993). Sex, money, and paternity: The evolutionary psychology of domestic violence. *Ethology and Sociobiology, 14,* 353–379.

Fox, J., & Levin, J. (2001). *The will to kill: Making sense of senseless murder.* Boston: Allyn & Bacon.

Giannangelo, S. (1996). *The psychopathology of serial murder: A theory of violence.* Westport, CT: Praeger.

Given, J. (1977). *Society and homicide in thirteenth-century England.* Stanford, CA: Stanford University Press.

Hacker, F. (1977). *Crusaders, criminals, crazies: Terror and terrorism in our time.* New York: Norton.

Hampton, R., Oliver, W., & Magarian, L. (2003). Domestic violence in the African American community. *Violence against Women, 9,* 533–557.

Hanawalt, B. (1979). *Crime and conflict in English communities, 1300–1348.* Cambridge, MA: Harvard University Press.

Harris, A., Thomas, S., Fisher, G., & Hirsch, D. (2002). Murder and medicine: The lethality of criminal assault 1960–1999. *Homicide Studies, 6,* 128–166.

Hickey, E. (2006). *Serial murderers and their victims* (4th ed.). Belmont, CA: Wadsworth.

Hoffman, C. (2002). Rethinking terrorism and counterterrorism since 9/11. *Studies in Conflict & Terrorism, 25,* 303–316.

Holmes, R., & DeBurger, J. (1998). Profiles in terror: The serial murderer. In R. Holmes & A. Holmes (Eds.), *Contemporary perspectives on serial murder* (pp. 1–16). Thousand Oaks, CA: Sage.

Hudson, R. (1999). *The sociology and psychology of terrorism: Who becomes a terrorist and why?* The Library of Congress, Federal Research Division, Washington, DC.

Hurd, M. (2003). *The psychology of junior sniper Lee Malvo.* Retrieved from http://www.capmag.com.

Johnson, P., & Feldman, T. (1992). Personality types and terrorism: Self-psychology perspectives. *Forensic Reports, 5,* 293–303.

Keeney, B., & Heide, K. (1995). Serial murder: A more accurate and inclusive definition. *International Journal of Offender Therapy and Comparative Criminology, 39,* 299–306.

Krug, E., Dahlberg, L., Mercy, J., Zwi, A., & Lozano, R. (2002). *World report on violence and health.* Geneva: World Health Organization.

Lepowsky, M. (1994). Women, men, and aggression in egalitarian societies. *Sex Roles,* 30, 199–211.

Leyton, E. (1986). *Hunting humans: Inside the minds of mass murderers.* New York: Pocket Books.

Mann, C. (1990). Black female homicides in the United States. *Journal of Interpersonal Violence, 5,* 176–201.

Mealey, L. (2003). Combating rape: Views of an evolutionary psychologist. In R. Bloom & N. Dess (Eds.), *Evolutionary psychology and violence* (pp. 83–113). Westport, CT: Praeger.

Messerschmidt, J. (1993). *Masculinities and crime.* Lanham, MD: Rowman & Littlefield.

Miller, J. (1998). Up it up: Gender and the accomplishment of street robbery. *Criminology, 36,* 37–65.

National Counterterrorism Center. (2005). *A chronology of significant international terrorism for 2004.* Washington, DC: National Counterterrorism Center.

Newton, M. (2000). *The encyclopedia of serial killers.* New York: Checkmark.

Perlman, D. (2002). Intersubjective dimensions of terrorism and its transcendence. In C. Stout (Ed.), *The psychology of terrorism* (Vol. III, pp. 57–81). Westport, CT: Praeger.

Reich, W. (1990). Understanding terrorist behavior: The limits and opportunities of psychological inquiry. In W. Reich (Ed.), *Origins of terrorism: Psychologies, ideologies, theologies, states of mind* (pp. 261–279). New York: Cambridge University Press.

Rennison, C. (2003). *Intimate partner violence, 1993–2003.* Bureau of Justice Statistics Report. Washington, DC: U.S. Department of Justice.

Rummel, R. (1992). Megamurderers. *Society, 29,* 47–52.

Schussler, M. (1992). German crime in the later middle ages: A statistical analysis of the Nuremberg outlawry books, 1285–1400. In L. Knafla (Ed.), *Criminal justice history: An international annual* (Vol. 13). Westport, CT: Greenwood Press.

Segrave, K. (1992). *Women serial and mass murderer: A worldwide reference, 1580 through 1990.* Jefferson, NC: McFarland.

Simonsen, C., & Spindlove, J. (2004). *Terrorism today: The past, the players, the future.* Upper Saddle River, NJ: Prentice Hall.

Smith, B. (1994). *Terrorism in America: Pipe bombs and pipe dreams.* Albany: State University of New York Press.

Thornhill, R., & Palmer, C. (2000). *A natural history of rape: Biological bases of sexual coercion.* Cambridge, MA: MIT Press.

Tolen, P., Gorman-Smith, D., & Henry, D. (2006). Family violence. *Annual Review of Psychology, 57,* 557–583.

United States Department of State. (2004). *Patterns of global terrorism: 2003.* Washington, DC: Author.

van Berlo, W., & Ensink, B. (2000). Problems with sexuality after sexual assault. *Annual Review of Sex Research, 11,* 235–257.

Walsh, A. (2005). African Americans and serial killing in the media: The myth and the reality. *Homicide Studies, 9,* 271–291.

Walsh, A., & Ellis, L. (2007). *Criminology: An interdisciplinary approach.* Thousand Oaks, CA: Sage.

Wright, R., & Decker, S. (1997). *Armed robbers in action.* Boston: Northeastern University Press.

Section 11

Baines, H. (1996). *The Nigerian scam masters: An expose of a modern international gang.* Hauppauge, NY: Nova Science.

Brett, A. (2004). "Kindling theory" in arson: How dangerous are firesetters? *Australian and New Zealand Journal of Psychiatry, 38,* 419–425.

Business Software Alliance. (2005). *2005 piracy study.* Retrieved from http://www.bsa.org/globalstudy.

Catalano, S. (2006). *Criminal victimization, 2005.* Washington, DC: Bureau of Justice Statistics.

Cornell Law School. (2005). *U.S. Code collection: Criminal infringement of a copyright.* Retrieved from http://www.law.cornell.edu/uscode/html/uscode18/usc_sec_18html

Department of Justice. (2004). *Nineteen individuals indicted in Internet "carding" conspiracy.* Retrieved from http://www.cybercrime.gov/montovaniIndict.html.

Federal Bureau of Investigation. (2005). *Uniform Crime Reports handbook.* Washington, DC: U.S. Government Printing Office.

Federal Bureau of Investigation. (2006). *Crime in the United States 2005.* Washington, DC: U.S. Government Printing Office.

Hakkanen, H., Puolakka, P., & Santilla, P. (2004). Crime scene actions and offender characteristics in arsons. *Legal and Criminological Psychology, 9,* 197–214.

Jacobs, B., Topalli, V., & Wright, R. (2003). Carjacking, streetlife and offender motivation. *British Journal of Criminology, 43,* 673–688.

Klaus, P. (2004). *Carjacking, 1993–2002.* Bureau of Justice Statistics Crime Data Briefs. Washington, DC: U.S. Department of Justice.

Kshetri, N. (2006, January/February). The simple economics of cybercrimes. *IEEE Security & Privacy,* 33–39.

Lamontagne, Y., Boyer, R., Hetu, C., & Lacerte-Lamontagne, C. (2000). Anxiety, significant losses, depression, and irrational beliefs in first-offense shoplifters. *Canadian Journal of Psychiatry, 45,* 63–66.

Levy, S, & Stone, B. (2005). *Grand theft identity.* Newsweek Business On-line. Retrieved from http://msnbc.com/id/835i692/site/newsweek/print/1/displaymode/1098/

Linden, R., & Chaturvedi, R. (2005). The need for comprehensive crime prevention planning: The case of motor vehicle theft. *Canadian Journal of Criminology and Criminal Justice, 47,* 251–270.

Mawby, R. (2001). *Burglary.* Colompton, Devon: Willan.

McGoey, C. (2005). *Carjacking facts: Robbery prevention advice.* Retrieved from http://www.crimedoctor.com.carjacking.htm.

Moore, R. (1984). Shoplifting in middle-America: Patterns and motivational correlates. *International Journal of Offender Therapy and Comparative Criminology, 23,* 29–40.

Mowbray, J. (2002). *Justice interrupted.* Retrieved from http://www.nationalreview.com

Mullins, C., & Wright, R. (2003). Gender, social networks, and residential burglary. *Criminology, 41,* 813–839.

National Insurance Crime Bureau. (2006). *2005 hot wheels.* Retrieved from http://www.nicb.org/public/newsroom/hotwheels/index.cfm.

Orlando Business Journal. (2005). *Survey: Shoplifting losses mount.* Retrieved from http://orlando.bizjournals.com/orlando/stories/2005/12/05/daily.html.

Power, R. (2000). *Tangled web: Tales of digital crime from the shadows of cyberspace.* Indianapolis, IN: Que Books.

Rengert, G., & Wasilchick, J. (2001). *Suburban burglary: A tale of two suburbs.* Springfield, IL: Charles C Thomas.

Rice, K., & Smith, W. (2002). Socioecological models of automotive theft: Integrating routine activities and social disorganization approaches. *Journal of Research in Crime and Delinquency, 39,* 304–336.

Rosoff, S., Pontell, H., & Tillman, R. (1998). *Profit without honor: White-collar crime and the looting of America.* Upper Saddle River, NJ: Prentice-Hall.

Ross, J. (1999, February 28). Suspect with Satanic impulses confesses to burning churches. *Associated Press.* Retrieved from http://www.rickross.com/reference/satanism36.html.

Seale, D., Polakowski, M., & Schneider, S. (1998). It's not really theft! Personal and workplace ethics that enable software piracy. *Behavior and Information Technology, 17,* 27–40.

Stafford, M. (2004). Identity theft: Laws, crimes, and victims. *The Journal of Consumer Affairs, 38,* 201–203.

Tonglet, M. (2001). Consumer misbehavior: An exploratory study of shoplifting. *Journal of Consumer Behaviour, 1,* 336–354.

Voiskounsky, A., & Smyslova, O. (2003). Flow-based model of computer hackers' motivation. *CyberPsychology & Behavior, 6,* 171–180.

Wright, R., & Decker, S. (1994). *Burglars on the job: Streetlife and residential break-ins.* Boston: Northeastern University Press.

Section 12

Bartol, C. (2002). *Criminal behavior: A psychosocial approach.* Englewood Cliffs, NJ: Prentice-Hall.

Buck, K., & Finn, D. (2000). Genetic factors in addiction: QTL mapping and candidate gene studies implicate GABAergic genes in alcohol and barbiturate withdrawal in mice. *Addiction, 96,* 139–149.

Bullough, B., & Bullough, V. (1994). Prostitution. In V. Bullough & B. Bullough (Ed.), *Human sexuality: An encyclopedia* (pp. 494–499). New York: Garland Press.

Burns, E. (2004). *The spirits of America: A social history of alcohol.* Philadelphia: Temple University Press.

Casey, E. (1978). *History of drug use and drug users in the United States.* Schaffer Library of Drug Policy. Retrieved from http://www.druglibrary.org/schaffer/History/CASEY1.htm.

Crabbe, J. (2002). Genetic contributions to addiction. *Annual Review of Psychology, 53,* 435–462.

Davenport-Hines, R. (2002). *The pursuit of oblivion: A global history of narcotics.* New York: W.W. Norton.

Demir, B., Ucar, G., Ulug, B., Ulosoy, S., Sevinc, I., & Batur, S. (2002). Platelet monoamine oxidase activity in alcoholism subtypes: Relationship to personality traits and executive functions. *Alcohol and Alcoholism, 37,* 597–602.

Drug Enforcement Administration. (2003). *Drugs of abuse.* Arlington, VA: U.S. Department of Justice.

DuPont, R. (1997). *The selfish brain: Learning from addiction.* Washington, DC: American Psychiatric Press.

Durant, W. (1939). *The life of Greece.* New York: Simon & Schuster.

Federal Bureau of Investigation. (2005). *Uniform Crime Reports handbook.* Washington, DC: U.S. Government Printing Office.

Fishbein, D. (1998). Differential susceptibility to comorbid drug abuse and violence. *Journal of Drug Issues, 28,* 859–891.

Fishbein, D. (2003). Neuropsychological and emotional regulatory processes in antisocial behavior. In A. Walsh & L. Ellis (Eds.), *Biosocial criminology: Challenging environmentalism's supremacy* (pp. 185–208). Hauppauge, NY: Nova Science.

Goldstein, P. (1985). The drugs/violence nexus: A tripartite conceptual framework. *Journal of Drug Issues, 15,* 493–506.

Goldstein, P., Browstein, H., Ryan, P., & Belluci, P. (1989, Winter). Crack and homicide in New York City 1988: A conceptually based event analysis. *Contemporary Drug Problems,* 651–687.

Goode, E. (2005). *Drugs in American society* (6th ed). Boston: McGraw Hill.

Gove, W., & Wilmoth, C. (2003). The neurophysiology of motivation and habitual criminal behavior. In A. Walsh & L. Ellis (Eds.), *Biosocial criminology: Challenging environmentalism's supremacy* (pp. 227–245). Hauppauge, NY: Nova Science.

Insurance Information Institute. (2005). *Drunk driving.* Retrieved from http://www.iii.org/media/hottopics/insurance/drunk/

Johnson, L., O'Malley, P., & Bachman, J. (2000). *Monitoring the future national survey results on drug use, 1975–1999.* Bethesda, MD: National Institute of Drug Abuse.

Kleber, H. (2003). Pharmacological treatments for heroin and cocaine dependence. *The American Journal on Addictions, 12,* S5–S18.

Kornblum, W., & Julian, J. (1995). *Social problems* (8th ed.). Englewood Cliffs, NJ: Prentice Hall.

Lo, C., & Stephens, R. (2002). The role of drugs in crime: Insights from a group of incoming prisoners. *Substance Use and Misuse, 37,* 121–131.

Martin, S. (2001). The links between alcohol, crime and the criminal justice system: Explanations, evidence and interventions. *The American Journal on Addictions, 10,* 136–158.

McBride, D., & McCoy, C. (1993). The drugs-crime relationship: An analytical framework. *The Prison Journal, 73,* 257–278.

McCaghy, C., & Cernkovich, S. (1987). *Crime in American society* (2nd ed). New York: Macmillan.

McDermott, P. A., Alterman, A. I., Cacciola, J. S., Rutherford, M. J., Newman, J. P., & Mulholland, E. M. (2000). Generality of psychopathy checklist-revised factors over prisoners and substance-dependent patients. *Journal of Consulting and Clinical Psychology, 68*(1), 181–186.

McGue, M. (1999). The behavioral genetics of alcoholism. *Current Directions in Psychological Science, 8,* 109–115.

McMurren, M. (2003). Alcohol and crime. *Criminal Behaviour and Mental Health, 13,* 1–4.

Menard, S., Mihalic, S., & Huizinga, D. (2001). Drugs and crime revisited. *Justice Quarterly, 18,* 269–299.

Maruschak, L. (1999). *DWI offenders under correctional supervision.* Washington, DC: Bureau of Justice Statistics.

Mustaine, E., & Tewksbury, R. (2004). Alcohol and violence. In S. Holmes & R. Holmes (Eds.), *Violence: A contemporary reader* (pp. 9–25). Upper Saddle River, NJ: Prentice-Hall.

National Highway Traffic Safety Commission. (2003). *Traffic safety facts.* Washington, DC: U.S. Department of Transportation.

National Institute of Justice. (2004). *Drug and alcohol use and related matters among arrestees, 2003.* Washington, DC: Department of Justice.

National Institute on Alcohol Abuse and Alcoholism. (1998). *Economic costs of alcohol and drug abuse estimated at $246 billion in the United States.* Retrieved from http://www.niaaa.nih.gov/press/1998/economics.htm.

Parker, N., & Auerhahn, K. (1998). Alcohol, drugs, and violence. *Annual Review of Sociology, 24,* 291–311.

Pridemore, W. (2004). Weekend effects on binge drinking and homicide: The social connection between alcohol and violence in Russia. *Addiction, 99,* 1034–1041.

Richardson, A., & Budd, T. (2003). Young adults, crime and disorder. *Criminal Behaviour and Mental Health, 13,* 5–17.

Robinson, M. (2004). *Why crime: An integrated systems theory of antisocial behavior.* Upper Saddle River, NJ: Prentice Hall.

Robinson, T., & Berridge, K. (2003). Addiction. *Annual Review of Psychology, 54,* 25–53.

Ruden, R. (1997). *The craving brain: The biobalance approach to controlling addictions.* New York: HarperCollins.

Tutty, L., & Nixon, K. (2003). Selling sex? It's really like selling your soul. Vulnerability to and the experience of exploitation through child prostitution. In K. Gorkoff & J. Runner (Eds.), *Being heard: The experience of young women in prostitution* (pp. 29–45). Black Point, Nova Scotia: Fernwood.

United States Department of Health and Human Services. (2005). *National household survey on drug abuse.* Washington, DC: Author.

Wanberg, K., & Milkman, H. (1998). *Criminal conduct and substance abuse treatment: Strategies for self-improvement.* Thousand Oaks, CA: Sage.

Section 13

Abadinsky, H. (2003). *Organized crime* (7th ed.). Belmont, CA: Wadsworth.

Albanese, J. (2000). The causes of organized crime. *Journal of Contemporary Criminal Justice, 16,* 409–432.

Bell D. (1962). *The end of ideology.* New York: Collier Books.

Benson, M., & Moore, E. (1992). Are white-collar and common offenders the same? An empirical and theoretical critique of a recently proposed general theory of crime. *Journal of Research in Crime and Delinquency, 29,* 251–272.

Browning, F., & Gerassi, J. (1980). *The American way of crime.* New York: G.P. Putnam.

Burr, M. (2004, December). SEC gains power, prestige in post Enron era. *Corporate Legal Times,* 10–13.

Bynum, T. (1987). Controversies in the study of organized crime. In T. Bynum (Ed.), *Organized crime in America: Concepts and controversies* (pp. 3–11). Monsey, NY: Willow Tree Press.

Calavita, K., & Pontell, H. (1994). "Head I win, tails you lose": Deregulation, crime, and crisis in the savings and loan industry. In D. Curran & C. Renzetti (Eds.), *Contemporary societies: Problems and prospects* (pp. 460–480). Englewood Cliffs, NJ: Prentice Hall.

Calavita, K., Pontell, H., & Tillman, R. (1999). *Big money game: Fraud and politics in the savings and loan crisis.* Berkeley: University of California Press.

Carter, D. (1994). International organized crime: Emerging trends in entrepreneurial crime. *Journal of Contemporary Criminal Justice, 10,* 239–266.

Clinnard, M., & Yeager, P. (1980). *Corporate crime.* New York: Free Press.

English, S. (2004). Enron legal bills will cost $780m. *Business.telegraph.* http://www.telegraph.co.uk/money.jhtml?xml=/money/2004/1

Finckenauer, J. (2004, July/August). The Russian "Mafia." *Society,* 61–64.

Firestone, T. (1997). Mafia memoirs: what they tell us about organized crime. In P. Ryan & G. Rush (Eds.), *Understanding organized crime in global perspective* (pp. 71–86). Thousand Oaks, CA: Sage.

Gutman, H. (2002). Dishonesty, greed, and hypocrisy in corporate America. *Statesman* (Kalkota, India), http://www.commondreams.org/cgi-bin/print.cgi?file=/views.

Hagan, F. (1994). *Introduction to criminology.* Chicago: Nelson-Hall.

Hill, P. (2003). *The Japanese mafia: Yakuza, law, and the state.* Oxford: Oxford University Press.

Hirschi, T., & Gottfredson, M. (1987). Causes of white-collar crime. *Criminology, 25,* 949–974.

Iwai, H. (1986). Organized crime in Japan. In R. Kelly (Ed.), *Organized crime: A global perspective* (pp. 208–233). Totowa, NJ: Rowman & Littlefield.

Johnson, E. (1990). Yakuza (criminal gangs) in Japan: Characteristics and management in prison. *Journal of Contemporary Criminal Justice, 6,* 113–126.

Kappeler, V., Blumberg, M., & Potter, G. (2000). *The mythology of crime and criminal justice* (3rd ed.). Prospect Heights, IL: Waveland.

Lupsha, P. (1987). La Cosa Nostra in drug trafficking. In T. Bynum (Ed.), *Organized crime in America: Concepts and controversies* (pp. 31–41). Monsey, NY: Willow Tree Press.

Lyman, M., & Potter, G. (2004). *Organized crime* (3rd ed.). Upper Saddle River, NJ: Prentice Hall.

McCabe, B., O'Reilly, C., & Pfeffer, J. (1991). Context, values and moral dilemmas: Comparing the choices of business and law school students. *Journal of Business Ethics, 10,* 951–960.

Parekh, R. (2004). Fraud by employees on the rise, survey finds. *Business Insurance, 38,* 4–6.

Pileggi, N. (1985). *Wiseguy: Life in a mafia family.* New York: Simon & Schuster.

President's Commission on Organized Crime. (1986). *The impact: Organized crime today.* Washington, DC: U.S. Government Printing Office.

Rosoff, S., Pontell, H., & Tillman, R. (1998). *Profit without honor: White-collar crime and the looting of America.* Upper Saddle River, NJ: Prentice Hall.

Rush, R., & Scarpitti, F. (2001). Russian organized crime: The continuation of an American tradition. *Deviant Behavior, 22,* 517–540.

Schmalleger, F. (2004). *Criminology today* (3rd ed.). Upper Saddle River, NJ: Prentice Hall.

Simon, D. (2002). *Elite deviance* (7th ed.). Boston: Allyn & Bacon.

Sutherland, E. (1940). White collar criminality. *American Sociological Review, 5,* 1–20.

Sutherland, E. (1956). *Crime of corporations* In A. Cohen, A. Lindesmith, & K. Schuessler (Eds.), *The Sutherland papers.* Bloomington: Indiana University Press.

Thorburn, K. (2004). Corporate governance and financial distress. In H. Sjogren & G. Skogh (Eds.), *New perspectives on economic crime* (pp. 76–94). Cheltenham, UK: Edward Elgar.

Useem, M. (1989). *Liberal education and the corporation.* Hawthorn, NY: Aldine de Gruyter.

Trevino, L., & Youngblood, S. (1990). Bad apples in bad barrels: A causal analysis of ethical decision-making behavior. *Journal of Applied Psychology, 78,* 378–385.

Walsh, A., & Hemmens, C. (2000). *From law to order: The theory and practice of law and justice.* Lanham, MD: American Correctional Association.

Weber, J. (1990). Managers' moral reasoning: Assessing their responses to three moral dilemmas. *Human Relations, 43,* 687–702.

Weisburd, D., Wheeler, S., Waring, E., & Bode, N. (1991). *Crimes of the middle classes: White-collar offenders in the federal courts.* New Haven, CT: Yale University Press.

Wells, R. (1995, June 16). Study finds fines don't deter Wall St. cheating. *Idaho Statesman,* 1e–2e.

Wilson, C. (1984). *A criminal history of mankind.* London: Panther Books.

Section 14

Amir, M. (1971). *Patterns of forcible rape.* Chicago University of Chicago Press.

Bartol, C. (2002). *Criminal behavior: A psychosocial approach* (6th ed.). Englewood Cliffs, NJ: Prentice Hall.

Bostaph, L. (2004). *Race and repeat victimization: Does the repetitive nature of police motor vehicle stops impact racially biased policing.* Unpublished doctoral dissertation, University of Cincinnati.

Catalano, S. (2006). *Criminal victimization, 2005.* Washington, DC: Bureau of Justice Statistics.

DeVoe, J., Peter, K., Kaufman, P., Ruddy, S., Miller, A., Planty, M., et al. (2003). *Indicators of school crime and safety.* Washington, DC.: U.S. Department of Education and U.S. Department of Justice.

Doerner, W., & Lab, S. (2002). *Victimology* (3rd ed.). Cincinnati, OH: Anderson.

Finkelhor, D. (1984). *Child sexual abuse: New theory and research.* New York: Free Press.

Fisher, B., Cullen, F., & Turner, M. (2001). *The sexual victimization of college women.* Washington, DC: National Institute of Justice.

Glaser, D., & Frosh, S. (1993). *Child sex abuse.* Toronto: University of Toronto Press.

Herman, J. (1990). Sex offenders: A feminist perspective. In W. Marshall, D. Laws, & H. Barbaree (Eds.), *Handbook of sexual assault: Issues, theories, and treatment of the offender* (pp. 177–193). New York: Plenum.

Karmen, A. (2004). *Crime victims: An introduction to victimology* (5th ed.). Belmont, CA: Wadsworth.

Knudsen, D. (1991). Child sexual coercion. In E. Grauerholz & M. Koralewski (Eds.), *Sexual coercion: A sourcebook on its nature, causes, and prevention* (pp. 17–28). Lexington, MA: D.C. Heath.

Loeber, R., Kalb, L., & Huizinga, D. (2001, August). Juvenile delinquency and serious injury victimization. *Juvenile Justice Bulletin.* U.S. Department of Justice.

Macmillan, R. (2001). Violence and the life course: The consequences of victimization for personal and social development. *Annual Review of Sociology, 27,* 1–22.

Mann, C. (1990). Black female homicides in the United States. *Journal of Interpersonal Violence, 5,* 176–201.

Mawby, R. (2001). *Burglary.* Colompton, Devon: Willan.

National Association of Crime Victim Compensation Board. (2005). *FY 2004 compensation to victims continues to increase.* Retrieved from http://www.nacvcb.org/

Talbot, T., Gilligan, L., Carter, M., & Matson, S. (2002). *An overview of sex offender management.* Washington, DC: Center for Sex Offender Management.

Tseloni, A., & Pease, K. (2003). Repeat personal victimization. *British Journal of Criminology, 43,* 196–212.

van Berlo, W., & Ensink, B. (2000). Problems with sexuality after sexual assault. *Annual Review of Sex Research, 11,* 235–257.

von Hentig, H. (1941). Remarks on the interaction of perpetrator and victim. *Journal of Criminal Law, Criminology, and Police Science, 31,* 303–309.

Walsh, A. (1988). Lessons and concerns from a case study of a "scientific" molester. *Corrective and Social Psychiatry, 34,* 18–23.

Walsh, A. (1994). Homosexual and heterosexual child molestation: Case characteristics and sentencing differentials. *International Journal of Offender Therapy and Comparative Criminology, 38,* 339–353.

Warchol, G. (1998). *Workplace violence, 1992–1996.* Bureau of Justice Statistics, U.S. Department of Justice special report.

Index

About the Authors

Anthony Walsh received his Ph.D. at Bowling Green State University. He teaches criminology, statistics, and criminal law at Boise State University in Idaho. He is the author, coauthor, or editor of 17 other books and over 100 published articles/essays. He has had field experience in both law enforcement and corrections. His proudest accomplishment is marrying the sweetest and most drop-dead gorgeous woman on the planet.

Craig Hemmens holds a J.D. from North Carolina Central University School of Law and a Ph.D. in Criminal Justice from Sam Houston State University. He is the Director of the Honors College and a Professor in the Department of Criminal Justice at Boise State University, where he has taught since 1996. He has previously served as Academic Director of the Paralegal Studies Program and Chair of the Department of Criminal Justice. Professor Hemmens has published 10 books and more than 100 articles on a variety of criminal justice-related topics. His primary research interests are criminal law and procedure and corrections. He has served as the editor of *Journal of Criminal Justice Education.* His publications have appeared in *Justice Quarterly, Journal of Criminal Justice, Crime and Delinquency, Criminal Law Bulletin,* and *The Prison Journal.*